Analysis of Visual Behavior

Analysis of Visual Behavior

edited by
David J. Ingle
Melvyn A. Goodale
Richard J. W. Mansfield

The MIT Press
Cambridge, Massachusetts
London, England

This book was set in VIP Palatino by Achorn Graphic Services, Incorporated, and printed and bound by Halliday Lithograph in the United States of America.

Library of Congress Cataloging in Publication Data
Main entry under title:

Analysis of visual behavior.

Articles first presented to the NATO Advanced Study Institute entitled New Advances in the Analysis of Visual Behavior, Brandeis University, 1978
    Includes bibliographies and index.
    1. Visual perception—Congresses. 2. Perceptual motor processes—Congresses. 3. Psychology, Comparative—Congresses. 4. Psychology, Physiological—Congresses. I. Ingle, David. II. Goodale, Melvyn A. III. Mansfield, Richard J. W. IV. NATO Advanced Study Institute (1978: Brandeis University)

BF241.A55   1982   596'.01823   81-15582
ISBN 0-262-09022-8   AACR2

Contents

Foreword
ix

**Identification and Localization
Processes in Nonmammalian
Vertebrates**

Introduction   David J. Ingle
3

1
Neuronal Basis of   Jörg-Peter Ewert
Configurational Prey Selection
in the Common Toad
7

2
Prey Selection in Salamanders   Werner Himstedt
47

3
Organization of Visuomotor   David J. Ingle
Behaviors in Vertebrates
67

4
Depth Vision in Animals   Thomas S. Collett
111   Lindesay I. K. Harkness

5
Mechanisms for Discriminating   Barrie J. Frost
Object Motion from Self-Induced
Motion in the Pigeon
177

6
Right-Left Asymmetry of
Response to Visual Stimuli
in the Domestic Chick
197

Richard J. Andrew
J. Mench
C. Rainey

7
Retinal Locus as a Factor in
Interocular Transfer in
the Pigeon
211

Melvyn A. Goodale
Jefferson A. Graves

8
Mechanisms of Concept
Formation in the Pigeon
241

John Cerella

**Visual Guidance of Motor
Patterns: The Role of Visual
Cortex and the Superior
Colliculus**

Introduction
263

Melvyn A. Goodale

9
Fractionating Orientation
Behavior in Rodents
267

Melvyn A. Goodale
A. David Milner

10
Contrasting Behavioral Methods
in the Analysis of Vision in
Monkeys with Lesions of the
Striate Cortex or the Superior
Colliculus
301

Charles M. Butter
Daniel Kurtz
Clare C. Leiby III
Alphonso Campbell, Jr.

11
Visual-Motor Transforms of the
Primate Tectum
335

E. Gregory Keating
John Dineen

12
The Contribution of Peripheral
and Central Vision to Visually
Guided Reaching
367

Jacques Paillard

13
Visuomotor Mechanisms in
Reaching within Extrapersonal
Space
387

Marc Jeannerod
B. Biguer

14
Vision in Action: The Control
of Locomotion
411

David N. Lee
James A. Thomson

**Recognition and Transfer
Processes**

Introduction
437

Richard J. W. Mansfield

15
Role of the Striate Cortex in
Pattern Perception in Primates
443

Richard J. W. Mansfield

16
Behavioral Measurement of
Normal and Abnormal
Development of Vision in
the Cat
483

Donald E. Mitchell
Brian Timney

17
Role of the Geniculocortical
System in Spatial Vision
525

Mark A. Berkley
James M. Sprague

18
Two Cortical Visual Systems
549

Leslie G. Ungerleider
Mortimer Mishkin

19
Analysis of Visual Behavior in
Monkeys with Inferotemporal
Lesions
587

Paul Dean

20
Visual Agnosia in Monkey and
in Man
629

David N. Levine

21
Visual Perception and Oculomotor   Richard Latto
Areas in the Primate Brain
671

22
Mechanisms of Interocular   Charles R. Hamilton
Equivalence
693

23
Interocular and Interhemispheric   Giovanni Berlucchi
Transfer of Visual Discrimination   Carlo Alberto Marzi
in Cats
719

24
Evaluation of Visual Performance   Paul Cornwell
in Cats with Posterior Extrastriate J. M. Warren
Lesions
751

25
Search for the Neural Mechanisms   Howard C. Hughes
Essential to Basic Figural
Synthesis in the Cat
771

26
Geometrical Approaches to Visual   Peter C. Dodwell
Processing
801

Contributors
826

Index
831

Foreword

This volume contains articles first presented to the NATO Advanced Study Institute entitled New Advances in the Analysis of Visual Behavior. The purpose of this conference (held at Brandeis University in June 1978) was to survey a variety of new experimental and theoretical approaches to the understanding of visual mechanisms in diverse vertebrate species, including man. The innovative reports presented dealt with physiological methods (brain lesions and unit recording), ethological approaches, and the analysis of visuomotor sequences (such as running, turning, tracking, and grasping). These three areas of interest are not mutually exclusive, but lend themselves to various experimental correlations. Thus, a behavioral analysis of stimulus determinants of snapping at prey by toads or salamanders (see the chapters by Ewert and by Himstedt) can become an exercise in psychophysics or in the study of specific motor sequences, or a problem in correlation of behavioral selectivity with the selectivity of single visual units. A fourth area represented was the acquisition and generalization of form discriminations by birds and mammals. Studies in this area aim to specify the features and the gestalt principles used to identify visual forms. They can have a psychophysical bent (Berkley and Sprague), a cognitive emphasis (Dean), or a naturalistic approach (Cerella). Perhaps the chief message of this volume is that analysis of the "real-life" behavior of humans and animals provides a wealth of data concerning stimulus control of behavior that should be taken into account when one is posing analytical and physiological questions in the laboratory.

The authors attempt to provide broad statements of the issues that prompted them to develop new techniques or better test paradigms. In many instances, they venture opinions as to where these methods may take us, and whether additional new methods may be required to solve the problems at hand. Thus, this volume looks into the future, emphasizing the many new opportunities for approaches to visual function, rather than pretending to be a systematic summary of information on the visual system. Perhaps the best measures of its success will be the number of new and better experiments generated by its readers and its encouragement of a wider acceptance of an ethological view of behavior.

The conference was dedicated to the memory of Hans-Lukas Teuber, who combined innovative research in clinical neuropsychology with international statesmanship in defining relationships between brain and behavior. There are several reasons for our wholehearted dedication of this book to Professor Teuber. First, many of the lecturers at the conference had much earlier been invited by Lukas to lecture at the weekly psychology colloquium at MIT, and it was there that I first gained their acquaintance. Second, his tireless travels across the Atlantic contributed to the recent internationalization of the brain-behavior sciences. His elegance of phrasing expressed itself equally well in English, French, or German, and his familiarity with the experimental and clinical literature of all three traditions was exemplary. Lukas would have been delighted to discover how well the participants in this conference were able to communicate their diverse methods and philosophies. Finally, the arrangement of topics at this meeting was in accord with his vigorous insistence that studies of behavior, physiology, and anatomy are convergent disciplines that must be kept in frequent seminal contact with one another. This principle became an article of faith for some of us who spent our formative years in or near the MIT psychology department. This view, unfortunately, has not yet become part of the conventional wisdom of our time. The "neurosciences" have become increasingly concerned with discoveries on the cellular and biochemical level, and have drawn away from an organismic view of the brain or its relation to real-life situations. Students of the visual system know little about the paradigms of pioneers such as Lashley, Klüver, and von Holst.

Unlike the spectacularly fast advances at the cellular level, advances in the study of brain-behavior relationships have come in fits and starts, and have resulted as much from alterations in theoretical viewpoints as from application of new techniques. The present volume reveals that both sources of discovery are with us today in the study of visually influenced behavior, and this fact conveys an excitement about the future. We are not naively waiting for new information about neuronal cells and molecules to fall into place and explain behavior. Instead, we are discovering new patterns in behavior itself, and we are using the techniques of brain sciences, old as well as new, to revise and sometimes to rework the functional anatomy of vision.

David J. Ingle

Identification and Localization Processes in Nonmammalian Vertebrates

Introduction

David J. Ingle

The 1978 NATO conference was one of the few meetings concerned with brain-behavior relations that gave equal consideration to the lower vertebrates and to rodents, cats, and monkeys. Although the mammalian forms provide most of the experimental knowledge of visual physiology and discrimination capacities, there has been a productive renaissance of interest in nonmammalian visual systems. A few laboratories have exploited new anatomical knowledge and have begun to analyze brain-behavior relations in certain species, particularly among the amphibians and the birds. These organisms seem likely to provide useful "models" of object recognition and spatial functions that can, to some extent, be generalized to mammals.

In part I, four papers deal with visual abilities of amphibians, and each emphasizes that their visual abilities are more complex than is usually assumed for "precortical" animals. The other four chapters discuss aspects of perceptual processing in birds, but three of these also emphasize the value of studying visuomotor systems. These various themes invite comparison with the studies of mammalian visuomotor mechanisms in part II.

In chapter 1, Ewert takes a neuroethological view of vision, proposing that the interaction of tectum and pretectum determines the rules for prey recognition by toads. He combines detailed behavioral studies with quantitative analysis of single-neuron visual responses within the toad's tectum and pretectum to establish a neural model for selective feeding upon "wormlike" shapes. Himstedt adds an important comparative note in chapter 2: Salamanders are even more sensitive than toads to the wormlike shape, and develop this discrimination during early months of life, but frogs are less sensitive to the parameter of "shape." The stage is now set for explanation of species differences in a common function (prey selection) in terms of specialized neural filters in the retina and within the primary retinofugal projection targets. Himstedt also notes that salamanders (unlike frogs and toads) will strike at stationary prey—an ability generated by visual structures that, at first glance, appear more "primitive" than those of anurans. Ewert's work on complex shape discrimination abilities of toads demonstrates that short-term memory for specific configurations can be analyzed without conventional training methods: Toads simply habituate to one class of stimuli and continue to respond to novel shapes. Generalization gradients can be obtained by ethologists as well as operant conditioners.

My chapter, 3, reports on the use of the frog to distinguish various classes of visually elicited orienting behavior. In addition to the previously known functions of optic tectum (pursuing prey and avoiding predators), I describe the role of the frog's pretectum in negotiation of barriers and apertures. This chapter adds the surprising observation that frogs behave differently in confronting horizontally and vertically striped surfaces, demonstrating an edge-orientation discrimination capacity usually associated with the evolution of neocortex. I also propose a model of interaction among these two neural subsystems that may explain the frog's ability to detour around barriers in pursuit of prey objects.

Mechanisms of depth vision are discussed in chapter 4 by Collett and Harkness, who use both vertebrates and insects to demonstrate remarkable functional specializations. Like Ewert's chapter and my own, this chapter reveals surprising abilities of toads to show size-distance constancy—a kind of visuospatial "abstraction" process that had been linked with neocortical function by neuropsychologists. The demonstration reported here that toads use binocular disparity cues for accurate depth estimation is another recent surprise for neuropsychologists. However, the work of Harkness shows that appearance of stereopsis among lower vertebrates is not ubiquitous: The more complex brain of the chamaelon does not possess stereopsis, but gains accurate depth vision by use of "accommodation cues." The question of use of binocular parallax is also raised, and examples are found among the insects where depth judgment follows side-to-side peering. Collett and Harkness add much new weight to Ewert's and my earlier proposals that animal psychophysics is entirely compatible with a naturalistic methodology.

In chapter 5, Frost continues to support the theme that specific behavioral tasks are controlled by specific visual events. He uses the pigeon to clarify differences between the elicitation of orienting responses by small objects moving independently of the environment and the elicitation of head-bobbing movements by whole-field motion produced by the bird's own locomotion. Units in the pigeon's tectum are shown to be well suited for detection of object motion but to ignore wide-field motion. Using the deoxyglucose method to map those neurons specifically excited by moving stimuli, Frost confirms his hypothesis that moving-spot detectors can be localized within the pigeon's tectum while large-field motion detectors (corresponding to self-induced motion) are found elsewhere in the pigeon's visual system.

In chapter 6, Andrew, Mench, and Rainey bring the pigeon into the limelight of social relevance: Not only do birds show significant specialization of hemispheric function (as do humans), but they too seem to show significant sex-linked differences in degree of cerebral asymmetry. Andrew has used relatively simple visuomotor tasks: feeding, aggressive threat to a model, and avoidance of simple visual objects to which the newly hatched chick is innately prepared to respond. Without time for acculturation, sex differences in behavior suddenly emerge. Together with the implications of Goodale and Graves that the two hemispheres keep much private information to themselves, the new data open basic questions regarding hemispheric dominance in specific tasks or at specific moments. The motivational, intentional, and attentional mechanisms that lie just back of visual perception may be studied in an interesting way by comparing the predilections of the hemispheres or by pitting each against the other in "conflict" tests. It is again clear that ethologists are insinuating themselves into a variety of topics unimagined by Lorenz and Tinbergen.

In chapter 7, Goodale and Graves focus on the surprising examples where visual learning does not transfer from one hemisphere to the other after a pigeon switches from the "trained" eye to the "naive" eye. These authors replicated earlier (controversial) findings that interocular transfer by intact pigeons is task-dependent. They argue that the portion of the retina used for viewing is critical for successful transfer. Unlike fishes and amphibians, birds place great emphasis on stereotyped viewing habits: the use of a posterior retinal zone for identifying shapes to be pecked, but a central retinal area for fixation of objects at a distance. The studies of Goodale and Graves imply that central visual pathways may be differently partitioned among the two (or more) kinds of viewing postures. The stereotypy of the pigeon may thus be exploited to discover specializations of visual encoding that begin within different retinal zones and continue along somewhat different central visual pathways.

Chapter 8, by Cerella, also begins with the notion that pigeons easily discriminate among natural classes of objects (for example, humans versus other animals) but generalize broadly within classes (such as "trees" or "bodies of water"). However, it is not clear whether pigeons search for a few very specific features, whose combinations determine strength of class-inclusiveness. Cerella's experiments with geometrical figures and with cartoon characters address this question. In fact, the birds seem to recognize local features rather than configurations per se. One wishes that the clever experiments of Cerella could be used by mammalian psychologists to reopen the old question of what brain structures must evolve to give competency in recognition of "form per se."

1 Neuronal Basis of Configurational Prey Selection in the Common Toad

Jörg-Peter Ewert

If we need a spanner for tightening a particular nut on our car, we select from a set of spanners the appropriate shape and the matching size. Here our visual recognition system has to perform two operations: discrimination of configurational cues and estimation of absolute size. Both operations must be invariant to changes in other visual parameters; for example, an appropriate red spanner solves the problem as well as a green one. What is the neural basis of these functions?

As first step, a retinal image is evaluated in terms of its angular size, its angular velocity, and its contrast with the background on the basis of well-known retinal mechanisms. In mammals, further analysis in terms of edge orientation and motion direction takes place in striate cortex (Hubel and Wiesel 1962, 1963, 1965). In extrastriate cortex, hypercomplex cells show sensitivity to particular configurations within a local area (Hubel and Wiesel 1965). However, the mechanism by which objects are recognized regardless of size and spatial orientation has not yet been revealed by single-unit recording methods. Some approaches to the problem of complex-shape recognition and the visual constancies are reviewed by others in this volume (Cornwell; Sprague; Hughes; Dodwell; Ungerleider).

Since the common toad (*Bufo bufo*) does not have a recognizable visual cortex, it seems likely that shape recognition in the toad obeys rules simpler than those attributed to mammals. However, for a physiological analysis of pattern recognition the use of lower vertebrates may be a real advantage. For one thing, their behavior is rather stereotyped: They tend to show species-specific behavioral patterns to particular objects (prey, predators, mates, rivals). By altering the stimulus characteristics of these key objects, it is often possible to discover the rules that determine their recognition and to generalize these to every member of the species. A second advantage is that the number of visual areas to be explored by physiological or surgical methods is much smaller than the number known for mammals. A third advantage is that more is known about the relationships of subcortical areas (such as tectum and thalamus-pretectum) to the motor system in the lower vertebrates. We have a realistic hope of tracing the flow of visual information from retina to each thalamic and midbrain target, and from these areas into brain regions where patterns of motor coordination are organized.

I shall review a series of behavioral experiments demonstrating that the prey-catching activity of the toad is reliably determined by a limited set of visual parameters, especially a particular interaction of spatial and movement variables that remain invariant with changes in background, type of contrast, and temporal variations in motion. The easily described rules for prey recognition allow investigation of the particular neurons within the optic tectum that help to detect prey and to initiate appropriate responses to prey. Our physiological investigations have shown that the prey-recognition system involves interactions between neurons of tectum and of thalamus-pretectum, and have thus provided a first step in understanding why most parts of the visual system are richly interconnected.

Gestalt Perception

Toads respond to moving objects of particular size and configuration with prey-catching behavior, which consists of a sequence of patterns: orienting toward prey, approaching, binocular fixation and snapping, and swallowing and gulping. Since frogs and toads have neither involuntary nor tracking eye movements (Autrum 1959), the retinal stimulus that elicits the response is relatively easy to describe. The prey/nonprey decision precedes the orienting motor response (Ewert 1968; Ewert et al. 1978). If a visual stimulus does not have prey features, the prey-catching orientation does not occur. (Of course, there are other orienting responses that are unrelated to the prey-catching behavior that I am reviewing here [Ingle 1976a; Ewert 1976].)

The orienting activity of the toad in response to different stimuli can be measured quantitatively by the following procedure (figure 1.1A). The corresponding values resemble the probability that a given stimulus fits the prey category (Ewert 1969). The toad sits in a diffusely illuminated area within a cylindrical glass vessel. The stimulus object can be moved mechanically at constant angular velocity v and constant distance d around the vessel. If the prey is far enough away, the orienting activity can be elicited in isolation from the fixation and snapping components of the prey-catching sequence. The toad follows the circling prey object continuously by successive turning movements. When the releasing value of the prey dummy increases, the average turning angle $\overline{T}$ decreases and the number of turning responses (R_b) per time interval t^* increases. In other words, the greater the efficacy of the prey stimulus, the smaller is the path in the visual field by which the

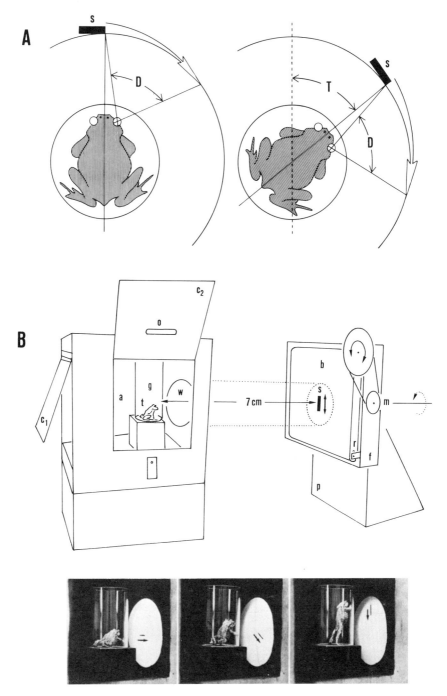

Figure 1.1 (A) Procedure for measurements of the prey-catching orienting activity in the common toad *Bufo bufo*. D: effective displacement of prey stimulus (s) in the visual field. T: angle of a turning movement by experimental animal. (B) Procedure by which stimulus can be moved in different directions through a frontal part of the toad's visual field (see examples below). a: arena. b: movable belt system. $c_{1,2}$: covers (closed during experiment). f: frame for belt system. g: glass vessel. m: electric motor. o: slit for observer. p: post. (During experiment, post was near arena, so distance between toad's eyes and stimulus was around 70 mm.) w: window in frontal arena wall. s: visual stimulus (prey dummy). Sources: Ewert 1969; Ewert et al. 1979a.

stimulus must be moved in order to elicit a prey-catching response and the smaller is the toad's turning angle T. The angle by which the stimulus must be displaced in order to elicit an orienting response is called effective displacement ($D°$). In the present procedure, $D > T$; that is, $\overline{D}° = k \cdot \overline{T}°$ with $k > 1$. Since $\overline{T} = \overline{D}/k$ and $\overline{D} \cdot R_b = v^*$ (degrees/min),

$$R_b = v^* \cdot \overline{D}^{-1} \text{ (turning responses/min)}. \tag{1}$$

For constant visual angular velocity, the relative effectiveness of a visual dummy as prey can be measured by R_b, the number orienting movements per minute. Experiments with different stimuli are separated by recovery pauses (for example, 60 seconds) in order to avoid habituation effects. The average value of $\overline{R}_b$ from different individuals is used as an index for the releasing value of a given visual stimulus.

Configurational Parameters

The investigation is concerned with the recognition of two-dimensional moving rectangular patterns (figure 1.2A) moved linearly at constant velocity. Changing the length (xl_2) perpendicular to the direction of movement modifies only the spatial parameters of the stimulus, whereas changing xl_1 in the direction of movement alters the spatial and temporal components. Configurational cues can be easily changed (figure 1.2A) by using stripes of constant width xl_2 ($x = 1$) and variable length xl_1 in the direction of their movement, stripes of constant width xl_1 ($x = 1$) and variable length xl_2 perpendicular to the direction of movement, or squares of variable edge length $xl_{1,2}$. The edges are elongated by steps with $x = 1,2,4,8,16$, and the prey-catching activity of the toad is measured in successive experiments.

The following stimulus parameters are held constant:

angular velocity v (degrees/sec) = 20,

movement direction m_d = 0° (horizontally),

minimal stimulus distance d = 70 mm,

background luminance L_b (candela/m²) = 40,

stimulus-background contrast $C = (L_b - L_s)(L_b + L_s)^{-1} = 0.9$, and

stimulus edge length $l_1 = l_2 = 2.5$ mm (1 mm $\approx$ 2° visual angle).

The stimulus edge magnification factor is changed in the range $1 \leq x \leq 16$.

The Worm/Antiworm Phenomenon

The effects of configuration on the toad's prey selection are shown in figure 1.2B. Stepwise elongation of the rectangle in the direction of movement (xl_1) progressively increases prey-catching activity in the range $1 \leq x \leq 8$. Increasing the height of the object (xl_2) perpendic-

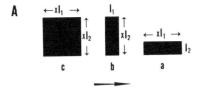

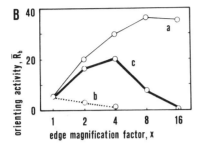

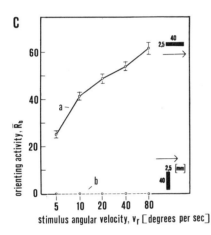

Figure 1.2 (A) Methods for changing configurational features of a visual moving contrast pattern (black stimuli moved on white background). From a $xl_1 \times xl_2$ square, (a) the edge xl_1 (xl_2 = constant), in the direction of movement, is stepwise elongated; (b) the edge xl_2 (xl_1 = constant), perpendicular to the direction of movement, is stepwise varied; and (c) both edges xl_1 and xl_2 are stepwise elongated in equal amounts. Sources: Ewert 1969, 1972. (B) Influence on configurational parameters xl_1 and xl_2 of patterns a–c on the prey-catching activity, $\overline{R}_b$ (turning movements per minute) ($N = 20$), of the common toad for $1 \leqslant x \leqslant 16$ and $l_1 = l_2 = 2.5$ mm. Stimulus-background contrast $C = \pm 0.9$; luminance of background $L_b = 40$ cd/m²; stimulus visual angular velocity $v = 20°$/sec; movement direction in x,y coordinates $m_d = 0°$ (horizontal); minimal distance between toad's eye and stimulus $d = 70$ mm. (Procedure of figure 1.1A.) (C) Example of velocity invariance in selection of a 2.5 $\times$ 40 mm² stripe moving in a wormlike (a) or an antiwormlike (b) fashion. Averages of 20 different animals *B. bufo*. Source: Burghagen, unpublished data.

ular to the direction of movement decreases the prey-catching activity for $x > 1$. Among squares of different sizes, those with $x = 4$ are optimal releasers for prey catching. The results of these experiments clearly show that a stripe moving in a *wormlike* manner (in the direction of its long axis) has a high probability of being categorized as prey. However, if the same stripe is moved perpendicular to its long axis, its probability of being classified as prey is very low. In order to emphasize the unnatural character of this second stimulus, I call it the *antiworm* configuration. The discrimination between worm and antiworm increases as x increases. Among all of the configurational moving stimuli tested, wormlike objects best fit the prey category: If $x = 8$, the square is a stronger releaser than the stripe moved as antiworm; however, moving in a wormlike manner the stripe is more attractive as prey than the square. The t-test probability that toads make a "misjudgement" between these stimuli ($x = 8$) is $P < 0.001$. We chose another example, and found that a 2.5×2.5 mm² square is a stronger releaser than a 2.5×20 mm² antiworm, but less effective than a 2.5×20 mm² worm (Ewert 1968).

Common toads display the worm/antiworm phenomenon immediately after metamorphosis with the transition to terrestrial life. Hence, this perceptual discrimination is innate. During ontogeny configurational selectivity in prey catching matures (Traud and Ewert, in prep.), and the selectivity pattern may also be subject to modification by individual experience. However, our toads could not be trained to prefer long antiworms over worms.

The toad not only discriminates among different rectangles, but is also sensitive to any extension in the direction perpendicular to axis of motion. A 2.5×2.5 mm² black square moved at 7 cm distance at constant angular velocity against a white background represents a prey stimulus that is just above threshold for the release of prey catching. If another square is added 2 mm away in the same plane, both moving simultaneously at the same velocity, the releasing value is increased, and if this chain is prolonged by further links the attractiveness as prey continues to increase. The releasing value is increased further if a fourth member next to the third is moved with the others, but the configuration loses its signal as a prey object almost completely if the fourth member is placed 10 mm above one of the other members.

It is interesting to note that a wormlike stripe with a vertical component (perpendicular to the direction of movement) reproduces the enemy image of a snake. As figure 1.3 shows, the toad responds to the dummy with the same specific defense posture with which it responds to the natural enemy: It blows itself up, assumes a stiff-legged avoidance posture, and offers its flank. Though the snake might seem to be similar to a big worm, its actual pattern of movement produces a very different configuration.

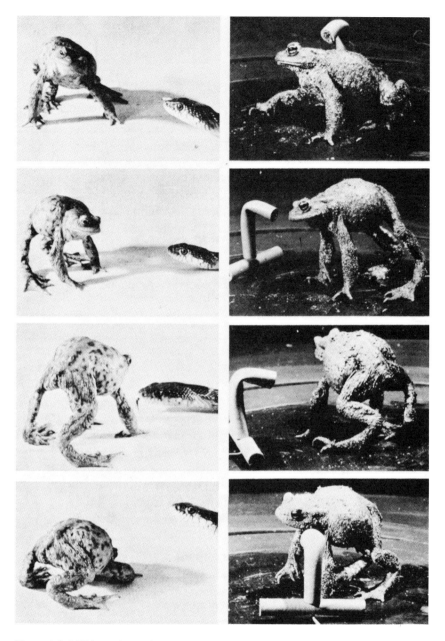

Figure 1.3 Stiff-legged avoidance postures of the toad *Bufo bufo* in response to a snake *Coluber hippocrepis* (left) and a simple dummy constructed from electric cable (right) moving around the toad horizontally. Source: Ewert and Traud 1979.

Invariants

Movement Direction

The question arises whether the ability of common toads to distinguish between wormlike and antiwormlike stripes is invariant for changes of the movement direction in the x, y coordinates. Another experimental procedure was designed for these experiments (figure 1.1B). The toad sits in a glass vessel in front of a movable belt of homogeneous white or black cloth which can be moved continuously by a motor. The movement direction of the belt within the vertical plane can be changed by a revolving frame, which is mounted on a cement post. The rectangular stimulus is attached to the white or black belt with an adhesive. The stimulus belt automatically reverses direction immediately after carrying the stimulus 15 cm within the frontal part of the toad's visual field. In this situation the toad either responds to the stimulus with orienting and approach responses or sits quietly. The number $\bar{R}_b^*$ of successive movements toward the stimulus object during a fixed time interval of t^* = 30 sec measures the releasing value of the stimulus as a prey object.

The stimulus parameters v, d, L, C, l are held constant; however, different kinds of stimuli are used: a 2.5×2.5 mm² square, a 2.5×30 mm² stripe moved as worm, and a 2.5×30 mm² stripe moved as antiworm. The stimulus movement is changed in the range $0° \leqslant m_d \leqslant 180°$.

Figure 1.4A shows the results. One major point is that the stimulus movement direction (plane orientation) does not influence the level of prey-catching activity in a statistically significant manner. However, within the vertical plane, upward movements of prey have a stronger releasing value than downward movements (Beck and Ewert 1979). The effect of movement direction becomes particularly evident when the shape of the stimulus does not fit the prey category so well. A second major result is that the ability of toads to distinguish a long stripe moved either as a worm or as an antiworm is invariant with respect to the movement direction in the x, y coordinates ($P < 0.001$, t-test) (Ewert et al. 1979a). Some toads tilted their heads slightly during pursuit of the diagonally moving wormlike object; however, this does not invalidate the conclusion that worm/antiworm discrimination is generally independent of the direction of the retinal image's motion.

In another version of the worm/antiworm experiment, the two stimuli either approached the toad while moving downward (z^+/y^- vector) or receded from the toad while moving upward (z^-/y^+ vector). In this situation toads still preferred the worm to the antiworm. Response to the receding worm was significantly greater than that to the approaching worm. Other studies have shown that "looming" is a factor in eliciting avoidance behavior (Beck and Ewert 1979). This may be a factor in decreasing the number of orientations toward an approaching worm.

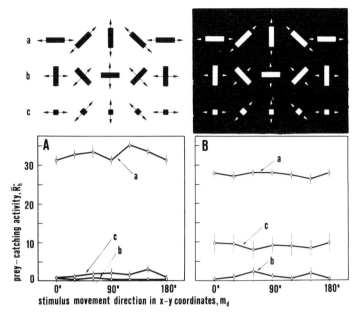

Figure 1.4 Influence of movement direction and contrast direction on configurational prey selection in common toads. Average prey-catching activity, $\bar{R}_b^*$ (orienting movements per 30 seconds) ($N = 20$) in response to different configurational stimuli (a–c) moved in different directions of the visual field x,y coordinates, to and fro (see double arrows) for $0° \leqslant m_d \leqslant 180°$. Different stimuli are used: (a) 2.5×30 mm² stripe oriented with its axis parallel to the direction of movement, (b) 2.5×30 mm² stripe oriented with its axis perpendicularly to the direction of movement, (c) 2.5×2.5 mm² square. (A) $C = +0.9$; (B) $C = -0.9$; $v = 20°$/sec; $d = 70$ mm. (Procedure of figure 1.1B.) Source: Ewert et al. 1979a.

Movement Dynamics

Is the worm/antiworm discrimination of the common toad influenced by certain movement dynamics of the stimuli? In order to investigate this question, a 6×6 mm square, a 3×30 mm "worm," and a 3×30 mm antiworm were moved linearly in the horizontal direction ($m_d = 0°$ and $180°$) at different velocities v' (mm/sec). In the investigated v' range of 2–36 mm/sec, toads preferred worms against antiworms (see figure 1.2C). Of course, this velocity invariance has been proved to be independent of the kind of prey with which they were familiar.

In another series of experiments, different kinds of movement patterns (step movements) were generated by square, triangular, and sinusoidal waves (figure 1.5, parts A–C). For all movement modes and frequencies tested, toads showed a clear preference for the worm configuration. The t-test probability that animals make a "misinterpretation"—tracking the antiworm—is $P < 0.001$. The prey-catching activity for the configurational neutral square averaged less than that obtained with the wormlike object. Surprisingly, activity

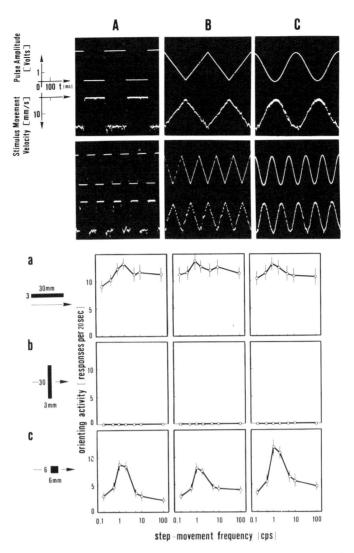

Figure 1.5 Influence of movement dynamics (step frequencies and functions, A–C) of stimuli with different configurations (a–c) on prey-catching activity $\bar{R}_t^*$ (orienting movements per 20 seconds) in the common toad. The stimulus belt could be driven continuously by an electric device consisting of a motor, a tachogenerator, and a controller. Step movements were generated by means of a function generator, which gave the reference input for the controller. Upper traces: Output voltage of the tachogenerator as an index of the movement velocity v' (mm/sec) of the stimulus. Records above: example of step movement frequencies $f = 2$ cycles per second. Records below: $f = 5$ cps. (General procedure of figure 1.1B.) Source: Borchers et al. 1979.

showed a statistically significant dependence on the step-movement frequency; 1–2 cycles per second were found to be best frequencies for all movement modes tested. Since the optimal frequency range did not change significantly in the higher or lower ranges for v' between 4 and 36 mm/sec, it can be concluded that the time pattern rather than the spatial dislocation pattern of the steps is the important feature with square stimuli (Borchers et al. 1978).

Sign of Stimulus-Background Contrast
Is the toad's ability to prefer a stripe moving as a worm over the same one moving as an antiworm invariant for changes in the sign of the stimulus-background contrast, $\pm C$? With stimulus parameters held constant as mentioned above, prey-catching activity was measured in response to a 2.5×2.5 mm^2 square, a 2.5×30 mm^2 "worm," and a 2.5×30 mm^2 "antiworm", for $0° \leqslant d_m \leqslant 180°$ and $C = -0.9$.

As figure 1.4 shows, toads also discriminate long worms from antiworms of the same length when the stimulus-background contrast is reversed. However, the stimulating effect of edge expansion in direction of movement was stronger for $C = +0.9$ than for $C = -0.9$ (see curves a and c in figure 1.4). The ability of the toad to distinguish configurational objects appears to be more selective for black stimuli moved against a white background than for white ones moved against a black background (Ewert et al. 1979a).

Background Structure
In the natural environment of the toad, prey objects very seldom move against homogeneous backgrounds. In the following experiments a computer-generated random process (Julesz 1965) was used to investigate whether the toad is able to distinguish configurational stimuli when they are moved on a structured background. Figure 1.6 shows the data of experiments in which black wormlike and antiwormlike stripes and squares of different sizes were moved against homogeneous white (A), grey (B), and structured (C–E) backgrounds. The average luminance in B–E was constant.

The results can be summarized in four major points:

• A stimulus moved against a homogeneous background was not detected by the toad when both stimulus and background had the same luminances. However, the stimulus was detected when the background was structured. The effect increased to some extent with structure width. Structure "compensates" missing luminance, which appears to be important for an animal active during twilight.

• The ability of toads to distinguish between black wormlike and an-

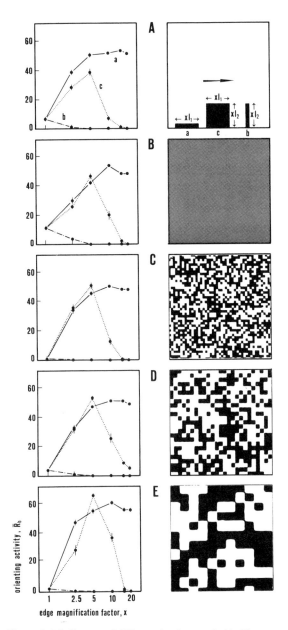

Figure 1.6 Influence of different backgrounds (A–E) on prey-catching activity $\bar{R}_b$ (turning movements per minute) ($N = 20$) of common toads in response to different black configurational stimuli (a–c) for $1 \leq x \leq 20$ and $l_1 = l_2 = 2.5$ mm. The minimal width W of the computer-generated random pattern ("Julesz pattern") of the background was (C) 1 mm, (D) 2 mm, (E) 4 mm; $v = 20°$/sec, $m_d = 0°$, $d = 70$ mm. The average luminance of the background, L_b, was (A) 62.7, (B–E) 36 cd/m². (Procedure of figure 1.1A.) Source: Ewert et al., in prep.

tiwormlike objects was not reduced by any of the structured backgrounds tested (figure 1.6, parts C–E).

• The discrimination between *small* wormlike stripes of different length appeared to increase when stimuli were moved on a structured background. (Compare values in curves a for $x = 1$ and $x = 2.5$ in parts B–E of figure 1.6). This effect increased with size of texture, as far as has been investigated for $w = 1$–4 mm. Small *white* stimuli moving against the same structured backgrounds were strongly masked.

• Square stimuli with area extensions *in* and *perpendicular to* the direction of movement around $x = 5$ had a greater efficacy when moved on structured backgrounds rather than on homogeneous ones. (Compare parts B and E of figure 1.6.) The effect increased with structure width in the investigated range, $w = 1$–4 mm (Ewert et al., unpublished).

The facilitation effect of background structure on black stimuli might be explained by spatial "flicker" effects generated across the black-white boundaries. Retinal class R3 ganglion cells, which respond best to on-off stimulation, may add excitation to the central neurons involved in pattern recognition, thus enhancing the entire information flow. Considering the different results of small black and white stimuli, we may postulate different processes linked to "white" and "dark" systems.

Nonmoving Stimuli
It is known that toads respond to moving objects of appropriate size and configuration with prey catching. However, shifting retinal images may be produced in other stimulus situations, such as the following:

• *The stimulus, a stripe of 2.5 × 30 mm², is stationary, and the toad passes it during walking.* If the stimulus is located on a homogeneous background and the stripe axis is oriented parallel to the movement direction of the toad, it may release orienting and snapping (Burghagen and Ewert, in prep.). However, toads fail to respond when the stationary stripe has an antiworm configuration in relation to the animal's own movement direction. In some cases such a stimulus is treated as an obstacle; it releases a detour movement by the toad (Ingle 1971). During the toad's own movement, stationary worms release no prey catching if they are attached to a structured background (as shown in parts C–E of figure 1.6).

• *The toad, facing a stationary stripe attached to a homogeneous background, is passively rotated in the horizontal plane by means of a modified procedure as described in figure 1.1A.* Even in this situation, the moving retinal image of a worm configuration (such as a 2.5 × 30 mm² stripe) releases prey catching whereas an antiworm configuration does not elicit any response. Generally the same results can be obtained when the toad is rotated stepwise at different accelerations in order to increase vestibular input (Burghagen and Ewert, in prep.). However, the worm stimulus does not release any response from the passively rotated toad when the stationary stimulus is attached to a structured stationary background.

The essential points of these experiments may be summarized as follows:

• Stationary prey objects located on homogeneous backgrounds elicit prey catching when appropriate moving retinal images are produced (for example, during active or passive movements of the toad).
• Potential prey objects are totally masked by a structured background as long as they do not move on it.
• A prey object is also masked if it is moved together with its background (Ewert and Härter 1969).

We may conclude from these results that the background plays an important role for the visual system of the toad in classifying moving retinal images according to the origin of movement (Burghagen and Ewert, in prep.). Obviously, this evaluation is not solely due to "reafference principles" in sensorimotor systems (von Holst and Mittelstaedt 1951).

Configurational Prey Selection by Individual Experience

As we have already seen, toads are able to discriminate between particular moving configurational prey stimuli (such as worm and antiworm configurations), and this ability obviously has an innate basis (Ewert and Burghagen 1979a). In light of the variety of natural prey objects' shapes and patterns, the question arises whether toads are able to distinguish between differently patterned and shaped prey objects on the basis of individual experience (Buytendijk 1941; Freisling 1948; Eibl-Eibesfeldt 1951).

Using the selective habituation method developed by Birukow and Meng (1955), we have demonstrated that the common toad is able to discriminate detailed patterns (for example, within the outline limits of a moving two-dimensional rectangular 5×20 mm^2 wormlike shape, as in figure 1.7C) (Ewert and Kehl 1978). The principle of the habituation method is as follows: If the orienting response of the toad is released in procedure of figure 1.1A (see figure 1.7A) in a long-term stimulus series, the number of prey-catching responses per minute (s) decreases approximately according to the formula

$$R_s = R_1 e^{-\sigma(s-1)} \text{ (turning responses/min)} \tag{2}$$

(see Ba in figure 1.7). That means the average effective dislocation, $\overline{D}$,

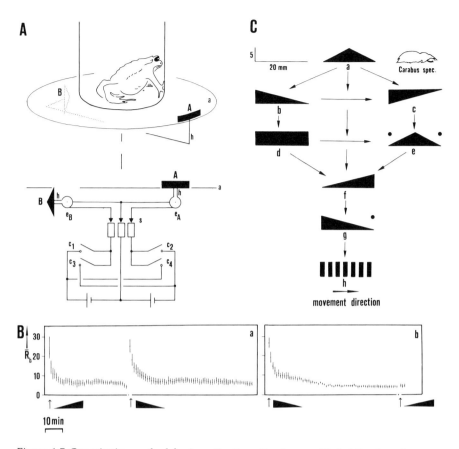

Figure 1.7 Quantitative method for investigating stimulus specific habituation of prey-catching orienting behavior in the common toad. (A) Procedure for automatic change of two different stimuli A and B, here shown for a black rectangular stripe and a black isosceles triangle. a: arena base. c_{1-4}: electrical contacts. e_A, e_B: electric devices for presentation of stimulus A or B. h: stimulus holder. s: sliding contacts. (B) Selection test between image and mirror image of a 5 mm × 20 mm right triangle, inverted about the vertical plane; $C = +0.9$, $v = 20°$/sec, $m_d = 0°$, $d = 70$ mm. Ordinate: Successive number of orienting responses per min, $\bar{R}_b$ (turning movements per minute) ($N = 15$); abscissa: habituation time t (minutes). (Ba) Habituation of the response first to the triangle with the small cathetus leading in the direction of movement and then, immediately afterwards, test of response (see vertical arrow) to the same triangle but moved with its tip leading in the direction of movement. (Bb) Reverse succession, in which a new group of 15 animals was tested. (C) "Hierarchical ordering" of features governing prey selection. Source: Ewert and Kehl 1978.

increases until the toad loses the stimulus object out of the binocular field of vision. Then, according to equation 1,

$$\overline{D} = k \cdot c(R_1 e^{-\sigma(s-1)})^{-1} \text{ (degrees of visual angle),} \qquad (3)$$

where R_s is the number of successive 1-minute intervals, R_1 refers to the number of turns made during the first minute, and c (= const) is the angular velocity of the stimulus related to the center of rotation. The criterion of habituation is that the orienting activity R_s has a value close to zero, i.e., the animal responds to the dummy in a given 1-minute interval fewer than five times. The following conclusions are allowed:

• If the toad after habituation of the orienting response to a prey dummy with the pattern A immediately shows orienting activity to one with pattern B (figure 1.7Ba), we can conclude that the animal discriminated both patterns if all other stimulus parameters were held at constant values.

• If in the reverse type of experiment the stimulus A releases no prey-catching responses after habituation of the response first to B, we can conclude that toads *prefer* pattern B as prey over pattern A (figure 1.7Bb).

• If in both habituating series the second dummy "renews" the toad's prey-catching activity by equal amounts, we can conclude that both dummies are distinguished by the toad and that they are almost equally attractive as prey.

The fact that the third outcome was often observed indicates that the toads "learned" to discriminate particular objects that were equally salient as prey when initially seen.

On the basis of paired tests of prey dummies with different shapes, a hierarchical ordering of significant features can be constructed (parts a–h of figure 1.7C). The main features governing prey selection during experience seem to be area components, tips leading the stimulus in the direction of movement, isolated dots, and striped patterns. These features are components of natural prey objects (Ewert and Kehl 1978). Patterns shown in figure 1.7C are not spontaneously preferred or neglected by the toad. Their releasing values, measured by the initial response rate (compare parts a and b of figure 1.7B) are almost the same. Thus, we need the habituation method in order to bring out the toad's discrimination ability.

Central Filtering of Configurational Cues

The activity of neurons from different levels of the visual system can be recorded in response to different configurational moving stimuli as described above. During the experiment the animal is immobilized after injection of succinyl choline and centered with one eye in a diffusely illuminated hemisphere (figure 1.8). The visual stimulus can be at-

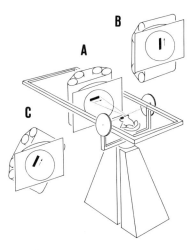

Figure 1.8 Perimeterlike device for stimulus movement and stimulus position in the x,y,z coordinates of the visual field of an immobilized toad. (A) Horizontal ($m_d = 0°$ or 180°), (B) vertical ($m_d = 90°$ or 270°), (C) oblique (e.g., $m_d = 135°$). As the stimulus traverses parts of the visual field, spike activity of single neurons from different levels of the central visual system can be recorded extracellularly. Source: Ewert et al. 1979b.

tached to a rotating belt, the visible area of which is restricted by a 40° diameter aperture placed in front of the toad. The excitatory receptive field (defined below) of the neuron under investigation has to be centered within this aperture. The distance between the toad's eye and the stimulus is held constant at 250 mm during all experiments. The belt consists of homogeneous white cloth which can be moved continuously in opposite directions and may also be rotated in the x, y plane about the center of the aperture (figure 1.8, parts A–C). For the quantitative recording studies a standard electronically controlled stimulation program is used, with a stimulus traverse of the center of the receptive field of the neuron under investigation, a recovery pause of 60 seconds, a reverse traverse of the receptive field, and a recovery pause of 60 seconds.

For the investigation of neuronal responses to configurational cues of a moving stimulus, the following parameters were held constant:

visual angular velocity $v = 7.6°$ (or 18°) per second,

movement direction $m_d = 0°$ (horizontally),

stimulus distance $d = 250$ mm,

background luminance $L_b = 40$ cd/m²,

stimulus background contrast $C = 0.9$, and

stimulus edge length $l_1 = l_2 = 2°$.

The stimulus edge magnification factor was changed in the range $1 \leqslant x \leqslant 10$.

Retinal Ganglion Cells

General Functional Properties

Three different types of ganglion cells project their axons to the surface layers of the optic tectum: classes 2, 3, and 4, as first described in the frog by Lettvin et al. (1959) (for summary see Grüsser and Grüsser-Cornehls 1976). In order to distinguish these retinal classes from neurons of tectum and pretectum, I have called them classes R2, R3, and R4. The region of the visual field from which a ganglion cell can be activated by a moving stimulus is called the *excitatory receptive field*, or ERF (figure 1.10, top left). The ERF shape is oval, on the average approximately radially symmetric (figure 1.9A). It is surrounded by an inhibitory receptive field (IRF). A contrasting stimulus moving in the IRF inhibits the activation of the ganglion cell in response to a second stimulus moving simultaneously through the ERF. The average diameter of the ERF increases from class R2 to R4: $ERF_2 \approx 4°$, $ERF_3 \approx 8°$, and $ERF_4 = 12–16°$. The strength of the IRF decreases from class R2 to classes R3 and R4. With respect to *on* and *off* areas, the receptive-field organization is different from that obtained in *on*- and *off*-center neurons in the mammalian retina (Kuffler 1953). Although class R2 neurons may show weak *on* response to diffuse brightening of the entire receptive field, R3 neurons respond with sharp bursts at *on* and *off*, and R4 neurons only with phasic-tonic *off* responses. Whereas class R3 and R4 neurons respond to moving objects, R2 neurons are termed movement-specific (Grüsser et al. 1967), since motion is always required for a strong unit response. Also, the amount of neuronal adaptation after a traverse of the receptive field by a stimulus is greatest for neurons of class R2.

Responses to Configurational Moving Stimuli

As an example, the activity of retinal R2 neurons in response to different configurational moving stimuli shall be described in detail (figure 1.10, R2). If the edge, xl_1 ($xl_2 = 2°$) of a stripe is extended in the direction of movement by steps of $x > 1$, the discharge rate is not remarkably changed. However, for class R2 neurons, when the edge xl_2 ($xl_1 = 2°$) of a stripe is elongated perpendicular to the direction of movement, there is an increase in the discharge frequency as the stripe length increases in the range of $1 \leqslant x \leqslant 2$. The frequency reaches a maximum when the length of the stripe equals the diameter of ERF_2 ($xl_2 = 4°$), and then decreases with further increase in stripe length owing to encroachment upon the inhibitory surround (IRF). Increasing the edge length $xl_{1,2}$ of a square has an effect on the discharge frequency similar to that of extending the xl_2 of an antiwormlike stripe.

Corresponding experiments with the two other neuronal classes show generally similar results (figure 1.10, R3,R4), with the difference that maximal activity to antiwormlike stripes and squares is seen in the range of their ERF sizes, $xl_2 = 8°$ (class R3) and $xl_2 = 16°$ (class R4) (Ewert and Hock 1972). In summary, retinal ganglion cells are sensitive to the

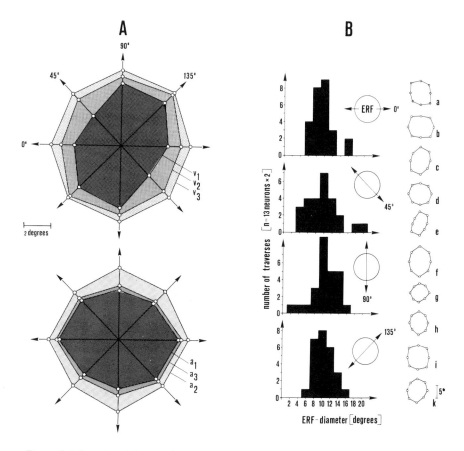

Figure 1.9 Functional shapes of excitatory receptive fields of retinal class R3 ganglion cells in the common toad. (A) Average functional ERF diameters ($N = 13$) from measurements with black test stimuli traversing the ERF centers on white background in different directions, $m_d = 0°$ and $180°$, $45°$ and $225°$, $90°$ and $270°$, $135°$ and $315°$. Above: test stimuli of constant size and configuration ($4° \times 4°$ black square) and different angular velocities: $v_1 = 1$, $v_2 = 8$, $v_3 = 20$ degrees per second. Below: test stimuli of constant angular velocity ($v = 7.6°/\text{sec}$) and different configurations ($a_1 = 2° \times 2°$, $a_2 = 2° \times 6°$ (movement in direction of long axis), $a_3 = 2° \times 6°$ (long axis oriented perpendicularly to direction of movement). (B) Left: sample distribution of ERF diameters of class R3 neurons ($N = 13$), calculated from duration of neuronal impulse trains elicited when a $4° \times 4°$ black square traversed the ERFs in the x,y plane at $v_2 = 8$ and $v_3 = 20$ degrees per second. Right: calculated ERF shapes of nine individual R3 neurons a–i; k represents the average for $N = 13$, Stimulus-background contrast was $C = +0.9$; $L_b = 40$ cd/m²; $d = 250$ mm. Source: Ewert et al. 1979b.

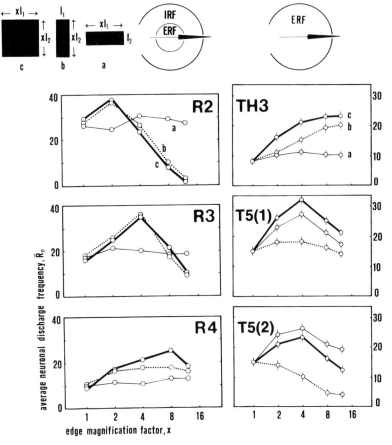

Figure 1.10 Average activity $\bar{R}_n$ (impulses per second) of single neurons of different levels of the common toad's visual system in response to different configurational moving stimuli, a–c, for $1 \leqslant x \leqslant 10$ and $l_1 = l_2 = 2°$; $C = +0.9$, $L_b = 40$ cd/m², $v = 7.6°$/sec, $m_d = 0°$, $d = 250$ mm. R2, R3, R4 ($N = 10$ each): responses of three different classes of retinal ganglion cells recorded from their fiber terminals in the optic tectum. TH3 ($N = 20$): movement-sensitive neurons recorded extracellularly from thalamic-pretectal region. T5(1), T5(2) ($N = 20$ each): two classes of movement specific neurons recorded extracellularly from the central layers of the optic tectum. Sources: Ewert and Hock 1972; Ewert and von Wietersheim 1974a.

stimulus area and, with regard to the configuration, mainly to lengthening of the stimulus perpendicular (xl_2) to the direction of movement.

Thalamic-Pretectal Neurons

The caudal dorsal thalamus and the pretectal region are anatomically distinct areas (for review see Fite and Scalia 1976). As a result of their close proximity, however, it was not possible to discriminate between these areas during physiological experiments. Therefore, this region is somewhat loosely called the thalamic-pretectal (TP) region.

Various different types of neurons have been recorded in the TP region, for example classes TH1–10 (Ewert 1971). Since toads are able to recognize configurational prey objects with one eye, the following quantitative neurophysiological investigations are performed with class TH3 neurons (TH stands for thalamus), which are monocularly driven. (These neurons were first described [Ewert 1971] in the toad *Bufo americanus* and termed type 3.) The average horizontal diameter of the ERF is about 46°. These neurons are lateral to the dorsal posterior thalamus and near the pretectal region (posterolateral cell group). The receptive fields of these neurons are topographically organized with respect to the retina as described in Ewert et al. 1974. Class TH3 neurons are sensitive to moving visual stimuli and show brief *off* responses to sudden darkening of the entire visual field.

The activity of class TH3 neurons was tested in response to various configurational moving stimuli (figure 1.10, TH3). If the edge xl_1 ($xl_2 = 2°$) of a stripe is extended stepwise by $x > 2$ in the direction of movement, the discharge rate shows no statistically significant change. However, there is an increase in the neuronal activity when the edge length xl_2 ($xl_1 = 2°$) of an antiwormlike stripe is extended perpendicular to the direction of movement. These effects are even stronger with squares of increasing size $xl_{1,2}$ (Ewert and von Wietersheim 1974a).

To summarize: Class TH3 neurons are sensitive to the entire area of a moving stimulus and, with regard to the configuration, mainly to elongation perpendicular (xl_2) to the direction of movement.

Neurons of the Optic Tectum

Many different types of neurons have been identified in the optic tectum (Grüsser and Grüsser-Cornehls 1976). Among those, the T5 neurons (T stands for optic tectum) are monocularly driven. They have relatively small excitatory receptive fields with an average horizontal diameter of about 26°. These neurons can be recorded extracellularly from the central layers of the optic tectum. They are topographically organized. With regard to the rostro-caudal axis, the retinal projection map in the optic tectum is a mirror image of the retinal projection to the TP region (Lázár and Székely 1969; Lázár 1971; Ewert et al. 1974; Scalia

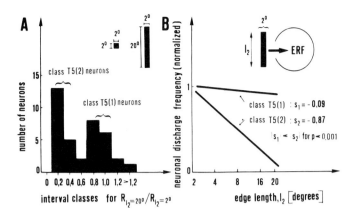

Figure 1.11 (A) Statistical analysis of T5 neurons recorded from the optic tectum in common toads. The criterion for this distinction is the response $\bar{R}_n$ to the configurational parameters of two black stimulus patterns moved at $v = 7.6°/\text{sec}$ and $m_d = 0°$ through the center of the ERF on a white background; $C = +0.9$, $d = 270$ mm. Quantitative criterion is the quotient Q of the discharge rates to a $2° \times 2°$ square and a $2° \times 20°$ stripe oriented with its axis perpendicular to the direction of movement. The histogram for the tectal T5 neurons clearly shows two maxima according to two neuronal populations, called T5(1) and T5(2) neurons. (B) The dependence of the neuronal discharge rate on the logarithm of the stripe length, l_2, can be described by the equation $\bar{R}_n = -s \log l_2 + k$ (impulses per second). The values of $\bar{R}_n$ are normalized. The difference in the amounts of s for T5(1) and T5(2) neurons is statistically significant at $P < 0.001$ by the t-test. Source: von Wietersheim and Ewert 1978.

and Fite 1974). T5 neurons in *B. bufo* show neither *on* nor *off* responses to sudden changes of diffuse light in the entire visual field.

The activity of T5 neurons was tested in response to the different configurational moving stimuli. In a sample of 38 single T5 neurons, two different response types—T5(1) and T5(2)—can be distinguished when configurational stimuli traverse the ERF centers in the horizontal direction (figure 1.11). In T5(1) neurons, stepwise elongation of the edge xl_1 ($xl_2 = 2°$) of a stripe in horizontal direction of movement leads to an increase of the discharge rate in the range of $1 \leqslant x \leqslant 4$ (see T5(1) in figure 1.10). This effect is even stronger for squares of increasing size $xl_{1,2}$. However, elongation xl_2 ($xl_1 = 2°$) of a stripe perpendicular to the movement direction does not cause a significant change in the neuronal activity. T5(2) neurons are different from T5(1) neurons in the property that stimulus extensions perpendicular to the direction of movement—xl_2—strongly reduce the discharge rate (see T5(2) in figure 1.10). For $x = 8$ their response to the square is weaker than that to the wormlike stripe, but greater than that to the antiwormlike stripe ($P < 0.01$, sign test).

To summarize: T5(1) neurons are sensitive to the entire area of a moving stimulus and, with regard to configurational cues, mainly to stimulus expansion in the horizontal direction of movement (xl_1). The response of T5(2) neurons is determined by configurational parameters

xl_1 and xl_2 in a stimulating (xl_1) or an inhibiting (xl_2) manner (Ewert and von Wietersheim 1974a).

Recently, neurons with response characteristic of TH3 were also recorded from the optic tectum (Ewert et al. 1979c; see figure 1.13D). At present we are not sure whether we recorded from terminals of TH3 fibers projecting to the optic tectum or from tectal cells.

Correlation Between Behavioral and Neurophysiological Results

Results from quantitative behavior experiments in the common toad show clearly that configurational parameters xl_1 and xl_2 of a moving stimulus play an important role in the decision-making process for the classification of objects in the prey category. These data from single neurons of different integration levels of the visual system provide some insight into how these configurational parameters are processed. By means of statistical methods we can ask how sharply these neurons may distinguish the parameters xl_1 and xl_2. Are any of these neurons—by their selective response properties—suitable for classification of prey stimuli?

Form Contrast
A quantitative measure of discrimination values for selection between wormlike (W) and antiwormlike (A) moving objects of equivalent sizes is given by the form-contrast formula (Ewert et al. 1978)

$$D_{W,A} = \frac{\overline{R}_W - \overline{R}_A}{\overline{R}_W + \overline{R}_A},$$

where $\overline{R}_W$ is the average response to a wormlike stripe and $\overline{R}_A$ the average response to the same stripe moved in an antiwormlike manner. In figure 1.12 the values of $D_{W,A}$ are plotted against the stimulus parameters xl_1 and xl_2. The values $D_{W,A}$ can be expected to be between $+1$ and -1. In the case of wormlike objects being preferable to the same ones moved in antiwormlike fashion, $D_{W,A}$ is positive. For $D_{WmA} = \pm 1$ both stimuli are distinguished by yes/no decision. There is no discrimination if $D_{W,A} = 0$.

Discrimination Analyses
The activities, $\overline{R}$, in response to wormlike (W) and antiwormlike (A) stimuli were compared for different behaviors and for each neuronal class. This procedure gives hints for answering the question of how similar transformations of the stimulus parameters xl_1 and xl_2 are performed by the corresponding system

$$\overline{R} = f_1(xl_1) \text{ compared with } \overline{R} = f_2(xl_2), \tag{4}$$

and thus discriminates between different configurational moving stimuli (Borchers and Ewert 1979).

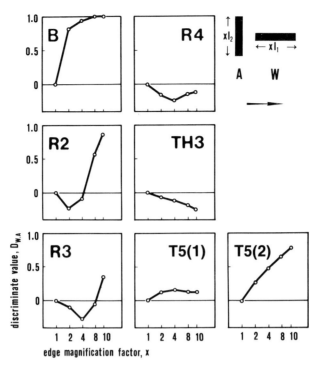

Figure 1.12 Discriminate values $D_{W,A} = (\bar{R}_W - \bar{R}_A)/(\bar{R}_W - \bar{R}_A)$ for selection between "worms" (W) and antiworms (A) of different length xl_1 or xl_2 respectively, calculated for the common toad's prey-catching behavior and for neurons belonging to different classes recorded from different levels of the visual system: R2, R3, and R4 retinal ganglion cells; TH3 neurons from the thalamic-pretectal region; and T5(1), T5(2) neurons from the central layers of the optic tectum. Values were measured for $1 \leqslant x \leqslant 10$ and $l_1 = l_2 = 2.5$ mm (B) or $2°$ (all others). Source: Ewert et al. 1978.

The experimental data expressed as curves can be compared by means of the Pearson waveform correlation (Sachs 1976). Continuity is approximated within each curve describing the stimulus-response relationships. Data were processed by means of a computer which gave directly the values $r_{W,A}$ of the Pearson waveform correlation. Data for behavioral experiments were from Ewert 1972 (compare Ewert 1976); data for retinal ganglion cells R2, R3, R4 were from Ewert and Hock 1972, and data for thalamic-pretectal neurons (class TH3) and tectal neurons (classes T5(1) and T5(2)) from von Wietersheim and Ewert 1978. Table 1.1 shows the correlation coefficients $r_{W,A}$.

Waveform Correlation
In this test the stimulus-response relationships measured for prey-catching behavior,

$$\bar{R}_b = f_b(xl_1), \qquad \bar{R}_b = f_b^*(xl_2), \tag{5}$$

were compared with the corresponding neurophysiological stimulus-response relationships

$$\bar{R}_n = f_n(xl_1), \qquad \bar{R}_n = f_n^*(xl_2) \tag{6}$$

for n = class R2 ($N = 10$ neurons), R3 ($N = 10$), R4 ($N = 10$), TH3 ($N = 21$), T5(1) ($N = 20$), or T5(2) ($N = 18$). The question of correlation between behavioral and neurophysiological response activities has to be analyzed for both parameters xl_1 and xl_2. Thus, the correlation coefficient is measured pairwise for stimuli with worm configuration (r_W) and antiworm configuration (r_A). In the case of perfect correlation, a pair $\{r_W; r_A\} = \{1; 1\}$ should be expected. The result is shown in table 1.2.

Conclusions

We can conclude from these analyses that the antiworm configuration is most decisive for determining the response characteristic of a neuron in the present context. Among retinal ganglion cells, the class R2 neurons are most sensitive to moving configurational stimuli (table 1.1). In visual projection fields beyond the retinal level, neuronal populations exist showing different kinds of sensitivity in response to wormlike and antiwormlike stimuli (tables 1.1 and 1.2). In class T5(1) the correlation coefficient for comparison between behavioral and neurophysiological data is positive for "worms" ($r_W = 0.8$), but there is no correlation for "antiworms" ($r_A = 0$). In class TH3 neurons the correlation coefficient for antiworms is negative ($r_A = -0.9$). Only class T5(2) neurons exhibit selective response to moving configurational stimuli (table 1.1). Relatively high values of $\{r_W; r_A\}$ for positive correlations with the prey-catching activity are found; that is, the response spectrum of T5(2) neurons reflects approximately the toad's discrimination among configurations of moving stimuli (figure 1.12). No neurons were found showing selective response to a prey object of only one particular configuration. Although the responsiveness of a single T5(2) neuron in this respect (compare T5(2) and B in figure 1.12) is not as selective as is seen in the prey-catching behavior, we may presume that the T5(2) neurons are participating in a system for the recognition of prey.

Table 1.1
Waveform correlation coefficient $r_{W,A}$ for comparison of responses to stripes moved in wormlike (W) and antiwormlike (A) manner, measured for prey-catching orienting behavior and for activity of neurons from retina (R), thalamic-pretectal region (TH), and optic tectum (T).

System						
Prey Catching	R2	R3	R4	TH3	T5(1)	T5(2)
−0.9	−0.8	0.6	0.5	0.8	0.6	−0.6

Source: Borchers and Ewert 1979.

Table 1.2
Pairs of waveform correlation coefficients $\{r_W, r_A\}$ for comparison between behavioral and neurophysiological responses to wormlike (W) and antiwormlike (A) moving stripes.

Neuronal Class	Correlation-Coefficient Pair	
	r_W	r_A
R2	0.5	0.6
R3	0.2	−0.7
R4	0.6	−0.9
TH3	0.9	−0.9
T5(1)	0.8	0.0
T5(2)	0.7	0.9

Source: Borchers and Ewert 1979.

The Question of Invariants

We now must ask whether the configurational sensitivity of these neurons is invariant with regard to changes of other stimulus parameters. Thus far, recording experiments with T5(2) neurons indicate that the worm/antiworm response differences are not basically altered during changes of movement velocity tested between 5° and 18° per second, changes in the stimulus-background relationship from black-on-white to white-on-black, variation of stimulus distance tested from 5 to 25 cm, or change from a homogeneous background to one with random visual noise (such as in figure 1.6). However, the effect of motion direction on responses of T5 neurons is a more complicated matter. Although the configurational selectivity of T5(2) neurons is largely independent of the stimulus direction in the x, y plane, the T5(1) neurons do show varying degrees of sensitivity to stimulus direction, as shown in parts B and C of figure 1.13. The relationship of these variations to behavioral decisions is not yet clear. However, taking the average $D_{W,A}$ values from 35 tectal T5 neurons, the greatest discrimination between worm and antiworm configurations is seen when the stimuli move along the horizontal meridian of the eye. This preference might explain the head-tilting behavior of toads in situations where prey stimuli traverse the visual field of oblique directions.

A Hypothesis Concerning the Decision-Making Process

In the present context, configurational pattern recognition can be defined as the classification of two-dimensional space- and time-dependent distributions of brightness from the environment into innate classes of functional significance (in the present study, prey and nonprey). This operation proceeds in two steps: extraction of behaviorally relevant gestalt features from the pattern to be perceived by neuronal "gestalt coding" systems, such as class T5(1) and TH3

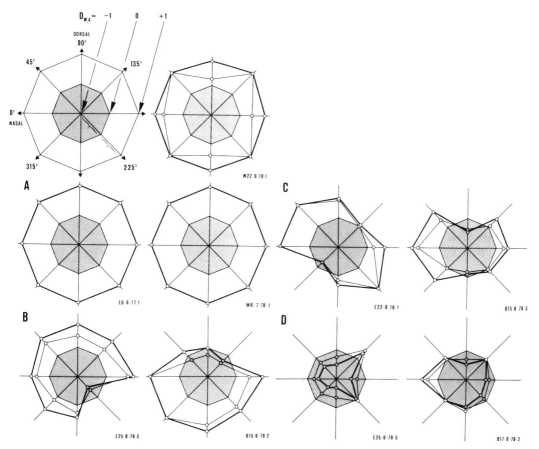

Figure 1.13 Influence of movement direction on configurational sensitivity of tectal T5 neurons in common toads. Top left: "standard field" for plotting functional properties of T5 neurons. ERFs were located in different regions of the visual field. The "standard fields" shown here correspond to the left eye (electrode position in the right tectum). Wormlike and antiwormlike stripes of $2° \times 8°$ (E, B) or $2° \times 20°$ (W) traversed the ERF center in different directions of the x,y coordinates for $0° \leq m_d \leq 360°$; $v = 7.6°/sec$, C = +0.9. The discriminate values $D_{W,A}$ (thin lines) were calculated from the neuronal discharge rates $R_{W,A} = n/t$, and $D^*_{W,A}$ (thick lines) from the product $nR_{W,A} = R^*_{W,A}$, which is related to the power of the output (n: number of spikes during stimulus traverse; t: time from first to last spike recorded). If the worm configuration is preferred, $D_{W,A}$ becomes positive and is plotted outside the "zero circle"; if the antiworm configuration is preferred, $D_{W,A}$ becomes negative and its values are plotted within the "zero circle." The origin of the coordinate system corresponds to $D_{W,A} = -1$. (A) Measurements of three single tectal T5(2) neurons. (B–C) Various subtypes of tectal T5(1) neurons, two examples each. Source: Ewert et al. 1979c.

neurons; and separation of the resulting feature vectors acquired by means of discriminating functions based on subtractive interaction between coding systems and weighted neuronal threshold value operations. The intensity of the corresponding behavior reaction (number of turning movements per time interval) would then depend on the number of threshold crossings.

The selective responsiveness of T5(2) gestalt decoding neurons might be explained by inhibitory inputs from class TH3 neurons and excitatory ones from T5(1) neurons (figure 1.17). That means that if the expansion of a visual object perpendicular to the direction of movement exceeds a certain amount, it is "categorized" as unsuitable for food and the prey-catching trigger system in the optic tectum is inhibited by TP neurons. In case of an enemy object, additional neuronal populations in the TP region may presumably be stimulated via excitatory pathways from the tectum and—together with activity in the tectum—trigger avoidance movements. These kinds of pretectal-tectal interactions (see figure 1.15A) are revealed by physiological evidence. First, after thalamic-pretectal lesions in toads (*Bufo bufo*) the T5 neurons from the central layers of the optic tectum recorded lose their configurational selectivity, or even their sensitivity (figure 1.14B). This might suggest that T5(1) neurons, as well, receive some inhibitory thalamic-pretectal input (see figure 1.17A) (Ewert and von Wietersheim 1974b). The immediate effect of thalamic-pretectal lesions (Ewert 1967, 1968) is to abolish the ability of toads and frogs to behaviorally distinguish prey from enemies (figure 1.15).

Anything moving will be categorized as prey by TP-lesioned toads for some time after the lesion. (Ingle [1980] reports, however, that frogs recover the tendency to avoid large objects by two weeks after a complete ablation of the TP region.) Second, after TP lesions in frogs (*Rana pipiens*), receptive fields of T5 neurons are enlarged and the neurons lose their habituation properties to repeated stimulation (Ingle 1973).

Further indications of thalamo-tectal interactions are given by experiments where a stimulating electrode is implanted in the TP region and a recording electrode is positioned in the central layers of the optic tectum (figure 1.16Aa), from which the response of a T5(2) neuron to a moving visual object is recorded (Ewert et al. 1974). The response to the same visual stimulus is strongly decreased after electrical stimulation of the thalamic-pretectal region (figure 1.16Ab), but recovers within about 30 seconds (figure 1.16Ac). The assumption that neurons of the TP area send axons directly to tectum is supported by anatomical studies (Trachtenberg and Ingle 1974; Wilczynski and Northcutt 1977; Székely and Lázár 1976). In the reverse type of experiment—when stimulation and recording positions are interchanged (figure 1.16B)—excitatory connections from the optic tectum to the TP region can be demonstrated. For a similar conclusion about frogs see Brown and Ingle 1973.

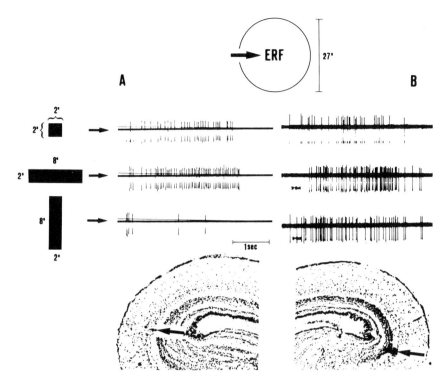

Figure 1.14 (A) Records of a T5(2) neuron from the central layers of toad optic tectum in response to three different configurational stimuli: square, "worm," and antiworm; $v =$ 7.6°/sec, $C = +0.9$. (B) Records of T5(1,2) neurons from same layer of optic tectum after ipsilateral lesions of thalamic-pretectal region. Bottom: staining of iron deposit (arrow) of the steel microelectrode after passing anodal DC current. Source: Ewert 1978.

Ewert and von Seelen (1974) have discussed a computer model of two-dimensional nerve nets. In this model neuronal structures from the retina, optic tectum, and thalamus-pretectum are associated with the experimentally determined functions of the system (see parts B and C of figure 1.17). The central point of the conceived model is a combined space-time filtering in neuronal networks and a weighted two-dimensional subtraction of stimulus distributions. The different evaluation of the extension of a pattern in the direction of movement (xl_1) and perpendicular to it (xl_2) occurs despite symmetrical coupling and is derived by an asymmetry in the time domain. The feature-characterizing patterns emerge quantitatively as the difference of two filter operations (space and time domain). Thus, the gestalt evaluation is here largely independent of the stimulus movement direction in the x, y plane and of the shape of the receptive fields (Ewert et al. 1979c). The principle of two-dimensional subtraction of patterns is simple and efficient, and may also exist as a basic principle in other systems. It seems likely that the fundamental problems of feature extraction and decision formation can be solved in the present relatively simple sys-

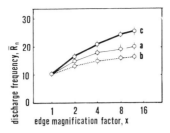

Figure 1.15 Influence of configurational stimulus parameters on average activity $\bar{R}_n$ of tectal T5(1,2) neurons ($N = 20$) after ipsilateral thalamic-pretectal lesions. Stimuli were tested as described in figures 1.2 and 1.9. Source: Ewert and von Wietersheim 1974b.

tem. This account, of course, represents only an initial crude simplification of what actually occurs in the neuronal networks. Nonlinear models are more likely to be adequate representations of reality; these are being modeled by Arbib (1981) and Lara et al. (1981).

Unsolved Problems

Wiring Patterns
We expect from our recent studies that T5(1) as well as T5(2) neurons are driven by thalamic-pretectal inhibitory inputs (Ewert 1980). These inputs are stronger in T5(2) than in T5(1), and they can be modulated, a phenomenon predominantly occurring in T5(1) neurons (Ewert et al. 1979c). We cannot decide whether T5(2) neurons are really derived from T5(1) (compare models B and C in figure 1.17). Of course, T5(1) neurons may participate in various other detection problems besides the analysis of prey and predator.

Rigidity of Wiring
The behaviorally meaningful worm/antiworm discrimination appears to be based on fixed wiring principles corresponding to an innate releasing mechanism. Sperry (1944) showed in frogs that after cutting of the optic nerve the animal is at first blind in the corresponding eye. After 180 days the retinotopic projection will be repaired according to genetically predetermined principles: The axon of a ganglion cell of a retinal area grows and makes synapses with one of a matched set of neurons in the optic tectum. Experiments using the worm/antiworm discrimination should be carried out to answer the question whether only the retinal map or also the connection patterns to thalamic-pretectal and tectal neurons are restored. Stimulation tests at different stages in regeneration will determine in which temporal succession pretectum and optic tectum are reinnervated.

Functional Plasticity
After unilateral TP lesions, the prey-catching behavior is disinhibited toward visual objects moving in the contralateral visual field. The size

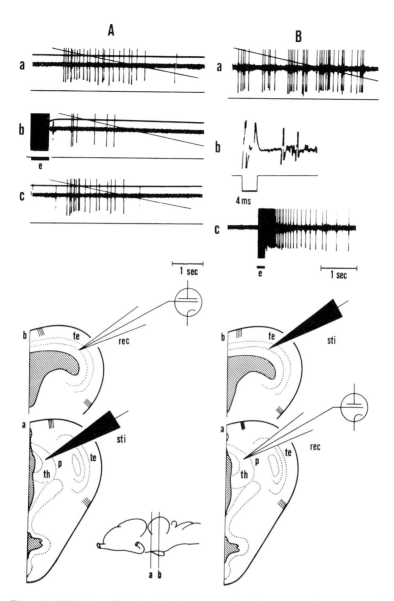

Figure 1.16 Evidence for physiological connections between optic tectum and thalamus-pretectum. (Aa) Response of a tectal "small-field unit" to a black 4° square moved at 7.6°/sec against white background. (Ab) Three minutes later: weak response to the same visual stimulus after previous electrical (*e*) point stimulation in TP region with a train of negative square-wave pulses of 50 cps, 5 msec pulse duration, and intensity of 30 μA. (Ac) Recovery of the visual response 30 seconds later. (Ba) Response of a thalamic-pretectal "large-field unit" (ERF ≈ 90°) to a 8° moving visual stimulus. (Bb) Activation of same unit by electrical point stimulation of the optic tectum with a pulse train. (Bc) Response to a single electrical square-wave pulse (latency ≈ 4 msec). Source: Ewert et al. 1974.

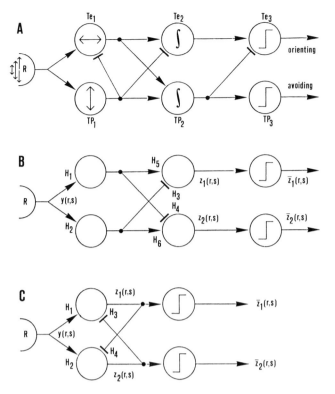

Figure 1.17 Models for configurational prey recognition in the common toad. (A) Model constructed on basis of experimental results. Arrows denote excitatory and dashed lines denote inhibitory interactions. The three vertical double arrows in R symbolize the main stimulus transformations of the three retinal ganglion cell classes on the basis of their different excitatory receptive field sizes. The horizontal double arrow in Te_1 symbolizes the configurational sensitivity of class T5(1) neurons; the vertical double arrow in TP_1 the configurational sensitivity of class TH3 neurons (shown for $m_d = 0°$ or $180°$). TP_2 (hypothetical thalamic-pretectal neurons) and Te_2 (class T5(2) neurons) may receive converging inputs from TE_1 and TP_1 and trigger prey-catching-orienting or enemy-avoiding movements via other neurons (TE_3, TP_3) after corresponding threshold operations. (B,C) Two system-theoretical models; $y(r,s,t)$ is the space- (r,s) and time- (t) dependent input signal, $z(r,s,t)$ is the output signal, and $H(r,s,t)$ is the coupling function of the neurons. Source: Ewert and von Seelen 1974.

of the "disinhibited" visual-field region increases, within limits, with the size of the lesion. The effect lasts after small lesions only hours to days, and after large lesions up to several weeks (Ewert 1968; Ingle 1980). However, in any case the "normalization" of the behavior—when quantitatively measured in response to configurational stimuli—is not the same as in the intact animal; it is only a rough "repair." Although fast functional recovery cannot be explained by regeneration (e.g., sprouting of axons), at least two explanations are possible: that other pathways from neighboring intact TP regions to the optic tectum become facilitated and that changes in intrinsic tectal-mechanism neuron thresholds restore some selective prey response. It is known that application of curare to the optic tectum also disinhibits prey catching (Stevens 1973; Ewert et al. 1974). Experiments must be designed to answer the question whether curare disinhibits thalamo-tectal connections in the tectum or (also) modifies intrinsic tectal circuits.

Habituation

The phenomenon of stimulus-specific habituation cannot be explained by the "two-stage filter" model. It requires storage processes in particular neuronal networks. The response preference of certain stimulus configurations, as shown in figure 1.7, cannot simply be explained by known retinal or tectal properties. Recording experiments must be performed from T5 and TH3 neurons and thalamic "memory cells" (Ewert 1971) in which the parameters from behavioral studies are precisely adapted.

Learning

There have been many experiments showing that toads are able to classify stimuli on the basis of experience (Schneider 1954; Ewert 1968; Brzoska and Schneider 1978). For example, toads associate the odor of mealworm excrement with feeding. After some kind of self-training, large objects—which are normally associated with threat—are classified as prey if presented simultaneously with the familiar odor. This effect must be linked to telencephalic mechanisms sensitive to olfactory input. There are various pathways by which telencephalic structures can influence the optic tectum indirectly via posterior thalamic or hypothalamic nuclei (Northcutt and Kicliter 1980). Recent recording studies from T5 neurons indicate that at least some T5(1) neurons may change their configurational selectivity over time (Ewert et al. 1979c). Recording experiments with TH3 and T5 neurons in response to configurational stimuli should be carried out in the presence of known and unknown prey odor or during electrical stimulation of the telencephalon or the TP region.

Motivation

The toad responds even to optimal prey objects only when it is in corresponding motivation, and that depends upon a variety of factors, such as season, time of the day, and hunger. Many T5 neurons display relatively long response latencies, or their activity appears to be temporarily inhibited. From our current observations it appears that the general response level, at least of T5(2) neurons, depends upon the factors mentioned above (Ewert 1980). An experiment should be designed in which the activity of T5(2) neurons in response to visual stimuli is recorded during short-term changes of the hunger state. For example, it may be possible to reduce blood glucose in the paralyzed toad by injecting adequate hormones.

Command Systems

The decision-making process "prey or nonprey" precedes the motor command for the orienting turn of the prey-catching sequence. Little is known of how such decisions are made and transformed into efferent commands. Most recently, recording experiments with freely moving toads have suggested that there are neurons whose activity precedes and predicts motor responses (Borchers 1979).

Electrical brain-stimulation experiments with chronic electrodes suggest a thalamo-pretectal avoidance system and a tectal prey-catching system (Ewert 1967, 1968). This is also in accordance with lesion and recording studies. However, since periventricular layers in posterior dorsal thalamus appear to show some physiological and morphological "continuity" with the periventricular layers of the optic tectum, the question arises whether an avoidance system is also present in the optic tectum. On the other hand, mechanisms for feeding are not restricted to the optic tectum: Snapping to tactile stimuli can be elicited in atectal frogs (Comer and Grobstein 1978).

The electrical brain-stimulation experiments suggest that there are prewired circuits for appropriate motor responses that can be called into play by corresponding command systems. These systems serve particular functions for stimulus recognition and localization. Recordings of single cells belonging to those systems during visually released behaviors will provide information about the mode of their function. These experiments may decide whether command systems fire at "all or nothing" or whether the strength of their output determines the latency of the motor response.

Species Differences

Although we are beginning to learn about configurational prey selection in the common toad, it is now important to investigate corresponding questions in other amphibians, such as frogs (Schürg-Pfeiffer and Ewert 1980), in salamanders (Roth and Himstedt 1978; Finkenstädt 1980; Himstedt, this volume), and in various other species. It seems likely that there is something resembling a common innate "wiring" pattern (time- and space-dependent filter operations) in all amphibian species. On the basis of different "emphases" among neuronal filter operations, variations of prey-selection systems may be derived in different amphibian groups and species according to their special ecological and behavioral adaptations.

Acknowledgments

The work was supported by the Deutsche Forschungsgemeinschaft and the Foundations' Fund for Research in Psychiatry.

References

Arbib, M. A. 1981. "*Rana computatrix:* An evolving model of visuomotor coordination in frog and toad." *COINS Tech. Report* 81-6: 1–27.

Arbib, M. A., and R. Lara. 1981. "A neural model of the interaction of tectal columns in prey-catching behavior." *COINS Tech. Report* 81-3: 1–37.

Autrum, H. 1959. "Das Fehlen unwillkürlicher Augenbewegungen beim Frosch." *Naturwissenschaften* 46: 435.

Beck, A., and J.-P. Ewert. 1979. "Prey selection by toads (*Bufo bufo* L.) in response to configurational stimuli moved in the visual field z,y-coordinates." *J. Comp. Physiol.* 129: 207–209.

Birukow, G., and M. Meng. 1955. "Eine neue Methode zur Prüfung des Gesichtssinnes bei Amphibien." *Naturwissenschaften* 42: 652–653.

Borchers, H.-W. 1979. "Single unit recordings from the optic tectum of freely moving toads *Bufo bufo* L." *Neurosci. Lett.* Suppl. 3: 291.

Borchers, H.-W., and J.-P. Ewert. 1979. "Correlation between behavioral and neuronal activities of toads *Bufo bufo* (L.) in response to moving configurational prey stimuli." *Behav. Processes* 4: 99–106.

Borchers, H.-W., H. Burghagen, and J.-P. Ewert. 1978. "Key stimuli of prey for toads (*Bufo bufo* L.): Configuration and movement patterns." *J. Comp. Physiol.* 128: 189–192.

Brown, W. T., and D. Ingle. 1973. "Receptive field changes produced in frog thalamic units by lesions of the optic tectum." *Brain Res.* 59: 405–409.

Brzoska, J., and H. Schneider. 1978. "Modification of prey-catching behavior by learning in the common toad (*Bufo bufo* (L.), Anura, Amphibia): Changes in responses to visual objects and effects of auditory stimuli." *Behav. Processes* 3: 125–136.

Buytendijk, F. J. 1941. *Wege zum Verständnis der Tiere.* Zürich.

Comer, C., and P. Grobstein. 1978. "Prey acquisition in atectal frogs." *Brain Res.* 153: 217–221.

Eibl-Eibesfeldt, I. 1951. "Nahrungserwerb und Beuteschema der Erdkröte (*Bufo bufo* L.)." *Behaviour* 4: 1–35.

Ewert, J.-P. 1967. "Aktivierung der Verhaltensfolge beim Beutefang der Erdkröte (*Bufo bufo* L.) durch elektrische Mittelhirnreizung." *Z. vergl. Physiol.* 54: 455–481.

————. 1968. "Der Einfluss von Zwischenhirndefekten auf die Visuomotorik im Beutefang- und Fluchtverhalten der Erdkröte (*Bufo bufo* L.)." *Z. vergl. Physiol.* 61: 41–70.

————. 1969. "Quantitative Analyse der Reiz-Reaktions-Beziehungen bei visuellem Auslösen der Beutefang-Wendereaktion der Erdkröte (*Bufo bufo* L.)." *Pflügers Arch.* 308: 225–243.

————. 1971. "Single unit response of the toad's (*Bufo americanus*) caudal thalamus to visual objects. *Z. vergl. Physiol.* 74: 81–102.

————. 1972. "Zentralnervöse Analyse und Verarbeitung visueller Sinnesreize." *Naturwiss. Rundsch.* 25: 1–11.

————. 1976. "The visual system of the toad: Behavioral and physiological studies of a pattern recognition system." In *The Amphibian Visual System*, K. V. Fite, ed. New York: Academic.

————. 1978. "Sensorische Erkennungssysteme." In *Die Psychologie des 20. Jahrhunderts, Bd. VI: Lorenz und die Folgen*, R. A. Stamm and H. Zeier, eds. Zürich: Kindler.

————. 1980. *Neuroethology: An Introduction to the Neurophysiological Fundamentals of Behavior.* Berlin: Springer.

Ewert, J.-P., and H. Burghagen. 1979a. "Configurational prey selection by *Bufo, Alytes, Bombina* and *Hyla*." *Brain, Behav., Evol.* 16: 157–175.

Ewert, J.-P., and H. Burghagen. 1979b. "Ontogenetic aspects of visual 'size constancy' phenomena in the midwife toad *Alytes obstetricians* (Laur.)." *Brain, Behav., Evol.* 16: 99–112.

Ewert, J.-P., and H.-A. Härter. 1969. "Der hemmende Einfluss gleichzeitig bewegter Beuteattrappen auf das Beutefangverhalten der Erdkröte (*Bufo bufo* L.). "*Z. vergl. Physiol.* 64: 135–153.

Ewert, J.-P., and F. J. Hock. 1972. "Movement sensitive neurons in the toad's retina." *Exp. Brain Res.* 16: 41–59.

Ewert, J.-P., and W. Kehl. 1978. "Configurational prey-selection by individual experience in the toad *Bufo bufo*." *J. Comp. Physiol.* 126: 105–114.

Ewert, J.-P., and Traud, R. 1979. "Releasing stimuli for antipredator behavior in the common toad *Bufo bufo* (L.)." *Behaviour* 68: 170–180.

Ewert, J.-P., and W. von Seelen. 1974. "Neurobiologie und System-Theorie eines visuellen Muster-Erkennungsmechanismus bei Kröten." *Kybernetik* 14: 167–183.

Ewert, J.-P., and A. von Wietersheim. 1974a. "Musterauswertung durch Tectum- und Thalamus/Praetectum-Neurone im visuellen System der Kröte (*Bufo bufo* L.)." *J. Comp. Physiol.* 92: 131–148.

————. 1974b. "Der Einfluss von Thalamus/Praetectum-Defekten auf die Antwort von Tectum-Neuronen gegenüber visuellen Mustern bei der Kröte (*Bufo bufo* L.)." *J. Comp. Physiol.* 92: 149–160.

Ewert, J.-P., F. J. Hock, and A. von Wietersheim. 1974. "Thalamus/Praetectum/Tectum: Retinale Topographie und physiologische Interaktionen bei der Kröte (*Bufo bufo* L.)." *J. Comp. Physiol.* 92: 343–356.

Ewert, J.-P., H.-W. Borchers, and A. von Wietersheim. 1978. "Question of prey feature detectors in the toad's (*Bufo bufo* (L.)) visual system." *J. Comp. Physiol.* 126: 43–47.

Ewert, J.-P., B. Arend, V. Becker, and H.-W. Borchers. 1979a. "Invariants in configurational prey selection by *Bufo bufo* (L.)." *Brain, Behav., Evol.* 16: 38–51.

Ewert, J.-P., H. Krug, and G. Schönitz. 1979b. "Activity of retinal class R3 ganglion cells in the toad *Bufo bufo* (L.) in response to moving configurational stimuli: Influence of the movement direction." *J. Comp. Physiol.* 129: 211–215.

Ewert, J.-P., H.-W. Borchers, and A. von Wietersheim. 1979c. "Directional sensitivity, invariance and variability of tectal T5 neurons in response to moving configurational stimuli in the toad *Bufo bufo* (L.)." *J. Comp. Physiol.* 132: 191–201.

Finkenstädt, Th. 1980. "Disinhibition of prey-catching in the salamander following thalamic-pretectal lesions." *Naturwissenschaften* 67: 471.

————. 1981. "Effects of forebrain lesions on visual discrimination in *Salamandra salamandra*." *Naturwissenschaften* .

Fite, K. V., and F. Scalia. 1976. "Central visual pathways in the frog." In *The Amphibian Visual System: A Multidisciplinary Approach*, K. V. Fite, ed. New York: Academic.

Freisling, J. 1948. "Studien zur Biologie und Physiologie der Wechselkröte (*Bufo viridis* Laur.)." *Österr. Zool. Z.* (Vienna) 1: 383–440.

Grüsser, O.-J., and U. Grüsser-Cornehls. 1976. "Neurophysiology of the anuran visual system." In *Frog neurobiology*, R. Llinas and W. Precht, eds. Berlin: Springer.

Grüsser, O.-J., U. Grüsser-Cornehls, D. Finkelstein, V. Henn, M. Patutschnik, and E. Butenandt. 1967. "A quantitative analysis of movement detecting neurons in the frog's retina." *Pflügers Arch. (Ges. Physiol. Menschen, Tiere)* 293: 100–106.

Hubel, D. H., and T. N. Wiesel. 1962. "Receptive fields, binocular interaction and functional architecture in the cat's visual cortex." *J. Physiol.* 160: 106–154.

————. 1963. "Shape and arrangement of columns in cat's striate cortex." *J. Physiol.* 165: 559–568.

————. 1965. "Receptive fields and functional architecture in two non-striate visual areas (18 and 19) of the cat." *J. Neurophysiol.* 28: 229–289.

Ingle, D. 1971. "Prey-catching behavior of anurans toward moving and stationary objects." *Vis. Res.* Suppl. 3: 447–456.

————. 1973. "Disinhibition of tectal neurons by pretectal lesions in the frog." *Science* 180: 422–424.

————. 1976a. "Spatial vision in anurans." In *The Amphibian Visual System: A Multidisciplinary Approach*, K. V. Fite, ed. New York: Academic.

————. 1976b. "Central visual mechanisms in anurans." In *Frog Neurobiology*, R. Llinas and W. Precht, eds. Berlin: Springer.

————. 1980. "Some effects of pretectum lesions on the frog's detection of stationary objects." *Behav. Brain Res.* 1: 139–163.

Julesz, B. 1965. "Texture and visual perception." *Sci. Amer.* 212: 38–48.

Kuffler, S. W. 1953. "Discharge patterns and functional organization of the mammalian retina." *J. Neurophysiol.* 16: 37–68.

Lara, R., and M. A. Arbib. 1981. "A neural model of interaction between pretectum and tectum in prey selection." *COINS Tech. Report* 81-4: 1–30.

Lara, R., M. A. Arbib, and A. S. Cromarty. 1981. "The role of the tectal column in facilitation of amphibian prey-catching behavior: A neural model." *COINS Tech. Report* 81-2: 1–54.

Lázár, G. 1971. "The projection of the retinal quadrants on the optic centers in the frog: A terminal degeneration study." *Acta Morphol. Acad. Sci. Hung.* 19: 325–334.

Lázár, G., and G. Székely. 1969. "Distribution of optic terminals in the different optic centers of the frog." *Brain Res.* 16: 1–14.

Lettvin, J. Y., H. R. Maturana, W. S. McCulloch, and W. H. Pitts. 1959. "What the frog's eye tells the frog's brain." *Proc. IRE* 47: 1940–1951.

Northcutt, R. G., and E. Kicliter. 1980. "Organization of the amphibian telencephalon." In *Comparative Neurology of the Telencephalon*, S.O.E. Ebbesson, ed. New York: Plenum.

Roth, G., and W. Himstedt. 1978. "Response characteristics of neurons in the tectum opticum of *Salamandra*." *Naturwissenschaften* 65: 657.

Sachs, L. 1976. *Statistische Methoden*. Berlin: Springer.

Scalia, F., and K. V. Fite. 1974. "A retinotopic analysis of the central connections of the optic nerve in the frog." *J. Comp. Neurol.* 158: 455–478.

Schneider, D. 1954. "Beitrag zu einer Analyse des Beute- und Fluchtverhaltens einheimischer Anuren." *Biol. Zbl.* 73: 225–282.

Schürg-Pfeiffer, E., and J.-P. Ewert. 1981. "Investigation of neurons involved in the analysis of Gestalt prey features in the frog *Rana temporaria*." *J. Comp. Physiol.* 141: 139–152.

Sperry, R. W. 1944. "Optic nerve regeneration with return of vision in anurans." *J. Neurophysiol.* 7: 57–70.

Stevens, R. J. 1973. "A cholinergic inhibitory system in the frog optic tectum: Its role in visual electrical responses and feeding behavior." *Brain Res.* 49: 309–323.

Székely, G., and G. Lázár. 1976. "Cellular and synaptic architecture of the optic tectum." In *Frog Neurobiology*, R. Llinas and W. Precht, eds. Berlin: Springer.

Trachtenberg, M. C., and D. Ingle. 1974. "Thalamo-tectal projections in the frog." *Brain Res.* 79: 419–430.

von Holst, E., and H. Mittelstaedt. 1951. "Das Reafferenzprinzip." *Naturwissenschaften* 37: 464–476.

von Wietersheim, A., and J.-P. Ewert. 1978. "Neurons of the toad's (*Bufo bufo* L.) visual system sensitive to moving configurational stimuli: A statistical analysis." *J. Comp. Physiol.* 126: 35–42.

Wilczynski, W., and R. G. Northcutt. 1977. "Afferents to the optic tectum of the leopard frog: An HRP study." *J. Comp. Neurol.* 173: 219–229.

2 Prey Selection in Salamanders

Werner Himstedt

A considerable amount of neurobiological and behavioral research has been done on nonmammalian vertebrates. However, only a few species have been studied in detail. Most investigations in this field use species that are easy to obtain and to maintain in the laboratory, such as the goldfish, the toad, the frog, and the pigeon. Our knowledge of visually guided behavior in amphibians is based mainly upon studies in anurans. Since the work of Lettvin et al. (1959), visual mechanisms have been studied in a few species of frogs and toads (see the chapters by Ewert and by Ingle in this volume). Studies of urodeles are still relatively new. Although the early anatomical study of the tiger salamander by Herrick (1948) provided detailed information on a typical urodele brain, electrophysiological and behavioral data have, until very recently, remained sparse. Because the differences in environmental adaptation and food selection require very different behavioral mechanisms in urodeles than in anurans, studies of the visual behavior of urodeles should give further insights into specializations of brain function among lower vertebrates.

This chapter deals with some results obtained in experiments with the fire salamander, *Salamandra salamandra,* an amphibian common in European forests. After metamorphosis these animals live on land. Mating takes place outside of the water, and the larvae develop in the maternal oviduct until a rather advanced stage. The female deposits about 20–70 larvae into the water, where they stay for about 1–3 months. Before and after metamorphosis *Salamandra salamandra* is carnivorous. In the guts of adults a great variety of prey animals can be found (Szabó 1962): slugs, earthworms, isopoda, millipedes, insects, spiders.

We have investigated which stimuli release the prey-capture behavior in *Salamandra* and what neural mechanisms guide this behavior. Some of the behavioral experiments reported here were carried out by Regina Anselmann, Ingeborg Hendel, and Hans-Jürgen Sperling. The electrophysiological recordings were performed together with Gerhard Roth at the University of Bremen.

Night Vision

Salamandra salamandra is active during the darkness, leaving its hiding place after sunset and returning before sunrise (Himstedt 1971). We asked whether these salamanders are able to detect their prey at night by vision. They might use other senses, such as olfaction. In the laboratory we tested the ability of salamanders to respond to visual prey stimuli at extremely low light intensities. It is known that prey-capture behavior in *Salamandra*, as in other amphibians, is released by moving objects in a certain range of velocity, size, and stimulus-background contrast (Himstedt 1967; Grüsser-Cornehls and Himstedt 1976). Thus, we used moving visual stimuli (figure 2.1). The illumination level could be reduced by neutral-density filters in logarithmic steps. In order to observe reactions in the dark, an infrared viewing apparatus was used. The infrared light did not affect the normal behavior of the animals.

Figure 2.2 shows the frequency of responses to moving objects at three different light intensities. The following reactions were scored: turning toward the prey object, approaching, and snapping. At 10^{-5} candela per m², stimuli of low ($C = 0.15$), medium ($C = 0.3$), and high (C

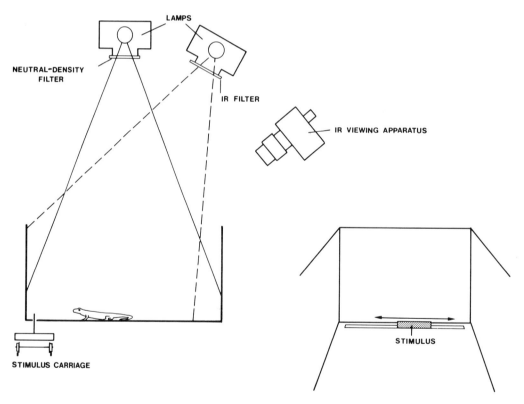

Figure 2.1 Experimental arrangement to investigate prey-capture behavior of salamanders at low illumination. The stimulus, a piece of gray cardboard, moves slowly in front of the salamander. The reactions of the animal are observed with an infrared viewing apparatus.

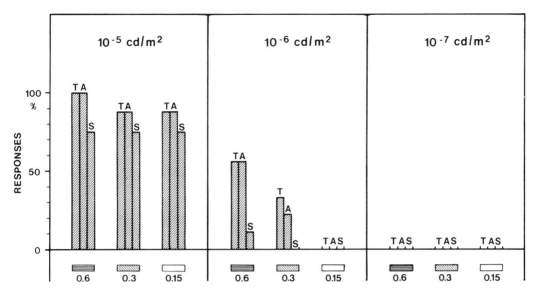

Figure 2.2 Responses of *Salamandra salamandra* to visual prey stimuli at three different illumination levels (measured in candela per m²). Three patterns of different stimulus-background contrast (0.6, 0.3, 0.15) were presented. Responses are (*T*) turning reaction, (*A*) approaching (at least four steps towards the stimulus), and (*S*) snapping. The responses of ten animals are added up.

= 0.6) contrast released almost the same number of responses. At 10^{-6} cd/m² the reactions decreased; snapping occurred only at high contrast, and the stimulus with low contrast did not elicit a single response. At 10^{-7} cd/m² the salamanders seemed unable to detect any visual stimulus.

In nature, the illumination level during a moonless night is about 10^{-4}–10^{-5} cd/m², so it is possible that *Salamandra* can see its prey even if the stars are covered by clouds (which may reduce the light by another logarithmic step).

To test the possible significance of olfactory stimuli, the prey models were scented with mealworm juice. Figure 2.3 shows the interaction of visual and olfactory stimuli. In the first series of experiments the models were moved as in the previous tests. There was no significant difference between the reactions to scented and to odorless objects. With increasing darkness the odor did not seem to act as an additional cue. This changed, however, if the prey objects were not moved continuously but were stopped after a short movement and then kept motionless. In this case olfactory stimuli were very effective in releasing snapping behavior, and it was evident that the significance of the odor increased with decreasing illumination level. So in the darkness motionless prey can be detected by olfaction. If there is, however, a small amount of illumination and some movement, the prey-catching behavior is guided mainly by vision. A similar interaction of visual and olfactory stimuli was found in the Italian lungless salamander *Hydromantes italicus* by Roth

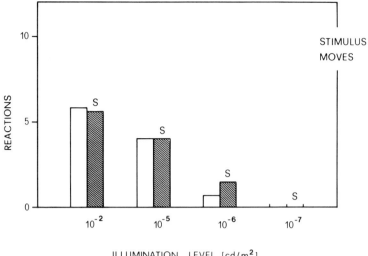

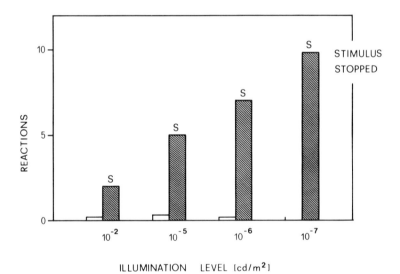

Figure 2.3 Snapping reactions of salamanders at different illumination levels to stimuli with or without prey odor. Reactions to smelling objects are indicated by S. Above: The stimuli moved continously. Below: The stimuli stopped after a short period of movement and then remained stationary. The responses of ten animals are added up.

(1976). These animals too respond preferably to visual patterns in light, and olfaction becomes more important in darkness.

The Significance of Prey Configuration

About 80 percent of the prey animals found in the gut of *Salamandra* (Szabo 1962) are elongated in form; they look like and move like worms. Ewert (1968 and this volume) showed that toads prefer visual patterns that are elongated in the direction of movement, and ignore objects that are elongated perpendicular to the direction of movement. We were interested in seeing whether there are similar mechanisms of pattern preference in salamanders.

Preference tests on prey-capture behavior of amphibians can be performed in a very simple manner, such as that used in field experiments on frogs. Two different black cardboard patterns are fixed on a fork of thin wire and presented to the animal at equal distance. The turning reactions to each pattern are scored. Since movement is an essential stimulus parameter, in quantitative experiments in the laboratory it is advisable to move the patterns not by hand but by a motorized system providing precise velocities.

In a choice situation such as that shown in Figure 2.4, *Salamandra* prefers the horizontal "wormlike" pattern. This is a very distinct decision if the stimuli are moved continuously and slowly. Amphibians that feed more on flying insects than on worms and slugs show a different pattern preference. In figure 2.5 the results of choice tests with frogs, salamanders, and toads are compared. The experiments with *Rana esculenta*, *Hyla arborea*, and *Bufo bufo* used forks of thin wire; the salamanders were tested in an apparatus that moved the models at a

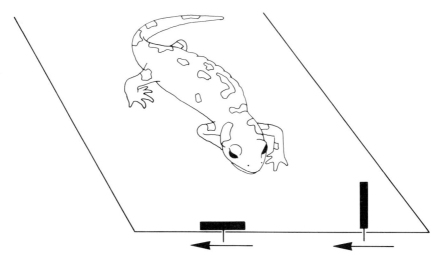

Figure 2.4 *Salamandra salamandra* turning toward a rectangular pattern with its long axis in the direction of movement.

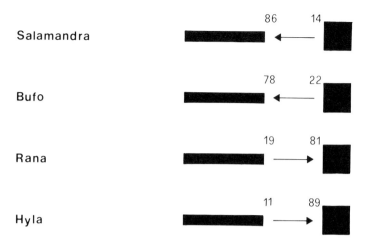

Figure 2.5 Preferences of four amphibians to patterns of equal area. Arrows point toward pattern chosen more frequently; numbers indicate scores of 100 choices.

constant velocity of 1 cm per second. Twenty subjects of each species carried out five choices, for a total of 100 tests per species. The rectangles measured 5 mm × 40 mm, and the edge of the square was 14 mm long. In the experiments with *Hyla* the dimensions of the stimuli were halved, since this tree frog has about half the body size of *Rana esculenta* and prefers smaller patterns. When these stimuli were presented, *Salamandra* and *Bufo* turned more frequently to the rectangle; *Hyla* and *Rana* chose the square. These results are from experiments with patterns of equal area. In *Rana esculenta,* too, a preference of the rectangle can be observed when the height of the square is the same. If the frogs had the choice between a square of 5 mm × 5 mm and a horizontal rectangle of 5 mm × 40 mm they turned with a frequency of 79% toward the rectangle. Also, an inhibitory effect of vertical edges was observed in the frog experiments: vertical rectangles more than 30 mm high were avoided. Thus, as in toads and salamanders, frog prey-catching behavior can be activated by increasing a pattern in the direction of movement and inhibited by an increase in the vertical direction. However, in frogs this effect is not as strong as in salamanders and toads.

Before metamorphosis, salamanders feed on prey which is quite different from that of adults. Larvae of *Salamandra* capture small crustaceans, such as *Amphipoda, Copepoda,* or *Cladocera,* and insect larvae that generally do not show the configuration and the movement of worms. It has been demonstrated that there is a distinct change in pattern preference during the development of *Salamandra* (Himstedt et al. 1976). In these experiments the animal was placed in a small glass tank, outside of which a stimulus was moved horizontally at an angular velocity of 10 degrees per second. Since the animals could not approach the stimulus, they responded by turning toward the objects. The number of responses in one minute was counted. This is the same method of mea-

Himstedt

suring the reactions as Ewert used in his behavioral experiments. Twenty larvae at an age of about 1 month after birth were used. Newly metamorphosed salamanders were divided into three groups of 20 animals. These groups were tested at different times after metamorphosis. To exclude effects of habituation to the experimental situation, each animal was presented only once to each of the different stimuli.

Figure 2.6 shows the findings of those experiments in which horizontally or vertically oriented patterns were presented. In one series the stimuli became more and more elongated in the direction of movement; in the other series they were elongated perpendicular to the direction of movement. In larvae and in juvenile salamanders during some weeks after metamorphosis, preference for the horizontal patterns was not seen. Only the longest rectangles (4° × 32°) caused some inhibition if they were oriented vertically. However, after about 6–8 months of terrestrial life the stimulus "filters" seem to become much narrower. In these experiments the animals at an age of 8 months responded exclusively to the long horizontal pattern of 32° × 4°. This change is probably not due to any learning processes. The salamanders were fed only with small heaps of tubifex, which outside the water are convoluted in a ball and never show the configuration or the motion of the preferred stimulus. We cannot explain the rather rapid change in behavior after the sixth month, and we are not sure whether this change always occurs precisely at this age. But in all of our salamanders this sharpening of preference was observed. As a control, all animals were tested once more at 10 months after metamorphosis, and all groups—those that had been tested as larvae as well as those tested at 2, 6, or 8 months— exhibited the same preference for the "wormlike" pattern. As yet, we cannot exclude the possibility that the consistent change in prey selection was due to a seasonal change between 6 and 8 months after metamorphosis. Seasonal changes in prey selection have been demonstrated in *Bufo bufo* by Ewert and Siefert (1974).

Luthardt and Roth (1979) assumed that the efficiency of horizontal patterns in *Salamandra* is not invariant with respect to changes in stimulus velocity. While at low velocities the animals preferred stimuli (size: 4 mm × 32 mm) oriented parallel to the direction of movement, at high velocities stimuli oriented perpendicular to the direction of movement seemed to be more effective. This phenomenon, however, cannot be observed in all salamanders, and presumably it depends on experience with certain prey. Animals in our laboratory that were fed mainly on mealworms preferred the horizontal stimulus at low and high velocities. The salamanders studied by Luthardt and Roth got mainly crickets, and those authors have more recently shown that experience of juvenile *Salamandra* with a certain food can effect differences in pattern preference (Luthardt and Roth 1980). Salamanders that had been fed only with stationary pieces of meat later preferred continuously moving "wormlike" patterns. Thus, there seems to be some kind of matura-

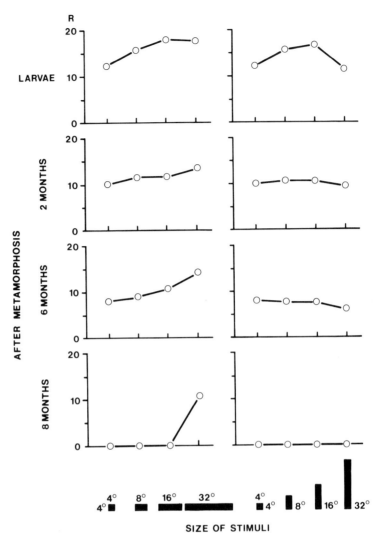

Figure 2.6 Change in pattern preference during development of *Salamandra salamandra*. Dependence of prey-capture responses (R = number of turning reactions during 1 minute) on length of horizontal or vertical edge of stimulus. Each point represents the mean value of 20 stimulus presentations.

tion process that leads to the selection of this type of prey, even though learning processes may modify the releasing mechanisms.

Size-Distance Discrimination

In the experiments described so far the size of an object was given either in millimeters or in degrees of visual angle. It was unknown until now which of these parameters is relevant in the salamander's prey selection. The question is whether or not salamanders are able to estimate an object's real size independent of the viewing distance. Ingle (1968) first showed that the preference of frogs for prey patterns is related to real size rather than to visual angle. Ewert and Gebauer (1973) and Ingle and Cook (1977) then carried out more systematic studies of the "size constancy" effect in toads and frogs, respectively.

We presented salamanders with two stimuli, which moved on circular paths (figure 2.7). The animal was placed on a small table that could be shifted by micromanipulator drives so that the head was positioned precisely in the center. The distance and the size of the stimuli was altered so that both objects either were of the same real size or subtended the same visual angle. The stimuli moved in two slits cut in the floor under the animal table. The floor was exchanged if other stimulus distances were to be presented. Two series of experiments were performed. During the first, the salamanders saw the patterns within the

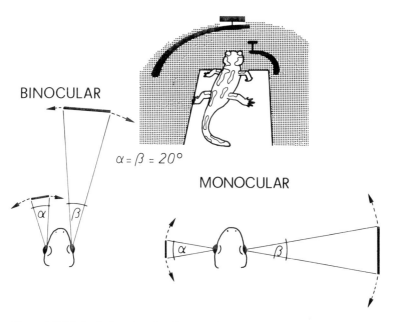

Figure 2.7 Method of testing size-distance discrimination in salamanders. Prey models of different real size but equal visual angle are presented at different distances. The stimuli are moving on circular paths either within the binocular visual field in front of the animal, or laterally within the monocular fields of each eye.

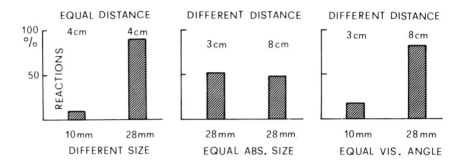

BINOCULAR

EQUAL DISTANCE — DIFFERENT DISTANCE — DIFFERENT DISTANCE

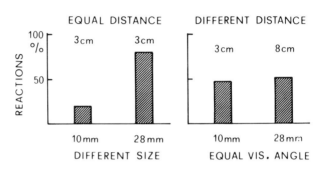

MONOCULAR

EQUAL DISTANCE — DIFFERENT DISTANCE

Figure 2.8 Size-distance discrimination in the frontal binocular and the lateral monocular field. Two stimuli of equal or different size were presented at equal or different distances. (Sizes of stimuli in mm, distances in cm.)

frontal binocular visual field; during the second the stimuli moved only in the lateral monocular zones.

Some of the results are shown in figure 2.8. If two rectangular patterns 28 mm and 10 mm in length (the height of both was 1/5 of the length) were presented at equal distances, the salamanders always preferred the larger stimulus. This was observed in both series. In the other experiments, however, differences occurred. In the frontal binocular field a size-constancy effect could be observed. Patterns of equal absolute size were chosen with equal frequency, independent of the distance. Patterns of equal visual angles at different distances were discriminated. The salamanders clearly perferred the 28-mm stimulus at 8 cm distance above the 10-mm stimulus at 3 cm when both subtended a visual angle of 20°. In *Salamandra* the discrimination of real size is possible only for rather short prey distances, as has been shown for frogs and toads (Ewert and Gebauer 1973; Ingle and Cook 1977). In fact, for distances of several meters, salamanders respond to oversized objects as prey; for example, a pattern 70 cm × 14 cm seen 2 m away (20° × 4°) will regularly elicit turning responses and several approach steps. However if these patterns were moved in a region about 90° from the salamander's midline, so that only one eye could see one stimulus, the animals did not discriminate objects of equal visual angle at different distances. These negative data do not rule out the existence of size constancy for the monocular field, since there can always be one point in common between two separate angle-preference curves although they peak at different angles. In fact, studies with frogs by Ingle (1968) and Ingle and Cook (1977) proved size-distance discrimination using monocular stimulus presentation. In toads, Ewert and Gebauer (1973) also showed that size constancy is independent of binocular vision. Although we need more data to assert that salamanders differ from anurans in this respect, we add the suggestion that salamanders are probably deficient in using monocular accommodation cues for depth estimation, as frogs (Ingle 1972) and toads (Collett 1977) can do. In the urodele eye the lens is shifted only by one vental protector muscle, whereas two accommodation muscles are present in the anuran eye. At present, however, we do not know whether or not monocular salamanders can use accommodation cues in the frontal visual field or whether they have stereoscopic vision. Perhaps the accommodation focuses in the forward direction but not in the lateral direction. This has to be investigated in further experiments. Although our knowledge is still incomplete regarding mechanisms of depth vision, I suggest from present data that salamanders are deficient in using monocular accommodation cues for lateral-field depth estimation as frogs are certainly able to do.

Neurophysiology of Prey Detection

In order to study the neuronal basis of prey selection in *Salamandra*, single-unit recordings from neurons in the tectum opticum were carried out (Roth and Himstedt 1978; Himstedt and Roth 1980). Figure 2.9 shows schematically the experimental arrangement. By means of a perimeter similar to those used by Grüsser and by Ewert, stimuli could be presented at any position of the visual field. In order to study problems of invariance in pattern preference, not only were the parameters of size and of orientation changed but also the velocity of the stimuli. The salamanders used in these experiments were caught as adults in Yugoslavia. During captivity they got a mixed diet; thus, experience with only one kind of prey can be excluded.

In the optic tectum of *Salamandra* three layers of retinal fibers were described by Grüsser-Cornehls and Himstedt (1973). The response characteristics of these retinal neurons cannot explain the prey-selection behavior. Tectal cells, which process the retinal inputs, are in many

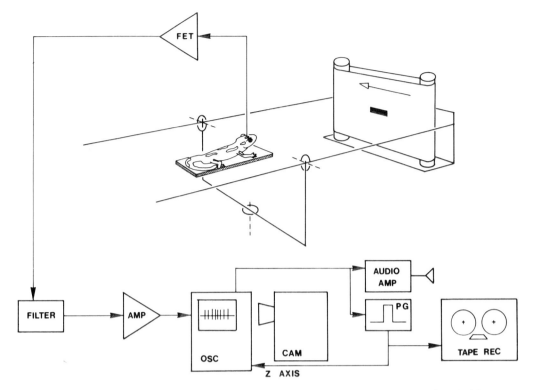

Figure 2.9 Experimental arrangement for the neurophysiological recordings. AMP: amplifier. CAM: camera for filming oscilloscope screen. FET: preamplifier with field-effect transistor. PG: pulse generator, triggered by action potentials (the pulses intensify the z axis of the oscilloscope and brighten the spikes, while the noise remains weak in intensity). OSC: oscilloscope. Placing the salamander in the center of a perimeter allows presentation of moving stimuli at any position in the visual field.

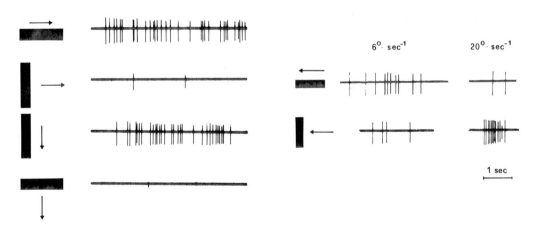

Figure 2.10 Responses of cells in the optic tectum of *Salamandra salamandra* to rectangular patterns oriented in or perpendicular to the direction of movement. Left: responses to constant velocity but different directions of movement. Right: responses of another cell to patterns of 2° × 8° moving at different velocities.

cases more selective to prey stimuli. Some tectal cells responded with a high impulse rate when a pattern of 2° × 8° visual angle was oriented in the direction of movement, but gave a low response if the pattern was oriented perpendicular to the direction of movement (figure 2.10). The receptive fields of these neurons measured about 20°–40° in diameter. The receptive fields are nearly circular, and pattern selectivity generally does not depend on the direction of movement. In several neurons, however, the responses were not independent of the velocity. The neuron shown at the right of figure 2.10 preferred the horizontal pattern at low velocity but the vertical pattern at higher velocity.

Figure 2.11 gives more data about this type of neurons. The direction of movement here always was horizontal, from right to left. With an increase in the length of the horizontal edge of the stimulus the impulse rate increased (figure 2.11A), whereas lengthening the vertical edge of the stimulus beyond 8° inhibited the response. This type of reaction is very similar to those of the "tectal type 2 neurons" Ewert and von Wietersheim (1974) recorded in toads (in recent publications these neurons are named class T5(2)).

When the stimulus velocity was varied, it was evident that these neurons in *Salamandra* are not a homogenous group. The velocities of 2°, 6°, and 20° per second, which we used here, are within the range of optimal prey movement (Himstedt 1967; Grüsser-Cornehls and Himstedt 1976). Some cells showed no change in their respective stimulus preference; that is, the impulse rates to the horizontal and vertical pattern increased proportional with an increase in stimulus velocity (figure 2.11B). Other neurons (2.11C) were characterized by a convergence of the velocity functions, while a third group of neurons (2.11D) showed a clear inversion of the mode of stimulus preference with an increase in stimulus velocity.

It certainly is difficult to study the neural basis of behavior in paralyzed animals. Perhaps the responses of tectal cells are different in freely moving salamanders which are motivated to catch prey. Some general conclusions, however, can be drawn from these experiments. Among the great variety of response characteristics, some cells in the salamander's tectum are selective for stimulus parameters that are preferred in prey-capture behavior. However, this selectivity seems to be not always invariant with respect to a change in stimulus velocity. Henn and Grüsser (1968) described that in frog class 2 ganglion cells the exponent of the velocity function changed if the stimuli became larger than the excitatory receptive field caused by different temporal frequency properties of excitatory and inhibitory mechanisms. Also, in frog class T5 tectal neurons, the exponent of velocity functions can be different, presenting different stimulus patterns (Grüsser and Grüsser-Cornehls 1976). A quantitative study on invariance and variability of T5(2) neurons has been carried out by Ewert et al. (1979) (see Ewert, this volume).

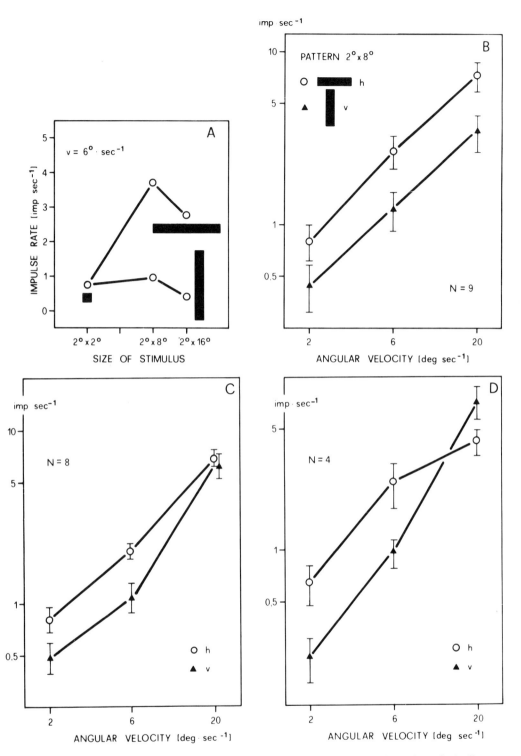

Figure 2.11 Response characteristics of cells in the optic tectum of *Salamandra salamandra*, which all gave a stronger response to the pattern oriented horizontally. (Movement direction was always horizontal, from right to left.) (A) Dependence of discharge rate of a single cell on length of horizontal or vertical edge. In the other experiments the horizontal (*h*) and the vertical (*v*) pattern were presented at different velocities. Average values and mean error from several neurons (*N* = number of cells) grouped according to the courses of the velocity functions. (B) Parallel curves. (C) Converging curves. (D) Inverted curves.

Detection of Stationary Objects

Usually, amphibians respond to visual prey stimuli only if these objects are moving (Ingle 1971a; Ewert 1974; Grüsser and Grüsser-Cornehls 1968; Himstedt 1967). It was uncertain whether amphibians, when not in motion, are able to detect stationary prey by vision. As shown in Figure 2.3, our salamanders in some cases snapped at a motionless pattern if it had moved previously and then stopped. Such behavior was also described for toads (*Bufo americanus*) by Ingle (1971b). Roth (1976) described the same response in the salamander *Hydromantes italicus.*

In a different situation (figure 2.12) the responses to stationary visual patterns were investigated in more detail. Two black patterns were fixed on a white wall, and a salamander was attracted toward this wall by moving a prey model (a piece of black card on a thin glass rod). When the salamander had followed the moving stimulus for about 6–8 cm and was standing about 3 cm away from the test patterns, the moving stimulus was raised away. When a moving stimulus suddenly disappears, a salamander remains motionless. In about 75% of the experiments the salamander subsequently turned toward one of the stationary

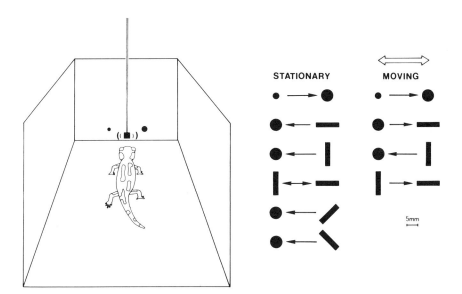

Figure 2.12 Responses of *Salamandra* to stationary patterns. Left: A salamander follows the moving prey dummy attached to a thin glass rod and is thus induced to approach the wall, where stationary patterns are fixed. Right: comparison of prey-catching reactions to stationary and moving patterns. Arrows in each case point toward the pattern preferred in a choice experiment. Arrow with two tips indicates that the two patterns were chosen equally often.

patterns. After the disappearance of the stimulus it took 12–87 seconds (37 seconds on the average) for a turning movement to occur. In one extreme case, 4 minutes elapsed before the animal turned toward a stationary pattern. Then the salamanders approached the pattern either rapidly, in about 2 seconds, or slowly, with long pauses after each step, within up to 90 seconds. When the animal was standing with its snout directly in front of the pattern, another period of 8–85 seconds passed until the salamander either turned away or snapped at it.

The choice between two dissimilar patterns was offered in order to test whether the salamanders prefer those forms that elicit the most prey-capture responses when they are used as moving stimuli. The results show some remarkable differences. If circles of different sizes were presented, one 5 mm in diameter was chosen more often than one with a diameter of 2.5 mm, regardless of whether the patterns were moving or stationary. But the situation was different with the horizontal rectangle: When presented as a moved dummy it was preferred to a circle of equal area, but as a motionless pattern it was chosen less often. When two moving rectangular dummies differed in orientation the horizontal rectangle was chosen more often than the vertical one, but when they were presented as still patterns the responses to the two rectangle were equally frequent. In general, one can say that the effectiveness of dummies moving horizontally is reduced the greater their vertical extent, and that the effectiveness increases with increasing horizontal extent. However, orientation does not affect the responses to motionless stimuli. When the choice was between a rectangle and a circle of equal area, the circular pattern was always chosen, regardless of the orientation—horizontal, vertical, or tilted 45° right or left—of the rectangle.

Detection of these motionless contours may occur in various ways. First, it could occur through sustained reactions of the retinal neurons. In the anuran retina, ganglion cells of classes 1 and 2 exhibit long-sustained reactions when a stimulus pattern is moved into the receptive field and stops there. As the salamander follows the moved dummy, the images of the stationary patterns shift on its retinae. When the animal stops moving, such neurons could continue to discharge for a time. However, in *Salamandra* the longest sustained reactions of ganglion cells observed last only 4–6 seconds after pattern movement (Grüsser-Cornehls and Himstedt 1973). As yet, we have not recorded other neurons in the visual centers of the salamander's brain that continue to fire for 30–60 seconds after stimulus movement. Second, retinal cells may be kept active via small movements of the head or of the eyes alone. Films of the animals show, however, that when the moved stimulus disappears the salamander stands motionless. Analyses of single frames of the films reveal no detectable head or body movements of any sort prior to the turning movement. When turning did occur, it was directed toward the stationary pattern from the outset. However, we could not

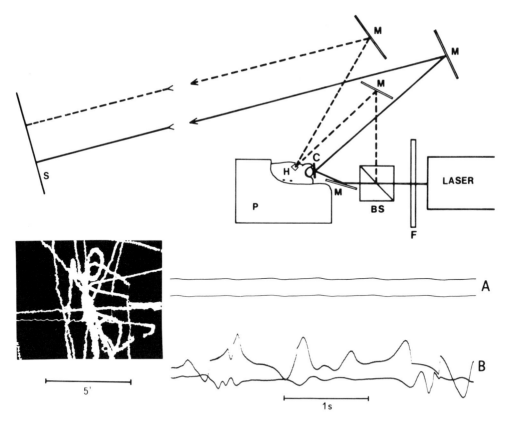

Figure 2.13 Measurement of eye movements in *Salamandra* by means of a reflected laser beam. BS: beam splitter. C: mirror on cornea. F: neutral-density filter. H: head of salamander. M: mirror. P: plaster block with fixes head. S: screen or position-sensitive photodetector. Below left: photograph of the screen (exposure time: a few seconds). Below right: signals from the photodetector recorded on an oscilloscope: upper beam represents deviations in x axis, lower beam deviations in y axis. (A) Movements of head mirror, (B) movements of cornea mirror.

analyze in the film movements with amplitudes lower than about 0.1 mm. Although the head did not appear to move in our animals, small eye movements may have occurred; Schipperheyn (1965) found in frogs movements of the eyeball produced by respiration movements. If these deviations are vertically oriented in the salamander, as in the frog, the most effective motionless rectangles should be those oriented vertically. But this was not the case.

To see whether undirected eye movements may exist, we recorded eye movements in *Salamandra* using reflection of a laser beam from a mirror on the cornea using the method of Manteuffel et al. (1977). As figure 2.13 shows, there are indeed undirected eye movements in this animal. Most of these rapid movements have amplitudes in a range of about 5–10 arc minutes, which is sufficient to displace the retinal image over one or two receptors. In addition to respiratory oscillations, rapid large-amplitude "flicks" occurred at intervals of some seconds. It seems that large shifts of the prey contour every 5–10 sec could regenerate the discharge of certain retinal ganglion cells. To get more insight into the significance of the eye movements in freely moving salamanders, we should try to record which kind of eye movement precedes a behavioral response.

There is another behavior in which detection of nonmoving contours is important: Homing salamanders use stationary visual patterns such as trees or stones as orientation marks (Himstedt and Plasa 1979). Perception of stationary barriers and obstacles likewise is important in guiding anuran behavior (Ingle 1971a, 1976, and in this volume), but in contrast to the salamanders no frog or toad was observed to react to stationary prey without facilitation by odor. Ingle provides data that pretectum and not the tectum is critical for barrier detection in the frog. Furthermore, Ewert (1971) found that "stationary edge detector" neurons are present in the toad's pretectal region but not in the optic tectum. Perhaps in the salamander, as well, the detection of inanimate objects and surfaces depends upon neurons of the pretectum. Thus, an important area for further research on pattern recognition by amphibians will be the analysis of pretectal functions.

Acknowledgment

The work was supported by the Deutsche Forschungsgemeinschaft.

References

Collett, T. 1977. "Stereopsis in toads." *Nature* 267: 349–351.

Ewert, J.-P. 1968. "Der Einfluss von Zwischenhirndefekten auf die Visuomotorik im Beute- und Fluchtverhalten der Erdkröte (*Bufo bufo* L.)." *Z. vergl. Physiol.* 61: 41–70.

Ewert, J.-P. 1974. "The neural basis of visually guided behavior." *Sci. Amer.* 230: 34–42.

Ewert, J.-P., and L. Gebauer. 1973. "Grössenkonstanzphänomene im Beutefangverhalten der Erdkröte (*Bufo bufo* L.)." *J. Comp. Physiol.* 85: 303–315.

Ewert, J.-P., and G. Seifert. 1974. "Seasonal change of contrast-detection in the toad's (*Bufo bufo* L.) visual system." *J. Comp. Physiol.* 94: 177–186.

Ewert, J.-P., and A. von Wietersheim. 1974. "Musterauswertung durch Tectum- und Thalamus/Praetectum-Neurone im visuellen System der Kröte *Bufo bufo* (L.)." *J. Comp. Physiol.* 92: 131–148.

Grüsser, O. J., and U. Grüsser-Cornehls. 1968. "Neurophysiologische Grundlagen visueller angeborener Auslösemechanismen beim Frosch." *Z. vergl. Physiol.* 59: 1–24.

———. 1976. "Neurophysiology of the anuran visual system." In *Frog Neurobiology,* R. Llinás and W. Precht, eds. Berlin: Springer.

Grüsser-Cornehls, U., and W. Himstedt. 1973. "Responses of retinal and tectal neurons of the salamander (*Salamandra salamandra* L.) to moving visual stimuli." *Brain, Behav., Evol.* 7: 145–168.

——. 1976. "The urodele visual system." In *The Amphibian Visual System*, K. V. Fite, ed. New York: Academic.

Henn, V., and O. J. Grüsser. 1968. "The summation of excitation in the receptive field of movement-sensitive neurons of the frog's retina. *Vision Res.* 9: 57–69.

Herrick, C. J. 1948. *The Brain of the Tiger Salamander*. University of Chicago Press.

Himstedt, W. 1967. "Experimentelle Analyse der optischen Sinnesleistungen im Beutefangverhalten einheimischer Urodelen." *Zool. Jahrb. Physiol.* 73: 281–320.

——. 1971. "Die Tagesperiodik von Salamandriden." *Oecologia* (Berlin) 8: 194–208.

Himstedt, W., and L. Plasa. 1979. "Home-site orientation by visual cues in salamanders." *Naturwissenschaften* 66: 372–373.

Himstedt, W., and G. Roth. 1980. "Neuronal responses in the tectum opticum of *Salamandra* to visual prey stimuli." *J. Comp. Physiol.* 135: 251–257.

Himstedt, W., U. Friedank, and E. Singer. 1976. "Die Veränderung eines Auslösemechanismus im Beutefangverhalten während der Entwicklung von *Salamandra salamandra* (L.)." *Z. Tierpsychol.* 41: 235–243.

Ingle, D. 1968. "Visual releasers of prey catching behavior in frogs and toads." *Brain, Behav., Evol.* 1: 500–518.

——. 1971a. "Discrimination of edge-orientation by frogs." *Vision Res.* 11: 1365–1367.

——. 1971b. "A possible behavioral correlate of delayed retinal discharge in anurans." *Vision Res.* 11: 167–168.

——. 1972. "Depth vision in monocular frogs." *Psychonom. Sci.* 29: 37–38.

——. 1976. "Behavioral correlates of central visual function in anurans." In *Frog Neurobiology*, R. Llinás and W. Precht, eds. Berlin: Springer.

Ingle, D., and J. Cook. 1977. "The effect of viewing distance upon size preference of frogs for prey." *Vision Res.* 17: 1009–1013.

Lettvin, J. Y., H. R. Maturana, W. S. McCulloch, and W. H. Pitts. 1959. "What the frog's eye tells the frog's brain." *Proc. IRE* 47: 1940–1951.

Luthardt, G., and G. Roth. 1979. "The role of stimulus movement patterns in the prey catching behavior of *Salamandra salamandra*." *Copeia*: 442–447.

Manteuffel, G., L. Plasa, T. J. Sommer, and O. Wess. 1977. "Involuntary eye movements in salamanders." *Naturwissenschaften* 64: 533.

Roth, G. 1976. "Experimental analysis of the prey catching behavior of *Hydromantes italicus* (Amphibia, Plethodontidae)." *J. Comp. Physiol.* 109: 47–58.

Roth, G., and W. Himstedt. 1978. "Response characteristics of neurons in the tectum opticum of *Salamandra*." *Naturwissenschaften* 65: 657–658.

Schipperheyn, J. J. 1965. "Contrast detection in frog's retina." *Acta Physiol. Pharmacol. Neerlandica* 13: 231–277.

Szabó, I. 1962. "Nahrungswahl und Nahrung des gefleckten Feuersalamanders (*Salamandra salamandra* L.)." *Acta Zool. Acad. Sci. Hung.* 8: 459–477

3 Organization of Visuomotor Behaviors in Vertebrates

David J. Ingle

Several of the authors in this volume support the idea that visual behavior is compounded from multiple discrimination and localization processes. A major concern of neurobehaviorists in this field is the functional dissection of these many visual systems and the matching of these apparently separate "channels" with particular anatomical units. This chapter reviews the multiple visuomotor systems in the common frog as a possible model for problems and solutions that experimenters may encounter in analyzing behaviors of more complicated vertebrates. Some of my own studies of orienting behaviors in the Mongolian gerbil are included to illustrate how comparative studies can be planned and how an evolutionary view of visuomotor organization can be sketched; that part of the chapter considers evidence that neocortex of mammals, as well as subcortical centers, participates in localization processes. Finally, I will try to show that these propositions concerning cortical-subcortical collaboration can form the basis of a broad theory of visuomotor organization within which phylogenetic and ecological variations in behavior can be understood.

Traditional studies of animal vision usually involve training animals in progressively more difficult tests of detection or discrimination, in order to probe their full capacities under various experimental conditions. Only recently have physiologists and psychologists begun to systematically analyze the practical uses of vision by which animals get around their everyday world—pursuing prey, negotiating barriers, jumping from rock to rock, avoiding holes, or avoiding threats. Animals as diverse as fishes, rodents, and birds face a number of common visuomotor problems, despite their obvious specializations. Appreciation of these common functions may point the way toward discovery of homologous subunits within vertebrate visual systems. Since several basic types of vertebrate behavior appear in fishes and amphibians that lack a definite neocortex, one can identify several basic mechanisms with visual subsystems of thalamus and midbrain. I have proposed that, for the frog at least, there are one-to-one correspondences between separate retinofugal targets and distinctive classes of visual behavior (Ingle 1976). Demonstration of this proposition can be an important step for mammalian neuropsychologists as well, since a substantial proportion of output fibers from visual cortex descend to these thalamic and midbrain visual centers. Thus, I shall consider the multiplicity of subcortical visuomotor systems as the essential *bauplann* for visual guidance of movement in all vertebrates. Whether or not this perspective on visuomotor process can also provide an approach to understanding other functions of the visual system, such as object recognition, is a speculative question I have approached elsewhere (Ingle 1978).

Visuomotor Classes in the Frog

The Visual Grasp Reflex

This term, as used initially by Hess et al. (1946) for the response elicited by tectal stimulation in the cat, implies that head and eyes are reoriented so as to bring a peripheral object to the area centralis of the cat's retina. In the monkey, the equivalent movements (eye saccade with head turn) are usually described as "foveation." The assumed function of such reorientations of gaze is "investigation" of an object that elicited attention as a peripheral stimulus. On the other hand, reorientation of head and body by a fish or amphibian typically precedes pursuit or grasping of prey with the mouth. In most amphibians and fishes there is no retinal region of high acuity that would justify such accurate reorientation just to obtain more acute visual information. However, the advantages of binocular vision (which involves a rather narrow zone for most nonmammals) may be sufficient to justify precise reorientation of the frontal midline to the target.

The similarity (or homology) of ocular reorientation in primates, head turning in birds, reptiles, and most mammals, and body turning in fishes and amphibians is more objectively demonstrated by showing the dependence of these functions on the optic tectum. For amphibians and fishes, the removal of tectum results in a total loss of the ability to turn toward or snap at small food objects (Ingle 1973a; Springer et al. 1977). In birds and reptiles, such evidence is less clear but in the same direction (Jarvis 1974; Bass et al. 1973). In lower mammals, deep tectal ablations (Casagrande and Diamond 1974) or deep undercuts of tectum (Schneider 1969) produce a "blindness" to peripheral food objects (seeds, worms) comparable to that reported for frogs. In cats or monkeys such dependence upon tectum can be demonstrated only after additional removal of primary visual cortex.

Another basic similarity between the ocular saccade of a monkey and the frog's prey orientation is that each has been shown to be an "open-loop" response, unmodified by visual feedback once initiated (Ingle 1973a; Robinson, 1968). It seems likely that this open-loop characteristic (as distinct from continuous tracking) is a defining character-

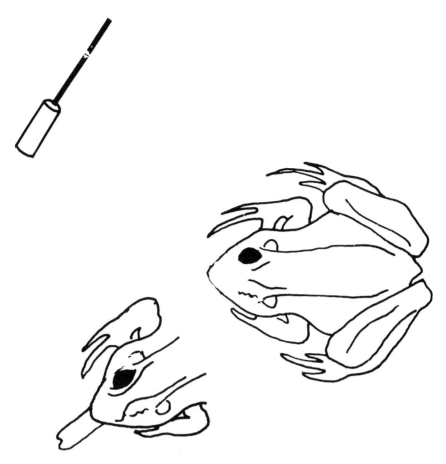

Figure 3.1 Example of a wrong-way prey strike by a frog with an abnormal retinal projection to the ipsilateral optic tectum. Each location of the dummy stimulus on the right side elicits a strike such that the frog's tongue hits the floor in a position on the left that is symmetrical to the target locus in respect to the frontal midline.

istic of the visual-grasp reflex. This is apparent in frogs in which the optic fibers have been rewired to the ipsilateral optic tectum: Such frogs strike precisely at loci that are mirror images of the actual position of a wiggling worm (figure 3.1), even though the worm image moves farther away from the frog's midline as the frog begins to turn. An experiment undertaken by Schneider (1973) indicates that similarly misdirected orienting movements by rewired hamsters are far short of the symmetrical locus. Scheider explains this discrepancy by the additional hypothesis that the slow orientation turn of the hamster is really composed of two or three "decisions" to aim for the target. We found more direct confirmation of this idea in our own cinematic studies of orientation by gerbils (Ingle et al. 1979), which provided direct evidence that the first portion of a gerbil's turn toward a seed target is an open loop, whereas the second portion can be modified by additional stimulus sampling at the end of the initial turn.

Striking versus Orienting Responses

The term "grasp reflex" was metaphorical for Hess—it implied a locking on of the fovea to the stimulus, rather than a grasp with jaws or claws. Yet frogs or fishes will lunge at and grasp prey falling within a critical distance (figure 3.2). Is this response also elicited by the tectum? In the frog, the answer is positive; frogs with ipsilateral retinotectal innervation will either orient toward or strike at prey in the mirror-image direction. In the frog, these responses have a distinctive "morphology": The snap involves throwing the body toward the object (even to the side); the orient mainly requires body rotation around a fixed point by a sequence of leg movements. Of course, orientation in the vertical plane is achieved by lowering or elevation of the head in addition to repositioning of the feet. Since a frog may either orient toward or snap at prey along a given line of sight (depending upon its distance), we must suppose that most regions of the frog tectum harbor two classes of efferent neuron with distinct tectofugal targets. Confirmation of this supposition comes from our recent analyses of frogs with transections of the ventral tegmental crossing of tectal efferent fibers. This disconnection produced a frog that could not turn its body toward prey objects moving in the lateral visual field, but would instead leap straight forward and extrude the tongue along the midline (figure 3.3).

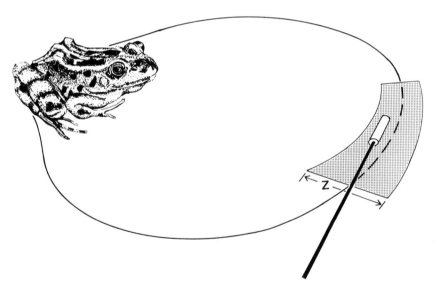

Figure 3.2 Snapping zone of a typical *Rana pipiens* frog, as derived from film records of feeding behavior elicited by wormlike dummy stimulus. When dummy moves along floor within outlined area, frog usually leaps forward and strikes by extending tongue. The same dummy moving outside this snap zone elicits a body reorientation or a short hop without the snap. The "zone of ambiguity" (Z) is that region where snaps and orienting responses occur equally often. This 1-inch-wide zone seems to reflect the limits of the frog's ability to discriminate absolute distances, as predicted by Collett and Harkness (this volume).

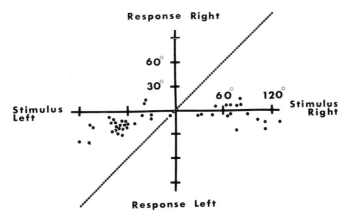

Figure 3.3 Graphic representation of the constricted snap zone of a split-tegmentum frog. All stimuli up to 120° from midline in either direction elicit snaps in vicinity of frontal midline. By contrast, responses of normal frogs would fall along the dashed line, where response angle equals stimulus angle.

Clearly, the control of jumping and of tongue extrusion are mediated by ipsilateral fibers, while turning toward prey depends upon a contralateral target of the tectum.

The dichotomy between investigative orienting and striking with the beak or jaws also appears in birds and in reptiles. In fact, the head-cocking movements of birds can involve turning the beak farther away from a frontal-field stimulus so as to bring the stimulus image onto the central retina (Erichsen 1979). The pecking response, on the other hand, is controlled by objects in the lower rostral visual field (Goodale, this volume), and may utilize another area of retinal specialization. As yet, there is no precise analysis of the dependence of either movement upon integrity of the optic tectum. The complex architecture of the avian tectum certainly fuels speculation that two or more movement patterns can be controlled from this area in response to selective visual filtering. A pioneering study by Friedman (1975) suggested that head-cocking and pecking are elicited by quite different regions of the visual field in doves; however, there was no critical analysis of the role of distance in the birds' decisions. The frequently heard statement that frogs orient toward lateral stimuli and snap at frontal objects proved too simple when the role of stimulus depth was carefully considered (Ingle 1968, 1976), and the parallel statement regarding birds may have to be similarly qualified. Just as we have described a snapping zone in the frog, we need to plot out pecking zones in birds, tongue-licking zones in iguanid lizards, and paw-grasping zones in gerbils reaching for food.

Head Rotation During Grasping
Although the striking responses of fishes and amphibians appear to be all-or-none fixed patterns, there is more subtlety in the grasping patterns of reptiles and birds. A newly hatched chick quickly learns how to

rotate its head to grasp a mealworm effectively. Indeed, it spends a conspicuous amount of time eyeing a worm via different head rotations, as if trying to develop a practical "description" of the same worm viewed from different angles. Reptiles show comparable abilities to rotate the head during food grasping or aggressive biting. We measured head rotation in a trained iguana while the animal was craning its neck to seize an elongated piece of liver. The head was quickly rotated by as much as 90° to match the orientation of the target. In our experience, nothing like this is seen in frogs and toads. Indeed, toads will persist in failing efforts to grasp a vertically climbing earthworm with a horizontal mouth position. Because reptiles have evolved a substantially larger telencephalic visual projection target (ectostriatum) than amphibians (Hall and Ebner 1970) and because electrical stimulation of reptile (but not amphibian) telecephalon elicits orienting movements of various kinds (Distel 1977), we suggest that one achievement of the telencephon in reptiles is the addition of this new dimension of visuomotor coordination. Since chicks with telencephalic ablation readily peck and grasp food, it should be possible to test this suggestion that loss of visually elicited head rotation can be obtained without a gross loss of localization accuracy for simple targets.

In the social behavior of lizards and birds, head movements have a distinctive signaling value. Observations of fighting between pairs of male *Anolis* lizards show that jaw fencing is an intricate visuomotor skill that is both "communicative" and potentially effective in combat. The timing and extent of mouth opening (in addition to head orientation) seem to be controlled by visual monitoring of head movements of the opponent. This behavior should also be appropriate for analysis by the lesion method. Indeed, Greenberg (1977) has already implicated telencephalic regions in the recognition of intruding males by *Anolis*. The anatomical relationship of "enemy recognition" systems to practical visuomotor systems for combat with the enemy should be an intriguing problem for the neuroethologist.

Orientation Away from Threat
The observations of Akert (1949) that "weak" stimulation of the trout's tectum could elicit eye movements toward the contralateral field and strong stimulation would produce the reverse saccade direction suggested that both approach and avoidance turns might be mediated by tectum. This is in line with the observation of Bechterev (1884) that ablation of tectum in the frog resulted in "blindness" to visual threat as well as to moving food objects. The dependence of visual threat-avoidance behaviors (sidestepping, ducking, jumping away) on tectum has been confirmed (Ingle 1973a; 1977), and we have added new data by filming the avoidance response to approaching black disks of frogs with regeneration of optic tract to the ipsilateral tectum. In these frogs, all

jumps from threat are now directed to the side of the threatening disk instead of away to safety. This new evidence indicates not only that tectum is necessary for activation of avoidance behavior, but that the spatial information inherent in the retinorectal projection is sufficient for direction of the spatial avoidance jumps. In coming to that strong conclusion, we must add that anatomical and physiological studies show that projections to other thalamic and midbrain centers remain normal, despite a reversed tectal projection.

Studies of the role of tectum in other vertebrate groups should reveal whether the dual role of tectum holds generally. The goldfish appears to show a complete loss of threat avoidance after tectum ablations as well (Springer et al. 1977). This is probably not due to surgical pressure upon more rostral diencephlic or pretectal structures; our unpublished studies show that even caudal tectum ablation in goldfish produces a clear loss of spontaneous avoidance of a red disk approaching from the rear. Nonetheless, it would be important to analyze avoidance movements in goldfish with wrong-way retinotectal projections.

The participation of the caudal thalamus in threat-avoidance behavior of toads was proposed by Ewert (1970) on the grounds that large ablations of this area abolished such behavior and that electrical stimulation of the general area activated ducking or sidestepping responses in free-moving *Bufo bufo* (Ewert 1967). Our unpublished studies have replicated such observations in *Bufo marinus*, but with the addition that directed escape behavior is no longer elicited by stimulating the caudal thalamus on the same side as a unilateral tectum ablation. From this initial evidence, one can imagine that diencephalic activation "motivates" the animal to escape but that a visuomotor link via the tectum tells the toad where to jump or run. The exact role of caudal thalamus is called into question by our more recent experiments (Ingle 1980) on frogs with large caudal thalamic ablations. Such frogs are initially unable to avoid large dark objects moved toward them. As Ewert (1970) described for toads, such lesions result in a "disinhibited" animal that actually *pursues* large stimuli as prey. However, within a week or two all of our lesioned frogs had recovered nearly normal avoidance behavior, despite absence of the entire caudal half of the thalamus in some cases. The efficiency of avoidance is shown the jump-direction histograms of figure 3.4. For this reason, I conclude that visual input to caudal thalamus is not absolutely necessary for activation of the tectal mechanism that directs jumps away from threats. However, we did not test this behavior with other stimulus configurations, as Ewert did. It remains possible that the toad's tendency to avoid a group of four spots placed about 40° apart is a specific "object recognition" function of the protectum. Thus, the residual avoidance capacity of the caudal-thalamus-lesioned frog or toad might include only a response to large dark objects, and not to more subtle configurations that frighten the normal toad (Ewert and Rehn 1969).

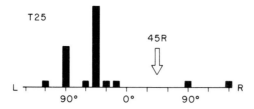

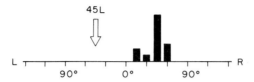

Figure 3.4 Avoidance jump-direction histogram for frog with pretectum ablated, showing normal topography. (Source: Ingle 1980.) When a 4-inch-wide black disk approaches at 45° from the midline on either right or left side, nearly all escape jumps are directed to the opposite side of the field. The same frog could not turn even 30° to avoid a grid barrier, as shown in figure 3.8.

Phototaxis

Fishes, amphibians, and reptiles often show the tendency to orient spontaneously towards a source of light within an otherwise dim environment. Frogs can be easily evaluated for phototactic tendencies by placing them within a dark enclosure, facing between symmetrical windows with different levels of back-illumination or different colors. Jaeger and Hailman (1973) conducted a mammoth study of more than 100 species of frog, nearly all of which showed positive phototaxis behavior. They found that this behavior was remarkably stable during laboratory tests: There was little variation in jump frequency versus intensity difference functions for a given species with variations in age, sex, season, or daily light cycle. A quantitative study of frog phototaxis by Muntz (1962a) showed that spectrally pure blue light was the most effective light source for attracting frogs. He followed up these demonstrations with the discovery that one class of optic-tract axons, the "sustained-on" fibers, were similarly blue-sensitive (Muntz 1962b). The finding that these units terminated only in the rostral dorsal thalamus suggested that this retinal-projection target is the only afferent channel that mediates color preference.

The hypothesis of Muntz was apparently confirmed by Kicliter (1973), who ablated frogs' rostral thalamus and produced a loss of color preference in a two-window preference test. However, the tendency of these frogs to orient toward a bright window was not totally abolished by these large thalamic lesions. The question thus remains as to whether the projection to rostral thalamus is the only channel that guides the animal toward light sources. Kicliter also showed that phototaxis survived splitting of the optic chiasm and thus could be mediated by ipsilateral fiber projections. In the species used by Kicliter (*Rana pipiens*) the ipsilateral projection to the pretectal area is rather sparse. The conclusion of Kicliter was confirmed by a similar method (Ingle, unpublished data): Frogs with both optic tract and optic nerve sectioned on the same side were able to jump normally toward a light source on either side of a dark compartment, despite their total loss of prey detection and threat avoidance by vision. From such data it might be assumed that orientation of frogs toward bright windows is mediated by a visuomotor mechanism quite separate from that used for orientation to prey. This assumption is premature without a further demonstration that ablation of tectum does not impair phototaxis, and is clouded by the observations of Kicliter (1973) that severing the optic input to tectum at its rostral margin does not impair phototaxis, but that removal of tectum itself plus incursions into pretectum and tegmentum does produce a substantial deficit in phototaxis. Whether the extratectal damage accounts for the observed deficit or whether a thalamotectal pathway (Trachtenberg and Ingle 1974) is required for orientation toward light is not known.

In tests of phototactic behavior, there are circumstances where the lighter aperture is not the preferred target. Our own studies showed that normal frogs (but not the ipsilateral group) almost always followed a jump toward the light panel with a subsequent turn and hop into the adjacent shadow of the opaque background next to the lighted panel. It seemed as if the frogs avoided too much light after drawing closer to the light source. An earlier study by Mrosovsky and Tress (1966) noted that frogs did not always approach a lighted aperture in preference to a dark aperture within a gray arena. One factor that exerted a definite bias on aperture choice was the degree of preexposure to a bright area. Frogs so exposed tended to choose the dark aperture, as if to hide. Mrosovsky and Tress carried out one experimental test of the hypothesis that the ambient light level was "measured" by the intracranial photoreceptors (the stirnorgan). By covering the frogs' heads with opaque "hats," they abolished the dark-aperture preference. These observations plus our own suggest that a dark-orienting (or light-avoiding) visuomotor channel coexists with the phototactic system. This channel must be independent of the system (involving optic tectum) that mediates threat avoidance in anurans, since this class of avoidance behavior is elicited by dark rather than light objects (Ewert and Rehn 1969). The recent discovery that the frog's stirnorgan projects to the pretectal region (Eldred et al. 1980) suggests new physiological experiments on the integration of optic information and the nonoptic light-sensitive channel.

Negotiation of Barriers and Apertures
At first thought, the tendency of an animal to jump away from an approaching threat resembles the reluctance of a moving organism to collide with obstacles in its path. However, the routes taken by frogs

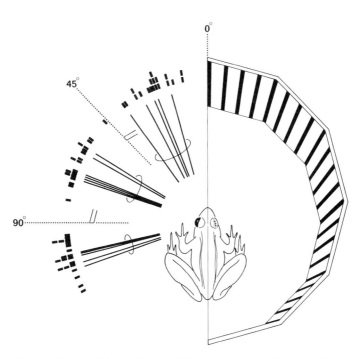

Figure 3.5 Normal jump-direction histogram derived from five frogs with ablation of optic tectum contralateral to open eye. (Source: Ingle 1973a.) In these tests three different barrier positions elicited three distributions of detour directions, as shown by the three encircled response clusters of one frog. In each case the animal turned just far enough to clear lateral edge of barrier. Similar turning accuracy of four other frogs is indicated by short vertical line segments located just to left of each barrier edge.

partially surrounded by a barrier (figure 3.5) do not resemble the avoidance directions elicited by visual threat (figure 3.4). Frogs clear the barrier edge by the minimal required distance, instead of maximizing the distance from the stimulus as in threat avoidance. Evidence that these two avoidance behaviors reflect separate visual subsystems came from experiments on frogs with rewired ipsilateral retinotectal innervation. These frogs responded to an approaching black disk by jumping toward the threat, yet they correctly negotiated barriers. Since I also found that frogs with bilateral tectal ablations could detour accurately around such barriers (Ingle 1977), it seems that barrier detection is not a tectal function. The survival of detours around barrier edges as far as 90° to the side provides two important controls for interpretation of the tectal deficit: First, it is difficult to argue that tectum has merely a nonspecific role in "energizing" approach or avoidance tendencies, since the tectumless frogs are able to respond briskly to mildly noxious skin stimulation and make sizable leaps to escape. Second, these animals have not lost the motor flexibility to make accurate turns; in fact, their orientation movement to clear the barrier edge closely resembles the turning sequence normally elicited by prey.

Frogs enclosed in an arena will usually orient and jump through apertures slightly wider than their own body size. The decision can usually be hastened by moving a dark threat object overhead. Unpublished studies show further that prompt approach to apertures usually depends on the frog's ability to see through the hole—that is, to judge depth disparity between the enclosing wall and the background behind the hole. Orientations toward dark squares taped to the wall surface can be elicited, but attempts to "jump through" these are infrequent. Since the animal aims directly for the hole, the behavior might seem to be a "targeting" behavior, like prey catching. On the other hand, one can interpret "aiming for apertures" as an avoidance of the enclosing solid surfaces. Studies with atectal frogs support the second view, since aperture approach (figure 3.6), like barrier detours, remains intact, while prey-catching is abolished (Ingle 1977). I propose to classify tectal-mediated behaviors of frogs as turning toward or away from moving (animate) objects, while both barriers and apertures can be classified together as stationary (inanimate) objects. Here it seems that stimulus category, rather than type of response, distinguishes tectal functions from those of other visual centers of the frog.

In order to further prove the independence of prey-induced and threat-induced behaviors from detection of barriers and apertures, I have studied the behavior of frogs with lesions of the so-called pretectal region, which receives an independent retinal projection (Scalia et al. 1968; Lázár and Székely 1969). Figure 3.7 gives a schematic view of the larger type of lesion, as reconstructed from stained sections through the thalamus and midbrain. Removing either the posterocentral cell group or the retinal-terminal zone plus the posterolateral cell group proved sufficient to produce a permanent loss of ability to avoid collision with the standard hemicylindrical grid barrier (Ingle 1980). Figure 3.8 (left) shows that such frogs jump randomly to escape mild tactile stimulation, even when the rostral edge of the barrier is on the frog's midline and a slight detour would suffice. These animals are able to strike accurately at prey, and show efficient and prompt jumps away from the approaching black disks after a week's postoperative recovery. Control lesions of the anterior thalamus (which receives a large independent retinal projection) did not produce a loss of barrier avoidance in the same tests (figure 3.8, right). Further tests with apertures showed that frogs with pretectal lesions were unable to turn toward the open window with better than chance frequency.

Ewert (1971) recorded units within the toad's caudal thalamus (the pretectal region) that gave a prolonged discharge to large stationary dark objects placed within the excitatory receptive field (ERF). Using frogs with bilateral tectum ablations, I have shown that such stationary-edge-detecting neurons are frequently recorded in the pretectal area (Ingle, unpublished). I have not found such neurons with

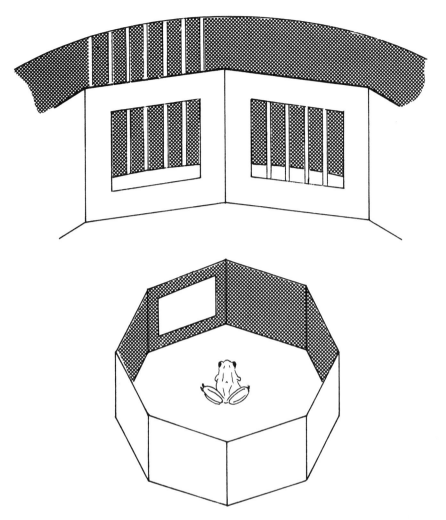

Figure 3.6 Aperture-detection test used by Ingle (1977) in experiments on frogs with bilateral tectum ablation. In the simpler version, the frog has only to turn about 30° in order to escape noxious stimulation by jumping through the open aperture. Both black and white enclosures were used in these tests. A more difficult test is shown at the top: discrimination between an "open" and a "closed" window, where white stripes equated for equal angular size are set behind each aperture. From the frog's perspective the only difference between the windows was the distance of the striped barrier from the aperture. Tectumless frogs showed normal depth discrimination in this test.

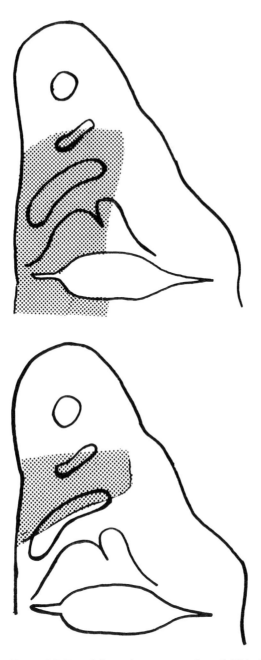

Figure 3.7 Top: Schematic representation of CT lesion in frog T-25. Shading indicates area of tissue removed. Note that the region included all of PC and PL as well as the retinal fiber projection to pretectum. The lesion extended downward to the ventral thalamus. Bottom: Representation of lateral tectal lesion in frog LT-5. In this animal, as was typical, all of the PC group and much of the PL group was spared, but the retinopretectal projection zone was ablated.

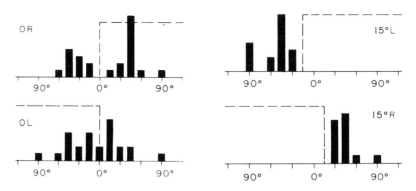

Figure 3.8 Left: jump distributions to the large 180°-wide barrier on OL and OR positions by frog with bilateral ablation of pretectal region (same as in figure 3.7, top). Note that the barrier is ignored in each case; the distributions remain symmetrical as they did with no barrier present. Right: distributions for an anterior thalamic lesioned control frog that consistently jumped to the open side of the barrier in either location (15° right and 15° left).

large (30°–60°) ERFs in the frog's optic tectum (see also Lettvin et al. 1961; Grüsser and Grüsser-Cornehls 1976). These data reinforce the conclusion that the stimulus-detection capacities of tectum and pretectum are substantially different: the tectum alone harbors neurons with selective response to small moving spots, and the pretectum alone shows specialization for detection of large stationary stripes or boundaries. This dichotomy is complicated by the fact that frog tectum receives some retinal axons (class 1 fibers) that are responsive to stationary edges. It remains unclear why this afferent-fiber characteristic has not been noticed among postsynaptic cells of the tectum yet is found among intrinsic neurons of pretectum even after tectum has been removed.

Stepping Onto Flat Surfaces
Frogs will not only treat stationary surfaces as obstacles to be avoided, but will also use them as places upon which to step in preference to falling off a high perch. In order to analyze this surface-approach tendency, I have adapted the visual-cliff method initiated by Walk and Gibson (1961). Figure 3.9 shows how one or two wedge-shaped surfaces can be placed 1 inch below a pedestal from which a frog is easily induced to escape. Normal frogs nearly always step down upon a patterned surface rather than falling 18 inches to the white floor below. However, when given the choice, frogs strongly prefer to step upon horizontal rather than vertical stripes. They have no apparent preference for horizontal stripes over a randomly splotched pattern with the same black-to-white ratio. This spontaneous discrimination of stripe orientation is comparable to an earlier observation (Ingle 1971a) that frogs readily detoured around a vertical grid barrier to pursue a worm seen between the stripes, but the same animals usually collided with an

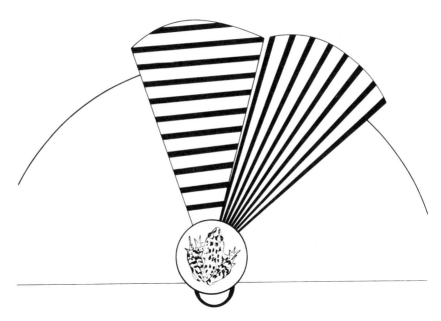

Figure 3.9 Step-down preference test demonstrating the normal frog's discrimination between horizontally and vertically striped surfaces. Frogs consistently step down to the horizontal platform (set 1 inch below) in a large majority of trials. Note that the vertical stripes converge toward the frog in this example; frogs avoided parallel vertical stripes in other tests as well.

equivalent horizontal grid. These demonstrations imply that the frog possesses a central mechanism (perhaps involving the pretectal system) for discriminating line orientation discrimination.

Although it makes sense that a frog would naturally avoid objects such as reeds rising vertically from the ground, it also seems likely that the frog would jump toward an empty space in preference to colliding with a horizontally striped barrier. Would the tendency to approach horizontal stripes hold true only for the lower visual field (where the alternative is falling), or would frogs also approach an eye-level array of horizontal stripes? A recent test of 15 normal frogs (*Rana pipiens*) provided a surprising result: 80% of 250 choices were directed toward an upright horizontally striped barrier in preference to escape toward the empty half of a large white arena. Additional unpublished experiments have shown that frogs have a strong tendency to jump toward dark areas when placed within a bright arena. The response toward the horizontal stripes was probably a manifestation of this dark-area preference, since in a gray area frogs seek out a panel with black and gray stripes but ignore one with white and gray stripes. Ewert (1971) noted that one class of pretectal neuron responds selectively to darkening of their receptive fields. Further recording and ablation studies will be necessary to localize this new subsystem, since current evidence does not exclude either tectum or visual thalamus from participation in this "dark-hole" localization behavior.

Optokinetic Nystagmus

When surrounded by a rotating vertically striped cylinder, frogs (like other vertebrates) turn the head slowly in the direction of stripe motion. As with many lower vertebrates, only the nasal direction of stripe motion "captures" the posture of the monocular frog. The natural function of this behavior is assumed to be the approximate stabilization of the retinal image of an animal undergoing passive rotation (a frog on a floating lily pad or a bird on a perch). Vision of a stable environment can even suppress the spontaneous ocular drift of a rabbit sitting still (Collewijn 1969). The frog should be best able to detect real object motion, such as that of prey or of predators, when the retinal image is stable.

Frog studies by Lázár (1973) indicated that ablation of tectum, telencephalon, and rostral thalamus had little effect on the optokinetic response. Incursions into pretectum produced a partial deficit, while lesions invading the basal optic tract completely abolished optokinetic nystagmus (OKN). However, no control experiments using other visually guided behaviors (feeding or avoidance) were carried out with the latter group of frogs to rule out the suggestion that they were too sluggish for any normal sequences of behavior. The finality of Lazar's interpretation is also questioned by our own observations of pretectal-lesioned frogs in which OKN was completely absent. Frogs with unilateral pretectal ablation showed a corresponding loss of OKN only on the side contralateral to the lesion. Histological examination of several animals has shown that the lesion does not enter ventral thalamus or tegmentum and does not approach the basal optic tract. One such frog with a unilateral pretectal lesion showed normal transport of horseradish peroxidase from the optic nerve to the terminals of the basal optic tract. Although our data suggest that innervation of the basal optic tract is not sufficient for mediation of OKN in the frog in absence of the pretectum, it remains possible that both projection targets of the retina are required for OKN.

Distinction Between Orienting Subsystems

Several of our observations, as reviewed above, support the idea that separate retinofugal protection targets can be distinguished in terms of visual stimulus selectivity rather than in terms of the motor typology of the elicited behavior. For example, orientations toward prey, past the edge of a barrier, and toward a lighted aperture involve the same essential motor sequence elicited by quite distinctive stimuli. The three sets of behaviors seem to be guided by different retinal inputs: class 2 motion-sensitive units in the tectum, a class of edge-sensitive units found in pretectum (Ingle, unpublished data), and blue-sensitive "on" units found in anterior thalamus. On the other hand, the same type of retinal input can trigger either snapping or orienting responses via the tectum, depending upon the stimulus distance. In distinction to these "sudden" orienting responses, OKN represents a relatively slow adjustment to a large moving surround. Here, both stimulus and response modes are unique. In the case of the frog, it seems likely that our main future task is not to distinguish more classes of visuomotor behavior, but to appreciate the variety and subtlety of cues used during object recognition, surface discrimination, and depth judgments (see, for example, Collett and Harkness, this volume). Indeed, it will probably take less experimental effort to trace connections from each retinofugal target to specific components of the motor system than to understand the integrative mechanisms within those visual centers that contribute to object vision (Ingle 1981).

Integration of Visual Subsystems: Prey-Barrier Interactions

Thus far we have been concerned with simple responses to a single visual object. In real life, responses to complex visual stimuli are likely to be interactive. A frog not only avoids a visual threat, but aims for a hiding place, taking care not to strike a tree root during his frantic dash to safety. As we consider the choices to be made when two or more objects influence an animal's decision, we move from a "reflexology" of behavior toward those decisions that demand "higher integrative functions." One such problem that lends itself to precise study is the pursuit of prey objects that move behind barriers or are otherwise out of reach. I have previously described the basic capacities of frogs and toads to make adaptive detour movements around semitransparent barriers to pursue food (Ingle 1970; 1971a). An interposed barrier has two effects: It inhibits a direct strike at a prey within the usual snap zone, and it often induces a sidestepping sequence by which the frog can peer around the barrier edge in position to jump forward at the prey. Thus, the two stimuli, with an appropriate spatial relationship, can induce a new movement sequence never elicited by either alone.

The perceptual integration of moving prey and stationary barrier must in some way reflect the interaction of tectum and pretectum. Since we find that a barrier does not elicit movement but only modifies the movements elicited by other visual or tactile stimuli, it seems likely that pretectal "barrier-detecting" neurons modify the discharge patterns of neurons in the optic tectum that direct the animal toward prey. The simplest mechanism here would be monosynaptic inhibition of efferent tectal neurons by cells of the pretectum. Wilczynski and Northcutt (1978), using transport of horseradish peroxidase, showed that cell groups of the pretectal region send axons directly into the tectum. These pretectal axons might directly inhibit neurons sensitive to prey stimuli located within that region of tectum corresponding to the location of the barrier. On the further assumption that prey-elicited excitation of tectal neurons can "spread" laterally within the tectum, weakly excited

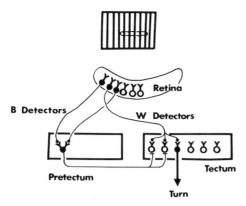

Figure 3.10 Schematic model to explain the ability of a frog to detour around a grid barrier to approach a moving worm. The retinal projections to the pretectum—"barrier (B) detectors"—encode the stationary object; the projection to tectum includes ganglion cells which encode the prey—"worm (W) detectors." Inhibitory fibers from pretectum block the response of tectal neurons to prey in that part of tectum corresponding to the barrier image. However, neurons whose receptive field centers are just lateral to the inhibited zone are able to discharge, initiating a turn beyond the edge of the barrier.

neurons located beyond the inhibited zone might direct the frog's orientation toward that region just beyond the barrier edge. This scheme is illustrated in figure 3.10.

This scheme demands some further assumptions. Lock and Collett (1979) reported that toads are able to reorient beyond the edge of a grid barrier in a direction at least 30° beyond the locus of the eliciting prey object. This fact implies that the critical tectal efferent neurons can summate excitation over a 30° range in one or both directions. In the latter case, the efferent-cell ERF would measure 60° across; yet cells with prey-detection properties have not been described as having such large ERFs. On the other hand, Ewert and Borchers (1971) and Ingle (1973b) reported that many of the prey-selective neurons in tectal layer 8 have 30° ERFs extending off center from the locus of retinal fibers. Thus, it may be possible that the response direction mediated by these prey-detector cells is not that associated with the center of the ERF but that of one side of the field. Therefore, an objective account of the ERF structure of these neurons and the frequencies of their rostral versus caudal deviations is essential to the formation of a realistic neural model of detour behavior.

A second assumption is that the rather weak excitation derived from stimulating the periphery of tectal ERFs must "build up" to threshold level over time. It is true that frogs or toads confronted with occluded prey require a few seconds to initiate a detour response, even when their snap toward an open prey is always elicited within 1 second. Furthermore, earlier studies (Ingle 1975) showed that readiness to snap at prey does build up over a few seconds during very brief successive prey movements. Another helpful addition to our model might be the assumption of mutual lateral inhibition between neighboring efferent neurons. As figure 3.10 shows, the effect of pretectal inhibition on a row of tectal neurons (corresponding to the barrier locus) is, in this case, to produce a "disinhibition" effect just beyond the boundary of pretectal modulation, so that this subset of efferent neurons will more easily respond to "nontopographic" retinal activation. The result of this shift in thresholds within the tectal map is that the frog will turn rather quickly beyond the edge of the occluding barrier.

Although tests of this particular model might be most decisively carried out by recording single tectal neurons in the freely moving animal during prey and barrier negotiations, a simpler experiment allows further tests of some assumptions of the model. We made the discovery (Ingle 1970) that distortions in stimulus-response mapping by frogs could be induced by partial undercuts of the lateral tectum. As figure 3.11 shows, cuts orthogonal to the presumed course of efferent tectal fibers (seen more directly in coronal sections) sever output connection from tectal neurons located medially to the cut. If these were activated by moving prey in the frog's upper visual field, one would expect large overshoot orientations toward the rear visual field. This is explained by the assumption that excitation of more caudal neurons with intact effer-

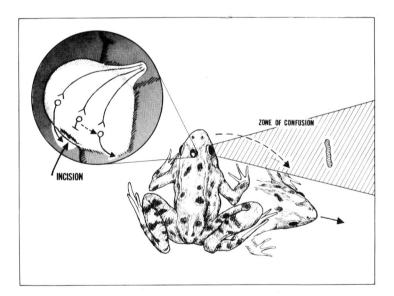

Figure 3.11 Schematic view of frog's left optic tectum where a localized horizontal cut has severed efferent fibers from the middle region of tectum. Incoming retinal fibers can activate neurons in all three regions, but only the anterior and posterior neurons still have efferent outflow intact. According to this model, when a prey object activates neurons of the middle region, the lateral spread of excitation (dashed line) activates neurons in neighboring regions, resulting in overshooting or undershooting of the prey depending upon exact prey location.

ent axons could now mediate orientation elicited by the prey. Using film records, such overshoots were observed in two frogs with partial undercuts of one tectum (Ingle 1970; see figure 3.11).

Our pretectal-inhibition model is somewhat complicated by the observation (Ingle 1971b) that toads detour even when the prey and the barrier are seen with opposite eyes. In this study, I placed a partially occluding eye cover on *Bufo marinus* toads and presented prey stimuli in the occluded portion of the visual field but in view of the contralateral eye. As figure 3.12 shows, toads made distinctive detour movements according to the positions of the prey and the barrier edge. The existence of an ipsilateral retinal input to pretectum might explain this ability, since new experiments have now revealed some barrier-detection capacity in frogs using ipsilateral input alone. However, there is good evidence that pretectal neurons project significantly to the contralateral tectum as well as to the ipsilateral side (Trachtenberg and Ingle 1974; Wilczynski and Northcutt 1978). The role of this projection system could be tested by interocular integration experiments in frogs or toads in which pretectal efferents crossing via the posterior commissure are severed.

Even if we knew the route of pretectal-tectal interaction, other conceptual problems in understanding the perceptual judgments underlying detour behavior would remain. Lock and Collett (1979) demonstrated that toads (*Bufo marinus*) are remarkably good in judging the depth disparity between a worm and a barrier while standing at a distance from the barrier. The worm must be a critical distance behind the barrier before detour attempts become frequent. In fact, the critical worm-barrier distance remains the same when the toad stands at various distances from the barrier. A student in my laboratory (Sopher) has repeated these observations on *Rana pipiens* with just the same results. This achievement raises the question of how the animal can judge distance disparity when each object is registered initially within a different visual projection. Although Collett (1977) showed that toads can measure prey depth via a stereoscopic mechanism, it is hard to imagine that pretectum-tectum connections preserve the fine-grain topography required for barrier-edge information to be integrated into a tectal stereoscopic system.

One possibility is that the prey-barrier distance is actually measured within the tectum, with direct retinal information used to represent "prey" and "barrier edge." Remember that class 1 units of the frog's tectum continue to discharge in response to a long stationary border in their receptive field, whereas the class 2 fibers do not (Lettvin et al. 1959). If the tectum can measure disparities between two images from opposite eyes, it may also note disparities between regions where only class 1 fibers discharge (the edge) and regions where both classes are active (the prey). A possible weakness in this argument is the failure of

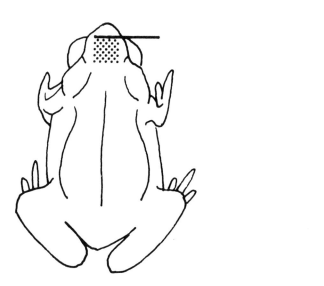

Figure 3.12 Cinematic reconstruction of orienting responses to wormlike stimuli in rostral visual field of toad wearing opaque white occluder in front of right eye. All worm loci (O) on left side resulted in direct locomotor approaches, whereas loci occluded from right eye by barrier (●) resulted in detour sidestep movements. This response derives from interocular integration of an edge stimulus (right eye) and worm location (left eye).

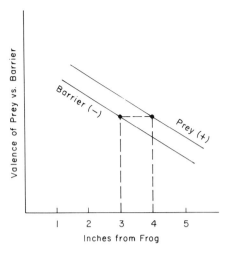

Figure 3.13 Schematic illustration of a "conflict model" to explain distance constancy in the frogs' detour behavior. Attractiveness (positive valence) of prey is assumed to decrease linearly with distance from the frog, while repelling effect of the barrier (negative valence) does likewise. At any given distance, positive value of prey exceeds negative value of the barrier, so frog will always snap directly at the prey. In this diagram, prey 1 inch behind barrier equals negative barrier valence; that is, frog will detour in about 50% of test trials. When barrier-worm distance exceeds 1 inch, frog will usually detour because barrier effect dominates decision. Because the two curves are parallel (an untested assumption), the critical worm-to-barrier distance for 50% detours is invariant with frog-to-barrier distance.

Ewert and Hock (1972) to find class 1 units in the tectum of *Bufo bufo*. However, the most superficial retinal terminals in the road tectum may be sufficiently responsive to stationary edges, if the facilitatory effect of small eye movements (induced by respiration) is considered.

A more parsimonious solution would be to treat discrimination of barrier distance and of prey distance as two independent visual processes, inducing competing response tendencies in the animal: *strike* and *avoid collision*. These conflicting tendencies might be resolved at the brain-stem level. It is known that the attractiveness of prey varies as a function of frog-worm distance (Ingle and Cook 1977), and it seems likely that the inhibitory (repelling) strength of a barrier also decreases with distance from the animal. As figure 3.13 demonstrates, adjustment of the *food attraction* versus distance function and the *barrier repulsion* versus distance function can yield the prediction that detour movements should begin at a fixed prey-barrier distance regardless of the absolute distance from the frog. Although the linearity of these functions seems unlikely in reality, this assumption is subject to independent experimental testing. Successful modification of this type of explanation would undercut the difficult problems of depth-disparity judgment mentioned in the preceding paragraph.

Size Constancy During Prey-Barrier Integration

Lock and Collett (1979) reported that toads show a kind of "size-distance constancy" when judging the width of an aperture in a striped barrier through which they can see moving prey (see figure 4.12 in the chapter of Collett and Harkness). Their study follows up an observation of ours (Ingle and Cook 1977) that frogs judge aperture width during escape behavior (figure 3.14) in terms of the visual angle of the gap. Our frogs used the same critical visual angle for deciding whether or not to jump through the gap at distances of 7.5–18.8 cm, but Lock and Collett found that actual gap size, rather than gap angle subtended at the toad's retina, determined the frequency of detour behavior. As those authors pointed out, it is important to explore a number of behavioral situations in asking whether a relatively stereotyped animal (the frog) is capable of "constancy" behavior. In this section, I will first attempt to examine practical differences in the test situations and then discuss issues of experimental design inherent in the "spontaneous preference" method used in our laboratory and that of Lock and Collett.

First, let us assume that toads do have the kind of perceptual constancy attributed to the higher mammals. Unknown species differences might account for our discrepant results; in an informal experiment in our laboratory, toads—but not frogs—quickly learned to detour in a particular direction for food reward. Though our unpublished experiments with lab-reared froglets tells us that detour behavior per se is unlearned in *Rana pipiens*, it is quite possible that the toad's aperture size judgments are modified by experience. Because the goal of our laboratory is to develop a predictive neural model of size-constancy behavior (see Ingle 1981 for discussion of one model), whether size constancy depends on experience remains an unanswered question.

However, I favor a second alternative explanation of the reported experimental differences between frogs and toads: that different visual cues and different behaviors were manifest in the two test situations. The frogs faced two stark black posts, 1 cm wide, standing symmetrically about the animals' frontal midline axis (figure 3.14). Unfortunately, we have no evidence that the frogs did accurately judge the depth of these posts under the test circumstances. A surface defined by multiple contours might have been better resolved in depth. Furthermore, we suddenly frightened the frog by moving dark objects overhead or by pinching the tail. It is possible that the animals did not have (or take) time for accommodation cues to operate before deciding where to jump. The toads tested by Locke and Collett had much more time and were not "frightened" by the situation (they calmly stalked their prey), and in moving slowly along they also obtained good motion-parallax information not available to the stationary frogs.

Another behavioral distinction is obvious: The frogs committed

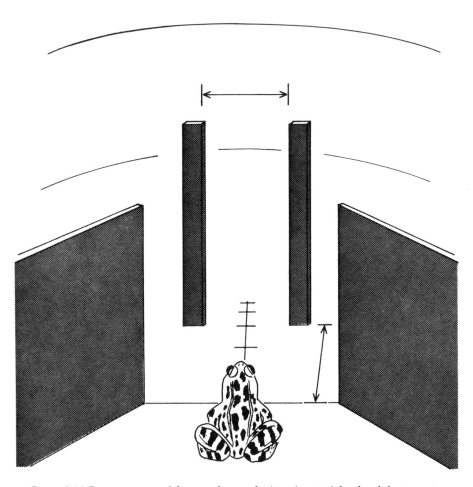

Figure 3.14 Frogs were tested for a preference for jumping straight ahead, between two vertical black posts, or for leaping off to one side. Since the distances between posts were varied, we determined the relationship between willingness to jump ahead to the inter-post distance (or visual angle). This critical distance was determined for the same subjects at each of four barrier distances from the frog's nose: 7.5, 11.3, 15.0, and 18.8 cm. The critical angular distance at which frogs were disinclined to jump between the posts (about 25°) was the same for each viewing distance. (Source: Ingle and Cook 1977.)

themselves to a long jump and possible collision, whereas the toads could change their minds at any step of the sequence. It seems possible that frogs would be less accurate in jumping through the more distant gap and would require more "clearance room" in proportion to the distance jumped. Thus, it might not be adaptive for a frog to follow a tight size-constancy rule when judging apertures to be negotiated by a single jump. Our frogs were realistic about clearing the gap at a 7.5 cm distance, since the gap accepted was only about 1 cm larger than their head width. In a later unpublished study, we found that small (4 cm long) and large (8 cm) *Rana pipiens* had very different (but realistic) criteria as to the smallest aperture that they could clear at a 7.5-cm distance. It would not be adaptive for frogs to behave as if objects were all at this distance, since following the same "law of visual angle" could result in a fatal mistake if the aperture were only 5 cm away.

While the tests by Lock and Collett (chapter 4) give results very different from our own, there remain some untested assumptions underlying their interpretation regarding size constancy. When testing for discrimination of gap width, Lock and Collett placed toads at a greater distance from the same barrier. The gaps were now framed by contours of different stripe intervals and a reduced height. If we view the toad's decision as a conflict between approaching food and avoiding a nearby pair of barrier edges, then the summation of the "repulsion" attributed to the barrier be reduced as gap height and stripe size are diminished. A near gap has greater edge height than a more distant gap, so its "repulsion" effect with an x-cm gap may be equivalent to the repulsion effect of a more distant gap whose angular width is $x/2$ cm. This kind of explanation is a simple alternative to the view that toads perceive gap size per se, as the human observer does. Further studies are necessary to show that the toad's behavior is controlled by the real size of the gap, regardless of the large variations in stripe height or width present in the experiment of Lock and Collett (1979). Our own studies of "distance constancy" in the frog (Ingle and Sopher, unpublished) did replicate the previous result of Lock and Collett when 6-inch-wide barriers were set at two distances from the animal, but we did not find the constancy effect when using barriers only 3 inches wide. Thus, we believe that the circumstances under which various "constancy" discriminations operate need to be better defined.

Visuomotor Mechanisms in Lower Mammals

A limited set of questions will be tackled in this section: How do orienting functions of rodents compare with those described for the frog? Can we link one or more of these classes of behavior to the function of optic tectum? What does visual cortex contribute to visually guided orienting behavior?

Classes of Visual Behavior in Lower Mammals

We must begin by noting that the classes of behavior described for frogs (above) can be recognized among rodents as well. We can begin with a fundamental subdivision of responses into object pursuit, dangerous-object avoidance, maneuvers with respect to stationary objects, and optokinetic responses. The third class has been studied little, but the chapter by Goodale reviews most of what we know about abilities of rats and gerbils to negotiate barriers and apertures. If we are to homologize this class of behavior (entering apertures and avoiding barriers) between frogs and mammals, our current prediction would be that pretectal structures play critical roles in these behaviors for rodents as well as for frogs. This hypothesis is just now being tested by Goodale.

The supposition that threat-avoidance behavior depends on the function of optic tectum in mammals was not definitively tested until very recently. Some positive evidence appeared in the study of Sprague and Meikle (1965), who noted that the cat's blink response to an approaching hand was absent for some time after tectum lesions. (Those authors noted some recovery of this response at later test periods.) Whether the animal would avert the head or actually flee an unexpected looming object in the periphery was not determined. Indeed, an attempt to elicit avoidance of looming objects in the hamster failed (Rosinski and Keselica 1977)—probably because the animals had habituated to numerous overhead visual events in the laboratory. We have noticed a progressive decrease (but not a complete loss) of this behavior in gerbils reared in glass-walled aquariums during the first months of life, but we have not determined whether the taming effect is maturational or experiential. In any case, brisk avoidance responses in normal hamsters can be restored by using a natural stimulus, such as a "bully" hamster (B. Finlay, pers. comm.) or by restricting the environment of the animal so that a sudden overhead movement is truly surprising. Under the latter circumstance, Merker (1980) found a consistent disappearance of threat-avoidance behavior in hamsters with large tectal lesions. This discovery is an important step in "homologizing" functions of tectum between amphibians and mammals.

Analysis of Orientation and Pursuit by Gerbils

The rest of this section is concerned with the functions of tectum and of visual cortex in pursuit of moving targets. Even casual observation indicates that gerbils, squirrels, and tree shrews resemble frogs in that respect: They merely turn the head or body to "inspect" a novel object appearing in certain parts of the peripheral field, but they turn *and* pursue objects under other conditions. The determinants of the decision to pursue are not known. Although familiarity with the stimulus is sometimes a factor, we have repeatedly observed that gerbils will immediately run toward an unexpected object striking the floor of their

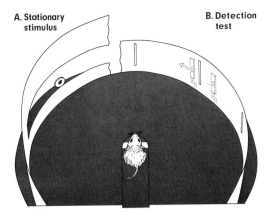

A. Stationary
stimulus

B. Detection
test

Figure 3.15 Cutaway sketch depicting apparatuses used for two experiments. Gerbils were trained to wait on end of platform for a mechanically controlled stimulus to appear. For experiment A the plastic disk was in view continuously for a brief time. For experiment B the stimulus (a seed), on its motorized holder, moved in the horizontal plane, carrying the stimulus quickly past one of seven vertical slits cut in the white wall. If the gerbil did not respond, the experimenter reactivated the motor after a 1.5–2.0-sec delay, so that a second seed was moved past the same aperture.

living tank but will only orient head and body toward a familiar object entering the visual field well above eye level. Thus, the elevation of an object may be a more important determinant of pursuit than familiarity.

Film records of gerbils pursuing familiar food objects (such as a raisin at the end of a wire) indicate that a nearby target will elicit a lunge and a paw grasp. We may be able to plot a "grasping zone" for the gerbil that is analogous to the snapping zone of the frog shown in figure 3.2. Species differences are apparent in some aspects of pursuit: We have consistently failed to teach gerbils to jump even 1 inch off the ground for a seed just out of reach, but I have observed chipmunks trained to leap more than 12 inches to grasp a dangling bread crust on a string. The failure of the gerbils to leap to catch food, as frogs do, is surprising in view of their spontaneous leaps of more than 12 inches to the tops of enclosures.

We carried out a fine-grain analysis of food pursuit sequences by gerbils (Ingle et al. 1979). In order to obtain data comparable to those obtained from frogs, we first had to induce gerbils to wait patiently in a fixed position before a food-baited target would appear. The test arena is shown in figure 3.15. For a stimulus, we chose to present a white disk ringed with black against a white background. A dark sunflower seed was affixed to the center of the disk. During training, the disk was presented first at the end of a stiff white wire and later by a spring-loaded lever which could be triggered with a camera shutter-release cable. This target would snap into position anywhere within a 120°-wide horizontal window and remain fixed in place until the gerbil had grasped the seed reward. We could also present "brief stimuli" by rapidly flipping

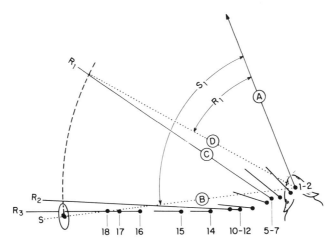

Figure 3.16 Detailed reconstruction made from single-frame film projections of a single pursuit sequence toward a stationary stimulus (S), appearing about 75° to the left of a gerbil's initial head direction (A). The stimulus appears on frame zero. The gerbil begins to move on frame 3, but pauses during frames 5–7 with head direction (C). Line C intersects an arc of points equidistant from the stimulus (S) to define the point of regard. A line drawn from the animal's head center to this point (line D) defines the initial head-turn direction. The angle between lines A and D is the first response angle, R_1 (the angular displacement of the point of regard along the arc during the first component of the turn). Here R_1 is only about half of the initial stimulus angle, S_1. As the animal moves forward (beginning with about frame 10), it completes the second component of the turn (R_2) and is pointing nearly at the stimulus. After another rotational pause, the animal makes a final corrective turn.

the disk in and out of view before the gerbil had time to complete its first head orientation. We could thus test the hypothesis that the whole orienting and pursuit sequence was an "open loop" program.

Using this "stationary" target, we found that normal gerbils show an orienting sequence more complicated than that shown by frogs. Typically, the gerbil responded to the appearance of the target within 200 msec by a sharp head turn. If the target appeared more than about 40° from the animal's frontal midline, the first turn would undershoot the target. The error was roughly proportional to the target eccentricity, such that a gerbil would turn only 55°–60° toward a target appearing 90° from the midline. By contrast, the frog's initial turn to such a stimulus was right on target. However, the gerbil paused only 80–160 msec before initiating a second turn that brought its nose direction nearly on target (figure 3.16). When we examined the accuracy of the second turn in response to brief stimuli, we found that the gerbil "guessed" at the exact stimulus location: The second turn overshot stimuli that had been seen close to the midline but badly undershot stimuli that had been eccentric. We concluded that the gerbil had preprogrammed a continuation of its pursuit turn, but required information during the brief pause to accurately control the turn direction.

Thus, the gerbil initiates two turns (and sometimes a third) to attain the same target position that the frog achieves in a single turn. We doubt that this undershooting is a simple consequence of skeletomuscular limitations. Why would the animal turn only 60° toward a 90° target, when it can turn 90° in response to a target seen at 140°? Why then would the gerbil intentionally undershoot, except that it runs a risk of losing the target by taking the more ambitious turn? Because natural objects a gerbil may pursue (insects or conspecifics) are as likely to move in either direction, an on-target turn would actually overshoot a moving quarry about half the time. We assume that a second corrective turn in the same direction is made more quickly than a reversal of turn direction. This hypothesis could be tested by arranging for the reappearance of the target in a new location at the end of the first turn.

Whatever the adaptive value of initial undershoots, the data obtained suggest that the visuomotor system for head turning in the gerbil does not contain a "motor map" for the rear visual field. In this respect, the normal gerbil behaves similarly to the frog, whose entire retinal projection is compressed upon a rostral "half-tectum" (Ingle 1979). When the gerbil wants to approach an object located 150° to the rear, three successive commands of 90°, 40°, and 20° may be required. We now must consider the assumption that this visuomotor system is located within the optic tectum.

Cinematic Analysis of Tectum-Lesioned Gerbils

In order to compare functions of the gerbil tectum with those well documented for the frog tectum, I carried out cinematic analysis of animals with the superficial and intermediate tectal laminae removed. To achieve reproducible, lamina-specific ablations of tectum, I elected to operate 1–3 days after birth, when the tectum is exposed and good visual inspection of lesion depth through a dissecting microscope is possible. Furthermore, the method of cooling the postnatal pup (introduced to us by Schneider) reduces bleeding to a minimum. As it turned out, our intended superficial lesions removed little more than the superficial gray layer. But our intended deep lesions always spared the lateral portions of the stratum profundum. In this respect, our lesions were not as deep as the deepest lesions reported by Schneider (1969) in hamsters and by Casagrande and Diamond (1974) in tree shrews.

When we studied the response of gerbils with superficial lesions to the standard ringed disks, we observed no significant deficits: These animals oriented accurately within all parts of the visual field. In two of the animals, proline-leucine mixture was injected into both eyes and the retinal projections were examined autoradiographically in Schneider's laboratory. As Schneider (1970) reported for hamsters with similar postnatal lesions, the optic fibers projected evenly across the residual tectum, ending deeper than usual within the intermediate gray. However, in five gerbils with postnatal deep lesions we did not find any such aberrant retinal projections penetrating to the deep gray lamina. Instead, there was only a very limited projection to a spared rim of superficial gray tissue at the rostral-lateral edge of the tectum. Outputs from these cells seemed to be undercut by the lesion.

On the basis of such histological data, we would have supposed that these gerbils would be unable to localize a small disk, except perhaps within the lower frontal visual field. During initial training the gerbils learned to wait on the platform for a food reward but did not seem to notice the visual target unless it was presented directly in front of them. However, this deficit did not remain for long. Within two or three weeks the animals had improved substantially in turning toward disks presented as much as 90° to the side. When film records were analyzed, we were surprised to find that all of the lesioned animals turned and ran fairly accurately toward the stimulus up to locations 60° or 70° from the midline. Beyond this point they tended to undershoot badly, as figure 3.17 indicates. Most animals tended to ignore stimuli appearing more than 100° laterally, but unfortunately too few tests were filmed with caudal field stimuli to make possible a conclusive statement on the peripheral limits of disk detection.

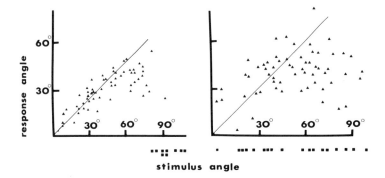

Figure 3.17 Left: Distribution of turn directions for a gerbil with infant ablation of superficial and intermediate tectal laminae plotted with respect to initial stimulus angle. Here, a "turn" actually constituted two rapid head rotations, with the second head movement superimposed on a running sequence. However, these lesioned animals typically stopped for several frames (40 msec/frame) at the end of this turn-and-run sequence. Unlike intact (or cortically damaged) gerbils, these deep-tectum-lesioned gerbils made large undershoot errors for stimuli more than 60°–70° from the frontal midline. Not only did error magnitude increase with further eccentricity, but a number of "wrong-side" responses (represented by ■ below abcissa) occurred when stimuli appeared at angles larger than 80°. Despite these significant deficits for acquiring more peripheral targets, all of these gerbils showed fairly good localization of disks presented rostrally. Right: After an additional lesion (as adults), largely confined to area 17, the same tectum-lesioned gerbil failed to localize baited disks at all. It usually ran off the start block to the correct side, but only "guessed" at locations between 30° and 60°. There was no significant tendency for the turn to be influenced by the angle of the stimulus, such that for small angles overshoots were common. In fact, these animals often ran in the wrong direction, and these errors occurred with equal likelihood for all stimulus loci. The data shown here were replicated when the same animals were tested "unexpectedly" in their large home tanks.

When we examined the detailed orienting pattern of the first two "deep tectum" subjects, we found that most of the undershoot error could be accounted for by failure to make a second turn while running off the platform. The number of runs straight forward much exceeded those obtained for two normal control animals. In other animals we did not measure the R_1 and R_2 components separately, but only the point at which the gerbil stopped turning for a large number of frames, or the point at which it ran past the stimulus. Because these animals did better than we had expected, we became concerned that they were simply using auditory cues from the mechanical release device to localize the disk. Control studies—one covering the aperture during activation of the pop-up stimulus and one using noiseless stimuli moved behind the aperture via long wire handles—showed that the lesioned animals did rely upon visual cues.

Finally, we attempted to compare orienting accuracy toward stimuli in upper versus lower visual fields, on the assumption that greater sparing of stratum profundum in lateral tectum would enhance performance in the lateral lower field. When stimuli were quickly lowered to a position about 20° below eye level, we found an increase in the number of accurate turns to objects within the 70°–90° range. Such turns were absent when the stimulus was lowered to a position about 20° above eye level. The variability in orienting accuracy elicited by the lower field stimulus may be accounted for by the assumption that gerbils sometimes made a decision to turn while the object was still descending through the upper field. This test would have been more informative had we redesigned the apparatus to allow entry of the disk from below.

Our conclusion that good visual orienting capacity can survive lesions that remove nearly all of the retinal innervation from the optic tectum is similar to that of Casagrande and Diamond (1974), who removed superficial layers postnatally from tree shrews. They also concluded, on the basis of Nauta degeneration staining methods, that no significant retinal projections to deeper lamina remained to explain the good functional development of food orienting by these animals. A major alternative source of visual input to the deep layers of the tectum comes from neocortex. In the gerbil, as in the rat, this projection originates from premotor cortex (area 6) and medial frontal cortex (areas 8 and 24), according to an experimental study by Sherman et al. (1979). I pursued this hypothesis by adding to our neonatal lesions of tectum further ablations of visual cortex in adults.

Interpretations of Double-Lesion Experiments
When gerbils with intended superficial lesions received large occipital lesions (removing striate cortex and most of adjacent area 18), there was no striking change in orientation accuracy within the rostral visual field, although more undershooting errors did appear for targets in the far lateral visual field. It seems likely that visual input from deep retinal

projections accounts for this performance, since equivalent lesions of visual cortex in each of six gerbils with deeper tectum ablations did abolish the capacity for directed head turning toward stimuli throughout the visual field (figure 3.17B). As figure 3.17 shows, these animals usually responded to the entry of the disk within the rostral and lateral visual field by running off the waiting platform, but they tended to run randomly within a zone extending from 30° to 70° regardless of where the stimulus actually appeared. In fact, about one-third of all runs were actually directed to the wrong side. This pattern indicates that the gerbil is only "guessing" where the disk might be located. We found the same failure to localize stimuli in two gerbils in which both tectum and cortex ablations had been made postnatally. The failure of the double-lesioned animals in these tests supports the assumption that nonretinal input to tectum can elicit orienting responses, provided that a sufficient amount of deep tectal tissue remains.

Our data plus those of Sherman et al. (1979) enhance speculation as to whether area 24 or a thin adjacent architectonic zone (area 8) may be homologous to the frontal eye fields of the monkey. This intriguing notion should be further encouraged by the finding that frontal-eye-field lesions added to tectal lesions in the monkey can cause a severe restriction of ocular scanning and response to lateral visual stimuli (Conway et al. 1979). We look forward to an experiment in which lesions of areas 8 and 24 are added to early tectum lesions in the gerbil. Would these combined lesions abolish orientation to disks?

Cortical Contributions to Orienting Functions
The visual cortex participates in a system that "compensates" the destruction of the retinotectal projection system, according to our studies of lesion-produced deficits in the ability to localize baited disks. Although the cortical mechanism was revealed only after ablation of most of the optic tectum, it must have unique functions that are not entirely redundant with those of the tectum. We already have one hint: The cortical system is focused on events in the rostral half of the visual field, and does not suffice for accurate turns toward caudal targets. Two categories of function are plausible for such a cortical system: modulation of turning frequency to visual objects, and modification of the temporal or spatial pattern of orienting responses. The effects of motivational state (fear versus hunger) and the effects of stimulus repetition (habituation versus dishabituation) are examples of the first class of factors, and may result in modulation of optic tectum. Though these factors have not been studied in lesioned mammals, short-term "attentional" effects on orienting have been demonstrated in our laboratory in normal gerbils (Ingle et al. 1979). We found that the detection of peripheral stimuli (figure 3.18) is enhanced during the delayed appearance of a second brief stimulus in the same place. However, this effect is not unique to the frontal visual field; on the contrary, it might be of greater use to the

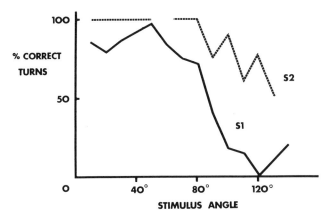

Figure 3.18 Mean percentage of turns to the correct slit as a function of stimulus angle for two gerbils. Detection rates were grouped for each 10° interval. Accurate response to brief (41 msec) stimuli was seen throughout the rostral 80° of the visual field, but not for more peripheral locations. When the gerbil failed to turn toward the stimulus on its first appearance (S_1) and the stimulus was presented again (2 sec later) there were significantly more correct responses for peripheral locations.

animal in aiding detection of rear-field stimuli. Our present evidence for cortical influence on orienting behavior derives from the second type of function: modification of topography between stimulus and response.

Our approach to the possible complexities of orienting behavior was conditioned by a "naturalistic" perspective, which forced us to ask "What are the objects to which gerbils would normally orient?" Two major classes of interesting stimuli involve animate objects: conspecifics and prey. We observed that male gerbils often pursued estrous females and other males carrying food, at high speeds. They also seemed adept at rapid pursuit of crickets after one or two encounters had revealed them to be tasty food objects. The gerbil's pursuit behavior seemed comparable to that of a football tackler, who must quickly estimate the direction and distance of his moving target. Therefore, we added to our standard disk-localization apparatus a moving-stimulus device that could challenge the pursuit capacity of the gerbil. In this test, white disks (without a black rim) emerged from behind one of three or four opaque barriers and moved at velocities of 30°–40° per second around the rim of the semicircular arena (figure 3.19). If the gerbil did not grasp the briskly moving target before it disappeared at one end of the arena, it was not rewarded on that trial. We were surprised that gerbils did not need special training to catch these moving stimuli on even the first test trials, after having been trained with stationary disks.

Cinematic records of normal gerbils revealed a new aspect of orientation elicited by moving targets: These animals typically turned and ran *ahead* of the stimulus, so as to approximate a collision course. On the

Visuomotor Behaviors in Vertebrates

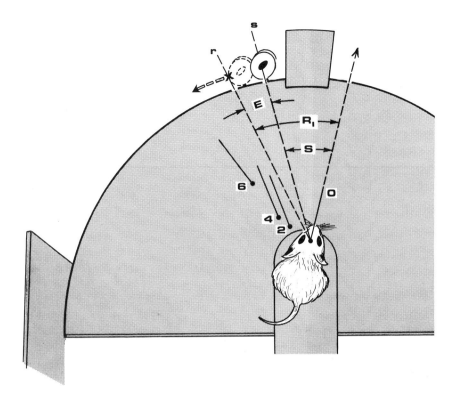

Figure 3.19 The gerbil was trained to pursue moving disks in an apparatus similar to that shown in figure 3.15. Here, the stimulus was moved via a rotating arm attached to a motor located beneath the semicircular platform. As before, the stimulus angle (S) is defined in terms of the seed location on the frame of film before the head turn is evident. Here, the gerbil's first head rotation has ceased by frame 2 (the head direction remains the same for three frames), so that the R_1 direction of regard is extrapolated from the head direction on frame 2. This direction has "overshot" the initial location of the stimulus (s) at the moment when it elicited the response. If the gerbil made the desicion to turn somewhat earlier than reckoned here, that would increase our estimation of response overshoot.

first head turn, the nose was pointed slightly ahead of a stimulus moving temporally (away from the animal's midline) and this overshoot was repeated on the second turn as the gerbil jumped off the waiting platform and dashed toward the target. When the stimulus moved nasally, the animals undershot the target on the first response by the same margin by which they would undershoot stationary disks, but on the second turn the undershoot significantly exceeded that elicited by stationary objects. Figure 3.20 shows that the populations of second-turn directions for nasal versus temporal objects do not even overlap. The orienting patterns of normal gerbils were plotted in terms of the discrepancy between initial target direction and head direction at the end of each response. As figure 3.21 shows, the tendency to overshoot stimuli moving temporally was sharply limited to the frontal visual field. For stimuli farther than 40° from the midline, undershoots were elicited by either stimulus direction. On the whole, there were no consistent dif-

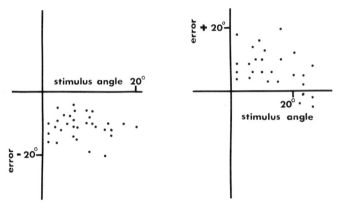

Figure 3.20 Scatter graphs contrasting direction of head turns for the R_2 component of a gerbil's pursuit of the moving disk. If the stimulus angle is calculated anew at the end of R_1 (see figures 3.15 and 3.19) and the change of direction of regard during R_2 is compared, a positive or negative error can be measured. When nasally moving stimuli elicited turns (see left graph), gerbils always showed negative errors (undershoot). However, on the right side it is seen that responses to temporally moving disks almost always resulted in positive errors (overshoot). The discrepancy between responses to the two kinds of moving stimuli was greater for the R_2 component than for the R_1 component (as seen in figure 3.21).

ferences in extent of undershoot toward peripheral targets among the three test situations (temporal motion, nasal motion, and stationary stimuli). Thus, turns toward peripheral objects were directed only in terms of their position, whereas with frontal stimuli the directionality was also important. Since we did not vary stimulus velocity sufficiently in these experiments, we do not yet know whether the degree of overshoot or undershoot varies with the velocity of frontal targets.

Comparison of the Gerbil with Other Species

The ability to anticipate the trajectory of a moving target is undoubtedly shared by other mammals, although I know of no experimental analysis of this behavior. The skill of a dog in pursuing and capturing airborne balls or sticks is an impressive example of such anticipation and deserves quantitative study. Similarly, the ability of a hawk to catch small birds on the wing seems to demand appreciation of both target velocity and direction in plotting the collision course. Whether the predator bird can make such calculations while swooping at an angle to the target bird's trajectory, or whether it tends to approach the quarry from the rear, would have to be known in order to appreciate the kind of visual information needed for success. Only two sources of evidence seem to be available for close comparison with the gerbil. One is our own series of films of prey-catching accuracy in frogs and toads. In these studies we have consistently observed that when a wormlike object moves before a frog or a toad at 10°–15° per second the predator strikes at a locus exactly predicted by the location of the worm's head on the frame before

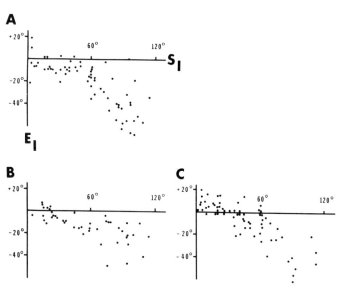

Figure 3.21 Plots of the first partial turn (R_1) during the orientation sequence for three stimulus conditions for one gerbil. For each response, the angle remaining after the first turn (E_1) was plotted as a function of the original stimulus angle (S_1). (A) The S_1 vs. E_1 plot for responses to stationary stimuli obtained in experiment 1. The first turns were relatively accurate for stimuli located up to 60° from the gerbil's midline. Very large errors were elicited by stimuli located beyond 70°. (B) An S_1 vs. E_1 plot for responses to stimuli moving toward the midline (nasally), obtained from experiment 3. Similar accuracy of response occurred for stimuli appearing within 60° of the midline. Although more peripheral stimuli seem to be associated with larger errors in the stationary condition than in the nasal-motion condition, our comparisons for three other animals showed no significant differences. (C) An S_1 vs. E_1 plot for responses to stimuli moving toward the periphery (temporally), obtained from experiment 3. This gerbil tended to overshoot temporally moving disks that appeared within 30° of the midline; for three other animals the tendency to overshoot occurred for stimuli up to 40° from the midline.

the lunge began. With faster movements, the strike misses the target altogether. The notion that a frog can calculate the trajectory of a fast-moving fly seems to be a myth! A detailed quantitative analysis of orientation toward falling food objects by a teleost fish (Lancaster and Mark 1975) supports a similar conclusion. The fish could not anticipate the path of food falling below its body axis, but turned intermittently to keep the retinal locus of the target constant as it closed in. Behavior toward food falling from above involved continuity in swimming horizontally, and was attributed to a delay in turning the body in response to the error signal. The findings thus far contribute to the hypothesis that evolution of predictive orientation is associated with development of a telencephalic visual system (the cortex in mammals, and possibly the striatal systems in birds). A simple version of this hypothesis predicts that predictive orientation will be abolished by ablation of visual cortex in mammals. We find some support for this idea in the report of Humphrey (1974) that a highly trained destriate monkey could rapidly fixate, but could not track, bits of food moved across the frontal midline (as judged from oculographic records).

Predictive Tracking is Abolished by Visual-Cortex Lesions in Gerbils
We have examined the visual orienting abilities of four adult gerbils after bilateral removal of primary visual cortex (including areas 17 and 18), and of one other gerbil with a unilateral visual-cortex ablation. All five gerbils no longer overshot temporally moving disks; orientations toward these targets were identical to turns made toward nasally moving or stationary targets. However, their accuracy and consistency of response toward stationary targets remained normal. These animals were not sloppy in their orienting behavior, but, having lost the ability to use motion direction cues, they now relied upon stimulus position alone. For the animal with a unilateral lesion, the deficit was confined to the contralateral visual field. (Three of the gerbils were lesioned 3 days postnatally, and showed good abilities in tests of pattern discrimination and depth detection.)

If we were inclined to play with words, we might conclude that our tests of cortically damaged gerbils revealed a deficit in one class of *localization* ability but none on a pattern-*identification* task—the reverse of Schneider's (1969) "two visual system" formulation. Goodale (this volume) describes another "localization" deficit in the same infant lesioned gerbils: the failure to localize apertures in a striped arena. It is important to emphasize the oddness of these results. Though our efforts to probe visual-cortex function failed when we used the standard "pattern discrimination" approach, we easily documented visual loss when animals were required to attend to certain dimensions of object motion. It is relevant that neurons with "direction selectivity" in the tectum of hamsters appear to derive this sensitivity from visual cortex (Chalupa and Rhoades 1977). This fits our current model: Visual cortex supplies to subcortical visuomotor centers new dimensions of sensitivity to complex stimuli (such as relative motion) or to parameters such as velocity and directionality.

We must seemingly choose between a model in which visual cortex supplies new information to the optic tectum to generate "overshoots" to moving targets and one in which visual cortex guides this new response via a different motor system. Visual input to frontal cortex or to caudate could either gain access to tectum via fronto-tectal or caudate-nigro-tectal routes or initiate overshoots via other pathways than tectum. If the second postulate were true, the known connections of frontal cortex and substantia nigra with tectum might be inhibitory so as to prevent conflict between the two guidance systems. At present, the most parsimonious hypothesis utilizes known topographic projections from visual cortex to deep tectum, where tecto-motor command neurons are presumed to exist. We have constructed a "sliding map" model (figure 3.22) to explain how corticotectal modulation can result in

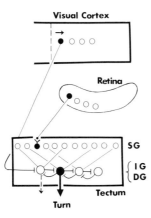

Visual Cortex

Retina

SG

IG
DG

Tectum

Turn

Figure 3.22 A schematic model to explain how visual cortex can modulate the optic tectum to produce overshoots of moving targets. The model bears a similarity with the pretecto-tectal inhibition hypothesis for the frog (figure 3.10). Here, only neurons of the deeper tectal lamina (IG: intermediate gray; DG: deep gray) are inhibited by corticofugal axons. Neurons in one extrastriate visual area projecting to the IG (e.g., area 18B or 19) are here activated only by temporally moving stimuli. Topographically restricted inhibition of deep tectal neurons results in asymmetrical disinhibition within this lamina, because they normally show reciprocal inhibition. Those neurons most excitable after inhibition of rostral tectum would normally mediate turns to slightly more caudal stimuli. Because deep neurons have rather large receptive fields, as depicted by diagonal lines funneling input from an array of superficial gray (SG) small-field neurons, the low-threshold neuron can be activated quite well by the off-center (dark) neuron of the SG layer, which receives unmodulated excitation from the retina. The asymmetrical functional link between the two (dark) cells at this instant result in the animal turning *beyond* the initial locus of the moving target. For a stationary target, the first neuron of the array (dashed arrow) would have made the greater contribution to the turn. This model explains events during the R_1 component where nasally moving targets are treated the same as stationary targets.

stimulus overshoot. The model is essentially the same as that proposed earlier (figure 3.10) to explain ability of the frog to overshoot the edge of a barrier in response to prey movement. In the present model, the modulating input is cortical rather than from pretectum. If we assume that neurons in extrastriate cortex detect temporal motion and inhibit the most rostral tectal zone, there will be a caudalward shift of excitation of the tectal efferent neurons due to unidirection disinhibition along the horizontal plane. This idea is testable, since partial lesions of nasal-field representation in the appropriate cortical region should produce asymmetries in corticotectal modulation and shift orienting behavior in the temporal direction. Just such a deficit was produced by an incomplete striate-cortex lesion in one of our gerbils. All responses to stationary and nasally-moving objects were overshoots for the side opposite the incomplete lesion, whereas no overshoots occurred in the other hemifield. The single case seems to demonstrate that small cortical lesions can produce greater distortion in orienting than the surgical removal of a larger cortical zone. The lesions needed to test the model rigorously could be best achieved after microelectrode mappings of the visual-field topography of striate and extrastriate areas. This becomes a plausible plan in view of the ability to evaluate each hemifield separately via cinematic records of elicited behavior.

Future Research on Cortical Functions

Two new research directions can be recommended on the basis of this review. First, the identification of visuomotor and attentional functions with particular cortical areas is a key goal for rodent visual research. Are there independent cortical systems for pattern recognition and spatial perception, as suggested in this volume by Ungerleider and Mishkin for the monkey? Second, we need to know much more about the particular kinds of information processing underlying spatial vision. For example, does the gerbil utilize motion parallax cues or expansion gradients to distinguish apertures or barriers or to gauge its own running speed? Do rodents—like primates—have effective binocular stereopsis? How is information on the layout of the environment picked up by rodents: via head-scanning movements, or also via small exploratory eye movements? Is there a "frontal eye field" in rodents that participates in strategic scanning patterns? We can remain within a naturalistic framework while pursuing a variety of new studies on the exact spatiotemporal changes in the retinal image that inform animals of detailed structure of their visual world.

References

Akert, K. 1949. "Der visuelle Greifreflex." *Helv. Physiol. Acta* 7: 112–134.

Bass, A. H., M. B. Pritz, and R. G. Northcutt. 1973. "Effects of telencephalic and tectal ablations on visual behavior in the side-necked turtle, *Podocnemis onfils*." *Brain Res.* 55: 455–460.

Bechterev, W. 1884. "Uber die Function der Vierhugel." *Arch. gesamte Physiol. Menschen und Tiere* 33: 413–439.

Casagrande, V. A., and I. T. Diamond. 1974. "Ablation study of the superior colliculus in the tree shrew (*Tupaia glis*)." *J. Comp. Neurol.* 156: 207–238.

Chalupa, L. M., and R. W. Rhoades. 1977. "Responses of visual somatosensory and auditory neurons in the golden hamster's superior colliculus." *J. Physiol.* 270: 595–626.

Collett, T. 1977. "Stereopsis in toads." *Nature* 267: 349–351.

Collewijn, H. 1969. "Optokinetic eye movements in the rabbit: Input-output relations." *Vision Res.* 9: 117–132.

Conway, J., P. H. Shiller, and S. True. 1979. "Eye movement control: The effects of paired superior colliculus and frontal eye field ablations." *Soc. Neurosci. Abstr.* 5: 366.

Distel, H.-J. 1977. "Behavioral responses to the electrical stimulation of the brain in the Green Iguana." In *Behavior and Neurology of Lizards,* N. Greenberg and P. D. MacLean, eds. Rockville, Md.: National Institute of Mental Health.

Eldred, W. P., T. E. Finger, and J. F. Nolte. 1980. "Neural projections of the frontal organ in the frog." *J. Cell Tissue Res.*

Erichsen, J. 1969. "How birds look at objects." D. Phil. thesis, Department of Zoology, Oxford University.

Ewert, J.-P. 1967. "Elektrische Reizung des retinalen Projektionsfeldes im Mittelhirn der Erdkröte (*Bufo bufo* L.)." *Pflügers Arch.* 295: 90–98.

———. 1970. "Neural mechanisms of prey-catching and avoidance behavior in the toad (*Bufo bufo* L.)." *Brain, Behav., Evol.* 3: 36–56.

———. 1971. "Single unit response of the toad's (*Bufo americanus*) caudal thalamus to visual objects." *Z. vergl. Physiol.* 74: 81–102.

Ewert, J.-P., and H.-W. Borchers. 1971. "Reaktionscharakteristik von Neuronen aus dem Tectum opticum and Subtectum der Erdkröte (*Bufo bufo* L.)." *Z. vergl. Physiol.* 71: 165–189.

Ewert, J.-P., and F. Hock. 1972. "Movement-sensitive neurones in the toad's retina." *Exp. Brain Res.* 16: 41–59.

Ewert, J.-P., and B. Rehn. 1969. "Quantitative Analysis der Reiz-Reaktionsbeziehungen bei visuellem Auslösen des Fluchtverhaltens der Wechselkröte (*Bufo viridis* Laur.)." *Behaviour* 35: 212–234.

Friedman, M. B. 1975. "How birds use their eyes." In *Neural and Endocrine Aspects of the Behaviour of Birds,* P. Wright et al., eds. Amsterdam: Elsevier.

Greenberg, N. 1977. "A neuroethological study of display behavior in the lizard *Anolis carolinenis.*" *Amer. Zool.* 17: 191–201.

Grüsser, O.-J., and U. Grüsser-Cornehls. 1976. "Physiology of the anuran visual system." In *Neurobiology of the Frog,* R. Llinas and W. Precht, eds. Berlin: Springer.

Hall, W. C., and F. F. Ebner. 1970. "Thalamotelencephalic projections in the turtle, *Pscadomys scripta.*" *J. Comp. Neurol.* 140: 101–122.

Hess, W. R., S. Burgi, and V. Bucher. 1946. "Motorische Funktion des Tektal- und Tegmentalgebietes." *Monatsschr. Psychiatr. Neurol.* 112: 1–52.

Humphrey, N. K. 1974. "Vision in a monkey without striate cortex: A case study." *Perception* 3: 241–255.

Ingle, D. 1968. "Visual releasers of prey catching behavior in frogs and toads." *Brain, Behav., Evol.* 1: 500–518.

———. 1970. "Visuomotor functions of the frog optic tectum." *Brain, Behav., Evol.* 3: 57–71.

———. 1971a. "Discrimination of edge-orientation by frogs." *Vision Res.* 11: 1365–1367.

———. 1971b. "Prey catching behavior of anurans toward moving and stationary objects." *Vision Res.* Suppl. 3: 447–456.

———. 1973a. "Two visual systems in the frog." *Science* 181: 1053–1055.

———. 1973b. "Disinhibition of tectal neurons by pretectal lesions in the frog." *Science* 180: 422–424.

———. 1975. "Selective visual attention in frogs." *Science* 188: 1033–1035.

———. 1976. "Spatial vision in anurans." In *The Amphibian Visual System*, K. Fite, ed. New York: Academic.

———. 1977. "Loss of anticipatory orientation following lesions of visual cortex in the gerbil." *Soc. Neurosci. Abstr.* 3: 68.

———. 1978. "Mechanisms of shape-recognition among vertebrates." In *Handbook of Sensory Psychology*, volume 7: Perception, R. Held et al., eds. Berlin: Springer.

———. 1979. "Behavioral evaluation: Abnormal retinotectal projections in the frog." *Neurosci. Res. Program Bull.* 17: 328–333.

———. 1980. "The frog's detection of stationary objects following lesions of the pretectum." *Behav. Brain Res.* 1: 139–163.

———. 1981. "Selectivity in feeding behavior of frogs and toads: A neurobehavioral model." In *Handbook of Behavioral Neurology*, P. Teitelbaum and E. Satinoff, eds. New York: Plenum.

Ingle, D., and J. Cook. 1977. "The effect of viewing distance upon size preference of frogs for prey." *Vision Res.* 17: 1009–1014.

Ingle, D., M. Cheal, and P. Dizio. 1979. "A cine analysis of visual orientation and pursuit by the Mongolian gerbil." *J. Comp. Physiol. Psychol.* 93: 919–928.

Jaeger, R. G., and J. P. Hailman. 1973. "Effects of intensity on the phototactic responses of adult anuran amphibians: A comparative survey." *Z. Tierpsychol.* 33: 352–407.

Jarvis, C. D. 1974. "Visual discrimination and spatial localization deficits after lesions of the tectofugal pathway in pigeons." *Brain, Behav., Evol.* 9: 195–228.

Kicliter, E. 1973. "Flux, wavelength and movement discrimination in frogs: Forebrain and midbrain contributions." *Brain, Behav., Evol.* 8: 340–365.

Lancaster, B. S., and R. F. Mark. 1975. "Pursuit and prediction in the tracking of moving food by a teleost fish (*Acanthaluteres spilomelanrus*)." *J. Exp. Biol.* 63: 627–645.

Lázár, G. 1973. "Role of the accessory optic system in the optokinetic nystagmus of the frog." *Brain, Behav., Evol.* 5: 443–460.

Lázár, G., and G. Székely. 1969. "Distribution of optic terminals in the different optic centres of the frog." *Brain Res.* 16: 1–14.

Lettvin, J. Y., H. R. Maturana, W. S. McCulloch, and W. H. Pitts. 1959. "What the frog's eye tells the frog's brain." *Proc. IRE* 47: 1940–1951.

Lettvin, J. Y., H. R. Maturana, W. H. Pitts, and W. S. McCulloch. 1961. "Two remarks on the visual system of the frog." In *Sensory Communication*, W. A. Rosenblith, ed. Cambridge, Mass.: MIT Press.

Lock, A., and T. Collett. 1979. "A toad's devious approach to its prey: A study of some complex uses of depth vision." *J. Comp. Physiol.* 131: 179–189.

Merker, B. H. 1980. "The sentinel hypothesis: A role for the mammalian superior colliculus." Ph.D. diss., Department of Psychology, Massachusetts Institute of Technology.

Mrosovsky, N., and K. H. Tress. 1966. "Plasticity of reactions to light in frogs and a possible role for the pineal eye." *Nature* 210: 1174–1175.

Muntz, W. R. A. 1962a. "Effectiveness of different colors of light in releasing positive phototaxic behavior of frogs and a possible function of the retinal projection of the diencephalon." *J. Neurophysiol.* 25: 712–720.

———. 1962b. "Microelectrode recordings from the diencephalon of the frog (*Rana pipiens*) and a blue-sensitive system." *J. Neurophysiol.* 25: 699–711.

Robinson, D. A. 1968. "Eye movement control in primates." *Science* 161: 1219–1224.

Rosinski, R. R., and J. J. Keselica. 1977. "Failure to avoid impending collision by the golden hamster, *Mesocricetus auratus*." *Bull. Psychonom. Soc.* 9: 53–54.

Scalia, F., and K. Fite. 1974. "A retinotopic analysis of the central connections of the optic nerve in the frog." *J. Comp. Neurol.* 158: 455–478.

Scalia, F., H. Knapp, M. Halpern, and W. Riss. 1968. "New observations on the retinal projection in the frog." *Brain, Behav., Evol.* 1: 324–353.

Schneider, G. E. 1969. "Two visual systems: Brain mechanisms for localization and discrimination are dissociated by tectal and cortical lesions." *Science* 163: 895–902.

———. 1970. "Mechanisms of functional recovery following lesions of visual cortex or superior colliculus in neonate and adult hamsters." *Brain, Behav., Evol.* 3: 295–323.

———. 1973. "Early lesions of superior colliculus: Factors affecting the formation of abnormal retinal projections." *Brain, Behav., Evol.* 8: 73–109.

Sherman, H. B., V. S. Caviness, and D. Ingle. 1979. "Corticotectal connections in the gerbil." *Soc. Neurosci. Abstr.* 5: 120.

Sprague, J. M., and T. H. Meikle, Jr. 1965. "The role of the superior colliculus in visually guided behavior." *Exp. Neurol.* 11: 114–146.

Springer, A. D., S. S. Easter, and B. W. Agranoff. 1977. "The role of the optic tectum in various visually mediated behaviors of goldfish." *Brain Res.* 128: 393–404.

Trachtenberg, M. C., and D. Ingle. 1974. "Thalamo-tectal projections in the frog." *Brain Res.* 79: 419–430.

Walk, R. D., and E. J. Gibson. 1961. "A comparative and analytical study of visual depth perception." *Psychol. Monogr.* 75 (no. 15): 1–44.

Wilczynski, W., and R. G. Northcutt. 1978. "Afferents to the Optic Tectum of the Leopard Frog: An HRP Study." *J. Comp. Neurol.* 173: 219–229.

4 Depth Vision in Animals

Thomas S. Collett
Lindesay I. K. Harkness

Animals engage in many activities that involve some knowledge of the disposition of objects in the space around them. They identify and locate things, they chase things, jump at things, catch things, and avoid things. They also plan their routes through cluttered and complicated environments, and learn the location of significant landmarks. Animals behaving in these ways can demonstrate to us the quality of their spatial vision. They tell us, for instance, how well or badly they judge the distance of an object, whether they appreciate spatial relations between objects, and how they integrate such information. The first part of this chapter is concerned with such issues: the different uses of distance vision and what we can learn from them about an animal's visual world. In the second half, we discuss the various cues that are potentially available for measuring distance, and how different species actually obtain distance formation. Each distance cue has its own requirements and limitations, and is not exploited in the same way by all species. Which methods a particular species has come to adopt depends both on simple physical properties, such as its size, and on remote historical decisions that have led it to emphasize one particular visual style at the expense of others. In this chapter we have chosen to stress insects, amphibians, and lizards, because these are the groups with which we are most familiar and because it is intriguing to discover how species so distantly related to man cope with the kinds of visuomotor problems that the world presents to all visually alert animals.

Why Animals Measure Distance

For Guiding Motor Programs
There are a variety of situations in which an animal must know how far it is from an object in order to organize a preprogrammed movement towards it. Animals commonly launch themselves, some part of themselves, or occasionally even a missile at a target, and to do this with precision they need to measure the distance of the target. Figure 4.1 illustrates this with two stroboscopic pictures of a tree frog (*Hyla cinerea*) jumping from one block of wood to another. The frog does not

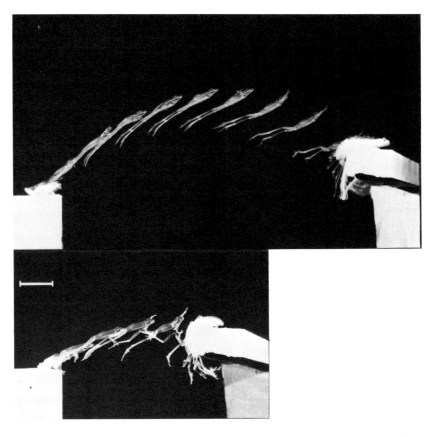

Figure 4.1 Tree frog jumping from one perch to another. Two jumps are shown. Velocity is greater and takeoff angle steeper in the longer jump, as corroborated by more extensive film records. Flash rate: 30 Hz. Bar: 5 cm.

simply fling itself as hard as it can directly at the block, but designs its trajectory ballistically, aiming upward to allow for the action of gravity. The more distant a horizontal target, the steeper the trajectory the frog adopts (to a maximum of about 45°) and the higher its takeoff speed. Similarly, a chameleon measures the distance of its insect prey before shooting out its tongue, and matches its tongue extension to the prey distance. As figure 4.2 shows, the tip normally reaches only just beyond the insect before being retracted. This improves the chances of capture, since the insect is less likely to be knocked away, and also prevents the tongue from whipping or coiling uncontrollably.

Distance vision is also used in the preprogramming of entire motor sequences. We humans, for instance, can observe our surroundings and then, with our eyes shut, walk around a series of obstacles to reach a predetermined goal (see Lee and Thomson, this volume). The advantages of such preprogramming are various. For example, segments of a movement can be prepared in advance, or sensory reflexes whose sign or gain may need to be altered during the course of a movement can be

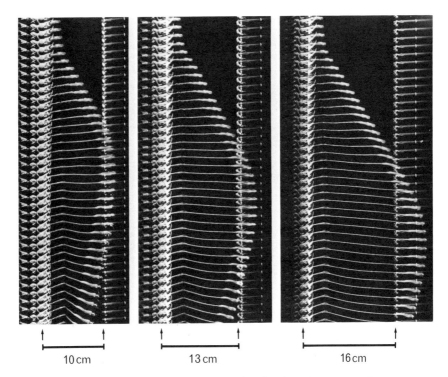

10 cm	13 cm	16 cm

Figure 4.2 Three multiple-exposure photographs of a chameleon attempting to capture insects at three different distances. Time moves down the page. In each photo the chameleon is on the left with just its head showing, and its tongue is flying out toward the insect on the end of the stick at the right. The positions of the chameleon's eye and the insect are indicated by arrows. The tongue does not hit the insect because the chameleon is wearing lateral-prism spectacles (see figure 4.36). Tongue tip starts to decelerate when it is a little beyond the plane of the insect, whatever its distance, and is brought to a halt a few centimeters beyond (a little farther beyond the insect) before being retracted. The chameleon's behavior varies with the target distance: The farther the insect, the farther (and faster) the chameleon shoots its tongue. Flash rate: 400 Hz.

set appropriately at the right time. Toads seem to preprogram their approach toward an insect for a rather different reason (Lock and Collett 1979). While walking, a toad seems to be incapable of using visual information to track its prey, which may change course, stop, or even disappear without the toad apparently noticing. If visual feedback is unavailable, the animal must decide before setting out how far and in what direction it must walk in order to reach its prey.

However, many animals use vision to make continuous and precise adjustments to their course, and distance information then plays an important role in the moment-to-moment control of many motor activities. Some of this variety is illustrated by the tracking behavior of a small hoverfly, *Syritta pipiens* (Collett and Land 1975). Male *Syritta* track other flies in order to find a receptive female. A male will "shadow" another fly it has sighted, keeping at a relatively constant distance (about 10 cm) from it and continuously turning to face it with a high-resolution region of the eye (a "fovea" subtending 10°) that is absent in females. This combination of keeping its distance from the other fly and pointing its long axis at it means that *Syritta* can remain inconspicuous until it selects a suitable moment to pounce, which is usually after the leading fly has settled on a flower to feed. When the distance between the flies grows, the tracking fly approaches; if it finds itself too close it retreats (figure 4.3). This "distance tracking" can be summarized by a simple rule relating the fly's forward velocity to the difference between the set distance the tracking fly wants to maintain and the actual distance.

Once the leading fly has settled, another use of distance vision comes into operation. The tracking male hovers, occasionally circling the leading fly to look at it from different vantage points, all the time turning its body to face the other fly (figure 4.4). This sideways circling happens too quickly and too precisely for the turning to be driven by visual feedback, as it is during normal tracking. It seems that while the fly is circling sideways, it predicts how much it must turn to keep the target on its fovea. The relation between sideways and angular velocity depends on the distance of the fly from its target, and the precision of the fly's behavior means that it must know this distance accurately and use this knowledge to adjust the linkage between its sideways and angular movements.

If the male fly is satisfied with the appearance of the fly on the flower, he will pounce, accelerating at a constant 5 m/sec² toward his target and showing no sign of braking before landing (figure 4.5a). The female is thus allowed little chance to escape. The male's terminal speed will simply depend on how far he was when he began his "rape," which implies that distance measurements play no part in this behavior. Such collisions need not be harmful, as *Syritta* is small. But many flying animals must land accurately on surfaces and avoid dangerous collisions. To do this they have to know when to decelerate or swerve. Figure 4.5b shows *Syritta* decelerating and hesitating as it approaches a flower on which it subsequently lands and feeds.

To control landing and obstacle avoidance, most animals other than toads can exploit visual information given by the shifting of the image of their surroundings across the retina as they move around, as Gibson (1950) emphasized. From this changing optic array or flow field, an animal can monitor its progress through its environment and locate objects in it. The subject of flow fields is beyond the scope of this chapter; all we wish to stress is that information for controlling motor programs need not be in terms of spatial properties such as object distance. Lee (1976) showed that the simplest quantity for an animal to compute from its flow field to avoid a collision is how much time will elapse before it collides with an object in its path. For an animal traveling at a uniform speed, the time to collision is given by the ratio of the angular separation of any two image points of the object to the rate of separation

Collett, Harkness

a

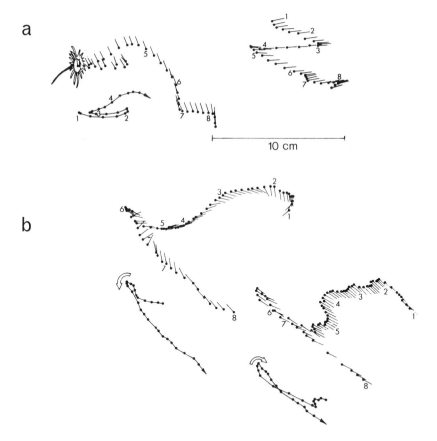

10 cm

b

Figure 4.3 Two examples of a male *Syritta* adjusting its forward velocity to keep a constant distance from a fly that it is tracking. In part a the leading fly was initially on a flower and the tracking fly hovered stably. After a while the leader flew off the flower, then returned briefly before again flying off. The leader followed these movements, flying away toward and then away from the flower. The dots with tails show the position of the leader (left) and pursuer (right) every 20 msec from the time the leading fly left the flower. In part b the tracking fly also copies the leader's movements; their positions (leader top left, pursuer bottom right) every 40 msec are shown. Source: Collett and Land 1975.

Depth Vision in Animals

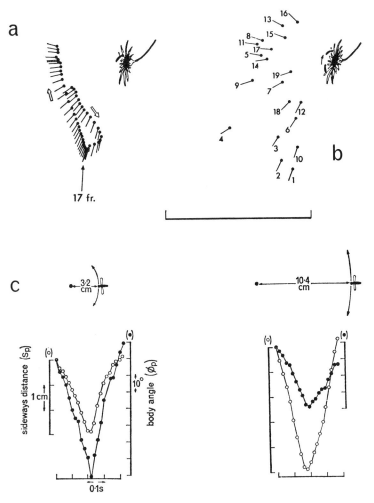

Figure 4.4 Circling in hoverflies. The fly moves sideways and at the same time turns so as to keep a stationary target fixated on its fovea. To do this, it uses its knowledge of the distance of the target to specify how much it must turn to compensate for a given sideways movement. (a) An example of circling, showing the position of the circling fly every 20 msec. (b) Successive stable hovering positions adopted by a circling fly during 25 sec. The positions are held for times varying between 0.2 and 2.5 sec. The leading fly moves slightly; its initial and final positions are shown. (c) Plots of angular and sideways movements during circling of two flies at different distances from the target. The shorter the distance, the greater the ratio of angular to sideways movement. Source: Collett and Land 1975.

a

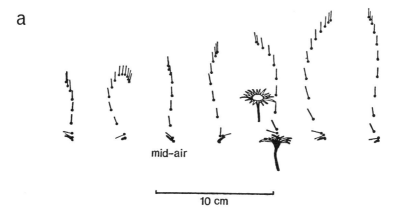

mid-air

10 cm

b

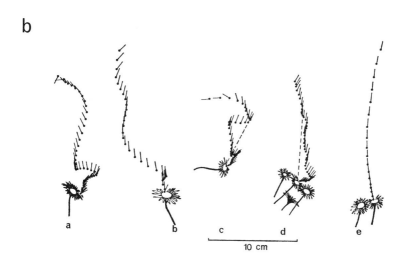

a   b   c   d   e

10 cm

Figure 4.5 (a) Seven "rapes". The male approaches at a constant acceleration, turning just before reaching the leading fly, to collide head to head and tail to tail. In all cases except one the leading fly is on a flower. (b) Flights towards flowers. The behavior is very variable, but in all cases the fly decelerates well before landing. Positions every 20 msec shown. Source: Collett and Land 1975.

of those image points. This ratio could easily be used to control landing and braking, and does not require extravisual information.

The importance of accurate distance measurement to a wide variety of species makes it seem likely that depth vision evolved to provide precise quantitative information about the location of objects in space, and not merely to help separate objects from the background by giving a qualitative impression of depth.

For Planning Routes
Animals are often so adept at choosing a route through a complex environment that it takes some effort to realize how impressive a skill it is. In this section, we describe some of the exploits of chameleons and toads that indicate the extent to which they are aware of the arrangement of objects in their immediate surroundings and that suggest that these animals can use distance information to build up an internal representation of their surroundings.

Chameleons are arboreal, and clamber rather slowly among branches. Their environment thus presents them with a natural maze where there is a real advantage to selecting the easiest and shortest path to their food supply or other goal and to avoiding blind ends or delicate twigs that cannot bear their weight. To solve such problems, chameleons need to make a variety of perceptual decisions: They must measure distances between branches in order to decide whether a gap can be spanned, and they should be able to recognize forks, to scan branches to their ends, and so forth.

If a young chameleon is placed on a vertical maze, as shown in figure 4.6a, it makes straight for the top, at each choice point selecting the limb that provides an unbroken path to the next segment of the maze. It will thus detect and avoid gaps when there is another equally accessible route. However, if the easiest path does involve spanning a gap, chameleons are quite ready to cross it. A chameleon presented with the maze illustrated in figure 4.6b will behave in one of two ways: It either spans the gap between one central vertical wire and the next, treating the maze as a ladder and reaching directly from rung to rung, or it detours around the rectangle. Whether it does one or the other depends on both the height of the gap and the length of the rectangle. With an 8-cm-long rectangle, it is equally likely to "span" or to detour around a gap of 2.5 cm (figure 4.6c), whereas with a 10-cm-long rectangle the gap has to be about 2.9 cm before the chameleon shows no preference for one solution to the problem over the other. Thus, the gap it is willing to span grows in proportion to the length of the detour it would otherwise have to make. This signifies that the chameleon's choice depends on an interaction between two distance measurements.

It is instructive to consider what this implies. Chameleons have turretlike eyes (see figure 4.35) with a relatively narrow field of view, and almost certainly have to scan the maze in order to measure the length

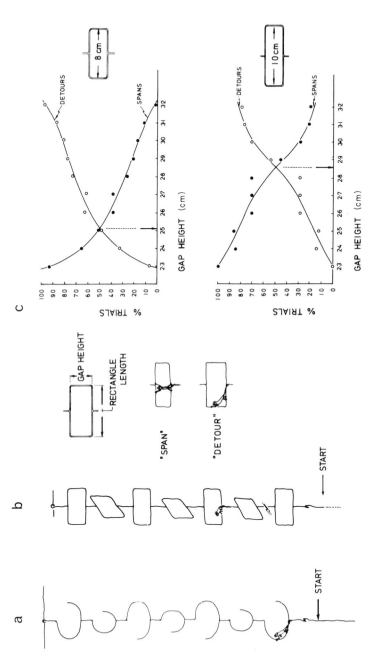

Figure 4.6 Behavior of young chameleons climbing on vertical mazes. (a) Maze presenting a choice between a "dead end" and a continuous route upward. Each maze unit is set at right angles to the next to ensure that animals do not take short cuts. Chameleons a few days old will climb up these mazes, avoiding dead ends about 80% of the time. (b) Maze presenting a choice between "spanning" across a rectangular unit and a less direct route "detouring" around. The dimensions of the maze units were varied in order to test whether the spanning distance (gap height) or the extra path length required to detour (rectangle length) affected the chameleons' choice of behavior. (c) Percentage of trials in which chameleons detoured (○) or spanned (●) around maze units of variable height and width of 8 cm (top) or 10 cm (bottom). The chameleons' choice of behavior is affected by the gap height; the probability of spanning decreases with increasing gap height in both plots. Detour length also affects how a chameleon behaves. The intersection of the curves on each plot defines a gap height that chameleons would be equally likely to span as to detour around. This gap height is about 2.5 cm when the maze units are 8 cm wide, but is about 2.86 cm for the 10-cm-wide units. The longer the detour, the greater the span a chameleon is prepared to make.

and width of the rectangle. Furthermore, since they probably rely on accommodative cues while climbing as well as when feeding, it is likely that they can only measure one distance at a time. A chameleon thus has to bring together information from a sequence of "snapshots," each limited in depth and extent; before it selects a route, it must assemble these snapshots into an internal representation of the position of objects within its environment.

Terrestrial toads also cope very effectively with the kinds of obstacles they are normally likely to encounter. They detour or sidestep around barriers, detect and make for gaps, and step up or down with precision to reach surfaces on a different horizontal level (Ingle 1976). And, like chameleons, toads make extensive use of distance information in selecting their route. Figure 4.7 summarizes an experiment in which a chasm, variable in width and depth, is interposed between a toad and its food. A toad's behavior in this situation shows that it measures depth and width and uses both distance measurements in deciding what to do. Typically, a toad notices its prey, turns toward it, approaches the edge of the chasm, and pauses. It then selects between three possible courses of action, depending on the width and depth of the chasm: to step down to the floor of the chasm and cross it, to leap across, or to turn away from the edge, as many animals do in the "visual cliff" situation. When the chasm is deep but narrow, a toad jumps across; but it will sensibly turn away if the width is increased (figure 4.7a). With shallower chasms, the toad uses the option of stepping down. The likelihood of stepping down drops as the chasm is made deeper, but it falls more gradually for wide than for narrow chasms (figure 4.7b). Not only do depth and width measurements contribute to the toad's choice of behavior, but once the option is selected, this information is also used to control the response itself.

Distance information again plays a role in a toad's decision whether to detour around a barrier to reach a prey on the other side. Two different forms of behavior can be displayed when a picket fence is placed between the toad and its food (figure 4.8). The toad, after orienting toward the prey, may pause and head directly for it. This is a reasonable strategy if the palings are sufficiently far apart and the prey sufficiently close to be caught through the fence. Alternatively, the toad after orienting towards the prey may make directly for the edge of the fence and detour around it. This is clearly only a useful maneuver if the worm actually is behind the fence. One might therefore expect the toad's choice between these alternatives to be determined in part by the distance between prey and fence, and this is indeed the case. If the prey is farther than 10–15 cm from the fence the toad prefers to make a detour, but when the prey is closer the toad is more likely to make a direct approach. Significantly, the changeover point is unaffected by the initial distance between toad and fence. The toad thus seems to measure

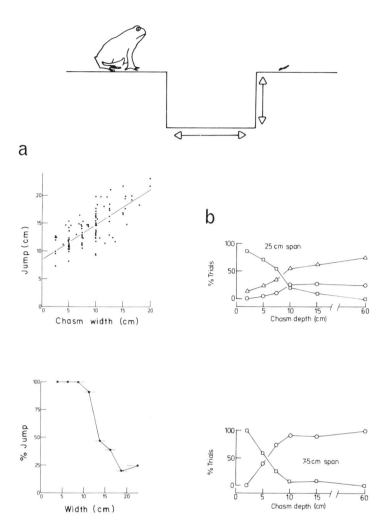

Figure 4.7 Behavior of *Bufo bufo* on approaching a 60-cm-deep chasm placed between it and its prey, as shown at top. Span of chasm is variable. (a) Upper graph shows relation between chasm width and distance jumped by the toad. Each point represents a single jump. Toad adjusts its jump so that it clears the chasm. Bottom graph plots relation between width of chasm and proportion of trials in which toad jumps chasm rather than turning away. Probability of jumping decreases as chasm width increases. *Bufo bufo*'s treatment of a chasm varies with both the chasm's depth and its span. (b) Percentage of trials in which a toad steps down into (□), leaps across (○), or turns away from (△) the chasm for different depths and widths. In both plots span is kept constant and depth varied. Top: 25-cm span with prey 5 cm from far end of chasm. Bottom: 7.5-cm span with prey 22.5 cm from far edge of chasm. In both cases probability of jumping across increases linearly with depth to reach an asymptote at 9 cm, indicating that the toad classifies depressions of more than 9 cm as chasms that need to be jumped. Source: Lock and Collett 1979.

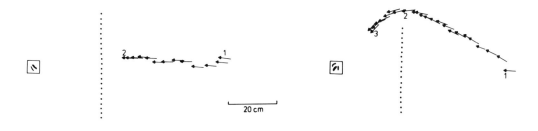

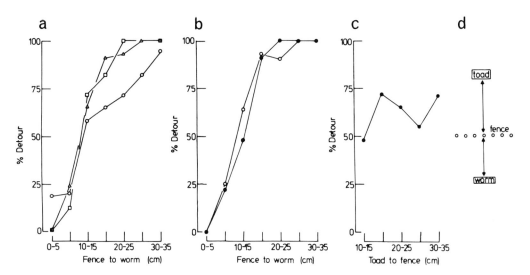

Figure 4.8 Toad's behavior when a picket fence is interposed between it and its prey. If distance between fence and worm is greater than 10–15 cm toad detours around fence; if this distance is shorter toad approaches worm directly. Diagram at top shows examples of a toad (*Bufo marinus*) approaching directly or detouring. Position of toad's head and midline are marked by an arrow every 0.2 seconds. Numerals indicate brief pauses between walks. Bottom: (a) Relation between percentage of trials in which toad detours and fence-to-worm distance with toad starting a constant 30 cm from fence for three different paling separations: (△) 2 cm, (□) 4 cm, and (○) 6 cm. Toad's choice depends strongly on distance between fence and worm, but not on paling separation. (b) Relation between detour percentage and toad-to-fence distance with toad starting either farther than 25 cm (○) or nearer than 15 cm (●) from fence. Paling separation 4 cm. (c) Relation between detour percentage and toad-to-fence distance when worm is kept at a constant 10–15 cm from fence. Toad can thus measure distance between fence and worm and use this information to guide its choice regardless of its own distance from fence. Source: Lock and Collett 1979.

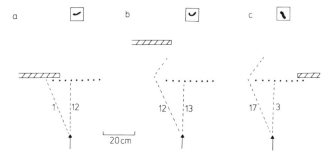

Figure 4.9 Toad's behavior when opaque barrier extends the picket fence. Toad's starting position is shown by arrow, and its choice of detour or direct approach by dashed lines. (a) Barrier close to left end of fence; toad approaches worm directly in 12 of 13 trials. (b) Barrier set back; toad detours in half the trials. (c) Barrier close to right end of fence; toad detours to left in almost all trials. Toad has strong side preference and never detours to right. Data indicate that toad can detect gap in situation b. Source: Lock and Collett 1979.

the distance between objects in the outside world irrespective of its own distance from them.

If the situation is made more complicated, the toad includes additional distance information in deciding whether to make a detour. Figure 4.9 demonstrates that if the fence is lengthened by an opaque barrier, so that a detour is impracticable, the toad consistently makes straight for the worm. However, when the barrier is set back from the fence to provide a passage, the animal detours on about half the trials. In this case, the toad's decision depends both on its measurement of the distance between fence and prey and on its realization that the barrier is set back.

These experiments and those illustrated in figure 4.12 suggest that toads have a good appreciation of the spatial relationships between stationary and moving objects and that they use such information in deciding how to negotiate obstacles in their surroundings. In the light of demonstrations of this kind, it seems appropriate to talk about a chameleon's or a toad's internal representation of its environment. Unfortunately, this does not mean we know anything about how such a representation may be organized. And, although we are prejudiced in favor of the first alternative, we cannot even say whether such a map is "perceptual" in the sense that the same assembled information is available to a variety of motor programs, or whether depth information is processed and integrated separately for directing each individual motor program.

For Identifying Objects

Distance information is used in a variety of tasks that are clearly perceptual in nature, of which the best known is judging the real size of an object. Since a small nearby object subtends the same visual angle as a larger, more distant one, an appreciation of an unfamiliar object's size is only possible if both its retinal image size and its distance from the viewer are taken into account. In man, there is much evidence that perceived distance strongly influences the perceived size of an object. The simplest demonstration of this is a phenomenon that has come to be known as Emmert's law: that an afterimage of constant retinal size is seen as larger the more distant the surface onto which it is projected. Direct manipulation of certain distance cues has also been shown to produce changes in perceived size. Thus, if an observer looks at an object while the angle of convergence of the two eyes is artificially increased to reduce the apparent distance, the object appears smaller than it really is (Heinemann et al. 1959). Under monocular viewing conditions, perceived size is more accurate if a viewer is allowed to move his head to obtain parallax information than it is if the viewer's head is clamped (Hell 1978).

It is obviously helpful to know the real size of an object when deciding, for instance, whether it is something that can be eaten or is better left alone, or whether a gap in a barrier can be negotiated. However, the curious forms of mimicry adopted by some insects (figure 4.10) in which they seem to copy with surprising realism the outward forms of much larger animals (Hinton 1973) suggest that perhaps some predators can be fooled, at least momentarily, into mistaking a bug for an alligator despite the enormous discrepancy in size.

There is relatively little information about the ability of nonmammalian species to perceive the real size of an object. Two different experimental approaches have been used to study this question. In one, an animal is trained to discriminate between two similarly shaped but differently sized objects by, for example, learning to approach one for food. During this training phase, the two objects are kept equidistant from the animal. The animal's ability to choose correctly is then tested when the distance of the objects is varied to make the larger one subtend a smaller visual angle than the smaller. The octopus (Boycott and Young 1956), the carp (Herter 1930, limited data), and the duckling (Pastore 1958) have all been found to be undisturbed by such manipulations, and so must automatically incorporate a distance cue into their judgments of size. In the other approach, an animal that normally shows a characteristic form of behavior toward particular objects in its environment is presented with a range of different object sizes and distances. The animal's size preference then reveals whether it perceives the object's real size or its retinal size at some assumed distance. Bulldog ants snapping at prey (Via 1977) and hoverflies (Collett and Land 1978) in their initial response to potential mates are governed by retinal-image size; that is, they respond in the same way to a large distant object as to a closer smaller one. Since they use image size as a cue to distance, this is hardly surprising.

Although size constancy generally involves the adjustment of the perceived size of an object according to its perceived distance, for certain

Figure 4.10 Insect mimic of an alligator: (a) head of the bug *Fulgora lucifera* from Brazil; (b) spectacled caiman, also from Brazil. Source: Hinton 1973.

restricted purposes even an insect can use a very simple strategy to furnish itself with a form of size constancy. The principle is well shown by the backswimmer *Notonecta*, which hangs in a stereotyped posture below the water surface of ponds and streams, waiting for small insects trapped just beneath the surface (figure 4.11). This prey is detected by the surface waves it generates, and is then probably tracked visually. Since *Notonecta*'s eye can be considered fixed in relation to the water surface, there is a one-to-one mapping of points on the water-air interface onto the surface of the eye. Distant points are seen by dorsal regions of retina, closer ones by more ventral retina. Consequently, all that is needed to recognize a prey of a constant real size at different distances is an appropriate gradation of receptive field sizes across the eye, diminishing from ventral to dorsal retina. Schwind (1978) found a visual interneurone in *Notonecta* that displays this property. The neurone has a large receptive field within which it responds to small

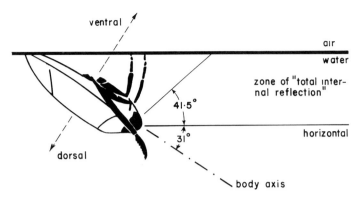

Figure 4.11 Typical posture adopted by *Notonecta* hanging beneath water surface. The "size constancy" neurone operates within the region of visual field defined by the zone of total internal reflection. Source: Schwind 1978.

stimuli moving vertically through the field. The optimal size of the stimulus varies across the receptive field in such a way that the unit is calibrated to react to an object of constant size moving toward the insect over the water surface. This trick works because *Notonecta*'s world is essentially a two-dimensional sheet and the insect only grapples with prey of a certain limited range of sizes. The same principle could be used by terrestrial animals looking for prey on the ground, although a rough or sloping surface will make the method rather inaccurate.

Ingle and Cook (1977) and Ewert and Gebauer (1973) showed that over a range of about 10–20 cm frogs and toads prefer to snap at prey on the ground that are of a constant real size (0.7 cm high). If size constancy in anurans relies on the same principle as that outlined for *Notonecta*, one would expect that for a constant, close stimulus distance animals would prefer smaller prey when the stimulus is raised above rather than on the ground. However, tests with *Rana pipiens* (Collett, unpublished data) provide no support for such an explanation of their appreciation of real prey size.

Although frogs and toads treat prey as though they are aware of its real size, size constancy is not always shown. Ingle and Cook (1977) found that frogs, when deciding whether to jump through an aperture to escape from a threatening stimulus, determine whether the aperture is sufficiently large solely on the basis of its retinal angle. Over the range tested (7.5–18.8 cm), the frog considers gaps of about 25° wide enough. This example points out very clearly the disadvantages of not knowing the real size of an obstacle that must be negotiated. If the frog uses this criterion when it is closer than about 5 cm, it may try to jump through a gap that is too narrow to clear; and conversely, at longer distances, it will avoid gaps that are easily negotiable.

However, anurans do show size constancy in other somewhat analogous situations. Toads (*Bufo viridis*), in planning their route around a barrier, demonstrate that they perceive the real size of two of its features. If there is a sufficiently wide gap in the barrier, a toad will aim its approach at the gap in preference to going around the end (figure 4.12). And, provided that the toad is no farther than 20 cm from the barrier (about the distance at which prey size constancy breaks down), its assessment of the width of the gap is in terms of the real and not the retinal size of the gap. If the barrier is made longer, the toad becomes more reluctant to detour around it, and again, as figure 4.12 illustrates, the toad's behavior is determined by the physical length of the barrier and not by the angle it subtends at the retina. In conclusion: It seems likely that selection pressures will have ensured that all animals equipped with objective and unambiguous methods of measuring distance use depth information in assessing the sizes of salient objects in their surroundings.

The Distance Cues Available to Animals

In the first part of this chapter, we considered various uses animals make of distance vision, and stressed that for all these activities (including capturing prey or mates, moving around the environment, and even making qualitative judgments such as identifying things) animals require *quantitative* depth information. Here we discuss the different cues that are available for measuring distance, and the range and accuracy they can offer to different animals.

Whatever cue an animal uses, the accuracy with which it can make a distance estimate depends on the resolving power of its eye. (The reasons will become clear in the discussion of particular distance cues.) The resolving power of an animal's eye, in turn, depends on its size, as shown empirically in figure 4.13: The larger the animal, the better the angular resolution of its eye.

The explanation of this empirical finding is that the size of an animal's eyes is roughly correlated with overall body size, but the size of its photoreceptors and their density per unit area are not. Photoreceptors cannot be scaled down infinitely, because the wave properties of light impose a limit on how narrow a receptor can be if it is to act as a light guide and absorb light efficiently. Also, to avoid crosstalk between them, photoreceptors must not be packed too tightly (Kirschfeld 1976). For this reason, whatever their size, animals with high-acuity vision have photoreceptors about 1–2 μm in diameter, with very similar packing density. Consequently, the smaller the eye, the grainier the picture it transmits. To achieve good visual resolution, then, small animals often have disproportionately large eyes. For example, within the primates, body weight varies over a 500-fold range from a tarsier to an orangutan, whereas eye size (volume) only varies by a factor of about 1.5 (Schultz 1940). But there are obviously physical limits to how large an eye can be accommodated in a small head, and, as figure 4.13 shows, visual acuity remains roughly proportional to body size. It follows,

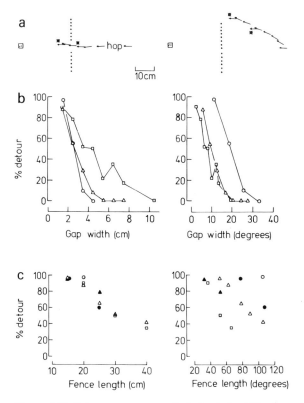

Figure 4.12 Data on measurement of physical width of gaps and barriers by toads (*Bufo viridis*). (a) Toad approaching mealworm when picket fence is interposed. Arrows represent toad's position every 20 msec. Asterisks indicate brief pauses during approach. Fence is composed of unpainted wooden dowels 0.5 cm in diameter and 30 cm high inserted into floor of 1-m² arena. The animal is lured to the desired position in front of barrier and its behavior recorded. It plans its approach before making any movement, and aims either directly at the gap or at the end of the fence. Toad tends to make for a gap more than about 3 cm wide (the width of its head). (b) Left: Choice between detour and direct approach plotted against gap width in cm for three different starting distances from the fence:(○) 7.5 cm, (△) 15 cm, and (□) 30 cm. Relation between percentage of trials in which toad detours and gap width is the same for starting distance of 7.5 and 15 cm, indicating that the toad measures real size over this range. However, if starting distance is increased to 30 cm, size constancy has evidently broken down and toad seems to underestimate gap width. Right: Same data plotted against gap width measured as angle subtended by gap at the toad's starting position. There is a significant difference between plots for 7.5-cm and 15-cm starting distances ($P < 0.001$), but plots for 15-cm and 30-cm starting distances are very similar. Best estimate from these data is that constancy holds for distances up to about 20 cm. (c) Dependence of toad's readiness to detour on length of fence. As length is increased, toad detours less. Relation between detour percentage and fence length in cm is unaffected by starting distance of toad from fence (over at least 30 cm), indicating that toad measures real width of barrier. (○, △, □) as in part b; (●) 9 cm, (▲) 25 cm. Note increased scatter when detour percentage is plotted against angle subtended by fence at starting position. Source: Lock and Collett 1980.

128   Collett, Harkness

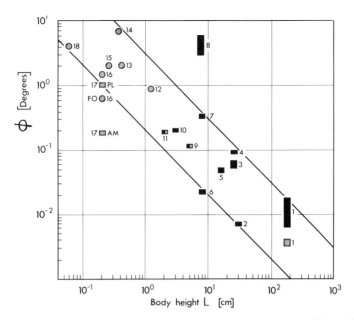

Figure 4.13 The relationship between an animal's body height (L) and the spatial resolution of its eye (φ). Black symbols represent resolution measured physiologically; shaded symbols represent resolution measured anatomically. (1) Man, (2) peregrine falcon, (3) hen, (4) cat, (5) pigeon, (6) chaffinch, (7) rat, (8) *Myotis*, (9) frog, (10) lizard, (11) minnow, (12) *Aeschna*, (13) *Apis*, (14) *Chlorophanus*, (15) *Musca*, (16) *Syritta*, (17) *Methaphidippus*, (18) *Drosophila*. Source: Kirschfeld 1976.

then, that the range and potential accuracy of distance cues will also tend to decrease proportionally with an animal's size. Small animals are beset by a variety of physical limitations to the distance information available to them, and are far more restricted than larger animals in their choice of cue.

Cues from Focus

Distance information is available from the basic optics of any image-forming eye. If the distance between the lens and receptor plane is fixed, the position of the image plane can be used as a measure of an object's distance from the eye, as image distance (v) and object distance (u) are directly related by the simple lens equation $1/f = 1/v - 1/u$, where f is the focal length of the eye. (Eyes cannot usually be treated as thin lens systems, so the definitions of focal length, image and object distance appropriate for thick lenses must be used.) The position of the image plane could be monitored, for instance, by a multiple-layered retina, in which case the position of the best focused image could be used by an animal as a direct measure of the object distance. Multiple-layered retinae are not common in the animal kingdom except, curiously, in the deep-sea fishes (Munk 1966; Locket 1977). It has been suggested that these animals may use the different receptor layers to establish the position of the image plane and so estimate object distances (Vilter 1954). However, there are reasons for thinking that this is unlikely and that there must be some other reason why this organization should be favored by deep-sea fishes (Munk 1966). It is possible that multiple-layered retinae in other animals—jumping spiders, for example (Land 1969)—provide the basis for distance estimates, but we have no evidence that any animal actually uses this information.

An animal with a fixed-focus eye and only a single layer of receptors cannot monitor the position of the image plane directly, but could conceivably use the degree of image blurring to place an object roughly in space. Many species have fixed-focus eyes, most notably the arthropods, but the very reasons that make it possible for a visually active animal to tolerate a fixed-focus eye make it unlikely that any aspect of focus would be useful to that animal in estimating distance. Very small animals have no need of focusing mechanisms, because the images formed by their eyes are effectively focused for objects lying anywhere from infinity to a few millimeters away. Changes in focus could only be detected and used by such an animal to estimate distance over a range too small to be biologically significant.

Consider the simple ray diagram of figure 4.14, which shows the geometric optics of a simplified "eye." Parallel light rays from a point source at infinity converge on the receptor plane to give a focused image; the image plane coincides exactly with the receptor plane. If the object were closer to the lens, the rays would be focused at a point be-

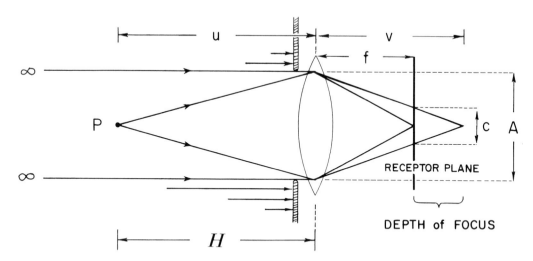

Figure 4.14 The geometric optics of a simplified eye, showing depth of field when lens is set to focus objects at infinity. Light focused by a thin lens is imaged on the "receptor plane," which coincides with the lens focus. A stop in front of the lens limits its aperture. Given the thin-lens equation ($1/f = 1/v - 1/u$), it can be shown from the diagram (see Rolls 1968, p. 121) that $H = f^2/nc$, where $n = f/A$ (the relative aperture, or f number). The depth of field of this "eye" of focal length f and aperture A extends from infinity to a distance H from the lens, when a point source of light at P forms an image of diameter c, the maximum allowable blur-circle diameter at the receptor plane.

Collett, Harkness

hind the receptor plane, forming a blur circle on the receptor surface. Whether or not this blurred image is *seen* as "in focus," however, will depend on the resolution of the receptors. This limited resolution is represented on the figure by c, which defines a maximum allowable blur-circle diameter. A point source can be brought as close as P before the blur circle formed in the receptor plane exceeds the size permitted, and all objects farther than a distance H from the eye will be seen as "in focus."

It is clear from the diagram that the hyperfocal distance H depends on both the lens aperture (A) and the diameter of the blur circle that is allowed (c); if A is increased or c is reduced in size, the hyperfocal distance will be increased. It can also be shown from the diagram that the hyperfocal distance depends on the square of the lens focal length (f). Thus, the size of the eye is important in determining its depth of field, or the range of distances in object space that will give images "in focus." Small eyes will have very much greater depth of field than large eyes. To give some idea of the absolute distances involved, figure 4.15 shows how hyperfocal distance varies with different focal lengths given constant but biologically reasonable values of relative aperture and maximum allowable blur-circle size.

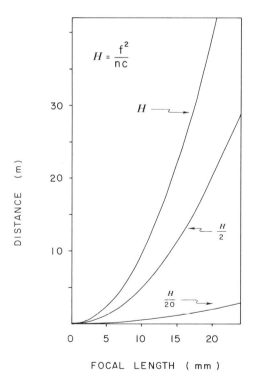

Figure 4.15 The relationship between H, the hyperfocal distance of a simplified "eye" (see figure 4.2), and its focal length f. H is calculated from $H = f^2/nc$ using a value of 2 μm for c, the maximum allowable blur-circle diameter, and a value of 5 for n, the relative aperture.

Depth Vision in Animals

Geometric optics is obviously only an approximation to the physiological optics of a real eye. However, the figures and calculations above illustrate why animals only need to accommodate their eyes over a limited range of near object distances, and how that range is disproportionately small for animals with small eyes. The depth of field of a fly's eye, for example, has been calculated to extend from infinity to a few millimeters from the eye (Collett and Land 1975). It is clear why such a small animal does not need to focus its eye, and also that distance cues from "focus" will be of no use to it whatever.

The larger the animal, the smaller will be the depth of field of its eye and the greater the range over which it could use some aspect of focus to estimate distance. It follows also that the larger the animal, the less well it will be able to see objects close by, as their images will be blurred. Many larger animals have overcome this problem by evolving mechanisms for focusing (accommodating) their eyes, by moving or deforming the lens to maintain the image in the plane of the receptors for a wide range of object distances (Walls 1942; Packard 1972). Within the range where accommodation is necessary, all these animals have distance information available from their focusing mechanisms, as different object distances will be represented by different focus settings. The accuracy with which a distance estimate can be made depends on the depth of field at the various focus settings. Unfortunately, it is not possible to predict exactly what that accuracy will be for real animals, as we do not know how or with what accuracy they can detect misfocus. However, simple geometric optics can again be used to illustrate the sort of constraints that are imposed by an animal's size and visual style.

Figure 4.14 illustrates the depth of field for a lens set to focus objects at infinity. The depth of field extended from infinity to a nearer point (P), the position at which the image of the point source first reached the arbitrarily defined permissible diameter. When an object lies closer to the lens than this point, and the lens is moved or deformed to refocus the image, there is a range of lens settings over which the image will be "in focus," and for any one lens setting there is a range of distances in space at which the object may be positioned and still give an "in focus" image. This latter range, the depth of field at a particular lens setting, is defined by two limiting conditions, as shown in figure 4.16. Given a maximum allowable blur-circle diameter (c), the image at the receptor plane will be seen as "in focus" if the image plane or "true" focus lies anywhere between A and B on the figure. It is also clear from figure 4.16 that the limiting positions A and B of the focused points will be altered by the same variables that affect the hyperfocal distance of the lens (recall figure 4.14). The "near" and "far" points of the accommodated lens, which are the points in object space that correspond to the two points A and B in image space, can be simply defined in terms of the hyperfocal distance (H):

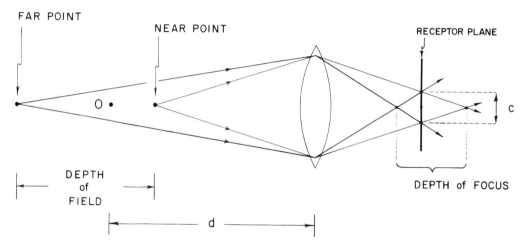

FAR POINT

NEAR POINT

RECEPTOR PLANE

O •

c

DEPTH
of
FIELD

DEPTH of FOCUS

d

Figure 4.16 The geometric optics of a simplified "eye," showing the depth of field about a point object O for which the eye is set to give a focused (point) image at the receptor plane. At this lens setting, objects positioned at the "far" and "near" points form images of diameter c, the maximum allowable diameter of blur circle at the receptor plane. Thus, the near and far points define the depth of field, that slice of object space within which an object can be positioned and still be seen as focused. The images of point objects outside this slice of space will be blur circles with diameters larger than c, and will therefore be judged as misfocused.

$$\text{Near point} = \frac{H \times d}{H + (d - f)},$$
$$\text{Far point} = \frac{H \times d}{H - (d - f)},$$

where d is the distance of the object plane for which the lens is set (i.e., the distance at which a point object, O, gives a point image at the receptor plane).

The first interesting thing these equations tell us is that when a lens is set to focus objects placed at the hyperfocal distance, the depth of field extends to a distance half the hyperfocal distance. (At far distances, $d - f \approx d$; so when $d = H$, far point $= H/O = \infty$ and near point $= H^2/2H = H/2$). An animal with a fixed-focus eye can thus optimize the performance of its eye by positioning its receptor surface a little farther than one focal length behind the lens so as to focus objects at H. The theoretical range over which an animal could estimate distance using focus cues is thus more likely to be related to $H/2$ than to H. For this reason, values of $H/2$ are also given in figure 4.15.

When a lens is set to focus on objects at distances closer than $H/2$, the accuracy of distance estimates will be limited by the depth of field (far point − near point). A convenient measure of the accuracy of focus cues is then given by the ratio of the depth of field to the object distance (d). For example, if the depth of field is 1/10 the object distance, then the error in estimating the distance of this object will be approximately ± 5%. (The near and far points are not exactly symmetrical about the ob-

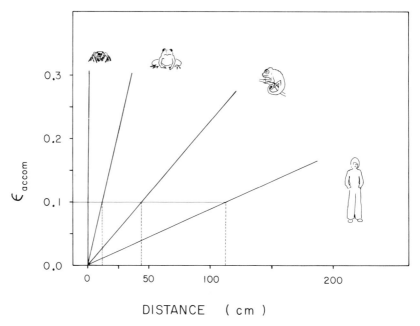

Figure 4.17 Relationship between relative accuracy of accommodation cues ($\epsilon_{accom} = (2nc/f^2)d = 2d/h$) and distance for jumping spider, frog, chameleon, and man. We have included the spider, even though it does not focus its eye, to illustrate how inaccurate focus cues would be for very small animals. Values used in calculated ϵ_{accom}: for jumping spider (*Metaphiddipus aeneolus*), $f = 0.5\ \mu m$, $n = 2.56$, $c = 1.7\ \mu m$ (Land 1969); for frog (*Rana esculenta*), $f = 4.08$ mm, $n = 0.91$ (duPont and de Groot 1976), $c = 8\ \mu m$ (calculated from Birukow's 1937 value for the spatial resolution of *Rana temporaria*); for chameleon (*Chameleo jacksoni*), $f = 7$ mm, $n = 3.8$ (Harkness 1977b), $c = 1.5\ \mu m$ (Rochon-Duvigneaud 1933, *Chameleo chameleo*); for man, $f = 17$ mm, $n = 5.6$ (from sources given in Davson 1972), $c = 2.25\ \mu m$ (mean of four values quoted in Pirenne 1967).

ject distances for which the lens is set.) Figure 4.17 shows how this ratio (ϵ_{accom}) changes with increasing distance for four simplified eyes representing animals of widely ranging sizes. ϵ_{accom} increases approximately linearly with object distance over the range illustrated, at a rate that depends on the focal length of the lens. The accuracy of focus cues thus deteriorates with distance, but it does so more rapidly for animals with small than with large eyes.

For relatively near distances, $\epsilon_{accom} \simeq 2ncd/f^2$, so the accuracy of estimating the distance of a certain object will improve as a function of f^2. It is also clear from this relationship that the range over which an animal can achieve a reasonably accurate estimate of depth is a fixed percentage of the hyperfocal distance. For $\pm 5\%$ accuracy, for example, ϵ must have a value of 0.1, and $d \simeq f^2\epsilon_{accom}/2nc$, so $d \simeq H/20$. (Values of $H/20$ are given in figure 4.15.) The distance at which the depth of field is 1/10 the object distance is also indicated for the various animals included in figure 4.17.

The accuracy with which an animal can actually estimate distance using accommodation cues, however, will depend on how closely the

Collett, Harkness

accuracy of its focusing system approaches the limits we have calculated using geometric optics. There are various reasons why we might expect an animal's accuracy to fall short of the maximum calculated in this way. One is that point objects cannot give point images on the retina. Even assuming that an eye has perfect optics, light is diffracted at the edges of the pupil, and this results in spreading of the light into a pattern of concentric light and dark rings (an "Airy disk"). Furthermore, eyes are not perfect optical systems, so further "blurring" of the ideal focus occurs as a result of lens aberration, lack of homogeneity in the eye's "transparent" media, and so on. It has been estimated that in the human eye, for example, the retinal image of a point source of light would have an intensity distribution with a half-width of about 2.5 μm (0.66 arc-minutes) when the pupil diameter is 3 mm, even when the eye is "perfectly" focused (Westheimer and Campbell 1962). Blurring of the image caused by diffraction and the eye's optical imperfections will increase the depth of field above the values calculated using geometric optics. The effect on the depth of field could in principle be estimated, but in practice the calculation is sufficiently complex that it has not been attempted even for humans.

Another reason is that the accuracy of the human focusing system, the only one we know much about, turns out to be very poor compared with what we might expect, even given the image degradation described above. Even under ideal viewing conditions, humans have been found to tolerate up to 0.3 diopters of focus error for small high-contrast objects at 50 cm (Campbell 1957; Campbell and Westheimer 1958). In geometric optics, this means that a blur circle up to 10.5 μm in diameter is "permitted," rather than one of 2.25 μm representing the center-to-center distance between foveal cones. Thus, humans seem to be about one-fifth as accurate at focusing as figure 4.17 would indicate. Other aspects of human accommodative response suggest that it would not provide a good basis for depth perception (Owens 1979), but this is perhaps not surprising, as the wealth of alternative depth information available to us makes it unlikely that our focusing system was designed with this function in mind.

Indeed, human accommodation probably does not normally contribute directly to distance perception, although there are rare instances of people whose subjective sensation of distance seems closely linked to the refractive state of their eyes and not to other cues (Carr and Allen 1906). The role of accommodation in normal human distance vision is probably, at most, to act with convergence as a subsidiary cue assisting stereopsis. The relative importance of convergence and accommodation in this role is still unclear, because the two systems are so interactive it is very difficult to isolate and assign specific effects to one or the other (Owens and Leibowitz 1980). For example, humans can make distance estimates monocularly when context and motion cues are excluded, presumably using the focusing system as the source of distance information (Baird 1903). However, it is not clear from Baird's experiment that the focusing system itself is providing depth information directly, as even in monocular viewing vergence of the two eyes still occurs because accommodation drives convergence. However, convergence alone has been shown to be an effective distance cue (Heineman et al. 1959), so it seems that even if accommodation is also used its role is not crucial for humans.

In contrast, accommodative cues may be very important to lower animals, and those that use them behave as if they are capable of better focus resolution than man. For example, frogs (*Rana pipiens*) have been shown to have a fairly sharply delineated "snapping zone" around them. If prey lies within this zone the frog snaps at it, but if the prey lies outside the zone "boundary" the frog only orients toward it (Ingle 1970). Individual animals have been shown to change their behavior from snapping to orienting when the prey lies within a "zone of ambiguity" about 1 inch wide at distances of 7 inches (Ingle, 1972 and pers. comm.); this indicates that they can estimate a distance of 7 inches with about 15% accuracy. This seems to correspond quite closely to the maximum accuracy we estimate that a frog could achieve at this distance (figure 4.17). Frogs, toads, and chameleons also seem remarkably accurate at using focus cues to estimate distance within the range over which they snap or shoot their tongues at prey (see figures 4.32, 4.36, 4.37). Snapping and tongue-shooting accuracy in these animals comes close to that expected on the assumption that the depth of field limits performance, even though it seems likely that error in the distance estimate will not be the only cause of behavioral variation. For example, some motor error is probably incurred in coordinating the snaps and tongue strikes. It is not clear why these animals appear to be more accurate than humans. It could be that the optics of their eyes are better or that the image characteristics they use to drive their focusing systems differ from our own (see, for example, Harkness and Bennet-Clark 1978). Alternatively, animals could have devised specially accurate ways of exploiting focus information, such as using the transition from focus to misfocus as the criterion from which distance is "read." Humans can actually detect this transition rather accurately—to within 0.02 diopters (Campbell and Westheimer 1958).

Whatever aspects of the image are used to detect focus error, an animal can improve the accuracy of its distance estimates by altering those characteristics of its eye that affect the depth of field defined in terms of geometric optics. For improved accuracy, eye size (and thus focal length, f) should be as great as the animal's body size will permit; the aperture should be large; and the photoreceptor density should be as high as possible to improve visual acuity, which will probably allow more accurate detection of misfocus. Chameleons, who rely on their focusing system to estimate the distance of their insect prey, seem ideally suited to exploit accommodation cues. The chameleon's eyes are very large compared with the size of its head (figure 4.34), it feeds with the pupils wide open (Harkness 1977b), and it has very high cone densities (Rochon-Duvigneaud 1933).

Chameleons, like many animals with high-acuity vision, restrict the necessary specializations to a small patch of retina (high receptor densities, high ratio of ganglion cells to receptors, etc.) and use head or body movements to bring the image in which they are interested onto that area best equipped to inspect the image closely. It is more important, then, for the animal to have the image in focus on this specialized region, the area centralis or fovea, than elsewhere on the retina; and it is reasonable to assume that foveate animals focus their eyes on the basis of image quality in that area, as humans do (Toates 1972). Thus, to use accommodation cues to estimate distance, such an animal would only have to calibrate its focusing system on one axis to give consistent readings. Recalibration would only be necessary as the animal grew.

In contrast, animals without foveas whose retinal structure is more uniform often behave as though they are able to accommodate their eyes to focus on objects seen by any part of the retina. If they are to make distance estimates based on their focusing systems, a single calibration will not give consistent distance estimates, as the focus setting required for an object at a certain distance will vary according to where on the retina the image falls. Frogs and toads have no fovea, but they do have a focusing system, and they are to a degree capable of using accommodation cues to estimate the distance of objects, either on the primary axis of their eyes or to one side when the image falls elsewhere in the periphery of the retina—a remarkable feat. There are retinal locations for which frogs do not seem capable of correcting their accommodative distance estimate. Monocular *Rana pipiens* undershoot prey in the field of view contralateral to the seeing eye (Ingle 1976), and it seems likely that error in the distance estimate may exist as a result of the offset between the image and the optic axis (along which the lens is moved to focus), as Ingle suggested. This error would clearly not be crucial to normal binocular animals, however, as stereopsis is an available cue, and Ingle's data imply that the other eye could provide an adequate accommodative distance estimate for this part of the visual field. At equivalent angles off axis in the ipsilateral (monocular) field, frogs only attempt to snap at prey at very close distances (Ingle 1976). Inaccuracy of the accommodative distance estimate may be one of the reasons that frogs restrict their snapping range at these extreme visual angles, where alternative depth information is unavailable.

Once calibrated correctly, however, any system of distance estimation based on accommodative cues has the advantage of being objective and only requires information from one eye. A disadvantage of accommodative cues is that only one distance estimate can be made at a time, as generally eyes can only be focused for one distance at a time. As explained at length above, a more serious restriction of the use of accommodative cues is the size of an animal, as both the range and the accuracy with which a distance estimate can be made using accommodative cues drops rapidly the smaller the animal's eye. However, over a range of close distances, cues from "focus" can provide precise quantitative distance estimates, which is on many occasions just what animals need.

Image Size

The visual angle subtended by an object is inversely proportional to its distance. Thus, given the retinal image size of an object at one distance, image size can be used to calculate any other object distance (figure 4.18). This very simple method requires either that the observer is genetically endowed with a knowledge of the size of the object or that it can learn its size, as when honeybees learn the image sizes of arbitrary landmarks in order to find a food source at some fixed distance from them (Cartwright and Collett 1979). The method will clearly fail if objects are misidentified or seen from an unfamiliar viewpoint. Furthermore, it can only be used for certain classes of natural objects that, like holometabolous insects that emerge fully grown from the pupa, are of a reasonably uniform size. Figure 4.19 plots the distribution of body length of workers from a single hive of honeybees and that of a sample of female flies (*Dilophis febrilis*) collected from sycamore leaves. Body length of these insects is unimodally distributed with a relatively narrow peak, and so could provide a usable cue to distance. On the other hand, body length within a population of juvenile and adult crickets, all of which look basically similar, varies widely. Depth judgments based on an assumed mean value for such a mixed population would often be seriously wrong.

The range and sensitivity of depth information that can be derived from image size depends on the accuracy with which visual angle can be measured, and so, as in the other cases we discuss, the precision of depth information will improve with body size. However, spatial resolution is the only parameter directly related to the size of the animal that has any effect on the accuracy of the cue. Its sensitivity will thus vary linearly with body size and not, like that of accommodation or

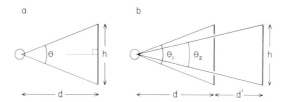

Figure 4.18 (a) Visual angle subtended by object of height H at distance d. $\tan(\theta/2) = H/2d$; therefore, $d = H/2 \tan(\theta/2)$. When θ is small, $\tan(\theta/2) \approx \theta/2$, so $d \approx H/\theta$. (b) Change in visual angle ($\delta\theta$) when an object of height H is moved a distance d' away from its position at distance d, changing visual angle from θ_1 to θ_2.

Collett, Harkness

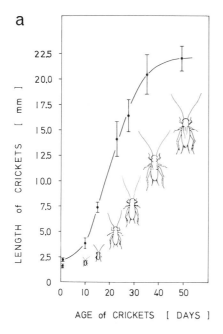

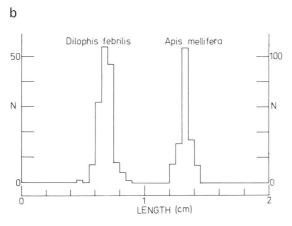

Figure 4.19 (a) Relation between body length and age in crickets (*Acheta domestica*). It would not be practical to use the retinal image size of one of these animals to estimate its distance, because all growth stages look basically similar (see sketches) but their body length varies over a tenfold range. (b) Histograms showing variation in body length of worker *Apis mellifera* from a single hive (length: head to abdomen) and of female *Dilophis febrilis* (length: head to wingtip). Honeybees provide good yardsticks for distance measurements. Swifts, for instance, are reputed to have sufficient confidence in their size to distinguish drones from workers on the wing (Lack 1956). Flies, although more useful than crickets, are less reliable than bees; their size varies over a twofold range.

Depth Vision in Animals

stereopsis, as a higher power. Consequently, in some situations image size is likely to be the preferred depth cue of very small animals. But the accuracy of depth information also varies with object size, and animals are often concerned with objects of about their own size (for instance, a mate).

The relation between acuity, object size, and depth sensitivity is easily derived from figure 4.18. If an object H cm across at distance d cm is moved away from an observer by d' cm, then the change in the visual angle of the object subtended at the observer ($\delta\theta$) will be given by

$$\delta\theta = d'H/(d^2 + d'd) \simeq d'H/d^2,$$

or

$$d' \simeq \delta\theta d^2/H.$$

If d' is the minimum discriminable depth interval at distance d, then the expression implies that depth sensitivity increases both with visual acuity (the accuracy with which $\delta\theta$ can be measured) and with object size. So, if an animal has any choice, it should use information from as large an object as it can.

In any real situation, though, curious difficulties are likely to arise. We mentioned earlier that male *Syritta pipiens* track other flies keeping a relatively constant distance from them. They probably do this by measuring some aspect of the image of the leading fly, retreating when the image is bigger than it should be and advancing when it is smaller. But what does the tracking fly measure? The leading fly holds its body axis more or less horizontal; its body length is its largest dimension and therefore at first sight the most suitable measure to take. However, flies change their angular orientation, and sometimes a tracking fly will see the leading fly end on and sometimes from the side (figure 4.3). This sudden change in the horizontal extent of the image (which is unrelated to the separation between the two) has little or no effect on the forward velocity of the tracking fly, who must in all probability measure either head or body height or possibly the envelope of the wing beat.

In principle, then, the apparent size of an object can provide a cue to distance—provided, of course, that the animal is correctly calibrated and knows for at least one pair of values the distance that corresponds to a given image size. In the few cases in which it has been examined, however, image size does not seem to serve this function. Rather, it is used to specify a desired position in space relative to some object. The animal knows no more than that a desired position is associated with a particular image size and how it needs to move to keep the image at that set size.

Stereopsis and Vergence
Binocular animals with partially or totally overlapping visual fields have another set of depth cues available to them. Because the eyes are sepa-

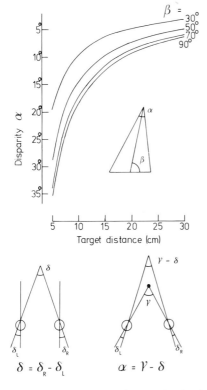

Figure 4.20 Geometry of binocular depth vision, showing how the distance of a target is specified by the angle α. Top: Relation between α and target distance for different values of β, assuming an interocular distance of 3.5 cm. Bottom left: An animal with fixed eyes can measure α from the difference in the horizontal position of the images on the two retinae ($\alpha = \delta = \delta_R - \delta_L$). Bottom right: An animal with vergence needs to know both the angle of vergence of its eyes (γ) and the horizontal disparity of the target (δ) in order to compute α ($\alpha = \gamma - \delta$, for uncrossed disparities).

rated in the head, each views the same region of space from a slightly differing vantage point, and the difference in the direction of the object as seen by the two eyes is related to its distance from the animal. The closer the object, the larger is the angular difference between the images on the two retinae. The geometry of this situation is shown in figure 4.20, which represents a binocular animal looking at a small target. The animal could gauge its distance from the target if it could measure the angle α. Binocular animals can compute α in two different ways. In one, the eyes are fixed in the head and α is calculated from the differences in the horizontal position of the images in the two retinae (the horizontal disparity, δ); in the other, the eyes are swiveled to position the target image onto a particular region of retina, such as the fovea, when α is given by the difference in the position of the two eyes (the angle of vergence, γ). In the first case the animal clearly needs a neural mechanism with which to compare the two images, whereas in the second the two eyes must agree to fixate the same target, which probably also requires machinery for matching the images on the two retinae.

The relation between α and target distance (d) depends on both the separation of the eyes (a) and the angle of the target to the midline (defined by β in figure 4.20), so that

$$\cos\alpha = \frac{4d^2 - a^2}{(16d^4 - 8a^2d^2 \cos2\beta + a^4)^{1/2}}.$$

When $\beta = 90°$, the expression reduces to

$$\tan(\alpha/2) = a/2d.$$

Thus, for accurate distance estimates, an animal must know its interpupillary separation and assess both α and β (or simplify the situation by turning its head to face the target, making $\beta = 90°$).

As with focus cues, we can use simple geometry to set limits to the precision with which animals of different sizes can use binocular cues to estimate depth. The shortest depth interval (d') that can be measured at a distance d from an animal's eye depends on two factors: the separation of the eyes and the smallest difference in the value of α that can be resolved (α'). It can be shown from figure 4.21 that

$$d'_f \simeq d^2\alpha'/(a - d\alpha') \qquad (\alpha \text{ in radians})$$

when d' is farther than d, and that

$$d'_n \simeq d^2\alpha'/(a + d\alpha')$$

when d' is closer. A convenient figure of merit for the relative sensitivity of binocular depth cues, which bears direct comparison with the accuracy of focus cues (figure 4.17), is given by

$$(d'_f + d'_n)/d.$$

We have called this ratio ϵ_{binoc}; it is approximately equal to $2d\alpha'/a$. In figure 4.22, ϵ_{binoc} is plotted against d for three representative animals with very different spatial resolution and interocular separation. The precision with which α can be measured is limited by an animal's acuity (ϕ) or by the accuracy with which it monitors the position of its

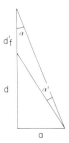

Figure 4.21 Illustration of relation between depth interval (d') and disparity difference (α'). By inspection, $a/(d + d'_f) = \alpha$ (where α is in radians). Substituting and rearranging gives $d'_f = d^2\alpha'/a - d\alpha'$. (After Ogle 1962.)

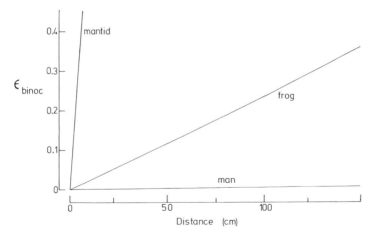

Figure 4.22 Relation between relative accuracy of binocular depth estimates (ϵ_{binoc}) and distance in mantid, frog, and man. For illustrative purposes have assumed that mantids use binocular cues and that humans do not have vergence. $\epsilon_{binoc} = 2d\alpha'/a$, where d = distance, a = interocular separation, and α' = smallest difference in α that can be resolved. When only disparity information is available, α' equals spatial resolution ϕ measured in radians (2 rad = 360°). Values of a and ϕ for mantid (*Ciulfina*) from Horridge 1978; value of ϕ for frog (*Rana temporaria*) from Birukow 1937. For comparison we have used human grating resolution rather than stereoacuity.

eyes. Since very little is known about how well vergence eye movements are monitored, we have taken the eyes to be fixed and assumed that the animals make use of disparity measurements.

It is clear from figure 4.22 that low acuity and close-set eyes provide a poor base from which to exploit binocular depth cues. Small animals, such as insects, labor under this double disadvantage. In vertebrates one finds a wide range of interocular separations. Figure 4.23 plots the distance between the centers of the two eyes against body length for adult specimens of different-sized species. For anurans and primates the two parameters are linearly related, showing that interocular distance is a constant proportion of body length irrespective of size and that within a single group small species do not attempt to improve their depth range by pushing their eyes farther apart. However, the ratio of eye separation to body length varies widely from group to group. It is about 1:5 in ranid frogs and 1:20 in primates, and one can only speculate why this might be so. To do this we must discuss the anatomy and physiology of binocular connections, which impose constraints of a very different kind. Disparity measurements inevitably require an animal to compare retinal images in the two eyes, and this means having binocular interactions at some level within the visual system. Were an animal to try to compare every point on one retina with every point on the other, the amount of wiring necessary would be hugely cumbersome for even the lowest-"grain" retina. An animal with fixed eyes that wishes to measure a wide range of distances over a wide range of angles

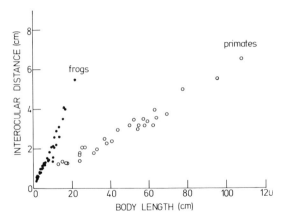

Figure 4.23 Relation between interocular separation and body length for a range of different-sized species of ranid frogs and old- and new-world primates. Each point represents a single specimen of a different species in the collection of the British Museum of Natural History. Body length was the distance between snout and vent (frog) or between crown and rump (primate). The ratio between interocular separation and body length is 1:4 in the frog, 1:20 in the primate, and about 1:10 in a series of rodents.

approaches this condition. The farthest distance it can estimate will be determined by the smallest disparity it can measure; the near point will be set by the maximum disparity to which it is sensitive. Thus, to measure far distances each point on one retina must be capable of precise interaction with a closely corresponding area in the other. But if the animal is to monitor near distances too, each of these positions on the retina must be able to talk to an extensive region of the other eye. Now, it seems to be a general rule that the higher the spatial resolution of an area of retina, the greater the area of visual cortex or optic tectum devoted to it. This means that the lateral spread of binocular interactions emanating from a small area of cortex will be determined by both the animal's spatial resolution and its disparity range, and will for our hypothetical animal be very extensive.

One might therefore expect a variety of compromises and economy measures to have developed which would limit the amount of wiring without sacrificing the range and accuracy of depth vision. One way a high-resolution animal can reduce its disparity range and still monitor short distances is by bringing its eyes closer together, and this may in part explain the difference in eye separation between primates, with their high-acuity vision, and frogs, whose resolution is significantly poorer (figure 4.13).

A rather different means of limiting the necessary amount of binocular interaction is to restrict depth vision to a narrow beam of directions within the visual field. If an animal always turns its head to face the target squarely and only attempts to measure the distance of objects straight ahead, it need merely arrange that a small patch of one retina

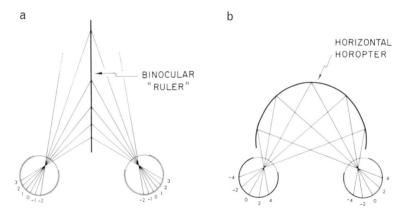

Figure 4.24 Two forms of binocular interactions. Both schemes assume that a region in one retina interacts with no more than a limited region in the other. (a) The animal measures distance along its midline. It has a form of binocular ruler extending outwards from between its eyes. To achieve this, points 1,2,... in the left eye are linked respectively with points 1,2,... in the right. Targets must be accurately aligned on the ruler, and to do this saccadically an independent method of depth measurement is needed. (b) The animal can measure a single distance along each cyclopean line of sight. Points marked with the same number are again linked between the two eyes, but in this case the "corresponding" points are arranged differently from those in part a. The loci of these binocularly interacting points are projected into space form the horizontal horopter. All vertebrates known to use binocular cues conform to this scheme and measure a small band of disparities on either side of the horopter.

converse with a small patch on the other, as shown in figure 4.24a. No examples of animals using this strategy are known.

Convergent eye movements provide another powerful device for reducing the necessary wiring. Instead of limiting the horizontal extent over which depth can be measured, vergence movements relieve the need to monitor a wide range of disparities. With changing vergence, a relatively narrow band of disparities shifts in depth with the fixation point. In this case, binocular interactions can be limited to what are termed corresponding points on the two retinae.

An animal with vergence movements has, however, given itself another complication: that of translating disparity values into distance. To measure the distance of a target that is nearer or farther than the fixation point, the disparity angle must be added to or subtracted from the vergence angle:

$\alpha = \gamma + \delta$ for objects nearer than fixation,

$\alpha = \gamma - \delta$ for more distant objects.

As is obvious from figure 4.20, the depth interval corresponding to a given disparity value will grow as the fixation point moves away from the animal. Conversely, if an animal wishes to monitor a certain depth interval on either side of its fixation point, say 20% of the fixation distance, the range of disparities it must watch will increase as the fixation point comes closer. Moreover, the rate of increase is proportional to the interocular separation. So it is worth keeping one's eyes set close together in order to reduce the shift in the range of disparities that must be monitored as fixation is switched from a distant to a nearby target.

Primates have relatively close-set eyes and convergent eye movements, and seem to have arrived at a satisfactory compromise that gives both good depth vision and a manageable amount of binocular interaction. Psychophysical and physiological evidence suggest that they only measure a few degrees of disparity. When a target is fixated binocularly, a series of points in space are seen by corresponding points on the two retinae (figure 4.24b). The locus of such points is known as the *horopter*, and in man it can be defined as those positions in space that appear to lie in the same direction when viewed through each eye separately (Ogle 1962). In the horizontal plane (at certain fixation distances) this turns out to be a slightly squashed circle passing through the fixation point and the nodal points of both eyes. Humans monitor disparities accurately over a relatively narrow region of space on either side of the horopter, so that the horopter indicates which regions of the two retinae are interconnected within the visual cortex. There is in addition a very coarse-grained, seemingly independent system for measuring large disparities imprecisely (Bishop and Henry 1971; Jones 1977) that is used primarily for driving vergence movements.

Cat and primate visual cortex contain single neurons sensitive to the horizontal disparity of binocular stimuli. Each neuron responds to a rather specific disparity value, and the whole population of neurons covers no more than a few degrees of disparity (cat: Barlow et al. 1967 and von der Heydt et al. 1978; monkey: Poggio and Fischer 1977). The spread of disparities monitored by these neurons presumably limits the animal's stereoscopic range, while the position of the mean of the distribution for a given direction of regard can be used to define the location of the horopter along that line of sight. In the cat, Joshua and Bishop (1970) showed that when the means of the distributions of the disparities were projected into space their positions described a curve resembling the human horizontal horopter. Thus, physiological data also imply that a point on one retina interacts with no more than a small corresponding region in the other.

Generally, the closer an object is to a terrestrial animal the lower it is in the visual field; this regularity gives some animals a way of measuring a wide range of distances without the necessity of extensive binocular connections or convergent eye movements. The principle is to fix corresponding retinal points so that the lower part of the visual field is specialized for close objects and the upper for more distant ones. For an animal with high acuity and fixed eyes the trick may be an essential economy; however, there are also substantial advantages for an animal with vergence, since the extent of the stereoscopic field at any particular fixation point will be increased. This result is achieved by tilting the vertical horopter. Helmholtz (1925) was the first to note that in man the "vertical horopter along the mid-saggital plane is not in fact vertical, but a straight line tilted away from the observer." Cooper and Pettigrew (1979) found in the burrowing owl (an essentially terrestrial predator) and the cat that the mean value of the disparities of binocular neurons sampled along the vertical meridian (recorded respectively in the Wulst and the visual cortex) shifts gradually from the upper to the lower field, but that the range of disparities remains constant (figure 4.25). When these disparities are projected into their appropriate position in space, they form a line that is tilted upward from the animal's feet toward the fixation point. Since the extent of binocular interactions within any small region of brain is determined by the range of disparity values that must be monitored and not the mean, the owl can judge distances on the ground from 20 cm outward without needing an implausible amount of wiring and without eye movements.

The owl and cat tilt their horopters by twisting the zero meridians of the two eyes with respect to each other so that they diverge at the top and approach at the bottom. The projections into space of these meridians will then intersect in a tilted line defining the corresponding points of the vertical horopter (figure 4.26).

Since the owl does not converge and bring its fixation point closer, there might seem to be a hole in its representation of three-dimensional space above the ground. However, it can fill this hole simply by tilting its head upward. Pettigrew tells us that this is just what a burrowing owl does when an object is brought toward it at eye level. It is using head movements as a substitute for vergence eye movements, altering the tilt of its head to place an object at a given distance in the correct position on the vertical horopter. Moreover, this behavior depends on binocular vision, and if one eye is masked the head is not tilted. Owls that have more aerial habits and cannot assume any fixed relationship between distance and position in the visual field do not behave in this way.

Toads are also terrestrial creatures with fixed eyes and stereopsis, but their lower visual acuity and behavior suggest a different binocular organization. When a toad orients to a distant prey on the ground it lifts its head; but it lowers it to look at a prey that is very close. However, as figure 4.27 shows, the head movements are not nearly large enough to keep the image on the same horizontal band of retina. Indeed, it would only seem to be a good strategy for a toad to fixate prey with a constant horizontal region of retina if its stereoscopic system were organized as in figure 4.24a. On the other hand, the toad does not behave like a burrowing owl either and reserve a specific part of the zero meridian for measuring a specific distance; when prey is presented at various distances along a 45° slope tilted upward away from the toad, it uses a different region of retina from that with which it views prey on the

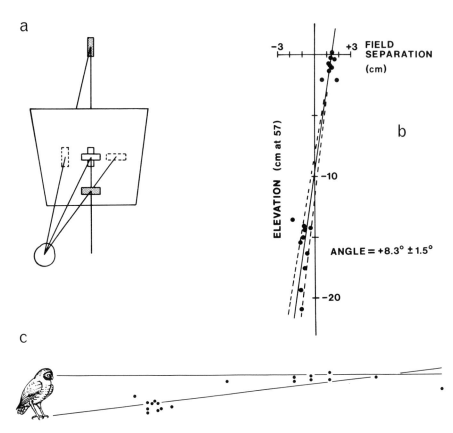

Figure 4.25 Neurophysiological demonstration of vertical horopter in burrowing owl. (a) Diagram illustrating the relation between the point in space monitored by a binocular neurone and the positions of its receptive fields (RF) on a tangent screen. RFs plotted through left and right eyes are shown by broken and unbroken lines, respectively. The point in space where the two RFs are superimposed is shown stippled. The horizontal unit has crossed RFs with respect to the fixation point on the screen (that is, the RF through left eye is to the right of the RF through right eye) and corresponds to a point in front of the screen. (b) Separations of RFs of single binocular units recorded in the Wulst of the burrowing owl. Each point shows the separation and elevation of one unit plotted on the tangent screen. Units with low elevations have negative (crossed) separations, units with high elevations positive (uncrossed) separations. (The slope of the line through the points gives the angle between the meridians of the two eyes.) Dotted lines represent 95% limits of regression line (see Cooper and Pettigrew 1979 for a fuller explanation of this point.) (c) RF positions of binocular units in part b projected into space. Each point shows the distance at which the RFs of a given binocular unit are superimposed. The line through these points is the lilted vertical horopter. Source: Cooper and Pettigrew 1979.

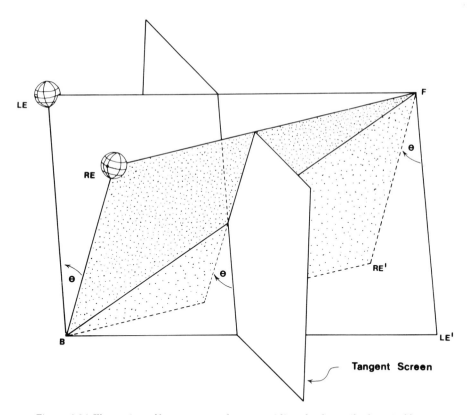

Figure 4.26 Illustration of how out-torted zero meridians lead to a tilted vertical horopter. If binocular neurons have RFs on the two zero meridians, then the points in space monitored by these neurons will lie where the projections of the two meridians intersect. Projections of zero meridians are represented by two inclined planes (LEFLE'B and RE-FRE'B). BF, the line of intersection, is the vertical horopter, and θ is the angle of shear between the zero meridians.

ground. *Rana pipiens* does the same, employing different areas of retina to fixate prey at different elevations in the visual field (figure 4.27). It appears that anurans have not gone to great pains to economize on the extent of their binocular interactions. They are able to measure a relatively wide range of disparities over a sizable region of retina, and this may be possible just because their acuity is relatively low and the distances they measure relatively short. This situation is not surprising, since anurans do not track objects with smooth head or eye movements, and are able to catch prey that appears within regions other than their frontal visual field. To do this they need to monitor the distance of objects over a wide area. Their binocular field is extensive, running vertically from the ground below the animal right over the top of its head (which gives almost backwards binocular vision) and horizontally over at least 70° (Fite 1973).

Until recently it was often thought that frontal eyes, partial decussation at the optic chiasm, and convergent eye movements were all essential for effective stereopsis (however, see Duke-Elder 1958 for a contrary

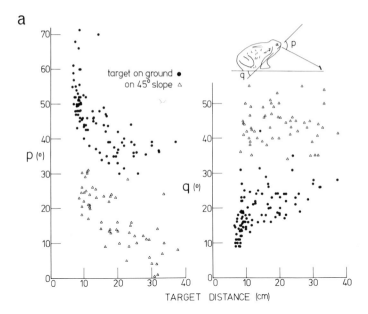

a

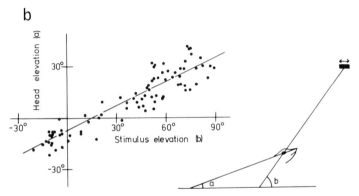

b

Figure 4.27 How the toad (*Bufo marinus*) and the frog (*Rana pipiens*) orient to a prey dummy at various elevations and distances. Films of pupil orientation show that when an animal looks up or down the eye maintains a constant position in the orbit so that head position can be used to define relative elevation of images on the retina. (a) Toad orients to mealworm or prey dummy either on the ground (●) or on a 45° slope tilted up from the animal's head (△). Left: Relative angular elevation of target after orientation (p) (measured as angle between jaw line and a line from the center of the pupil to the target) plotted against distance from target to eye. The more distant the target, the higher in the visual field (the lower on the retina) it is imaged. Different regions of retina are used to view targets on the ground and on a 45° slope. Right: Plot of angle of head relative to horizontal, *q*, against distance of target from eye. The animal lowers its head to look at nearby targets and raises it slightly for distant ones. It raises it higher for targets on a 45° slope than for those on the ground. (b) Head position plotted against elevation of target when frog orients to prey dummy 10–15 cm distant. Head position is angle made by horizontal and line passing through tip of nose and center of pupil. Elevation of target is angle between horizontal and line passing through target and center of pupil. Frog tips its head to compensate in part for elevation of target. However, as in *Bufo marinus*, different regions of retina are used to view the target at different elevations.

view). But the demonstration of stereopsis in birds and amphibians has shown that none of these properties are really necessary.

Pettigrew (1978) argued that frontal eyes evolved to cope, not with stereopsis, but with the problems of nocturnal vision. If an eye is to work adequately at low light levels it must have a small f ratio, which means a relatively large aperture. The penalty to be paid for this is that off-axis rays suffer serious optical aberration, and good resolution is only possible for directions close to the optic axis. Consequently, if a nocturnal animal wishes to make precise disparity measurements, the optic axes of the two eyes must be approximately parallel. Thus, the owl's eyes are truly frontal and their optic axes point in the same direction, whereas the optic axes of the diurnal kestrel, which also has stereopsis, diverge by about 90°.

Various pathways are used to combine information from the two eyes. In birds (Karten et al. 1973; Pettigrew and Konishi 1976) each thalamus receives input only from the contralateral retina, and information from both retinae is first brought together in the Wulst. In anurans each optic tectum receives a direct projection, primarily from the contralateral eye. Binocular interactions occur by way of a circuitous pathway between the two tecta that connects corresponding areas (Keating and Gaze 1970; Glasser and Ingle 1977; Gruberg and Udin 1978).

Insects do not appear to have anatomical pathways capable of mediating extensive interactions between the two eyes. Yet some predatory species, such as mantids and robber flies, have a large binocular field and may require binocular vision to catch prey successfully. It has often been supposed that such insects use both eyes to decide whether their prey is within catching distance, but how it might be done was a mystery until Rossel (1980) came up with a very plausible and simple suggestion. In the mantid *Tenodera australasiae*, the foveas of the two eyes have parallel lines of sight and are separated by some 5 mm. When both eyes view a distant peripheral target, saccadic movements of the head bring the image of the target onto the two foveas. Similarly, if vision is restricted to one eye, a head saccade will carry the image to the fovea of that eye. Each eye thus measures the angular position of a target with respect to its own fovea and generates an appropriate command signal. Should a target be brought close to the insect, then—because of the separation of the two foveas—the command signals from the two eyes will differ. This means that the mantid will be able to obtain a measure of the distance of a target simply by taking the difference between the saccadic signals generated by the two eyes. The underlying geometry is, of course, identical to that of conventional stereopsis. The difference is that, whereas the latter relies on local binocular interactions, the mantid, Rossel suggests, has merely to compare output signals from both its eyes. Such a method will work well for situations in which there is just one target, but may lead to confusion should there be several.

Image Transformations as Animals Move
Any translational movement made by an animal will result in retinal image changes related to the distance of objects in the world.

Parallax Component
When an animal looks at an object and moves perpendicular to its line of sight there is a shift in the position of the image on the retina, which can provide absolute or relative distance information in a manner directly analogous to the binocular cues discussed above. So long as the animal can estimate its own displacement (or velocity), translational movements will supply the base to the triangle that in the binocular case is given by its interocular distance. For an animal with fixed eyes, image displacement (or velocity) is then equivalent to retinal disparity (figure 4.28). For an animal with mobile eyes that keeps a target fixated on its fovea, the rotation (or angular velocity) of the eyes themselves could provide the same information as binocular convergence.

As with binocular cues, the accuracy with which distance can be judged using parallax will depend on the animal's visual acuity, and will also decrease with object distance. However, accuracy will not be further restricted by eye separation, but by how far (or fast) the animal moves. The precision of motion parallax as a distance cue and the range

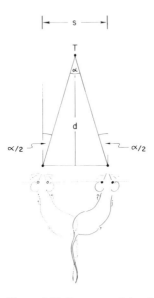

Figure 4.28 Geometry of the simplest way an animal can estimate distance using motion parallax. The animal moves a known distance s at right angles to the target T, and reads α from the change in visual angle at which it sees T (or the change in eye position if it has mobile eyes). Knowing both s and α, the animal can compute d, the target distance. The relation is $d = s/2 \tan(\alpha/2)$.

Collett, Harkness

over which it can be useful is thus not as constrained by an animal's size as the precision and range of binocular cues. A small animal can, in some cases, compensate for the poor resolution of its eyes by making a translational head or body movement of sufficient size or speed to give it the accuracy it requires.

Looming Component

Movement along the line of sight can also provide an animal with distance information. Moving directly toward or away from an object causes its retinal image to expand or contract, and the change (or rate of change) in size caused by a given displacement is related to the distance of the object independent of its size (see figure 4.29). Several factors determine the precision of this cue: The accuracy with which an animal can estimate looming depends upon how well it measures retinal size, which is in turn limited by its visual acuity. The accuracy of looming will also increase with the size of the object, since the larger an object the greater is the absolute change in image size for a given movement toward or away. Like all other cues, the precision of looming drops as object distance increases, because for a given head movement the further away an object is the smaller the change in its retinal size. Finally, the bigger (or faster) the animal's movement, the greater the change in retinal size; so, as with parallax, an animal can control the precision and accuracy of this cue by adjusting the amount or speed of its movement.

For other directions of motion, the distance of objects can be specified by a combination of their image shift (or velocity) and magnification (or rate of looming), so long as the animal does not rotate. Figure 4.30 shows the trigonometry involved. If an animal can measure both motion parallax and looming components, it can retain perfect freedom of translational movement and still be able to estimate the distance of any

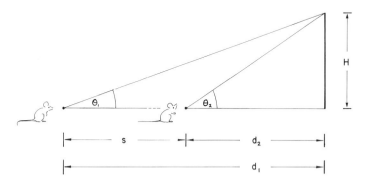

Figure 4.29 Geometry of looming as a distance cue. The animal is initially positioned at distance d from a target of height H, which subtends a visual angle of θ on the animal's retina. Animal moves forward a known distance s directly toward target, causing retinal image of target to expand to a visual angle of θ_2. If animal knows s, θ_1, and θ_2 it can compute both d_1 (its initial distance) and d_2 (its current distance from the target); $d_2 = s/(\tan\theta_2/\tan\theta_1 - 1)$.

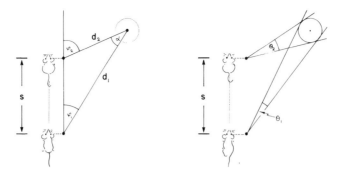

Figure 4.30 Geometry showing how an animal can estimate distance using a combination of parallax and looming. The animal, initially at distance d from a target, makes a translational movement of known distance s. This movement causes the retinal image of the target to move through $\alpha°$ on the retina (or causes the eyes to rotate $\alpha°$), and also causes the target image to expand from an angular subtense of $\theta_1°$ to $\theta_2°$. Knowing s, α, θ_1, and θ_2, the animal can compute d_2, its current distance from the target. $\alpha = \phi_2 - \phi_1$; $d_1/d_2 = \tan(\theta_2/2)/\tan(\theta_1/2) = \lambda$; $d_2{}^2 = s^2/(\lambda^2 - 2\lambda \cos\alpha + 1)$.

object within range. Furthermore, the animal could simultaneously assess the distance of many objects, whatever their position in the visual field. Image transformations can thus provide distance information to animals that make rather stereotyped side-to-side or back-and-forth movements directed toward some objects of interest, but they can also allow any animal wandering about its environment to interpret the spatial relationships of objects around it.

Rotation (yaw, pitch, or roll) will also cause movements of the retinal image, but these are not dependent on distance and would thus distort the simple relationship between distance and image displacement caused by pure translational movement. Thus, as well as all the variables that effect the accuracy of parallax and looming cues, an animal's ability to maintain constant angular head orientation may affect the precision of its distance estimates derived from image transformations.

It is often assumed that animals are attempting to estimate distance when they make stereotyped "head bobbing" or "peering" movements, particularly if they do so when confronted by a visual cliff (Walk 1965, p. 119). However, there is only one animal for which there is any evidence that this assumption is correct: It seems that this is what the locust may be doing when it makes characteristic side-to-side peering movements (Wallace 1959). Furthermore, these movements turn out to be well designed to obtain parallax information (Collett 1978) (figure 4.31). Head movement is actively restricted to translation; visual feedback from the lateral field of view prevents any head rotation which would decrease the accuracy of the distance estimate. Also, the locust increases the amplitude of peering when object distance is increased, and so to some extent is able to compensate for the decline in accuracy with distance. The facets that cover the binocular field of a locust

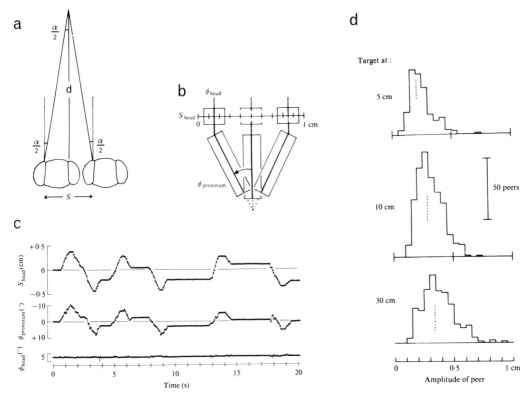

Figure 4.31 Peering in the locust. (a) Geometry of parallax. If locust knows s and α, it can compute d, which equals $s/2 \tan(\theta/2)$. (b) Lateral head movements (s_{head}) result from leg movements which rotate the animal's body (ϕ_{pronotum}). Head orientation (ϕ_{head}) is nonetheless kept constant. (c) Plot against time of orientation and lateral position of a locust's head and orientation of its body during a series of peers. Anticlockwise angles are taken as positive. The pronotum pivots through about 20°, shifting the head through 7 mm; head orientation, however, remains constant throughout. The rotation of the head with respect to the body, which results in constant head orientation, is given by ($\phi_{\text{head}} - \phi_{\text{pronotum}}$). (d) Effect of target distance on peering amplitude. Mean peering amplitude increases slightly but significantly ($P < 0.01$) with distance (0.21, 0.29, and 0.35 cm at target distances of 5, 10, and 30 cm, respectively). Vertical calibration is 50 peers. Source: Collett 1978.

155   Depth Vision in Animals

(*Schistocerca gregaria*, fourth instar) are about 2.25 mm apart, so by moving its head about 7 mm the animal has achieved a gain of about three times the accuracy it could achieve using retinal disparity.

Locusts may be forced to use image transformation to estimate distance because, being so small, they have no viable alternative. Larger animals may also have to use parallax and looming information when other cues fail. For example, it is not possible for fast-moving animals to stabilize their visual field by appropriate eye or head movements, and visual acuity, at least in humans, drops rapidly with increasing angular velocity of the retinal image (Ludvigh 1949; Brown 1972). This loss of acuity would seriously affect the accuracy of retinal image size, vergence, and accommodative cues to distance. Furthermore, in humans, the vergence and accommodation systems respond and adjust to depth changes rather slowly (Westheimer and Mitchell 1956; Campbell and Westheimer 1960) and would thus be inappropriate to guide maneuvers at speed. Stereo acuity in man also drops precipitously when objects move in depth (Westheimer and McKee 1978). Indeed, man cannot "see" very rapidly moving images at all in the conventional sense, although neural motion detectors still function. Under such extreme conditions, the only available distance information is from optical flow, and it seems likely that many animals are forced to use optical flow patterns for navigation through complex environments at speed. Although we tend to think of "motion parallax" as a cue that does not provide distance information instantaneously but rather represents a sequential analog of disparity, this need not be the case. All the relevant information is available at any instant from the velocities in the flow field, and a fast-moving animal could probably guide itself using the same geometrical principles as the locust, which sits and peers deliberately.

Interaction of Cues
Combining Depth Information
Often distance information is available from several independent cues, and one might expect an animal to improve the overall precision of its depth estimates by combining information from different sources. But one would also expect an animal to take more notice of a very sensitive cue than one that is relatively inaccurate, and it is worth considering how the information should best be pooled. Suppose that a long series of independent judgments of the same distance are made using a single cue. The series forms a distribution whose mean will be the best available estimate of target distance. The value of the variance, on the other hand, will depend on exactly how the information is extracted, but in part it will be determined by the physical limits to the accuracy of the cue. Assume now that two cues are used to measure the same distance, and that their distributions have the same mean (μ) but different variances (σ_1^2 and σ_2^2). It can then be shown that if each cue provides just

Collett, Harkness

one judgment of target distance (x_1 and x_2), as might normally be the case, the most accurate estimate of μ is given by

$$X_1 \left(\frac{\sigma_2{}^2}{\sigma_1{}^2 + \sigma_2{}^2} \right) + X_2 \left(\frac{\sigma_1{}^2}{\sigma_1{}^2 + \sigma_2{}^2} \right).$$

This expression simply means that an animal's best strategy is to use all the available information, but to weight it according to the precision of the estimate each cue supplies.

Despite some uncertainties, the use of accommodation and stereopsis by frogs and toads seems to fit this model. Monocular anurans probably measure depth using accommodative cues; as figure 4.32 shows, they can do so reasonably accurately. Nonetheless, when vision is binocular, cues of focus are largely ignored and the animal relies primarily upon stereopsis. If the two cues are pitted against each other experimentally (as when prey is viewed binocularly through spectacles), depth estimates in *Bufo marinus* can be predicted by the linear expression

$$D = 0.06A + 0.94S,$$

where A and S are the depth values given by accommodation and stereopsis respectively and D is the overall estimate. Thus, there is a 16-to-1 preference for disparity information. If the toad has adopted this weighting in order to maximize its accuracy, it follows that the range of uncertainty of estimates based on accommodation should be greater than that of those based on disparity by a factor of about 4 (or that $\epsilon_{accom}/\epsilon_{stereo} \simeq 4$). Figure 4.33 plots ϵ_{accom} and ϵ_{stereo} against a target distance for frogs of the size of *Rana pipiens* and behave in a qualitatively similar way to *Bufo marinus*. The ratio $\epsilon_{accom}/\epsilon_{stereo}$, about 3.6, supports the hypothesis that anurans exploit the two cues available to them to obtain the most accurate depth information possible.

There are, of course, doubts about the numbers we have used to derive our prediction. For instance, ϵ_{stereo} and ϵ_{accom} are estimates of the greatest stereo acuity and focusing accuracy an animal can be expected to achieve, and the frog's actual resolution using stereopsis or accommodation may well be less. Also, we have assumed that, like humans, anurans focus their eyes in tandem and so only derive a single depth estimate from accommodation. If they have independently focused eyes, accommodation cues will provide two estimates and the predicted ratio $\epsilon_{accom}/\epsilon_{stereo}$ will be $4/\sqrt{2}$, i.e., 3. Nonetheless, there is no doubt about our general conclusion that stereopsis is much more accurate than accommodation and that it is therefore appropriate for anurans to rely very heavily on binocular cues when these are available.

One can also ask what increase in accuracy is gained by incorporating a small accommodative component into the toad's overall estimate. It can be shown (Gelb 1974, p. 4) that

$$\sigma_2/\sigma_1 = [f^2/(1 - f^2)]^{1/2},$$

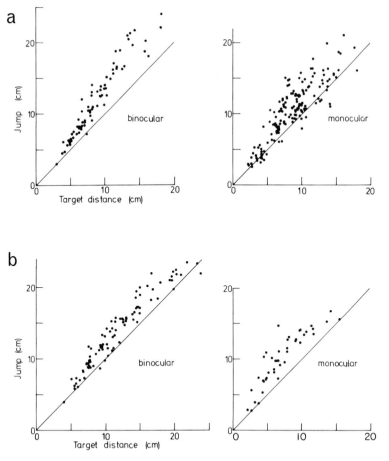

Figure 4.32 Accuracy of jumping at prey in monocular and binocular frogs. Films were taken of *Rana pipiens* jumping at mealworms suspended above them. Distance between eye and target and height jumped were measured from film. Each data point represents one jump. Frogs were first tested binocularly; one optic nerve was then severed and testing continued for several weeks. The behaviors of two individuals are plotted separately.

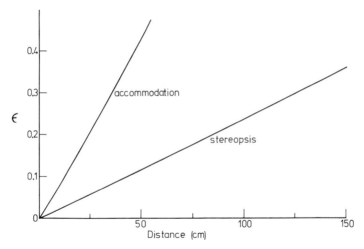

Figure 4.33 Comparison of relative sensitivities of accommodation and stereopsis in grass frogs. ϵ_{accom} (= $2d(nc/f^2)$) and ϵ_{stereo} (= $2d(\phi'/a)$) are plotted against distance, using data from figures 4.17 and 4.21. $\epsilon_{accom}/\epsilon_{stereo} = a/H\phi' = 3.6$, where $a = 1.5$ cm, $H = 240$, and $\phi' = 0.1°$.

where f is the ratio of the accuracy using two cues to the accuracy using one. This means, for example, that if accommodation is to improve the precision of the overall estimate by 10% (i.e., $f = 0.9$), then $\sigma_{accom}/\sigma_{stereo} \simeq 2$, so that the accommodation would need to be at least half as accurate as stereopsis. However, the data suggest that accommodation is much worse than this and that $\sigma_{accom}/\sigma_{stereo}$ lies between 3 and 4. The value of f will then be between 0.97 and 0.95, so the animal's accuracy would improve by no more than 3%–5%. This is a small gain, and one might ask why toads should bother to include focus cues along with their binocular estimates. But then it is probably no more difficult to reduce the contribution of accommodation to the overall estimate than it is to eliminate it entirely.

Some animals employ only one depth cue, and the same kind of reasoning can tell us how precise a second cue would have to be for it to add significantly to the available information. The chameleon, for instance, relies on accommodation to judge the distance of its prey. Although its eyes converge accurately onto the prey, convergence angle plays no part in its estimate (Harkness 1977a,b). This is not surprising when one calculates that for convergence to be as accurate as accommodation eye position would need to be known with a precision of at least $\pm 0.15°$. Furthermore, since the chameleon almost certainly has accommodative information available independently from the two eyes, the accuracy would only be improved by a factor of $\sqrt{3}/\sqrt{2}$.

Choosing Between Cues
If the accuracy of distance estimates is not significantly improved by the less accurate of the two cues, why should animals trouble at all with the

less precise one? There are two straightforward answers to this question: Backup cues can become invaluable (for example, it is not uncommon to come across toads with impaired vision in one eye, for whom accommodation will have changed from a subsidiary to an essential cue), and also a given cue may not be available in all situations. Without accommodation, anurans would have no depth information available within their extensive monocular field of view. In man, stereoscopic acuity varies with the configuration of the target. It is best for vertical targets and very low for horizontal ones, when depth sensitivity falls to the level of monocular vision (Blake et al. 1976). Similarly, motion parallax is not an ideal method for measuring the distance of moving objects; so an animal would have to move alternately in one and then the opposite direction and average the resulting image motion. The hoverfly, *Syritta pipiens*, does not perform maneuvers of this kind, and probably uses image size rather than parallax to estimate the distance of a moving conspecific. However, once the leading fly settles, the situation changes. Image size, which earlier provided an accurate cue, is no longer reliable, since the leading fly may now adopt any orientation with respect to the horizontal and the tracking fly would do well to shift to motion parallax. Beersma et al. (1977) found that the mean distance of male *Syritta* looking at a fly settled on a flower is the same whether the fly is a conspecific or a considerably larger *Eristalis*; this suggests that distance is no longer primarily determined by a cue that relies on image size.

These few examples are unlikely to be special cases, and it seems reasonable to suppose that many animals have ways of assessing rather flexibly how much credence should be placed upon the information provided by a particular cue in a given situation.

Using Cues to Assist Each Other

It is especially important for animals with stereopsis to be capable of using other depth cues, because without subsidiary depth information the interpretation of retinal disparities becomes very difficult. Recall that for animals with mobile eyes, disparities alone can give no more than relative depth information. Absolute estimates require additional information. When target and animal are stationary, the animal needs to know the distance of the fixation point through vergence or accommodation cues. However, if there is relative motion between target and observer, changes of image size provide an alternative supplementary cue. Regan and Beverley (1979) pointed out that the absolute size of an object can be determined from changes in its retinal disparity and image size without any knowledge of the object's distance. To show how changing image size can supplement disparity cues to give absolute distance, we need only formulate their argument in a slightly different way. For a constant vergence angle, the change in horizontal disparity as an object at distance d_1 approaches by a distance d' to distance d_2 is given approximately by

$$\delta_2 - \delta_1 = \frac{ad'}{d_1{}^2},$$

where δ_1 and δ_2 are the horizontal retinal disparities of the object before and after the movement and a is the interocular separation. The ratio of the image size after the movement (θ_2) to that before (θ_1) is given by

$$\frac{\theta_2}{\theta_1} = \frac{d_1 - d'}{d_1}.$$

Rearranging, we have

$$d' = d_1 (1 - \theta_2/\theta_1),$$

and by substitution

$$\delta_2 - \delta_1 = \frac{a}{d_1} (1 - \theta_2/\theta_1),$$

so that

$$d_1 = \frac{a(1 - \theta_2/\theta_1)}{\delta_2 - \delta_1}.$$

In this case, by using visual information from two independent cues the animal is able to arrive at a computation of the absolute distance of a target without any information about the position of its eyes (so long as the vergence angle is constant) or how far or fast the target has moved.

Our discussion of the ways in which cues interact suggests that the neural representation of sensory qualities such as location in depth should not only be coded in terms of separate cues like accommodation, or parallax, or disparity. Ultimately these need to be brought together, and, as discussed by Marr and Nishihara (1978), one might expect to find within the visual system a neural representation of three-dimensional space incorporating information from all the various available depth cues.

Conclusion

The conclusion from this section that we stress is that generally the depth information available to a small animal is very poor compared with that available to a larger one. Table 4.1 summarizes the various geometrical factors determining the relative sensitivities of the cues we have considered. The precision of all the distance cues depends on visual acuity, which as we pointed out earlier depends on an animal's size. Both accommodation and binocular cues are especially inaccurate for small animals—accommodation because the short focal length of their eyes results in an effectively infinite depth of field; binocular cues because small animals have both low acuity and close-set eyes. Small size thus means a more restricted choice of depth cue, a less accurate

Table 4.1
Relative sensitivities of depth cues and their scaling with body length.

Cue	Geometric Factors Determining Relative Sensitivity of Cue $\left(\dfrac{d'_n + d'_f}{d}\right)$	Scaling of Sensitivity with Body Length for Dimensionally Similar Animals
Accommodation	$2dnc/f^2$	$1/L^2$
Stereopsis	$2d\boldsymbol{\phi}'/\boldsymbol{a}$	$1/L^2$
Image size	$2d\boldsymbol{\phi}'/\boldsymbol{H}$	$1/L$
Motion parallax	$2d\boldsymbol{\phi}'/\boldsymbol{S}$	$1/L$
Looming	$\dfrac{\boldsymbol{\phi}'(2 - S/d)}{\tan(\theta_2 - \theta_1)}$	$1/L$

Notes. d is distance between animal and target; n is F ratio of lens; c is receptor separation; f is focal length, ϕ' is spatial resolution, a is interocular separation; H is height of target; S is distance animal moves to obtain parallax or looming information; d_2 is distance from target to animal after it has moved toward target; L is linear body dimension. Boldface indicates factors that are proportional to body size.

estimate of distance, and a shorter range over which depth discrimination is possible. This is not necessarily an impediment, since for many purposes the space in which an animal operates is also size-dependent. Manipulative space, for example, is limited by an animal's reach; similarly, the range over which prey can be caught is likely to be related to an animal's size, as it depends on length of tongue or forelimb. Also, animals with small mass do not need to respond to obstacles until they are very close, because, having low inertia, they can correct their course very rapidly and accidental collisions will be relatively harmless. Thus, although insects fly fast and over large distances, they probably only need relatively close distance vision.

However, the distances that animals may need to estimate are not always proportional to their body size. For example, small animals leap relatively large distances. A locust can jump 18 times its body length (Bennet-Clark 1975), whereas a man can only jump about two body lengths from a standstill (the Olympic record for the standing broad jump is 11 ft, 4⅞ in. [Long 1969]). In these situations, small animals may need special behavioral strategies. One option is to devise ways of coping without distance information; another is to move enough to obtain adequate parallax information.

The Distance Cues That Animals Have Been Shown to Use

Despite an enormous literature on the use of depth cues by man, which we make no attempt to cover, information about the remainder of the animal kingdom is surprisingly sparse. A full picture of an animal's spatial vision requires some knowledge of the range of cues available to it and how different cues are employed in a variety of behavioral situations. However, questions are usually posed in the form "Can such and such an animal exploit a particular cue?", and, by and large, this information has come from a study of simple motor responses such as jumping and prey capture. The aim of this section is simply to review demonstrations of the use of different depth cues by various animals. Our seemingly haphazard choice of examples has been dictated by what people have chosen to work with, and this is presumably the result of historical accidents. Four essentially different approaches have been used to investigate how animals see depth.

The first and simplest approach is to deduce what cues are theoretically available, and by this means candidate depth cues can often be eliminated. For example, when a male *Syritta* "shadows" another fly (figure 4.3), it maintains a distance of about 10 cm to within about 1 cm. For reasons outlined earlier, arthropods do not have focus cues, and binocular cues can only give them depth information over a short range. Furthermore, *Syritta* is unlikely to be using parallax information while the leading fly is moving. Thus, by a process of elimination, it seems that the distance of the fly ahead is probably calculated from the size of the retinal image.

The second approach is to deprive an animal of one or more cues and then observe the accuracy of its distance estimates. A standard first step is to compare monocular and binocular depth judgments. In man, it is a commonplace that casual depth judgments are much less accurate if one eye is closed. In insects and lower vertebrates, on the other hand, the effects of limiting vision to one eye are nonexistent or far less dramatic (Canella 1936; Wallace 1959; Ingle 1972; Harkness 1977a,b; Collett 1977). Figure 4.32 shows the results of an experiment in which binocular and monocular frogs leaped for mealworms suspended from a thin transparent nylon cord above them. The distance between the frog's eye and the worm is plotted against the height the frog jumps when attempting to reach the worm. Individual *Rana pipiens* were first tested binocularly; then the optic nerve was cut on one side and testing was continued for several weeks. Despite some increase in scatter in the monocular frog (which may have resulted from a motor imbalance and not from a perceptual deficit), the predominant impression was how well the frogs did with just one eye. From tests of this kind one can deduce that animals have monocular mechanisms of judging depth, but it is not legitimate to conclude that they lack binocular mechanisms.

To decide what cues are used, it is necessary to turn to the third approach, that of manipulating cues. An elegant example is Wallace's (1959) demonstration that locusts use motion parallax information to gauge the distance of an object onto which they are about to jump. Before they jump, they often perform a series of side-to-side head

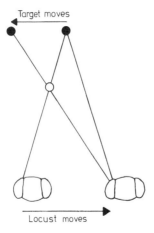

Target moves

Locust moves

Figure 4.34 Principle of Wallace's 1959 experiment to show that the locust obtains distance information from its side-to-side head movements. As locust moves from left to right, changing the angular position of the target on retina, experimenter moves target (●) from right to left, increasing movement of target across retina. If locust uses parallax information, target will appear closer than it really is, as shown by ○.

movements. If the object is moved in synchrony with the head movements horizontally in a fronto-parallel plane, the locust misjudges the distance. When the object is shifted in the opposite direction from the head movement, increasing the amount by which the image of the object moves across the retina, the locust jumps short (figure 4.34). This manipulation changes parallax information without affecting any other depth cue, so the observation that the locust undershoots is sufficient to establish that parallax information plays an essential role in controlling the size of the jump. To determine whether a distance estimate is based on a single cue alone, however, it is necessary to show that when that cue is altered the animal's behavior changes correspondingly in a quantitatively predictable way.

For example, a quantitative study by Wehrahn (in prep.) has established that flies rely exclusively on retinal image size to control their distance from a conspecific that they are tracking. He has shown that in muscid flies there is a relationship between distance and forward velocity such that the farther the fly is from its target, the faster it flies. By using targets of different sizes, he has been able to prove that this relationship is determined solely by the size of the target image on the retina.

If an animal is big enough, a simple and informative method of manipulating depth cues is to persuade it to make distance judgments while wearing spectacles containing lenses and prisms of various powers (figure 4.35). A target seen through a lens will be blurred unless the animal compensates for the spectacle lens by adjusting its own accommodative setting. To allow for a negative lens, the accommodative setting must be changed as though the target were nearer than it really is,

Figure 4.35 Chameleon wearing spectacles. A chameleon's turretlike eyes normally move independently, each eye scanning (saccadically) over more than a hemisphere. When taking aim at an insect, however, both eyes are swiveled directly toward the target, and chameleons always aim and shoot their tongues straight ahead. Thus, spectacles can be positioned so that the animal must look through them while taking aim.

Depth Vision in Animals

and conversely for the positive lens. Accordingly, if spectacle lenses lead an animal to misjudge the target by an appropriate amount, it follows that the judgment was made using accommodative cues. Chameleons viewing prey monocularly or binocularly behave in exactly this way (Harkness 1977). They are consistently fooled by lens spectacles, negative lenses causing them to undershoot and positive lenses to overshoot their prey by amounts very close to those predicted from lens strengths (figure 4.36). Like muscid flies tracking, chameleons seem to rely on one distance cue only while feeding—in this case, accommodation.

Similar tests performed on toads lead to somewhat different results (Collett 1977) suggesting that they use accommodative cues if viewing targets monocularly but that they rely on stereopsis when vision is binocular (figure 4.37). A toad with the use of one eye will undershoot its prey when wearing a negative lens by about the predicted amount, but will judge depth almost correctly when viewing prey through lenses placed over both eyes. This result implies that when binocular cues are available they are preferred to monocular ones. And indeed, tests with prisms indicate that toads do have a form of stereopsis. Appropriately aligned prisms will alter the apparent binocular disparity of a target. Consequently, the behavior of animals wearing prism spectacles should change if stereopsis is used to judge distance. Unlike chameleons, toads are affected by prism spectacles. Base-out prisms, which increase the apparent value of α, cause toads to undershoot their prey; base-in prisms decrease α, and toads equipped with such prisms are found to overshoot their prey. Because a toad's eyes are fixed in its head, binocular convergence is not an available cue. Thus, toads must be using stereopsis.

The last way of showing that an animal exploits a particular cue is to devise a perceptual task that can only be solved by the use of that cue. The best known of such methods is that devised by Julesz (1971) for work on stereopsis. Almost the same random microdot pattern is presented to each eye. Some segment of the field of dots (for instance, a triangular area in the center) is displaced slightly to one side in one of these fields. If the two patterns are then fused binocularly, the displaced segments will have a different disparity value from the background and will be seen in front of or behind the rest of the pattern. The triangular shape is not seen in either pattern separately; it only exists as a difference between the two fields. Accordingly, animals show whether they have stereopsis by whether they notice such an embedded pattern. In most cases, animals have been trained to give a discriminative response to a real stimulus and then tested with a random-dot version. By this means, stereopsis has been demonstrated in falcons (Fox et al. 1977) and monkeys (Bough 1970). A variant avoids the necessity of training animals, and uses a "Julesz" version of an optomotor drum (Fox et al. 1978). Rotating stripes are only present as disparity differences. It will

Collett, Harkness

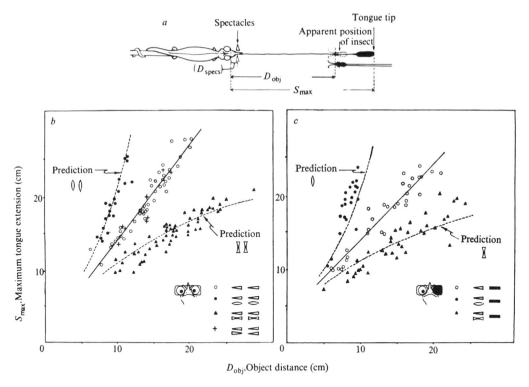

Figure 4.36 Effect of lenses and prisms on depth vision in the chameleon. Chameleons (*Chameleo jacksoni*) were filmed while shooting at illusory targets in order that the distance reached by the tongue would not be affected by impact with the insect. Part a shows how this was achieved. The chameleon (viewed from above) is wearing thin prism spectacles that displace the image of the insect laterally, and in consequence the trajectory of its tongue is to the side of the real insect. Target distance D_{obj} (distance between near side of insect and chameleon's pupil just before tongue is fired) and the maximum tongue extension S_{max} (distance between pupil and tip of tongue when it is fully extended) were measured from multiple-exposure photos of animals attempting to capture insects while wearing these glasses. Parts b and c show the relationship between D_{obj} and S_{max} for chameleons wearing lateral prisms, as in a, combined with various other lens and prism spectacles. (b) Binocular animals, (c) monocular animals (view of one eye obscured by a black patch). Each point represents one shot by one of the three animals tested. (O) Shots by animals wearing lateral prism spectacles only. The tongue reaches a few centimeters beyond the insect whatever its distance. The farther the target, the farther the chameleon shoots its tongue, whether it is feeding binocularly or monocularly. Dashed lines predict overshoots and undershoots for chameleons wearing lens spectacles if they use accommodative cues. Shots by chameleons wearing positive lenses (●) and negative lenses (▲) fit the predictions quite closely in both plots, suggesting that both monocular and binocular animals do indeed use accommodative cues, estimating prey distance as a function of the focal plane. In contrast, note the crosses (+). These points were obtained from binocular animals wearing convergent, base-out prisms of sufficient angular deviation to predict dramatic undershoots if the animals were estimating distance as a function of the convergence angle of the eyes. Chameleons do not seem to triangulate to estimate distance. Source: Harkness 1977a.

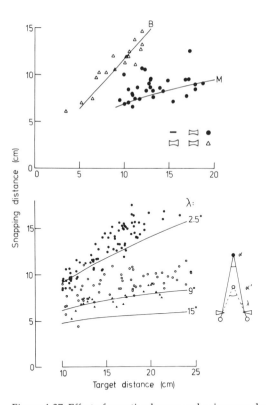

Figure 4.37 Effect of negative lenses and prisms on depth vision in the toad *Bufo marinus*. Toads were filmed wearing spectacles containing either −136-mm negative lenses or base-out prisms. Target distance (T) and snapping distance (S) were measured. T is distance from a point midway between the eyes, before toad moves, to middle of prey. S is distance from the same point between the eyes to tip of tongue at its fullest extent. Without any spectacles monocular toads shoot their tongue accurately at the prey, with the tip extending about 1 cm beyond it. Top: Effect of −136-mm lenses on S. (●) Lens placed in front of one eye while the other eye is occluded. The line labeled M is S predicted on the assumption that toads employ accommodative cues. (△) Lenses over both eyes. Line B is the predicted S assuming toads depend on stereopsis and ignore monocular cues. Each data point represents a single snap. When vision is monocular, toad undershoots by an amount that implies it uses accommodative cues, but when vision is binocular toad no longer undershoots, suggesting that disparity measurements are now more important. Bottom: Effect of base-out prisms on S. Diagram shows why base-out prisms should cause a toad using disparity to undershoot its prey. The prisms will increase the apparent disparity of the target (α') by an amount equal to the summed angular deviation produced by the two prisms at the eyes (2λ). Thus, $\alpha' = \alpha + 2\lambda$. As the prey is moved away, $\alpha \to 0$ and $\alpha' \sim 2\lambda$. Scattergram shows relation between S and T for prisms of various powers. (●) Minimum angular deviation caused by each prism is 2.5° ($2\lambda = 5°$). (○) $2\lambda = 18°$; (▲) $2\lambda = 30°$. Solid lines show predicted S assuming that toads employ stereopsis alone. Source: Collett 1977.

Table 4.2
The distance cues animals have been found to use.

Species	Image Size	Motion Parallax	Accommodation	Convergence	Stereopsis
Insects					
Housefly	√		×	×	
Hoverfly	√	√ ?a	×	×	
Locust		√	×	×	
Honeybee	√	√ ?	×	×	
Bulldog ant			×	×	×
Amphibians					
Grass frog			√	×	√
Toad			√	×	√
Reptiles					
Chameleon			√	×	
Birds					
Owl				×	√
Falcon					√
Mammals					
Cat					√
Monkey					√
Man	√	√	√ ?	√	√

a. Evidence is not equally strong for all entries to the table. A question mark is used to qualify the more doubtful entries.

be intriguing to see whether animals found to have stereopsis by more laborious techniques will exhibit stereopsis by this ingenious method as well.

Table 4.2 serves as a summary of this section, listing the cues different species have been shown to use. By and large, it seems that animals behave as they should. Insects exploit motion parallax and retinal image size, the most useful cues for small animals. Although insects manage remarkably well with them, motion parallax can be cumbersome, as the animal must move, whereas retinal image size is inherently ambiguous unless objects are of a known fixed size. It thus makes good sense that when an animal is big enough it should turn to accommodation and stereopsis, both of which will only work over a reasonable range for larger animals.

Accommodation mechanisms are common within the animal kingdom, and many animals have eyes set laterally with wide monocular fields of view where binocular cues are unavailable. Consequently, it would be surprising if focusing systems were not exploited to give monocular distance vision, and it seems likely that many visually active animals other than those mentioned in table 4.2 will turn out to use accommodative cues. However, stereopsis should generally have greater precision and range, and it is striking that animals from such diverse groups as amphibians, birds, and mammals have all evolved ways of interpreting retinal disparities.

There is as yet no evidence that any insects use binocular cues, although stereopsis could in principle supply valuable information over short distances. It has in fact often been suggested that predatory insects such as mantids triangulate to estimate the distance of their prey, thus avoiding the ambiguity of image size and the conspicuity of motion parallax. But the only case where the hypothesis has been tested (Via 1977) turns out to be an example of an animal apparently coping without distance vision. The bulldog ant (*Myrmecia gulosa*) captures prey between its pincerlike jaws, working within a range of distances that could be specified by triangulation. As the ant approaches its prey, the pincers must be opened and then, at the appropriate moment, snapped shut to grip the victim. If a prey dummy is moved towards an isolated head (which is easier to manage than an aggressive whole animal), the jaws open at a distance that is linearly related to the size of the prey object and then close (figure 4.38). The data seem to imply that there are two trigger zones, one in each eye, onto which the edge of the prey stimulus must fall for jaw opening to occur. If one eye is masked, the jaws still open, though much less readily, when one edge of the stimulus falls within the remaining trigger zone. Effectively, then, the ant responds to an image of constant size but uses two eyes to make its measurements. This arrangement ensures that the jaws will always open before the ant collides with its prey, and also filters out prey that are too large, since the pincers will have both opened and closed before an oversize prey is reached.

It is curious that although there are many examples of animals having some form of stereopsis, there is still no evidence that any animal apart from man uses the apparently simpler cue of convergence. But the chameleon is the only subprimate with mobile eyes that has been tested; it may well turn out that convergence is a much commoner cue than table 4.2 suggests.

Perhaps the most conspicuous feature of table 4.2 is the small number of entries. This is a pity, because depth vision provides a good tool for learning how the nervous systems of a variety of animals handle complex sensory information, and because studies of the ways depth vision is used are revealing unexpected subtleties in the spatial knowledge different animals have about their immediate surroundings. A comparative study of depth vision would seem to be a good starting point for understanding how our own visual world has evolved.

Acknowledgments

We are very grateful to Jack Pettigrew for sending data on the behavior of the burrowing owl, to Harold Edgerton for advice and the loan of stroboscopic equipment, and to Jim Commerford, David Ingle, Mike

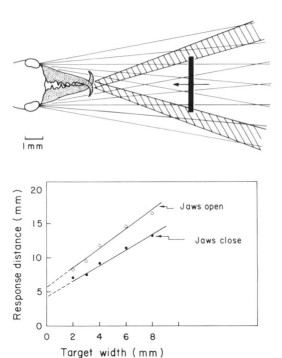

Figure 4.38 Size-distance ambiguity in the bulldog ant. Top: Sketch of ant's head and approaching stimulus. Jaws open when stimulus falls within trigger zones (hatched areas of visual field) and close again as stimulus approaches still closer. Bottom: Relation between stimulus size and distance from ant at which jaws open and close. It is clear that the ant's decision to respond to the stimulus depends on its size; the smaller the target the closer it can approach before the ant snaps its jaws. The distance at which the ant responds is proportional to the target width, which implies that the ant reacts when the target subtends some critical angle. The ant is not responding to a fixed retinal image size (fixed visual angle) as we normally understand it, however; the lines drawn through the data, when extrapolated to intercept the *y* axis, indicate that the ant responds when the target subtends a fixed angle not to the *eye,* but to points about 6 mm (for opening the jaws) and 4 mm (for closing) in front of the eyes. There are ommatidia within the binocular field of the bulldog ant whose lines of sight intercept at about these distances in front of the animal. Thus, it seems likely that opening and closing of the jaws is triggered by simultaneous stimulation of sets of such ommatidia as the object enters their fields of view—the "trigger zones," as indicated for jaw opening in the sketch. After Via 1977.

Land, Fred Owens, and Greg Zacharias for comments on part or all of this chapter. Some of the work described here was carried out in David Ingle's laboratory at McLean Hospital while T. S. C. was supported by an MRC traveling fellowship and a McLean Hospital Biological Research Support Grant (FR 05484). T. S. C. thanks the SRC and L. I. K. H. thanks Fight for Sight, Inc. for support during the preparation of this chapter. L. I. K. H. also thanks John Dowling for his generous hospitality at Harvard.

References

Baird, J. W. 1903. "The influence of accommodation and convergence upon the perception of depth." *Amer. J. Psychol.* 14: 150–200.

Barlow, H. B., C. Blakemore, and J. D. Pettigrew. 1967. "The neural mechanism of binocular depth discrimination." *J. Physiol.* 193: 327–342.

Beersma, D. G. M., D. G. Stavenga, and J. W. Kuiper. 1977. "Retinal lattice, visual field, and binocularities in flies: Dependence on species and sex." *J. Comp. Physiol.* 119: 207–220.

Bennet-Clark, H. C. 1975. "The energetics of the jump of the locust *Schistocerca gregaria*." *J. Exp. Biol.* 63: 53–84.

Birukow, G. 1937. "Untersüchungen über den optischen Drehnystagmus und über die Sehschärfe des Grasfrosches (*R. temporaria*)." *Z. vergl. Physiol.* 25: 92–142.

Bishop, P. O., and G. H. Henry. 1971. "Spatial vision." *Annu. Rev. Psychol.* 22: 119–161.

Blake, R., J. M. Camisa, and D. N. Antoinetti. 1976. "Binocular depth discrimination depends on orientation." *Percept. Psychophys.* 20: 113–118.

Bough, E. W. 1970. "Stereoscopic vision in macaque monkeys: A behavioural demonstration." *Nature* 225: 42–45.

Boycott, B. B., and J. Z. Young. 1956. "Reactions to shape in *Octopus vulgaris* Lamarck." *Proc. Zool. Soc.* (Lond.) 126: 491–547.

Brown, B. 1972. "Resolution thresholds for moving targets at the fovea and in the peripheral retina." *Vision Res.* 12: 293–304.

Campbell, F. W. 1957. "The depth of field of the human eye." *Optica Acta* 4: 157–164.

Campbell, F. W., and G. Westheimer. 1958. "Sensitivity of the eye to differences in focus." *J. Physiol.* 143: 18P.

———. 1960. "Dynamics of accommodation responses of the human eye." *J. Physiol.* 151: 285–295.

Canella, F. 1936. "Quelques recherches sur la vision monoculaire." *C. R. Soc. Biol.* (Paris) 122: 1221–1224.

Carr, H. A., and J. B. Allen. 1906. "A study of certain relations of accommodation and convergence to the judgment of the third dimension." *Psychol. Rev.* 13: 258–275.

Cartwright, B. A., and T. S. Collett. 1979. "How honey-bees know their distance from a nearby landmark." *J. Exp. Biol.* 82: 367–372.

Collett, T. 1977. "Stereopsis in toads." *Nature* 267: 349–351.

———. 1978. "Peering: A locust behavior pattern for obtaining motion parallax information." *J. Exp. Biol.* 76: 237–241.

Collett, T. S., and M. F. Land. 1975. "Visual control of flight behaviour in the hoverfly (*Syritta pipiens* L.)." *J. Comp. Physiol.* 99: 1–66.

———. 1978. "How hoverflies compute interception courses." *J. Comp. Physiol.* 125: 191–204.

Cooper, M. L., and J. D. Pettigrew. 1979. "A neurophysiological determination of the vertical horopter in the cat and owl." *J. Comp. Neurol.* 184: 1–26.

Davson, J. 1972. *Physiology of the Eye.* Edinburgh: Churchill Livingstone.

Duke-Elder, S. 1957. *System of Ophthalmology,* vol. 1: The Eye in Evolution. London: Kimpton.

du Pont, J. S., and P. J. de Groot. 1976. "A schematic dioptric apparatus for the frog's eye (*Rana esculenta*)." *Vision Res.* 16: 803–810.

Ewert, J.-P., and L. Gebauer. 1973. "Grössenkonstanzphänomene im Beutefangverhalten der Erdkröte (*Bufo bufo* L.)." *J. Comp. Physiol.* 85: 303–315.

Fite, K. V. 1973. "The visual fields of the frog and the toad: A comparative study." *Behav. Biol.* 9: 707–718.

Fite, K. V., and F. Scalia. 1976. "Central visual pathways in the frog." In *The Amphibian Visual System: A Multidisciplinary Approach,* K. V. Fite, ed. New York: Academic.

Fox, R., S. Lehmkuhle, and R. C. Bush. 1977. "Stereopsis in the falcon." *Science* 497: 79–81.

Fox, R., S. Lehmkuhle, and L. E. Leguire. 1978. "Stereoscopic contours induce optokinetic nystagmus." *Vision Res.* 18: 1189–1192.

Gelb, A. 1974. *Applied Optimal Estimation.* Cambridge, Mass.: MIT Press.

Gibson, J. J. 1950. *The Perception of the Visual World.* Boston: Houghton Mifflin.

Glasser, S., and D. Ingle. 1977. "The nucleus isthmus as a relay station to the ipsilateral visual projection to the frog's optic tectum." *Brain Res.* 159: 214–218.

Gruberg, E. R., and S. B. Udin. 1978. "Topographic projections between the nucleus isthmus and the tectum of the frog *Rana pipiens.*" *J. Comp. Neurol.* 179: 487–500.

Harkness, L. 1977a. "Chameleons use accommodation cues to judge distance." *Nature* 267: 346–349.

———. 1977b. A Behavioural Study of Chameleons. D. Phil. thesis, Oxford University.

Harkness, L., and H. C. Bennet-Clark. 1978. "The deep fovea as a focus indicator." *Nature* 272: 814–816.

Heinemann, E. G., E. Tulving, and J. Nachmias. 1959. "The effect of oculomotor adjustments on apparent size." *Amer. J. Psychol.* 72: 32–45.

Hell, W. 1978. "Movement parallax: An asymptotic function of amplitude and velocity of head motion." *Vision Res.* 18: 629–635.

Herter, K. 1930. "Weitere Dressurversuche an Fischen." *Z. vergl. Physiol.* 11: 730–748.

Helmholtz, H. 1925. *Treatise on Physiological Optics*, vol. III, J. P. C. Southall, ed. Translation of third German edition (1909–1911). New York: Columbia University Press (for Optical Society of America).

Hinton, H. E. 1973. "Natural deception." In *Illusion in Nature and Art*, R. L. Gregory and E. H. Gombrich, eds. New York: Scribner's.

Horridge, G. A. 1978. "Separation of visual axes in apposition compound eyes." *Phil. Trans. Roy. Soc.* B 258: 1–59.

Ingle, D. 1970. "Visuomotor functions of the frog optic tectum." *Brain, Behav., Evol.* 3: 57–71.

———. 1972. "Depth vision in monocular frogs." *Psychonom. Sci.* 29, 37–38.

———. 1976. "Spatial vision in anurans." In *The Amphibian Visual System: A Multidisciplinary Approach*, K. V. Fite, ed. New York: Academic.

Ingle, D., and J. Cook. 1977. "The effect of viewing distance upon size preference of frogs for prey." *Vision Res.* 17: 1009–1013.

Jones, R. 1977. "Anomalies of disparity detection in the human visual system." *J. Physiol.* 264: 621–640.

Joshua, D. E., and P. O. Bishop. 1970. "Binocular single vision and depth discrimination: Receptive field disparities for central and peripheral vision and binocular interaction on peripheral single units in cat striate cortex." *Exp. Brain Res.* 10: 389–416.

Julesz, B. 1971. *The Foundations of Cyclopean Perception.* University of Chicago Press.

Karten, H. J., W. Hodos, W. J. H. Nauta, and A. M. Revzin. 1973. "Neural connections of the "visual Wulst" of the avian telencephalon: Experimental studies in the pigeon (*Columba livia*) and owl (*Speotyto cunicularia*)." *J. Comp. Neurol.* 150: 253–278.

Keating, M. J., and R. M. Gaze. 1970. "The ipsilateral retinotectal pathway in the frog." *Q. J. Exp. Physiol.* 55: 284–292.

Kirschfeld, K. 1976. "The resolution of lens and compound eyes." In *Neural Principles in Vision*, F. Zettler, and R. Weiler, eds. Berlin: Springer.

Lack, D. 1956. *Swifts in a Tower*. London: Methuen.

Land, M. F. 1969. "Structure of the retinae of the principle eyes of jumping spiders (*Salticidae: dendryphantinae*) in relation to visual optics." *J. Exp. Biol.* 51: 443–470.

Lee, D. N. 1976. "A theory of visual control of braking based on information about time-to-collision." *Perception* 5: 437–459.

Lock, A., and T. Collett. 1979. "A toad's devious approach to its prey: A study of some complex uses of depth vision." *J. Comp. Physiol.* 131: 179–189.

———. 1980. "The 3 dimensional world of a toad." *Proc. Roy. Soc.* B 206: 481–487.

Locket, N. A. 1977. "Adaptations to the deep-sea environment." In *Handbook of Sensory Physiology*, vol. VII/5, *The Visual System in Vertebrates*, F. Crescitelli, ed. Berlin: Springer.

Long, L. H. 1969. *The 1970 World Almanac*. New York: Newspaper Enterprise Association.

Ludvigh, E. J. 1949. "Visual acuity while one is viewing a moving object." *Arch. Ophthalmol.* 42: 14–22.

Marr, D., and H. K. Nishihara. 1978. "Visual information processing: Artificial intelligence and the sensorium of sight." *Technol. Rev.* 81: 28–49.

Marr, D., and T. Poggio. 1979. "A computational theory of human stereo vision." *Proc. Roy. Soc.* B 204: 301–328.

Munk, O. 1966. "Ocular anatomy of some deep-sea teleosts." *Dana Report* 70: 1–62.

Ogle, K. N. 1962. "The optical space sense." In *The Eye*, vol. 4, part II, H. Davson, ed. New York: Academic.

Owens, D. A. 1979. "A comparison of accommodative responsiveness and contrast sensitivity for sinusoidal gratings." *Vision Res.* 20: 159–167.

Owens, D. A., and H. W. Leibowitz. 1980. "Accommodation, convergence, and distance perception in low illumination." *Amer. J. Opt. Physiol. Opt.* 57: 540–550.

Pastore, N. 1958. "Form perception and size constancy in the duckling." *J. Psychol.* 45: 259–261.

Packard, A. 1972. "Cephalopods and fish: The limits of convergence." *Biol. Rev.* 47: 241–307.

Pettigrew, J. D. 1978. "Comparison of the retinotopic organisation of the visual Wulst in noctural and diurnal raptors, with a note on the evolution of frontal vision." In *Frontiers of Visual Science*, S. J. Cool and E. L. Smith, eds. Berlin: Springer.

Pettigrew, J. D., and M. Konishi. 1976. "Binocular neurones selective for orientation and disparity in the visual Wulst of the barn owl (*Tyto alba*)." *Science* 193: 675–678.

Pirenne, M. H. 1967. *Vision and the Eye*. London: Chapman and Hall.

Poggio, G. F., and B. Fischer. 1977. "Binocular interaction and depth sensitivity in striate and prestriate cortex of behaving *Rhesus* monkey." *J. Neurophysiol.* 40: 1392–1405.

Regan, D., and K. I. Beverley. 1979. "Binocular and monocular stimuli for motion in depth: Changing-disparity and changing-size feed the same motion in depth stage." *Vision Res.* 19: 1331–1342.

Rochon-Duvigneaud, A. 1933. "Le chaméleon et son oeil." *Ann. d'ocul.* 170: 177.

Rolls, P. 1968. "Photographic optics." In *Photography for the Scientist*, C. E. Engel, ed. New York: Academic.

Rossel, S. 1980. "Foveal fixation and tracking in the praying mantis." *J. Comp. Physiol.* 139: 307–331.

Schultz, A. H. 1940. "The size of the orbit and of the eye in primates." *Amer. J. Phys. Anthropol.* 26: 389–403.

Schwind, R. 1978. "Visual system of *Notonecta glauca*: A neuron sensitive to movement in the binocular visual field." *J. Comp. Physiol.* 123: 315–328.

Toates, F. M. 1972. "Accommodation function of the human eye." *Physiol. Rev.* 52: 828–863.

Via, S. E. 1977. "Visually mediated snapping in the bulldog ant: A perceptual ambiguity between size and distance. *J. Comp. Physiol.* 121: 33–51.

Vilter, V. 1954. "Interprétation biologiques des trames photorécptrices superposées de la rétine du *Bathylagus benedicti*." *C. R. Soc. Biol.* (Paris) 148: 327–330.

von der Heydt, R., C. Adorjani, P. Hanny, and G. Baumgartner. 1978. "Disparity sensitivity and receptive field incongruity of units in the cat striate cortex." *Exp. Brain Res.* 31: 523–545.

Walk, R. D. 1965. "The study of visual depth and distance perception in animals." In *Advances in the Study of Behavior*, vol. 1, D. S. Lehrman et al., eds. New York: Academic.

Wallace, G. K. 1959. "Visual scanning in the desert locust *Schistocerca gregaria* (Forskål)." *J. Exp. Biol.* 36: 512–525.

Walls, G. L. 1942. *The Vertebrate Eye and its Adaptive Radiation*. New York: Hafner.

Westheimer, G., and A. M. Mitchell. 1956. "Eye movement responses to convergence stimuli." *Arch. Ophthalmol.* 55: 848–856.

Westheimer, G., and F. W. Campbell. 1962. "Light distribution in the image formed by the living human eye." *J. Opt. Soc. Am.* 52: 140–145.

Westheimer, G., and S. P. McKee. 1978. "Stereoscopic acuity for moving retinal images." *J. Opt. Soc. Am.* 68: 450–455.

5

Mechanisms for Discriminating Object Motion from Self-Induced Motion in the Pigeon

Barrie J. Frost

It has long been recognized that the detection of visual motion has immense survival value for predator and prey alike. However, recent experiments have shown that different classes of stimulus motion and motion-form compounds may provide the organism with very different types of information about the visual world and may therefore precipitate quite diverse patterns of behavior and perception. Some of the other chapters in this volume provide excellent examples of how particular patterns of motion are tightly correlated with certain behavioral sequences and phenomenal impressions. For example, the following have long been known:

• Movement of the entire visual array, or large segments of it, produces a striking impression of self-motion, where the observer himself feels he is moving while the visual world appears to remain stationary (Mach 1906; Helmholtz 1911; Thalman 1921; Dichgans and Brandt 1972; Wong and Frost 1978). It has also been shown experimentally that the peripheral retinal image is more effective than the central retinal image in producing this effect (Brandt et al. 1973). As one might expect, motion of the whole visual field has also been shown to be intimately associated with the control of posture and balance (Gibson 1966; Lee and Lishman 1977; Lee, this volume), and even the extension phase of reaching behavior is under the control of peripheral retinal stimulation (Paillard, this volume).

• Movement of large objects in an otherwise stationary visual environment produces avoidance behavior and autonomic reactions indicative of fear. The shadow reflex displayed by fish (Springer et al. 1977) and the avoidance response displayed by frogs and toads (Ewert 1970; Ingle 1970) are prime examples of this visually evoked behavior.

• In contrast, the movement of small objects (of the appropriate shape and orientation) often triggers an orienting response toward these objects followed by prey-catching behavior (Ewert 1970; Ingle 1970; Friedman 1975; Himstedt, this volume).

• Large patterns of relative motion in the visual array produced by either head movements or locomotion past stationary objects provide depth and distance information through motion parallax (Wallace 1959; Collett 1978).

• Certain other patterns of relative motion, usually in small subregions of the visual field, give rise to striking perceptions of perspective transformations of solid objects (Johansson 1977).

Though not an exhaustive list or a formal taxonomy of visual-motion phenomena, this rough classification does seem useful. Since different patterns of stimulus movement give rise to functionally different perceptual and motor responses, it seems reasonable to ask whether these patterns are processed in different parts of the brain and to look for specific adaptations in populations of motion-detecting neurons to meet the demands of each class of stimulus motion. The "two visual system" notion proposed by Schneider (1969) to account for differences between the retino-geniculo-striate system and the retinocollicular system was an important first step toward understanding how parallel processing produces the subdivision of labor in the visual system. With the great increase in anatomical, electrophysiological, and behavioral experiments on the visual system over the past decade, the time now seems ripe to expand the "two visual system" notion to incorporate other functionally separated subsystems.

In terms of analysis of motion information, one of the first and most basic functions that has to be performed is to differentiate movement in the visual image that is caused by moving objects from that which is caused by the animal's own body, head, and eye movements. In this chapter an attempt will be made to show how certain groups of neurons in the pigeon's optic tectum respond selectively to object motion, and to relate the findings from single-cell recording studies to head movement.

The problem of how self-induced visual motion is distinguished from object motion is an old one. It was described clearly by Helmholtz (1911), who felt that occulomotor commands played an important role in neutralizing the visual motion resulting from the actual execution of the eye movement. In a modified version of Helmholtz's notion, von Holst and Mittelstaedt (1950) postulated that motor centers of the brain might send corollary discharges to visual structures to cancel out the reafferent pattern of movement sensation produced by the animal's movement. Support for this formulation was found in observations of human subjects with paralyzed or anesthetized extraocular muscles (Kornmuller 1931). When such individuals attempt to shift their gaze to the left, their entire visual world appears to jump to the left. In this case they "see" only the "efference copy" produced by the corollary discharge. When a normal eye movement to the left is executed, no image motion is perceived, because the "reafferent" movement of the retinal image to the right is canceled by the efference copy to the left. Several single-cell studies indicate that some units in area 17 of the cat cortex (Noda et al.

1972) and in the superior colliculus of the monkey (Richmond and Wurtz, 1977) may be performing a function like this, since they are either excited or inhibited during saccadic eye movements, even when these eye movements occur in complete darkness.

However, it seems clear that there are several other situations in which body, head, and locomotor movement patterns may produce visual motion that might not be appropriately canceled or inhibited by corollary discharges. Take for example the gliding flight of birds. Here it is difficult to imagine where a phasic signal might arise to provide the source of a corollary discharge signal. Also, it should be kept in mind that whereas self-induced visual motion might need to be eliminated in some behavioral systems (for example, the visual subsystem involved in orienting toward visual movement), it is undoubtedly the prime source of information for other subsystems (such as the accessory optic system, which apparently uses such information for visual control of posture and stabilization of the head and eyes).

What will be described below will be a number of single-cell recording studies of the pigeon tectum and some behavioral observations showing that there may be several alternative mechanisms by which self-induced visual motion might be distinguished from object motion.

Most of our electrophysiological experiments have been concerned with the processing of motion in the optic tectum of the pigeon. Abundant evidence indicates that the tectum, like its homologous mammalian structure the superior colliculus, is intimately associated with "orienting" to novel objects (Hess et al. 1946; Skultety 1962); therefore it seems reasonable to conclude that it might be processing the motion information about objects required to elicit the "visual grasp" reflex (Robinson 1972; Goldberg and Wurtz 1972; Schaefer and Schneider 1969). Several years ago Diane DiFranco and I found that most directionally specific cells in the pigeon tectum preferred motion in the forward direction (posterior to anterior) in the visual field and were even more specific in terms of *not* responding to backward motion (Frost and Thomson 1972; Frost and DiFranco 1976). It can be seen in figure 5.1 that these directionally specific cells are rather broadly tuned and respond to most directions of motion through their receptive fields, but are very precise in vetoing backward (anterior to posterior) motion in the visual field. For convenience we have called these *backward notch* units. This directional bias is clearly evident when preferred and null directions of a sample of tectal units are plotted as in figure 5.2. It should be emphasized that in the tectum there are some neurons that do prefer backward motion in the visual field, but they are found quite infrequently.

When we first discovered this bias in directionally specific tectal neurons, we began to search for a behavioral manifestation of this directional bias and found that monocularly driven optokinetic head movements displayed a similar asymmetry to backward and forward

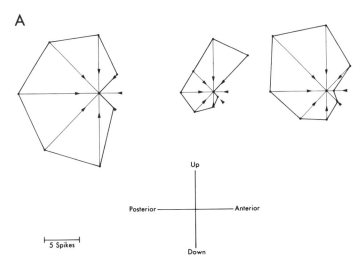

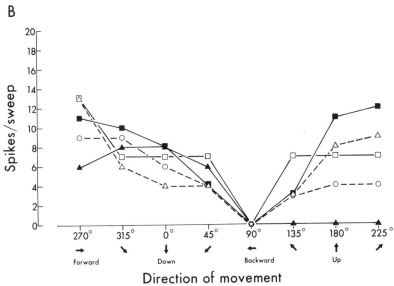

Figure 5.1 Directional response characteristics of broadly tuned tectal units. (A) Polar plots of three units. Left: unit 129, for which the stimulus consisted of a 1° spot swept at 8°/sec through the receptive field. Middle: unit 130; stimulus 1° spot swept at 14°/sec. Right: unit 138; stimulus 1° spot swept at 14°/sec. Points represent average of three stimulus sweeps for 129 and two for units 130 and 138. (B) Directional tuning curves of another five backward notch directionally specific neurons. Note that the minimum response for these broadly tuned units is always produced by backward motion in the visual field.

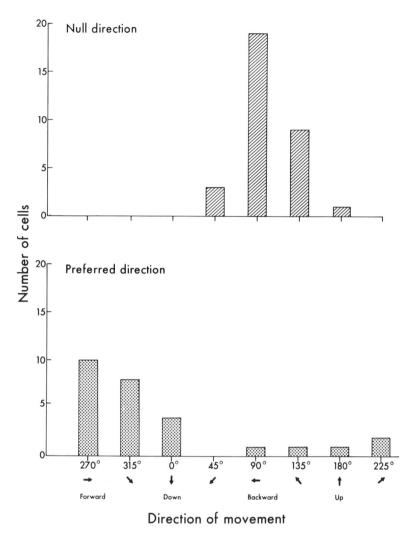

Figure 5.2 Histograms showing preferred and null directions of 32 broadly tuned directionally specific tectal neurons. In some cases when a cell responded equally to several directions of motion it was not possible to specify a single preferred direction. In these cases no entry was made for the preferred direction of the cell. χ^2 tests reveal that both null and preferred directions differ significantly from chance expectations.

Motion Discrimination in the Pigeon

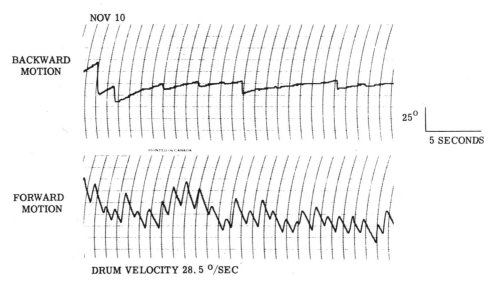

RIGHT EYE OPEN

NOV 10

BACKWARD
MOTION

25°

5 SECONDS

FORWARD
MOTION

DRUM VELOCITY 28. 5 °/SEC

Figure 5.3 Monocularly driven optokinetic head movements of a pigeon viewing a vertically striped drum (spatial frequency 0.25 cycles/degree) rotating at 28.5°/sec. Only the
right eye of the bird was open. Top trace: optokinetic head movements produced by
clockwise rotation, which stimulated the right eye with backward moving stripes (anterior to posterior in the visual field). Bottom trace: optokinetic head movements produced by forward motion at same velocity. Note that forward motion produces normal
optokinetic head movements, where head velocity closely matches stimulus velocity,
whereas backward motion produces inappropriate optokinetic head movements that are
insufficient to stabilize the moving pattern.

motion in the visual field. Optokinetic head movements (recorded with
a microtorque potentiometer coupled to the bird's head through a universal joint while the bird was inside a revolving striped cylinder)
showed symmetrical responses to both directions of rotation with both
eyes open. When one eye was covered, however, directional asymmetries similar to that shown in figure 5.3 appeared. What appears to be
happening in this illustration is that with counterclockwise rotation of
the drum, which produces forward motion across the visual field of the
right eye, resulting optokinetic head movements are quite normal and
show good matching between stimulus velocity and head velocity.
However, when the direction of the drum was reversed to present the
same eye with a backward flow of motion across the visual field, optokinetic head movements were reduced and there was no longer a tight
relationship between stimulus and head velocity. Similar biases in directionally specific neurons have been observed in the optic tecta of fish
(Cronly-Dillon 1964) and the retinae of rabbits (Oyster 1968; Daw and
Wyatt 1974). Although we at first thought this response might have
been mediated by the tectum, it now seems clear it is not and that it
may instead be mediated by the accessory optic system and pretectal

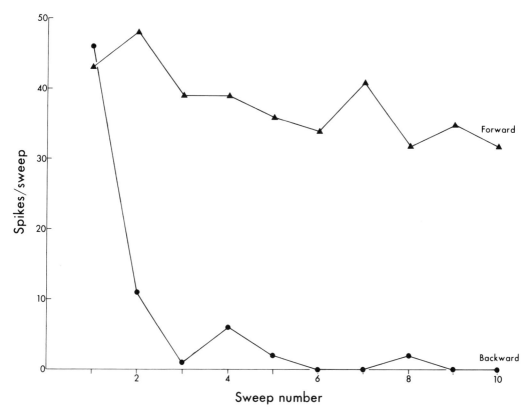

Figure 5.4 Graph showing differential rates of adaptation of backward and forward motion in a broadly tuned directional tectal unit. The unit was allowed to recover for 20 minutes, and then a 1° spot was alternately swept forward and backward at 8°/sec through its receptive field. Note how rapid adaptation is to backward motion compared to forward motion.

nuclei (Fite et al. 1977; Springer et al. 1977; Collewijn 1975; Simpson et al. 1979).

In a related series of experiments we looked at the habituation and adaptational characteristics of a sample of backward notch cells (Frost and DiFranco 1976; Woods and Frost 1977). After isolating these cells and determining that they responded to most directions of motion (except backward) through their receptive field, we then let them recover for periods ranging from 5 to 15 minutes and then retested them with forward and backward motion of a test stimulus. The data presented in figure 5.4 are typical of our adaptation experiments and show that initially this cell responded as vigorously to backward motion through the receptive field as to forward. However, with successive sweeps of the test stimulus there was rapid adaptation to backward motion, which contrasted with a relatively small degree of adaptation to forward motion.

Another set of experiments showed that many of the cells located in deeper tectal structures (stratum griseum and fibrosum centrale and

Motion Discrimination in the Pigeon

stratum griseum and fibrosum periventriculare) also exhibit characteristics of habituation to repeated stimulus presentations, since their responsivity can be immediately restored by altering a single parameter
of stimulation (Woods and Frost 1977).

Our interpretation of the response characteristics of these tectal cells
at this stage was as follows. It was suggested (Frost and DiFranco 1976)
that the backward notch units might be specially adapted to respond to
object motion, since they would specifically veto the self-induced
backward flow of motion produced by the bird's flight. The rapid differential adaption of these cells to backward motion was consistent with
this interpretation, for when the pigeon was stationary these units
would recover and thus respond to any direction of motion through
their receptive fields. When the bird locomoted forward past stationary
objects again, the backward flow of motion would rapidly desensitize
these cells to this class of self-induced motion, yet preserve their sensitivity to other directions of motion.

One of the many problems associated with this interpretation was
that directionally specific units such as these might also be maladap-
tively desensitized to backward motion while the bird is walking.
However, when pigeons and many other species of birds walk they
produce a characteristic bobbing of the head that at first sight bears a
resemblance to optokinetic nystagmus. We therefore decided to study
head bobbing in detail to see if its nature might throw any light on our
interpretation of the function of backward notch units found in the
tectum and on the differentiation of self-induced and object motion.

At the outset, we used high-speed motion photography to film the
walking of feral pigeons and analyzed their head, leg, and body positions frame by frame. As can be seen in figure 5.5, head bobbing consists of two well-defined phases: one in which the head appears to be
locked in space and one where it is rapidly thrust forward to a new
position. Thus, the head's illusory backward and forward motion during head bobbing is only in relation to the body, whereas its movement
through space is clearly of a "hold and thrust" nature. The stroboscopic
photograph shown in figure 5.6 makes this clear.

Although head bobbing looked very much like optokinetic head
movements, I was at that stage still not certain it was under visual control. For example, an alternative account might suggest that head bobbing was an equilibratory response to help maintain the center of gravity over the legs. If this is the case, head bobbing should occur when
pigeons walk on treadmills; but if it is a response to stabilize the visual
world, head-bobbing should not occur. Figure 5.7, which shows that
head bobbing does not occur during walking on a treadmill, supports
the notion that it is serving a visual function (Frost 1978a). A similar
conclusion was reached by Friedman (1975), whose elegant experiments
on head bobbing in the ring dove were published while our experiments were in progress.

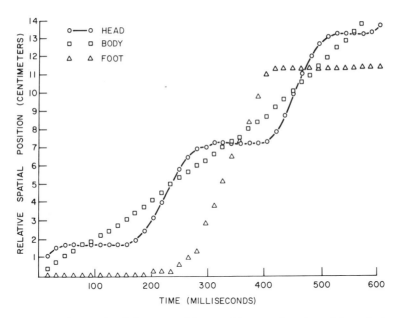

Figure 5.5 Pigeon head, breast, and foot position as a function of time, during normal walking. Data points were obtained from frame-by-frame analysis of movie film shot at 64 frames/sec. Note the characteristic "hold and thrust" pattern of head movements, where the head is apparently locked in space for a period of time and then rapidly moved forward to a new "hold" position.

However, a problem still remained: If (as our studies seemed to show) head bobbing was under visual control, and if the head was truly stabilized in space during the "hold" phase, how could vision provide an error signal to produce the compensatory backward movement of the head on the body to achieve stabilization? We found the error signal serendipitously. After filming a bird on the treadmill one day we accidently turned the treadmill to Very Slow instead of Off. After a few seconds we noticed that the bird's head was being slowly extended farther and farther forward until eventually the bird toppled over. What appeared to be happening was that the very low treadmill speed, although insufficient to elicit walking, still caused the bird to attempt to stabilize his visual world. When we filmed this behavior we saw the slow slippage of the head during stabilization (see figure 5.8), which must be the source of the visual error signal to produce a compensatory movement of the head to effect stabilization. If the slippage rate were approximately the same during normal head bobbing, it would amount to a change in head position of approximately 0.37 mm per hold or head bob, which would be difficult to detect by analyzing film records.

Having concluded that the pigeon's head bobbing while walking was under visual control and consisted of a "hold and thrust" ratcheting sequence, we attempted to relate it to our single-cell recording data. Since our analysis of head-bobbing records indicated that the head was

Figure 5.6 (Top) Stroboscopic photograph of pigeon walking showing the typical "hold and thrust" action of head bobbing. (Bottom) Stroboscopic photograph of pigeon landing, showing smooth trajectory of head although head bobbing is often reported during this maneuver.

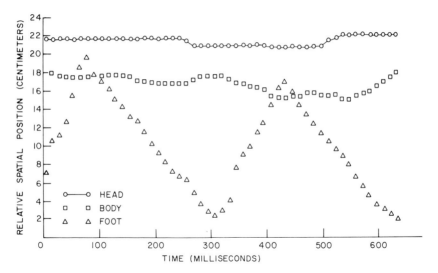

Figure 5.7 Head, breast, and foot position as a function of time for pigeon walking on a treadmill moving at 60 cm/sec. Film shot at 64 frames/sec. The "hold and thrust" head movements typical of normal walking have stopped.

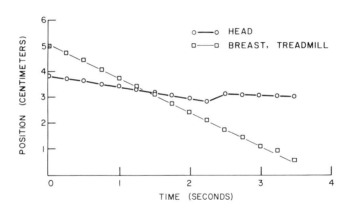

Figure 5.8 Head and breast position as a function of time for pigeon moved very slowly backward on a treadmill so that no walking was induced. Belt speed of treadmill was 1.2 cm/sec; film shot at 16 frames/sec; position in every fourth frame plotted. The bird exhibits attempted head stabilization, but slow slippage is apparent.

held still approximately 65% of the time during walking, it could be argued that this would keep the backward notch units in a recovered or semirecovered (that is, a nonadapted) state, in which they would be more likely to detect object motion in the backward direction. Also, the forward thrust would result in relatively fast backward motion of the visual array across the retina, which would produce velocities much greater than the optimum for tectal cells. Perhaps more important, the velocity and direction of movement of the visual array during a forward head thrust would place it within the range of backward image flow produced during flight. This would allow a single neural mechanism to process motion-parallactic information during flight and walking. As I have stated elsewhere (Frost 1978a), if these joint constraints of relative stabilization of the visual world to facilitate motion detection and head movement to provide motion-parallactic depth information are considered, head bobbing seems an elegant compromise. An obvious implication of these ideas is that head-bobbing birds might be alternately processing motion and depth information, phase-locked respectively with the "hold" and "thrust" phases of head bobbing during walking.

One of the obvious ways visual motion produced by moving objects differs from that produced by body, head, or eye movements is in the size or area of stimulation. Usually, moving objects stimulate a relatively small area of the visual field, whereas self-induced visual motion often fills the entire visual world or large segments of it. In fact, a compelling illusion of self-motion is produced when large regions of the visual field are filled with moving contours (Brandt et al. 1973; Wong and Frost 1978). In another series of single-cell recording experiments, we showed how the directionally specific responses of pigeon tectal cells to small moving test stimuli can be altered dramatically by the addition of separate moving background patterns (Frost 1978b; Frost et al. 1978). When these large background patterns are moved in phase with or at the same velocity and direction as the test stimulus, they completely abolish the directional response. However, movement of the background in the opposite direction to the test stimulus (out of phase) often enhances or facilitates the directional response. An example of this effect, which is exhibited by a large majority of the directional neurons below a depth of 400 μm in the tectum, is shown in figure 5.9. It can be seen that when a small light spot was moved through the receptive field in the preferred direction (forward) this tectal cell responded (A). However, a randomly textured background pattern (with edges masked) moved in the same direction at the same speed (B) abolished the response completely. In contrast, when the background pattern was moved out of phase (C) the cells' response was even greater than that produced by the test spot alone. We have yet to determine whether these interactive effects are completely relational or result from two separate directional mechanisms.

The in-phase inhibitory effect may even be obtained with a second

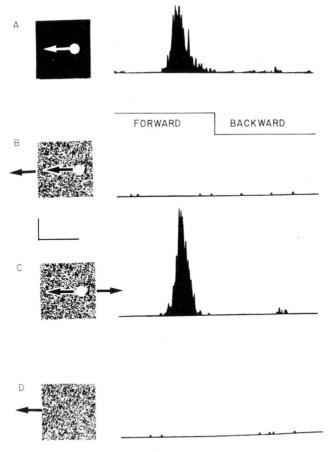

Figure 5.9 Peristimulus-time histograms (PSTHs) obtained from a directionally specific unit located at a depth of 256 μm in the pigeon optic tectum. This cell preferred a 1.3°-diameter light spot swept at 23.25°/sec in a forward (posterior to anterior) direction through its receptive field. (A) PSTH produced by movement of the test stimulus (optimal stimulus) swept forward and back through the receptive field. Contrast of test spot was 0.63 (($L_{max} - L_{min})/(L_{max} + L_{min})$). (B) PSTH produced by same stimulus as in A but with a 29.0° × 19.5° Julesz-type random-dot field moved in phase with the test stimulus. The probability of the 0.23 square elements in the background pattern being light or dark was 0.5. Contrast of background was 0.26. (C) PSTH produced by the test stimulus when the background pattern was moved out of phase. (D) PSTH produced by movement of the background field alone. Each PSTH represents the sum of 16 sweeps. Calibration markers: 10 spikes and 500 msec.

spot instead of the large moving background pattern. Here again, the directional response to a test stimulus is abolished if the second spot moves in the same direction (or has an in-phase component vector), but not if it moves in the opposite direction. The fact that this effect can be produced by a second spot stimulus allows us to map the boundaries of the directionally specific inhibitory receptive field of these cells, as illustrated in figure 5.10. When the second in-phase spot's path is close to the test stimulus no inhibition occurs, but as it is moved farther away inhibition develops and can be produced for very eccentric paths until eventually the directional response returns. At this point it can be concluded the second inhibitory in-phase spot is no longer falling within the directionally specific inhibitory receptive field (IRF) for that cell. In this fashion the outer boundary of the IRF can be mapped with reasonable accuracy. Other investigators have also shown that a second remote spot inhibits directionally specific cells, but have assumed that the inhibition is independent of the directional relation between the test spot and the remote spot (Rizzolatti et al. 1973). Our studies clearly indicate that the surround inhibition for these cells is directionally specific also. However, the important point to note is that these IRFs are large and in several instances occupy the entire visual field. We have not attempted to locate the site of origin of this surround inhibition. However, since units near the stratum opticum—the main retinal input area—do not exhibit these effects, it seems reasonable to conclude that the interaction takes place in the tectum, even though the inhibition could conceivably have a remote origin.

The relevance of these cells exhibiting in-phase inhibition for discriminating object motion from self-induced visual motion is as follows: If a small visual stimulus moves through the excitatory receptive field in the preferred direction, these cells will respond vigorously. If, however, the same stimulus moves through the receptive field together with other surrounding stimuli moving along similar paths—as would occur with translation of the visual field produced by body, head, or eye movement—then these cells would not be activated. They would of course also be triggered by stimuli moving through the receptive field in the appropriate direction with stationary background patterns, or when backgrounds were moved in different directions. It seems reasonable to conclude that tectal units such as these are well suited to respond to object motion but would ignore most classes of self-induced motion.

An exciting new technique allows us to visualize regions of the brain that have been activated while the animal is processing specified types of sensory information or engaging in certain types of behavioral response. In this process (termed [^{14}C]2-deoxyglucose autoradiography), [^{14}C]2-DG is injected intravenously into an animal, whose sensory or behavioral experience is then carefully controlled for the next 45 minutes. The deoxyglucose is taken up by neurons and metabolized in a

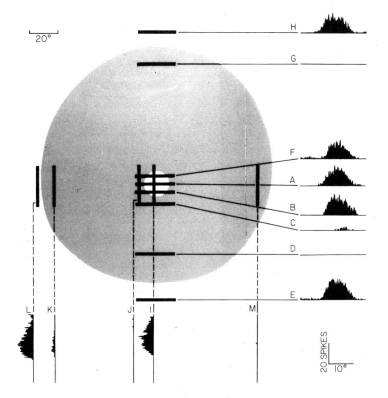

Figure 5.10 Receptive field map for a directionally specific neuron located at a depth of 1,562 μm from the tectal surface in the stratum griseuum and album periventriculare. The peristimulus-time histogram labeled A was obtained by moving a 2.7° test spot forward through the center of the excitatory receptive field (white area on reconstructed receptive field), along the path indicated, at 21.7 cm/sec. The PSTHs labeled B–H were obtained by moving the test spot along precisely the same path used to produce A, but in conjunction with a second 1.6° spot moved in the same direction and velocity along the paths indicated below (B–E) and above (F–H) the test stimulus. All the PSTHs in the figure were recorded from the test spot as it moved through the center of the ERF on the path shown in A. The addition of the second stimulus produced complete inhibition of the excitatory response over a large area of the visual field. A similar procedure was used with both the test stimulus and a secondary stimulus rotated through 90° and swept down through the receptive field. PSTHs labeled I–M were produced with the second stimulus moved in phase with the test stimulus along the paths indicated on the receptive field map. Each PSTH represents the sum of 32 sweeps.

similar manner to glucose, except that 2-deoxyglucose-6-phosphate is not readily removed from the neurons. The neurons that have been most active accumulate relatively large concentrations of 2-DG-6-phosphate. Immediately after the period of controlled sensory experience or behavior the animal is killed and its brain removed, frozen, and sectioned. The sections are then placed on x-ray film and kept in the dark until the radioactivity exposes the film sufficiently to produce useful autoradiographs. When the film is developed one is left with images of sections of the brain in which the densities indicate the relative rates of glucose utilization, and presumably the relative levels of neuronal activity in various brain structures and sites.

We have used [^{14}C]2-DG autoradiography in an attempt to converge on our notion that the tectum is primarily concerned with object motion and vetos most cases of self-induced motion (Frost and Ramm 1979). Pigeons were lightly anesthetized, injected intravenously with [^{14}C]2-DG, and placed in a stereotaxic instrument. One eye was opened to view a tangent screen, upon which stimuli were projected. One group of pigeons viewed a single light spot, less than 1° in visual angle, which was repeatedly swept forward along the same 70°-long path across the visual field for 45 minutes. The other group received a large textured pattern similar to the background patterns used in the single-cell studies described above. In each bird the unstimulated eye and tectum served as controls. The autoradiography revealed no discernible increase in activity in the region of the tectum stimulated with the large moving patterns. However, as can be seen in figure 5.11, the small spot

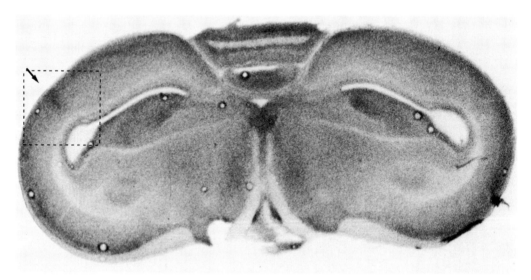

Figure 5.11 [^{14}C]2-Deoxyglucose autoradiograph of pigeon exposed to a small (< 1°) spot swept repeatedly across the same 70°-long horizontal path centered on the bird's right fovea. Note the increased density in the left optic tectum just dorsal to the ventricle. Similar effect in serial sections permit reconstruction of the path of activity traced out by the moving spot. When large textured patterns are moved over the same visual areas, no increased density is produced.

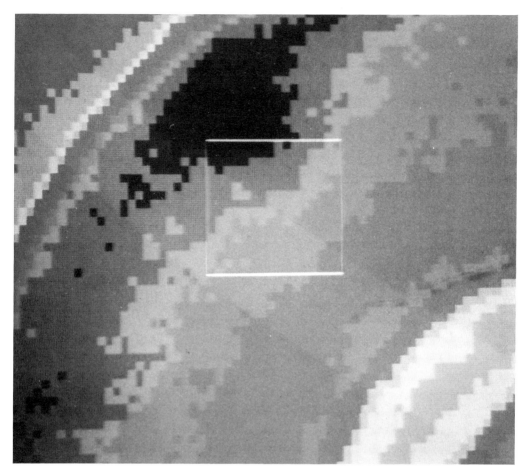

Figure 5.12 Black and white conversion of computer-digitized and processed color map of the tectal region containing the stimulus path shown in figure 5.11. The map portrays the area of increased density, which presumably results from increased activity of neurons in this region, as the red "hot spot." Blue region at top left indicates the clear area outside the surface of the tectum; the blue region at bottom right represents the ventricle. Different densities are represented as different spectral hues.

increased the activity of neurons in the appropriate retinotopic region of the tectum. Serial sections show the path traced by the stimulus.

Autoradiographic images similar to that shown in figure 5.11 can be computer-digitized and processed to quantify these effects. The differences in grain are translated into color maps, which enhance features of interest such as the "hot spot" produced in the tectum by forward movement of a small spot (figure 5.12). One important point to be drawn from this experiment is that these results are entirely consistent with our single-cell recording experiments. Further, they suggest that the tectum is particularly interested in small moving objects, and not in the movement of patterns covering large areas of the visual field.

References

Brandt, T., J. Dichgans, and E. Koenig. 1973. "Differential effects of central versus peripheral vision on egocentric and exocentric motion perception." *Exp. Brain Res.* 16: 476–491.

Collett, T. S. 1978. "Peering—A locust behavior pattern for obtaining motion parallax information." *J. Exp. Biol.* 76: 237–241.

Collewijn, H. 1975. "Oculomotor areas in the rabbit's midbrain and pretectum." *J. Neurophysiol.* 6: 3–22.

Cronly-Dillon, J. 1964. "Units sensitive to direction of movement in goldfish optic tectum." *Nature* 203: 214–215.

Daw, N. W., and H. J. Wyatt. 1974. "Raising rabbits in a moving visual environment: An attempt to modify directional sensitivity in the retina." *J. Physiol.* 240: 309–330.

Dichgans, J., and T. Brandt. 1972. "Visual-vestibular interactions and motion perception." In *Cerebral Control of Eye Movements and Motion Perception,* J. Dichgans and E. Bizzi, eds. Basel: Karger.

Ewert, J.-P. 1970. "Neural mechanisms of prey-catching and avoidance behavior in the toad (*Bufo bufo* L.)." *Brain, Behav., Evol.* 3: 36–56.

Fite, K., T. Reiner, and S. Hunt. 1977. "The accessory optic system and optokinetic nystagmus" *Soc. Neurosci. Abstr.* 1: 1771.

Friedman, M. B. 1975. "How birds use their eyes." In *Neural and Endocrine Aspects of Behaviour in Birds,* P. Wright et al., eds. Amsterdam: Elsevier.

Frost, B. J. 1978a. "The optokinetic basis of head-bobbing in the pigeon." *J. Exp. Biol.* 74: 187–195.

———. 1978b. "Moving background patterns alter directionally specific responses of pigeon tectal neurons." *Brain Res.* 151: 599–603.

Frost, B. J., and D. E. DiFranco. 1976. "Motion specific units in the pigeon optic tectum." *Vision Res.* 16: 1229–1234.

Frost, B. J., and P. Ramm. 1979. "Functional activity of pigeon optic tectum revealed by ^{14}C-2-deoxyglucose autoradiography." *Soc. Neurosci. Abstr.* V: 33.31.

Frost, B. J., and D. E. Thomson. 1972. "Directional asymmetries of cells in the pigeon tectum." Paper presented to Canadian Psychological Association, Montreal.

Frost, B. J., S. C. P. Wong, and P. L. Brooks. 1978. "Surround antagonism and opponent processes in directionally specific pigeon tectal neurons." *Soc. Neurosci. Abstr.* 2: 627.

Gibson, J. J. 1966. *The Senses Considered as Perceptual Systems.* Boston: Houghton Mifflin.

Goldberg, M. E., and R. H. Wurtz. 1972. "Activity of superior colliculus in behaving monkey. 1. Visual receptive field of single neurons." *J. Neurophysiol.* 35: 542–559.

Helmholtz, H. von. 1911. *Physiological Optics* (translation edited by J. C. P. Southall). New York: Dover, 1962.

Hess, W. R., S. Burgi, and V. Bucher. 1946. "Motorische Funktion des Tekalund Tegmentalgobieles." *Monatsschrift Psychiatr. Neurol.* 112: 1–52.

Ingle, D. 1970. "Visuomotor functions of the frog optic tectum." *Brain, Behav., Evol.* 3: 57–71.

Johansson, G. 1977. "Studies on visual perception of locomotion." *Perception* 6: 365–376.

Kornmuller, A. E. 1931. "Eine experimentelle Anaesthesis der ausseren Augenmuskeln am Menschen und ihre Auswirkungen." *Z. Psychol. Neurol.* 41: 354–366.

Lee, D. N., and J. R. Lishman. 1977. "Visual control of locomotion." *Scand. J. Psychol.* 18(3): 224.

Mach, E. 1906. *The Analysis of Sensation* (translation by S. Waterlow, based on fifth edition). New York: Dover, 1959.

Noda, H., R. B. Freeman, and O. D. Creutzfeldt. 1972. "Neuronal correlates of eye movements in the visual cortex of the cat." *Science* 175: 661–664.

Oyster, C. W. 1968. "The analysis of motion by the rabbit retina." *J. Physiol.* 199: 613–635.

Richmond, B. J., and R. H. Wurtz. 1977. "Visual responses during saccadic eye movements. A corollary discharge to superior colliculus." *Soc. Neurosci. Abstr.* 1: 1834.

Rizzolatti, G., R. Camarda, L. A. Grupp, and M. Pisa. 1973. "Inhibition of visual responses of single units in the cat superior colliculus by the introduction of a second visual stimulus." *Brain Res.* 61: 390–394.

Robinson, D. A. 1972. "On the nature of visual oculomotor connections." *Invest. Ophthalmol.* 11: 497.

Schaefer, K. P., and H. Schneider. 1969. "Stimulation experiments and neurone correlations in optic tectum." *Electroencephalogr. Clin. Neurophysiol.* 26: 433.

Schneider, G. E. 1969. "Two visual systems." *Science* 163: 895.

Simpson, J. J., R. E. Soodak, and R. Hess. 1979. "The accessory optic system and its relation to the vestibulocerebellum." In *Reflex Control of Posture and Movement*, R. Granit and O. Pompeiano, eds. (*Progr. Brain Res.* 50: 715–724).

Skultety, F. M. 1962. "Circus movements in cats following midbrain stimulation through chronically implanted electrodes." *J. Neurophysiol.* 25: 152–164.

Springer, A. D., S. S. Easter, and B. W. Agranoff. 1977. "The role of the optic tectum in various visually mediated behaviors of goldfish." *Brain Res.* 128(3): 393.

Thalman, W. A. 1921. "The after-effect of seen movement when the whole visual field is filled by a moving stimulus." *Amer. J. Psychol.* 32: 429–441.

von Holst, E., and H. Mittelstaedt. 1950. "Das Reafferenzprinzip." *Naturwissenschaften* 37: 464.

Wallace, H. 1959. "Visual scanning in the desert locust (*Schistocera Gregaria* Forskål)." *J. Exp. Biol.* 36: 512.

Wong, S. C. P., and B. J. Frost. 1978. "Subjective motion and acceleration induced by the movement of the observer's entire visual field." *Percept. Psychophys.* 24(2): 115–120.

Woods, E. J., and B. J. Frost. 1977. "Adaptation and habituation characteristics of tectal neurons in the pigeon." *Exp. Brain Res.* 27: 347–354.

6

Right-Left Asymmetry of Response to Visual Stimuli in the Domestic Chick

R. J. Andrew
J. Mench
C. Rainey

Differences between right and left systems in the central nervous system, which hold for the majority of individuals in a species, are now known to be present in animals other than man. Nottebohm and Nottebohm (1976) and Nottebohm et al. (1976) demonstrated clear lateralization of this sort in the motor control of song in passerine birds. However, there appears to be no evidence of such right-left differences for simple visually evoked responses outside man.

We have recently found in the young domestic chick marked differences between the effects of presenting stimuli to the right and to the left eye. These differences were superimposed upon differences between the sexes, which made it necessary to consider male and female chicks separately. The visual system fed by the right eye habituated to novelty faster and was better able to withhold response to irrelevant stimuli. Conversely, there was some indication that the systems fed by the left eye, which were more persistently affected by novel or striking stimuli, were more likely to be used to view a stimulus while it was evoking fear responses. We will argue below that the chick's condition has some suggestive basic resemblances to that found in man, and that lateralization of function may be ancient and widespread in the vertebrates. However, before speculating we should set out the evidence that differences of some sort exist at all.

In all the experiments which we shall discuss, chicks (Warren sex-link) were singly housed, having been separated on arrival from the hatchery; at test they were between 2 and 6 days old. The cages were 20 × 20 × 25 cm, open at the top (so that stimuli could readily be introduced) and with a transparent front allowing a close view of the chick when this was necessary. Results will be presented fully elsewhere.

Habituation Experiments

The first evidence of right-left differences in the behavior of normal domestic fowl chicks was obtained during investigation of a sex difference—itself unexpected—in response to the repeated presentation of a small white bead in the binocular and lateral visual fields. The

Asymmetrical Response in the Chick

bead (5.0 mm in diameter and mounted on a stiff thin wire) was introduced in each test in a series of eight presentations, each 5 seconds in length and separated from the next by 5 seconds during which the bead was withdrawn from the cage. Successive presentations were immediately in front of the beak (binocular: B) and to one eye (right: R or left: L) in the order BRBLBRBL or BLBRBLBR (the chicks were counterbalanced in this respect). When the presentation was made into the field of one eye only, the bead was introduced from behind the head. When the chick caught sight of the bead appearing from behind, it might ignore it or turn and fixate it binocularly; if it then pecked, this was recorded as a peck due to a monocular presentation.

Once preliminary experiments had suggested that females habituated pecking to monocular presentations more quickly than males, two groups, one male and one female, were tested under identical conditions in six sessions twice daily for three days. Since we had no expectation that left and right eyes would differ, in this experiment (and this experiment only) all chicks were tested in all sessions in a single order (BLBRBLBR). Habituation did indeed prove to develop sooner in females (figure 6.1). This was true within sessions as well as between them; the sex difference appeared only in the fourth and subsequent sessions. However, the most striking feature of the data was that there was only a clear and consistent difference in the case of presentations to the right eye; otherwise it appeared only in the later binocular presentations of two of the sessions. Left-eye rates were very similar in males and females, and much the same as male right-eye rates.

No valid simple comparison could be made between right- and left-eye rates in this experiment, since the order of presentation had not been counterbalanced. However, there was a strong suggestion that there might be a right-left difference in females, and indeed in males also: A seventh session, in which a new (blue) bead was used in an attempt to restore interest among the chicks, resulted in higher left-eye rates in the males (the right-left difference was significant at $P < 0.02$ by the Wilcoxon two-tailed test when the second left presentation was compared with the first right, to take into account the slight tendency for rates to fall progressively through each session).

Subsequent experiments of this type used female chicks only, since it seemed that they more readily developed the habituation necessary to reveal right-left differences. The results from the habituation sessions, which will not be considered here, showed that even under apparently identical conditions habituation of pecking proceeded at very different rates in different groups of chicks. The final sessions were all given under the standard counterbalanced conditions already described. In experiments 1 and 2 the systems fed by the right and by the left eye had similar counterbalanced experience in each individual during the habituation sessions (table 6.1). The results presented in the table are for the final or the final two test sessions (distinguished as a and b).

Andrew, Mench, Rainey

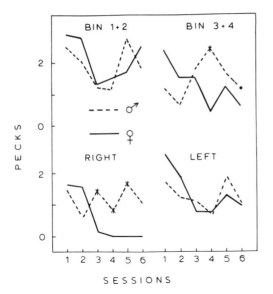

Figure 6.1 Pecking rates per presentation are shown for the first two (BIN 1 + 2) and second two (BIN 3 + 4) binocular presentations of the bead, and for the two right-field (RIGHT) and left-field (LEFT) presentations in each session. Male pecking rates fall significantly below female rates for right-field presentations only, as habituation develops in the later sessions. Asterisk indicates $P < 0.01$; dot indicates $P > 0.1$ (Mann-Whitney two-tailed test). Sex differences in binocular-field presentations occur only late in each session, and probably reflect the dominance of right-eye mechanisms in such presentations.

Table 6.1
Habituation of right-eye mechanisms.

Type of Presentation	Experimental Session								
	1	3b	2a	4aR[a]	3a	4bR	2b	4bL[a]	4aL
	Pecks per Presentation								
Binocular	0.7	1.1	1.6	1.9	1.9	2.2	2.6	3.1	3.9
Right	0.1[b]	0.5[b]	1.0	1.6	<u>1.6</u>[c]	1.1	<u>1.9</u>[d]	<u>2.2</u>	<u>3.3</u>
Left	<u>1.8</u>	<u>1.1</u>	<u>1.4</u>	<u>2.8</u>	1.2	<u>1.6</u>	0.7	2.0	1.7

a. R (right) and L (left) indicate which eye was covered.
b. $P < 0.01$; Wilcoxon two-tailed test.
c. Underline indicates the higher of the lateral-field rates.
d. $P < 0.05$; Wilcoxon two-tailed test.

Asymmetrical Response in the Chick

Pecks per presentation are shown for the second two presentations in the binocular visual field (B) and for the two presentations in the right (R) and in the left (L) visual field; the order of presentations was BRBLBRBL or BLBR---, and was fully counterbalanced. Two types of data are shown: In experiments 1 and 2 in each individual the right and left eyes had comparable counterbalanced experience during the habituation sessions; the order was BRRRRBRR or BLLLLBLL given in alternating sessions. Right-left differences appeared in the test sessions nevertheless. In experiments 3 and 4 each bird had one eye covered in each of the habituation sessions; the covered eye had seen the bead for the first time in the test session. The birds were divided into two subgroups, one receiving BLBLBLBL and one BRBRBRBR in the habituation sessions (which were four in number in experiment 3 and six in experiment 4). In experiment 3 the two groups were nevertheless very similar and have been lumped; in experiment 4 they differed and are shown separately (R and L indicate covered eye). The sessions are ordered according to binocular rates, and the right-field rates are significantly below the left only when binocular rates show marked habituation. Irrespective of prior experience, rates in right-field and in binocular-field presentations vary closely together, whereas left-field rates vary almost at random and considerably less.

In the final test session of experiment 1, right-eye rates were markedly and significantly lower than left-eye rates. However, in experiment 2, where there was little or no habituation (as can be seen from the binocular rates), there was no obvious left-right difference in the first test session; a second test session produced even higher rates and a significant left-right difference in the direction opposite to that predicted.

In order to examine the relationship between overall pecking rates and the direction of right-left differences, data from five other final test sessions with different groups of chicks were examined. (The procedures were the same.) There was a clear relationship between overall pecking rates in a particular session and the direction of right-left differences (compare table 6.1 and figure 6.1). Right-eye pecking rates fell markedly and significantly below left-eye rates in sessions in which habituated binocular pecking rates were less than 1–1.5 per presentation. On the other hand, in those experiments in which habituation did not occur and rates remained at or rose to 2–2.5 per presentation, there was a consistent trend for right-eye rates to be the higher.

Pecking rates in binocular and in right-eye presentations were markedly and significantly correlated ($r_s = + 0.95, P < 0.01$) across sessions, whereas there was effectively no correlation ($r_s = + 0.12$) between rates in binocular and left-eye presentations. Therefore, it seems probable that—at least when presentations in left, right, and binocular fields alternate rapidly—right-eye mechanisms dominate responses to stimuli appearing in the binocular field.

Andrew, Mench, Rainey

The difference between right-eye and left-eye rates also correlates strongly and significantly with right-eye rates ($r_s = +0.80$, $P < 0.01$), whereas it does not correlate significantly with left-eye rates ($r_s = -0.38$). It is likely, then, that right-left differences here reflect systematic changes in right-eye mechanisms as a result of prior experience, which affect both right-eye and binocular rates, whereas left-eye mechanisms in the same bird at the same time show lesser and (as far as can be seen from group means) random variation.

One possible interpretation is that pecking is affected by two different visual mechanisms. One (probably situated in the midbrain; see Andrew 1974) is directly responsible for pecking, and does not show right-left differences. The other is higher in level and can inhibit or facilitate pecking by its partner lower-level mechanism. If so, then the mechanism fed by the right eye is either better able to control partner lower-level mechanisms or better able to learn from prior experience and so to acquire information making it appropriate to withhold pecks.

It is impossible at present to exclude right-left differences in ability to learn. However, Bell and Gibbs (1977), who trained chicks monocularly in a discrimination task involving small beads presented binocularly (very much as in the habituation experiments reported here), found that both right- and left-eyed chicks learned equally rapidly and completely. We have found no consistent right-left differences with this test in our strain of chicks.

Differences Between the Sexes, and Further Tests

In the habituation experiments no systematic attempt was made to compare lateralization in male and female chicks, since it seemed likely that right-left differences would only appear after very different periods of prior experience in the two sexes.

A range of other tests were used in an attempt to explore further both the sex difference and the properties of lateralization itself. Three will be considered here. In all of them, birds were tested with one eye temporarily covered. (A small square of masking tape pinched lightly onto the down in front of and behind the eye was found to disturb the chicks little, if at all.)

The first test measured the rate at which chicks chose food grains in preference to pebbles of similar size and appearance, among which food grains were scattered in approximately equal number (Andrew 1972; Rogers et al. 1974). The pebbles were stuck down firmly, whereas the grains were loose. The chicks were accustomed to eating food grains of the type offered, but had not encountered pebbles before. A series of 80 choices were scored; repeated pecks at the same object were treated as a single choice. Half the chicks had the right eye and half the left eye covered throughout.

Male chicks began to avoid pebbles considerably sooner when using

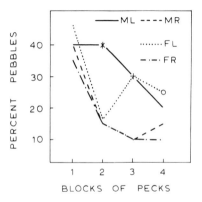

Figure 6.2 Percentage of pecks at pebbles rather than food grains for successive blocks of 20 pecks over the course of a test at the beginning of which the chicks had not before seen pebbles. The left-eyed males (ML) differ significantly from the right-eyed males (MR); asterisk indicates $P < 0.02$, + indicates $P < 0.05$. FL: left-eyed females. FR: right-eyed females. Open dot indicates $P < 0.05$, for comparison with FL and FR. ML and FL differ significantly in the second block ($P < 0.05$). All statistical comparisons are by the Mann-Whitney two-tailed test.

the right eye than when using the left (figure 6.2). There was no such difference in females, at least over the first 40 choices: Both groups avoided pebbles as soon as the right-eyed males. As a result, there was a significant difference in early choices between males and females using the left eye ($P < 0.02$ by the Mann-Whitney two-tailed test). A reappearance of pecks at pebbles in the left-eyed females produced a belated (but significant) right-left difference in the female also. However, this was not replicated in a second experiment, in which the other effects were again obtained.

Measures based on the pebble-floor test might be affected by the ability to learn to discriminate food and pebbles, by degree of interest in the novel pebbles, and by ability to inhibit pecks at pebbles. In view of the habituation data, perhaps the most economical explanation for the moment is that left-eye mechanisms are poorer at inhibiting pebble pecks. It is worth noting in passing that all groups readily acquired some degree of discrimination; even left-eyed males markedly and significantly decreased the proportion of pebble chosen over the course of the test ($P < 0.016$ by the Mann-Whitney two-tailed test).

In the second type of test, a single large mealworm was presented to a chick whose right or left eye was covered. Chicks give a complex series of responses on first encountering a mealworm (Hogan 1966): After initial startle, they fixate and usually peck the mealworm, and then pick it up and run with it, often dropping it and then picking it up again before finally swallowing it. Only two measures will be considered here: total time of holding the mealworm in the bill and latency to swallow. Two tests were given. The sexes differed markedly in that right-left differences were present in males only at the second test, and in females

Andrew, Mench, Rainey

Table 6.2
Response of monocular chicks to a mealworm.

	First Test		Second Test	
	Left-eyed	Right-eyed	Left-eyed	Right-eyed
Duration of hold (sec)[a]				
Males	10.4	12.1	16.6	10.6
Females	15.1[b]	8.9	8.0	9.6
Swallow latency (sec)				
Males	30.3	30.9	25.6[c]	13.0
Females	30.0[d]	13.0	13.7	15.2

Note: Any bird that did not swallow the mealworm in both tests, given on successive days, was excluded; the proportion involved was small and did not differ markedly between sexes or eyes.
a. Total time spent in holding worm (usually while running).
b. $P < 0.05$; two-tailed t-test (left vs. right).
c. $P < 0.02$; two-tailed t-test (left vs. right).
d. $P < 0.01$; two-tailed t-test (left vs. right).

only at the first (table 6.2). This was because at the first test right-eyed females showed little preliminary excitement and swallowed the mealworm as quickly as at the second test; left-eyed females held the mealworm much longer at first test, but by the second test were coping with it rapidly and efficiently, much like the right-eyed females. Both groups of males showed prolonged excitement at first test. At second test the right-eyed males behaved like the females and swallowed the mealworm quickly, whereas the left-eyed males still showed prolonged periods of holding.

The chicks throughout appeared to be dominated by the sight of the mealworm. Even when the mealworm was held in the bill, one end of its long body was likely to be still visible. They showed few overt signs of fear—in particular, peep calls (Andrew 1973) were unusual. Rather as in the habituation experiments, it seems that right-left differences appear only when the chick is neither too excited nor too indifferent to the stimulus, and that this point is reached sooner in females.

The third type of test has so far been administered only to male chicks. In this test, in contrast with the previous ones, a stimulus was chosen that was known to be likely to evoke marked escape behavior. This was a brightly colored toy ladder, introduced from above, and placed standing vertically against the cage wall. Care was taken to counterbalance between chicks the point of introduction, so that it was equally often to the right and to the left of the chick. After an initial startle response, almost all chicks became still and fixated the object with one eye; care was taken to present the object equally often to the right or left of the chick in its original position. Birds that fixated with the left eye gave peep calls (which are evoked in frightening and painful situations), whereas those that fixated with the right eye did not ($P <$ 0.02; two-tailed test). Here the choice of which eye to use was that of the chick; this suggests that right-eye and left-eye mechanisms may be used in different ways in normal behavior.

Summary

When right-left differences appeared in these tests, the right-eye system appeared better able to withhold a number of responses that are very effectively evoked in a naive chick by the visual stimuli presented. These are pecking at targets like bead or pebble, excited response to the sight of a mealworm, and the evocation of fear responses (peep calls) by a conspicuous novel stimulus. All of these responses, as performed at full intensity by a naive chick, could well be mediated by relatively automatic lower-level mechanisms. Thus, the hypothesis that the main right-left asymmetry is between higher-level mechanisms, and that the higher-level mechanism fed by the right eye is better able to control its partner lower-level mechanisms, would fit all of the results described.

Other explanations cannot be excluded but seem less likely. Thus, right-eye mechanisms appear entirely capable of performing all of the responses at full intensity. There might be quantitative right-left differences in the likelihood of performance by motor mechanisms, which are revealed only when that likelihood is falling; it is not clear why this should be true of a number of different motor mechanisms. Another possibility already considered is that left-eye mechanisms might learn more slowly and therefore be slower to habituate to the bead, or to distinguish pebbles and food grain, or to come to treat objects like the mealworm or the frightening ladder as no longer novel. Against this are the failure to demonstrate clear right-left differences in bead discrimination tasks and the evident ability of female chicks to learn the discrimination between food grain and pebbles at the same high speed with either eye.

The relevant difference between the sexes may be of the same sort as the difference between right-eye and left-eye systems. If that difference is in the ability to control lower-level visual mechanisms, then this might be highest in the right-eye systems in females, followed by the right-eye system in males, the left-eye system in females, and finally the left-eye system in males.

Evolution of Left-Right Asymmetry

The tests described above were concerned with differences between mechanisms fed by the right eye and by the left eye, but provided no evidence as to the location of these mechanisms. If they were to be situated in the forebrain this would be consistent with the hypothesis so far developed, calling as it does for asymmetry between higher-level visual mechanisms. Rogers (1979) provided direct evidence for interhemispheric differences of this sort. Permanent disturbance of visual response can be produced by injection of cycloheximide into the chick forebrain (Rogers et al. 1974). Such injection into the left hemisphere greatly facilitates both copulation and attack (which are responses that can be evoked in normal male chicks, but only with carefully presented stimuli, and often only at low intensity; see Andrew 1966). Injection into the right hemisphere has no such effect. Since optic decussation is typically complete in birds, and since the projections from the tectum and thalamus to visual areas of the forebrain are largely ipsilateral (Cohen and Karten 1974), the left hemisphere will be directly supplied by the right eye. It is interesting that it is the right-eye mechanisms that appear to tonically inhibit responses such as copulation and attack. Such an effect would be consistent with the hypothesis developed here of more effective control by higher-level right-eye mechanisms of the evocation of instinctive responses.

Any interpretation of the chick data is complicated by the fact that we are dealing with a young (although highly precocial) animal. It seems very unlikely, however, that such marked left-right differences exist only as a means towards the development of an entirely symmetrical adult condition. It can be argued, in fact, that asymmetry of the chick type might be expected, not only in adult fowl, but in birds in general. Such an argument commences with the observation that the two eyes commonly scan independently in a number of vertebrate classes, including birds (but not mammals).

Rochon-Duvigneaud (1943) and Walls (1942) both noted that all birds with perceptible eye movements move their eyes independently when scanning; conjugate movements occur only during compensation for head-body movement relative to the environment. Both of those authors recorded independent eye movements as typical for reptiles (for example, chameleons, agamid and iguanid lizards, chelonians) and some teleost fish, and emphasized that conjugate scanning is typical of mammals. Thus, the "binocular" fields of each eye would coincide only during binocular fixation in a bird like the chick in which convergence is marked. During independent scanning two "binocular" fields—both presumably feeding mechanisms mediating responses such as prey catching—would search independently for possible targets. Making one hemisphere better able to control such responses and dominant during binocular fixation (as in habituation experiments) seems an excellent way to prevent the simultaneous initiation of incompatible responses. Ingle (1973) noted that frogs are able to select between two prey objects seen with opposite eyes, and suggested that interhemispheric commissures play a role in reducing the potential competition. Such problems would be most acute in forms with laterally placed eyes and without a mobile neck; by this argument, lateralization of the chick type might have been evoked in the aquatic ancestors of the tetrapods.

Asymmetry of this sort would have interesting consequences that deserve setting out, even though its existence outside the domestic chick is entirely speculative. The right hemisphere, being fed by the left eye (which we have argued is more likely than the right eye to be moved by lower visual mechanisms free of higher control), would tend to act as a passive observer. Its visual experience would therefore be somewhat different from that of the left hemisphere. For example, it might be easier for the right hemisphere to build up a representation of the spatial organization of the environment than for the left hemisphere, when the latter is functioning as an active observer, selecting for fixation only those stimuli that are of interest to its own immediate aims.

Any theory that suggests that lateralization may have evolved early in the vertebrates also suggests that homologous organization may underlie those cases of lateralization already described. Some degree of correspondence may indeed exist. Nottebohm and Nottebohm (1976) showed that the lateralization of song to the left hemisphere in songbirds may depend more directly upon the lateralization of specialized motor mechanisms than on that of auditory mechanisms. During development, interference with left motor outflow can cause control to shift to the right hemisphere, but cutting off the direct auditory input to the left hemisphere does not. Song involves complex and precise control of elaborate sequences of responses; in order to allow this, the left hemisphere has acquired a surprising degree of direct control of brain-stem mechanisms (Nottebohm and Nottebohm 1976). A left hemisphere better able to control such mechanisms than its partner (as has been suggested for the chick) might be preadapted for such a role.

Correspondences with man are easier to find (or to imagine), since a variety of human abilities have been shown to exhibit some degree of lateralization. Nevertheless, it is important to the present argument that parallels can be drawn. Witelson (1976, 1977) described the human left hemisphere as analytical, isolating and identifying stimuli one at a time, and concerned with the control of sequences of movement. The right hemisphere is better at the analysis of patterns in space and time (Kimura 1967). The human right hemisphere is also more likely than the left to judge an experience horrific or shocking (Dimond et al. 1976), and right-hemisphere damage or dysfunction is associated with attenuation or disorder of emotional and affective behaviors (Gianetti 1972; Flor-Henry 1976).

The association in the chick of left-eye fixation with fear responses offers a parallel, which is extended by the fact that the human right hemisphere appears to become dominant at the onset of anxiety (Gur and Gur 1977). Such dominance in man is also suggested by the greater intensity of emotional expressions on the left side of the face (Sackeim et al. 1978).

All of these features of human lateralization could derive from a basic vertebrate asymmetry. Set out once again (this time fully), the hypothe-

Andrew, Mench, Rainey

sis is that the left hemisphere is better able to control brain-stem mechanisms, and so tends to be dominant when a course of action (that is, sequential or sustained response to selected stimuli) has to be carried out. However, when the right hemisphere is dominant it leaves lower-level mechanisms relatively free to act, but continues to give access to learned information and to allow learning to continue. A special ability to extract patterns might be associated with acting as a passive observer of the results of scanning, rather than focusing on particular relevant stimuli.

If lateralization is retained in man from a condition present before the evolution of conjugate eye movements in the mammalian line, some evidence of lateralization would be expected in other mammals. The available evidence is equivocal. On the one hand, visual discrimination and other learning tasks have revealed little evidence of right-left differences of the human type (that is, differences in the same direction in the great majority of individuals) in animals as diverse as the rat, the cat, and the rhesus monkey (Glick et al. 1977; Hamilton 1977; Warren 1977). On the other hand, Dewson (1977) reported asymmetry in auditory discrimination in the rhesus monkey and Petersen et al. (1978) found an advantage of the right ear in discriminating species-specific vocalizations in the Japanese macaque. Most interesting of all, removal of the right cortex in young rats that have had varied experience during rearing markedly disturbs behavior in novel surroundings, but removal of the left cortex has no such effect (Denenberg et al. 1978); the parallel with human disturbances of emotional behavior after right-hemisphere lesions is suggestive.

Conclusion

Two final points seem worth making. The first is that any theory postulating an early origin of lateralization in the vertebrates has the advantage of providing a possible explanation for right-left asymmetry in the same direction in all or most members of a species. Whatever the functions or the properties ascribed to lateralization, it would seem likely to work just as well if its direction varied at random between individuals (as is true of handedness in mammals other than man and the great apes). However, Jefferies and Lewis (1978) argued that the most primitive chordates (calcichordates; Cambrian to lower Devonian) were extremely asymmetrical in their head region, which was innervated from a structure homologous with the vertebrate forebrain. Cornute calcichordates lay on their right side, so the head region had gill slits on the left side only, as is true of *Amphioxus* today early in larval life. The more advanced mitrate calcichordates, with a body plan foreshadowing that of heterostracan vertebrates, had become almost symmetrical externally, but still retained marked internal asymmetry.

Brain and cranial nerves were largely embedded in the head skeleton, which allows their reconstruction.

The second point is that if our speculation turns out to be unjustified, and the chick condition is a specialized one not homologous with lateralization in other vertebrates, then its further study will be justified on quite different grounds. It will offer a useful model for understanding what functions right-left asymmetries coupled with sex differences might serve, and under what special selective pressures they tend to evolve.

References

Andrew, R. J. 1966. "Precocious adult behavior in the young chick." *Animal Behav.* 14: 485–500.

———. 1972. "Changes in search behaviour in male and female chicks, following different doses of testosterone." *Animal Behav.* 20: 741–750.

———. 1973. "The evocation of calls by diencephalic stimulation in the conscious chick." *Brain, Behav., Evol.* 7: 424–446.

———. 1974. "Changes in visual responsiveness following intercollicular lesions and their effects on avoidance and attack." *Brain, Behav., Evol.* 10: 400–424.

Bell, G. A., and M. E. Gibbs. 1977. "Unilateral storage of monocular engram in day-old chick." *Brain Res.* 124: 263–270.

Cohen, D. H., and H. J. Karten. 1974. "The structural organization of the avian brain: An overview." In *Birds: Brain and Behaviour*, I. J. Goodman and M. W. Schein, eds. New York: Academic.

Denenberg, V. H., J. Garbanati, G. Sherman, D. A. Yutzer, and R. Kaplan. 1978. "Infantile stimulation induces brain lateralization in rats." *Science* 201: 1150–1152.

Dewson, J. H. 1977. "Preliminary evidence of hemispheric asymmetry of auditory function in monkeys." In *Lateralization in the Nervous System*, S. Harnad et al., eds. New York: Academic.

Dimond, S. J., L. Farrington, and P. Johnson. 1976. "Differing emotional response from right and left hemispheres." *Nature* 261: 690–692.

Flor-Henry, P. 1976. "Lateralized temporal-limbic dysfunction and psychopathology." *Ann. N.Y. Acad. Sci.* 280: 777–795.

Gainetti, G. 1972. "Emotional behaviour and hemispheric side of the lesion." *Cortex* 8: 41–55.

Glick, S. D., T. P. Jerussi, and B. Zimmerberg. 1977. "Behavioral and neuropharmacological correlates of nigrostriatal asymmetry in rats." In *Lateralization in the Nervous System*, S. Harnad et al., eds. New York: Academic.

Gur, R., and R. Gur. 1977. "Correlates of conjugate lateral eye movements in man." In *Lateralization in the Nervous System*, S. Harnad et al., eds. New York: Academic.

Hamilton, C. R. 1977. "Investigations of perceptual and mnemonic lateralization in monkeys." In *Lateralization in the Nervous System*, S. Harnad et al., eds. New York: Academic.

Hogan, J. A. 1966. "An experimental study of conflict and fear: An analysis of behaviour of young chicks towards a mealworm. The behaviour of chicks which eat the mealworm." *Behaviour* 27: 273.

Ingle, D. 1973. "Selective choice between double prey objects by frogs." *Brain, Behav., Evol.* 7: 127–144.

Jefferies, R. P. S., and D. N. Lewis. 1978. "The English Silurian Fossil *Placocystites forbesianus* and the Ancestry of the Vertebrates." *Phil. Trans. Roy. Soc.* B 282: 207–321.

Kimura, D. 1967. "Functional asymmetry of the brain in dichotic listening." *Cortex* 3: 163–178.

Nottebohm, F., and M. E. Nottebohm. 1976. "Left hypoglossal dominance in the control of canary and white-crowned sparrow song." *J. Comp. Physiol.* 108: 171–192.

Nottebohm, F., T. M. Stokes, and C. M. Leonard. 1976. "Central control of song in the canary, *Serinus canarius*." *J. Comp. Neurol.* 165: 457–486.

Peterson, M. R., M. D. Beecher, S. R. Zoloth, D. B. Moody, and W. C. Stebbins. 1978. "Neural lateralization of species-specific vocalizations by Japanese macaques (*Macaca fuscata*)." *Science* 202: 324–327.

Rochon-Duvigneaud, A. 1943. *Les yeux et la vision des vertébrés*. Paris: Masson.

Rogers, L. J. 1979. "Lateralization of function in the chicken forebrain; cyclohexamide produces attentional persistence and slowed learning in chickens." *Pharmacol. Biochem. Behav.* 10: 679–686.

Rogers, L. J., H. D. Drennen, and R. F. Mark. 1974. "Inhibition of memory formation in the imprinting period: Irreversible action of cycloheximide in young chickens." *Brain Res.* 79: 213–233.

Sackeim, H. A., R. C. Gur, and M. C. Saucy. 1978. "Emotions are expressed more intensely on the left side of the face." *Science* 202: 434–436.

Walls, G. L. 1942. *The Vertebrate Eye and its Adaptive Radiation*. New York: Hafner.

Warren, J. M. 1977. "Handedness and cerebral dominance in monkeys." In *Lateralization in the Nervous System*, S. Harnad et al., eds. New York: Academic.

Witelson, S. F. 1976. "Sex and the single hemisphere: Specialization of the right hemisphere for spatial processing." *Science* 193: 425–427.

———. 1977. "Early hemisphere specialization and interhemisphere plasticity: An empirical and theoretical review." In *Language Development and Neurological Theory*, S. J. Segalowitz and F. A. Gruber, eds. New York: Academic.

7 Retinal Locus as a Factor in Interocular Transfer in the Pigeon

Melvyn A. Goodale
Jefferson A. Graves

It is often assumed that if an animal has learned to modify its behavior in response to stimuli that were present in a particular part of its visual field, it will continue to respond appropriately to the same stimuli if they now occur at a new locus in the visual field. In the case of bilaterally symmetrical animals such as vertebrates, such generalization is assumed to occur not only within the visual field of one eye but also between eyes. Thus, if a vertebrate has learned to discriminate between visual stimuli presented to only one eye, it will usually continue to discriminate accurately if the stimuli are presented to the other eye. Intra- and interocular "transfer of training" has obvious adaptive significance, since it is often important for an animal to react appropriately to visual stimuli no matter where they are located in the visual field. However, we will argue here that most demonstrations of interocular transfer of training (in birds, at least) may be an "artifact" of binocular convergence.

In mammals, visual information from one eye can reach both hemispheres directly via contralateral and ipsilateral projections from the retina. Projections from the binocular portions of the two retinas have converging input upon the same neuronal population in a single hemisphere. But by sectioning the crossed projections at the chiasm, direct input from the retina can be confined to one hemisphere so that it can reach the opposite hemisphere only via commissural pathways. However, in certain classes of vertebrates, such as birds, the fibers arising from the retina of one eye project entirely to the opposite half of the brain. Therefore, if information derived from one eye is to reach both hemispheres, even from the binocular fields, it must be transmitted to the ipsilateral hemisphere via the interconnecting commissural and decussational pathways. For this reason, a bird resembles an ideal "split-chiasm" preparation in which it is possible to study the conditions that determine the transmission of visual information from one hemisphere to the other as well as the pathways by which such information is transmitted.

Early Work on Interocular Transfer in Birds

In the laboratory, interocular transfer (IOT) is commonly studied by training animals on a visually guided task with one eye covered by a blindfold. Once the animal has reached an arbitrary criterion of successful performance on the task, the blindfold is switched to the other eye and the performance of the animal using the "naive" eye is assessed. Measures of IOT have varied from simple savings scores on retraining to reversal-learning scores or stimulus-preference tests (particularly in the case of visual-discrimination tasks).

The earliest work on IOT in birds was begun in 1917 by Köhler (cited in Levine 1945a) as part of his classic study on relative-brightness judgments in chickens. He found that domestic hens that had been trained monocularly to select and peck one of two sheets of gray paper differing in brightness showed excellent IOT when tested with the other eye. This observation was used by Köhler as an argument for stimulus equivalence at different retinal loci, an important principle in his general gestalt approach to perception. Some years later, Diebschlag (1940) showed that pigeons trained monocularly to select a platform by its color for a food reward demonstrated excellent IOT when tested with the other eye.

However, in contrast with these clear demonstrations of IOT, Beritov and Chichinadze (1936) found that pigeons trained monocularly on go/ no-go pattern or color discriminations failed to show any evidence of IOT when tested with the eye that had been covered during training. In these experiments the birds were trained to approach a food box from some distance when the positive stimulus was projected onto the wall above the food box. When tested with the naive eye, they showed no ability to differentiate between positive and negative stimuli.

The discrepancy between these two sets of results was extensively investigated and apparently resolved by Levine (1945a, 1945b, 1952), a graduate student of Lashley's. Using a modified Lashley jumping stand, Levine found that he could reliably produce either perfect IOT or none at all by varying the position of the stimuli with respect to the head of the bird. If the discriminanda were presented vertically in front of the pigeon, there was no evidence of IOT on either pattern-, color-, or brightness-discrimination tasks. The failure of IOT was complete by all the measures of transfer Levine employed: There was no evidence of transfer when stimulus-preference tests were given to the naive eye, nor were there any savings when the pigeon was retrained on the same task using the naive eye, nor any retardation in learning when the naive eye was used to learn the reversal of the first task. The complete absence of IOT in this situation was in good agreement with the earlier observations of Beritov and Chichinadze (1936).

However, when Levine trained pigeons with the stimuli located horizontally in a plane below the pigeon's head, there was excellent IOT of all the discrimination tasks that had failed to transfer when the stimuli were located in the vertical position. Levine also replicated Köhler's earlier observations of good IOT in a task where the bird was required to select one of two adjacent feeding dishes marked with different-colored paper. However, if the pigeon had to select and then peck through a hole in the same colored paper placed vertically in front of the feeding dish there was no evidence of transfer. It appeared that a necessary condition for IOT in either jumping-stand or pecking tasks was that the stimuli be located below the head of the bird.

Levine (1945a) used this distinction between the two positions of the stimuli to demonstrate a failure of *intra*ocular transfer. If a pigeon was trained monocularly on the jumping stand with the stimuli placed in a horizontal plane below the head and then tested with the same eye but with the stimuli placed vertically in front of the head, performance on the discrimination fell to chance.

On the basis of the results of his extensive series of experiments, Levine concluded that in the pigeon there are at least two functionally independent areas of vision in the eye. He went on to argue that the retinal locus that corresponds to the subrostral position in the visual field (below the head) has neural connections that permit communication between the hemispheres and thereby mediate IOT, whereas the locus that corresponds to the anterostral position (in front of the head) does not have such interhemispheric connections.

These conclusions were challenged by Catania (1965), who found complete IOT in pigeons of brightness, color, and pattern discriminations with the stimuli located vertically in front of the bird—the same position that had yielded no IOT in Levine's experiments. Catania trained his birds on successive discriminations where the response was pecking at transilluminated keys in a Skinner box. His findings have been replicated by a large number of investigators for both successive and simultaneous presentations of stimuli (Mello et al. 1963; Mello 1966; Ogawa and Ohinata 1966; Cuénod and Zeier 1967; Green et al. 1978; Cheney and Tam 1972). Mello (1968) also demonstrated good IOT of an up-down movement discrimination using a key-pecking task.

Catania (1965) suggested two possible explanations for the discrepancy between his own results and those of Levine. The first of these was derived from the characteristics of the pigeon's visual system. He argued that the pigeon is laterally far-sighted and anteriorly near-sighted (Catania 1964), so that in order to view the stimuli at the distance required in the jumping stand the pigeons had to cock their heads to one side and so bring the stimuli into the lateral visual field. Since the direction in which the head was cocked would depend on which eye was covered, the jumping response in each case could also have been affected. Thus, switching the blindfold would necessitate a change in posture for the bird, and learning to make this change would have obscured any IOT that was present. Catania also hypothesized that the discrepancy in results could have been due to the difference in the number of responses demanded in performing the discrimination in the two situations. Levine's jumping-stand experiments required only a few hundred responses, whereas Catania's key-pecking task required several thousand. The resultant overtraining might have facilitated transfer, although this explanation does not speak to the clear difference in IOT that Levine found between subrostral and anterostral stimuli on the jumping stand.

These suggested explanations for the difference between Catania's results and those of Levine were accepted by a large number of investigators (Sheridan 1965; Ogawa and Ohinata 1966; Cuénod 1972; Zeier 1976). Nevertheless, it is still possible that Levine's jumping-stand findings do represent a genuine failure of transfer. None of his critics have attempted to replicate his original observations on the jumping stand, and almost all the subsequent investigations have followed Catania's lead and used transilluminated keys and the pecking response.

Recent Experiments Comparing IOT in Jumping Stand and Skinner Box

Jumping-Stand Experiments
Several years ago we began a series of experiments (Goodale and Graves 1979; Graves and Goodale 1977, 1979) designed to reexamine Levine's original observations and investigate directly Catania's criticisms. We attempted to reproduce the jumping stand used by Levine (see Halstead and Yacorzynski 1938). As figure 7.1 illustrates, the final reconstruction contained a rotating perch 45 cm in length and 2.5 cm in diameter, covered with rough cloth so that the pigeon could easily grip it with its feet. Two collapsible platforms measuring 19 × 14 cm were located 7 cm in front of and 5 cm below the perch. Each platform could be made secure or allowed to collapse when a bird landed on it. The stimulus cards, measuring 19 × 14 cm, were suspended vertically behind the platforms. A 10-cm partition separated the two platforms, allowing a pigeon to jump to one or the other but not to both platforms at once. In all the jumping-stand experiments the pigeons were trained and tested monocularly with one eye covered with a cloth blindfold. Every training day, a pigeon was given 20 noncorrection trials in which it was placed on the perch and the perch was rotated, forcing the pigeon to jump onto one of the two platforms. If it jumped onto the correct platform it was allowed to remain there for 15 seconds; if it landed on the incorrect platform, the platform would collapse and the bird would fall 20 cm into a pile of hay, whereupon it would be picked up immediately by the experimenter. The position of the stimulus cards was switched from left to right according to a pseudorandom sequence, and each bird was

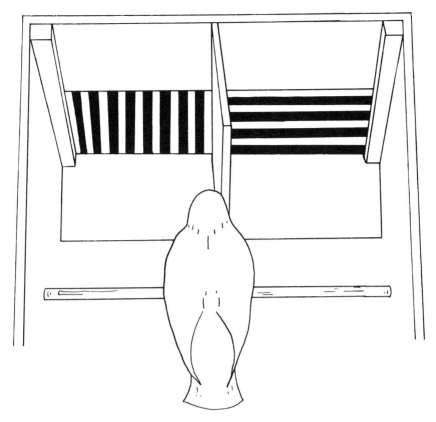

Figure 7.1 Illustration of jumping stand, showing relative position of platforms, perch, and stimuli.

trained until it reached a criterion performance of 18 correct trials in one day.

In the first experiment, six pigeons were tested for IOT of a triangle-circle discrimination. In this and subsequent experiments, training was carried out in four stages. As figure 7.2 illustrates, during stage 1 this particular bird was trained with the right eye covered to jump toward the triangle until it reached criterion. During stage 2, the blindfold was shifted to the left eye and the bird was again trained with the triangle positive. After reaching criterion, the bird was moved on to stage 3 of training, in which the left eye was still occluded but the circle was the positive stimulus. In stage 4, the blindfold was switched back to the right eye and the bird was trained with the circle positive once more. In addition, a retention test was administered between stages 3 and 4. This test consisted of two daily sessions of 20 trials during which neither platform would collapse. The blindfold was switched from one eye to the other every five trials. In this way, the stimulus preference of

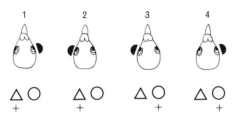

Figure 7.2 Experimental design, showing the four stages of training. Solid crescent indicates position of blindfold; + indicates positive stimulus at each stage.

Table 7.1
Triangle-circle discrimination performance on jumping stand.

Pigeon	Stage 1		Stage 2		Stage 3		Stage 4	
	Errors	Days	Errors	Days	Errors	Days	Errors	Days
7	292	36	194	24	295	33	283	37
8	197	25	196	22	257	32	187	25
9	259	33	256	33	380	43	368	39
10	190	22	196	24	223	25	286	32
11	173	21	87	15	310	33	341	38
12	189	22	202	24	230	29	267	28
$\bar{X}$	216.7	26.5	188.5	23.7	282.5	32.5	288.7	33.2

the bird when it used each eye could be measured. This rather tedious design allowed us to measure IOT in a number of different ways, a feature which is important in a situation where IOT is purported to be absent.

The results of the first experiment are summarized in table 7.1 and figure 7.3. In agreement with Levine's findings, there was no significant difference between stages 1 and 2 in either error scores or days required to reach criterion performance. The pigeons took as long to learn the task with the second eye as with the first. Furthermore, they failed to learn the reversal of the original discrimination any faster in stage 4 than in stage 3, although in both these stages they made many more errors and required far more days to reach criterion than they had in stages 1 and 2. These results provide a strong confirmation of Levine's earlier reports of failure of IOT in pigeons trained on a jumping stand. Furthermore, the failure of transfer between stages 3 and 4 is inconsistent with Catania's (1964, 1965) proposal that the subsequent change in posture after switching of blindfolds was obscuring IOT. During stage 4, the pigeons were using the eye that they had already used on a large number of trials during stage 1, and were well accustomed to the apparatus. Moreover, they had used this eye for 20 trials during the retention test administered immediately before stage 4. Nevertheless, con-

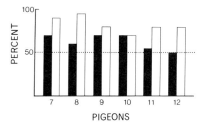

Figure 7.3 Stimulus-preference scores on interpolated retention test administered between stages 3 and 4 on triangle-circle discrimination task with jumping stand. Height of each bar indicates percentage of trials in which stimulus that had most recently been correct for that eye was chosen. Performance of each pigeon using eye that had been last trained in stage 1 is represented by a solid bar; performance with eye last trained in stage 3 is represented by an open bar.

trary to what Catania might have predicted, the birds took as long to learn the task in stage 4 as they had in stage 3. Indeed, the performances in both these stages were highly similar, and for a large number of trials in the early portions of both stages all the pigeons persisted in jumping toward the stimulus card that had been correct in stages 1 and 2 before they either began to jump to the correct platform or formed a position habit.

The results of the retention test interpolated between stages 3 and 4 were also consonant with a failure of IOT (figure 7.3). On most test trials the birds chose the stimulus that had been correct in stage 3 when they were allowed to use the eye that had been uncovered in stage 3; however, when forced to use the other eye, they tended to jump toward the stimulus card that had been correct in stages 1 and 2, even though they had not been trained with that eye for an average of 1,120 trials. In fact, in the last 660 of those trials, they had been trained with the other eye *not* to jump toward that stimulus, but toward the other one.

It could be argued that the retention test had simply revealed the existence of a conditional discrimination; that the pigeons jumped toward either stimulus depending on which eye was uncovered. Certainly, Catania (1965) and Konnerman (1966) have demonstrated that it is possible to train birds on conflicting discriminations using each eye separately in situations where there is normally good interocular transfer. However, Catania alternated the training sessions for each eye from day to day, whereas in our experiment the pigeons were trained to criterion on one eye before being switched to the other. Konnerman failed to establish conflicting discriminations in domestic geese when one eye was trained to criterion first, even though he was able to do so when the geese were given alternate sessions for the left and right eyes in the same day. This suggests that the pigeons' behavior in the interpolated retention test of our experiment was not the result of their having learned a conditional and conflicting discrimination, but was a reflection of the absence of interocular transfer in the situation.

Triangle-circle discrimination proved to be a rather difficult discrimination for the birds to learn on the jumping-stand, requiring on average nearly a month of daily testing before they reached criterion on stage 1. It has been suggested that more difficult discriminations transfer less readily than do easier discriminations for pigeons (Zeier 1976) and goldfish (Ingle 1963, 1968). Ingle showed that an easy 0°-versus-52° orientation discrimination transferred successfully in the goldfish whereas a more difficult 0°-versus-23° orientation discrimination did not. Zeier found that pigeons trained monocularly on a complicated sequential learning task did not show any evidence of IOT. Accordingly, we trained pigeons on two much easier discrimination problems on the jumping stand in order to determine whether IOT could be improved (Graves and Goodale 1979). Four pigeons were trained on a red-green discrimination and four on a blue-yellow (using the same four-stage design as before). In the red-green task, the stimulus cards were painted either solid red or green but no attempt was made to equate the brightness of the two cards. In the blue-yellow task, the stimulus cards were transilluminated from behind by yellow and blue filters mounted in a projector and the two colors were equated for intensity by using spectral sensitivity functions for the pigeon established by D. Blough (1957). The pigeons learned both these discriminations rapidly, taking an average of 5.5 days to learn the red-green task on stage 1 and 4 days to learn the blue-yellow. But despite the fact that these two problems were much easier to learn than the triangle-circle discrimination, there was still no evidence of IOT as measured by either savings scores or performance on the retention test (figures 7.4, 7.5). It appeared that even simple discrimination tasks fail to show transfer from one eye to the other on the jumping stand.

One observation that troubled us in these experiments was the relatively poor performance of the pigeons in the retention test when they were using the eye that had last been trained during stage 1. In the three experiments, the pigeons chose the "correct" stimulus card 62.5%,

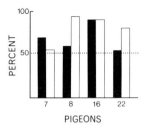

Figure 7.4 Stimulus-preference scores on interpolated retention test administered between stages 3 and 4 on red-green discrimination task with jumping stand. Height of each bar indicates percentage of trials in which stimulus that had most recently been correct for that eye was chosen. Performance of each pigeon using eye that had been last trained in stage 1 is represented by a solid bar; performance with eye last trained in stage 3 is represented by an open bar.

Goodale, Graves

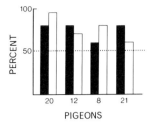

Figure 7.5 Stimulus-preference scores on interpolated retention test administered between stages 3 and 4 on yellow-blue discrimination task with jumping stand. Height of each bar indicates percentage of trials in which stimulus that had most recently been correct for that eye was chosen. Performance of each pigeon using eye that had been last trained in stage 1 is represented by a solid bar; performance with eye last trained in stage 3 is represented by an open bar.

68.7%, and 75.0% of the time when they were using this eye, compared with 82.5%, 80.0%, and 76.2% when they were using the eye that had just finished learning the reversal in stage 3. It could be argued the poor performance with one eye was due to interhemispheric transfer of conflicting information from the other eye-hemisphere system, rather than being a function of the length of time elapsed between the first stage of training and testing. The simple "decay" hypothesis receives some support from the observation that the difference between the bird's performance using the two eyes gets smaller as the interval between stage 1 and the retention test gets shorter. Nevertheless, a proper control experiment seemed necessary.

Four pigeons were therefore tested in parallel with the four trained on the blue-yellow discrimination described above. However, after being trained in stage 1 on the blue-yellow problem, they were then trained with the second eye on an irrelevant horizontal-vertical discrimination for the same number of trials as stages 2 and 3 took for the experimental group. They were then given the same retention test as the experimental group (figure 7.6). Not only was there no significant difference be-

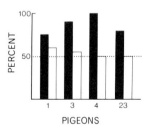

Figure 7.6 Stimulus-preference scores on yellow-blue discrimination task for those pigeons given irrelevant training in horizontal-vertical discrimination with other eye. Height of each bar indicates percentage of trials in which the stimulus that had been correct in stage 1 was chosen. Performance of each pigeon using eye that had been trained in stage 1 in yellow-blue discrimination is indicated by a solid bar; performance with eye that had been trained on horizontal-vertical is represented by an open bar.

tween the two groups when they were using the eye that had been trained on the blue-yellow discrimination in stage 1, but the four birds that had been trained on the irrelevant horizontal-vertical problem with the other eye showed no evidence whatsoever of IOT when they were tested with this eye on the blue-yellow problem. With the eye given irrelevant training, these birds chose the color that had been positive during stage 1 only 53.7% of the time on average, a score that does not differ significantly from chance performance. Thus, despite the fact that these birds had not been trained on conflicting discriminations with each eye, they failed to exhibit IOT. These results suggest that the poor retention-test performance of the pigeons using the eye trained in stage 1 in the earlier experiments cannot be interpreted in terms of interhemispheric transfer of a conflicting habit. Instead, the scores probably reflect a simple deterioration of habit retention over time.

However, in a situation where *failure* of IOT is apparent, it becomes important to optimize the possibility of demonstrating its existence. A number of authors (Menkhaus 1957; Shulte 1957; Ingle 1968) have suggested that information transferred from one hemisphere to the other may be much "weaker," less complete in some way, or more labile than the information received directly from the trained eye. Sheridan (1965) argued that the probability of such attenuated information being transferred from one hemisphere to the other, at least in the albino rat, can be increased by overtraining. For these reasons, we decided to overtrain pigeons on a discrimination using one eye, and then test immediately for IOT. If IOT was at all possible on the jumping stand, then such a procedure should maximize the chances of measuring it. Accordingly, the four birds that had already been trained with one eye on the horizontal-vertical task in the last experiment were overtrained on the same problem using the same eye, before being tested in the usual fashion for retention and transfer of the horizontal-vertical task. As figure 7.7 illustrates, with the "overtrained" eye the pigeons scored very well, jumping an average of 96.2% of the time to

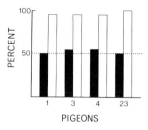

Figure 7.7 Stimulus-preference scores on horizontal-vertical discrimination task after overtraining. Height of each bar indicates percentage of trials in which the "correct" horizontal-vertical stimulus was chosen. Performance of each pigeon using eye that had been trained in stage 1 in yellow-blue discrimination is indicated by a solid bar; performance with eye that had been trained in horizontal-vertical discrimination is represented by an open bar.

the platform that had been rewarded during training. However, with the other eye, which had been used to learn only the blue-yellow discrimination, they performed at chance level. Thus, in spite of overtraining with one eye, they failed to show any suggestion of transfer to the other eye.

Summary

The results of the preceding experiments provide firm support for Levine's original conclusion that interocular transfer fails to occur in a monocularly trained pigeon on the jumping stand when the discriminative stimuli are located vertically in front of the pigeon's head. Moreover, the results are inconsistent with Catania's (1965) hypothesis that transfer exists but is obscured by the necessity of a change in posture when the occluder is transferred from one eye to the other. The failure of IOT appears to be a genuine phenomenon. Indeed, even birds trained on the jumping stand with both eyes open often show evidence of learning with only one eye when tested monocularly (Levine 1945a; Goodale and Graves 1980; Graves and Goodale 1979). But if IOT is so difficult to obtain in this situation, why is it so easily demonstrated in the key-pecking task employed by Catania (1964, 1965)? In order to determine what factors might be involved, we began a series of investigations of IOT in the Skinner box.

Skinner-Box Experiments

Although the stimulus keys in a Skinner box are apparently located in the same vertical position in front of the bird's head as they are in a jumping stand, most investigators have found nearly perfect IOT of a wide variety of visual discrimination habits in this situation (Mello et al. 1963; Catania 1965; Mello 1966, 1968; Ogawa and Ohinata 1966; Cuénod and Zeier 1967; Cheney and Tam 1972; Green et al. 1978). The results of the jumping-stand experiments described in the previous section show quite clearly that an explanation couched in terms of learned postural readjustments cannot account for the striking difference in IOT in the two situations. However, as Catania (1965) emphasized, the response demands of the two tasks are quite different: Whereas only a few hundred responses (at the most) were required to learn a discrimination problem on the jumping stand, birds often made "tens of thousands" of responses in the Skinner box before reaching criterion. According to Catania (1965), the resultant overtraining might have facilitated transfer.

In order to test this hypothesis, we designed a key-pecking task in which the number of responses required to reach criterion was made comparable to the number of responses demanded in the jumping-stand situation (Graves and Goodale 1979). This was accomplished by requiring only a single peck for reinforcement, and by using a correction procedure rather than the 3-minute VI schedule used by Catania

(1965). The situation was made even more comparable to the jumping-stand by using a two-choice simultaneous discrimination instead of the successive discrimination procedure used by Catania. The same four-stage training paradigm (figure 7.2) employed in the jumping-stand experiments was used to evaluate IOT.

As table 7.2 shows, pigeons trained to discriminate a red trans-illuminated key from a green one made as few errors in reaching criterion as other pigeons did in learning the same task on the jumping stand. Yet the five birds in this experiment quite plainly showed excellent IOT between stages 1 and 2 and between stages 3 and 4.

In a second experiment, the similarity between the Skinner-box and jumping-stand situations was further increased by adding a 10-cm partition between the two keys and using a noncorrection training procedure. Despite these extensive alterations to Catania's basic key-pecking task, five birds trained on a triangle-circle discrimination continued to show evidence of good IOT between stages 1 and 2 and between stages 3 and 4 (table 7.3).

The findings of these two experiments together with the results ob-

Table 7-2
Red-green discrimination performance in Skinner box (correction training).

Pigeon	Stage 1		Stage 2		Stage 3		Stage 4	
	Persev-erative Errors	Days	Persev-erative Errors	Days	Persev-erative Errors	Days	Persev-erative Errors	Days
1	21	1	6	1	67	2	22	2
2	49	2	3	1	197	4	42	3
3	84	2	5	1	91	2	9	1
6	37	2	12	1	171	2	8	1
21	62	2	25	2	217	4	5	1
$\overline{X}$	50.6	1.8	10.2	1.2	148.6	2.8	17.2	1.6

Table 7.3
Triangle-circle discrimination performance in the Skinner box (non-correction training).

Pigeon	Stage 1		Stage 2		Stage 3		Stage 4	
	Errors	Days	Errors	Days	Errors	Days	Errors	Days
2	285	7	10	1	354	11	133[a]	5[a]
17	623	15	60	3	389	10	38	3
23	211	8	106	3	238	8	132	4
24	134	4	30	2	229	5	20	2
$\overline{X}$	313.25	8.5	51.5	2.25	302.5	8.5	80.75	3.5

a. There was an unavoidable 10-day hiatus during the stage 4 training of this pigeon.

tained from the jumping stand strongly suggest that neither of Catania's (1965) proposed explanations (postural habits or difference in the number of responses) can account for the failure of transfer in the jumping stand or its success in the Skinner box. It would appear that a nontrivial difference in transfer exists and that monocularly learned visual discriminations do not transfer to the other eye on the jumping stand, whereas they do in the Skinner box.

Retinal Locus as a Factor in Interocular Transfer

A major difference between the Skinner-box and jumping-stand experiments may lie in the part of the retina subtended by the discriminative stimuli. Before discussing the nature of the difference, it is necessary to review briefly the evidence for a distinction between the frontal and lateral visual fields.

Accommodation Differences Between Lateral and Frontal Fields
It has been generally observed that many birds appear to scrutinize distant objects monocularly with one of their lateral fields (Pumphrey 1948). This observation has led a number of investigators to examine avian optics to see whether the frontal and lateral visual fields differ with respect to the near point of accommodation.

Catania (1964) put forward a variety of indirect evidence suggesting that the frontal field is myopic and the lateral fields hypermetropic in the pigeon. He found that pigeons failed to learn a pattern discrimination when the stimuli were presented either laterally very close to the eye, or frontally but at a distance behind transparent stimulus keys. He then speculated that the division of the visual field by accommodation distance into two monocular lateral fields and one binocular frontal field might affect the posture of the pigeon in different discrimination tasks.

The determination of the near point of accommodation for the lateral fields of the pigeon had been the subject of several earlier experiments. Using a jumping stand, Hamilton and Goldstein (1933) and Chard (1939) concluded that acuity decreased for targets nearer than 30–40 cm. These experiments probably involved only lateral viewing.

When P. Blough (1971) measured visual acuity at 73 cm, the pigeons adopted a very pronounced sideways stance, which permitted them to fixate the stimuli with the lateral fields. Blough remarked that "pigeons are commonly observed to look at nearby stimuli (such as food) with frontal, binocular regard and distant stimuli with lateral, monocular regard" (Blough 1971, p. 65).

Blough (1973) went on to investigate the effect of increasing and decreasing viewing distance on the visual acuity of pigeons. She found that acuity was worst at intermediate distances of about 26 cm. At either lesser or greater distances, acuity improved. However, when the pigeons were fitted with goggles that restricted their viewing to the frontal field, they did not show increasing acuity with increasing distance. This last finding provides some confirmation for Catania's (1964) hypothesis.

The difference between lateral and frontal fields has also been examined retinoscopically and by analysis of other optical data. Mello (1966 and pers. comm.), using retinoscopic examination, found no difference in accommodative power between the lateral and frontal fields. On the other hand, Millodot and P. Blough (1971) reported retinoscopic evidence for nonuniformity in refractive characteristics at different retinal locations, although their data suggest that the difference between the two areas is slight. Using ray-tracing computations based on measurements from frozen eyes, Nye (1973) calculated that the near point of the pigeon eye is at a greater distance in a lateral direction than it is in a frontal direction (or, for that matter, in a posterior direction). From these calculations, he concluded that the greatest distance at which objects would be precisely in focus is 2–3 cm in the anterior field and 4–5 cm in the lateral. Unfortunately, the accommodative state of these eyes at the time of freezing was unknown. Indeed, this degree of myopia seems unreasonable, since the distance from the end of the beak to the anterior surface of the cornea is about 4 cm in most pigeons.

Using high-speed motion-picture film to analyze the location of the pigeon's head and eyes when it inspects and pecks an illuminated key, Hodos et al. (1976) estimated the near point in the frontal field to be 5.85 cm from the anterior surface of the cornea.

Summary
It seems likely that Catania's suggestion that the pigeon is relatively myopic in the anterior field and hypermetropic in the lateral fields is substantially correct. At the same time, it seems equally unlikely that this relative difference could fully account either for the failure of IOT in Levine's or our studies or for the poor performance of pigeons using their lateral fields in Catania's studies.

Red and Yellow Fields
In contrast with Catania's studies, Nye's (1973) experiments suggest that the deficient discrimination performance of pigeons using their lateral visual fields arises, not because of poor optics at the distances used, but rather from limitations in the way visual information is used or organized at more "central" levels of the nervous system. In a series of experiments, Nye was unable to train his pigeons to sustain a discrimination learned in the anterior field when the stimuli were moved to the lateral field. This decline in performance was observed even when the stimuli were moved laterally in small steps and training was continued to asymptote at each step. More important, the loss in performance occurred with both a color discrimination and suprathreshold luminance discrimination, neither of which would have re-

quired precise focusing by the pigeon. These observations led Nye to propose that "pigeons do not possess the neural capability required to learn to use information contained in laterally located stimuli to *directly* control pecking behavior" (Nye 1973, p. 570).

Furthermore, although the film analysis of the scanning behavior on the jumping stand in our own experiments revealed that the birds often viewed the discriminative stimuli with their lateral visual fields, on many trials they jumped toward the correct platform even though the correct stimulus had not fallen on any part of the visual fields beyond 40° lateral from the axis of the beak. Thus, on many occasions the stimuli would be falling well outside either of the central foveas, which are located on the optic axes at approximately 75° lateral from the beak. However, the central fovea may not be as important for visual acuity in the pigeon as it is in primates. Although pigeons certainly use information from the lateral visual fields in go/no-go tasks of visual acuity, there appears to be no difference in the threshold performance of intact birds and birds whose foveas have been coagulated with a laser (Blough 1971, 1973). Nonetheless, when the film records of scanning behavior on the jumping stand were compared with those taken of birds in a key-pecking task, lateral viewing was observed more often in the jumping stand than in the Skinner box. But the most striking difference between the two tasks was related not to the lateral position of the stimuli, but to the location of the stimuli along the dorsal-ventral axis of the visual fields.

On the jumping stand, the pigeon sits on the perch with its head in the usual resting position, so that the angle between the long axis of the beak and the horizon is about 25° down (Duijm 1951; Goodale and Graves, unpublished observations). With the head held in this position, the discriminative stimuli would fall on the lower quadrants of the retina, in the so-called yellow field (named after the color of the oil droplets in the cones of this field). On the other hand, in the Skinner box there is some evidence to suggest that the pigeon views the stimulus keys with another part of the visual fields outside the yellow area. In a recent experiment, we filmed pigeons while they were pecking at keys in a simultaneous red-green discrimination. A frame-by-frame analysis of the film record showed that the pigeons either raised or tilted their heads in such a way that the stimulus key would often fall on the dorsal posterior quadrant of the retina, in the so-called red area, where the cones contain mainly red oil droplets. This area, which subserves the binocular portion of the lower frontal field of vision, has a relatively high tectal magnification factor and a retinal ganglion-cell density comparable to that of the central fovea (Whitteridge 1965; Galifret 1968; Binggelli and Paule 1969; Nye 1973; Yazulla 1974; Clarke and Whitteridge 1976), which suggest an area specialized for acute vision, comparable to the fovea itself.

The existence of two areas of specialization in the pigeon retina and the evidence that the areas differ in near point of accommodation suggest the possibility that the pigeon has two visual systems for controlling different patterns of behavior. Nye's (1973) experiments showing that laterally presented visual information cannot be used to control pecking directly are in good agreement with such a distinction, and suggest even further that information arriving from these two areas of the retina may be processed differently at higher levels. In this context, Levine's (1945b; 1952) long-ignored distinction between stimuli presented in front of the beak and stimuli presented below the beak assumes great relevance. Certainly stimuli presented below the beak would stimulate the red area, whereas stimuli presented in front of the beak would fall outside this area (if the bird adopts a normal resting posture with the axis of its beak pointing 25° below the horizon). Similarly, in Catania's (1965) key-pecking task, even though the stimuli were presented vertically in front of the pigeons, they may nevertheless have raised or tilted their heads in such a way that the stimuli fell on the red field.

It is of considerable interest that the visual field that projects onto the red area includes the tip of the beak as well as the field below the beak. By using relatively small stimuli at which the pigeon pecked directly, Romeskie and Yager (1976a,b) obtained spectral-sensitivity curves that differed significantly from those obtained by Blough (1957), who had used stimuli located several centimeters *above* the key the pigeons were required to peck. The curves generated from the experiment of Romeskie and Yager are very similar to those plotted from the average spectral sensitivities of units in the pigeon's optic tectum whose receptive fields lie in the red area, whereas Blough's curves correspond to the functions obtained from units with receptive fields in the yellow area of the retina (King-Smith, unpublished observation cited in Muntz 1972). Thus, even though the birds were free to move their heads in both experiments, Romeskie and Yager argued that stimuli in their experiments were projected onto the red field whereas in Blough's experiment the stimuli were probably confined to the yellow field. More recently, Martin and Muntz (1978) obtained two separate photopic spectral-sensitivity curves from the same pigeons by projecting the test stimuli onto either the red or the yellow fields by means of fiber-optic light guides. Their two curves correspond to Romeskie and Yager's (1976a,b) and Blough's (1957) results, respectively.

Failure of Interocular Transfer in the Skinner Box

On the basis of our own film records and by extrapolation from the psychophysical literature reviewed above, it could be argued that in all situations in which interocular transfer has been clearly demonstrated—such as Köhler's (cited in Levine 1945a) and Levine's (1945b) pecking tasks and Catania's (1965) Skinner-box experiments, as well as our own—the image of the discriminative stimuli fell upon the

Table 7.4
Performance in Skinner box on yellow-blue discrimination with 10-cm stimulus-response separation.

Pigeon	Stage 1		Stage 2	
	Days	Errors	Days	Errors
6	31	1155	25	960
17	35	1467	17	491
18	14	481	12	476
22	11	342	33	1430
$\overline{X}$	22.75	861.25	21.75	839.25

red area of the retina, whereas in the jumping-stand experiments carried out by Levine and us in which no IOT was found, the red area was not stimulated, but rather a portion of the yellow areas located more ventrally. It follows that a necessary condition for IOT in the pigeon may be that the image of the discriminative stimuli fall on the red area.

In order to evaluate this hypothesis directly, we devised a key-pecking task in which the discriminative stimuli were presented on small panels located 10 cm above the response keys. Thus, the stimuli were still presented in the vertical plane and the birds were still required to peck at the response keys, but now, unless the birds tilted their heads high in the air, the stimuli would fall well outside the field of the red area and in the yellow fields. The results from four pigeons trained to criterion monocularly on a simultaneous blue-yellow discrimination and then retrained using the other eye are presented in table 7.4.

The 10-cm stimulus-response separation made the task extremely difficult for the pigeons. This difficulty may be related to Nye's (1973) suggestion that pigeons could not learn to use information presented to the lateral fields to control pecking. In both cases, the stimuli would have been projected onto the yellow area, a part of the retina that may not send projections to those visuomotor pathways directly involved in the control of pecking.

In days and in errors to criterion there was no significant difference between stage 1 and stage 2. In distinct contrast with the usual finding in the Skinner box, there was no evidence of IOT. This conclusion is strengthened by an additional observation: One bird accidentally received 50 trials on day 4 of stage 2 with the wrong eye open (the eye that had already been used in stage 1). With this eye, the pigeon performed at a level of 88% correct. When the mistake was discovered, the cover was switched to the correct eye, and the training was continued, the pigeon's performance fell immediately to 50% correct for the remaining 50 trials that day. Thus, even the experience of using the "trained" eye to control pecking after several days of using the less-experienced eye produced no evidence of transfer whatsoever when the eye cover was immediately switched.

These results lend further support to Levine's original contention (and the argument developed in this chapter) that the area of the retina stimulated is an important determinant of the success or failure of IOT.

Successful Interocular Transfer on the Jumping Stand

By placing the stimuli on the platforms, so that they were located well below the beak of the pigeon, Levine was able to demonstrate successful IOT on the jumping stand. These findings were replicated some years later by Siegel (1953a,b), who trained ring doves on a similar sort of jumping stand with the stimuli presented horizontally below the doves on the platforms.

In our own jumping-stand experiments in which the stimuli were presented in the vertical plane in front of the animal, individual pigeons would occasionally show evidence of IOT, even though in the vast majority of pigeons no IOT was demonstrated. It is possible that those rare individuals that exhibited some transfer may have scanned the array in such a way that the stimuli fell on the red area for a significant proportion of the time. Although there is no systematic evidence to support such a proposal, several observations provide at least some confirmation. The few available film records show that for one pigeon with an unusually high error-savings score of 52% between stages 1 and 2, the stimuli were within the red field a remarkable 77% of the time before the bird jumped. In contrast, another pigeon, which showed no evidence of transfer at all, scanned the stimulus cards in such a way that the discriminative stimuli were within the red field on only 1.5% of the frames of film that we analyzed.

In these experiments we had forced the pigeon to jump onto one of the platforms by rotating the perch in such a way that the pigeon was carried under its own momentum toward the platforms and the stimulus cards. One very simple way to change the angle at which the head is held while the bird is on the perch, and thereby to increase the chances of the discriminative stimuli falling within the red field, is to reverse the direction of rotation of the perch. Four pigeons were trained in this manner on a blue-yellow discrimination using exactly the same jumping-stand apparatus that had already yielded no evidence of IOT in a number of previous experiments. Whereas the average number of errors to criterion on stage 1 of the previous blue-yellow discrimination had been 18.25, the average number for these pigeons experiencing reverse rotation was 35.5. Although the mean error-savings score between stages 1 and 2 in the previous experiment had been only 5.8%, merely changing the direction of perch rotation increased the savings score to 65.7%. Unfortunately, the film records of head position during this last experiment were incomplete. Nevertheless, the rank-order correlation between the amount of time the discriminative stimuli fell within the red field and the error-savings scores was +0.6. Although this experiment is far from satisfactory, it does provide some additional validation for the suggestion that the red area and its projections to higher levels of the nervous system play a crucial role in IOT.

Physiological Considerations

Why should a visual-discrimination task in which the stimuli fall within the red field of one eye transfer so easily to the other eye? Perhaps the answer to this question lies in the binocularity of the red fields.

Even though the retinofugal projections in the pigeon are completely crossed, binocular convergence on the same neuronal population in a single hemisphere is possible via postchiasmal decussations and commissures. For example, the thalamic target of the optic tract, the nucleus opticus principalis thalami, has bilateral projections to the "visual Wulst" of the telencephalon in both the pigeon and the owl (Karten et al. 1973). Single neurons in the Wulst of the owl have been shown to have a high degree of binocular interaction, and many of these units are highly tuned for retinal disparity (Pettigrew and Konishi 1976a). In the pigeon, binocular units have also been recorded in the visual Wulst (Perišić et al. 1971). Thus, neuronal populations in both hemispheres would be activated by stimulation of one eye if that stimulation includes the binocular field. Since the red area receives the image of the binocular field in the region of the beak, binocular convergence of input from this retinal area in the Wulst would provide a neural substrate for stereopsis, an important depth cue for the control of pecking.

As Hamilton points out in this volume, IOT in an intact animal can be explained most parsimoniously in terms of these binocular convergence mechanisms. In the case of the monocular pigeon, discriminative stimuli falling within the normally binocular red field of the one eye available to the bird would activate binocular cells in both hemispheres. In this way, both hemispheres would participate in the learning of the visual-discrimination habit even though only one eye was exposed to the stimuli. The switching of the blindfold to the "trained" eye would pose no special problem for the bird, since the same binocular neurons in both hemispheres could be activated by the "naive" eye. This explanation of IOT in the pigeon is compelling, as it relies on known anatomical connections and does not depend on sophisticated (and unknown) mnemonic transfer mechanisms. Transfer simply results from the normal mechanisms underlying binocular perception. Some additional support for this hypothesis can be derived from the results of several lesion experiments in birds.

The binocular cells in the Wulst receive input from the contralateral nucleus opticus principalis thalami (and thus from the ipsilateral eye) via the dorsal supraoptic decussation (Karten et al. 1973). Although sectioning the tectal, posterior, anterior, or pallial commissures has been found to have no effect on IOT in pigeons, transection of the dorsal supraoptic decussation in both juvenile and adult pigeons prevents the IOT of color and pattern discriminations learned subsequently (Chichinadze 1940; Beritov 1965; R. Meier 1971; R. Meier et al. 1972; V. Meier 1976, 1979; Burkhalter and Cuénod 1978). O'Connell (1979) has demonstrated that transection of the dorsal supraoptic decussation *after* a domestic chick has learned a color discrimination monocularly does not interfere with the successful transfer of the task to the other eye. These experiments suggest that interference with binocular convergence in the Wulst by removing input from the ipsilateral eye eliminates IOT of tasks learned after surgery, and that once the information has been acquired by both hemispheres via monocular exposure, subsequent interruption of converging input does not interfere with IOT.

Burkhalter and Cuénod (1978) have also shown in a series of monocular-deprivation experiments that the dorsal supraoptic decussation is implicated in both IOT and the development of normal binocular integration of input from the two eyes. They found that monocular deprivation during the first few months of life not only interfered with the pigeon's later ability to learn pattern discriminations with the deprived eye, but also severely disrupted transfer of the pattern discrimination from the deprived to the experienced eye. Since transection of the dorsal supraoptic decussation in the newly hatched squab later resulted in normal discrimination performance with monocular deprivation, Burkhalter and Cuénod argued that the impaired visual discrimination and IOT in intact monocularly deprived pigeons can be explained by an asymmetrical development of the retino-thalamo-hyperstriatal projections due to binocular competition in the Wulst. This hypothesis receives further support from the finding of Pettigrew and Konishi (1976b) that owls that had been monocularly deprived as young animals lacked binocular cells in the Wulst. The results of these experiments are in close agreement with the hypothesis that binocular convergence of input from the red field mediates transfer of discrimination training to the untrained eye in monocularly trained pigeons.

Other Behavioral Investigations of Interocular Transfer
Several studies appear to suggest that projection of the discriminative stimuli onto the red area of the retina is not sufficient to ensure IOT. Zeier (1976) has shown that pigeons trained monocularly on a sequential key-pecking task in which they were required to peck the left of two keys first and then the right did not show any evidence of IOT. He argued that task difficulty was the principal reason IOT was impaired. This contrasts with the results of our experiments, in which manipulating task difficulty had no effect on the degree of transfer.

More recently, Green et al. (1978) found that although pigeons showed excellent IOT of a simultaneous visual discrimination, they did not show any transfer of a spatial conditional habit in which the pigeons were trained monocularly to peck the left key when a particular stimulus (color or pattern) was presented on both keys and to peck the right key when a different stimulus was presented. Those authors suggested that the nature of the task itself—independent of either task difficulty or the location of the stimulus in the bird's visual field—was yet another important variable in IOT.

Although it is possible that task difficulty as well as other more subtle characteristics of the visual-discrimination problem may influence the amount of transfer from one eye to the other, the two experiments described above share a common problem in design that makes this interpretation of the results questionable: In both the sequential task and the spatial conditional problem the pigeons were required to discriminate left from right, not on the basis of differential cues present on the keys themselves, but on the basis of the position of the keys with respect to the pigeon and/or the apparatus. For a pigeon with one eye covered, the left-right distinction may be coded in terms of the initial location of the key in the visual field before the bird turns toward it and pecks at it. If this were the case, then moving the blindfold to the other eye would create not only a very different visual field but also a serious recoding problem for the bird. Since the pigeon has very laterally placed eyes, the visual field of one eye for any given head position is very different from that of the other eye. Thus, the apparently poor IOT of a spatial conditional problem may have arisen for exactly the same reason IOT seemed deficient in birds trained on a sequential pecking task—the birds would have been forced to relearn the distinction between left and right when the blindfold was switched. In addition, it is possible that a differential head turn toward the left or the right might have been initiated by the image of one of the keys falling quite laterally within the yellow field. The problem of left-right coding, with its inherent mirror-image symmetry (a quality that makes the distinction often a difficult one for bilaterally symmetrical animals [Corballis and Beale 1970]), could have been avoided had the keys been positioned one above the other. Since the distinction between up and down remains unchanged (in terms of relative visual-field position) when the blindfold is switched, IOT scores can be more easily and unambiguously interpreted.

Recently, Green (pers. comm.) has attempted to meet these criticisms directly by training pigeons on a spatial conditional problem in which the choice keys are positioned one above the other. Again, he finds no evidence of IOT. However, even in this "improved" situation, the choice keys are both located to the right (or left) of an observing key, instead of being positioned above and below this central key. Thus, it is still possible for the "correct" responses to be coded asymmetrically in the visual fields of the two eyes and to require recoding when the blindfold is switched from one eye to the other.

On the basis of a series of rather complex experiments (Palmers and Zeier 1974; Zeier 1976), it has been suggested that conflicting information presented simultaneously to both hemispheres can affect the nature of interhemispheric communication. Palmers and Zeier have speculated that in certain situations one hemisphere can "dominate" the other, so that visual information presented directly to that hemisphere will suppress information arriving from the other. By using goggles fitted with colored filters, they were able to train pigeons on simultaneous or successive color discriminations in a Skinner box in such a way that each stimulus would be seen by only one eye. In the simultaneous condition, Palmers and Zeier argued, the conflicting information in the two hemispheres resulted in one hemisphere "dominating" the other. Using the same goggle-and-filter arrangement, Graf and Zeier (1979) showed that pigeons could learn to integrate information presented to each eye separately in order to solve a discrimination problem. Though these experiments illustrate that the relationship between the two hemispheres is complex and poorly understood, they do not speak directly to the hypothesis that stimulation of the binocular red field is an important prerequisite for successful IOT in the pigeon.

A number of authors have investigated IOT in chicks and ducklings, using tasks very different from the familiar two-choice visual discrimination tasks commonly employed to measure IOT in adult birds.

Chicks reared on the deep side of a visual cliff lose the typical depth-avoidance response (Tallarico and Farrell 1964). Although chicks reared in this situation with one eye covered showed habituation with the trained eye, this modification of the response did not transfer—with the naive eye they still showed depth avoidance (Zeier 1970).

Newly hatched chicks presented with a small target will peck at it without prior experience, but this response is abolished in only one trial if the target (a small metallic bead) has been coated with a highly distasteful liquid (Lee-Teng and Sherman 1966). Chicks that had undergone this training with one eye occluded avoided pecking at the target even when the occluder was transferred to the other eye (Cherkin 1970). However, a chick exposed monocularly to repeated presentations of an uncoated target until the passive avoidance response had extinguished would still avoid pecking at the target with the other eye (Benowitz 1974).

In recent years, Bell and his colleagues have suggested that the "engram" for this kind of one-trial passive avoidance learning in the chick may be stored unilaterally. Using intracranial injections of ouabain or cycloheximide (which interfere with the sodium pump and protein synthesis, respectively), Bell and Gibbs (1977) found that injections contralateral to the trained eye immediately after learning suppressed

IOT, and that injections to the other hemisphere did not. Sectioning the dorsal supraoptic decussation 4 hours after monocular passive avoidance training interfered with the performance of the chicks when they were subsequently tested with the trained eye, although the sham-operated control group showed excellent IOT (Bell and Ehrlich 1979). These results, together with the findings of a recent investigation of time-dependent factors affecting the induced "amnesia" (Bell and Gibbs 1979), strongly suggest that the hemisphere contralateral to the trained eye may be more important for one-trial passive avoidance learning in the chick than the other. However, the relationship between these observations and possible binocular convergence mechanisms is not well understood. Bell, Ehrlich, and Howard (1979) have begun to investigate what areas of the retina are used by chicks pecking at the small targets in the one-trial passive avoidance task, and as these investigations are continued the role of the different retinal areas in IOT of this particular task will become increasingly clear.

The imprinting literature is confused with respect to IOT. Klopfer (1973), using a two-choice approach test with ducklings that had previously been imprinted monocularly to a model, found little evidence for transfer. On the other hand, Moltz and Stettner (1962), who used a following-of-model measure, did report finding IOT of imprinting. Bateson (cited in Klopfer 1973) has suggested that length of exposure is the critical factor; that the longer the exposure time, the greater the amount of transfer.

Demarest et al. (1977) argued that duck decoys have visual elements that elicit innate preferences from ducklings, and that it is therefore better to use more "neutral" objects as imprinting stimuli when measuring transfer. These authors found IOT of imprinting to neutral models in a runway, but reported that when using the trained eye the ducklings were better at maintaining proximity to the models. They concluded, as Zeier (1970) had, that tasks for which there is a strong inherent disposition, such as imprinting, are transferred from one eye to the other more easily than reactions that counter to such predispositions.

In the course of investigating biochemical changes in the central nervous system consequent on imprinting, Horn and co-workers (Horn et al. 1971, 1973; Horn 1977) found that monocularly imprinted chicks showed similar incorporation of tritiated uracil into RNA in the forebrain roof in both hemispheres. However, transection of the supraoptic decussation restricted these changes to the hemisphere contralateral to the imprinted eye; this suggested that binocular convergence mechanisms may also explain IOT of imprinting.

The studies described in this section indicate that a number of different factors, such as task difficulty, exposure time, the innate response tendencies of the animal, and the nature and amount of visual "information" presented to each hemisphere, may play crucial roles in IOT.

However, it is not always clear whether those birds that show good evidence of IOT are simply scanning the visual stimuli differently than those that show little or no evidence of IOT. Since our work and that of many other investigators reviewed in previous sections strongly suggests that retinal locus is a powerful determinant of transfer in the pigeon, it becomes necessary to control or provide direct evidence about the location of visual stimuli within a bird's visual field before invoking speculation about other "higher-order" explanations of transfer or its failure.

The Problem of the Yellow Fields

Although convergence of input from binocular fields can account for IOT when the discriminative stimuli fall within the normally overlapping portion of the visual field, IOT is possible in some animals when no opportunity for low-order binocular convergence exists. In split-chiasm monkeys, IOT can still be explained by binocular convergence of input in a single hemisphere via callosal connections in the splenium (Hamilton, this volume). However, monkeys with transections of both the chiasm and splenium still show IOT of object learning sets, but not individual problems, if the rostral corpus callosum and anterior commissures are intact (Noble 1973). Thus, it appears that interhemispheric transfer of some sort of higher-order mnemonic function may take place by means of a transfer mechanism that is different in some fundamental way from simple binocular convergence (Hamilton, this volume).

The pigeon is very different from the monkey in that the largest portion of the visual field of each eye is monocular. Thus, throughout vast areas of the representation of each eye's field in the central nervous system, convergence of input of the sort demanded from the binocular fields for perceptual analysis is unnecessary. Since binocular convergence of input from the monocular yellow fields is absent in the pigeon, it perhaps is not surprising that IOT of visual-discrimination problems learned monocularly using this portion of the field is also absent. It is possible that interhemispheric transfer of a higher-order nature similar to that observed in monkeys (Noble 1973; Hamilton, this volume) could take place in the pigeon, even for visually guided behavior patterns learned using the monocular yellow fields. However, in our own experiments using the jumping stand and in Levine's (1945a,b; 1952) earlier studies there is no evidence for IOT of specific visual-discrimination problems.

Even pigeons that are learning a visual-discrimination task on the jumping stand with both eyes open often scan the stimulus cards with one eye more than the other, and consequently learn the discrimination with only one eye-hemisphere system (Goodale and Graves 1979). Similar findings have been reported for at least one mammalian species with large monocular fields: In rabbits, which normally show poor IOT (Van Hof 1970), some individuals that had been trained on a discrimination problem with both eyes uncovered and then retrained to the same criterion monocularly sometimes showed retention of the task with one eye but not with the other (Van Hof and Van der Mark 1976).

In animals with panoramic vision, such as the pigeon and rabbit, the lateral visual fields normally represent two very different views of the world. Perhaps if only one of these fields includes stimuli that are controlling behavior, some sort of selective attention and/or unilateral storage of information operates. In other words, the failure of IOT in the pigeon when stimuli fall outside the red field may be due not simply to the absence of binocular convergence but also to an active mechanism that allows input to one hemisphere to inhibit activity in the other. Certainly, there is physiological evidence for strong interhemispheric inhibition in the pigeon (Robert and Cuénod 1969). However, any further speculation about transfer mechanisms and selective attention mechanisms superordinate to explanations expressed in terms of direct convergence of input from overlapping fields would be premature.

Summary and Conclusions

Throughout this chapter we have developed the argument that interocular transfer of a visual-discrimination habit in the pigeon can only occur if the discriminative stimuli fall within the red fields. We have also suggested that IOT in a monocularly trained bird is therefore a straightforward consequence of the stimulation of pathways to both hemispheres from the binocular portion of the open eye's field— pathways that normally carry converging binocular information for the perception of depth. The absence of IOT in certain situations, such as on the jumping stand, is a consequence of the discriminative stimuli falling within the monocular field, where there is no necessity for later convergence of input from the two eyes. It is also possible that active inhibition mechanisms serve to lateralize input from one monocular field in animals, such as the pigeon, that have wide panoramic vision. Although it is very likely that factors other than retinal locus are important determinants of IOT, far too little attention has been paid to this variable in discussions of interocular and interhemispheric interactions in birds.

References

Bell, G. A., and D. Ehrlich. 1979. "Engram lateralization in the intact and split-brain chick." *Soc. Neurosci. Abstr.* 5: 313.

Bell, G. A., and M. E. Gibbs. 1977. "Unilateral storage of monocular engram in day-old chick." *Brain Res.* 124: 263–270.

———. 1979. "Interhemispheric engram transfer in chicks." *Neurosci. Lett.* 13: 163–168.

Bell, G. A., D. Ehrlich, and J. Howard. "Retinal input in the chick and its possible role in interocular transfer." In *Proceedings of the Australian Experimental Psychology Conference, 1979.*

Benowitz, L. 1974. "Conditions for the bilateral transfer of monocular learning in chicks." *Brain Res.* 65: 203–213.

Beritov, J. S. 1965. *Neural Mechanisms of Higher Vertebrate Behaviour,* W. L. Liberson, transl. Boston: Little, Brown.

Beritov, J. S., and N. Chichinadze. 1936. "Localization of visual perception in the pigeon." *Bull. Biol. Med. Exp.,* 2: 105–107.

Binggeli, R. L., and W. J. Paule. 1969. "The pigeon retina: Quantitative aspects of the optic nerve and ganglion cell layer." *J. Comp. Neurol.* 137: 1–18.

Blough, D. S. 1957. "Spectral sensitivity in the pigeon." *J. Opt. Soc. Am.* 47: 827–833.

Blough, P. M. 1971. "The visual acuity of the pigeon for distant targets." *J. Exp. Anal. Behav.* 15: 57–67.

———. 1973. "Visual acuity in the pigeon. II. Effects of target distance and retinal lesions." *J. Exp. Anal. Behav.* 20: 333–343.

Burkhalter, A., and M. Cuénod. 1978. "Changes in pattern discrimination learning induced by visual deprivation in normal and commisurotomized pigeons." *Exp. Brain Res.* 31: 369–385.

Catania, A. C. 1964. "On the visual acuity of the pigeon." *J. Exp. Anal. Behav.* 7: 361–366.

———. 1965. "Interocular transfer of discriminations in the pigeon." *J. Exp. Anal. Behav.* 8: 147–155.

Chard, R. D. 1939. "Visual acuity in the pigeon." *J. Exp. Psych.* 24: 588–608.

Cheney, C. D., and V. Tam. 1972. "Interocular transfer of a line tilt discrimination without mirror-image reversal using fading in pigeons." *J. Biol. Psychol.* 14: 17–20.

Cherkin, A. 1970. "Eye to eye transfer of an early response modification in chicks." *Nature* 227: 1153.

Chichinadze, N. 1940. "Das Problem der Localization kortikaler Prozesse, welche durch optische Reize hervorgerufen werden." *Mitt. Akad. Wiss. Georg. S.S.R.* 1: 609–614.

Clarke, P. G. H., and D. Whitteridge. 1976. "The projection of the retina, including the "red area," onto the optic tectum of the pigeon." *Q. J. Exp. Physiol.* 61: 351–358.

Corballis, M. C., and I. L. Beale. 1970. "Bilateral symmetry and behavior." *Psychol. Rev.* 77: 451–464.

Cuénod, M. 1972. "Split-brain studies: Functional interaction between bilateral central nervous structures." In *The Structure and Function of Nervous Tissue,* volume 5, G. H. Bourne, ed. New York: Academic.

Cuénod, M., and H. Zeier. 1967. "Transfer interhemispherique et commissurotomie chez le pigeon." *Schweiz. Arch. Neurol. Neurochir. Psychiat.* 100: 365–380.

Demarest, J., N. Brecha, and R. Bronstein. 1977. "Interhemispheric mediation of the following response in ducklings." *Physiol. Psychol.* 5: 378–382.

Diebschlag, E. 1940. "Über den Lernvorgang bei der Haustaube." *Z. vergl. Physiol.* 28: 67–104.

Duijm, M. 1951. "On the head position in birds and its relation to some anatomical features. I." *Proc. koninklijke Nederlandse akadamie van wetenschappen C* 54: 202–211.

Galifret, Y. 1968. "Les diverses aires fonctionelles de la rétine du pigeon." *Z. Zellforsch.* 86: 535–545.

Goodale, M. A., and J. A. Graves. 1980. "The relationship between scanning patterns and monocular discrimination learning in the pigeon." *Physiol. Behav.* 25: 39–43.

Graf, M., and H. Zeier. 1979. "Interhemispheric integration of visual information in the pigeon." In *Structure and Function of the Cerebral Commissures*, I. Steele-Russell et al., eds. London: Macmillan.

Graves, J. A., and M. A. Goodale. 1977. "Failure of interocular transfer in the pigeon (*Columba livia*)." *Physiol. Behav.* 19: 425–428.

———. 1979. "Do training conditions affect interocular transfer in the pigeon?" In *Structure and Function of the Cerebral Commissures*, I. Steele-Russell et al., eds. London: Macmillan.

Green, L., N. Brecha, and M. S. Gazzaniga. 1978. "Interocular transfer of simultaneous but not successive discriminations in the pigeon." *Animal Learning Behav.* 6: 261–264.

Halstead, W., and G. Yacorzynski. 1938. "A jumping method for establishing differential responses in pigeons." *J. Genet. Psychol.* 52: 227–231.

Hamilton, W. F., and J. L. Goldstein. 1933. "Visual acuity and accommodation in the pigeon." *J. Comp. Psychol.* 15: 193–197.

Hodos, W., R. Leibowitz, and J. C. Bonbright. 1976. "Near-field visual acuity of pigeons: Effects of head location and stimulus luminance." *J. Exp. Anal. Behav.* 25: 129–141.

Horn, G. 1977. "Neural studies of imprinting in the chick." Paper presented at European Brain Behavior Society Workshop on Structure and Function of the Cerebral Commissures, Rotterdam.

Horn, G., A. L. D. Horn, P. P. G. Bateson, and S. P. R. Rose. 1971. "Effects of imprinting on uracil incorporation into brain RNA in the "split-brain" chick." *Nature* 229: 131–132.

Horn, G., S. P. R. Rose, and P. P. G. Bateson. 1973. "Monocular imprinting and regional incorporation of tritiated uracil into the brains of intact and "split-brain" chicks." *Brain Res.* 56: 227–237.

Ingle, D. 1963. "Limits of visual transfer in goldfish." Ph.D. diss., University of Chicago.

———. 1968. "Interocular integration of visual learning by goldfish." *Brain, Behav., Evol.* 1: 58–85.

Karten, H. J., W. Hodos, W. J. H. Nauta, and A. M. Revzin. 1973. "Neural connections of the 'visual Wulst' of the avian telencephalon: Experimental studies in the pigeon (*Columba livia*) and owl (*Speotyto curicularia*)." *J. Comp. Neurol.* 150: 253–278.

Klopfer, P. H. 1973. "Imprinting: Monocular and binocular cues in object discrimination." *J. Comp. Physiol. Psychol.* 84: 482–487.

Konnerman, G. 1966. "Monokulare Dressur von Hausgänsen, z. t. mit entgegengesetzter Merkmalsbedeutung für beide Augen." *Z. Tierpsychol.* 23: 555–580.

Lee-Teng, E., and S. M. Sherman. 1966. "Memory consolidation of one-trial learning in chicks." *Proc. Nat. Acad. Sci.* 56: 926–931.

Levine, J. 1945a. "Studies in interrelations of central nervous structures in binocular vision. I. The lack of binocular transfer of visual discriminative habits acquired monocularly by the pigeon." *J. Genet. Psychol.* 67: 105–129.

———. 1945b. "Studies in the interrelations of central nervous structures in binocular vision. II. The conditions under which interocular transfer of discriminative habits takes place in the pigeon." *J. Genet. Psychol.* 67: 131–142.

———. 1952. "Studies in the interrelations of central nervous structures in binocular vision. III. Localization of the memory trace as evidenced by the lack of inter- and intraocular habit transfer in the pigeon." *J. Genet. Psychol.* 81: 19–27.

Maier, V. 1976. "Effekte unilateraler telencephaler und thalamischer Läsionen auf die monokulare masterdiskriminations Fähigkeit kommissurotomierter Tauben." *Rev. Suisse Zool.* 83: 59–82.

———. 1979. "Behavioral and anatomical aspects of monocular vision in birds." In *Structure and Function of the Cerebral Commissures*, I. Steele-Russell et al., eds. London: Macmillan.

Martin, G. R., and W. R. A. Muntz. 1978. "Spectral sensitivity of the red and yellow oil droplet fields of the pigeon (*Columba livia*)." *Nature* 274: 620–621.

Meier, R. E. 1971. "Interhemisphärischer Transfer visueller Zweifachwahlen bei kommissurotomierten Tauben." *Psychol. Forsch.* 34: 220–245.

Meier, R. E., M. J. Mihailovic̆, M. Peris̆ić, and M. Cuénod. 1972. "The dorsal thalamus as a relay in the visual pathways of the pigeon" (abstract). *Experentia* 28: 730.

Mello, N. 1966. "Concerning the interhemispheric transfer of mirror image pattern in pigeon." *Physiol. Behav.* 1: 293–300.

———. 1968. "The effect of unilateral lesions of the optic tectum on interhemispheric transfer of monocularly trained color and pattern discrimination in pigeon." *Physiol. Behav.* 3: 725–734.

Mello, N., F. R. Erwin, and S. Cobb. 1963. "Intertectal integration of visual information in pigeon: Electrophysiological and behavioural observation." *Bol. Inst. Estud. Mèd. Biol.* (Mexico) 21: 519–533.

Menkhaus, I. 1957. "Versuche über einäugiges Lernen und Transponieren beim Haushuhn." *Z. Tierpsychol.* 14: 210–230.

Millodot, M., and P. M. Blough. 1971. "The refractive state of the pigeon eye." *Vision Res.* 11: 1010–1022.

Moltz, H., and L. J. Stettner. 1962. "Interocular mediation of the following response after patterned-light deprivation." *J. Comp. Physiol. Psychol.* 55: 626–632.

Muntz, W. R. A. 1972. "Inert absorbing and reflecting pigments." In *Handbook of Sensory Physiology*, H. J. A. Dartnall, ed. Berlin: Springer.

Noble, J. 1973. "Interocular transfer in the monkey: Rostral corpus callosum mediates transfer of object learning set but not of single problem learning." *Brain Res.* 50: 147–162.

Nye, P. W. 1973. "On the functional differences between frontal and lateral visual fields of the pigeon." *Vision Res.* 13: 559–574.

O'Connell, N. 1979. "Role of the supraoptic decussation in interocular transfer in the chick: Age dependence?" Ph.D. diss., University of Rochester.

Ogawa, T., and S. Ohinata. 1966. "Interocular transfer of color discriminations in a pigeon." *Ann. Anim. Psychol.* 16: 1–9.

Palmers, C., and H. Zeier. 1974. "Hemispheric dominance and transfer in the pigeon." *Brain Res.* 76: 537–541.

Perišić, M., J. Mihailović, and M. Cuénod. 1971. "Electrophysiology of contralateral and ipsilateral visual projections to the Wulst in the pigeon *(Columba livia)*." *Intern. J. Neurosci.* 2: 7–14.

Pettigrew, J. D., and M. Konishi. 1976a. "Neurons selective for orientation and binocular disparity in the visual Wulst of the barn owl *(Tyto alba)*." *Science* 193: 675–678.

———. 1976b. "Effect of monocular deprivation on binocular neurons in the owl's visual Wulst." *Nature* 264: 753–754.

Pumphrey, R. J. 1948. "The theory of the fovea." *J. Exp. Biol.* 25: 299–312.

Robert, F., and M. Cuénod. 1969. "Electrophysiology of intertectal commissures in the pigeon. II. Inhibitory interaction." *Exp. Brain Res.* 9: 116–136.

Romeskie, M., and D. Yager. 1976a. "Psychophysical studies of pigeon color vision. I. Photopic spectral sensitivity." *Vision Res.* 16: 501–505.

———. 1976b. "Psychophysical studies of pigeon color vision. II. The spectral photochromatic interval function." *Vision Res.* 16: 507–512.

Schulte, A. 1957. "Transfer und transpositionsversuche mit monokulardressierten Fischen." *Z. Vergl. Physiol.* 39: 432–476.

Sheridan, C. L. 1965. "Interocular interaction of conflicting discrimination habits in the albino rat: A preliminary report." *Psychonom. Sci.* 3: 303–304.

Siegal, A. I. 1953a. "Deprivation of visual form definition in the ring dove. I. Discriminatory learning." *J. Comp. Physiol. Psychol.* 46: 115–119.

————. 1953b. "Deprivation of visual form definition in the ring dove. II. Perceptual-motor transfer. *J. Comp. Physiol. Psychol.* 46: 249–252.

Tallarico, R. B., and W. M. Farrell. 1964. "Studies of visual depth perception: An effect of early experience of chicks on a visual cliff." *J. Comp. Physiol. Psychol.* 57: 94–96.

Van Hof, M. W. 1970. "Interocular transfer in the rabbit." *Exp. Neurol.* 26: 103–108.

Van Hof, M. W., and F. Van der Mark. 1976. "A quantitative study on interocular transfer in the rabbit." *Physiol. Behav.* 16: 775–782.

Whitteridge, D. 1965. "Geometrical relations between the retina and the visual cortex." In *Mathematics and Computer Science in Biology and Medicine*. London: Medical Research Council.

Yazulla, S. 1974. "Intraretinal differentiation in the synaptic organization of the inner plexiform layer of the pigeon retina." *J. Comp. Neurol.* 153: 309–324.

Zeier, H. 1970. "Lack of eye-to-eye transfer of an early response modification." *Nature* 225: 708–709.

————. 1976. "Interhemispheric interactions." In *Neural and Endocrine Aspects of Behaviour in Birds*, P. Wright et al., eds. Amsterdam: Elsevier.

8 Mechanisms of Concept Formation in the Pigeon

John Cerella

It is easy to train a laboratory pigeon to distinguish one pattern from a set of patterns: The patterns are projected onto a small screen on the wall of an operant chamber, and pecking a key when a "positive" pattern is present is reinforced whereas pecking when "negative" patterns are showing is not. After a few sessions the pigeon's response comes to be controlled by the image on the screen: Pecking to positives is rapid, and pecking to negatives ceases. Positive patterns are thus "recognized" by the pigeon. We review here studies of how this is accomplished.

Abstract Patterns

We start with a series of studies employing a single geometric form as the positive pattern. These studies derive from the classical work of Klüver (1933) on the monkey and Lashley (1938) on the rat, in which, once a discrimination was established, test patterns were devised by altering characteristics of the positive. It was reasoned that if the altered characteristic was unnoticed or was overshadowed by unaltered characteristics, the test pattern would be treated as positive. If the altered characteristic was noticed and was prominent in comparison to unaltered characteristics, the test pattern would be treated as negative. Therefore, depending on whether responding extended to a test pattern or not, inferences could be made concerning the method by which one pattern was compared with another.

There have been technical advances since this work of the 1930s. Attneave and Arnoult (1956) published a method of generating indefinite numbers of geometric shapes by placing points randomly in the plane and connecting them to form irregular polygons. Posner developed a method of distorting such a shape by displacing each vertex a small variable amount to generate populations of shapes of varying degrees of similarity to the original (see Posner et al. 1967). Sources of positive, negative, and test patterns were thus readily available.

These methods were applied to the pigeon by Ferraro and Grisham (1972). Pigeons trained on the random polygon shown in figure 8.1

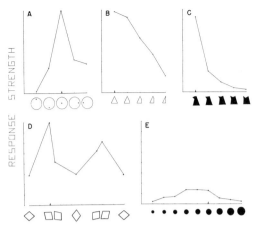

Figure 8.1 Pigeons' response to (A) translations, (B) truncations, (C) deformations, (D) rotations, and (E) dilations of a previously reinforced stimulus, indicated by mark. Data are from the first presentation(s) of the transformed stimuli; for C and B, pigeons were exposed to other unreinforced stimuli prior to those shown here. Stimulus transformations are measured by vertex displacement (C), degrees rotation (D), radius length (E), quadrant (A), and fraction of whole (B); response strength is measured by percent total responses (C, E), fraction of S+ responses (D), and probability of response (A, B). These are group curves, and in every case are representative of individual curves.

were tested on five transformations of this shape. Part C shows that response fell off sharply as the degree of distortion increased. Vetter and Hearst (1968) performed a similar experiment, transforming the training pattern by rotation in the plane rather than by vertex displacement in order to generate test patterns, and found that response to rotated figures was reduced (part D). Wildemann and Holland (1973) used a more regular training figure (a circle), and altered its size in testing; response to slightly smaller and larger circles was reduced (part E). Heinemann and Kadison (1976) used a spot as a training pattern, and varied its location during testing. Pigeons failed, for the most part, to respond to displaced spots (part A).

Taken together, these four studies suggest that the pigeon is sensitive to any alteration of a graphic configuration. I reached the same conclusion after analyzing responses to thousands of computer-generated distortions of a training figure (Cerella 1975). The best response predictor of all the pattern characteristics examined was simple shape congruity. (See Brown and Owens 1967 for a fascinating though incomplete account of 117 features defined on polygons.)

The data are suggestive of the performance of a template matching system. In such a system, the positive pattern is stored in the form of an internal template, and test figures are evaluated by fitting them to the template. Any discrepancy will reduce the match, and the output of the system will be lowered correspondingly.

We have established that the pigeon can detect the difference between two figures that differ by a deformation, a rotation, a dilation, or a translation. Can the pigeon also perceive the relation between such figures? The answer to this question requires a more complex procedure than that employed so far. In addition to a prototype and a set of transformations of the prototype, a third set of patterns unrelated to the prototype is needed. If both the prototype and its transforms are made positive and the unrelated patterns are made negative, the pigeon is required to differentiate not a fixed pattern but a class of patterns from a further set of patterns. Successful discrimination of the class would imply that the relationship between instances was recognized. This logic was elaborated by a contemporary of Klüver and Lashley, Fields (1932), who studied the rat, and more clearly by Andrew and Harlow (1948), who used monkeys.

In the case of pigeons, only a few experiments of this type have been performed. The first was by Jenkins and Sainsbury (1969), who followed Heinneman and Kadison in using a spot as a positive feature. In their experiment, however, there were four, not one, positive patterns, as the spot could appear in any quadrant. The unrelated negatives were spot-sized stars. (There were actually two stars and one spot in every positive pattern and three stars and no spots in every negative pattern, but this difference is immaterial.) Jenkins and Sainsbury found that the four positives were readily discriminated from the negatives. Thus, although the pigeon is aware of the difference between translated but otherwise unaltered images (as in the work of Heinemann and Kadison), it appears also to be aware of their equivalence.

This conclusion is not wholly warranted. It is possible that the pigeons of Jenkins and Sainsbury learned each of the four positives individually, without recognizing the relationship among them. We should like to know if learning would be as rapid if there were 24 or 104 positives rather than only four, or if after four were learned transfer would be obtained to others.

An answer is provided in more recent data of D. Blough (1979), who trained pigeons to identify the letter O in contrast to the letter X across 16 possible stimulus locations. Blough's success confirms, I think, my previous conclusion that pattern recognition in the pigeon is, or can be, position-invariant.

Only one other image transformation has been tested at the level of class learning, and that is shape distortion. Morgan et al. (1976) successfully trained pigeons to differentiate the letter A from the digit 2 across 18 different Letraset typefaces (see figure 8.2). These authors used a transfer test to rule out discrete-case learning: After a criterion was met on the original patterns, 22 additional typefaces were discriminated without further training. Thus, although successive positives were in general incongruent, their resemblance was perceived. Somewhat unconventionally, I will call this tolerance to "random" deformation *shape invariance*.

To what extent do these results on class learning force us to give up

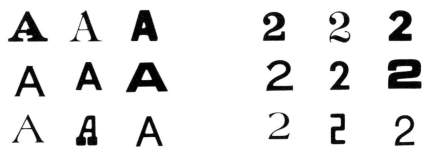

Figure 8.2 Patterns of the sort used by Morgan et al. (1976). In their experiment, pigeons successfully distinguished A's from 2's.

the template-matching model derived from single-stimulus learning? The model can, I think, be modified to accommodate both position invariance and shape invariance. Position invariance is readily obtained by postulating cross-correlation as the mechanism underlying template matching. This operation entails sweeping the template across the image (or vice versa) in a raster scan pattern, so that a match is picked up at any location. (See McLachlan 1962 for a tutorial on the two-dimensional cross-correlation.)

At first sight it may appear that shape invariance offers no further problem. Although the A's of Morgan et al. do not coincide exactly, they appear to overlap each other to a greater extent than any of the 2's, so that a conventional template would be sufficient to separate the classes. This explanation will not work, however, for the set of hand-printed, thin-lined characters used by Morgan et al. in an additional test. These A's intersect at only a few random points. Moreover, this explanation glosses over the question of how a template for a class of variable shapes originates in the first place.

Let us deal with the second question first. Posner and Keele (1968) have pointed out that the signal-averaging techniques used to extract a signal from a series of randomly distorted waveforms can be applied to two-dimensional patterns. The variation in A's can be thought of as noise perturbing a standard pattern. This standard, or prototype, can be recovered by averaging over successive images, and may then be used as a template to identify unlabeled exemplars. Posner and Keele present evidence that human subjects perform such averaging, as demonstrated by enhanced response to the prototype after exposure to a set of distorted patterns. More generally, response to novel exemplars is found to vary with their degree of distortion, which suggests subsequent use of the prototype as a template (Aiken and Brown 1971; Hyman and Frost 1975).

I have obtained similar results with pigeons (Cerella, unpublished). Pigeons were required to discriminate patterns generated by the random distortion of two prototypes, a square and a chevron (figure 8.3). The discrimination, as measured by overall response rates, was readily

Figure 8.3 Distortions of two prototypes. (Bottom row) Small distortions; (middle row) medium distortions; (top row) large distortions. In an experiment involving 40 distortions of each prototype, pigeons' accuracy of classification varied inversely with degree of distortion (Cerella, unpublished data).

achieved. Further analysis showed that response to individual patterns indeed varied with their distortion from the class prototypes.

In suggesting that the pigeon abstracts and employs the prototypes as templates, we are brought back to the first question of how line patterns not in exact correspondence can be matched to a template. Where cross-correlations are negligible within a class, conventional template matching cannot account for the classification of instances. But in each of the classes—A's, 2's, squares, and chevrons—a template might be made to work if only it could be "stretched" somewhat from instance to instance. The notion of a flexible template has been useful in pattern-recognition research (Fischler and Elschlager 1973; Widrow 1973). To make it more concrete, consider a physical model (adapted from Julesz 1971) of how a flexible template might be realized. Imagine an array of corks bobbing in the cells of a net floating loosely on a surface of water. Each cork is pierced by a magnetic needle. An image is stored in this array by setting the corks corresponding to black picture elements ("pixels") north side up and those corresponding to white pixels south side up. To match this magnetic template to a test image, the test image is represented by a second array of needles fixed rigidly in a sheet of cardboard. The needles in the cardboard are reversed— north end up for black pixels, south end up for white pixels. The cardboard is then flipped and lowered over the corks. The corks will be driven by the magnetic forces to align themselves with complementary image pixels, to the extent permitted by the constraints of the loose net. The overall force of attraction on the test image exerted by the corks below may then be measured, and will indicate the "similarity" of image and template. The extent of the deformation may also be measured by summing the deviations of the corks from their home positions, and will indicate another aspect of image-template similarity.

Concept Formation in the Pigeon

However realized, it seems that a flexible template could accommodate shape variants such as those illustrated in figures 8.2 and 8.3. Therefore, let us postulate the flexible template as the mechanism underlying shape invariance in the pigeon.

This completes the present survey of studies employing abstract images. We have ascribed to the pigeon the capacity to formulate position-independent, flexible templates, by means of which critical visual input may be fixed and subsequently reidentified. Whether these templates are also orientation-invariant or size-invariant is not known, as the experiments have not been done (but see Cerella 1977 and the section on synthetic patterns in this chapter).

Natural Patterns

Studies with geometric patterns have suggested that a single mechanism underlies the pigeon's recognition of patterns: the flexible template. To what extent can this mechanism account for results obtained with natural patterns? Two experiments that Anish (1978) did while an undergraduate at Harvard provide a point of comparison. In the first study, pigeons were trained to identify a single positive pattern: a 35-mm color slide of a tree. The negative patterns were a diverse collection of outdoor views not containing trees. After the discrimination was formed, the pigeons were tested on a variety of other slides, all of them containing trees. None of the test slides elicited a significant response. This result disappointed Anish, who was hoping to obtain evidence of "conceptual" encoding. What Anish had obtained was just the result of Ferraro and Grisham, Vetter and Hearst, Wildemann and Holland, and Heinemann and Kadison; in all these studies, response to test patterns after exposure to a training pattern was governed by purely geometric congruity. Anish's second study took the approach of Morgan et al. (1976) and employed a small set of positive patterns, each slightly different. For one group of pigeons, the positive patterns were four full-length views of a medium-sized tree, each view shot within a few feet of the others. For a second group of pigeons, the positives were four close-up views of leaves and twigs, each also taken from a slightly different angle. As in the first study, the negative patterns were outdoor scenes not containing trees. Again, after the discrimination was formed, test slides were introduced. This time, significant response was obtained to some of the test slides. In every case, these were slides physically similar to the training slides—either full-length views of medium-sized trees or close-up views of leaves and twigs.

Anish's results agree closely with the geometric studies. In both cases, images seem to be encoded in a templatelike fashion, so that generalization is obtained to only physically similar instances.

If we accept this conclusion, what are we to make of the classical "concept formation" studies performed on the pigeon? In the earliest of these studies (Herrnstein and Loveland 1964), pigeons were trained on a large number of positive patterns, each containing a person, and a large number of negatives without people. Pigeons readily discriminated positives and negatives, and subsequently generalized almost without error to large numbers of additional images, some with people and some without. This experiment has since been repeated, with similar success, using human images again (Siegel and Honig 1970) and using several other natural concepts: Poole and Lander (1971) used pigeons, and Herrnstein et al. (1976) used trees and water.

Can these results be understood in terms of template learning? Let us make an attempt and propose that the class "pictures containing people" may be defined as a set of "focal instances" (such as "male, front view, full figure, middle ground" and "female, profile view, face only, close up") and the subsets of physically similar instances surrounding each focal instance reachable by flexible template matching. The apparently unitary concept "person," then, would actually embrace a disjunction of miniconcepts. Each miniconcept would have to be learned independently, as the pigeon would have no way of knowing that every focal instance represented a different view of the same type of object.

How plausible is this theory? Notice first that it makes sense of Anish's results; the category resulting from exposure to a limited number of instances will itself be limited. It also makes sense of results of Malott and Siddall (1972), who introduced the concept "person" in stages. First, pigeons were trained on close-up faces only (plus non-faces, of course) until they generalized to novel faces without error. We interpret this as indicating the formation of the template(s) "human face." Malott and Siddall found that the pigeons transferred to the next stage of the problem, torsos (head plus chest), with very little further training. We interpret this to mean that the face template stretched sufficiently to fit the smaller heads in combination with shoulders and chests. Two of the three pigeons, however, failed to transfer to the third stage: crowds. Crowds required as much additional training as had faces in the first stage. We interpret this to mean that the face or torso template could not match the tiny faces and torsos in crowds; a new template had to be formed. (The third pigeon, though, succeeded.) Finally, after this separate stage training, the birds were able to discriminate the general class, people (stage 1 plus stage 2 plus stage 3), without further difficulty. We interpret this to mean that a full ensemble of templates had been formed and were being used interchangeably. Of course, each stage may have required more than one template (although it is possible that Malott and Siddall, unlike Herrnstein, restricted the range of variation in their exemplars).

Although the results of Anish and those of Malott and Siddall can be viewed in terms of a template theory, other data on concept formation in pigeons cannot be so readily interpreted. We shall consider three apparent conflicts between theory and data:

• The classical experiments (for example, those of Herrnstein et al. [1964, 1976]) employed thousands of images; ideally, no image was shown twice. These constitute the strongest demonstrations of concept formation. But much has been learned from experiments employing only 100 or so images presented repeatedly. For example, in Herrnstein 1979 pigeons were shown the same 80 slides daily: 40 containing trees (positive) and 40 without trees (negative). (These were the same slides used by Anish.) Herrnstein was thereby able to track the response to individual tree slides session by session. The reinforcement schedule was probabilistic: therefore, some positives passed through the first four or five sessions without a single reinforcement; other positives accumulated four or five reinforcements over the same sessions. Template theory postulates that many slides must be learned individually. We would expect as a consequence that acquisition rates for slides would in general be a function of the number of reinforcements each received. Herrnstein's data show no sign of such a relationship—those tree slides that by the fifth session had never been reinforced elicited as strong a response as those that had been reinforced on every occurrence. (By the fifth session trees were readily discriminated from nontrees overall.)

• Exposure to thousands of randomly selected exemplars would have a normalizing effect on the boundaries of a developing concept. Exposure to only 40 exemplars might on the other hand induce an incomplete or idiosyncratic generalization class, as key focal instances may be missing. Herrnstein and deVilliers (1980) trained one group of pigeons on 40 slides of fish (positive) and 40 underwater slides without fish (negative). A second group of pigeons were trained on a different and larger set of fish (120 positive, 120 negative). After training was completed, the resulting concepts were compared in a test session in which both groups received the first set of 40 positive and 40 negative slides. The two groups agreed on their ranking of these 80 images. Thus, the limited experience of the first group was not reflected in the resulting concept.

• As conceived by template theory, concept formation entails learning many focal discriminations independently and concurrently. The realization of a concept in this manner might be difficult, possibly far beyond the retentive capacity of the pigeon. Herrnstein and deVilliers put this to a test. In their "fish" experiment, a third group of pigeons received the same 80 slides as the first group every day, but the positive or negative status of a slide was fixed randomly, independent of its content. The only way to learn such an unrelated set of images, presumably, is to memorize each one individually. Surprisingly, the pigeons succeeded in this. However, the quasiconcept was more difficult, requiring over twice as many sessions to learn as the fish concept.

These three difficulties are, it seems to me, putative. The same issue seems to underlie each of them, namely the number of images a single template is able to reach in relation to the totality of the class. If the "capture ratio" of templates were small, many many templates would be needed—in the limit, one for every image. If this were the case, learning a natural concept would be extraordinarily difficult (which it isn't), the concept acquired would be highly dependent on the particular instances encountered (which it isn't), and responding to a given image would only follow reinforcement of that image (which it doesn't). If, on the other hand, the capture ratio of templates were large, only a few templates might be needed. Concepts would be readily learned, small subsets of instances might cover the full class, and any one image would benefit from reinforcements applied to many other images.

The plausibility of a template theory of natural concept formation therefore seems to hinge on the capture ratio of the templates. How can capture ratios be determined? We have considered one approach: limiting the pigeon's exposure to a category and mapping the resulting generalization class. The two studies taking this approach seem to give opposite answers. Malott and Siddall's work suggests that two or three templates may be sufficient to cover the class "people," and Anish's work suggests that more than 40 templates may be needed to cover "trees." But here we run out of studies. In the next section, a different approach is taken to further determine the characteristics of templates.

Synthetic Patterns

This section reviews a series of studies performed with patterns that, although loosely geometric, were designed to simulate aspects of natural patterns. These studies, done in Herrnstein's lab at Harvard, probe further the pigeon's representation of natural categories.

The first study (Cerella 1980) established a synthetic natural class and assessed the pigeon's response to various pathological instances. This involved projecting hundreds of frames from cartoons; pigeons were trained to identify Charlie Brown in these frames as positive, in contrast to the other characters (Linus, Snoopy, Marcia, et al.). The graphic variation here was extraordinary—running, sitting, lying, crawling; snowsuits, swimsuits, sweaters, and caps. Nonetheless, the discrimination was readily achieved (with one proviso, that backgrounds were inked out of the earliest frames). Let us adopt as a tentative explanation of this discrimination that a library of templates was abstracted from the *Peanuts* corpus that was sufficient to detect Charlie Brown in any of his attitudes and attires.

Figure 8.4 Truncated, scrambled, and upside-down Charlie Brown. Pigeons treated 30 truncated images and 18 scrambled images as equivalent to intact Charlie Brown images. Response to upside-down images was inconsistent. Data from Cerella 1980.

Thirty novel Charlie Brown frames were then selected and duplicated for a test. One set of frames was unaltered; in the other set, portions of Charlie Brown were inked out, so that only a fraction of the full figure remained (see figure 8.4). Both sets of frames were embedded in otherwise normal sessions a few at a time; test frames were never reinforced. Not surprisingly, the intact images were identified as Charlie Brown without difficulty. More surprisingly, the incomplete images were found to elicit comparable levels of response.

This combination of results is hard to reconcile with our understanding of template action. Full response to an intact image demonstrates the availability of a matching template. Applied to the truncated image, this template would fit only partially, and responding ought to be correspondingly reduced. This outcome was obtained consistently in the studies discussed in the first section of this chapter after alteration of a positive pattern. A particularly relevant example is that presented in part B of figure 8.1 (also from Cerella 1980). Pigeons were trained to identify an isosceles triangle as positive and other geometric figures as negative. As seen in the figure, subsequent response to abridged triangles was significantly lowered. In the case of Charlie Brown, however, response to truncated figures was sustained.

Eighteen other Charlie Brown frames were selected and duplicated for a second test. In these frames, Charlie Brown was cut into three parts: head, trunk, and legs. One set of frames was reassembled exactly as it had been: head/trunk/legs. The duplicate set was reassembled in haphazard orders: trunk/head/legs, legs/trunk/head, etc. (figure 8.4). Unreinforced test images were embedded in regular sessions, as in the first test. In the second test, intact test images were again identified without difficulty, and again, altered test images were found to elicit comparable levels of response. This result too conflicts with template theory: a template ABC clearly will not fit the pattern BCA or CBA. Notice, however, how easily a template theory could be modified to accommodate the new data. We need only think of fitting multiple local templates to the test patterns—rather than single, global templates—and the results fall into place. A discrimination mediated by local features would be sustained over incomplete patterns if enough features remained to trigger a response; furthermore, rearranging features would not matter if the feature detectors were position-invariant.

A shift from global to local templates is not as ad hoc as it may first appear. Consider the application of signal averaging (postulated in the first section above to underlie template formation) not to a single letter, A, as in the study of Morgan et al., but to a pair of letters, AX, where X represents a variable letter. (Negative patterns would be any letter pairs—BC, WX, etc.—not including A.) When summed, successive A's will reinforce one another, whereas successive X's, being always different, will cancel. The result, plausibly enough, might be a "local" template, encoding the fixed and repeated portions of the positives, (the A's) while the variable portions are simply lost.

In the example the constant pattern elements were always in register. Now consider a more complicated set of positives, ABX, where not only the identity of X varies, but also the location in the frame of all three letters varies from pattern to pattern. The above method of simple superposition will fail to uncover the regularity in these patterns, because corresponding elements will only by chance be aligned. Recall that template matching was assumed to be accomplished through cross-correlation, in order to provide for position invariance. It is reasonable to suppose that pattern averaging involves the same operation, and that a pattern and the developing template are first aligned to maximal overlap before summation. Now the correspondence between elements in the ABX patterns will be signaled by a series of peaks in the cross-correlation—one when the two A's are in register, and another when the two B's are in register. If each peak triggers a summation of the patterns as they are superimposed at that instant, two local templates, A and B, will be formed.

The graphic invariants (repeated subpatterns) in a variable set of patterns can therefore be abstracted, through a straightforward extension of the methods inferred in the first section above, to more complex patterns. In the case of Charlie Brown, we conjecture that such invariants were supplied by the artist, Schulz (cowlick, zigzag sweater stripe, etc.), and are adequate to account for the observed discrimination over both intact and fragmented images.

We have reached the conclusion that the pigeon decomposes line projections into sets of local features. Is the pigeon also aware of relations between features—specifically, those relations imposed by the projected, three-dimensional target? The *Peanuts* data are indecisive here. Although image integrity at a local level was sufficient to enable the identification of positives, it is possible that three-dimensional structure was also computed. The pigeon would thus have been aware that the altered test images were invalid projections of Charlie Brown, but chose to respond anyway. (It is unlikely that response rates would be totally unaffected if this were the case; see Biederman 1972 and 1974

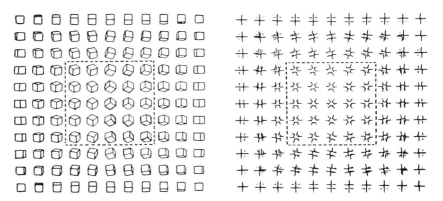

Figure 8.5 (Top) Derivation of a negative pattern by random distortion of a positive pattern. (Bottom left) A sample of 21 out of 9,261 projections of a cube. When pigeons were trained on the centermost projection, they generalized to only those other projections enclosed within the dotted line. (Bottom right) A set of abstract patterns of the same mathematical complexity as the set of cube projections. Pigeons trained on these patterns generalized as widely. Source: Cerella 1977.

for related data on humans.) The question, therefore, is whether the pigeon could distinguish coherent from anomalous projections of a target if expressly reinforced to do so. If the two sets of patterns had the same local features, the distinction would be possible only if relations between features were encoded. I used thousands of computer-generated images to test this possibility (Cerella 1977). Positives were the line projections of a cube; negatives were randomly distorted projections (figure 8.5, top). Positives and negatives were thus composed from more or less the same sets of line segments, vertices, and enclosed regions (see Perkins 1968 for a gratifying discussion of these features). Positives and negatives were trivially easy for people to distinguish (unlike the patterns devised by Shepard and Metzler [1971], which embody a similar logic), but these patterns proved impossible for pigeons, despite extensive training. In a second attempt, I trained with a single projection first (the center one in the bottom left part of figure 8.5), and introduced additional projections gradually by increasing the angle through which the cube was allowed to rotate around its home position. In this way pigeons were successfully trained to identify a small subset of positive projections, but the discrimination collapsed as the number of projections increased. (The size of the learnable subset is marked by the dotted line in figure 8.5. It is tempting to think of the enclosed region as the area in pattern space that could be reached by a flexible

Figure 8.6 (Top) White oak leaves. (Bottom) Leaves from other tree genera. Pigeons' tendency to generalize from the center oak to 40 other oaks was so strong that they were unable to distinguish this oak from the 40 others. Source: Cerella 1979.

template centered on the middle pattern.) Pigeons therefore proved unable to distinguish two classes of patterns that shared the same local features and differed only in three-dimensional structure.

The abstraction of three-dimensional structure from a coordinated series of line projections is thus beyond the capacity of the pigeon (see Ullman 1978 for a formal solution to this problem, and Guzman 1967 for a famous informal solution). A lesser question is "Can the pigeon encode at least the two-dimensional relations between features?" A pair of studies suggest not (Cerella 1979). In the first, pigeons were trained to identify a particular oak-leaf silhouette as positive and leaf silhouettes of other tree species as negative (figure 8.6). The pigeons were then tested on a variety of other leaves. The oaks among the test leaves were accepted as also positive, while the nonoaks were rejected as negative. This result suggested that the training oak leaf was encoded in terms of its local features—lobes, which it shared with other oaks—rather than in terms of its overall shape, which was unique. This can be understood as a consequence of averaging a pattern containing repetition symmetry: sliding the pattern AAA (think of this pattern as an oak leaf composed of three lobes) across the pattern AAA will isolate A as a constituent as well as AA and AAA. In the second study, pigeons were required to identify this same oak leaf as positive and other oak leaves as negative. This could be done only by encoding the overall shape of the positive, or at least a fair-sized fragment of that shape, which might be represented as a set of local features plus the spatial relations among

Concept Formation in the Pigeon

them. This I could do (with concentration); but the pigeons could not, despite extensive training. Again, pigeons proved incapable of separating two classes of patterns that shared the same local features and differed only in relations between features.

The experimental record is now exhausted. From it we have compiled the following picture of the pigeon's representation of natural categories.

In the training phase of a concept-formation problem, positive patterns are scanned for repeated subpatterns, which are abstracted in the form of local, flexible templates. In the testing phase of the problem, patterns of unknown status are scanned for positive subpatterns. If some threshold number of subpatterns is found, the pattern is accepted as positive; otherwise it is rejected as negative.

A category like Charlie Brown is therefore defined by the pigeon as a redundant set of distinctive features. A particular instance of Charlie Brown samples some subset of those features. Insensitivity to scrambling indicates that relations between features are not encoded, and insensitivity to truncation indicates that no specific features are necessary to establish an instance.

The second section concluded with a question on the capture ratio or extension of a template. We may now suggest as answer: If templates were global, a condition of their application would be a test image that matched completely a training image, to the tolerance allowed by flexible matching. If this were the requirement, nearly as many templates might be needed as there were images in the class. Local templates allow us to relax this condition considerably. For a template to be applied now requires only that some part of a test image match some part of a training image. A feature abstracted from one surround might apply to untold instances in different surrounds. Indeed, a small set of features might recur in a large set of otherwise different instances. Studies by Rosch and Mervis (1975) and Rosch et al. (1976) suggest that this is just the internal structure that characterizes natural categories (although the features cataloged are not defined graphically).

Problems and Limitations

We have argued that the pigeon is sensitive to only the low-order invariants in a set of images (that is, the repeated subpatterns), and that image classification is possible only at this level. Higher-order invariants, involving relations between subpatterns, are not encoded; hence, the pigeon is unaware of the structure of more complex patterns. These are profound limitations to visit on an animal whose visual categories—"people," "trees," "water," and the like—are comparable to our own. Let us therefore turn from the limitations of the pigeon to the limitations of our argument.

• Broad evidence suggests that there are two visual systems, one subserving movement through space and the other the identification or recognition of objects (Ingle et al. 1967–1968). In the pigeon these systems may be expressed in the lateral and temporal "foveae," respectively (P. Blough 1973). The images examined here have been delivered to the near-field frontal system. Corresponding limitations may not apply to the lateral system. Notice, for example, the curious insensitivity to slow movement in the near field reported by Hodos et al. (1975).

• Synthetic images were scaled so that their details approximated in extent those of Herrnstein's natural-concept studies. Sainsbury (1971) and Morris (1977) have collected data suggesting that different methods of analysis may be applied at different scales (see also Ingle 1978, on fish). Thus, the limitations found here may not extend to very much larger or smaller stimuli.

• The studies reviewed here involve images, not objects. Limitations demonstrated on line projections may not apply to halftones, or color slides, or the objects themselves. At each level of schematization, information is lost; this information may be crucial for comprehension. In particular, two pigeon studies suggest that whereas halftones and color slides may be equivalent to the corresponding objects, line projections may not be (Cabe 1976; Looney and Cohen 1974).

• The images were static. There is dramatic evidence that dynamic images more effectively engage three-dimensional processes in man (Johansson 1975). Limitations on static line projections may not apply to line projections rotating in real time.

• Granted that these studies involve small, static, nearby line patterns, limitations yet remain. I have distinguished between local and global predicates, in the spirit of Minsky and Papert (1972). But of the latter, only three-dimensional predicates have been tested. The pigeon may be able to compute other global predicates, such as Connected, Inside-of, or Symmetric.

• Local template models are comforting because they seem to lie within the range of Hubel-Wiesel detection units. A result on Connected, or Symmetric, or the like would be devastating. But we are threatened elsewhere as well. The first section concluded that feature detection was position-invariant and distortion-invariant, but drew no conclusion on two-dimensional rotational invariance (or size invariance). The third section went on to demonstrate the absence of rotational invariance in three dimensions. The two-dimensional case continues to haunt us. Inspect the patterns inside the dotted line in figure 8.5 closely. These patterns were readily identified by the pigeon; that is, after the centermost pattern was learned, generalization to the remaining patterns in the dotted region was pretty much immediate (see Cerella 1977). But if a pattern fell within the dotted region its 180° rotation did also; thus, these data seem to show rotation invariance in two dimensions. Figure 8.5 (bottom left) presents only 1.3% of all the patterns, so the actual extent of generalization was even more impressive. Pigeons were also trained with two-dimensional patterns equated to the three-dimensional cubes in information content. Figure 8.5 (bottom right) shows that the two-dimensional problem was of the same difficulty for the pigeon as the three-dimensional problem (Cerella 1977). The point here is that the pigeon exhibited two-dimensional rotational invariance on abstract patterns as well.

A test on upside-down patterns was also conducted in the *Peanuts* series. Immediately after the scrambled-image test reported above, the 18 intact test images were rotated 180° and reinserted—unreinforced, a few at a time—in otherwise normal sessions. The results of this test were inconsistent. Bird 44 responded to upside-down Charlie Browns at the same level as to right-side-up Charlie Browns; bird 47 responded very little to upside-down Charlie Browns (t-tests, repeated measures; unpublished data).

Thus there is some "anecdotal" evidence for rotational constancy. If this is borne out in more systematic experiments, our sense of the adequacy of current neurophysiology will be shattered. (We are glossing over neural-network problems raised by translation invariance, as discussed by Minsky and Papert [1972].)

• The patterns examined here have for the most part been presented on blank backgrounds. How might filled backgrounds alter a pattern-recognition problem? Cross-correlations between positives might be expected to yield spurious features due to the chance alignment of background elements. These features would occur in negatives as well as positives, and would have to be deleted from the positive feature list (perhaps through some form of "subtraction," as opposed to "summation"). Working through false features is likely to take some time, so we might predict that the addition of visual noise will first retard acquisition of a pattern-recognition problem. I have observed this effect in the *Peanuts* studies. In fact when *Peanuts* frames were used intact, with backgrounds, there was no learning whatsoever (four birds, 60 sessions).

On the other hand, with backgrounds deleted, pigeons showed significant discrimination on the very first session and reached virtually perfect discrimination by the fifth or sixth session (two birds). Once distinctive features have been isolated, however, one might expect that cross-correlations will pick them up in noisy slides as well as in noise-free slides. This too was observed in the *Peanuts* studies. After the discrimination was established in a three-bird study on frames without backgrounds, it transferred to frames with backgrounds without difficulty. But herein lies a puzzle. Herrnstein (1977), too, obtained single-session acquisition (trees positive, nontrees negative), but his frames included backgrounds. Indeed, all of the natural-concept-formation problems discussion in the second section above involved backgrounds, and all were solvable. Thus, backgrounds in cartoon frames make discrimination prohibitively difficult, whereas backgrounds in naturalistic slides do not seem to matter. I can think of two explanations of this difference. One is benign: Perhaps targets are more prominent in color slides (brighter, or larger, or the like), and thus physically more likely to resist fading and be carried through the cross-correlation process. The other explanation is more disturbing: Perhaps naturalistic images engage perceptual processes not elicited by line projections. This has been the case in artificial intelligence; for example, Horn (1975) demonstrated a computer algorithm for the recovery of curved surfaces whose domain is the halftone projection. Similarly, Waltz (1975) showed that the addition of shadows drastically reduces the three-dimensional ambiguities inherent in line projections of polyhedra. Again, therefore, I must warn that limitations found on line projections may not extend to other types of images. More research is needed.

Acknowledgments

The author acknowledges the support of Professor Richard Herrnstein. Work was performed under NIMH predoctoral fellowship MH48189 and NIMH grant MH15494 to Harvard University.

References

Aiken, L. S., and D. R. Brown. 1971. "A feature utilization analysis of the perception of class structure." *Percept. Psychophys.* 9: 279–283.

Andrew, G., and H. F. Harlow. 1948. "Performance of macaque monkeys on a test of the concept of generalized triangularity." *Comp. Psych. Monogr.* 19 (no. 3).

Anish, D. S. 1978. "The natural concept tree: A study on learning in pigeons." Undergraduate honors thesis, Harvard College.

Attneave, F., and M. D. Arnoult. 1956. "The quantitative study of shape and pattern perception." *Psychol. Bull.* 53: 452–471.

Biederman, I. 1972. "Perceiving real world scenes." *Science* 144: 424–426.

———. 1974. Paper presented at Psychonomic Science meeting, Boston.

Blough, D. 1979. "Effects of number and form of stimuli on visual search in the pigeon." *J. Exp. Psychol.: Animal Behav. Processes* 5: 211–223.

Blough, P. M. 1973. "Visual acuity in the pigeon. II: Effects of target distance and retinal lesions." *J. Exp. Anal. Behav.* 20: 333–343.

Brown, D. R., and D. H. Owens. 1967. "The metrics of visual form: Methodological dyspepsia." *Psychol. Bull.* 68: 243–259.

Cabe, P. 1976. "Transfer of discrimination from solid objects to pictures by pigeons." *Percept. Psychophys.* 19: 545–550.

Cerella, J. 1975. "Studies on concept formation in the pigeon." Ph.D. diss., Harvard University.

———. 1977. "Absence of perspective processing in the pigeon." *Pattern Recognition* 9: 65–68.

———. 1979. "Visual classes and natural categories in the pigeon." *J. Exp. Psychol.: Human Percept. Perf.* 5: 68–77.

———. 1980. "The pigeon's analysis of pictures." *Pattern Recognition* 12: 1–6.

Ferraro, D., and M. Grisham. 1972. "Discrimination and generalization of complex shape variations in pigeons." *Percept. Motor Skills* 35: 915–927.

Fields, P. E. 1932. "Studies in concept formation. I. The development of the concept of triangularity by the white rat." *Comp. Psych. Monogr.* 9 (no. 2).

Fischler, M., and R. Elschlager. 1973. "The representation and matching of pictorial structures." *IEEE Trans. Comput.* C 22: 67–92.

Guzman, A. 1967. "Decomposition of a visual scene into bodies." MIT Artificial Intelligence memo 139.

Heinemann, E. G., and K. Kadison. 1976. "Control of pigeon's choice behavior by position and luminance of a spot of light." *Bull. Psychonom. Soc.* 7: 522–524.

Herrnstein, R. J. 1979. "Acquisition, generalization, and discrimination of a natural concept." *J. Exp. Psych.: Animal Behav. Processes* 5: 116–129.

Herrnstein, R. J., and P. A. deVilliers. 1980. "Fish as a natural category for people and pigeons." In *The Psychology of Learning and Motivation*, G. H. Bower, ed. New York: Academic.

Herrnstein, R. J., and D. Loveland. 1964. "Complex visual concept in the pigeon." *Science* 46: 549–551.

Herrnstein, R. J., D. Loveland, and C. Cable. 1976. "Natural concepts in pigeons." *J. Exp. Psych.: Animal Behav. Processes* 2: 285–311.

Hodos, W., L. Smith, and J. Bonbright. 1975. "Detection of the velocity of movement of visual stimuli by pigeons." *J. Exp. Anal. Behav.* 25: 143–156.

Horn, B. 1975. "Obtaining shape from shading information." In *The Psychology of Computer Vision*, P. H. Winston, ed. New York: McGraw-Hill.

Hyman, R., and N. Frost. 1975. "Gradients and schema in pattern recognition." In *Attention and Performance*, vol. 5, M. A. Rabbitt and S. Dornic, eds. London: Academic.

Ingle, D. 1978. "Mechanisms of shape recognition among vertebrates." In *Handbook of Sensory Physiology*, vol. 8, R. M. Held et al., eds. Berlin: Springer.

Ingle, D., G. Schneider, C. Trevarthan, and R. Held. 1967–1968. "Locating and identifying." *Psych. Forsch.* 31: nos. 1, 4.

Jenkins, H. M., and R. S. Sainsbury. 1969. "The development of stimulus control through differential reinforcement." In *Fundamental Issues in Associative Learning*, N. J. MacKintosh and W. K. Honig, eds. Halifax: Dalhousie University Press.

Johansson, G. 1975. "Visual motion perception." *Sci. Amer.* 232 (no. 6): 76–88.

Julesz, B. 1971. *Foundations of Cyclopean Perception.* University of Chicago Press.

Klüver, H. 1933. *Behavior Mechanisms in Monkeys.* University of Chicago Press.

Lashley, K. S. 1938. "The mechanisms of vision. XV. Preliminary studies of the rat's capacity for detail vision." *J. Genet. Psych.* 18: 123–193.

Looney, T., and P. Cohen. 1974. "Pictorial target control of schedule induced attack in white Carneaux pigeons." *J. Exp. Anal. Behav.* 21: 571–584.

Malott, R., and J. Siddall. 1972. "Acquisition of the people concept in pigeons." *Psych. Rep.* 31: 3–13.

McLachlan, D. 1962. "The role of optics in applying correlation functions to pattern recognition." *J. Opt. Soc. Am.* 52: 454–459.

Minsky, M., and S. Papert. 1972. *Perceptrons,* 2nd edition. Cambridge, Mass.: MIT Press.

Morgan, M., M. Fitch, J. Holman, and S. Lea. 1976. "Pigeons learn the concept of an 'A'." *Perception* 5: 57–66.

Morris, R. C. 1977. "Spatial variables and the feature positive effect." *Learning Motivation* 8: 194–212.

Perkins, D. 1968. "Cubic corners." *Quarterly Progress Report* (MIT Research Lab of Electronics) 89: 207–214.

Poole, J., and D. Lander. 1971. "The pigeon's concept of pigeon." *Psychonom. Sci.* 25: 153–158.

Posner, M., and S. Keele. 1968. "On the genesis of abstract ideas." *J. Exp. Psych.* 77: 353–363.

Posner, M. I., R. Goldsmith, and K. E. Welton. 1967. "Perceived distance and the classification of distorted patterns." *J. Exp. Psych.* 73: 28–38.

Rosch, E., and C. Mervis. 1975. "Family resemblances: Studies in the internal structure of categories." *Cogn. Psychol.* 7: 573–605.

Rosch, E., C. Mervis, W. Gray, D. Johnson, and P. Boyes-Braem. 1976. "Basic objects in natural categories." *Cogn. Psychol.* 8: 382–439.

Sainsbury, R. S. 1971. "Effect of proximity of elements on the feature positive effect." *J. Exp. Anal. Behav.* 16: 315–325.

Shepard, R. N., and J. Metzler. 1971. "Mental rotation of three-dimensional objects." *Science* 171: 701–703.

Siegel, R., and W. Honig. 1970. "Pigeon concept formation: Successive and simultaneous acquisition." *J. Exp. Anal. Behav.* 13: 385–390.

Ullman, S. 1978. "The interpretation of structure from motion." MIT Artificial Intelligence memo 476.

Vetter, G., and E. Hearst. 1968. "Generalization and discrimination of shape orientation in pigeons." *J. Exp. Anal. Behav.* 11: 753–765.

Waltz, D. 1975. "Understanding line drawings of scenes with shadows." In *The Psychology of Computer Vision*, P. H. Winston, ed. New York: McGraw-Hill.

Widrow, B. 1973. "The 'rubber-mask' technique." *Pattern Recognition* 5: 175–211.

Wildemann, D. G., and J. G. Holland. 1973. "The effect of the blackout method on acquisition and generalization." *J. Exp. Anal. Behav.* 19: 73–80.

Visual Guidance of Motor Patterns: The Role of Visual Cortex and the Superior Colliculus

Introduction

Melvyn A. Goodale

A tradition in the visual neurosciences dating back to the latter half of the nineteenth century has associated visual projections to the cerebral cortex with "higher" perceptual functions, and the phylogenetically older projections to the visual midbrain with visually guided motor behavior. Thus, pattern or form vision has been thought to be the function of the geniculostriate pathway, whereas the control of eye movements and other visuomotor behaviors has been ascribed to the superior colliculus. Though the two-visual-systems hypothesis has been a useful heuristic for studying visual behavior in the mammal, much of the recent work on visually guided behavior suggests that this dichotomous view of the visuomotor network is an oversimplification.

This section reexamines the cortical and collicular contribution to the organization of visually guided behavior in the light of this recent evidence. The first three chapters focus directly on the behavior of laboratory animals with lesions in the geniculostriate pathway or the superior colliculus, the last three on the visual guidance of motor patterns in human subjects. Throughout the section, the authors emphasize the subtle interaction between various visuomotor channels in the control of different patterns of movement and show how the old distinction between visual perception and orientation behavior is becoming blurred.

Chapter 9, by Goodale and Milner, looks at the role of the superior colliculus in the visual control of orientation behavior in rodents. The nature of the orientation behavior examined in their experimental animals is spelled out rather precisely. As the authors point out, the term has been used in a wide variety of contexts in the past, and can be found to refer to any one of a number of behavior patterns, including saccadic eye movements, head movements, tracking, reaching with the forelimbs, visually guided locomotion toward targets and around barriers, and the maintenance of upright posture. Using frame-by-frame film analysis and several different behavioral situations, they and a number of other investigators have been able to show that colliculectomized rats and gerbils can locomote as efficiently as normal animals toward visual targets and around barriers even though the same animals are often unable to detect or orient their head and eyes toward novel stimuli presented in the visual periphery. Moreover, some of their data indicate that the pretectum and areas of visual cortex may be more important for the visual guidance of locomotion around barriers than is the superior colliculus. By using a single-frame analysis of behavior within a traditional lesion-study framework, these investigators have been able to tease apart orientation behavior into a number of separate visually guided movements and are beginning to show that these different patterns of visuomotor behavior depend on different neural substrates.

The next two chapters, by Butter, Kurtz, Leiby, and Campbell and by Keating and Dineen, both look at the effect of collicular and striate-cortex lesions on visual behavior in the monkey. Butter and coauthors provide a thorough review of the deficits (and residual abilities) that have been described in monkeys with extensive lesions of superior colliculus or complete striatectomies. Throughout their chapter they emphasize the opposing strategies that investigators have had to employ in order to sort out the contributions of these two visual-projection areas to visually guided behavior. The problem with studying colliculectomized monkeys is that their deficits are so subtle that sophisticated testing methods must be used to detect these impairments. In sharp contrast, the visual performance of monkeys is so devastated by lesions of striate cortex that equally sophisticated testing methods must be employed in order to detect any visually guided behavior at all. Keating and Dineen continue this theme in chapter 11, and review several experiments from their laboratory that were designed to evaluate collicular function by analyzing both the deficits in colliculectomized monkeys and the residual visuomotor abilities of monkeys with lesions of striate cortex. They argue that much of the visually guided behavior surviving striate lesions is likely to be mediated by retinal projections to the superior colliculus. In order to test this hypothesis, they made additional lesions of either colliculus or its cortical targets (via nuclei in the posterior thalamus) in striatectomized monkeys. The behavioral experiments on the monkey described in both these chapters have produced results which are remarkably consistent with those obtained in rodents. As the results from these careful, analytical studies accumulate, the two-visual-systems model is giving way to a multichannel view of visuomotor organization in mammals. In fact, the neural substrate of the mammalian visuomotor programs shows striking parallels with that of lower vertebrates, such as the frog and the toad. The homologies in the anatomy of the visual system that can be seen through the vertebrate line, from goldfish to rhesus monkey, are reflected (perhaps not surprisingly) in behavioral homologies. It is equally apparent that, within the basic vertebrate plan, individual classes and indeed individual species have evolved visuomotor solutions to problems presented by their own particular environments.

But what about the special visuomotor skills of human beings? Does the organization of their visually guided movements also conform to the basic vertebrate plan? The remaining three chapters in this section examine an area of human visuomotor behavior that has been sadly neglected in the past: the visual guidance of reaching, locomotion, and posture. The chapter of Paillard and that of Jeannerod and Biguer both deal with the problem of the visual control of reaching, and although they attack the problem in different ways they arrive at similar conclusions about the possible neural organization of this complex skill. Both suggest that a reaching movement consists of several different motor programs, and that visual information controlling the movement arrives over separate channels. Moreover, both argue that the exquisite precision of this human visuomotor skill involves cortical as well as collicular control mechanisms. The experiments supporting these conclusions involve direct measurement in space and time of the dynamic movements of the arms, the fingers, and the eyes, as well as systematic manipulation of the amount of available visual information. Again, it seems that the more analytical the behavioral methods the less useful is the dichotomous view of the visual system suggested by the two-visual-systems hypothesis.

In chapter 14, Lee and Thomson do not offer any direct suggestions about the neural mechanisms underlying visuomotor behavior, but instead suggest a way of looking at the fundamental function of vision that may provide new directions for neurobehavioral research in the future. They argue that vision provides three basic types of information for the control of movement: exteroceptive information about the layout of objects and events in the environment; proprioceptive information about the position, orientation, and movement of parts of the body relative to each other; and "exproprioceptive" information about the position, orientation, and movement of the body, or parts of the body, relative to the environment. The experimental support for such a division of labor within the visuomotor pathways is provided in large part by observations of the visual control of static and dynamic balance, the guiding of the feet and direction of movement during locomotion, and a number of other poorly understood visuomotor programs in human beings. The behavioral methods and subsequent analysis brought to bear on these problems can be adapted for use with both laboratory animals and neurological patients. Such a research program could provide important insights into the neural basis of these complex visuomotor programs.

In sum, the six chapters in this section arrive in different ways at much the same conclusion: that mammals, like the lower vertebrates, have evolved a number of separate visuomotor programs, some of which share the same neural substrate and some of which do not. Finally, the division of labor implied by the two-visual-systems hypothesis is not sufficient to account for the differences in the patterns of orientation behavior observed in animals with collicular and cortical lesions. It appears that there are many more than two visual systems at work in the visually active animal.

9 Fractionating Orientation Behavior in Rodents

Melvyn A. Goodale
A. David Milner

The last two decades have seen an enormous growth in the amount of information available about the anatomy and electrophysiology of the visual system in all major vertebrate classes. These comparative data have revealed a basic pattern of retinofugal projections in vertebrate species from the goldfish to the rhesus monkey (Ebbesson 1970; Riss and Jakway 1970). Unfortunately, our understanding of the relationship between the structure of the visual system and the control of visually guided behavior in different species is much more limited. In a 1975 review, Ingle and Sprague pointed out that behavioral studies have produced comparative data for only one retinofugal target, the optic tectum. Even here, the results from experiments on different classes of vertebrates are not always directly comparable, simply because very different measures have been used to study the behavior of each verte-brate class. These differences in experimental approach can be clearly seen if one compares the work on anurans (frogs and toads) with work on familiar mammalian species. Ingle's (1973, 1976, and this volume) investigations of visual behavior in the frog and Ewert's (1970 and this volume) studies of the toad have looked directly at the spatial and tem-poral characteristics of actual movements made by these animals in re-sponse to a variety of different visual stimuli. Moreover, they have examined the effect of ablating various retinofugal targets on behavior patterns normally displayed by these creatures in their natural habitats. Thus, for example, the behavioral distinction between visually guided prey catching and visually guided barrier avoidance has been found to be reflected in a dissociation of their respective neural substrates. In fact, at least five separate (but interactive) visuomotor systems have been postulated to exist in the frog (Ingle 1976).

In contrast with this "neuroethological" approach, most investigators studying mammalian visual behavior have focused on the performance of brain-damaged animals in visual-discrimination tasks. In a typical experiment, animals with lesions in the visual system have been trained (or retrained) on a two-choice visual-discrimination test and their performance compared with that of operated control animals. Such experiments have relied heavily on inferential measures derived from error scores or the number of trials required to reach criterion performance. Deficits have usually been interpreted in terms of a disturbance in the animal's ability to extract appropriate information from the stimulus array or to code and store this information. Thus, almost all the theoretical statements have tended to follow Sherrington (1906) and have assumed that vision functions primarily as an exteroceptive sense, providing information about the layout of the external world and the objects and events within it. Little attention has been paid to the role of vision in the orientation and movement of the body (or parts of the body) relative to the environment. As a consequence, these investigations have produced a wealth of information about the role of various neural structures (principally the cortical elaboration of the geniculostriate pathway) in pattern vision and "higher" visual perception in mammals, but little insight into the visual control of posture, locomotion, and other behaviors that these animals might use in the natural world. Indeed, with the notable exceptions of optokinetic and vestibular nystagmus, saccadic eye movements, and various other pursuit and reflexive movements of the eyes and head (see Robinson 1972; Wurtz and Goldberg 1972; Collewijn 1975), there are very few studies in which detailed measurements have been made of visually controlled movements in mammals with brain lesions. Such observations have consisted of informal tests of visual placing, avoidance of obstacles, tracking of food, responses to visual "threat," and accuracy of reaching. However, only rarely has the behavior of animals in these tests been carefully monitored and described (see, for example, Sprague and Meikle 1965). Therefore, only a small amount of information about the neural circuitry mediating visually oriented movements in mammals is comparable to existing work on anurans and other submammalian species.

In this chapter, we review some of the initial attempts to analyze the neural substrates of different kinds of visuomotor behavior in three rodent species: the Syrian hamster, the rat, and the Mongolian gerbil. In most of the experiments we discuss, the visually controlled movements of these animals have been recorded on film or videotape. Thus, it has been possible to look in fine detail at the spatial and temporal characteristics of different visuomotor sequences in normal animals and compare these parametric data with visuomotor deficits observed in animals with lesions in different retinofugal targets. Information yielded by this kind of approach is directly comparable to that obtained from the study of visual mechanisms in nonmammalian vertebrates and suggests that the functional organization of the visuomotor system may be highly similar in all vertebrates.

The particular question we focus upon in the following pages is the role of the rodent superior colliculus in orientation and locomotion toward stationary visual stimuli. The effects of lesions in this structure and in the pretectum on avoidance of barriers and obstructions is also discussed. The results of these experiments are compared with the rapidly accumulating information about visual mechanisms in non-mammalian vertebrates, particularly the work of Ingle on the frog and Ewert on the toad reported in this volume. Much of the material we present is closely related to experiments on the orientation and pursuit of moving targets by the gerbil that are discussed by Ingle at the end of chapter 3.

Two Visual Systems

Despite the shortcomings of the discrimination task as a research tool, pioneer investigators such as Lashley (1931, 1935) and Klüver (1937, 1941, 1942) used this paradigm to great effect in developing the classical description of striate cortex as the primary site for pattern or form vision in the rat and monkey. It was clear that animals with complete ablations of area 17 had great difficulty in making discriminations between different visual patterns even though they could readily discriminate between stimuli differing in the total amount of light emitted or reflected. Although these classical notions of striate function have been modified in recent years, there is general agreement that interruption of the geniculostriate pathway results in a profound disturbance of the ability to perform pattern discriminations and a large reduction in both contrast sensitivity and acuity in rats, monkeys, and man (see the review of Butter et al. in this volume).

The effect of lesions of the superior colliculus on discrimination performance is not nearly so clear, and the literature abounds with contradictory results. Although some investigators have reported acquisition or retention deficits in brightness and/or pattern discrimination tasks in animals with collicular damage (Freeman and Papez 1930; Blake 1959; Anderson and Williams 1971; Berlucchi et al. 1972; Casagrande et al. 1972; Milner et al. 1979), a number of workers have found that collicular lesions had no effect or produced only slight impairments on similar discrimination tasks (Layman 1936; Ghiselli 1937; Rosvold et al. 1958; Myers 1964; Fischman and Meikle 1965; Horel 1968; Urbaitis and Meikle 1968; Anderson and Symmes 1969; Thompson 1969).

Despite the confused state of the empirical literature, Schneider (1966, 1967, 1969) has argued persuasively that the role of the superior colliculus within the mammalian visual system can be clearly differentiated from that of visual cortex. According to the "two-visual-systems" hypothesis, the collicular system is directly involved in orientation behavior, enabling the animal to localize a stimulus in visual space, whereas the geniculostriate system participates in the identification of the stimulus. The notion of collicular mediation of orientation behavior grew out of the earlier concept of the "visual grasp" reflex (Hess et al. 1946), whereby objects in the peripheral visual fields were said to be relocated on the central field by movements of the head and eyes. Further support for a collicular role in orientation was provided by Schneider's observations that Syrian hamsters with under-cut superior colliculi were unable to locate food objects, such as sunflower seeds, presented by the experimenter in various parts of their visual fields. In addition, Schneider reported that such animals did not guide themselves accurately toward visual stimuli in a two-choice dis-crimination apparatus even though in their door-pushing behavior they discriminated the correct door from the incorrect. This last obser-vation, in particular, has been used in Schneider's arguments that the apparent contradictions in the earlier literature can be resolved by ex-amining the response demands of the visual-discrimination tasks employed by different investigators. According to Schneider (1969), dis-crimination tasks that require locomotor responses to be differentially oriented towards the visual discriminanda will produce deficits in col-liculectomized animals, whereas tasks that do not require orientation will not.

Schneider's distinction between "identification" and "orientation" is important, since it clearly implies that the visual pathways can no longer be considered a unitary system, but are a network of relatively independent visuomotor "channels." It is equally clear, however, that the two functional categories postulated in the two-visual-systems model may themselves consist of a number of distinct visuomotor se-quences. Certainly, the term "orientation behavior" has been used to refer to a variety of different behavior patterns, including saccadic eye movements, head movements, reaching with the forelimbs, tracking, visually guided locomotion towards targets and around barriers, and maintenance of upright posture. In different contexts, Schneider (1969) has himself used this term in reference to both (1) the head and eye movements and other postural changes constituting an orienting re-sponse toward a stimulus presented in the peripheral visual field and (2) the visually guided locomotor movements made toward a target, such as the correct door in a discrimination apparatus. In Schneider's original experiments, only the head turns involved in orienting towards sunflower seeds were measured directly; whereas his conclusions about deficits in visually guided approach responses were, for the most part, inferred indirectly from errors to criterion on visual-discrimination tasks. The possibility remained that the control of these very different patterns of behavior might depend on separate neural pathways. Therefore, as a further step in the "fractionation" of visual behavior into distinct visuomotor sequences we carried out a series of experi-ments in which we filmed orienting responses and visually guided locomotor behavior in rats and gerbils with different visual-system lesions.

Visual Orientation in the Rat: A Dissociation of Deficits Following Cortical and Collicular Lesions

Visually Guided Locomotion

In the first set of experiments (Goodale and Murison 1975; Goodale et al. 1978), we filmed the locomotor behavior of rats trained to run toward a small illuminated target in the wall of a large open area. The basic experimental setup (figure 9.1) was very simple. It consisted of two large open square boxes, each with a floor space measuring about 90 × 90 cm with walls 45 cm high. The boxes were interconnected by a short tunnel intersected by a photobeam halfway along its length. Five circular holes 5 cm in diameter and 15 cm apart were located in each box at floor level in the wall opposite the tunnel opening. Behind each hole was a white translucent plexiglas door hinged at the top, which, when pushed open, gave access to a small feeding hole through which a condensed-milk solution could be presented by means of a dipper mechanism. Also behind each door was a small light bulb that could be independently

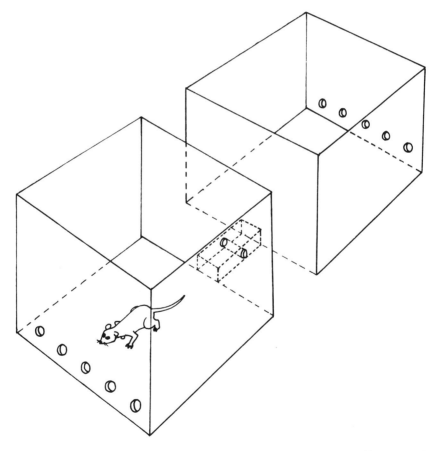

Figure 9.1 Schematic drawing of apparatus used to study visually guided locomotion in the rat.

illuminated. General illumination of the entire apparatus (which was painted matte gray) was provided by a single 100-watt light bulb suspended 1.5 m above the floor. Thus, the ambient light level in the situation was relatively high. A 16-mm movie camera fitted with a wide-angle lens was mounted above one of the boxes such that the field of view of the camera included the entire floor of the box, with the tunnel exit and the five circular openings in the opposite wall clearly visible. The entire apparatus was automatically controlled by solid-state programming equipment.

As far as the rat was concerned, the normal course of events was as follows: When the rat entered the tunnel and broke the photobeam, one of the five lights in the box ahead of it was automatically switched on. If the rat then ran over and pushed open the illuminated door, it gained access to the milk solution for 2 seconds. If the rat pushed open a darkened door instead, no reward was presented, an error was recorded, and the rat was allowed to correct and push open the illuminated door. At the end of the reinforcement period, the light behind the correct door went out, and no light would come on again until the rat reentered the tunnel and broke the photobeam, in which case one of the five doors in the other box was illuminated. In this way, one of the five doors in either box was illuminated in an alternating fashion, and to obtain access to the milk solution a hungry rat had to run from one end of the apparatus to the other. This running behavior was easily recorded on film, which enabled us to compare directly the visually guided locomotion of rats with collicular or cortical lesions against that of normal rats. In most cases these comparisons could be quite detailed, since a frame-by-frame analysis of the film records permitted us to relate the spatial and temporal characteristics of a rat's movements to the time of presentation and the position of a visual stimulus in the rat's visual field.

In this situation normal rats (or rats with sham operations) ran directly toward the illuminated target door almost as soon as they left the tunnel and entered the arena (Goodale and Murison 1975; Goodale et al. 1978). The high degree of visual control over their locomotor behavior is evident both in the plots of individual runs (figure 9.2) and in the rank-order correlations between the relative position of the paths the rats followed on different trials and the relative position of the doors to which they were running on those trials (figure 9.3).

Rats with large bilateral lesions of the superior colliculus were not impaired on this task. Even rats with complete lesions in which the damage extended slightly into the subjacent tegmental area ran toward the illuminated doors as efficiently and as rapidly as normal animals (figures 9.2, 9.3). It did not matter whether the rats had received preoperative training or not; the learning or relearning of this visuomotor skill did not appear to depend on the integrity of the superior colliculus. On some occasions, the running of the colliculectomized rats seemed

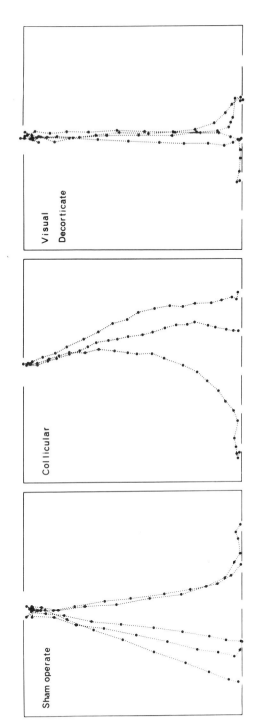

Figure 9.2 Routes followed by three different rats on all the correct trials of the last ten on the first day of postoperative training. None of these rats had been trained before operation. Large dots represent tip of rat's nose on each frame of film; dotted lines join together successive frames (10 frames/sec). Opening at top of outlined square represents tunnel entrance; spaces at bottom represent the five doors in the wall of the arena.

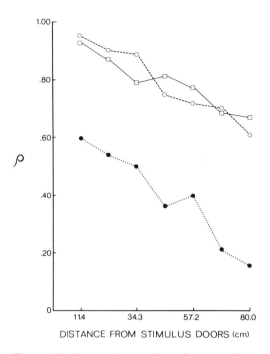

Figure 9.3   Rank-order correlation between relative path positions and position of correct doors on ten postcriterion trials for the three different groups of rats: (●) Rats with lesions of visual cortex; (○) rats with lesions of superior colliculus; (□) rats with sham operations only.

even more efficient than that of normal rats. They almost never paused en route from the tunnel exit to the correct door, and as a consequence their paths were more uniform and often more direct than those of normals, who sometimes appeared to be influenced by other stimuli in the situation. Indeed, it was in part this apparently supernormal performance that led us to study possible "distraction" deficits accompanying collicular lesions.

The fact that the rats with collicular lesions showed no impairment in their visual control of locomotion when approaching a small target door from some distance seems to be in direct conflict with Schneider's (1966, 1967, 1969) original observations. None of the "peculiar path habits" described by Schneider (1967) in his hamsters with undercut superior colliculi were ever observed in the colliculectomized rats. Although the apparent discrepancy in the results of the two experiments could be due to species differences or differences in the discrimination apparatus and surgical techniques, it is interesting to note that of the four hamsters in Schneider's experiment, only one of them (M-7) failed completely to learn to approach the correct door under visual control, and that was only during a horizontal-vertical discrimination. Yet this same animal had made only slightly more of these "approach errors" than normal animals on an earlier black-white discrimination problem. Thus, the deficit in visual control of approach responses was not absolute and, at least in one animal, appeared to be stimulus-dependent. Even normal hamsters make many approach errors in an apparatus of only slightly different design (Schneider 1969), and similar sorts of approach errors have been observed in normal rats in a comparable two-choice discrimination apparatus (Goodale and Milner, unpublished observations). Moreover, Dyer et al. (1976) failed to find any indication of an increase in approach errors in a two-choice pattern-discrimination task after collicular lesions in the rat. Like the rats in our experiments, their operated rats appeared to locomote toward the correct stimuli as accurately as their control rats.

More recently, the visually guided behavior of hamsters has been reassessed—this time in an apparatus similar to the one we used with the rats (Mort et al. 1980). In these experiments, the behavior of each hamster was recorded on videotape. A subsequent analysis of these records failed to reveal any group deficit in visual control of locomotion toward target doors after the hamsters had received collicular undercuts in a manner similar to that described by Schneider (1967). It is true that two hamsters did show rather profound, though not permanent, deficits in visually guided locomotion; but the undercuts in these animals included part of the posterior thalamus and pretectum. It may well be that some part of this nuclear complex is important for the visual guidance of locomotion, and that in Schneider's original experiment (1966, 1967, 1969) this critical structure was damaged. Certainly the one animal (M-7) that failed to learn to approach the discriminanda correctly in Schneider's investigation was the only one with a large amount of damage to the pretectal nuclei.

The absence of a deficit in visually guided locomotion in rats and hamsters with collicular lesions is consistent with the observations of other workers who have been looking at similar kinds of behavior in other nonrodent species. Casagrande and Diamond (1974) reported that tree shrews (*Tupaia glis*) with deep collicular ablations could nevertheless guide themselves accurately through a barrier maze presumably under visual control. Voneida (1970) has made similar observations in cats with midline sections of the mesencephalic tegmentum, an operation that effectively interrupts outflow from the colliculus via the predorsal bundle. Tunkl and Berkley (1974) reported that cats with collicular lesions could accurately localize and approach auditory or visual stimuli of long duration, although there were deficits when the stimuli were present only very briefly.

All these different experiments suggest strongly that, whatever its role in visuomotor behavior, the superior colliculus is not an essential component in the neural circuitry underlying the ability of an animal to locomote toward discrete visible landmarks.

In contrast with the excellent performance of the colliculectomized animals in our experiments, the rats with large lesions in posterior neocortex (including area 17 with some damage to neighboring areas 18 and 18a) were distinctly impaired in running accurately toward the illuminated target door (Goodale and Murison 1975; Goodale et al. 1978). In fact, almost all of the early runs of naive decorticates in this situation showed no sign of differential guidance by the lighted door on a particular trial. As the film record depicted in figure 9.2 illustrates, a decorticate rat would often run straight forward from the tunnel exit toward the center door and then push it open or make a corrective deviation toward the appropriate side. Some improvement in visual control did occur with repeated training, but the correlations that were computed between relative positions of the routes and the correct doors remained lower than those of either the normal or the colliculectomized group (figure 9.3).

Marks and Jane (1974) reported even more dramatic deficits in locomotor orientation toward an illuminated goal in both cats and squirrel monkeys with large ablations of areas 17, 18, and 19. Moreover, it has long been observed that the avoidance of unexpected obstacles while walking and visually guided placing are both severely impaired after ablation of area 17 in the macaque and areas 17, 18, and 19 in the cat, although a variable degree of recovery occurs with long postoperative survival (Denny-Brown and Chambers 1955; Doty 1961; Weiskrantz 1963; Fischman and Meikle 1965; Humphrey and Weiskrantz 1967; Spear and Braun 1969). The nature of these deficits remains unclear.

In our experiments, the possibility exists that the rats with large lesions of visual cortex simply could not disembed the small target from the surround in the moderate to high ambient illumination present in the apparatus. In other words, their deficit may have been a reflection of the impairment in visual acuity that accompanies area-17 lesions rather than a disturbance in visual feedback mechanisms controlling locomotion toward targets. This explanation receives some support from the work of Ferrier and Cooper (1976), who reported that rats with large ablations of posterior neocortex could nevertheless jump accurately toward visual targets. In their experiment, the targets were brightly illuminated and ambient illumination was very low—a situation where the acuity demands were not so great. More recently, gerbils with large lesions of posterior neocortex were trained in an arena where the contrast between the target door and the surround was much greater than in the earlier rat experiments (Goodale and Purvis, unpublished). In this situation, the gerbils with visual-cortex lesions ran as efficiently and as directly toward the target door as gerbils with sham operations. Of course, if the illumination gradient from target to surround is sufficiently steep, then even a simple "phototactic" strategy should enable an animal to approach the target directly. Whatever the contribution of visual cortex to the control of locomotion might be, the differing results obtained in the experiments described above suggest that the mechanisms involved in the visuospatial guidance of locomotor movements are not likely to be simple or unitary.

Orienting Responses to Novel Visual Stimuli

As mentioned in the preceding subsection, rats with bilateral lesions of the superior colliculus were often more efficient than normal animals in running toward target doors. Moreover, on the very few occasions when one of us had to approach the apparatus in order to adjust some of the equipment while the animals were performing, the colliculectomized rats seemed far less distracted by our sudden presence than the normal or visual-decorticate animals. These rather informal observations led us to design a series of "distraction" experiments in which novel visual stimuli were suddenly presented to an animal while it was running toward one of the target doors.

In the first of these experiments (Goodale and Murison 1975), a small light was located on top of the left or right side wall of the apparatus halfway between the tunnel exit and the target doors. On test trials, which followed a series of "normal" trials, this light was turned on 400 msec after the rat had broken the photobeam in the tunnel and was flashed twice a second for 5 sec. On these test trials, films were made of the rat's responses to this novel stimulus. In addition, the time taken to reach the correct door after leaving the tunnel were recorded on both normal and test trials. The film records, some of which are reproduced in figure 9.4, clearly show that both normal rats and rats with lesions of visual cortex exhibited pronounced head-raising, turning, rearing, freezing, and approach responses to the flashing light. This disruption of ongoing behavior is also reflected in the increased running times of these two groups on those trials when the distracting stimulus was presented (figure 9.5). Indeed, one of the normal animals in this experiment was so disturbed by the introduction of the flashing light that it immediately ran back into the tunnel and could not be induced to reenter the arena.

In contrast with the behavior of normal and visual-decorticate rats, the rats with collicular lesions showed no responses at all to the flashing light, but simply ran directly from the tunnel opening to the correct door (figures 9.4, 9.5). It is important to emphasize that the colliculectomized group did not make inappropriate or incomplete orienting movements toward the flashing light, nor did they show any evidence of freezing or slowing as they ran toward the target door. In other words, the deficit in orienting did not appear to be simply one of motor expression, but instead a much more fundamental disruption of the neural circuitry underlying the initiation of the sequence of head movements and postural changes making up an orienting response to a novel visual stimulus.

In a subsequent experiment (Goodale et al. 1978) we examined the possibility that the deficit is field-dependent; that is, that the colliculectomized animals might be responsive to novel visual stimuli in some parts of their visual field but not to stimuli presented in other parts. Rats were again trained to run toward target doors in the arena,

VC   NOR   SC

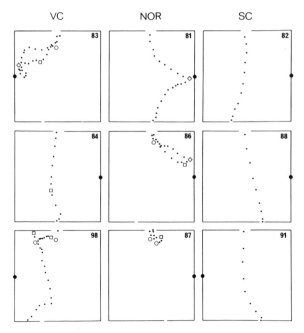

Figure 9.4 Behavior of each rat on first trial in which a novel visual stimulus was presented. Tunnel opening is indicated by gap at top of each outline of floor of box; correct door on that trial is indicated by smaller opening at bottom of outlined square. Position of the flashing light is indicated by ● on side of each outlined square. Points represent tip of rat's nose on successive frames of film. (○) Freezing responses; (□) head raising; (◇) rearing. SC: colliculectomized rats. NOR: sham operates. VC: visual decorticates.

but now instead of just two distracting lights to the left and right, small lights were mounted in 16 different positions around the perimeter of the apparatus, four in each wall. On test trials, which were randomly embedded within training trials, a flashing light would come on briefly in one of the 16 different positions when the rat interrupted an infrared photobeam located halfway between the tunnel exit and the target door. Eventually, each rat in this experiment was presented with a distracting light in each of the 16 locations. The latency between breaking the infrared photobeam and reaching the target door was recorded on both training and test trials. In addition, the position of the distracting light in the rat's visual field was calculated from the film record of each test trial. On the basis of such information, test-trial latency could be plotted as a function of the horizontal eccentricity of the flashing light in the visual field. Of course, this technique was limited by the fact that we had no way of measuring eye movement or calculating the eye's position in the orbit. Despite these shortcomings, it was readily apparent that the behavior of colliculectomized rats was strikingly different from that of the sham-operated or visual-decorticate animals, and that these differences were dependent on where in the visual field the distracting stimulus was presented. Both the sham operates and the visual decorticates showed an elevation of running time in the presence of flashing

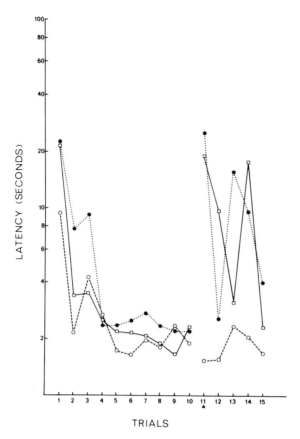

Figure 9.5 Mean running time (latency) for the three groups of rats in the first distraction experiment. The novel visual stimulus was first presented on trial 11. (□) Sham operates; (●) visual-decorticate group; (○) colliculectomized group.

lights located anywhere from directly ahead out to 160° into the peripheral visual field (figure 9.6). Only when the stimulus was presented almost directly behind them did they fail to respond. Most often, the rats in both groups showed oriented head turns toward the flashing light as well as other behaviors such as freezing and rearing. In contrast, the four colliculectomized rats in this study responded to the flashing light only when it was located within 40° of the midline, and were not at all disturbed by stimuli presented from 40° to 160°, where strong responses were elicited in the other two groups.

The response of the colliculectomized group to these stimuli cannot be simply explained by a systematic sparing of that portion of the superior colliculus subserving the central visual field. One animal, which had a complete bilateral lesion including all the anterior portion of the colliculus and extending slightly into the pretectum, continued to show head turning and freezing when stimuli were presented near the midline, but no response to stimuli located elsewhere in the field.

Thus, it appears that rats with large bilateral collicular lesions do not show orienting responses to novel visual stimuli throughout most of

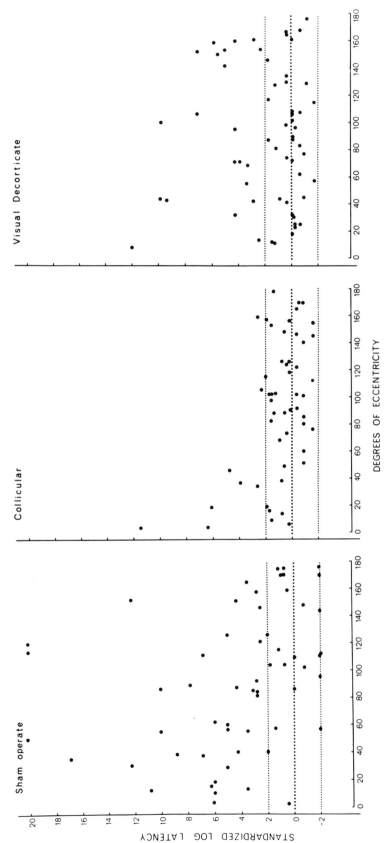

Figure 9.6 Latencies (running times) on all test trials of second distraction experiment. Times are plotted as logarithms standardized as a function of mean and standard deviation of 10 control runs. Position of light right or left of vertical midline of rat's head is indicated on abscissa. Points falling between the two lines of fine dots were within ± 2 SD of the control distribution of times.

their visual field, but will respond occasionally to stimuli occurring within 40° of the midline. The failure to respond to peripheral stimuli is consistent with descriptions of "attentional" deficits following collicular damage in a number of different mammalian species (Denny-Brown 1962; Sprague and Meikle 1965; Anderson and Symmes 1969; Casagrande et al. 1972; Sprague et al. 1973), although most of these accounts are based on rather gross observations of the neurological status of such animals. In a more carefully monitored situation (Milner et al. 1978), we found a failure to respond to novel stimuli presented in the visual periphery in macaques with incomplete collicular lesions, a deficit reminiscent of that observed in the rat. In that experiment, the stimuli missed most often were located in those parts of the peripheral visual field subserved by the ablated portion of the superior colliculus. More recently, Mort et al. (1980) found that hamsters, like rats, were distracted by the sudden presentation of novel visual stimuli in their peripheral visual field and often showed orientation towards the distractor. Nevertheless, like the colliculectomized rats in our experiments, hamsters with undercut colliculi were not as distracted as sham operates when a bright light was turned on in their far peripheral field as they ran toward a target door. There is no information, however, about the field dependency of the effect in these animals.

Although all these data suggest that the superior colliculus might play an important role in the neural mechanisms underlying rapid orientation toward potentially important visual stimuli, at least two important questions remain: What neural structures are mediating the "residual" visuomotor capacity or enabling an animal to orient toward novel stimuli in the central portion of the field after complete ablation of the superior colliculus? Are there any circumstances in which animals with large bilateral collicular lesions can be influenced by stimuli located beyond 40° from the midline (in other words, is it possible to show that colliculectomized animals are not simply peripherally blind)? There are, of course, several other retinofugal projections, including the geniculostriate pathway, which (theoretically, at least) should be intact after colliculectomy; some of these projections have representations of the entire visual field.

Recent work by Ingle (chapter 3, this volume) provides a tentative answer to the first question posed above. In a food perimetry test, gerbils with neonatal ablations of the superior colliculus extending into the deeper laminae either ran directly ahead or failed to respond at all when a small black disk baited with a sunflower seed was presented beyond some 90° from the vertical midline. Nonetheless, like the rats in our own distraction experiments, the gerbils would respond with appropriate orienting movements when the disk was introduced into their central visual fields. However, after subsequent ablations of visual cortex, the animals lost even this residual ability; in later perimetric testing they oriented randomly with respect to the baited disk. These results suggest that cortical mechanisms may underlie some of the visuomotor behavior surviving collicular lesions, although it is not clear what corticofugal pathways might be involved. The results also suggest that cortical mechanisms might normally play an important (and, to some extent, independent) role in mediating orienting responses toward stimuli occurring in the central portion of the visual field. Although Ingle's food-perimetry test is not directly comparable to the distraction paradigm we employed, it is possible that cortical mechanisms were responsible for the oriented movements toward flashing lights presented within 40° on either other side of the midline. This possibility has not been tested empirically.

In the next section we review some very preliminary work that provides a rather cautious affirmative answer to the second question asked above whether there are any circumstances in which colliculectomized animals will respond to peripheral stimuli). The animal used in these studies was not the rat but a much more "visually active" rodent: the Mongolian gerbil (*Meriones unguiculatus*).

Visual Orientation in the Gerbil

In our experiments on visually guided locomotion in the rat, the target doors, which were located on the wall directly opposite the tunnel exit, would almost always fall within 20° or so from the midline of the rat's visual field as it entered the arena. In other words, the target doors were located in the same part of the rostral field in which the flashing light would occasionally evoke an orienting movement in a colliculectomized rat. Thus, we had no information about the ability of animals with collicular lesions to detect and locomote toward targets located in the visual periphery in that part of the field where they would not respond to novel visual stimuli.

Some very recent investigations of visually guided behavior in gerbils (Goodale and Purvis, unpublished) have been looking at their ability to locate a target hole placed up to 90° from their visual midline. The simple procedure used in these investigations was similar in many respects to the earlier rat studies. Gerbils were trained to run into a small dark hole (5 cm in diameter) that could be located in any one of a number of different positions in the wall of a semicircular arena with a radius of 35 cm (figure 9.7). Upon entering the hole, they were presented with a sunflower seed. By filming the behavior of the gerbils in this situation through a 45° mirror, we were able to plot the routes the animals followed from the arena entrance to the hole. Since the camera was operating at 64 frames per second, it was possible to obtain detailed records of any orienting movements the gerbils made as they approached the hole.

As figure 9.8 illustrates, normal gerbils learned rapidly to detect and locomote toward the hole no matter where it was located, and the same

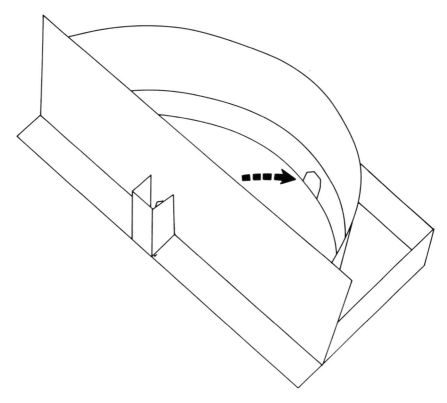

Figure 9.7 Schematic drawing of apparatus used to study visually guided locomotion in the gerbil.

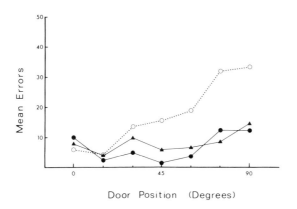

Door Position (Degrees)

Figure 9.8 Mean number of errors made in achieving criterion performance plotted as a function of position of opening in wall arena. Opening at 0° was located directly ahead of gerbil as it entered arena. Openings located to right and left of this central position are indicated on absissa. (O) Gerbils with collicular lesions; (▲) gerbils with visual-cortex lesions; (●) sham operates.

was true for gerbils with large aspiration lesions of posterior neocortex (including in most cases all of area 17 and extending into surrounding peristriate areas). However, gerbils with bilateral radio-frequency lesions of the superior colliculus made many more errors than the other two groups in learning this task; that is, they often ran toward another part of the arena wall before entering the hole—and the more peripherally the hole was located, the more often they did this. Nevertheless, extensive training eventually permitted them to locate doors well out in their peripheral visual field. This was true not only for those gerbils whose collicular lesions included only the superficial and intermediate laminae, but also for those with very deep lesions extending slightly into the dorsal tegmentum and periaqueductal gray. The behavior of one of these deep-lesion gerbils is illustrated in figure 9.9. The runs by this animal toward holes located in the rostral field were as efficient and as rapid as those of normal gerbils. However, as figure 9.9 shows, as the hole was moved beyond 40° from the midline into the visual periphery, the routes became markedly different from those of normal animals. On these occasions the animal apparently ran forward and then paused for a long period of time (sometimes well over a second) before turning in the correct direction and running promptly toward the hole, which was by now located in its rostral field. But despite the fact that the gerbils with collicular damage were much less efficient than normal or visual-decorticate animals in making the initial orientation toward the peripherally located holes, even those animals whose lesions included the deeper laminae were able to detect the hole and turn in the correct direction over 80% of the time after a considerable amount of post-operative training. Thus, it is likely that visuomotor pathways other than the superior colliculus were mediating the behavior initiated and directed by stimuli located in the peripheral visual field. This behavior was far from "normal." Particularly in the early stages of learning, the colliculectomized gerbils failed to detect the hole when it was located outside the rostral field. This initial difficulty in learning to orient toward peripherally located holes may again reflect the involvement of collicular mechanisms in mediating orientation toward important but peripheral visual stimuli in new situations.

In a food-perimetry test (figure 9.10) similar to that employed by Ingle (chapter 3, this volume), there were again clear differences between the behavior of colliculectomized gerbils and that of sham operates and visual decorticates. The gerbils with large lesions of neocortex (including area 17 and parts of areas 18, 18a, and 19) were as capable as sham operates of responding briskly and accurately to a stationary disk baited with a sunflower seed throughout their entire visual field up to about 150° from the midline (figure 9.11). Nevertheless, gerbils with deep collicular lesions often failed to respond to seeds located beyond 70° from the midline, and even when they did make oriented movements the angle of their final turn most often undershot the position of the disk by

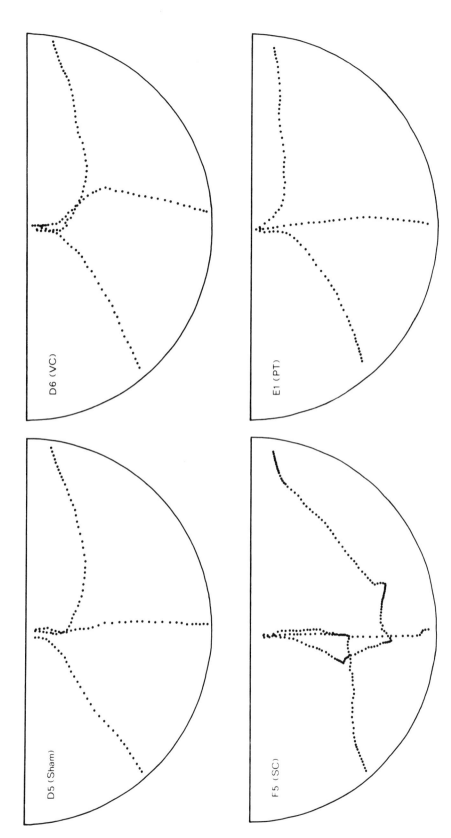

Figure 9.9 Routes followed in arena by representative gerbils from four different groups. Three different positions of opening are illustrated. Dots represent center of gerbil's head on successive frames of film (64 frames/sec). Sham: sham operate. VC: visual decorticate. SD: colliculectomized gerbil. PT: gerbil with pretectal lesion. Note that F5(SC) runs as directly to central opening as the other three gerbils but has trouble orienting toward opening when it is located more peripherally.

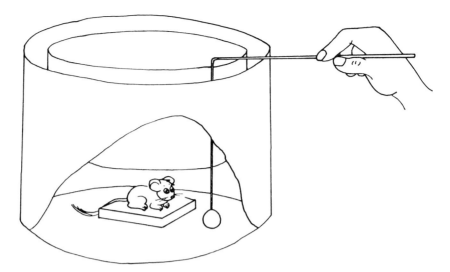

Figure 9.10 Schematic drawing of perimetry apparatus used to plot orienting responses to baited disks.

a considerable amount. In fact, as figure 9.11 illustrates, they rarely made final turns larger than 90°. These results are similar to those obtained by Ingle, who also examined the orientation ability of gerbils with deep collicular ablations (although the animals in his study received the lesions as neonates).

In a very preliminary experiment, we have assessed the ability of gerbils with different lesions of the visual system to negotiate a transparent grid barrier placed between the entrance to the arena and the hole in the arena wall (figure 9.12). The barrier extended horizontally through approximately 80° of the gerbil's visual field (at the entrance to the arena), and consisted of 16 vertical black stripes each occupying 2.5° of visual angle and spaced at equal intervals across a piece of transparent plexiglas. With this arrangement, the hole was clearly visible through the vertical grid of the barrier. As figure 9.12 illustrates, the barrier was located in such a way that on different trials the hole was nearer one end of the barrier than the other. When the barrier was introduced into the situation, sham-operated gerbils would often run up to it, sniff around its edges, rear up, and then eventually run around it toward the hole. On over half of the trials, however, they immediately ran around the barrier and approached the hole without touching either the barrier or the perimeter of the arena. Moreover, on one-third of the trials, they chose the more efficient route; that is, they ran around the edge of the barrier that was closer to the hole. As can be seen in figure 9.13, gerbils with superficial collicular lesions and those with deeper ablations could negotiate the barrier and approach the target hole as efficiently as the sham operates. However, the visual-decorticate animals were clearly impaired on this task. Even though in the absence of the barrier this group could locomote toward the hole in the arena as

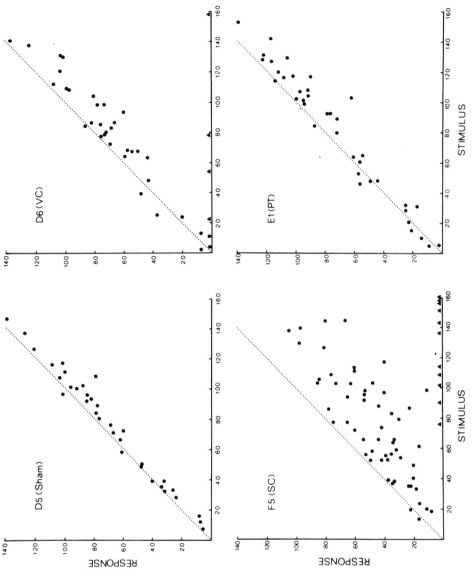

Figure 9.11 Angle of final turn toward disk, plotted as a function of initial position of disk in gerbil's visual field. Data for representative gerbils from four different groups. $x = y$ axis is indicated by dotted line. Failures to detect disk are indicated by triangles on abscissa. Note the undershoots and misses of F5(SC).

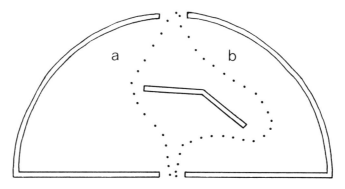

Figure 9.12 Schematic drawing illustrating position of grid barrier in arena. Two possible routes around barrier toward opening in arena wall are shown as dotted lines. Route *a* is clearly more efficient than route *b*.

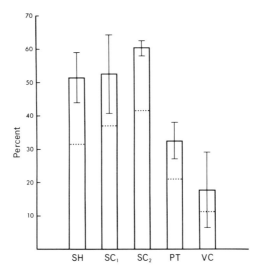

Figure 9.13 Mean percentage of trials in which each group successfully avoided barrier. Percentage of trials in which the more efficient route was chosen is indicated by dotted line. SH represents the six sham operates, SC_1 the seven gerbils with superficial collicular lesions, SC_2 the two with deeper collicular lesions, VC the four with pretectal lesions, and PT the six with visual-cortex lesions.

well as the sham operates, on most of the barrier trials they ran directly into the barrier. The performance of this group did not improve on other trials in which the hole was in plain view and not behind the barrier. On these trials, they ran toward the barrier (and thus away from the hole) as often as they ran directly toward the hole. This again suggests that when the acuity demands of the task are high, visual-decorticate animals often show deficits in their visually guided behavior. In other words, the visual-decorticate group did as well as the sham operates when no barrier was present, but when a barrier was introduced that either obscured the view of the hole or could be confused with the hole, their performance deteriorated.

A small group of gerbils with pretectal lesions were also tested on this task. As figure 9.13 illustrates, the performance of this group was significantly worse than that of the sham-operate or colliculectomized groups. On most trials, they were not successful at avoiding the barrier, but instead ran directly into it after entering the arena. However, when the barrier was not obscuring the view of the hole for the pretectal group, their performance (in contrast to that of the visual decorticates) improved dramatically, and, like the sham-operate and colliculectomized animals, they approached the hole directly on the majority of trials. The precise nature of the deficits responsible for these differences in performance is not clear. In frogs and toads (Ingle 1973, 1977, chapter 3 in this volume), there is a striking loss of barrier avoidance after large posterior thalamic lesions in the area considered homologous with the pretectal projections of higher vertebrate classes (Ebbesson 1970; Riss and Jakway 1970). Nonetheless, the same animals continue to demonstrate vigorous prey catching. Conversely, frogs with lesions in the optic tectum (homologous with the mammalian superior colliculus) can avoid barriers and jump correctly through apertures even though they are unable to make any overt response to either moving prey objects or threatening stimuli. This double dissociation of deficits is similar to that observed in the gerbil experiments described above. Gerbils with collicular lesions were able to negotiate barriers efficiently even though in the perimetry task they had difficulty in detecting small stimuli located in their visual periphery. In contrast, those with pretectal lesions were much less efficient in avoiding the barriers, even though in the perimetry test they were comparable to sham operates in detecting and orienting toward small stimuli anywhere in their visual fields. Not all these situations are directly comparable to the experiments carried out on frogs and toads, and precise descriptions of the deficits accompanying pretectal lesions in mammals await further experimentation.

The Nature of the Collicular Deficit

By now it should be apparent that a common thread runs through the results of all the experiments on colliculectomized rats, hamsters, and gerbils described above: Animals with bilateral collicular lesions extending into the deeper laminae often fail to detect stimuli in their peripheral visual fields. Even in those situations where they do respond to stimuli outside the central portion of the visual field, their oriented movements toward these stimuli are less efficient, less complete, and less frequent than those of normal animals. Finally, in addition to the sharp decrease in the probability of responding that occurs around 40° or so from the midline, there is a further decline in response probability as the stimulus is located more and more peripherally.

Although it is clear that animals with complete collicular ablations are not peripherally blind, it is equally clear that their "residual" vision is affected by more than the retinal eccentricity of the stimulus. For example, a large visual stimulus seems to be more effective in evoking a response than a smaller one, and an "anticipated" stimulus more than a novel one. But the same is undoubtedly true for normal animals in most situations, even though their "base rate" of responding is often much higher. It can still be argued with some force, however, that complex interactions will be found between the characteristics and retinal eccentricity of the stimulus, the nature of the lesion, and the response demands of the particular situation. Recent experiments by Midgley and Tees (1980) have shown that variations in stimulus parameters such as brightness and movement will dramatically affect the probability of eliciting an orienting response in colliculectomized rats. In our own investigations, collicular lesions seemed to affect brisk orientation toward small novel stimuli (such as the flashing lights in the distraction experiment) much more than they affected responses to larger, more prominent stimuli associated with food (such as the holes in the wall of the arena). This might mean that the superior colliculus, unlike other retinofugal targets, contains neuronal networks that are particularly sensitive to small novel stimuli and may participate in the initiation and sequencing of the motor commands that result in rapid and accurate orientation toward such stimuli. Unfortunately, not much more can be said with any certainty about the nature of collicular function in mammals until experiments are carried out in which the dimensions of stimulus "salience" and retinal eccentricity have been systematically varied across situations with different response demands. By generating a series of "detection" curves for normal and colliculectomized animals in a number of different tasks, such experiments would provide detailed information about the stimulus and response characteristics of the visuomotor channel that is disrupted by collicular lesions. In chapter 3 above, Ingle describes in some detail how such experiments might enable us to test specific hypotheses about the sensorimotor characteristics of collicular networks.

Despite the reservations and caveats put forward in the preceding paragraphs, some rather clear conclusions can be drawn from the results of the experiments described in the preceding two sections of this chapter. Schneider (1966, 1967, 1969) was correct in his assertion that the superior colliculus is involved in orientation behavior. But not all classes of orientation behavior require the participation of the superior colliculus. Contrary to Schneider's arguments, the weight of evidence now suggests that the superior colliculus in mammals, like its non-mammalian homolog the optic tectum (see Ingle, chapter 3), is not an essential part of the neural circuitry involved in the visual control of locomotion either around barriers or toward targets. In fact, the preliminary evidence from mammals supports the view that the pretectum may play a much more crucial role in these behaviors than does the colliculus. This observation is also consistent with the earlier work on frogs and toads.

The class of orientation behavior in which the superior colliculus seems to play an indispensable role is detection of and orientation toward novel visual stimuli suddenly introduced into the peripheral visual field. A parametric description of the sensory and motor characteristics of this orienting response has yet to be worked out for any mammalian species, but it is impressive how much this putative collicular function in mammals resembles the descriptions of tectal function in frogs and toads (Ewert 1970 and chapter 1 above; Ingle 1973 and chapter 3 above). The prey-catching behavior of anurans, which consists of a rapid reflexive orientation to a small visual target, could be regarded as an evolutionary antecedent of the orienting response to peripheral visual stimuli observed in mammals. Even though collicular mechanisms in mammals may be considerably modulated by the extensive corticotectal projections that have arisen in this vertebrate class, the operating characteristics of the visuomotor channel in which the superior colliculus (or optic tectum) is an important component may be highly similar throughout the vertebrate line. Thus, it could be argued that during the course of vertebrate evolution the principal neural systems underlying different classes of visually oriented behaviors have remained anatomically distinct despite the development of extensive interconnections.

Collicular Lesions and Discrimination-Task Deficits

Schneider (1969) argued that the poor performance of colliculectomized animals on discrimination tasks is a direct result of their failure to approach the stimulus doors correctly, and those investigators who have required their animals to orient toward the discriminanda have found deficits following collicular lesions whereas those who have not required orientation have therefore failed to observe such deficits. Yet in this chapter we have developed the argument that although the superior colliculus is an important component of at least one class of orientation behavior, it is not essential to visually guided locomotor behavior and hence to visually guided approach responses toward the correct stimulus doors in a discrimination task. But how then can the poor performance of colliculectomized animals on a number of different visual discrimination tasks be explained? We argue that their difficulty may be a direct consequence of their failure to orient to salient stimuli in locations displaced from the focus of their current behavior.

It has been demonstrated repeatedly that spatial factors are of considerable importance in the visual-discrimination performance of normal animals. For example, when the location of the stimulus and the site of the response are spatially separated, monkeys learn color and object discriminations less rapidly than when the required response involves touching the stimulus area (Cowey 1968). Moreover, a separation as small as 2 cm can retard learning quite noticeably (Schuck 1960), and one of 15 cm or more can result in no learning at all (Murphy and Miller 1955). Similarly, if a discontiguity of stimulus and response is introduced in a discrimination problem that was previously acquired under conditions in which the stimulus and the response site were contiguous, the performance of normal monkeys is often disrupted (Murphy and Miller 1955; Polidora and Fletcher 1964). In monkeys, the effect has been attributed to a "visual sampling gradient" (Schuck 1960) in which the area toward which the eyes and head are oriented—usually, the area touched in responding—is the region most adequately sampled for discriminative cues (Meyer et al. 1965). It has been suggested that in order to effectively sample the visual cues, particularly if these are displaced some distance away from the response site, an "orienting movement" is necessary (Polidora and Fletcher 1964). In rats, similar stimulus-response separation effects have been described. In experiments on pattern-discrimination learning, Lashley (1938) demonstrated that rats will show positive transfer from the complete stimulus cards on a jumping stand to the lower halves presented alone, but not to the upper halves. Ehrenfreund (1948) reported that when the stimuli were located above the level of the rats' eyes, as opposed to on the same level, the learning of a discrimination task was less efficient.

The difficulty normal monkeys and rats have in "sampling" visual cues some distance from the response site has been found to be exaggerated by lesions of the superior colliculus. Although colliculectomized monkeys did not differ from control animals when the stimuli were located at the response site, their performance deteriorated when the stimuli were moved to a different location (Butter 1974, 1979). Kurtz and Butter (1976) reported that monkeys with collicular lesions performed less accurately than they had before the lesions on a discrimination task when, and only when, the discriminative stimuli were separated from the response panels. (The impairment was most severe at large separations.) More recently, we have shown that the performance of rats with bilateral lesions of the superior colliculus was disrupted on a horizontal-vertical task when the stimuli were raised above the place where the rats pushed at the stimulus doors with their head and

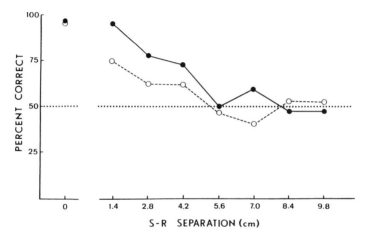

Figure 9.14 Mean performance levels during test sessions in first experiment of Milner et al. 1979. (O) Superior-colliculus-lesion group; (●) operated control group; (S-R) stimulus-response.

forepaws (Milner et al. 1979). As figure 9.14 shows, although the performance of sham-operate and colliculectomized rats were both disturbed by the separation of the stimulus from the response site, the performance of the collicular group deteriorated much more rapidly than that of the sham-operate controls as the distance between the stimulus and the response site was increased. In fact, the collicular impairment was apparent at the smallest separation distance used. In a second experiment (figure 9.15) we showed that these findings were not simply due to an increase in task difficulty. When contradictory stripes were introduced into the upper portions of the stimulus doors, the rats with collicular lesions were less disturbed than the control animals. Taken together, these two experiments confirm that colliculectomy narrows stimulus sampling in rats in much the same way as has been found in monkeys in Butter's laboratory. We would like to suggest, as did Butter (1974, 1979) and others before us, that all these results on monkeys and rats can be explained by arguing that when stimulus and response sites are discontiguous, animals must make an appropriate orienting response in order to sample the visual stimuli effectively, and that lesions of the superior colliculus alter performance by interfering with this orienting behavior. If indeed the superior colliculus has a special role in orientation toward potentially relevant stimuli some distance from where the animal is responding or manipulating devices within the apparatus, then it is not surprising that sometimes collicular lesions will produce deficits in visual-discrimination performance— particularly in situations where, inadvertently, a task has been used in which the relevant cues are spatially separated from the response site.

This kind of argument can be extended to explain the observation that animals with collicular lesions often fail to learn new visual-

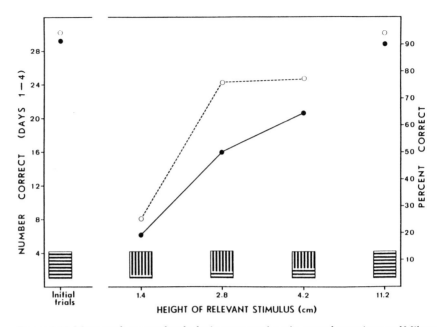

Figure 9.15 Mean performance levels during test sessions in second experiment of Milner et al. 1979. (O) Superior-colliculus-lesion group; (●) sham-operated group. For the first eight trials of each test session, animals were overtrained on the training task, and eight more such control trials were mixed with eight each of the three types of test trial.

discrimination problems even though they may show no retention deficits in problems acquired before operation. When faced with a new discrimination problem, an animal may have to scan the stimulus array—to make efficient orienting movements of the head and eyes from one part of the stimulus card to another—in order to bring potentially important aspects of the discriminative stimuli within the central portion of the visual field. In the later stages of training, the animal may concentrate its gaze on that portion of the stimulus array that contains relevant cues for accurate discrimination and is nearest the response site. If collicular lesions interfere with the capacity to make orienting movements to potentially relevant stimuli located in the visual periphery, then colliculectomized animals might be expected to show deficits in learning new discrimination problems as well as problems on which they received only a small amount of preoperative training. Berlucchi et al. (1972) and Anderson and Williams (1971) showed that the learning of an unfamiliar discrimination is impaired in cats as a result of collicular damage, while postoperative retention of previously learned tasks is scarcely affected. In our own experiment (Milner et al. 1979) we did find a deficit in the relearning of the horizontal-vertical discrimination problem after collicular lesions, but this could be explained by the fact that our preoperative criterion of learning (90% correct in a 40-trial session) was less stringent than the criteria used by most investigators. Previous workers (Ghiselli 1937; Thompson 1969) who used rats found no collicular deficit in the relearning of a horizontal-vertical discrimination, but they overtrained their animals prior to surgery. Their animals, in contrast to our own, may have come before operation to direct their gaze almost entirely on a small area of the stimulus cards, much as the rats in Lashley's 1938 experiment had done. In a very recent experiment (Milner and Bowen, in prep.), rats overtrained on a horizontal-vertical discrimination problem before surgery showed almost no deficit in relearning after collicular lesions. Other workers who have also employed a strict preoperative criterion have found good postcolliculectomy savings on pattern-discrimination problems in cats (Anderson and Williams 1971; Berlucchi et al. 1972; Myers 1964) and monkeys (Anderson and Symmes 1969). In sharp contrast, a study in which preoperative training was terminated upon attainment of perhaps the least stringent criterion in the relevant literature (19 correct in 20 trials) revealed a clear relearning impairment in a group of colliculectomized tree shrews (Casagrande and Diamond 1974).

Thus, the deficit in discrimination learning that follows lesions of the superior colliculus may have nothing to do with the "sensory" or "analytical" demands imposed by the task, or the ability of the animal to approach the stimuli under visual control, but instead may simply result from a failure to orient the gaze away from a primary goal toward any source of potentially useful information located elsewhere in the visual field and to "scan" the visual array as efficiently as normals. Of course, scanning movements and rapid orienting responses to discrete portions of the stimulus array may not involve exactly the same visuomotor pathways. To a certain extent, scanning movements seem to be generated spontaneously, whereas other orienting responses are clearly dependent on eliciting stimuli. Nevertheless, both kinds of orienting movements may be necessary in the early stages of discrimination learning and dependent on the integrity of the superior colliculus. But the experiments reviewed above provide only indirect evidence for a scanning deficit. What direct evidence is there that colliculectomized animals make fewer scanning movements in discrimination tasks than normals? Thus far, a deficit in scanning movements of the head after collicular damage has been documented for only one animal: the Syrian hamster (Mort et al. 1980). Although normal hamsters showed spontaneous scanning movements of the head as they learned a three-choice brightness discrimination, this motor pattern was absent in hamsters with collicular undercuts. This deficit may be related to differences in the discrimination learning of normal and colliculectomized animals. A clear demonstration of this relationship, however, will require far more observations of this kind.

References

Anderson, K., and D. Symmes. 1969. "The superior colliculus and higher visual functions in the monkey." *Brain Res.* 13: 37–52.

Fractionating Orientation in Rodents

Anderson, K., and M. R. Williams. 1971. "Visual pattern discrimination in cats after removal of superior colliculi." *Psychonom. Sci.* 24: 125.

Berlucchi, G., J. M. Sprague, J. Levy, and A. C. DiBerardino. 1972. "Pretectum and superior colliculus in visually guided behavior and in flux and form discrimination in the cat." *J. Comp. Physiol. Psychol.* 78: 123–172.

Blake, L. 1959. "The effect of lesions of the superior colliculus on brightness and pattern discrimination in the cat." *J. Comp. Physiol. Psychol.* 52: 272–278.

Butter, C. M. 1974. "Effect of superior colliculus, striate, and prestriate lesions on visual sampling in rhesus monkeys." *J. Comp. Physiol. Psychol.* 87: 905–917.

———. 1979. "Contrasting effects of lateral striate and superior colliculus lesions on visual discrimination." *J. Comp. Physiol. Psychol.* 93: 522–537.

Casagrande, J. A., and I. T. Diamond. 1974. "Ablation study of the superior colliculus in the tree shrew (*Tupaia glis*)." *J. Comp. Neurol.* 156: 207–237.

Casagrande, J. A., J. K. Harting, W. C. Hall, I. T. Diamond, and G. Martin. 1972. "Superior colliculus of the tree shrew: A structural and functional subdivision into superficial and deep layers." *Science* 177: 444–447.

Collewijn, H. 1975. "Oculomotor areas in the rabbit's midbrain and pretectum." *J. Neurobiol.* 6: 3–22.

Cowey, A. 1968. "Discrimination." In *Analysis of Behavioral Change*, L. Weiskrantz, ed. New York: Harper & Row.

Denny-Brown, D. 1962. "The midbrain and motor integration." *Proc. R. Soc. Med.* 55: 527–538.

Denny-Brown, D., and R. A. Chambers. 1955. "Visuomotor function in the cerebral cortex." *Arch. Neurol. Psychiatr.* (Chicago) 73: 566.

Doty, R. W. 1961. "Functional significance of the topographical aspects of the retinocortical projections." In *The Visual System: Neurophysiology and Psychophysics*, R. Jung, ed. Berlin: Springer.

Dyer, R. S., M. F. Marino, C. Johnson, and T. Kruggel. 1976. "Superior colliculus lesions do not impair orientation to pattern." *Brain Res.* 112: 176–179.

Ebbesson, S. 1970. "On the organization of central visual pathways in vertebrates." *Brain, Behav., Evol.* 3: 178–194.

Ehrenfreund, D. 1948. "An experimental test of the continuity theory of discrimination learning with pattern vision." *J. Comp. Physiol. Psychol.* 41: 408–422.

Ewert, J.-P. 1970. "Neural mechanisms of prey-catching and avoidance behavior in the toad (*Bufo bufo* L.)," *Brain, Behav., Evol.* 3: 36–56.

Ferrier, R. J., and R. M. Cooper. 1976. "Striate cortex ablation and spatial vision." *Brain Res.* 106: 71–85.

Fischman, M. W., and T. H. Meikle. Jr. 1965. "Visual training discrimination in cats after serial tectal and cortical lesions." *J. Comp. Physiol. Psychol.* 59: 193–201.

Freeman, G. L., and J. W. Papez. 1930. "The effects of subcortical lesions on the visual discrimination of rats." *J. Comp. Psychol.* 11: 185–191.

Ghiselli, E. E. 1937. "The superior colliculus in vision." *J. Comp. Neurol.* 67: 451–467.

Goodale, M. A., and R. C. C. Murison. 1975. "The effects of lesions of the superior colliculus on locomotor orientation and the orienting reflex in the rat." *Brain Res.* 88: 243–255.

Goodale, M. A., N. P. Foreman, and A. D. Milner. 1978. "Visual orientation in the rat: A dissociation of deficits following cortical and collicular lesions." *Exp. Brain Res.* 31: 445–457.

Hess, S., S. Burgi, and J. Bucher. 1946. "Motor-function of tectal and tegmental area." *Monatsschrift Psychiatr. Neurol.* 112: 1–52.

Horel, J. A. 1968. "Effects of subcortical lesions on brightness discrimination acquired by rats without visual cortex." *J. Comp. Physiol. Psychol.* 651: 103–109.

Humphrey, N. K., and L. Weiskrantz. 1967. "Vision in monkeys after removal of the striate cortex." *Nature* 215: 595–597.

Ingle, D. 1973. "Two visual systems in the frog." *Science* 181: 1053–1055.

———. 1976. "Central visual mechanisms in anurans." In *Neurobiology of the Frog*, R. Llinas and W. Precht, eds. Berlin: Springer.

———. 1977. "Detection of stationary objects by frogs (*Rana pipiens*) after ablation of optic tectum." *J. Comp. Physiol. Psychol.* 91: 1359–1364.

Ingle, D., and J. M. Sprague. 1975. "Sensorimotor function of the midbrain tectum." *Neurosci. Res. Program Bull.* 13: 169–288.

Klüver, H. 1937. "Certain effects of lesions of the occipital lobes in macaques." *J. Psychol.* 4: 383–401.

———. 1941. "Visual function after removal of the occipital lobes in monkeys." *J. Psychol.* 11: 23–45.

———. 1942. "Functional significance of the geniculo-striate system." In *Visual Mechanisms*, H. Klüver, ed. Lancaster, Pa.: Jaques Cattell.

Kurtz, D., and C. M. Butter. 1976. "Deficits in visual discrimination performance and eye movements following superior colliculus ablations in rhesus monkeys." *Neurosci. Abstr.* 2: 1122.

Lashley, K. S. 1931. "The mechanism of vision. IV. The cerebral areas necessary for pattern vision in the rat." *J. Comp. Neurol.* 53: 419–478.

———. 1935. "The mechanism of vision. XII. Nervous structures concerned in the acquisition of habits based on reactions to light." *Comp. Psychol. Monogr.* 11: 43–79.

————. 1938. "The mechanisms of vision. XV. Preliminary studies of the rat's capacity for detail vision." *J. Gen. Psychol.* 18: 123–193.

Layman, J. D. 1936. "Functions of the superior colliculi in vision." *J. Genet. Psychol.* 49: 33–47.

Marks, K. E., and J. A. Jane. 1974. "Effects of visual cortical lesions upon ambulatory and static localization of light in space." *Exp. Neurol.* 42: 707–710.

Meyer, D. R., R. R. Treichler, and P. M. Meyer. 1965. "Discrete training techniques and stimulus variables." In *Behavior of Nonhuman Primates: Modern Research Trends,* A. M. Schrier et al., eds. New York: Academic.

Midgley, G., and R. C. Tees. 1980. "Orienting behavior by rats with visual cortex and subcortex lesions." *Exp. Brain Res.* 41: 316–328.

Milner, A. D., N. P. Foreman, and M. A. Goodale. 1978. "Go-left go-right discrimination performance and distractibility following lesions of prefrontal cortex or superior colliculus in stumptail macaques." *Neuropsychologia* 16: 381–390.

Milner, A. D., M. A. Goodale, and M. C. Morton. 1979. "Visual sampling after lesions of the superior colliculus in rats." *J. Comp. Physiol. Psychol.* 93: 1015–1023.

Mort, E., S. Cairns, H. Hersch, and B. Finlay. 1980. "The role of the superior colliculus in visually guided locomotion and visual orienting in the hamster." *Physiol. Psychol.* 8: 20–28.

Murphy, J. V., and R. E. Miller. 1955. "The effect of spatial contiguity of cue and reward in the object-quality learning of rhesus monkeys." *J. Comp. Physiol. Psychol.* 48: 221–224.

Myers, R. E. 1964. "Visual deficits after lesions of brainstem tegmentum in cats." *Arch. Neurol.* 11: 73–90.

Polidora, V. J., and H. J. Fletcher. 1964. "An analysis of the importance of S-R spatial contiguity for proficient primate discrimination performance." *J. Comp. Physiol. Psychol.* 57: 224–230.

Riss, W., and J. S. Jakway. 1970. "A perspective on the fundamental retinal projections of vertebrates." *Brain, Behav., Evol.* 3: 30–35.

Robinson, D. A. 1972. "Eye movements evoked by collicular stimulation in the alert monkey." *Vision Res.* 12: 1795–1808.

Rosvold, H. E., M. Mishkin, and M. K. Szwarcbart. 1958. "Effects of subcortical lesions in monkeys on visual discrimination and single alternation performance." *J. Comp. Physiol. Psychol.* 51: 437–444.

Schneider, G. E. 1966. "Superior colliculus and visual cortex: Contrasting behavioral effects of their ablation in the hamster." Ph.D. diss., Massachusetts Institute of Technology.

————. 1967. "Contrasting visuomotor functions of tectum and cortex in the golden hamster." *Psychol. Forsch.* 31: 52–62.

————. 1969. "Two visual systems: Brain mechanisms for localization and discrimination are dissociated by tectal and cortical lesions." *Science* 163: 895–902.

Schuck, J. R. 1960. "Pattern discrimination and visual sampling by the monkey." *J. Comp. Physiol. Psychol.* 53: 251–255.

Sherrington, C. S. 1906. *The Integrative Action of the Nervous System.* Cambridge University Press.

Spear, P. D., and J. J. Braun. 1969. "Pattern discrimination following removal of visual cortex in the cat." *Exp. Neurol.* 25: 331–348.

Sprague, J. M., and T. H. Meikle, Jr. 1965. "The role of the superior colliculus in visually guided behavior." *Exp. Neurol.* 11: 115–146.

Sprague, J. M., G. Berlucchi, and G. Rizzolatti. 1973. "The role of the superior colliculus and pretectum in vision and visually guided behavior." In *Handbook of Sensory Physiology,* vol. VII/3B, R. Jung, ed. Berlin: Springer.

Thompson, R. 1969. "Localization of the 'visual memory system' in the white rat." *J. Comp. Physiol. Psychol. Monogr.* 69 (no. 4, part 2): 1–29.

Tunkl, J. E., and M. A. Berkley. 1974. "Form discrimination and localization performance in cats with superior colliculus ablations." *Soc. Neurosci. Abstr.,* p. 454.

Urbaitis, J. C., and T. H. Meikle, Jr. 1968. "Relearning a dark-light discrimination by cats after cortical and collicular lesions." *Exp. Neurol.* 20: 295–311.

Voneida, T. J. 1970. "Behavioral changes following midline section of the mesencephalic tegmentum in the cat and monkey." *Brain, Behav., Evol.* 3: 241–260.

Weiskrantz, L. 1963. "Contour discrimination in a young monkey with striate cortex ablation." *Neuropsychologia* 1: 145–164.

Wurtz, R. H., and M. E. Goldberg. 1972. "The primate superior colliculus and the shift of visual attention." *Invest. Ophthalmol.* 11: 441–450.

10 Contrasting Behavioral Methods in the Analysis of Vision in Monkeys with Lesions of the Striate Cortex or the Superior Colliculus

Charles M. Butter
Daniel Kurtz
Clare C. Leiby III
Alphonso Campbell, Jr.

We shall describe and evaluate the methods used to assess visually guided behavior in monkeys with ablations of two brain structures that contribute to vision: the striate cortex and the superior colliculus. Studies of several nonprimate species suggest that these two structures control different aspects of visually guided behavior. Direct observations of hamsters (Schneider 1967) and cats (Sprague and Meikle 1965) with superior-colliculus lesions indicate that they are impaired in orienting to visual stimuli presented in the periphery of the visual field. Similarly, when presented with novel visual or auditory stimuli, rats with superior-colliculus lesions, unlike control rats, continue to locomote toward a site where food is present and fail to orient to the novel stimuli (Goodale and Murison 1975). In contrast, after visual areas of cortex including striate cortex are removed, the animals studied in the experiments cited above showed only slight orientation deficits or none at all. However, when tested in two-choice pattern discriminations, hamsters show deficits after visual cortex ablations but not after collicular ablations (Schneider 1967). A similar dissociation of pattern-discrimination deficits is found in rats with collicular (Layman 1936; Thompson 1969) and cortical (Horel et al. 1966; Lashley and Frank 1934; Thompson 1969) lesions. These and related findings have led to the view, first stated explicitly by Schneider (1969), that the superior colliculus is involved in localizing and orienting to visual stimuli, whereas striate cortex contributes to identifying the features of visual stimuli.

The role attributed to striate cortex by this view has been amply confirmed by electrophysiological analyses too numerous to cite. Results of other electrophysiological studies involving unit recordings (Schiller and Koerner 1971; Goldberg and Wurtz, 1972; Wurtz and Goldberg 1972a) and electrical stimulation (Robinson 1972; Schiller and Stryker 1972) in monkeys support the notion that the primate superior colliculus is involved in orienting the eyes to extrafoveal targets. Since foveal vision is so crucial for identifying the detailed features of objects, it seems paradoxical that, until quite recently, most investigators reported no disturbances or only transient disturbances of visually guided behavior after superior-colliculus lesions in monkeys. Conversely, total removal of striate cortex in monkeys leads to visual impairments so devastating that until recently only very primitive visual capacities have been demonstrated in striatectomized monkeys.

Thus, different strategies—indeed opposing ones—have been used to analyze the contributions of the superior colliculus and striate cortex to visually guided behavior in monkeys: On the one hand, the effects of collicular lesions appear so subtle and transient that specialized behavioral methods must be employed to detect impairments. On the other hand, after total removal of striate cortex in monkeys, the visual impairments are so severe that specialized behavioral methods are needed to demonstrate any residual visual capacities.

Review of Visual-Discrimination Procedures and Related Methods Used with Monkeys

Because the visual capacities of monkeys have been assessed for the most part with two-choice visual discrimination tests, it is appropriate to briefly describe these tests before turning to their use in ablation studies of the superior colliculus and striate cortex.

Two-choice discrimination tests have been customarily administered to monkeys in the Wisconsin General Test Apparatus (WGTA), automated discrimination apparatuses, or (less frequently) the pull-in apparatus. Visual stimuli are presented on the front surfaces of the stimulus boxes of the pull-in apparatus, facing the monkey, and are illuminated by reflected or transmitted light (figure 10.1). The monkey indicates its choice by pulling a stimulus box until it is within reach. If its choice is correct, the monkey can obtain a food reward located inside the box (or, in the apparatus shown in figure 10.1, from a recessed platform behind the front surface of the box).

The WGTA (see figure 10.2) consists of a cage in which the monkey is separated by a sliding opaque screen from the stimulus compartment. The stimulus compartment contains a stimulus tray, on which two objects differing in one or more visual dimensions are customarily presented over food wells. The monkey indicates its choice by displacing one of the objects from the food well; if its choice is correct, it is permitted to retrieve a piece of food that is exposed in the uncovered well. Thus, the WGTA, like the pull-in apparatus, is designed to assess discrimination performance, either in initial learning or in relearning, by the monkey's choice response. Furthermore, the features of the stimuli that the monkey utilizes for discrimination performance can be evaluated by presenting stimuli varying systematically along certain dimensions in test trials, which are usually administered after a criterion of

Butter, Kurtz, Leiby, Campbell

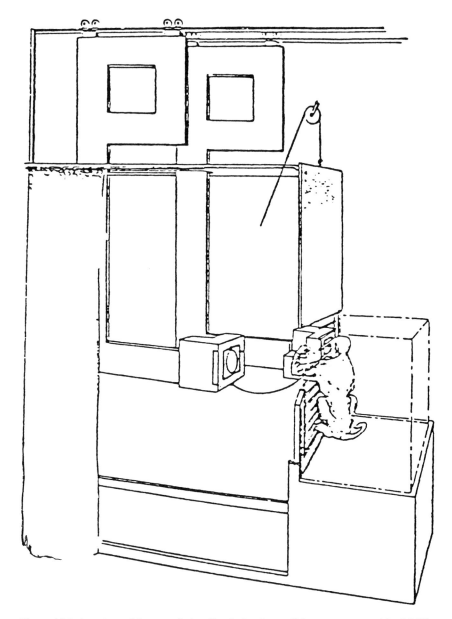

Figure 10.1 A variant of the two-choice discrimination pull-in apparatus used by Schilder et al. (1971) (figure adapted from that reference).

Vision in Monkeys with Lesions

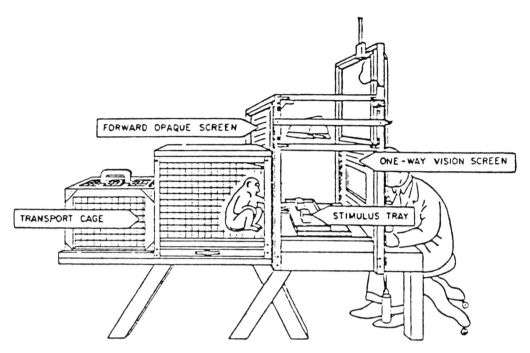

Figure 10.2 A variant of the Wisconsin General Test Apparatus used by Harlow (1951). Figure adapted from Meyer et al. 1965.

learning is attained. (See Cornwell's chapter for a detailed discussion of equivalence tests.) In addition, the monkey's ability to shift attention from one set of features to another can be evaluated in the WGTA by changing the contingencies between reward and different features of the stimuli (see, for example, Soper et al. 1975).

Much of the popularity of the WGTA for studying the visual-discrimination capacities of monkeys is due to the rapid rate at which they acquire discriminations in this situation. The rapidity of learning appears to be largely due to the close proximity of stimulus, reward, and choice-response site in the WGTA (Cowey 1968; Meyer et al. 1965). The extensive use of the WGTA is also attributable to its flexibility: The WGTA can be used to present objects with a wide range of stimulus dimensions (although in this regard it has certain limitations). It can also be used to present a variety of complex tasks, such as oddity and matching to sample, as well as simpler two-choice discriminations. Since monkeys are capable of learning the abstract rules associated with these complex tasks and applying them to unfamiliar stimuli, they can be used to assess capacities such as perceptual matching (by simultaneous matching to sample or oddity learning) and short-term visual memory (by delayed matching to sample).

However, there are several limitations in the use of the WGTA and similar discrimination apparatuses for evaluating visual functions. Vi-

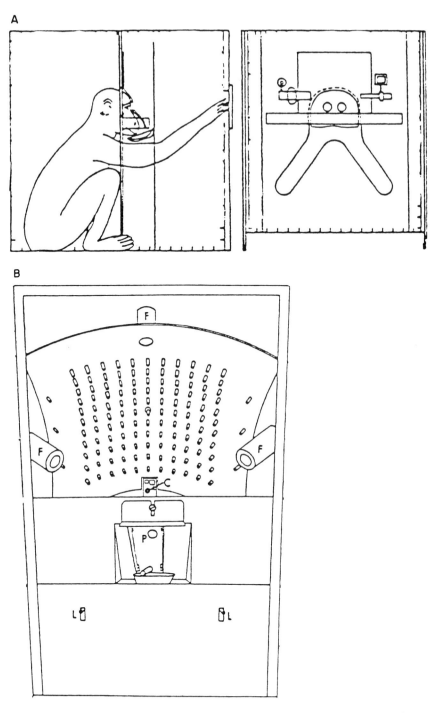

Figure 10.3 Two kinds of automated apparatuses for testing vision in primates. (A) Variant of the Ohio State Apparatus. Adapted from Otteson et al. 1962. (B) Apparatus for perimetry in monkeys. Wall above peephole (P) has been removed to reveal curved perimeter with stimulus bulbs set in it. Large floodlights (F) at sides of perimeter provide corneal reflections recorded by movie camera (C) mounted at bottom of perimeter. Foodwell is located below peephole. Response levers (L) are located on both sides of foodwell. Adapted from Weiskrantz and Cowey 1970.

sual acuity is difficult to assess in these situations, as the animal's freedom of movement results in unwanted variation in the retinal size of the stimuli. For the same reason, it is difficult to precisely control the effective duration at which the stimuli are exposed to the subject and thus to obtain reliable measures of response latencies in apparatuses like the WGTA. Furthermore, because the direction of gaze is not easily controlled or measured in manually operated discrimination apparatuses, they are not suitable for evaluating either orientation of the eyes or visual capacities in specific parts of the visual field. Consequently, monkeys can easily compensate for partial visual-field defects by making compensatory movements when tested in this kind of situation, as Cowey and Weiskrantz (1963) observed. It should also be pointed out that discrimination tests conducted in the customary manner in the WGTA do not strongly tax the capacity to spatially shift attention or gaze. Monkeys apparently have a strong tendency to orient (and presumably attend) to the place where they direct the choice response (Meyer et al. 1965), which is customarily the place where the stimulus is also located.

Some of these disadvantages are overcome or lessened by administering discrimination tests in automated apparatuses such as those shown in figure 10.3. In the apparatus shown in part A, the monkey must make an observing response (in this case by placing its head in a mask) in order to initiate presentation of the discriminative stimuli, which are rear-projected on panels facing the monkey. Thus, the retinal size and effective exposure duration of the stimuli are more precisely controlled in this kind of apparatus than in the WGTA.

The monkey can also be required to initiate presentation of the discriminative stimuli in automated apparatuses by first performing an observing response to a particular cue, such as the dimming of a small light, which must be fixated in order to be detected. Eye position can be monitored during testing to confirm that stimulus fixation is achieved (Wurtz 1969). Requiring fixation of a stimulus prior to brief presentation of discriminative stimuli makes it possible to control the position of the discriminative stimuli in the visual field. Alternatively, corneal reflections can be photographed to determine the direction of gaze at the time when brief stimuli are presented (Cowey and Weiskrantz 1963). Furthermore, by altering the task requirements in automated apparatuses so that the monkey must respond differentially to the presence versus the absence of one stimulus, it is possible to evaluate detection of visual stimuli (Bender 1973) and, with the aid of eye photography, to determine how detection varies in different parts of the visual field (Cowey and Weiskrantz 1963) (see part B of figure 10.3). The usual penalty for more precise controls in automated apparatuses is much slower discrimination learning than in the WGTA (Meyer et al. 1965).

Effects of Superior-Colliculus Lesions on Visually Guided Behavior and Eye Movements in Monkeys

We pointed out that two-choice discrimination tests administered by customary procedures in the WGTA were not suitable for evaluating sensory functions in particular parts of the visual field or the ability to orient to visual stimuli. In view of these limitations, it is not surprising that monkeys with lesions of the superior colliculus (SC), which is thought to be involved in detecting or orienting to peripheral stimuli, rarely show deficits in the WGTA. Thus, SC lesions in monkeys do not adversely affect the performance of discriminations involving patterns (Rosvold et al. 1958; Butter 1974a), color, brightness, size, or orientation of stripes (Butter 1974a) in the WGTA. Similarly, when monkeys with SC lesions are tested in automated apparatuses, which are also insensitive to sensory or orienting deficits, they are not impaired in discriminating between luminance differences and patterns (Keating 1976a), flickering versus steady light, or rotating versus stationary stimuli (Anderson and Symmes 1969). However, when investigators have employed testing procedures more sensitive to deficits in peripheral vision or deficits in orientation to visual stimuli than those customarily employed in the WGTA, they have found impairments in visually guided behaviors and in eye movements after SC lesions in monkeys.

Latto (1978) tested monkeys in a two-choice discrimination task in which they were rewarded for pressing only the panel on which a small target pattern appeared. Varying numbers of irrelevant patterns also appeared on both panels. Latto assumed that the task required searching the array of patterns for the relevant target, and that search time would be reflected in response latencies (at least on trials when the subject performed correctly). These assumptions were supported by the finding that the response latencies of unoperated monkeys increased monotonically as the number of irrelevant patterns increased. One monkey showed even greater latency increases with increasing numbers of irrelevant patterns after SC lesions. This monkey apparently required more search time in the task after surgery, possibly because of deficits in shifting fixation. Since eye movements were not recorded, this question cannot be answered. The other monkey that received SC lesions in Latto's study, unlike the first one, did not show latency changes after surgery; rather, its discrimination performance declined as the number of irrelevant patterns increased. It is not clear whether these differences in performance were due to differences that were found in the location of their lesions, or to possible differences in behavioral strategies, or to both factors.

A different approach to the analysis of orienting behavior was employed by Butter (1974a, 1979). In these studies, monkeys were tested in two-choice color-discrimination tasks in a WGTA. The testing procedure differed from that customarily employed in that the stimuli were spatially displaced from the response sites in some phases of testing. In the other phases of testing, the stimuli were located in the customary position: at the response sites. It was assumed that if monkeys with SC lesions are impaired in orienting to visual stimuli, their discrimination performance might be deficient when the stimuli are displaced from the response sites, for monkeys apparently direct their gaze to the response sites (Meyer et al. 1965). In the WGTA, the response, as mentioned, involves displacing a plaque or other object from over a food well. The farther the stimuli are displaced from the response sites (within an effective range), the more the performance of normal monkeys declines, presumably because increasing stimulus-response (S-R) separation makes it less likely that the monkeys identify the stimuli without reorienting their gaze from the response sites (Stollnitz 1963).

Some of the findings from Butter's studies involving S-R separation are shown in figure 10.4. The unoperated monkeys showed the expected decline in discrimination performance as S-R separation increased. Note that the monkeys with partial striate-cortex lesions were unimpaired in this test, as were monkeys with cortical lesions outside the visual system (Butter 1974a). The colliculus-lesioned monkeys performed as well as their controls did when there was no S-R separation, but they were impaired in discriminating between the same stimuli when they were displaced at intermediate distances from the response sites. All the monkeys performed poorly when the stimuli were displaced even farther from the response sites. The results of further ex-

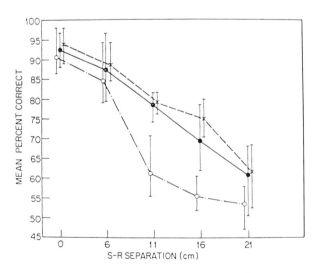

Figure 10.4 Mean percent correct performance of (O) monkeys with superior-colliculus lesions, (●) monkeys with lateral striate lesions, and (x) unoperated control monkeys in color-discrimination test with varying degrees of stimulus-response (S-R) separation. Source: Butter 1979.

periments indicate that the deficit in performing color discriminations involving S-R separation is not transient and is independent of the direction in which the stimuli are displaced from the response sites (Butter 1979).

These findings, together with those of Latto, suggest that the primate superior colliculus may participate in the control of orienting movements to visual stimuli. However, these results offer only indirect evidence for this conclusion. More direct evidence is provided by studies in which orienting movements of the eyes to visual targets were recorded in monkeys with SC lesions.

Pasik and Pasik (1964) reported that total ablation of the SC in monkeys did not impair nystagmus induced optokinetically or by caloric stimulation. However, there is more recent evidence that voluntary saccades to visual targets are impaired after SC lesions in monkeys.

Wurtz and Goldberg (1972b) reported that, after unilateral SC lesions, monkeys show a transient increase in the latencies of saccadic movements (recorded by external electro-oculogram electrodes) that they were trained to perform to extrafoveal visual targets presented in the field contralateral to the lesion. In all other respects, the saccades of the operated monkeys appeared to be normal. Using more sensitive methods of eye-movement recording in monkeys (miniature EOG electrodes implanted in the bone surrounding the orbit), Mohler and Wurtz (1977) found not only increases in saccade latencies but also a marked increase in the frequency of undersaccades to visual targets in the field contralateral to unilateral SC lesions. These abnormally short-amplitude saccades were followed by corrective saccades by which the monkeys foveated the target.

It appears that spontaneous saccades of monkeys are even more severely impaired after unilateral SC lesions. In a preliminary report, Albano and Wurtz (1978) described marked reductions in spontaneous saccades directed to the side of space opposite the lesion, whereas saccades toward the side ipsilateral to the lesion appeared normal. A monkey that received a bilateral SC lesion showed a selective reduction in short-amplitude spontaneous saccades after surgery.

Kurtz and Butter (1976) recorded eye movements of SC-lesioned monkeys performing a visual-discrimination task in which the stimuli were displaced varying distances from the response sites. This experiment was undertaken to determine whether the impaired performance of monkeys with SC lesions in this situation is accompanied by deficits in orienting the eyes to the stimuli. During testing, the monkey's head was immobilized and directed toward the observing-response (OR) panel located in the center of a display screen in the testing apparatus (see figure 10.5). The monkey initiated each trial by pressing the OR panel when a small, dim light appeared on it. When this observing response was made, the two discriminative stimuli, which differed in color, were projected on one of five pairs of stimulus panels located at

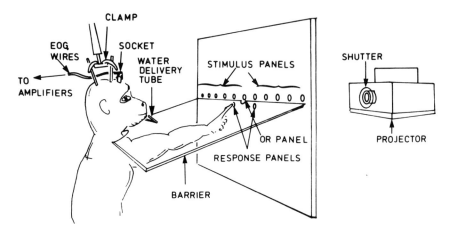

Figure 10.5 Apparatus used by Kurtz and Butter (1976) for studying visual-discrimination performance with stimuli projected varying distances from response panels.

varying distances from the OR panel. The two stimulus panels in each pair were located on either side of the OR panel, equidistant from it. The monkey was required to press the response panel closer to the correct stimulus in order to receive water reward from a tube. Since the discriminative stimuli were small and dim, the monkey had to reorient its gaze from the OR panel in the center of the display screen toward at least one of the discriminative stimuli in order to decide which response panel to press.

The two stimuli were apparently not equally informative to the monkeys. Before and after surgery, they performed the discrimination well only when they oriented their eyes toward the correct stimulus (S^+). Before surgery, both the monkeys' discrimination performance and the frequency with which they oriented to the S^+ declined as stimulus eccentricity increased; both performance and S^+-orientation scores also increased with training. After SC lesions, the monkeys' discrimination performance declined as stimulus eccentricity increased, whereas the operated control monkeys' performance remained high (see figure 10.6). Likewise, the colliculus-lesioned monkeys, in contrast with the controls, oriented less to the S^+ as stimulus eccentricity increased. Rather, as seen in figure 10.7, they continued to direct their gaze toward the center of the display screen to a greater extent than did the control monkeys. It appears that the monkeys with SC lesions had no difficulty in identifying the S^+ and associating it with the appropriate response panel; like the control monkeys, they performed the discrimination well in those few trials in which they did orient to the displaced S^+. These findings suggest that the monkeys with SC lesions were impaired in discriminating between stimuli presented eccentrically because they were deficient in shifting their gaze to them. Presumably, monkeys with SC lesions are also impaired in performing visual discriminations involving S-R separation in the WGTA because in that situation they

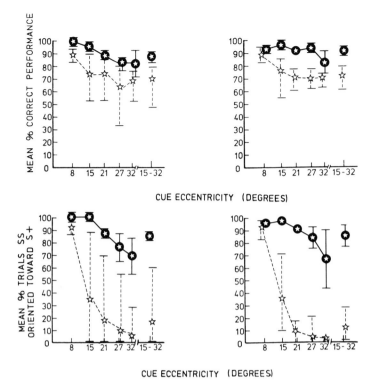

Figure 10.6 Top: Mean percent correct performance in a color discrimination task as a function of stimulus (cue) eccentricity after (●) superior colliculus or (☆) control surgery. Bottom: Mean percent trials in which subjects oriented to the S+ as a function of stimulus eccentricity after same types of surgery. Left graphs represent first postoperative session; right graphs represent last three postoperative sessions.

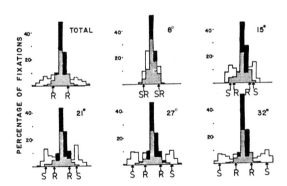

Figure 10.7 Distributions of fixations (in 5° bins) across the screen containing stimulus and response panels during color-discrimination performance of monkeys with superior-colliculus lesions (filled bars) and of operated control monkeys (open bars). Overlap between fixation distributions of the two groups is indicated by the dotted areas. Only horizontal coordinates of fixations were used in this analysis, since they accounted for 95% or more of all fixations within trials. Distributions of fixations associated with each degree of stimulus eccentricity (8°–32°) are shown separately, and are pooled in the distributions labeled TOTAL. R: positions of choice-response panels. S: positions of stimuli.

Vision in Monkeys with Lesions

have difficulty in shifting their gaze to the stimuli from the response sites.

Why were the eye-movement deficits found in this study more severe than those reported in the studies of Wurtz and Goldberg (1972b) and those of Mohler and Wurtz (1977), in which monkeys were also required to fixate visual targets? The discrepancy between these findings may have been due to differences in the visual targets or in the task requirements. In addition, it is possible that variations in the extent of the SC lesions may have been responsible for the different effects on eye movements: The lesions in the monkeys of Kurtz and Butter, unlike the lesions in the monkeys of Wurtz and his colleagues, were bilateral and consistently included the deepest layers of the SC.

Despite differences in the testing situations and uncertainties about the locus of SC lesions associated with alterations in eye movements, the findings described above indicate that under some conditions monkeys with SC lesions show severe deficits in eye movements to visual stimuli. These deficits may account for the deficient performance of monkeys with SC lesions in visual tasks requiring shifts of orientation (Butter 1974a, 1979; Kurtz and Butter 1976; Latto 1978).

In addition to these investigations of orienting behavior in monkeys, the results of several other recent studies implicate the primate SC in the control of processes underlying orientation to visual stimuli.

Orienting behavior, which involves directing the eyes, head, or body to a localized stimulus, depends upon several processes, including detection and localization of stimuli as well as selective attention to them. Motor control or orienting behavior and visuomotor coordination are obviously also important. Thus, a deficit in orienting behavior could be due to a disturbance of any one or combination of these processes.

Stimulus detection may be defined as the ability to discriminate between the presence and the absence of a sensory event. In addition to stimulus parameters such as intensity and duration, the persistence of a stimulus trace or "icon" is an important factor in detection, as shown by backward-masking studies (Turvey 1973).

Two kinds of localization are customarily distinguished: *allocentric* (localization of a stimulus relative to another one, independent of the subject's position) and *egocentric* (localization of a stimulus relative to the body). Egocentric localization is obviously necessary to orient to a stimulus; it requires information concerning not only the retinal position of the stimulus, but also the position of the eyes in the head and the position of the head relative to the body.

Selective attention to a stimulus refers to a central process that increases the likelihood of detecting or identifying a stimulus relative to other stimuli, independent of sensitivity increases due to stimulus changes (such as increased intensity) or changes in the position of the stimulus on the retina. Shifts of attention to a stimulus may be reflexive or voluntary. We define a *reflexive shift* of attention as one that is reliably elicited by a novel stimulus and habituates as that stimulus is repeatedly presented. A *voluntary shift* of attention we define as one that is elicited by a stimulus because of significance that the stimulus has acquired from experience. Attention may be directed not only to particular stimuli, but also to particular parts of the visual field, as demonstrated by a tradeoff in target detection between two portions of the visual field (Sperling and Melchner 1978). This tradeoff can be described by the "attention operating characteristic" of the subject (Kinchla 1969).

Stimulus detection, localization, and selective attention are probably not independent of one another; rather, it is likely that they interact, which complicates experimental analysis. Thus, as mentioned above, selective attention to a stimulus appears to decrease its threshold of detection or to enhance the likelihood of identifying it. Selective attention may also be linked to efferent processes underlying orienting behavior. There is some evidence that the probability of both stimulus detection (Singer et al. 1977; Nissen et al. 1979; Remington 1979) and identification (Bryden 1961; Crovitz and Daves 1962) are increased prior to a shift of gaze to the stimulus. Furthermore, the coordination of sensory and motor events underlying orienting behavior implies that these two sets of processes interact.

We now turn to investigations of the role of the primate SC in stimulus localization. The results of several studies indicate that lesions of the SC produce deficits in localizing briefly presented visual targets in monkeys. Keating (1974) tested monkeys in a task in which two brief light flashes differing in intensity were simultaneously presented on a pair of panels within a large array of panels; the two panels on which the flashes were presented varied from trial to trial. After receiving SC lesions, the monkeys were impaired in performing the correct response (pressing the panel on which the dimmer of the two light flashes appeared). Most of the errors that the monkeys with SC lesions committed were directed not to the panel on which the brighter of the two flashes appeared, but to panels near the correct one. This implied that the monkeys with SC lesions were impaired in localizing the positive stimulus.

In an experiment similar to that of Keating (1974), MacKinnon et al. (1976) reported that monkeys with SC lesions were impaired in responding to 1-second illuminations of sites that were shifted from trial to trial. However, when the duration of illumination was extended to 5 seconds, they performed as well as control animals did. In a further analysis of this impairment, Butter et al. (1978) tested monkeys for localization of brief light flashes presented while the direction of gaze was controlled so that the light flashes could be presented in particular part of the visual field. The monkeys' direction of gaze was controlled at the beginning of each trial by requiring them to successfully perform an observing response that involved pressing the panel in the center of the

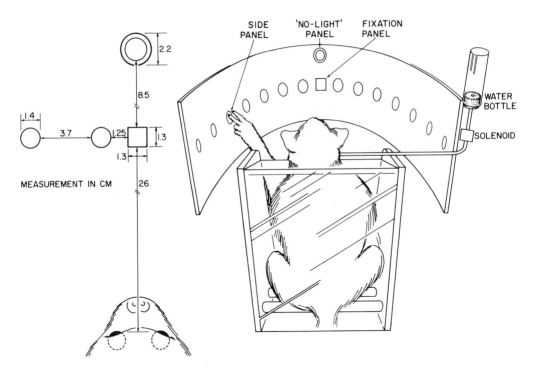

Figure 10.8 Apparatus for testing localization and detection of a light flash presented on one of the fourteen side panels when monkey performs an observing response, which requires fixating and pressing the center fixation panel. Water reward for correct responses is delivered from a tube connected to a solenoid. Adapted from Butter et al. 1978.

display screen within a short period of time after a small, dim spot on the panel changed color. The parameters of this light spot, including the varying durations between its appearance and its change of color, were adjusted so that the monkey had to fixate within 2° of the stimulus in order to respond correctly, as shown by EOG recordings of monkeys performing this task in an earlier study (C. M. Butter, R. Maciewicz, and C. G. Gross, unpublished). When the fixation task was performed correctly, a brief (50 milliseconds) light flash was presented on one of the side panels shown in figure 10.8. The monkey received water reward only for pressing the side panel on which the flash appeared.

After the monkeys were trained in the task, several of them received large, electrolytic lesions that destroyed all but a small anterior portion of the SC. The pre- and postoperative performance of these monkeys and of their controls on light trials is represented in figure 10.9. It is clear that the monkeys with large SC lesions were only impaired in responding correctly to the light flashes when they were presented on stimulus panels 40° or more from the center of the screen. This impairment was due in part to elevated localization errors, most of which resulted from pressing the panel adjacent and medial to the correct panel. Furthermore, two of the SC-lesioned monkeys showed deficient per-

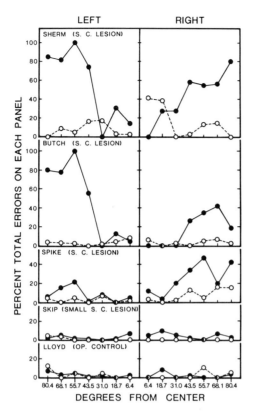

Figure 10.9 Percent total errors made by each monkey when brief flashes were presented on each of the side panels, 6.4°–80.4° from center fixation, before and after surgery. (O) Preoperative results; (●) postoperative results. Sherm, Butch, and Spike were monkeys with large superior-colliculus lesions; Skip and Lloyd were control monkeys. Adapted from Butter et al. 1978.

formance with peripheral flashes throughout the 6 months in which they were tested. In contrast, none of the lesioned monkeys were severely impaired when the light flash was extended to 1 second. It is apparent, then, that monkeys with large SC lesions show a long-lasting and severe localization deficit in this task, but one that is apparent only when brief light flashes are presented in the periphery of the visual fields.

This restriction of the deficit to peripheral vision might be expected to result from lesions limited to the posterior portion of the SC, which represents the periphery of the visual fields. However, this was clearly not the case: The lesions in the impaired monkeys were extensive and involved portions of the SC representing all the visual field except for the central 5°. One possible explanation for the discrepancy between the retinotopic locus of the lesions and the locus of impairment in the visual field is that the visual cortex takes over the functions of central vision normally mediated by the anterior part of the SC. Alternatively, it is possible that an impairment was not found in the central portion of the

Vision in Monkeys with Lesions

visual field because the task was not sufficiently difficult. The monkeys with SC lesions might have been impaired in localizing central as well as peripheral light flashes if the flashes had been near threshold.

Other findings from the studies of stimulus localization described previously are also relevant to the detection capacities of monkeys with SC lesions. MacKinnon et al. (1976) reported that, after SC lesions, monkeys made more errors of omission (failing to respond to an illuminated site) as well as more errors of commission (responding to an unilluminated site), which suggested that they were impaired in detecting the lights as well as in localizing them. The task used by Butter et al. (1978) was designed to investigate not only localization of light flashes, but also their detection. In that experiment, light flashes were presented on the side panels shown in figure 10.8 on only 50% of the trials. On the remaining trials, the monkeys received water reward only for pressing the "no light" panel, located above the center panel. If a trained monkey pressed the "no light" panel when a light flash was presented, it was assumed that it had failed to detect the flash. After large bilateral SC lesions, the monkeys showed increased detection errors as well as the increased localization errors mentioned previously. Furthermore, their detection deficit, like their localization deficit, was evident only when brief light flashes were presented on side panels located 40° or more from the center of the display screen. However, it is possible that deficits in detecting more centrally located flashes may have been found if flashes had been presented near threshold. This possibility gains credence from the finding that monkeys with bilateral SC lesions show a transient rise in thresholds for detecting flashes in a situation where they apparently fixate the stimuli (Latto 1977).

Furthermore, after small unilateral lesions of the superficial layers of the SC, a monkey with frontal-eye-field lesions showed transient deficits in detecting light flashes presented approximately 20°–35° from the fovea (Latto and Cowey 1971). However, Mohler and Wurtz (1977), using a different method of evaluating detection (releasing a lever when a light flash appeared in the visual field), failed to find deficits in monkeys after unilateral colliculus lesions similar in extent to those made by Latto and Cowey. This discrepancy may be due to brighter flashes used by Mohler and Wurtz or to the frontal-eye-field lesions in the monkey of Latto and Cowey.

The findings described above suggest that deficits in orienting to visual stimuli after SC lesions may be due to impairments in localizing and detecting stimuli. Alternatively, or in addition, these orientation deficits may be due to impairments in selectively attending to stimuli to which orienting movements are directed. This possibility is supported by the finding that the responses of many superficial SC neurons to visual stimuli are selectively enhanced shortly before the monkey directs its eyes to these stimuli (Wurtz and Goldberg 1972b; Wurtz and Mohler 1976).

Milner et al. (1978) examined the distracting effects of peripheral light flashes on the performance of a discrimination task by monkeys, and found that monkeys with SC lesions were less distracted by the flashes than control monkeys. This impaired distractibility may have been due to a disturbance in selective attention; however, it may have been due to a disturbance in detecting the peripheral flashes.

Kurtz et al. (1980) attempted to determine whether the SC-lesioned monkey's deficit in performing visual discriminations involving S-R separation is due to a disturbance in shifting attention to the discriminative stimuli. In order to investigate this, it was necessary to rule out the possibility that an impairment in orienting to the stimuli (see Kurtz and Butter 1976) might account for discrimination deficits that might be found. Orienting to the stimuli was prevented by presenting stimuli at a duration (100 msec) briefer than the latency of eye movements to visual stimuli, while the monkeys' fixation was controlled.

The monkeys in this experiment were trained in a two-choice color-discrimination task similar to the one employed in the experiment of Kurtz and Butter (1976). However, this task differed from the previous one in several respects:

• The monkeys were required to respond quickly to the dimming of a small spot located straight ahead in order to initiate presentation of the discriminative stimuli; this observing response requires fixating the stimulus (Wurtz 1969).

• When the observing response was made, the two discriminative stimuli were presented lateral to the central fixation spot, either for 2 sec or for 100 msec.

• In one-fourth of the trials the discriminative stimuli were not presented; the monkeys were rewarded only for pressing the "no-light" panel, so that detection errors on "light trials" could be evaluated as in the experiment of Butter et al. (1978).

• In alternate sessions, the monkeys were tested with choice-response panels located centrally (8° from center), as in the prior experiment, and peripherally (32° from center).

Figure 10.10 shows the performance of monkeys in initial preoperative testing. When the response panels were centrally located, the monkeys showed the expected decrement in performance as the eccentricity of the long-duration stimuli increased. This decrement in performance was due primarily to increased errors in detecting the peripheral stimuli and to a lesser extent to increased errors in discriminating between them. Conversely, when the response panels were located peripherally, performance with the nearby peripheral stimuli was very high and declined as the stimuli were presented more centrally. This decrement in performance with stimuli presented centrally was due mainly to increased discrimination errors; detection errors increased to a lesser ex-

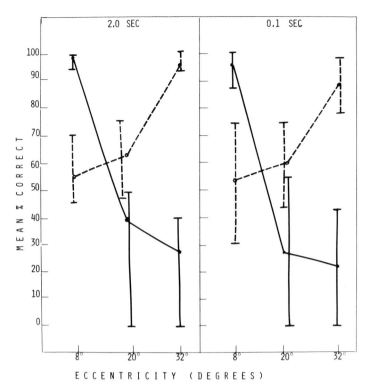

Figure 10.10 Mean percent correct performance of monkeys as a function of stimulus eccentricity in preoperative color-discrimination tests with (●) central and (○) peripheral response sites. Vertical bars indicate ranges of individual scores.

tent. The same pattern of results was obtained when the discriminative stimuli were presented for 100 msec, a duration at which the monkeys could not fixate them (figure 10.10). Furthermore, the monkeys committed discrimination errors and detection errors in approximately the same proportion as when the stimuli were longer in duration. The finding that S-R separation markedly affects discrimination performance independently of the opportunity to fixate the stimuli suggests that S-R separation worsens performance in this task by making it difficult for the monkeys to shift attention to the stimuli from the response panels.

After surgery, the monkeys with SC lesions, like those in the experiment of Kurtz and Butter (1976), were impaired when tested with long-duration stimuli presented peripherally (32° eccentricity) and central response panels (figure 10.11). Moreover, this impairment was entirely attributable to elevated discrimination errors; like their controls, the monkeys with SC lesions made few detection errors. However, when the response panels as well as the stimuli were located peripherally, the monkeys with SC lesions were unimpaired. Similarly, when brief stimuli were presented, the SC-lesioned monkeys were only impaired when the response panels were centrally located and the stimuli

Butter, Kurtz, Leiby, Campbell

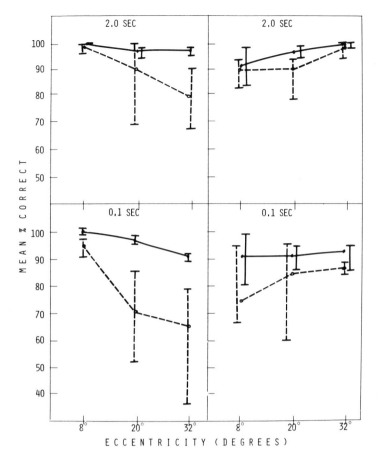

Figure 10.11 Mean percent correct performance of (○) monkeys with superior-colliculus lesions and of (●) operated controls as a function of stimulus eccentricity in preoperative color-discrimination tests with central and with peripheral response sites. Vertical bars indicate ranges of individual scores.

were presented peripherally. In fact, they were even more severely impaired than with long-duration stimuli. Their performance was deficient relative to the control animals when brief stimuli were presented not only at 32° eccentricity, but also at 20° eccentricity; they even showed a slight impairment with 8°-eccentric stimuli. These performance decrements were due to increases in both discrimination and detection errors.

It is difficult to account for these deficits in terms of sensory losses in peripheral vision, since they were not present when the response panels were located peripherally. It is more likely that the deficits were due to the disturbance of an attentional process facilitating peripheral vision, perhaps an attentional process associated with shifts of gaze. Although shifting the gaze from the center OR panel to the periphery would not aid the monkeys in identifying peripheral stimuli presented for 100 msec, a shift of attention accompanying such gaze shifts may aid

Vision in Monkeys with Lesions

stimulus detection and identification, as suggested by the studies of human perception cited above. It is likely that the control animals shifted their gaze to the peripheral stimulus panels even when the stimuli there were presented too briefly for fixation, as the monkeys could not anticipate whether a long-duration stimulus (which they probably did fixate) or a brief stimulus would appear in any trial. Finally, electrophysiological findings support the view that the primate SC is involved in shifts of attention accompanying gaze shifts. Many neurons in the superficial layers of the monkey's SC increase their firing rates to a visual stimulus in their receptive field prior to a saccade directed to the stimulus (Wurtz and Goldberg 1972b; Wurtz and Mohler 1976). However, the monkeys with SC lesions could apparently shift their attention to the periphery when they made a manual response in the periphery of the display screen, as shown by their unimpaired performance in this condition. Thus, it appears that attention directed to a part of space where the monkey reaches was intact after SC lesions. This conclusion is supported by the observation that the monkeys with SC lesions accurately grasped bits of food presented in the periphery of the display screen even when, on some occasions, they failed to orient their eyes to the food.

An impairment in shifting attention to a target in the periphery of the visual field prior to a saccade directed to the target might aid in understanding why SC lesions produced more severe eye-movement impairments in the discrimination task of Kurtz and Butter (1976) than in the task employed by Mohler and Wurtz (1977). In the testing situation used by Kurtz and Butter, two discriminative stimuli were presented simultaneously on each trial. The stimuli were sufficiently small and dim that, on most of the trials, the monkeys had to shift their gaze several times before directing it to the "correct" stimulus. In contrast, in the Mohler-Wurtz "saccade task," only one extrafoveal target was presented on each trial; their monkeys had been trained before surgery to shift their gaze to the target with one saccade. Thus, the greater demand placed on target selection in the Kurtz-Butter study than in that of Mohler and Wurtz might account for the more severe eye-movement deficits found in the former study.

Summary
Several recent ablation studies provide evidence that the primate SC, like the SC of other mammalian species, is involved in orienting activity. Some of this evidence is indirect, in that it derives from observations of discrimination deficits of monkeys in situations demanding shifts of orientation. Other, more direct evidence is based on observations of eye movements to visual targets in monkeys with SC lesions. However, there is clearly a need for further analyses of the role of the primate tectum in orientation, especially since the few relevant findings disagree over the effects of SC lesions on eye movements in monkeys.

Specifically, the particular parameters of saccadic eye movements that are impaired, the severity and permanence of the impairments, the effective lesions, and the particular testing conditions in which these impairments are found have yet to be established. There is also need for analysis of orienting movements involving the head and other parts of the body in monkeys with SC lesions, perhaps by means of motion-picture photography (see the chapters by Goodale and Ingle in this volume).

There is even less certainty about the nature of the impairment underlying orientation deficits in monkeys with SC lesions than there is about the orientation deficits themselves. Although recent findings suggest that impairments in detecting and localizing visual stimuli may follow SC lesions in monkeys, it is not clear whether they are present only in the periphery or throughout the visual field. This issue could be settled by determining thresholds for detecting and localizing stimuli presented in different parts of the visual field while fixation is controlled.

In addition, shifts of attention in monkeys need to be analyzed further in order to verify the conclusions based on the study of Kurtz et al. (1979). It is possible that methods for evaluating shifts of attention in monkeys may be adapted from those used to study spatial shifts of attention in humans (see, for example, Eriksen and Collins 1969; Nissen et al. 1979). These tests involve briefly presenting a spatial cue that indicates where in the visual field a stimulus that the subject is required to detect or identify will shortly appear. The performance of monkeys in this kind of test might reveal whether SC lesions decrease the speed with which attention shifts from one part of the visual field to another, as well as the likelihood of these shifts. Furthermore, this kind of test may make it possible to distinguish between impairments in shifting attention and impairments in detecting or localizing extrafoveal stimuli, a distinction that is not possible from previous findings (Butter et al. 1978; Keating 1974; Kurtz and Butter 1976). Additional studies would be required to determine whether and to what extent sensory or attentional deficits to extrafoveal stimuli might account for disturbances in orienting to these stimuli.

Effects of Striate-Cortex Lesions on Visually Guided Behavior in Monkeys

As we pointed out in the last section, efforts to understand the effects of superior-colliculus lesions on vision in monkeys have been directed toward devising tests that reveal visual impairments not apparent in customary tests. In contrast, the effects of striate-cortex ablations on visually guided behavior of monkeys are so devastating that most recent investigations in this area have attempted to define the residual visual capacities in striatectomized monkeys.

Klüver (1942) was the first to investigate in a thorough and systematic manner the visual capacities of monkeys after removal of the striate cortex. He concluded, primarily on the basis of two-choice discrimination tests in the pull-in apparatus, that monkeys with striate-cortex ablations are unable to differentiate stimuli on the basis of form, color, or brightness. Furthermore, Klüver found that the discrimination performance of these monkeys was not disrupted when values of area, intensity, distance, and duration of stimuli were interchanged so that the luminous flux was not altered. Hence, he also concluded that after removal of striate cortex monkeys could distinguish visual stimuli only in the basis of the total amount of light.

One might infer that Klüver's monkeys retained another visual capacity—the ability to localize visual stimuli—since they learned to respond in the appropriate direction to one of two simultaneously presented stimuli differing in flux. However, Klüver presented findings suggesting that his striatectomized monkeys did not localize stimuli in the normal manner, but rather oriented to the side of the apparatus where they received more (or less) light and then made the choice response on that side. Thus, Klüver asserted, all the dimensions of visual space, including the location of objects, were eliminated by striate-cortex ablation. Furthermore, since Klüver's striatectomized monkeys demonstrated a striking equivalence between visual and auditory stimuli, it appears that they may not have distinguished visual from nonvisual events.

More recently, investigators have claimed a greater sparing of visual capacities in striatectomized monkeys than Klüver's findings indicated. In this section we shall examine these claims and describe the behavioral methods employed in these experiments. We shall consider only studies involving total and nearly total striate cortex removals, for only after such large ablations is the visually guided behavior of monkeys altered drastically.

Whereas Klüver's monkeys failed to reliably orient toward visual stimuli, several investigators have reported that this capacity is still present after total or nearly total striate-cortex removal in monkeys. Denny-Brown and Chambers (1955) observed striatectomized monkeys orienting to and following moving targets. Humphrey and Weiskrantz (1967) also reported that this orienting behavior occurred in their monkeys (one with total and the other with nearly total striate-cortex removal). They then found that these monkeys could learn to reach out and grasp a moving light for food reward after orienting the head and eyes to it. Some time later, these monkeys demonstrated the ability to reach toward a stationary light. Initially, the monkeys groped in a clumsy manner toward the light and required some practice before reaching to it accurately. In addition, unlike intact monkeys, they immediately transferred the reaching response from a light to a buzzer; like Klüver's observations, this suggested a failure to identify visual

Butter, Kurtz, Leiby, Campbell

events as such. The ability to reach toward a stationary light was replicated by Feinberg et al. (1978) in totally striatectomized monkeys, which learned to accurately reach for targets of decreasing size (down to approximately 9° of visual angle). In our laboratory, we have also trained striatectomized monkeys to reach to a moving light and later to a stationary light in order to obtain food reward (A. Campbell, C. C. Leiby, and C. M. Butter, unpublished). Like the monkeys of Humphrey and Weiskrantz, ours initially reached in a groping and inaccurate manner and progressed only after many hundreds of trials. Their slow progress was partly due to a tendency to perseverate reaching to the position where the target had previously appeared. These perseverative errors slowly declined in frequency as accuracy increased. Furthermore, our monkeys, like those in the studies described, oriented the head and eyes to the light before reaching to it.

On the basis of these findings, one might be tempted to conclude that, contrary to Klüver's findings, localization of objects in two-dimensional space survives destruction of the geniculo-striate pathway and, presumably, survives the loss of many retinal ganglion cells, which undergo transneuronal, retrograde degeneration after removal of striate cortex (Van Buren 1963; Cowey 1968). Furthermore, supporters of the two-visual-systems hypothesis might conclude from these findings, as well as from similar findings in humans with visual-field defects due to cerebral lesions (Perenin and Jeannerod 1975; Weiskrantz 1978), that spatial localization is normally mediated by subcortical mechanisms, presumably involving the superior colliculus. However, before accepting these conclusions, one would have to rule out an alternative interpretation of these findings: that striatectomized monkeys, unlike normal monkeys, may scan their environment with head or eye movements and use the resulting temporal pattern of retinal stimulation to guide reaching. Doubt is cast on this interpretation by the finding that monkeys with striate-cortex ablations can reach accurately to a target illuminated too briefly to be localized by this strategy (Weiskrantz et al. 1977). Commenting on this finding, Weiskrantz (1978) speculated that the successful reaching demonstrated in that study may have been aided by illuminated keys that were protruding and thus easy to detect and localize by touch. Nevertheless, the demonstration that monkeys can reach to briefly illuminated targets after striate cortex is removed does not necessarily imply that these monkeys use the same mechanisms that normal monkeys use to guide reaching for objects. Humphrey (1970) suggested that striatectomized monkeys may make use of another mechanism involving the following stages: An object, which is initially differentiated from the background, triggers an orienting response that involves directing the eyes and head toward the object; information from efferent signals to the eyes (or proprioceptive information) is then used to guide the arm to the target. It is conceivable that a monkey could make use of such positional information even when it is generated by very brief visual stimuli. If striatectomized monkeys rely on such positional information to guide reaching, they should be able to reach accurately only to targets to which they orient. Furthermore, reaching errors should be closely related to where the monkey looks. These predictions have not been tested experimentally, but we have the impression that our striatectomized monkeys tend to misreach for targets in the same direction in which their head and eyes are pointed (Campbell, Leiby, and Butter, unpublished). At the least, then, one must entertain the possibility that monkeys with striate-cortex ablations differ from intact monkeys in that they rely on indirect cues to guide their reaching. Presumably, according to this view, the role of direct visual cues is limited to triggering and building their orienting responses.

On the basis of other findings, it has been claimed that striatectomized monkeys retain the capacity to distinguish between spatial attributes, color, brightness, and other dimensions of visual stimuli. Before describing these findings, we mention Humphrey's (1970) observations of Helen, a monkey with a large striate-cortex ablation that had demonstrated reaching to stationary objects. Helen consistently preferred to reach for one of two simultaneously presented objects on the basis of size or degree of contrast from the background. Even more surprising was this monkey's ability to differentiate rather small differences in these dimensions. In addition, Humphrey observed that this monkey no longer preferred a solid object when it was broken into fragments. The importance of these stimulus parameters, plus Helen's failure to recognize objects, led Humphrey to conclude that only stimulus salience—that is, the degree to which an object is perceptually differentiated from the background—was critical in determining Helen's preferences. In a subsequent series of tests, Helen was differentially rewarded for reaching to one of two objects. In these tests, Helen discriminated between two stimuli, one of which she had preferred over the other in prior preference tests, irrespective of which stimulus was rewarded. On the other hand, Helen failed to discriminate between a pair of stimuli for which she had shown no differential preference in prior tests. These findings are not only of interest in themselves; they may also have implications for interpreting other visual-discrimination capacities that have been demonstrated in striatectomized monkeys.

Weiskrantz (1963) reported that a striatectomized monkey discriminated between two patterns (speckles and regular stripes) matched for total flux and brightness. The discrimination could have been performed on the basis of either differences in the total amount of contour or differences in the number of elements in each stimulus. The monkey's success in performing this discrimination may have been due at least in part to the unique method of testing, which involved reaching through holes in the centers of the stimuli to obtain food reward for the correct choice. Keating (1976b) reported that totally striatectomized monkeys can relearn color and velocity discriminations; however, the testing procedures were not described.

Using a modified version of Klüver's two-choice pull-in apparatus, Pasik et al. (1969) showed that striatectomized monkeys can discriminate between transilluminated stimuli matched for total luminous flux. Since in a later experiment their striatectomized monkeys maintained the discrimination even when the stimuli were also matched for overall width and amount of contour, they concluded that their monkeys might have been discriminating on the basis of brightness differences (Schilder et al. 1971). They also concluded that striatectomized monkeys can discriminate between a hue and an achromatic stimulus (Pasik and Pasik 1971) and between two hues (Schilder et al. 1972), independent of variations in luminous flux. Furthermore, Schilder et al. (1972) demonstrated that monkeys with striate-cortex ablations can discriminate between two transilluminated forms equated for flux and brightness. The results of equivalence tests performed after the monkeys had mastered the form discrimination suggested that they may have been using configurational cues in performing the discrimination. The Pasiks have been cautious in interpreting these findings, and have acknowledged the possibility that their monkeys may have used movement-generated temporal cues to perform the brightness and pattern discriminations, since the stimuli were continuously exposed until the monkeys responded. The Pasiks' monkeys were severely retarded in relearning discriminations after striate-cortex removal. However, the differences between their results and those of Klüver are so striking that one is led to speculate about their cause. As pointed out by Schilder et al. (1971), their training apparatus differs from Klüver's in that, after pulling in the stimulus box, the monkey must reach behind its front surface, where the transilluminated stimulus is displayed, in order to retrieve the food reward. In doing so, the monkey directs its eyes at the stimulus (see figure 10.1). In Klüver's apparatus, the monkey opens the front surface of the stimulus box to retrieve the food reward inside. Thus, the Pasiks' apparatus may have provided more opportunities for the monkey to be exposed to the stimuli than did Klüver's. Furthermore, there are grounds for suspecting that Klüver's striate-cortex lesions, for which no histology was reported, may have included large amounts of prestriate cortex (Pasik and Pasik 1971). The Pasiks' lesions, in contrast, left most of the prestriate cortex intact. The importance of prestriate cortex in the discrimination performance of striatectomized monkeys is shown by the finding that when much of prestriate cortex, in addition to all of striate cortex, is removed, monkeys fail to discriminate between flux-matched stimuli; whereas when the ablation is limited to striate cortex, this discrimination can be relearned (Pasik and Pasik 1971).

What, then, can one validly conclude from studies of residual visual capacities of monkeys in the absence of the geniculostriate system? The findings presented in this section suggest, on the one hand, that striatectomized monkeys may show visual capacities not revealed by standard procedures when they are tested by different procedures. On the other hand, there are also good reasons for treating such conclusions with caution and considering alternative interpretations.

Thus, with regard to the ability to localize objects in space, there are some grounds for believing that, after striate-cortex removal, monkeys may use indirect, proprioceptive cues rather than direct, visual ones. In order to experimentally evaluate this interpretation, it would be necessary to control the orienting movements generating proprioceptive cues and relate these movements to accuracy of reaching.

There are several factors to be considered in interpreting findings that suggest that striatectomized monkeys can distinguish the total amount of contour, brightness, color, and form of stimuli. Striatectomized monkeys may make use of temporal cues resulting from scanning stimuli that differ in their spatial organization. This possibility could only be ruled out by a demonstration that discrimination performance is not affected by reducing the stimulus duration until scanning movements are no longer effective. Of course, one must be cautious in interpreting a failure to discriminate between briefly presented stimuli; Klüver (1942) found that in his testing apparatus striatectomized monkeys required exposure times of more than 1 second to discriminate between simultaneously (but not successively) presented stimuli varying in luminous flux. His control monkeys, in contrast, discriminated between these stimuli at much shorter durations. This finding suggests that striate-ablated monkeys may need to scan two spatially separated stimuli in order to localize the correct one and thus make the appropriate choice response in the pull-in apparatus.

Also, it is possible that striatectomized monkeys learn some discriminations on the basis of the differential stimulus salience, which may control their orienting movements toward the stimuli, as suggested by Humphrey's analysis of Helen's visually guided behavior. For example, stimuli used in brightness discriminations vary in degree of contrast from the background, which was an important determinant of Helen's preferences. Another example of the possible operation of salience is in the stripes-versus-speckles discrimination that Weiskrantz's striate-ablated monkey learned; this monkey, like Humphrey's, may have had a preference for orienting to the more articulated pattern (the stripes). Besides the stimulus features that Humphrey studied, other stimulus determinants of salience, such as particular wavelengths or curvatures of contours, might account for some other discriminative capacities of striatectomized monkeys. The potential role of salience in discrimination learning could be investigated by conducting combined preference-discrimination tests in the manner of Humphrey. Such tests should include stimuli that striatectomized monkeys successfully discriminated between in other experiments as well as those that they failed to differentiate.

One problem with the view that striatectomized monkeys learn visual discriminations on the basis of salience detection is that it fails to account for the failure of Klüver's monkeys to learn certain kinds of discriminations that the Pasiks' monkeys were able to perform. It is highly unlikely that the degree of differential salience differed systematically in these two series of experiments. However, it is possible that prestriate cortex, which may have been largely destroyed in Klüver's monkeys, may be critical for salience to operate as a cue. Prestriate cortex may be involved in detection of stimulus salience on the basis of visual inputs from the superior colliculus via the inferior pulvinar (Benevento and Rezak 1976; Ogren and Hendrickson 1977), a route by which visual information may reach these visual cortical areas independent of striate cortex.

In addition, interpretations of striatectomized monkeys' discriminative capacities do not sufficiently stress the point that these animals are severely retarded in learning or relearning visual discriminations (even those that are rapidly acquired preoperatively), and require many months of training. Moreover, considerable time may be required to retrain the striatectomized monkey to perform the choice response and retrieve food rewards in the testing apparatus. The need for prolonged training raises a question: To what extent does the discriminative ability of the striatectomized monkey reflect reorganization of the remaining visual system and possibly of nonvisual structures, and to what extent does it reflect residual capacities, normally mediated by the remaining visual system? The possibility of reorganization is suggested by the finding that lesions of the superior colliculus abolish the brightness-discrimination performance of previously striatectomized monkeys (Pasik et al. 1973) whereas superior-colliculus lesions in otherwise intact monkeys do not interfere with brightness-discrimination performance (Butter 1974b; Keating 1976a). These findings need not imply reorganization; an alternative interpretation is that the striatectomized monkey, unlike the intact monkey, relearns a brightness discrimination by a unique strategy (such as scanning) that involves the superior colliculus.

It is still far from clear whether and to what extent striatectomized monkeys can use the kinds of stimulus dimensions that normal monkeys use in visual discrimination, but they appear to retain the capacity to orient to objects that capture their attention—objects that are highly distinct from the background and may be characterized as possessing the property of "good articulation." This conclusion is based on limited observations of a single monkey, in which the striate cortex was not totally removed. Thus, it is clear that the analysis of stimulus characteristics that "catch the striatectomized monkey's eye" needs to be pursued systematically.

Summary and Conclusions

The two-visual-systems hypothesis has generated a number of investigations of visually guided behavior in various mammalian species. For the most part, the findings of these studies have tended to confirm the notion that striate cortex and superior colliculus contribute to qualitatively different aspects of visual perception. However, when similar ablation experiments have been carried out with monkeys, two related problems have retarded progress: difficulties in detecting visual impairments after collicular lesions and difficulties in identifying residual visual functions after striate-cortex lesions. We have presented evidence that the resolution of these problems appears to depend in large part on the development of new behavioral testing procedures that are more appropriate than standard tests for answering the particular questions we want to ask about the lesioned monkey's abilities.

At present, it appears that this strategy of devising more appropriate tests has yielded more consistent results when applied to the analysis of superior-colliculus lesions than when applied to the analysis of striate-cortex lesions. For example, whether the success of recent investigations in revealing discrimination capacities that Klüver failed to find in striate-ablated monkeys is due to differences in the extent of ablations or in training methods, or both, is still not clear. However, there is agreement that the striatectomized monkey retains the capacity to orient and reach toward objects. Thus, it would be appropriate to exploit these capacities by using them as indicators of discriminative ability, as Weiskrantz and Humphrey (1970) have done. The use of orienting or reaching as measures of discrimination ability may not only reveal capacities heretofore unrecognized in striatectomized monkeys; it may also reduce the time required for discriminations to be mastered after striate-cortex ablation in comparison with tests conducted in standard apparatuses, as suggested by Humphrey's findings. In addition to the practical advantages of reducing the number of testing sessions necessary to demonstrate discrimination ability, reducing testing time has the advantage of decreasing potential effects on discrimination learning of any slowly developing reorganization of the remaining visual system.

Whereas orienting ability appears to be retained after striate-cortex ablation, recent findings suggest that it is impaired after superior-colliculus lesions. Some of the recent tests in which monkeys with collicular lesions show altered behavior involve orientation to peripheral stimuli (Butter 1974a; Keating 1974; Latto 1978; Milner et al. 1978). Furthermore, in the few studies in which orientation to visual stimuli was directly measured, deficits have been found after collicular lesions (Wurtz and Goldberg 1972b; Mohler and Wurtz 1974; Kurtz and Butter 1976). The only measure of orientation employed in these studies was eye direction; consequently, other kinds of orienting behavior, involving the head and body, require investigation in colliculus-lesioned monkeys, perhaps by means of cinematography. Another problem requiring resolution is that a deficit in orienting behavior to extrafoveal stimuli can be interpreted in various ways: It may be due to one or a combination of impairments in sensory functions, including localization and detection in the periphery, shifting attention, or visuomotor coordination. Although monkeys with collicular lesions are deficient in detecting and localizing peripheral stimuli (Butter et al. 1978), these alterations may be due at least in part to a disorder in shifting attention, which may underlie the collicular-lesioned monkeys' deficit in discriminating between peripheral stimuli (Kurtz et al. 1979). Dissecting out these various functions in experimental situations and determining which are impaired after brain lesions remains a formidable task.

Why has there been so little concern with the role of orientation in the discrimination performance of animals, with the exceptions of Tolman's (1938) concept of vicarious trial and error and Klüver's (1933) description of comparison behavior? One reason, no doubt, is that measuring orienting behavior is more difficult than measuring choice behavior, the particular form of which can be arbitrarily defined at the experimenter's convenience. In addition to concerns with ease of measurement, there may be another reason why the analysis of orienting activity has developed so slowly: The tests of visual capacities that we devise and use tend to reflect our thinking about the processes underlying visual perception. In other words, one might say that we incorporate into our visual (as well as other) tests the assumptions underlying accepted models of perception. For a long time, the generally accepted model of perception was what might be called an input-output model. According to this view, perception involves the passive reception of environmental stimuli, which are conveyed to cortical sensory areas and then joined together by associative bonds (or, to use a more modern term, "processed") to form perceptions. While many modern investigators would reject this view, our tests still reflect one of its assumptions: that the perceiver, animal or human, is a passive recipient of environmental stimuli. The acquisition of information, by this view, is not directly relevant to visual perception, which depends exclusively on afferent activity. Perhaps it is because of this assumption that our standard tests of vision not only fail to measure orienting activity but are designed to ensure that the stimuli are displayed in such a way as to require a minimum of orientation shifts. Although under certain conditions animals will actively compare two simultaneously presented stimuli in a discrimination test (as noted by Tolman and Klüver), such shifts of orientation are an unanticipated consequence of the two-choice method and are ignored by most investigators. Of course, procedures for reducing the likelihood of shifts in orientation may be defended on the grounds that they promote rapidity of learning and thus make it easier to study the roles of stimuli, rewards, and other factors in discrimination learning. However, deliberate attempts to reduce the need for orienting also reflect the attitude that such activity is not crucial to visual perception. When the role of orienting activity in perception is recognized and systematically analyzed, a more adequate account of perception than is now available may be formulated.

Acknowledgments

The research reported in this chapter was supported by grant BMS74-16298 from the National Science Foundation and grant MH26489 from the National Institute of Mental Health. The authors thank Judith Lehman for her editorial assistance and Dr. Alan Cowey for his comments.

References

Albano, J. E., and R. H. Wurtz. 1978. "Modification of the pattern of saccadic eye movements following ablation of monkey superior colliculus." *Neurosci. Abstr.* 8: 161.

Anderson, K., and D. Symmes. 1969. "The superior colliculus and higher visual functions in the monkey." *Brain Res.* 13: 37–52.

Bender, D. B. 1973. "Visual sensitivity following inferotemporal and foveal prestriate lesions in the rhesus monkey." *J. Comp. Physiol. Psychol.* 84: 613–621.

Benevento, L. A., and M. Rezak. 1976. "The cortical projections of the inferior pulvinar in the rhesus monkey (*Macaca mulatta*): An autoradiographic study." *Brain Res.* 108: 1–24.

Bryden, M. P. 1961. "The role of post-exposural eye movements in tachistoscopic perception." *Canad. J. Psychol.* 15: 220–225.

Butter, C. M. 1974a. "Effect of superior colliculus, striate, and prestriate lesions on visual sampling in rhesus monkeys." *J. Comp. Physiol. Psychol.* 87: 905–917.

———. 1974b. "Visual discrimination impairments in rhesus monkeys with combined lesions of superior colliculus and striate cortex." *J. Comp. Physiol. Psychol.* 87: 918–929.

———. 1979. "Contrasting effects of lateral striate and superior colliculus lesions on visual discrimination performance in rhesus monkeys." *J. Comp. Physiol. Psychol.* 93: 522–537.

Butter, C. M., C. Weinstein, D. B. Bender, and C. G. Gross. 1978. "Localization and detection of visual stimuli following superior colliculus lesions in rhesus monkeys." *Brain Res.* 156: 33–49.

Cowey, A. 1968. "Discrimination." In *The Analysis of Behavioral Change*, L. Weiskrantz, ed. New York: Harper and Row.

———. 1974. "Atrophy of retinal ganglion cells after removal of striate cortex in a rhesus monkey." *Perception* 3: 257–260.

Cowey, A., and L. Weiskrantz. 1963. "A perimetric study of visual field defects in monkeys." *Q. J. Exp. Psychol.* 15: 91–115.

Crovitz, H. F., and W. Daves. 1962. "Tendencies to eye movements and perceptual accuracy." *J. Exp. Psychol.* 63: 495–498.

Denny-Brown, D., and R. A. Chambers. 1955. "Visuo-motor function in the cerebral cortex." *J. Nerv. Ment. Dis.* 121: 228–289.

Eriksen, C. W., and J. F. Collins. 1969. "Temporal course of selective attention." *J. Exp. Psychol.* 80: 254–261.

Feinberg, T. E., T. Pasik, and P. Pasik. 1978. "Extrageniculostriate vision in the monkey. VI. Visually guided accurate reaching behavior." *Brain Res.* 152: 422–428.

Goldberg, M., and R. Wurtz. 1972. "Activity of superior colliculus in behaving monkeys. I. Visual receptive fields in single neurons." *J. Neurophysiol.* 35: 542–559.

Goodale, M., and R. Murison. 1975. "The effects of lesions of the superior colliculus on locomotor orientation and the orienting reflex in the rat." *Brain Res.* 88: 243–261.

Harlow, H. F. 1951. "Primate learning." In *Comparative Psychology*, C. P. Stone, ed. (3rd edition). New York: Prentice-Hall.

Horel, J. A., L. A. Bettinger, G. J. Royce, and D. R. Meyer. 1966. "Role of neocortex in the learning and relearning of two visual habits by the rat." *J. Comp. Physiol. Psychol.* 61: 66–78.

Humphrey, N. K. 1970. "What the frog's eye tells the monkey's brain." *Brain, Behav., Evol.* 3: 324–337.

Humphrey, N. K., and L. Weiskrantz. 1967. "Vision in monkeys after removal of the striate cortex." *Nature* 215: 595–597.

Keating, E. G. 1974. "Impaired orientation after primate tectal lesions." *Brain Res.* 67: 538–541.

———. 1976a. "Effects of tectal lesions on peripheral field vision in the monkey." *Brain Res.* 104: 316–320.

———. 1976b. "The role of prestriate cortex in the recovery of vision after area 17 lesions." *Soc. Neurosci. Abstr.* 2: 1551.

Kinchla, R. A. 1969. Unpublished address at Attention and Performance III meeting, Soesterberg, The Netherlands; cited in Sperling and Melchner 1978.

Klüver, H. 1933. *Behavior Mechanisms in Monkeys*. University of Chicago Press.

———. 1942. "Functional significance of the geniculostriate system." *Biol. Symp.* 7: 253–299.

Kurtz, D., and C. M. Butter. 1976. "Deficits in visual discrimination performance and eye movements following superior colliculus ablations in rhesus monkeys." *Soc. Neurosci. Abstr.* 2: 1122.

Kurtz, D., C. C. Leiby, and C. M. Butter. 1980. "Reaching to the periphery abolishes deficits in peripheral vision of monkeys with superior colliculus lesions." *Soc. Neurosci. Abstr.* 6: 481.

Lashley, K. S., and M. Frank. 1934. "The mechanism of vision. X. Postoperative disturbances of habits based on detail vision in the rat after lesions in the cerebral visual areas." *J. Comp. Psychol.* 17: 355–391.

Latto, R. 1977. "The effects of bilateral frontal eye-field, posterior parietal or superior collicular lesions on brightness thresholds in the rhesus monkey." *Neuropsychologia* 15: 507–516.

———. 1978. "The effects of bilateral frontal eye-field, posterior parietal or superior colliculus lesions on visual search in the rhesus monkey." *Brain Res.* 146: 35–50.

Latto, R., and A. Cowey. 1971. "Visual field defects after frontal eye-field lesions in monkeys." *Brain Res.* 30: 1–24.

Layman, J. D. 1936. "Functions of the superior colliculus in vision." *J. Gen. Psychol.* 49: 33–47.

MacKinnon, D. A., C. G. Gross, and D. B. Bender. 1976. "A visual defect after superior colliculus lesions in monkeys." *Acta Neurobiol. Exp.* 36: 169–180.

Meyer, D., F. Treichler, and P. Meyer. 1965. "Discrete-trial training techniques and stimulus variables." In *Behavior of Non-Human Primates*, vol. 1, A. Schrier et al., eds. New York: Academic.

Milner, A. D., N. P. Foreman, and M. A. Goodale. 1978. "Go-left go-right discrimination performance and distractibility following lesions of prefrontal cortex or superior colliculus in stumptail macaques." *Neuropsychologia* 16: 381–390.

Mohler, C. W., and R. H. Wurtz. 1977. "Role of striate cortex and superior colliculus in visual guidance of saccadic eye movements in monkeys." *J. Neurophysiol.* 40: 74–94.

Nissen, M. J., M. I. Posner, and C. R. R. Snyder. 1979. Unpublished experiment cited by M. I. Posner in "Orienting of attention" (VIIth Sir Frederick Bartlett Lecture, Oxford).

Ogren, M. P., and A. Hendrickson. 1977. "The distribution of pulvinar terminals in visual areas 17 and 18 of the monkey." *Brain Res.* 137: 343–350.

Otteson, M. I., C. L. Sheridan, and D. R. Meyer. 1962. "Pattern discrimination." *J. Comp. Physiol. Psychol.* 55: 935–938.

Pasik, P., and T. Pasik. 1964. "Oculomotor functions with lesions of the cerebrum and superior colliculi." In *The Oculomotor System*, M. Bender, ed. New York: Harper and Row.

Pasik, T., and P. Pasik. 1971. "The visual world of monkeys deprived of striate cortex: Effective stimulus parameters and the importance of the accessory optic system." *Vision Res.* Suppl. 3: 419–435.

Pasik, P., T. Pasik, and P. Schilder. 1969. "Extrageniculostriate vision in the monkey: Discrimination of luminous flux-equated figures." *Exp. Neurol.* 24: 421–437.

Pasik, T., P. Pasik, P. Schilder, and J. Wininger. 1973. "Extrageniculostriate vision in the monkey: Effect of circumstriate cortex or superior colliculi ablations." *Excerpta Medica, Int. Congr. Series* no. 286: 202.

Perenin, M. T., and M. Jeannerod. 1975. "Residual vision in cortically blind hemifields." *Neuropsychologia* 13: 1–7.

Remington, R. 1979. Cited by M. I. Posner in "Orienting of attention" (VIIth Sir Frederick Bartlett Lecture, Oxford).

Robinson, D. A. 1972. "Eye movements evoked by collicular stimulation in the alert monkey." *Vision Res.* 12: 1795–1808.

Rosvold, H. E., M. Mishkin, and M. Szwarcbart. 1958. "Effects of subcortical lesions in monkeys on visual discrimination and single-alternation performance." *J. Comp. Physiol. Psychol.* 51: 437–444.

Schilder, P., T. Pasik, and P. Pasik. 1971. "Extrageniculostriate vision in the monkey. II. Demonstration of brightness discrimination." *Brain Res.* 32: 383–398.

Schilder, P., P. Pasik, and T. Pasik. 1972. "Extrageniculostriate vision in the monkey. III. Circle vs. triangle and 'red vs. green' discrimination." *Exp. Brain Res.* 14: 436–448.

Schiller, P., and F. Koerner. 1971. "Discharge characteristics of single units in the superior colliculus of alert rhesus monkeys." *J. Neurophysiol.* 34: 920–936.

Schiller, P., and M. Stryker. 1972. "Single-unit recording and stimulation in superior colliculus of alert rhesus monkeys." *J. Neurophysiol.* 35: 915–924.

Schneider, G. E. 1967. "Contrasting visuomotor functions of tectum and cortex in the golden hamster." *Psychol. Forsch.* 31: 52–62.

———. 1969. "Two visual systems." *Science* 163: 895–902.

Singer, W., J. Zihl, and E. Pöppel. 1977. "Subcortical control of visual thresholds in humans: Evidence for modality specific and retinotopically organized mechanisms of selective attention." *Exp. Brain Res.* 29: 173–190.

Soper, H. V., I. T. Diamond, and M. Wilson. 1975. "Visual attention and inferotemporal cortex in rhesus monkeys." *Neuropsychologia* 13: 409–419.

Sperling, G., and M. H. Melchner. 1978. "The attention operating characteristic: Examples from visual search." *Science* 202: 315–318.

Sprague, J., and T. Meikle. 1965. "The role of the superior colliculus in visually guided behavior." *Exp. Neurol.* 11: 115–146.

Stollnitz, F. 1963. "Spatial gradient of performance of monkeys (*Macaca mulatta*) on a single-color pattern discrimination problem." *Amer. Psychol.* 18: 473.

Thompson, R. 1969. "Localization of the 'visual memory system' in the white rat." *J. Comp. Physiol. Psychol. Monogr.* 69: 1–29.

Tolman, E. C. 1938. "The determiners of behavior at a choice point." *Psychol. Rev.* 45: 1–41.

Turvey, M. T. 1973. "On peripheral and central processes in vision: Inferences from an information-processing analysis of masking with patterned stimuli." *Psychol. Rev.* 80: 1–52.

Van Buren, J. M. 1963. "Trans-synaptic retrograde degeneration in the visual system of primates." *J. Neurol. Neurosurg. Psychiatr.* 26: 402–409.

Weiskrantz, L. 1963. "Contour discrimination in a young monkey with striate cortex ablation." *Neuropsychologia* 1: 145–164.

———. 1978. "Some aspects of visual capacity in monkeys and man following striate cortex lesions." *Arch. Ital. Biol.* 116: 318–323.

Weiskrantz, L., and A. Cowey. 1970. "Filling in the scotoma: A study of residual vision after striate cortex lesions in monkeys." In *Progress in Physiological Psychology*, vol. 3, E. Stellar and J. M. Sprague, eds. New York: Academic.

Weiskrantz, L., A. Cowey, and C. Passingham. 1977. "Spatial responses to brief stimuli by monkeys with striate cortex ablations." *Brain* 100: 655–670.

Wurtz, R. H. 1969. "Visual receptive fields of striate cortex neurons in awake monkeys." *J. Neurophysiol.* 32: 975–986.

Wurtz, R., and M. Goldberg. 1972a. "Activity of superior colliculus in behaving monkey. III. Cells discharging before eye movements." *J. Neurophysiol.* 35: 575–586.

———. 1972b. "Activity of superior colliculus in behaving monkey. IV. Effects of lesions on eye movements." *J. Neurophysiol.* 35: 587–596.

Wurtz, R. H., and Mohler, C. W. 1976. "Organization of monkey superior colliculus: Enhanced visual response of superficial layer cells." *J. Neurophysiol.* 39: 745–765.

11 Visuomotor Transforms of the Primate Tectum

E. Gregory Keating
John Dineen

The superior colliculus in the primate must be considered part of the visual system. It receives information directly from the eye and from other visual structures in the brain. It seems to have a motor function as well. Stimulation of the superior colliculus (SC or "tectum") in the monkey elicits saccadic eye movements. Lesions of the tectum in several nonprimate mammals causes visual and orienting deficits. Thus, by reason of its anatomy, physiology and ancestry, the primate superior colliculus might be thought to be the neural machinery controlling a "visual grasp" reflex: an arc between sensing and movement that allows the animal to orient its head and eyes toward targets falling outside the center of gaze.

Anatomy and Physiology

Direct projections from the eye come to the tectum from all sectors of the retina (Lund 1972; Hubel et al. 1975; Magnuson et al. 1979). There are inputs to the tectum from other primary retinal targets, such as the pregeniculate nucleus, the ventral part of the lateral geniculate nucleus, and the pretectal area (Carpenter and Pierson 1973; Edwards et al. 1974; Benevento et al. 1977). Projections of auditory and somatosensory origin, ascending from the lower brain stem and the spinal cord, also terminate in the superior colliculus (Moore and Goldberg 1966; Mehler 1969). The other main class of afferents to the SC is a descending influence from diffuse areas of the neocortex, with the major input coming from the occipital and frontal lobes (Myers 1963b; Kuypers and Lawrence 1967; Spatz et al. 1970; Wilson and Toyne 1970).

Some laminar segregation of these projections occurs within the superior colliculus (figure 11.1). Afferents from visual structures such as the retina and the occipital lobe terminate in the more superficial laminae, while cortical and brain-stem inputs of nonvisual origin synapse more deeply. The output of the SC follows a similar laminar pattern. Most of the axons exiting from the superficial laminae ascend toward more rostral portions of the brain stem and the thalamus en

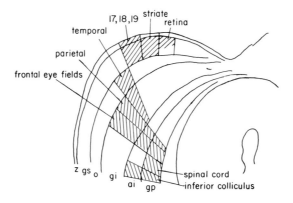

Figure 11.1 Laminar distribution of some afferent projections to the primate superior colliculus. Cross-hatching depicts laminae receiving the major portion of terminations. Sources are referenced in text. Abbreviations for strata: (z) zonale; (gs) griseum superficiale; (o) opticum; (gi) griseum intermediale; (ai) album intermediale; (gp) griseum profundum.

route to vision-related structures such as the pretectum, the inferior pulvinar nucleus, and the lateral geniculate nuclei. From the deeper laminae, axons both ascend and descend in the brain stem. Ascending projections from the deeper laminae are to diverse nuclei in the midbrain and the thalamus of nonvisual or multimodal function (Benevento and Fallon 1975). The descending axons terminate in the neighboring inferior colliculus and in several areas in the pons, the medulla, and the cervical spinal cord. These descending pathways of deep tectal origin project to regions involved in eye movements, or to areas projecting to the cerebellum or to cervical levels of the spinal cord (Myers 1963a; Benevento and Fallon 1975; Harting 1977).

By physiological measurement the entire topography of the retina is mapped onto the surface of the tectum, but not evenly. Neurons across two-thirds of the tectal surface respond to stimulation of the central 30° of the visual field (Cynader and Berman 1972; Lane et al. 1973). Tectal units have, on the average, larger receptive fields than those in the lateral geniculate nucleus (Schiller and Koerner 1971; Goldberg and Wurtz 1972a). They are less fastidious about the shape of a stimulus than are neurons of the visual cortex, and respond rather similarly to a light of any size falling within the central portion of their receptive field. A light outside this area weakly inhibits the cell, but the center-surround antagonism is not as sharp as in typical units recorded in the lateral geniculate nucleus. Most tectal visual units respond better to transient or moving stimuli. In the monkey, but not in other species, their responses are generally not specific to a particular direction of movement. Tectal neurons are not color-opponent, but are broadly tuned to a particular wavelength (Kadoya et al. 1971). Laminar variation in the responses of cells is consistent with the pattern of anatomical inputs. Neurons in the superficial laminae are fired by visual stimuli. In deeper laminae, the cells respond less strictly to visual events and can be triggered as well by noises or by somatosensory stimuli, including stretch information from muscles of the eye, the head, and the neck (Cynader and Berman 1972; Abrahams and Rose 1975).

Neurons that fire in some relationship to saccadic eye movements can be found in all laminae of the tectum (Goldberg and Wurtz 1972b; Schiller and Stryker 1972; Sparks 1978). Those in the superficial laminae are least tightly bound by this association. They respond to visual stimuli whether or not eye movements occur, but for many cells the responses are enhanced if the monkey makes a foveating saccade into the cell's receptive field (Goldberg and Wurtz 1972b; Wurtz and Mohler 1976). The enhancement increases over several trials in which the monkey foveates the cell's receptive field, and gradually decays in the repeated absence of such eye movements. Cells in the deep layers of the SC are related quite strictly to eye movements. These units are triggered only when the monkey moves its eyes, and they may have no visual receptive field. Cells in the intermediate laminae fall between these two extremes, in that either a visual stimulus or an eye movement, or the conjunction of the two, will fire them (Schiller and Stryker 1972; Mohler and Wurtz 1976; Sparks 1978).

Electrical stimulation of the SC induces eye movements (Robinson 1972; Schiller and Stryker 1972). The eye movements elicited are always conjugate and saccadic. Associated with stimulation of each point of the tectum is a movement of a particular vector (distance and direction). The direction of the movement is toward the point of the retinal map that received the stimulation, and in the monkey the distance and direction of the induced saccade is independent of the original position of the monkey's eyes. Electrical stimulation of many other brain structures than the SC, notably the occipital lobe and the frontal eye fields, also elicits eye movements (Schiller 1972; Bender and Shanzer 1964; Robinson and Fuchs 1969). However, the superior colliculus seems somewhat closer than they to the final pathway for eye movements, as the threshold currents for eliciting eye movements are lower in the SC than in these other structures (Schiller 1977).

If the superior colliculus is thought of as a reflex arc for orienting the animal to sensory events, then the above review suggests that in the monkey the motor or output side of the arc is primarily concerned with the production of saccadic eye movements. Its controlling input is of several modes (both visual and nonvisual) and of several origins (ascending sensory pathways from the eye, brain stem, and spinal cord, but also descending control from the neocortex).

For comprehensive reviews, see Sparks and Pollack 1977 and Goldberg and Robinson 1978.

Effects of Tectal Lesions in Primates

Most research has examined just the visuomotor function of the superior colliculus, and has risked (as will the rest of this chapter) ignoring its nonvisual function. The main challenge to deriving the function of the tectum in the primate from its anatomy and physiology is that vision, eye movements, and visual grasp proceed rather well without the superior colliculus. Ferrier and Turner (1901) first tried bilaterally removing the SC, and observed no change in either visual or oculomotor function. A much later study reported that temporary blindness resulted from lesions involving the tectum (Denny-Brown 1962). However, most laboratories have not found tectal-lesioned monkeys to be so drastically impaired. Typically, these monkeys show little deficit on the usual laboratory measures of vision. They perform pattern and color discriminations normally, and can distinguish moving from stationary targets (Rosvold et al. 1958; Anderson and Symmes 1969; Keating 1976). They can resolve luminance differences, although their difference threshold is temporarily elevated (Latto 1977). They can, if not pressed for time, reach accurately toward targets (Keating 1974; MacKinnon et al. 1976). After tectal lesion, monkeys make fewer spontaneous eye movements (Albano and Wurtz 1978) but can be induced to saccade toward any quadrant of the visual field. Nystagmus induced by optokinetic and vestibular stimulation appears to be normal within a few days of surgery, and there is no forced deviation of gaze (Pasik et al. 1966). Saccadic movements begin about 200 msec more slowly than normal, but foveate targets accurately (Wurtz and Goldberg 1972).

Tectal ablation in the monkey does produce deficits on certain behavioral tasks. Operated monkeys are temporarily poorer than normal at making a color discrimination if the salient stimulus, the patch of color, is set apart from where the monkey reaches to respond (Butter 1974). Tectally ablated monkeys have difficulty distinguishing the relative velocity of two moving objects (Anderson and Symmes 1969). Long-lasting deficits appear only when the monkeys are pressed for time. For example, SC-lesioned monkeys can locate a target hidden in a field of visual noise, but their reaction times for finding it are slower than normal (Latto 1978). In other studies, the monkeys had to reach toward lights that flashed for a fraction of a second in various sectors of the field (Keating 1974, 1976; MacKinnon et al. 1976). Tectal-lesioned monkeys required a long time to recover consistently good performance on these tasks. At their worst, however, they still reached accurately on many trials, and they were as good as normal if the stimulus lights were left on for more than 1 second.

Much of the impairment on these tasks could be explained by the reduction of exploratory saccades, but the deficit is more than merely a motor deficit. Butter et al. (1978) had monkeys view an array of targets and push the one that lighted briefly. A separate "no light" panel was to be pushed on those occasional blank trials when no light appeared. If the target flashed, the animal might err either by reaching inaccurately toward the light or by signaling that it had detected no light at all. Butter was able to plot the topography of the deficit after tectal lesions. He found that after operation the monkeys were poorer than normal at responding to a flashed light if it appeared beyond 40° from the central fixation point. On 36% of the incorrect trials the monkeys pushed the "no light" panel, signaling that they had failed to detect the peripheral target and were not simply misreaching toward it. Butter et al. argued that although even this deficit might be explained by an oculomotor impairment, it is better interpreted as a sensory neglect of stimuli appearing in the peripheral field. Their interpretation is supported by the results of tectal-lesion experiments in other species. Cats, rats, and hamsters fail to orient their heads toward the visual field contralateral to a tectal lesion, and they are not otherwise slowed, stopped, or distracted from their pursuits by events occurring in the affected field (Sprague and Meikle 1965; Kirvel et al. 1974; Goodale and Murison 1975; Finlay et al. 1978). Thus, one component of the tectal deficit seems to be a neglect of stimuli in the peripheral field.

Summary of Ablation Studies to Date

After tectal lesions, monkeys are slower to initiate saccades, and make fewer spontaneous exploratory eye movements. This deficit may be partly due to a reduced sensitivity for detecting peripheral events, and is not absolute. The sensory neglect may be an elevation of threshold for detecting a stimulus; the oculomotor deficit may be a relative paucity of saccades into the periphery.

The superior colliculus can be thought of as a reflex arc, a structure that provides some definable transform between sensory inputs and motor outputs. Our recent research began with the decision to accept that the function of tectal outputs was the production of conjugate saccadic eye movements. Described below is our attempt to determine by behavioral methods the nature of the SC's sensory inputs, in hopes of eventually defining the input-output transform between these and eye movements.

The intent of the first experiments was to determine the visual capacity of those inputs to the SC ascending directly from the retina and from brain-stem retinal targets. A second series of studies is pointed toward defining the nature of descending cortico-tectal control over saccadic eye movements.

Vision Surviving Removal of the Striate Cortex

The strategy of the first studies was influenced by our own and others' failure to capture the significance of the primate superior colliculus by the method of lesioning it directly and measuring the resulting deficit.

We therefore resorted to the converse maneuver. Ideally, the paradigm was to remove all of the monkey's visual system except the "ascending" inputs to the tectum from retina and other primary retinal targets, and then measure the visual control over eye movements exercised by these remaining pathways. The actual experiments departed somewhat from this ideal. We were not then equipped to record eye movements in monkeys, and instead measured their capacity to visually guide their arms in a variety of sensory tasks.

The research method of removing portions of the visual system and of testing the capacity of the remaining circuits follows a long tradition. By the end of the last century the pathway from the retina to the lateral geniculate nucleus and the radiations from there to the occipital cortex had already been described (see Polyak 1957). It was thought that damage to the occipital lobe blinded primates, with the exception that the pupils remained reactive to light. It was also known by then that the optic nerve projected to other targets besides the thalamus. Removing the superior colliculus, the largest of these targets, had negligible effect on vision (Ferrier and Turner 1901); this confirmed that the geniculostriate system ought to be the focus for studying the more interesting aspects of vision. Klüver (1941) more formally measured the vision remaining in a monkey after removal of the geniculostriate pathway, and found that destriated monkeys were not blind. He borrowed the model of a "photocell" to describe their remaining vision (Klüver 1942). The surviving visual system could, like a single-channel photocell, integrate the total luminous energy falling on the whole retina; however, local contrasts and their position on the retina could not be resolved.

The model explained why destriate monkeys could regulate their pupillary response to light and why they could discriminate stimuli of different luminous energy, and even explained other visual talents described by subsequent researchers. Humphrey and Weiskrantz (1967; see also Humphrey 1974) noted that destriate monkeys could maneuver about familiar environments, judge distances somewhat, and reach accurately for food. A single-channel photocell might accomplish such feats if it could be scanned across the visual field or if it could instead allow the environment to parade across its receptor angle and record the change in light over time. Temporal contrast and eye/head position could thus provide rough information about the visual shapes of things.

The photocell model was modified somewhat by the finding that a destriate monkey could distinguish a pair of patterns of equal luminosity but differing amounts of contour. Weiskrantz (1963) argued that the retina is "wired" to be acutely sensitive to changes in contrast (edges). A stimulus with more edges produced greater activity in the retinal ganglion cells, and the edgier stimulus thus appeared as brighter. He demonstrated that reducing the brightness of the edgier stimulus with a gray filter disrupted the discrimination. Thus, while the retina of the destriate monkey remained the mosaic of local contrasts that it was in the normal animal, the extrastriate system was only able to integrate all of the activity coming from the retina without preserving information about the location of retinal events. Recent experiments, however, have indicated that extrastriate structures have visual talents more sophisticated than that of a photocell. Destriate monkeys and humans are now known to possess true spatial vision in the sense that they can use the position of a light on the retina as a cue for reaching accurately toward targets (Sanders et al. 1974; Weiskrantz et al. 1977; Perenin and Jeannerod 1978). They also possess some rudimentary color vision (Schilder et al. 1972).

Recovery of Pattern Vision in Destriated Monkeys

In perhaps the most impressive demonstration of extrastriate vision in a primate, Schilder et al. (1972) were able to retrain monkeys without striate cortex to distinguish the shape of a triangle from that of a circle. We were interested in replicating this important finding. It was the sole demonstration of form vision in completely destriate primates. In other studies from the same laboratory, monkeys without striate cortex had been able to relearn a luminance discrimination (Schilder et al. 1971). The recovery in three such animals was later disrupted permanently by removal of the superior colliculus (Wininger et al. 1972). There was thus some reason to believe that recovery after striate lesions might serve as a useful index of the visual capacity of ascending inputs to the superior colliculus. We wished to explore the exact sense in which extrastriate pathways could process visual forms. Patterns can differ along a variety of dimensions. For example, a circle and a triangle differ not only in overall shape but also in some local features, such as length of contour and number of corners. Were the destriate monkeys attending to the overall shape of the stimuli, the way normal mammals (Hughes, this volume; see Zusne 1970 for review) and adult humans (Wertheimer 1938) discriminate such forms? Or were they ignoring the wide-field spatial relationships and becoming captured by local featural differences in the patterns, as do normal human babies (Salapatek 1975) and cats with large occipital lesions (Ritchie et al. 1976)?

Five adult monkeys (*Macaca mulatta* and *M. fascicularis*) were trained before operation on a series of two-choice pattern discriminations of the type shown in figures 11.2 and 11.3. Testing took place in a general primate test apparatus using the two-choice pull-in procedure (Schilder et al. 1971). At the start of a trial a guillotine door was raised, allowing the monkey to view two patterns which appeared to hang side by side in the dark with no visible surround. The patterns were projected from two light boxes suspended at arm's reach from the monkey. Each pattern in the pair subtended 31° and was separated from the other by 36°. If the monkey pulled in the box containing the correct pattern, it would find a food reward in a bin below and in front of the stimulus. The position of the correct pattern varied from left to right on an unpredict-

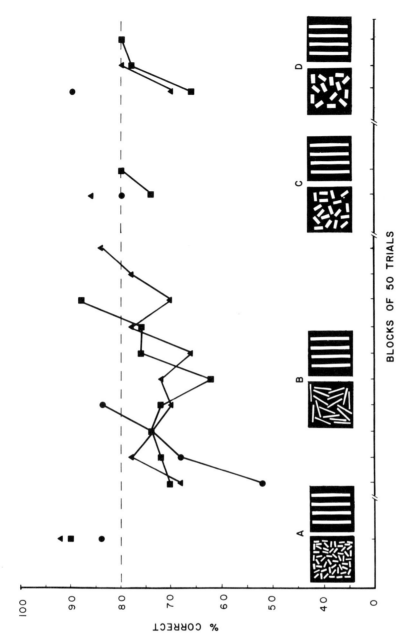

BLOCKS OF 50 TRIALS

Figure 11.2 Results of contour experiment. In task A–D monkeys were required to distinguish between members of a pair of flux-equated stimuli. The positive stimulus is shown on the left. In tasks A–C stimuli of each pair differ in a number of ways, including total amount of contour. The contour difference is gradually reduced in B and C, and in task D the stimuli are equated for contour. Symbols indicate performance of each of three monkeys on these tasks. The cumulative binomial probability of achieving a score above the dashed line in 50 trials is $P < 0.001$.

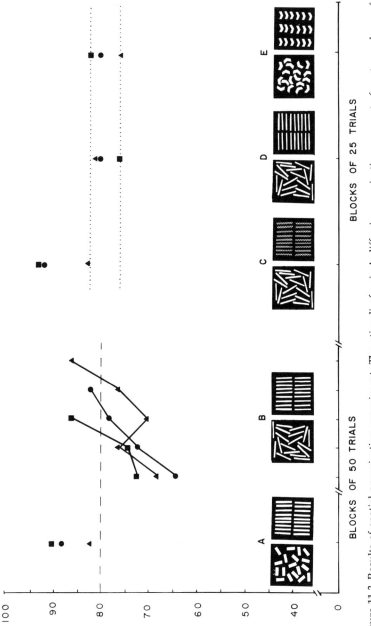

Figure 11.3 Results of spatial-organization experiment. The stimuli of pair A differ in organization, amount of contour, shape of subelements, and number of corners. After achieving criterion performance on this task, monkeys were transferred to task B, in which stimuli have identical local features but differ in organization of their elements. Each monkey eventually learned this discrimination and showed excellent transfer to tasks C, D, and E. Shading in C represents 0.5 log filter placed in front of right light box. (---) Training criterion (cumulative probability of $P < 0.001$) for the 50 block trials; ($\cdots$) confidence interval ($P < 0.008$–0.001) for the 25-trial transfer tasks.

able schedule. An error was counted and the trial was terminated if the animal pulled in the box with the wrong pattern. A monkey was judged to have learned to distinguish between a pair of stimuli if it reached correctly in 40 of 50 trials in a session. Once it achieved criterion on a pair of patterns, we switched the monkey to another pair. A monkey was said to have transferred good performance to the second pair if it maintained above-chance performance ($P < 0.008$ cumulative binomial probability) on the first 25 trials with the new stimuli.

The patterns were etched on plexiglas plaques, which were covered with opaque vinyl plastic and placed in front of a light box whose surface was evenly illuminated to 1.06 candela/m^2 with Aristo grid lamps and a diffusing plate. Considerable care was taken to ensure that all patterns had the same transilluminated surface area (54 cm^2). Stimulus patterns were changed by sliding a different plaque into the front of each of two light boxes. To disrupt the monkey's attention from local flux cues, each plaque was made reversible so that on a trial either a pattern or its mirror image might be presented to the animal.

Although the patterns of a pair were equal in luminance and luminous flux, they differed along other dimensions. They could have different local features, such as unequal amounts of contour, numbers of elements, or numbers of corners. They might also have different arrangements of the elements (organization), or they could differ in both organization and one or more local features. For example, a subject might use the scattered versus the regular organization to distinguish pair A in figure 11.2. However, one of the patterns of pair A also has different local features: more contour, more elements, and more corners than the other stimulus. In contrast, pairs B–E in figure 11.3 have equal local features and differ only in how the elements are arrayed on the slide. Prior to operation, monkeys were led through a series of such discriminations. The results confirm that normal mammals attend more readily to the overall organization of such arrays and are relatively impervious to changes in local features.

After preoperative testing, all five monkeys received bilateral lesions of striate cortex. Care was taken to remove all of area 17. After operation, the monkeys appeared, on casual examination, to be blind. However, after a week or so they began to orient to a penlight. They at first appeared confused in the testing apparatus, and required two weeks of shaping to become reacquainted with the training procedure. After that, formal testing began. Each monkey was tested on a series of visual discriminations. The entire series (not described here) was chosen to gradually wean them from using a flux cue and force them to make more complex pattern discriminations. The results of two experiments are described.

The animals were first required to distinguish between the patterns represented in part A of figure 11.2, but with the aid of a flux cue provided by a 2 log filter placed in front of the unrewarded stimulus. Eventually four of the five monkeys mastered this flux discrimination, requiring an average of 1,400 trials to do so. The flux difference was then gradually reduced and finally removed. Three of the monkeys were able to master pattern A in the absence of a flux cue. They went on to distinguish several patterns without the benefit of flux difference. The graph in figure 11.2 begins with the final performance of these three animals on pattern A.

The patterns of this pair differ in a number of ways, including amount of contour (length of black-white boundary). At this point, the three animals were performing similarly to the monkey described by Weiskrantz (1963) that could distinguish stimuli that differed in amount of contour or "edginess." Next, we attempted to determine whether destriate monkeys could be trained to distinguish patterns that were equal in contour. Each monkey was shifted from test A through test D in figure 11.2, which reduced the contour difference until in test D the amounts of contour were equal. Part D of figure 11.2 shows that the three monkeys mastered this equal-contour task rather quickly.

The results thus far confirm Weiskrantz's report that destriate animals could learn a contour difference equated for flux, and also indicate that destriated monkeys distinguished stimuli of equal flux and contour. Several possibilities now remained to be tested. For example, the monkeys could have been detecting differences in the amount of some other local feature, such as the number of elements in the array or the number of corners. The monkeys might also have been able, as are normal animals, to perceive the difference in the organization of the elements of the two stimuli.

The next test assessed this last possibility. Each monkey learned to discriminate test A of figure 11.3. The graph begins with the final performance of the three monkeys on this pair. These stimuli differed along several dimensions. The next pair (task B of figure 11.3) was designed to differ only in the organization or the spatial relationships of the elements; the number of elements, the length of contour, the form of each element, and the number of corners were the same. As figure 11.3 shows, within 250 trials all three monkeys had relearned this task. This was evidence that organization alone could serve as a discriminable cue for the destriate monkey.

However, several control tests were needed to determine whether the animals might not be using some other cue on task B of figure 11.3. In test C of figure 11.3, a 0.5 log filter was introduced in front of the right light box to control for minimal differences in total flux that might have eluded our efforts to equate the stimuli. Test D should have disrupted performance if the animals had been using local flux differences by attending to just the left edge or the bottom half of the stimulus. The animals maintained excellent performance through both of these controls. Finally, test E indicated that monkeys would generalize their responses to a new pair of stimuli that maintained the same difference in organization but introduced novel local features.

At the end of testing, one of the three monkeys (represented by ●——● in figures 11.2 and 11.3) was dissected to verify the completeness of the striate lesion. Gross examination of the brain suggested that the occipital cortex had been removed completely, with some invasion of area 18 along the banks of the lunate and calcarine sulci. Microscopic examination revealed a complete absence of cell clusters throughout both lateral geniculate nuclei. There were occasional single neurons, although they never appeared in clusters. Similar findings have been reported for other monkeys in which the striate lesions were judged complete (Schilder et al. 1971).

We view these data first as a confirmation of the important finding of Schilder et al. (1972) that monkeys with area 17 removed completely can distinguish flux-equated shapes. Second, our data indicate that the destriate monkey, when pushed, can process visual pattern information in some sense like a normal animal. It is not a single-channel photometer that resorts to eye-fixation strategies to isolate some local flux difference between the patterns. Nor does it simply sum the number of local features appearing over the entire retinal field.

However, these data do not argue that some extrastriate circuit, such as the tectal-pulvinar-prestriate pathway, is normally a major contributor to the perception of pattern in the normal primate. Although there is limited support for this idea with respect to other mammals (Berlucchi et al. 1972), particularly the tree shrew (Casagrande and Diamond 1974), it seems an unlikely possibility in the primate, given the long retraining required to elicit pattern vision in these destriate monkeys and the minimal effects of tectal lesions on pattern vision. We would argue that in the primate the geniculostriate pathway still remains the principal conduit of pattern information to the cortex.

These data do say that the extrastriate pathways have some ability to resolve local spatial contrast and preserve information about its position on the retina. Information about the spatial relationships of local contrasts can still be assembled from wide areas of the retina, and a summary of this information can still be extracted to allow for some generalization to other similar stimuli.

Pathways Involved in Recovery
Which of the extrastriate pathways are responsible for the recovery of vision in the destriate monkeys? Primary retinal targets remain intact in these monkeys, as do vision-related cortical structures such as preoccipital and inferotemporal cortex. Only a few experiments have sought out the structures critical to the recovery. Pasik and Pasik (1971) determined that the only portion of the visual system necessary for merely detecting the presence of a light is an intact projection from retina to the

accessory optic nucleus in the midbrain. For the recovery of more complex vision the superior colliculus appears critical, since in one study subsequent tectal ablation disrupted the destriate monkey's performance of a luminance discrimination (Wininger et al. 1972). However, although information processed at the level of the SC is critical, it is not necessarily sufficient for sustaining the residual vision. Those authors felt that the recovery of complex vision after removal of area 17 depended as well on the integrity of cortical structures, particularly areas 18 and 19 of preoccipital cortex. If they included large amounts of areas 18 and 19 in their lesions of striate cortex, this too prevented their monkeys from recovering a luminance discrimination (Pasik and Pasik 1971).

In a pilot experiment to seek out the pathways of recovery, two out of three of our destriate monkeys who had relearned the pattern tasks subsequently received bilateral lesions of the inferotemporal cortex. A large body of literature implicates the preoccipital and inferotemporal cortex in the control of pattern vision, and either of these areas might have been important to our monkeys' recovery. The preferred experiment would have been to remove the preoccipital target of the tectal–inferior pulvinar pathway (Benevento and Rezak 1976). However, in these two subjects we wished to avoid any lesion that would compromise the later histological analysis of the striate-cortex ablations. For that reason also, the inferotemporal lesions were small and anteriorly placed (figure 11.4). Thus, these preliminary results should be interpreted cautiously. The monkeys were retested on all the patterns. Surprisingly, their performance on the pattern problems was unimpaired by the inferotemporal lesions (figure 11.5).

From these preliminary results, it seems that the inferotemporal cortex is not greatly involved in the recovery of destriate pattern vision. The preoccipital cortex may be, but a recent experiment indicates that at least some spatial and color vision survives removal of both striate and preoccipital cortex (Keating 1979, 1980). In this experiment, five normal rhesus monkeys were trained on four visual tasks. Two of the tasks

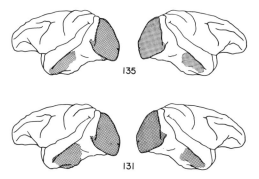

Figure 11.4 Lateral view of striate-cortex ablation in two monkeys and subsequent lesions of inferotemporal cortex. Monkeys mastered pattern discriminations after both stages of surgery.

Visuomotor Transforms of Primate Tectum

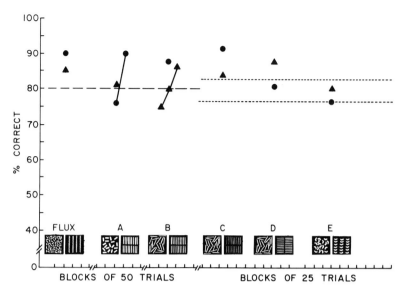

Figure 11.5 Performance of pattern discriminations after inferotemporal lesions in previously destriated monkeys. In flux test the positive stimulus, shown on the left, was 0.5 log units brighter than the negative pattern. Legend otherwise as in figure 11.3.

were velocity discriminations. In the absolute-velocity test the monkey had to distinguish a stimulus rotating at 26 rpm from a stationary one. In the relative velocity test both stimuli rotated, but at different angular velocities (26 and 44 rpm), and for a reward the monkey had to reach toward the slower stimulus. Figure 11.6 portrays two versions of the velocity tests. In one (part A) shadows were cast onto a projection screen by star-shaped cutouts mounted on drive pulleys and interposed between a light projector and the screen. The average luminance of the screen surrounding the shadows was 541 cd/m^2, and that of the shadows was 14 cd/m^2. The position of the correct shadow varied unpredictably according to a schedule that placed it over the left food well on half of the 50 trials of a daily session. The monkey was judged competent at this task when it performed 45 out of the 50 trials correctly. All five monkeys learned this version of the velocity tests before operation. In most of the monkeys the use of shadows seemed to retard the learning of the velocity difference; subject 126 had particular difficulty with the task. A second version (part B of figure 11.6) was introduced for testing the last subject in the study, 126, after surgery. It was similar to the first version except that instead of shadows the monkey viewed two black plexiglas disks (1 cd/m^2) mounted on an opaque black screen. Each disk contained a white spot surrounding a nail. The apparatus was lit from above by an ultraviolet tube which fluoresced the white spot (8 cd/m^2) but kept other stray light to a minimum. A raisin could be fixed to the nail of the correct disk. Control trials assured us that a destriate monkey could not distinguish the raisin from a small irregular black

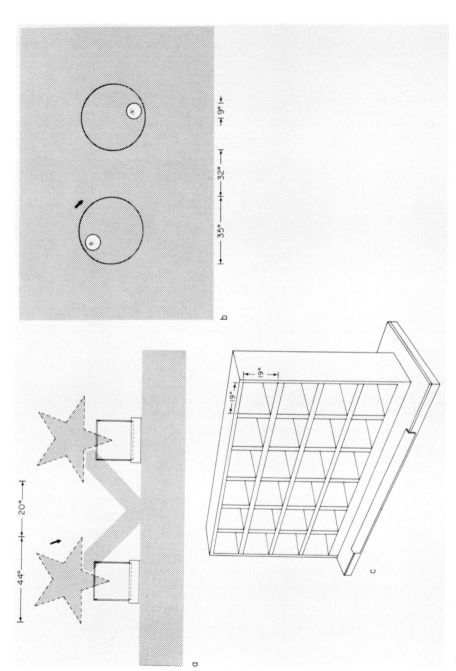

Figure 11.6 (a,b) First and second versions of the apparatus used in the velocity discriminations. Shading in part a represents shadows cast on stimulus screen by backlighted stars and occluders masking drive train. Shading in part b represents black plexiglas disks rotating in front of opaque black background. (c) Matrix of food wells used in target location test. Measurements are given in degrees of visual angle subtended by the apparatus at arm's length (20 cm) from monkey.

spot painted around the nail on both disks. The same angular velocities were used as in the earlier version.

A third test, the target location task (part C of figure 11.6) measured the monkeys' accuracy at reaching into the correct bin in a matrix of 24 food wells. A 1400-cd/m² light appeared for a few seconds in one of the wells, signaling the presence of a raisin. An error was scored if the monkey reached into the wrong well. Criterion performance was set at 80% correct reaches in 50 trials.

The fourth test was a blue-yellow color discrimination. At the start o a trial, the monkey faced a plexiglas projection screen. A blue field appeared on the screen over one of the food wells, a yellow field over the other. The monkey reached through the door under the blue stimulus for a food reward. The blue/yellow pair of stimuli was made from Kodak CC50 yellow and blue filters and a number of neutral-density filters sandwiched into a single slide. The exact makeup of a set of 20 such slides and the rationale of the wavelength test followed the description of Kicliter and Loop (1976). In general, the method involved measuring the spectral transmittance curves of the color filters at 25-nm intervals through the visible range of the spectrum. Then neutral-density filters were added behind either the blue or the yellow half of the slide until a set of 20 slides was created such that on 10 the optical density of the blue half was greater than that of the yellow half at every point along the visible spectrum. The relationship was reversed on the other 10 slides. The brightness of a stimulus is a function of the spectral sensitivity of the viewing animal and the absolute spectral distribution of the stimulus. With the procedure described above, it is not necessary to know the exact spectral sensitivity of the animal. For every animal whose sensitivity to light lies within the measured range of 400–700 nm, the blue portion of the slide will appear brighter on half of the slides and dimmer on the other half. If the animal uses brightness differences to distinguish the slides, it should perform at chance. Better-than-average performance is evidence of wavelength discrimination.

The monkeys received 50 trials a day selected from the 20 blue-yellow slides. The left-right position of the blue stimulus and its relative brightness varied from trial to trial in an unpredictable manner. After a monkey reached a criterion of 45 out of 50 correct responses in a session, it was switched to a similar set of blue-green slides. In normal animals, stable performance of a wavelength discrimination is generally indicated by good transfer to a second pair of stimuli containing a novel unrewarded hue.

After reaching criterion on all the tests, the monkeys received lesions intended to remove all of striate cortex bilaterally. They were again retested over several months after this surgery until they had recovered criterion performance on all four of the tasks. Then the monkeys received a second lesion; bilateral removal of "preoccipital" cortex. At their most extensive the combined lesions included all of striate cortex;

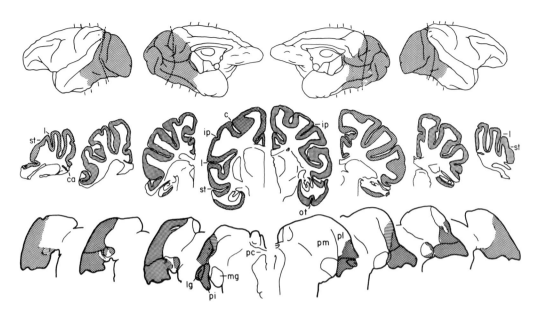

Figure 11.7 Neuropathology of the subject with the most extensive lesions (monkey 126). Top: Lesions have been shaded onto standard hemisphere drawings. All trace of tissue has been removed posterior to the dotted line. Hatchmarks indicate the level of representative cortical cross-sections. Middle: Anterior sections through the lesion are presented in the center and extend posteriorly toward either side of the figure. Shading represents intact cortex. Abbreviations of sulci: (c) central, (ca) calcarine, (ip) intraparietal, (l) lateral, (ot) occipitotemporal, (st) superior temporal. Bottom: Cross-sections through the posterior thalamus. Anterior sections are presented centrally. Stipple represents regions of complete retrograde degeneration (70% cell loss); hatched areas represent partial (30%–70%) cell loss. Abbreviations: (lg) dorsal lateral geniculate nucleus, (mg) medial geniculate n., (pc) posterior commissure, (pi) inferior pulvinar n., (pl) lateral pulvinar n., (pm) medial pulvinar.

preoccipital areas OB and transitional OA-TE were entirely removed. Area OA was also mostly destroyed except for a spared portion on the floor of the superior temporal sulcus corresponding to the STS-movement area. Degeneration of the inferior pulvinar was extensive but incomplete (figure 11.7).

Figure 11.8 shows the results from monkey 126, the subject with the most extensive lesions. The numbers in the middle of each graph indicate that several months and several hundred trials were required for this monkey to recover after area-17 lesions. This was the typical effect of the striate-cortex lesions. Subsequent preoccipital lesions generally disrupted recovery. It appeared to blind the monkeys for a few days and altered their performance on the formal tests. However, all of the animals eventually relearned all of the tests.

After the monkeys had returned to criterion we began control sessions, which assured us that they were not using sounds or other nonvisual cues coming from the apparatus to solve the discriminations. Additional control sessions with subject 126 (Keating 1980a) convinced us that this monkey, which had sustained the most extensive striate-preoccipital lesions, was capable of true spatial vision; that is, it could

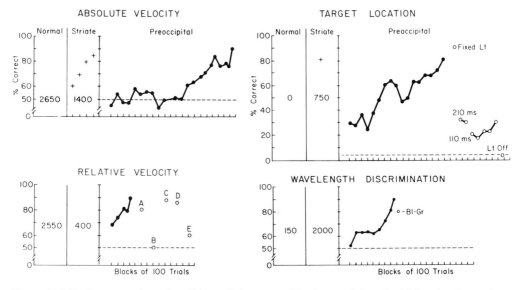

Figure 11.8 Performance of monkey 126 on all four tests. Numbers at left and middle of each graph are numbers of trials to criterion before and after striate-cortex lesion. At right are the relearning curves after subsequent preoccipital surgery. +: Performance over 50 review trials immediately prior to preoccipital surgery.

resolve the position of a light on the retina to find the food well in the target-location test and was using rate of movement across the retina to solve the velocity tests.

With sufficient time or retraining and under optimal conditions, primates can be shown to maintain a number of visual skills in the absence of the geniculostriate system. They can still resolve local contrast on the retina and use this information to reach accurately and even discriminate some patterns. They can perform a simple color discrimination. Some of the residual vision seems to represent the capacity of the brain-stem visual structures, since it survives additional removal of the pulvinar-preoccipital pathway, the remaining major route of visual information to the cortex.

The superior colliculus appears to be the most important of these brain-stem structures for sustaining the vision required by these tasks. After recovering from the combined striate-preoccipital surgery, one of the monkeys in the study received yet a third lesion: bilateral ablation of the superior colliculus and pretectum. The monkey appeared blind for 3 weeks after surgery. After that, its pupils began to react sluggishly to light. The monkey was observed informally for 3 more weeks and never gave any other signs of seeing. It performed at chance on 500 trials of the absolute-velocity test. From physiological evidence, neurons in the superior colliculus seem to possess all of the information required to carry out the tasks in this study. They have discrete receptive fields which would allow for resolution of separate points on the retina. They are sensitive to movement and in some cases are broadly tuned to a preferred wavelength.

These data do not argue that the retinal and brain-stem visual inputs to the tectum are normally important for velocity discriminations, pattern, or color vision. The long retraining required of destriate monkeys suggests that the tectum has in these studies been pressed into a service it does not normally perform. After all, research into prosthetic devices in humans indicates that the blind can learn to "see" patterns using spatially coded information from a variety of unusual sources. For what, then, is such visual information to the tectum normally used? Returning to the idea that the output of the SC is mainly concerned with the control of eye movements, we find that these experiments partly specify a "peripheral" visual grasp reflex: the dimensions of control over foveation exercised by ascending inputs to the superior colliculus. Retinal and brain-stem afferents will influence whether a saccadic eye movement occurs, and will control its distance and direction on the basis of the target's location on the retina, its velocity, and perhaps its size and brightness. However, the peripheral grasp reflex will be only crudely sensitive to the shape and color of the target, and not at all "attentive" to other kinds of information.

Descending Cortical Control of Eye Movements

Saccadic eye movements are certainly controlled by other sorts of information than the above narrow list. Loftus and Mackworth (1978) showed human subjects pairs of drawings like those in figure 11.9 and measured their eye scan paths. One of each pair of photos had an inappropriate figure near the picture's center. The central figure in the drawing was fixated sooner and more often when, like the octopus in the farmyard, it was inappropriate to the context of the drawing. The sequence of fixations might in this case also be called a visual grasp reflex. Its motor output, a saccade of some vector, is the same as when a monkey orients to a light flashing in its peripheral field, but its controlling input is from different sources. The control over eye movements in this case must be at least partially of cortical origin. The fine contrast sensitivity of the geniculostriate system and other cortical areas involved in pattern perception very likely participates in resolving the figures of the drawing. The eye movements are also driven by internal information, memory, and other forebrain systems that allow us to consider the octopus inappropriate to a farmyard.

We are developing analogous visual tasks for monkeys. Our intent is to capture the function of the cortical-tectal pathway and other avenues of forebrain control over eye movements. The basic mechanics of a single saccade (its latency, speed, and accuracy) seem to be driven from brain-stem structures outside the SC. Our hypothesis is that the superior colliculus is instead part of a pathway by which the cortex

Figure 11.9 Pair of pictures consisting of a coherent scene (top) and one with an inappropriate central figure (bottom).

controls the *sequence* of fixations (scan path) that is appropriate to some viewing situation.

Our first effort has been to define what constitutes an "appropriate" scan path in a normal monkey. We have so far evolved two situations where the monkey's fixations, besides appearing sensible to a human observer, have the more objective qualities of being nonrandom, reliable, and manipulable. The infrared corneal reflection method is used for recording eye position in the monkey. The animal sits in a chair facing a television screen that subtends a 40° × 50° area of its visual field. A mount screwed to its skull prevents head movement. The monkey first learns to press a lever to bring a small spot of light onto the screen. The spot of light remains on for an unpredictable amount of time, up to several seconds. The light then dims for a half-second. If the monkey releases the lever during this brief dim period, it receives a bit of orange juice. Releasing the lever at any other time causes a buzzer to sound and delays the next trial. The dimming is so slight that the monkey must foveate the light to see it occur. The light thus entrains the monkey's gaze and serves as a fixation point for calibrating the eye-tracking apparatus. Eye position is recorded with the Bio-Trac Eye Movement recorder (Gulf and Western Labs). A light-emitting diode surrounded by two photodiodes is placed directly in front of the monkey's left eye (figure 11.10). The LED bathes the cornea with invisible infrared light, and the light reflected from the cornea is detected by the photodiodes, transduced to a voltage, and amplified. A change in the

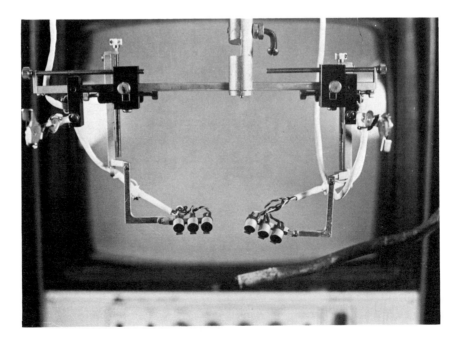

Figure 11.10 Infrared spectacles for recording eye movement. In front of each eye is a cluster of two photodiodes surrounding a light-emitting diode.

voltage represents a shift in the horizontal position of the eye. A similar array placed under the right eye monitors the lower lid, which shifts along with vertical eye movement. The advantages of this method of recording eye movements are that it is commercially available and relatively inexpensive, requires no surgical implantations, and (with rigid head fixation) does not drift over a 1-hour session. The method will detect an eye movement of 1° of visual angle. When used to measure eye position over a 50° diameter field, it can distinguish eye positions of greater than 5°. The disadvantages of the method are its limited accuracy, its frequent nonlinearity under practical conditions, and its partial obstruction of the view of one eye. The eye-position signals are now being processed by a PDP 11/03 microprocessor. However, the data reported here were analyzed manually by first taping them on an FM recorder, then playing them back onto an x,y plotter at one-fourth their original speed.

The first method of manipulating scan patterns in monkeys is a version of a human visual-search test. After the monkey learns to fixate a spot of light and wait for it to dim, the light is embedded in a field of visual noise like the one in figure 11.11. On each trial the target could appear at any one of nine possible positions in the noise. The monkey must find and fixate the target before it dims. A theoretically ideal searching strategy cannot be formally defined in this situation, but one can judge scan paths to be more or less suited to variations of this task. For example, the salience of the target can be altered by changing the amount and shape of the noise. With less noise and a conspicuous target, large areas can be grasped with each single fixation; this allows for greater spacing between fixations. However, with very noisy backgrounds the subject would be wise to make longer and more closely grouped fixations, processing smaller sectors at a glance (Engel 1977). The subject's prior information should also alter the search strategy (Williams 1966). If the target tends to appear with equal frequency at each of the nine possible locations, then the scan path ought to sample widely through the field in search of the light. If the target tends instead to repeat itself in the same position, accomplished human searchers quickly shift their scans to that point and risk ignoring the rest of the array (Llewellyn-Thomas 1963).

As it turns out, the monkeys also alter their search strategies to fit their recent experience with the task. We first presented a block of trials in which the target had an equal chance of appearing at any one of nine positions in the field. Then five blank trials occurred in which the visual noise was presented but without a target. The top half of figure 11.11 shows two examples of the fixations of a monkey searching for the target on those five blank trials. The fixations are superimposed on a grid of symbols (+) which mark the position of the nine possible target locations. In contrast, the lower half of figure 11.11 provides two examples of fixation patterns that resulted when the monkey was given a

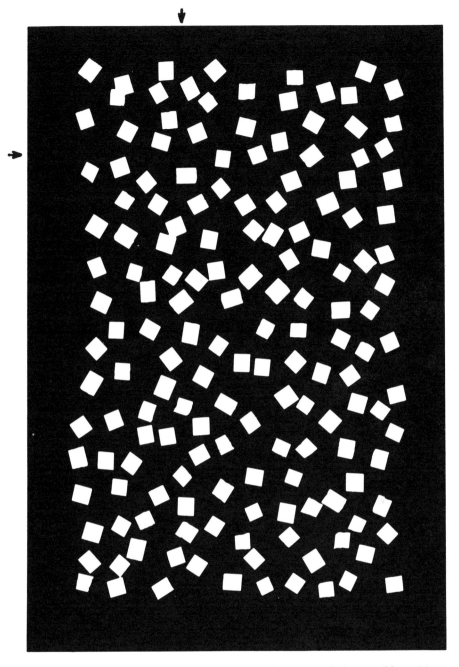

Figure 11.11 Visual noise with a light spot embedded at one of nine possible positions (arrows). The monkey must find and fixate the spot in time to see it dim.

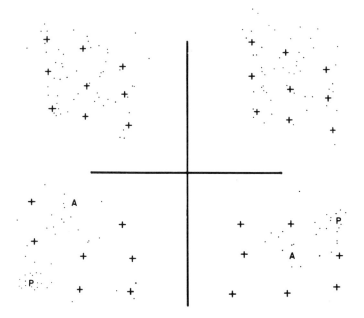

Figure 11.12 Each panel shows the fixations of a monkey searching for a target of five blank trials in which no target is present. Symbols +, P, and A mark the nine possible positions in which the target might appear on regular trials. The search strategy in the top two panels resulted after 36 trials in which the target appeared equally often in each of nine positions. In the bottom panels the target appeared most recently at A, but appeared most frequently over the last 36 trials at P.

sequence of 36 trials on which the target appeared with a high probability in the corner of the array marked P. Then for one trial the target shifted to an alternate location (A), followed by five blank trials. The search in this case is biased toward one quadrant of the array. If the chronological order of the fixations is considered, the monkey's scan path can be shown to be the sum of two separable strategies. First, there is a tendency to look to where the spot appeared in the immediately previous trial (A). This bias fades after 5–10 seconds and is followed by a search toward where the longer-term probabilities point (P).

Monkeys also show reliable, stereotypic scan paths in certain circumstances where viewing is freer. In a study carried out together with Dr. Caroline Keating we allowed monkeys to view photographs of faces. (Monkeys recognize photographic representations of objects [Humphrey and Keeble 1976], and are quite attentive to faces [Redican 1975], although no one to our knowledge has measured eye fixations to determine what facial features capture the monkeys' attention.) Two rhesus monkeys were presented with a series of photographs of faces of other rhesus monkeys, chimpanzees, humans, and schematic cartoons of a face. The faces fell on varied portions of the TV screen and subtended 20°–30° of the visual field. Figure 11.13 shows the resulting cluster of fixations made by one monkey over the 5-second viewing time for each

Keating, Dineen

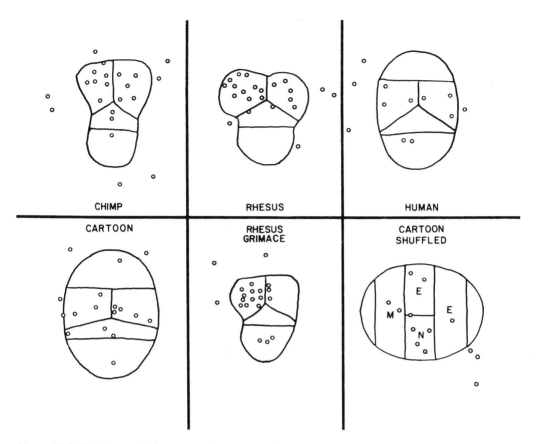

CHIMP   RHESUS   HUMAN
CARTOON   RHESUS
GRIMACE   CARTOON
SHUFFLED

Figure 11.13 Fixations made by one monkey viewing photographs of faces for 5 seconds each. Fixations are plotted onto outlines which divide faces into regions of eyes, nose, and mouth. Position of fixation within each compartment is not necessarily accurate. Outline at bottom right is a cartoon face whose features were rearranged into an unfacelike array.

face. The limits to the resolution of the eye-tracking apparatus permitted us to confidently assign eye movements to large compartments of the face (for example, the region of the eyes, the nose, or the mouth), but the fixations are not necessarily accurately placed within each compartment. There was considerable variability between the viewing patterns of the two subjects, particularly in the amount of time each cared to spend looking directly at the face. However, their on-face fixations were fairly similar and definitely not random. They attended mainly to the eyes (usually one eye more than the other), and spent little time fixating the rest of the face. This preoccupation with the eyes held for rhesus, chimp, and human faces. It was true for rhesus faces gesturing with mouth threats or grimaces. The concentration on the eyes was somewhat weaker for schematic cartoon faces. At bottom right in figure 11.13 are the fixations that resulted when the features of a cartoon were scrambled into an unfacelike array.

For human subjects viewing photographs, appropriate eye movements are those that fixate earliest and most often regions of the photo that, by other criteria, are judged high in information content (Mackworth and Morandi 1967). As another attempt to elicit cortically driven saccadic eye movements, we are beginning to measure the scan paths of monkeys discriminating simple geometric forms. For example, in distinguishing a hexagon from a circle the corners of the hexagon contain most of the information about the difference between the forms and should capture the bulk of the animal's fixations.

We speculate that the commands about where the monkey should look in these situations are of cortical origin and, therefore, depend for their execution on corticofugal pathways of control over the brain-stem oculomotor machinery. There is more than one pathway of cortical control over saccadic eye movements. At least one of these pathways is routed through the superior colliculus. Conway et al. (1979) observed a monkey's eye movements while it searched for a slice of apple on a tray. Normally the monkey scanned the entire tray. After lesion of either the superior colliculus or the frontal eye fields (FEF) there was little change in its searching pattern. After lesion of both FEF and SC, searching saccades stopped altogether. Electrical stimulation of either the FEF or visual cortex elicits saccadic eye movements in a monkey. However, after ablation of the superior colliculus eye movements can still be driven by electrical stimulation of the frontal lobe but no longer by stimulation of visual cortex (Schiller 1977).

The existence of more than one avenue of cortical influence over the oculomotor complex suggests why tectal lesions do not cause frank oculomotor deficits. Any one of several circuits between the visual and oculomotor system might be able to mediate the relatively simple task of foveating a light flashing in the visual field. Tectal lesion would cause only a partial disconnection of sensory control over eye movements. The partial disconnection should be reflected not as a paralysis of eye movements, but perhaps as an oculomotor apraxia; as inappropriate rather than inaccurate fixations on tasks in which eye movements are driven by information routed through the superior colliculus. Defining the terms of the apraxia should clarify that particular input-output transform between vision and eye movements that represents the function of the primate superior colliculus.

Acknowledgments

This work was supported by the U.S. Public Health Service under grants NS 10576 and EY 02941 and by the Veterans' Administration Research Service.

References

Abrahams, V. C., and P. K. Rose. 1975. "Projections to extraocular neck muscles and retinal afferents to superior colliculus in the cat: Their connections to the cells of origin of the tectospinal tract." *J. Neurophysiol.* 38: 10–18.

Albano, J. E., and R. H. Wurtz. 1978. "Modification of the pattern of saccadic eye movements following ablation of the monkey superior colliculus." *Soc. Neuroscience Abstr.* 4: 161.

Anderson, K. V., and D. Symmes. 1969. "The superior colliculus and higher visual functions in the monkey." *Brain Res.* 13: 37–52.

Bender, M. B., and S. Shanzer. 1964. "Oculomotor pathways defined by electrical stimulation and lesions in the brainstem of monkeys." In *The Oculomotor System*, M. B. Bender, ed. New York: Harper and Row.

Benevento, L. A., and J. H. Fallon. 1975. "The ascending projections of the superior colliculus in the rhesus monkey (*Macaca mulatta*)." *J. Comp. Neurol.* 160: 339–361.

Benevento, L. A., and M. Rezak. 1976. "The cortical projections of the inferior pulvinar and adjacent lateral pulvinar in the rhesus monkey (*Macaca mulatta*): An autoradiographic study." *Brain Res.* 108: 1–24.

Benevento, L. A., M. Rezak, and R. Santos-Anderson. 1977. "An autoradiographic study of the projections of the pretectum in the rhesus monkey (*Macaca mulatta*): Evidence for sensorimotor links to the thalamus and oculomotor nuclei." *Brain Res.* 127: 197–218.

Berlucchi, G., J. M. Sprague, J. Levy, and A. C. DiBerardino. 1972. "Pretectum and superior colliculus in visually guided behavior and in flux and form discrimination in the cat." *J. Comp. Physiol. Psych. Monogr.* 78: 123–172.

Butter, C. M. 1974. "Effect of superior colliculus, striate, and prestriate lesions on visual sampling in rhesus monkeys." *J. Comp. Physiol. Psychol.* 87: 905–917.

Butter, C., C. Weinstein, D. B. Bender, and C. B. Gross. 1978. "Localization and detection of visual stimuli following superior colliculus lesions in rhesus monkeys." *Brain Res.* 156: 33–49.

Carpenter, M. B., and R. J. Pierson. 1973. "Pretectal region and the pupillary light reflex: An anatomical analysis in the monkey." *J. Comp. Neurol.* 149: 271–300.

Casagrande, V. A., and I. T. Diamond. 1974. "Ablation study of the superior colliculus in the tree shrew." *J. Comp. Neurol.* 156: 207–238.

Conway, J., P. H. Schiller, and S. True. 1979. "Eye movement control: The effects of paired superior colliculus and frontal eye field ablations." *Soc. Neurosci. Abstr.* 5: 366.

Cynader, M., and N. Berman. 1972. "Receptive-field organization of monkey superior colliculus." *J. Neurophysiol.* 35: 187–201.

Denny-Brown, D. 1962. "The midbrain and motor integration." *Proc. R. Soc. Med.* 55: 527–538.

Edwards, S. B., A. C. Rosenquist, and L. A. Palmer. 1974. "An autoradiographic study of ventral lateral geniculate projections in the cat." *Brain Res.* 72: 282–287.

Engel, F. L. 1977. "Visual conspicuity, visual search and fixation tendencies of the eye." *Vision Res.* 17: 95–108.

Ferrier, D., and W. A. Turner. 1901. "Experimental lesions of the corpora quadragemina in monkeys." *Brain* 24: 27–46.

Finlay, B. L., D. Cordon, and K. Marder. 1978. "Ontogeny of visuomotor behavior in normal and neonatally colluculectomized hamsters." *Soc. Neurosci. Abstr.* 4: 111.

Goodale, M. A., and R. C. Murison. 1975. "Effects of lesions of superior colliculus on locomotor orientation and the orienting reflex in the rat." *Brain Res.* 88: 243–261.

Goldberg, M. E., and D. L. Robinson. 1978. "Visual system: Superior colliculus." In *Handbook of Behavioral Neurobiology*, vol. 1: *Sensory Integration*, B. Masterton, ed. New York: Plenum.

Goldberg, M. E., and R. H. Wurtz. 1972a. "Activity of superior colliculus in behaving monkey. I. Visual receptive fields of single neurons." *J. Neurophysiol.* 35: 542–559.

———. 1972b. "Activity of superior colliculus in behaving monkey. II. Effect of attention on neuronal responses." *J. Neurophysiol.* 35: 560–574.

Harting, J. K. 1977. "Descending pathways from the superior colliculus: An autoradiographic analysis in the rhesus monkey (*Macaca mulatta*)." *J. Comp. Neurol.* 173: 583–612.

Hubel, D. H., S. LeVay, and T. N. Wiesel. 1975. "Mode of termination of retinotectal fibers in Macaque monkey: An autoradiographic study." *Brain Res.* 96: 25–40.

Humphrey, N. K. 1974. "Vision in a monkey without striate cortex: A case study." *Perception* 3: 241–255.

Humphrey, N. K., and G. R. Keeble. 1976. "How monkeys acquire a new way of seeing." *Perception* 5: 51–56.

Humphrey, N. K., and L. Weiskrantz. 1967. "Vision in monkeys after removal of the striate cortex." *Nature* 215: 595–597.

Kadoya, S., L. R. Wolin, and L. C. Massopust. 1971. "Collicular unit responses to monochromatic stimulation in squirrel monkey." *Brain Res.* 32: 251–254.

Keating, E. G. 1974. "Impaired orientation after primate tectal lesions." *Brain Res.* 67: 538–541.

———. 1976. "Effects of tectal lesions on peripheral field vision in the monkey." *Brain Res.* 104: 316–320.

———. 1979. "Rudimentary color vision in the monkey after removal of striate and preoccipital cortex." *Brain Res.* 179: 379–384.

———. 1980. "Residual vision in the monkey after removal of striate and preoccipital cortex." *Brain Res.* 187: 271–290.

Kicliter, E., and M. S. Loop. 1976. "A test of wavelength discrimination." *Vision Res.* 16: 951–956.

Kirvel, R. D., R. A. Greenfield, and D. R. Meyer. 1974. "Multimodal sensory neglect in rats with radical unilateral posterior isocortical and superior collicular ablations." *J. Comp. Physiol. Psychol.* 87: 156–162.

Klüver, H. 1941. "Visual functions after removal of the occipital lobes." *J. Psychol.* 11: 23–45.

————. 1942. "Functional significance of the geniculo-striate system." *Biol. Symp.* 7: 253–299.

Kuypers, H. G. J. M., and D. G. Lawrence. 1967. "Cortical projections to the red nucleus and the brain stem in the rhesus monkey." *Brain Res.* 4: 151–188.

Lane, R. H., J. M. Allman, J. H. Kaas, and F. M. Miezin. 1973. "The visuotopic organization of the superior colliculus of the owl monkey (*Aotus trivirgatus*) and the bush baby (*Galago senegalensis*)." *Brain Res.* 60: 335–349.

Latto, R. 1977. "The effects of bilateral frontal eye-field, posterior parietal or superior collicular lesions on brightness thresholds in the rhesus monkey." *Neuropsychologia* 15: 507–516.

————. 1978. "The effects of bilateral frontal eye-field, posterior parietal or superior collicular lesions on visual search in the rhesus monkey." *Brain Res.* 146: 35–50.

Llewellyn-Thomas, E. 1963. "Visual search patterns of radiologists in training." *Radiology* 81: 288–292.

Loftus, G. R., and N. H. Mackworth. 1978. "Cognitive determinants of fixation location during picture viewing." *J. Exp. Psychol.* 4: 565–572.

Lund, R. D. 1972. "Synaptic patterns in the superficial layers of the superior colliculus of the monkey, *Macaca mulatta*." *Exp. Brain Res.* 15: 194–211.

MacKinnon, D. A., C. G. Gross, and D. B. Bender. 1976. "A visual deficit after superior colliculus lesions in monkeys." *Acta Neurobiol. Exp.* 36: 169–180.

Mackworth, N. H., and A. J. Morandi. 1967. "The gaze selects informative details within pictures." *Percept. Psychophys.* 2: 547–552.

Magnuson, D. J., M. Rezak, and L. A. Benevento. 1979. "Some observations on the organization of the retinal projections to the pretectum and superior colliculus in the macaque monkey as demonstrated by the combined use of laser beam lesions of the retina and autoradiography." *Soc. Neurosci. Abstr.* 5: 794.

Mehler, W. R. 1969. "Some neurological species differences." *Ann. N.Y. Acad. Sci.* 167: 424–468.

Mohler, C. W., and R. H. Wurtz. 1976. "Organization of the monkey superior colliculus: Intermediate layer cells discharging before eye movements." *J. Neurophysiol.* 39: 722–744.

Moore, R.Y., and J. M. Goldberg. 1966. "Projections of the inferior colliculus in the monkey." *Exp. Neurol.* 14: 429–438.

Myers, R. E. 1963a. "Projections of superior colliculus in monkey." *Anat. Rec.* 145: 264.

————. 1963b. "Cortical projections to midbrain in monkey." *Anat. Rec.* 145: 337–338.

Pasik, T., and P. Pasik. 1971. "The visual world of monkeys deprived of striate cortex: Effective stimulus parameters and the importance of the accessory optic system." *Vis. Res. Suppl.* 3: 419–435.

Pasik, T., P. Pasik, and M. B. Bender. 1966. "The superior colliculi and eye movements." *Arch. Neurol.* 15: 420–436.

Perenin, M. T., and M. Jeannerod. 1978. "Visual function within the hemianopic field following early cerebral hemidecortication in man. I. Spatial localization." *Neuropsychologia* 16: 1–13.

Polyak, S. L. 1957. *The Vertebrate Visual System.* University of Chicago Press.

Redican, W. K. 1975. "Facial expressions in non-human primates." In *Primate Behavior Developments in Field and Laboratory Research,* vol. 4, L. A. Rosenblum, ed. New York: Academic.

Ritchie, G. D., P. M. Meyer, and D. R. Meyer. 1976. "Residual spatial vision of cats with lesions of the visual cortex." *Exp. Neurol.* 53: 227–253.

Robinson, D. A. 1972. "Eye movements evoked by collicular stimulation in the alert monkey." *Vision Res.* 12: 1795–1808.

Robinson, D. A., and A. F. Fuchs. 1969. "Eye movements evoked by stimulation of frontal eye fields." *J. Neurophysiol.* 32: 637–648.

Rosvold, H., M. Mishkin, and M. K. Szwarcbart. 1958. "Effects of subcortical lesions on visual-discrimination and single alternation performance." *J. Comp. Physiol. Psychol.* 51: 437–444.

Salapatek, P. 1975. "Pattern perception in early infancy." In *Infant Perception: From Sensation to Cognition,* L. B. Cohen and P. Salapatek, eds. New York: Academic.

Sanders, M. D., E. K. Warrington, J. Marshall, and L. Weiskrantz. 1974. " 'Blind-Sight': Vision in a field defect." *Lancet* I: 707–708.

Schilder, P., T. Pasik, and P. Pasik. 1971. "Extrageniculo-striate vision in the monkey. II. Demonstration of brightness discrimination." *Brain Res.* 32: 383–398.

Schilder, P., P. Pasik, and T. Pasik. 1972. "Extrageniculo-striate vision in the monkey. III. Circle vs. triangle and "red vs. green" discrimination." *Exp. Brain Res.* 14: 436–448.

Schiller, P. H. 1972. "The role of the monkey superior colliculus in eye movement and vision." *Invest. Ophthalmol.* 11: 451–460.

———. 1977. "The effect of superior colliculus ablation on saccades elicited by cortical stimulation." *Brain Res.* 122: 154–156.

Schiller, P. H., and F. Koerner. 1971. "Discharge characteristics of single units in superior colliculus of the alert rhesus monkey." *J. Neurophysiol.* 34: 920–936.

Schiller, P. H., and M. Stryker. 1972. "Single-unit recording and stimulation in superior colliculus of the alert rhesus monkey." *J. Neurophysiol.* 35: 915–924.

Sparks, D. L. 1978. "Functional properties of neurons in the monkey superior colliculus: Coupling of neuronal activity and saccade onset." *Brain Res.* 156: 1–16.

Sparks, D. L., and J. G. Pollack. 1977. "The neural control of saccadic eye movements: The role of the superior colliculus." In *Eye Movements,* B. A. Brooks and F. J. Bajandas, eds. New York: Plenum.

Spatz, W. B., J. Tigges, and M. Tigges. 1970. "Subcortical projections, cortical associations and some intrinsic intralaminar connections of the striate cortex in the squirrel monkey (*Saimiri*)." *J. Comp. Neurol.* 140: 155–174.

Sprague, J. M., and T. H. Meikle, Jr. 1965. "The role of the superior colliculus in visually guided behavior." *Exp. Neurol.* 11: 115–146.

Weiskrantz, L. 1963. "Contour discrimination in a young monkey with striate cortex ablation." *Neuropsychologia* 1: 145–164.

Weiskrantz, L., A. Cowey, and C. Passingham. 1977. "Spatial responses to brief stimuli by monkeys with striate cortex ablations." *Brain* 100: 655–670.

Wertheimer, M. 1938. "Laws of organization in perceptual forms." In *A Source Book of Gestalt Psychology,* W. E. Willis, ed. New York: Harcourt, Brace.

Williams, L. G. 1966. "The effect of target specification on objects fixated during visual search." *Percept. Psychophys.* 1: 315–318.

Wilson, M. E., and M. J. Toyne. 1970. "Retino-tectal and cortico-tectal projections in *Macaca mulatta*." *Brain Res.* 24: 395–406.

Wininger, J., P. Pasik, and T. Pasik. 1972. "Effect of superior colliculi lesions on discrimination of luminous flux-equated figures by monkeys deprived of striate cortex." *Soc. Neurosci. Abstr.* 100.

Wurtz, R. W., and M. E. Goldberg. 1972. "Activity of superior colliculus in behaving monkey. IV. Effects of lesions on eye movements." *J. Neurophysiol.* 35: 587–596.

Wurtz, R. H., and C. W. Mohler. 1976. "Organization of monkey superior colliculus: Enhanced visual response of superficial layer cells." *J. Neurophysiol.* 39: 745–765.

Zusne, L. 1970. *Visual Perception of Form.* New York: Academic.

12 The Contribution of Peripheral and Central Vision to Visually Guided Reaching

Jacques Paillard

Studies of the neural mechanisms sustaining visual functions are limited by the experimental constraints of conventional psychophysical and neurophysiological approaches. Basic knowledge gathered in those fields has to be related to the use of visual functions in the real world to gain full significance. The physiology of vision will remain incomplete until attention is paid to the behavioral context in which the particular visual stimulus is delivered to the organism and accepted by its sensory filter for further processing. Further insights into the input-output relationships that are of behavioral significance for the organism are therefore needed.

We are still at an early stage in describing, operationally, the classes of visually guided behavior that the nervous machinery mediates. Ingle (chapter 3, this volume) has discussed this problem of visuomotor "taxonomy" for various vertebrate species. However, the adaptive motor patterns of reaching, as one of the most basic elements of the behavioral repertoire of mobile species, are of particular interest in this context. The reaching paradigm lends itself to a clear analysis of its functional segments; it has specific visual requirements and it provides an opportunity to separate and categorize the visual channels that are required for the triggering and the guidance of visuomotor behavior.

I shall first attempt to specify a number of distinct classes of behavioral subroutines that can be operationally defined as components of the whole performance and that can be related to their established neural counterparts. I shall then describe research carried out by a team of workers in our laboratory that is derived from this psychophysiological approach. Two groups of studies focused on the visual processes that subserve the reaching response in man and in the split-brain monkey. A third study focused on the problem of visuomotor recalibration in man after prismatic deviation of the visual field.

In my view, all these experiments have provided clear evidence of the distinctive role of central and peripheral vision in providing, through separate channels, different kinds of error signals to guide the trajectory of the hand toward the target. To some extent, peripheral vision can be regarded as assisting the "navigator" in charge of the transport of the hand from its initial position toward the target with the computing of the appropriate trajectory, and central vision might be considered as providing the "pilot" with the cues necessary to achieve the precise and smooth landing of the hand on the target. We shall try to show how navigator and pilot use different types of cue—movement versus position, direction versus distance—and how they have to refer their appreciation of spatial relationships to different systems of spatial coordinates.

Functional Segmentation of Reaching

First I propose a schema for the analysis of reaching into a number of distinctive behavioral components, as shown in figure 12.1. The organism, after positioning the head and eyes, is assumed to engage in

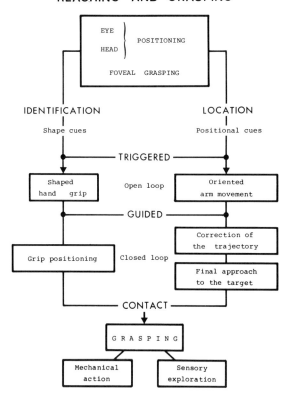

Figure 12.1 Functional segmentation of reaching behavior into two parallel visual channels subserving, respectively, identification and location functions. Visually triggered programs presetting the hand grip and orienting the arm movement might be assisted by visual guidance of grip positioning and of trajectory in its initial and final stages (closed-loop condition). Tactile cues then become prominent to assist manual grasping after contact with the object is established.

parallel processing of shape and location cues in order to trigger and then guide an appropriate reaching response. The target has to be localized in direction and distance and identified in shape.

Localization

The first step is the anchoring of the fovea on the visual target. This requires a measure of the retinal eccentricity of the target and the involvement of the classical visual grasp reflex. This step is usually accompanied by the centering of the head in the direction of gaze, involving the vestibulo-ocular reflex. Information regarding the position of the eye in the head (calculated principally from the efferent signals that monitor eye movement) and the position of the head relative to the trunk (signaled by muscular and joint afferents from the neck) is essential for the calculation of the direction of the target in relation to the body. The anchoring of the gaze on the target is probably one of most highly elaborated visuomotor mechanisms involved in efficient reaching. Direction cues for the location of the target have to be referred to the position of head and body. They therefore depend for their precision on the accuracy of proprioceptive cues derived from the relative positions of eye, head, and trunk (Paillard and Beaubaton 1978).

Factors determining the precision of pointing can be studied in two experimental conditions: open and closed visual loop. In the open-loop condition (that is, without vision of the moving arm), there is no visual feedback available to guide the movement in its course. Therefore, the accuracy of pointing may depend entirely on feedforward information about the location of the target relative to body position provided that, when plotted onto the hypothetical internal map, this information releases the appropriate motor program. This experimental condition has been widely used to study the remapping of the correspondence between the internal plotting of visual space and the release of direction-specific movements in the "motor space." It has also been used to analyze the information content of location cues, especially those concerned with direction.

Distance cues may also be studied in the open-loop condition. The respective roles in pointing of stereopsis, accommodation processes, and patterning of the visual array have not until now attracted the same attention as the study of direction cues, despite experimental opportunities offered by the reaching paradigm to investigate them (Welch 1974).

In the closed-loop condition, precision of reaching is greatly improved by the addition of feedback information that allows for the correction of the trajectory in its first (often considered ballistic) part and, as previously noted by Woodworth (1899), the adjustment of the final approach to the target. The factors contributing to this improvement were the main focus of the present studies.

Identification of the Target

The initial stage of foveal grasp, with the resulting stabilization of gaze (also involved in localization), allows the organism to process and analyze shape cues. The incoming information is then used to trigger a program for preshaping the handgrip, in accordance with previous experiences dealing with the features of the target, such as its size, shape, and orientation. This preshaping does not require any direct visual control of the hand. It can be achieved even in the open-loop condition. Hence, it clearly relies on feedforward information and on stored motor programs.

The next stage concerns the positioning of the grip in relation to the orientation of the target shape. This stage clearly benefits from feedback information in the closed-loop condition. It could be argued that the positioning of the grip relative to the orientation of the target shape could be determined to some extent by the location processes, as described above. I have discussed elsewhere (Paillard 1971) the possibility that two different systems of spatial reference operate: *"l'espace des lieux,"* in which the locus of an external target is plotted onto a body-centered map of external space, and *"l'espace des formes,"* in which the position of the handgrip has to be referred to a local "intrafigural" space-coordinate system in which the relative positions of the elements are taken into account. Although this theoretical position will not be further developed here, some experimental evidence from previous studies with split-brain monkeys (Paillard and Beaubaton 1975) will be noted because of its relevance for subsequent discussion. In the ipsilateral eye-hand combination, a failure to position the grip (control of the distal segments) according to the orientation of the target shape in reaching to an object is observed, as is the lack of correction of the arm position (control of the proximal segment) in the final approach of a pointing task. These findings suggest the possibility of a common origin for this defect in visual guidance. In addition, we may note that the preshaping of the handgrip is also lacking in this condition. As already mentioned, the preshaping of the grip is not dependent on visual guidance, and the defect can clearly be attributed to the known absence of individual control of finger movement by ipsilateral cortico-spinal motor commands (Kuypers 1964).

Location and Identification

The parallel processes of visual location and identification end with the effective contact of hand and target. Then a new phase develops in which tactile guidance of the grasping movements of the hand comes into prominence. This in turn may lead to sensory exploration and/or mechanical action.

The present studies employ a pointing task that minimizes information related to the processes of identification and to the final tactile guidance of grasp. It therefore concerns mainly the visual triggering

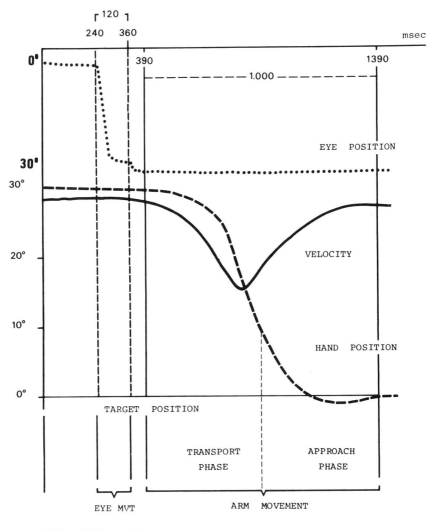

Figure 12.2 Cinematic record of eye and hand movements in reaching. (After Jeannerod and Carblanc 1978.)

and guidance of the pointing movement and places the major emphasis on location processes. More specifically, it is the guided component of reaching that has been investigated. Two aspects of guiding have been distinguished: the correction of the trajectory in its initial phase and the final approach to the target preceding contact. As shown in figure 12.2, the movement starts about 150 msec after foveal grasp of the target is achieved. The initial position of the hand is then at a retinal eccentricity of 30°. The approach phase is assumed to begin when the velocity curve presents discontinuous characteristics of the final adjustment (about 300 msec before contact, corresponding to a retinal eccentricity of 10°). The duration of the transport phase is then about 300 msec and corre-

sponds to a movement of the hand image on the retina starting at 30°
eccentricity and directed toward the fovea; the peak angular velocity is
about 100° per second.

The question of the respective contribution of positional and movement cues and the role of central and peripheral vision in processing
these cues arises. The important role of central vision, particularly in
the final phase of the movement, has already been well documented.
What is less clear is the possible role of peripheral vision in correcting
the trajectory.

Experimental Data

Evidence for an Early Visual Correction in the Peripheral Field
The first set of experiments dealt with the pointing of human subjects
when the timing and duration of visual feedback were controlled. The
apparatus is shown in figure 12.3. Precision of pointing was measured
by the use of an electronic grid that directly plotted responses on an x, y
coordinate system for on-line computation. Six normal adult subjects

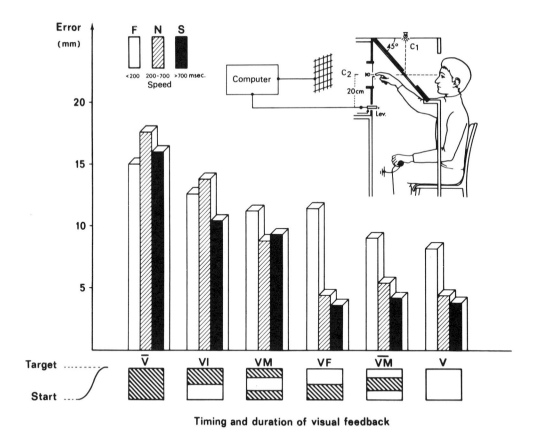

Timing and duration of visual feedback

Figure 12.3 Controlled visual feedback in pointing to a visual target. Movements have been sorted into three
classes of speed: fast (white column), normal (shaded column), and slow (black column). (After Conti and
Beaubaton 1976.)

were examined under conditions balanced for speed of response. Three classes of response speeds were studied: fast (movement duration < 200 msec), normal (200–700 msec), and slow (> 700 msec). Timing and duration of visual feedback were controlled by adjustable occluders. Further details of the experimental procedure have been published elsewhere (Conti and Beaubaton 1976).

The six patterns of visual feedback were, as indicated in figure 12.3, the following:

- $\overline{V}$: without visual feedback (open loop),
- VI: feedback during the first half of the trajectory,
- VM: feedback during the middle third of the trajectory,
- VF: feedback during the second half of the trajectory,
- V$\overline{M}$: feedback during the first and final thirds of the trajectory, and
- V: full visual feedback.

The salient findings were as follow. As far as slow movements were concerned, accuracy was best improved under the conditions that provided visual feedback in the final phase of the movement. If feedback was limited to the first or to the intermediary phase, it nevertheless produced a clear improvement over performance in the open-loop condition.

A systematic deterioration in accuracy of responses was no longer observed when movement time was less than 200 msec. There was, however, as shown in the figure 12.3, a clear difference in precision in the open-loop and closed-loop conditions, which suggests either the possibility of a progressive improvement of performance based on the knowledge of results available in closed-loop conditions or the availability of movement cues processed at a speed higher than that usually considered necessary for visual feedback to operate (Keele and Posner 1968). The important point is that any visual feedback, even if it is not provided during the critical final phase, can improve performance. If one takes into account the chronology of the different phases of a reaching performance, as shown in figure 12.2, and assumes that the gaze remains anchored on the target until final contact, it appears that the first part of the trajectory provides visual information that is processed in the peripheral visual field and contributes to the improvement of performance by using cues that could not involve the direct evaluation of the relative position of hand and target position. This hypothesis was further explored in the second set of experiments, with split-brain monkeys.

Evidence for Separate Contribution of Positional and Movement Cues to Visuomotor Reaching

Split-brain surgery (sectioning of the corpus callosum and of the anterior commissure, combined with midline sectioning of the optic chiasm) allows for the separate control of visual inflow to either eye and for visual information to be available in either hemisphere. This experiment involved two adult baboons performing a pointing task without tactile information about the position of the visual target (see Beaubaton and Chapuis 1974). Precision in pointing, as in the preceding experiment in man, was automatically recorded in rectangular coordinates and submitted to on-line computation. The room was illuminated by a monochromatic green light, and the target consisted of a red electroluminescent diode. Vision of the moving limb could be eliminated by fitting the head mask attached to the experimental cage with green filters, preserving vision of target but precluding vision of the limb (open-loop condition). Moreover, by placing either a green or a red filter in front of each eye, selective visual information about target location and about limb movement could be distributed separately to either eye. Therefore, the hand and the panel illuminated by the green ambient light were only seen by one hemisphere (contralateral or ipsilateral in relation to the moving hand); the target was perceived by the other.

Different experimental conditions were selected according to the two visual conditions: vision of only the hand, or only the target, or both hand and target (ipsilateral or contralateral). The left hand was used by the two monkeys in all sessions. Accuracy in pointing was compared in the split-brain and the control animals under the conditions shown in figure 12.4. By combining conditions it was thus possible to compare binocular versus monocular vision and contralateral versus ipsilateral control of the left arm by the seeing hemisphere. The main results, discussed at length in Beaubaton et al. 1978, can be summarized as follows:

• Comparison of the binocular condition and the monocular steering of the contralateral hand confirmed that the occlusion of one eye does not affect the precision of a visuomotor pointing task, either in the split-brain animals or in the control animals, whose performance was comparable.

• The suppression of the vision of the hand (open-loop condition) resulted, as expected, in less accurate responses in both monkeys for all combinations. Performance in ipsilateral and contralateral eye-hand combinations did not differ significantly. In contrast, and as reported previously (Paillard and Beaubaton 1975), when vision of the hand was allowed (closed-loop condition) performance in ipsilateral eye-hand combination remained significantly impaired, as if the ipsilateral hemisphere controlling the reaching hand was not able to process visual cues about position and/or movement of the moving limb in order to adjust the reaching program. The reaching program, however, was normally triggered on the basis of location cues about the position of the target.

• The salient feature of these results was the surprising improvement in accuracy observed in the pointing response associated with the ipsilat-

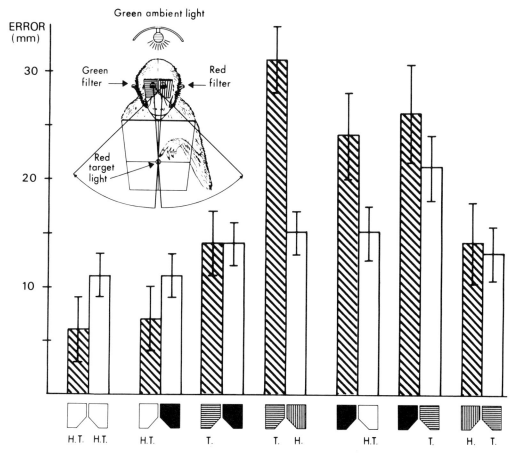

Green ambient light

ERROR (mm)

Green filter →

Red filter ←

Red target light →

H.T. H.T.   H.T.   T.   T. H.   H.T.   T.   H.   T.

Figure 12.4 Pointing performances in split-brain (shaded columns) and control monkeys with a separate control of target and hand vision through colored filter. One eye is occluded by a black occluder. Only the hand (H) is visible through the red filter (indicated by vertical stripes); only the target (T) through the green (horizontal stripes). Both hand and target are visible when no filter is present. (After Beaubaton et al. 1978.)

eral "target-seeing" eye when vision of the moving hand (but not of the target position) was simultaneously provided to the contralateral "hand-seeing" eye. The reverse was not true; a target-seeing hemisphere did not improve its reaching with the contralateral hand when visual information about limb movement was provided at the same time to the other hemisphere (ipsilateral to the reaching arm). Let us assume for the interpretation of these data that at least three main sources of visual information are involved in reaching: location cues about target position, which seem to be used in either contralateral or ipsilateral eye-hand combination to trigger the program of reaching (open-loop condition); position cues about the relative position of hand and target, which are present in the contralateral eye-hand combination but not in the ipsilateral (Paillard and Beaubaton 1975, 1978); and movement cues about the transport of the limb across the visual fields, which are only processed by the contralateral non-target-seeing hemisphere. This may account for the lack of precise reaching in the ipsilateral eye-hand combination.

The question of what mechanisms are involved then arises. Reaching to a visual target first requires the encoding of parameters related to the direction and amplitude of the trajectory. It then needs, for precise adjustment, detection of error signals and corrective information.

The triggering of the reaching program requires the use of positional cues about the location of the visual target. Extraretinal signals (oculomotor efference), generated by foveal grasping of the target, may give the position of the eye in the head; proprioceptive signals provide information relating the position of head and trunk (Beaubaton et al. 1978). These nonvisual signals about target location in egocentered space are presumably available to both hemispheres, and thus account for the known capacity of a split-brain monkey to steer adequately his contralateral as well as his ipsilateral hand toward a visual target seen by only one eye and one hemisphere.

We have observed that the corrective feedback mechanisms used to guide visually the movement once started, and to adjust position in the vicinity of the target, seem to operate efficiently only in the contralateral eye-hand combination and are lacking in the ipsilateral one. This defect could be explained as a consequence of the midline section of the optic chiasm, which suppresses the function of the nasal retinal field in the split-brain monkey, preventing any possibility of visual guidance of the ipsilateral hand moving across the blind nasal retinal field. However, the so-called macular sparing known to occur in hemianopic patients (Teuber et al. 1969) may allow the error-correcting system to operate when the hand enters the foveal field in the terminal part of the trajectory. Indeed, precise corrected reaching seems to be preserved in patients with hemianopic field defect or with tubular vision. Moreover, a definite overlap of cortical connections from the nasal and temporal hemiretinae nearest the vertical midline is now well documented in split-brain monkeys by both anatomical (Bunt et al. 1977) and electrophysiological (Van Essen and Zeki 1978) evidence. The 3° overlap of retinal ganglion cells projecting to the geniculocortical system that has been shown to exist in monkeys around the fovea suggests that a terminal correction of the trajectory is possible if visual acuity in this parafoveal zone is adequate for monitoring the final appearance of the hand. This kind of adjustment is, however, clearly lacking in the ipsilateral eye-hand combination in the split-brain monkey. The same amount of inaccuracy is observed whether the monkey is allowed to see its ipsilateral moving hand or not.

How, then, can we explain the efficiency of visual cues about hand movement available to the contralateral hemisphere (which is otherwise deprived of information about the position of the target) to improve dramatically the ability of the ipsilateral target-seeing hemisphere to steer the limb? The contralateral eye seems to be able to process adequately movement cues about the sweep of the trajectory of the hand across its functional temporal field in order to guide the movement toward the target and to improve significantly the accuracy of reaching (combination H-T, figure 12.4).

The same explanation could account for the fact that when vision of the ipsilateral eye is restricted to hand movement, whereas the contralateral eye receives information only about target location, no improvement and even an impairment of performance is observed (combination T-H, figure 12.4). In this condition, visual cues about hand movement clearly cannot be processed by the peripherally blind nasal field of the ipsilateral eye, and visual cues about the relative position of hand and target cannot be processed by the contralateral eye, which cannot see the hand.

This line of interpretation leads us to hypothesize the existence of two different visual mechanisms involved in the visual guidance of reaching: one operating predominantly in central vision and using position cues and extracting error signals from the relative positions of hand and target, and the other operating predominantly in peripheral vision and using speed and direction-of-movement cues. Direction cues are related to the clamping of the axis of the "non-target-seeing" eye to the target through the mechanism of binocular convergence, which is preserved in the split-brain monkey.

The hypothesis of distinctive roles for peripheral and central vision in processing two different types of visual cue (movement versus positional) for the visual guidance of movement has received additional support from a third set of experiments, reported below.

Evidence for Distinctive Roles of Peripheral and Central Vision in Visuomotor Coordination

Considerable effort has been devoted to determining the conditions under which accurate reaching to a visual target is restored after the dramatic impairment that follows prismatic displacement. There is little agreement as to the nature of the adaptive processes that may intervene and contribute, in an additive or supplementary fashion, to a full or partial recovery of normal function. Where is the compensating mechanism located—at the visual, the oculomotor, the proprioceptive, or even the cognitive level? What allows for the presumptive recalibration of eye position in the head? What determines the predominance of vision over other sensory channels in the detection of error signals or discrepencies, which is necessary for the adaptive process to occur? Without going into the controversial literature of the past two decades, I shall simply describe some recent studies of Brouchon-Viton and Jordan (1978) in our laboratory, which were based on the assumption that central and peripheral vision might contribute differently to the mechanisms involved in visuomotor recoordination after prismatic deviation of the visual field.

This experiment stemmed from the divergent results obtained by many authors using, among other variations, two main exposure conditions. The first, introduced by Held and Freedman (1963), allows the (human) subject to see his arm oscillating freely in front of him on an homogeneous visual background. This procedure leads to a certain amount of compensation in the self-moving condition, whereas passive displacement of the arm does not produce this effect. The second, introduced by Howard (1963), stresses the importance of the presence of a stationary target aimed at by the subject during the exposure condition. It leads to partial compensation that often is greater in amount than that obtained under the Held-Freedman condition, even when vision of the moving arm is restricted to the ultimate phase of terminal adjustment. Moreover, the results seem to dismiss the emphasis placed by Held on the role of active movement in the process of visuomotor recoordination (Howard and Templeton 1966). A mechanical occluder restricting the visual field to central vision and a procedure using a fixation point above the plane of the moving hand were used in this experiment to restrict visual input during the exposure of the moving arm to either central or peripheral retinal fields.

Four exposure conditions were compared in eight right-handed subjects wearing (20 diopters) prismatic goggles: vision of the arm moving to and fro and pointing to a stationary target, vision of the arm moving against a homogeneous background, vision of the final phase of hand movement in pointing to a stationary target, and vision of the arm precluded (with the subject looking at a fixation point in front of him).

Each exposure was tested under three visual conditions: whole-field vision, peripheral vision (the subject fixated a luminescent target in front of him and above the arm that he saw moving in the superior peripheral hemifield of the retina), and central vision (the subject looked at his moving hand through a mechanical occluder restricting vision to a conic central field of 8°).

Adaptation was expressed as a percentage of subjective displacement that was defined by the comparison of performance in preexposure, normal and prismatic conditions. This explains the relatively high values of adaptation obtained in this experiment in comparison with those usually quoted in the literature, where the percentage of adaptation is measured with reference to the physical displacement expected from prismatic deviation.

The main results, presented in figure 12.5, were the following:

• In whole-field vision, a 10-minute exposure with vision of the arm pointing to a stationary target was sufficient to produce a 90% correction of pointing performance. The percentage of adaptation dropped to

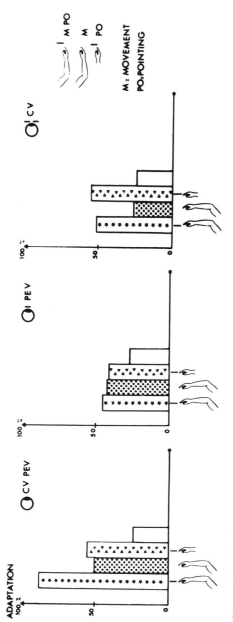

Figure 12.5 Adaptation to an 11° prismatic displacement of the visual field, following four exposure conditions described in the text, when whole-field vision is allowed (CVPEV) or when vision is restricted to peripheral vision (PEV) or to central vision (CV). (After Brouchon-Viton and Jordan 1978.)

Peripheral and Central Vision and Reaching

about 50% under both conditions where the arm was seen moving freely and where vision was restricted to final adjustment of the hand when pointing to a stationary target. The control condition without vision of the hand led to about 20% adaptation (McLaughlin et al. 1966).

• The striking finding was the reduction of the amount of adaptation to control level when vision of the freely moving arm was restricted to central vision, whereas adaptation was preserved when vision of the arm was confined to peripheral vision. In contrast, adaptation based on reaching to a stationary target was totally preserved in central vision and only partially preserved in peripheral vision.

These results lead to the conclusions that the "terminal feedback" procedure is mainly concerned with central vision and that the efficiency of the Held-Freedman procedure (free-moving arm) is entirely dependent upon peripheral visual cues. Previous results obtained by Graybiel and Held (1970) showed that two separate processes of adaptation are involved in scotopic and photopic conditions (which represent peripheral and central vision, respectively).

Adaptation, produced by photopic exposure, was significantly reduced under scotopic conditions. Scotopically induced adaption, however, could be demonstrated under both scotopic and photopic conditions. Graybiel and Held hypothesized the differential role of cones and rods as the origin of these phenomena.

Thus, it is tempting to attribute the mechanism of adaptation in central vision to the processing of position cues involved in the terminal adjustment of pointing to a stationary target, and the mechanism of adaptation in peripheral vision to the processing of movement cues. In regard to the use of movement cues, the contribution of an acceleration component of reafferent stimulation to prismatic adaptation was stressed by McCarter and Mikaelian (1978).

Two preliminary results obtained by Jordan and Brouchon-Viton provide additional evidence of the respective roles of central and peripheral vision:

• A stroboscopic procedure was used to suppress movement cues and to bring stationary cues into prominence. Under this condition, using the free-moving-arm procedure, the adaptation observed in the peripheral field dropped to control level whereas adaptation in the pointing-to-target procedure was totally preserved. The two processes were additive under the whole-field condition.

• Movement cues encoded by peripheral vision led to readaptation only when associated with self-induced movement, and not when associated with passive displacement in the peripheral visual field. In contrast, position cues encoded in central vision were equally efficient for adaptation when related to displacement, either active or passive.

If confirmed, these results may lead to a reconsideration of the mechanisms underlying the adaptive processes involved in the recalibration of reaching after prismatic displacement. They may, therefore, contribute to a clarification of some of the conflicting issues that have obscured this field for the last two decades.

Discussion

Compelling evidence has been obtained from these experiments that two separate channels may contribute to the visual guidance of movement. Movement cues, mainly processed in the peripheral field, are used to control the direction of the trajectory relative to the visual axis clamped on the visual target by the mechanism of foveal grasp.

Central vision, highly sensitive to position cues, subserves the error-detecting mechanism involved in the late phase of adjustment of a reaching movement. It encodes the position of the moving hand relative to the stationary target and allows corrective feedback to operate for an accurate positioning of the hand on the target.

Neurophysiological and psychophysical studies have provided evidence for the channeling of visual information into parallel selective subsystems. The results of these studies are summarized in table 12.1. These channels seem to operate as independent detectors of characteristic features of visual information. A multistage model of processing has been postulated (see Bonnet 1977 for a review). It is assumed that the filtering of spatial and temporal frequency characteristics occurs at the peripheral level. The processing of this filtered information would proceed by feature-analyzing structures. Many of these structures have

Table 12.1
Main characteristics of the two visual channels for motion and displacement analysis.

Filters	
TF	SF
Analyzers	
Motion	Position change
	Distance
Prevalent mode	
Peripheral vision	Central vision
Scotopic	Photopic
No target dependency	Target dependency
High velocity	Low velocity
Moving stimulus	Stationary position
Selectivity	
High TF	Low SF
Velocity direction	Distance direction
Linked to self-motion	Change of position, either active or passive
Impaired by strobe	Unimpaired by strobe

been identified for shape analysis. For movement detection, it has been proposed that two separate analyzers may operate. The first would be fed with the low-spatial-frequency components of the stimulus, selective for change of position and for relatively low velocity of change. Distance between initial and final position seems to be the parameter coded in this analysis. Interesting for our problems is the demonstration of a "target dependency" phenomenon involved in this detection (Tyler and Torres 1972). The presence of a stationary target greatly lowered detection thresholds for change of position in the central field of vision. This effect decreased monotonically with eccentricity of the stimulus. We may speculate that the error feedback, operating in central vision for the final adjustment of reaching, provides relevant visual information on the position of the hand relative to the stationary target that can be selectively processed by such a channel. The resistance of this feedback operation to stroboscopic conditions that was observed in our experiment would fit well with this interpretation.

The second channel would be fed with the high-frequency temporal components of the stimulus, which are selective for movement cues of relatively high velocity. A velocity signal would be extracted by the central analyzer and used together with a directional specification to account for the phenomena observed in peripheral vision. The dramatic impairment observed in stroboscopic conditions is consistent with this interpretation. Moreover, psychophysical studies show that thresholds for detection of movement, at their highest in the peripheral field, are not target-dependent. It is accepted that the so-called transient system of Y optic fibers, sensitive to movement, has higher temporal acuity than the "sustained" system of X optic fibers, which primarily carry information about spatial shape, pattern, and contour and have higher spatial acuity.

Space does not permit a lengthy discussion of the significance of this functional distinction between these two modes of processing visual information, but it is of considerable interest for contemporary neurophysiological research (see Grüsser and Grüsser-Cornehls 1973). Many problems remain to be solved before the substantial and ever-increasing corpus of data can be integrated within a coherent functional picture (see Bonnet 1977). Whether the two channels interact in a cooperative or an antagonistic way (see Sekuler and Levinson 1974; MacKay and MacKay 1976) during normal vision—and, specifically, in visuomotor guidance—must be determined by further experimentation.

What, then, are the implications of the selective sensitivity of the peripheral retinal field to self-moving stimuli that we have observed? How does it relate to the presumptive contribution of peripheral vision to a kind of "proprioceptive" mode of visual information processing, based on "reafference" generated by the movement of our own body? A Gibsonian approach is of interest in this context, as reflected in Lee's

Paillard

(1977) proposal to distinguish two kinds of proprioceptive visual cues: those provided by movement of a body segment in relation to another body part and those (defined as "exproprioceptive") supplied by changes in the visual array related to displacement of the head in space due to body movement.

It will certainly be rewarding to relate our experimental findings to studies of the differential effects of central versus peripheral vision on visually induced sensations of movement in the perception of self-motion and in the stabilization of body posture (Brandt et al. 1973). Such studies point to the interesting distinction between egocentric and exocentric modes of movement perception. They also raise the crucial problem of the neural counterparts of the space coordinate systems that enable the steering of goal-directed motor actions. They are more likely to be productive if they are based on a behavioral taxonomy that is directly related to the known functional components of nervous activity.

References

Beaubaton, D., and N. Chapuis. 1974. "Rôle des informations tactiles dans la précision du pointage chez le singe 'split-brain'." *Neuropsychologia* 12: 151–155.

Beaubaton, D., A. Grangetto, and J. Paillard. 1978. "Contribution of positional and movement cues to visuo-motor reaching in split-brain monkey." In *The Cerebral Commissures*, P. Van Hof et al., eds. Amsterdam: Elsevier.

Bonnet, C. 1977. "Visual motion detection models: Feature and frequency filters." *Perception* 6: 491–500.

Brouchon-Viton, M., and P. Jordan. 1978. "Relevant cues in visuo-motor rearrangement." *Neurosci. Lett.* Suppl. 1: S. 386.

Brandt, T., J. Dichgans, and E. Koenig. 1973. "Differential effects of central versus peripheral vision on egocentric and exocentric motion perception." *Exp. Brain Res.* 16: 476–491.

Bunt, A. H., D. S. Minckler, and C. W. Johansen. 1977. "Demonstration of bilateral projection of the central retina of the monkey with horseradish peroxidase neuronography." *J. Comp. Neurol.* 171: 619–630.

Conti, P., and D. Beaubaton. 1976. "Utilisation des informations visuelles dans le contrôle du mouvement: Étude de la précision des pointages chez l'homme." *Travail Humain* 39: 19–32.

Graybiel, A. M., and R. Held. 1970. "Prismatic adaptation under scotopic and photopic conditions." *J. Exp. Psychol.* 85: 16–22.

Grüsser, O. J., and V. Grüsser-Cornehls. 1973. "Neuronal mechanisms of visual movement: Perception and some psychophysical and behavioral correlations." In *Handbook of Sensory Physiology*, vol. VII/3, R. Jung, ed. Berlin: Springer.

Held, R., and S. J. Freedman. 1963. "Plasticity in human sensory-motor control." *Science* 142: 445–462.

Howard, I. P. 1963. "Displacing the optical array." In *The Neurophysiology of Spatially Oriented Behavior*, S. J. Freedman, ed. Homewood, Ill.: Dorsey.

Howard, I. P., and W. B. Templeton. 1966. *Human Spatial Orientation*. New York: Wiley.

Jeannerod, M., and C. Carblanc. 1978. "Résolution et plasticité de la coordination oeil-main." In *Du contrôle moteur à l'organisation du geste*, H. Hécaen and M. Jeannerod, eds. Paris: Masson.

Keele, S. W., and M. I. Posner. 1968. "Processing of visual feedback in rapid movements." *J. Exp. Psychol.* 77: 155–158.

Kuypers, H. G. J. M. 1964. "The descending pathways to the spinal cord, their anatomy and function." *Progr. Brain Res.* 11: 178–202.

Lee, D. N. 1977. "The functions of vision." In *Modes of Perceiving and Processing Information*, H. L. Pick and E. Salbman, eds. Hillsdale, N.J.: Erlbaum.

MacKay, D. M., and V. Mackay. 1976. "Antagonism between visual channels for pattern and movement." *Nature* 263: 312–314.

McCarter, A., and H. H. Mikaelian. 1978. "The role of acceleration information in prism adaptation." *Percept. Psychophys.* 23: 21–26.

McLaughlin, S. C., K. I. Rifkin, and R. Webster. 1966. "Oculomotor adaptation to wedge prisms with no part of the body seen." *Percept. Psychophys.* 1: 452–458.

Paillard, J. 1971. "Les déterminants moteurs de l'organisation de l'espace." *Cahiers Psychol.* 14: 261–316.

Paillard, J., and D. Beaubaton. 1975. "Problèmes posés par le contrôle visuel de la motricité proximale et distale après disconnexion hémisphérique chez le singe." In *Les Syndromes de disconnexion calleuse chez l'homme*, B. Schott and F. Michel, eds. Presses de l'Université de Lyon.

———. 1978. "De la coordination visuo-motrice à l'organisation de la saisie manuelle." In *Du contrôle moteur à la coordination du geste*, H. Hécaen and M. Jeannerod, eds. Paris: Masson.

Sekuler, R., and E. Levinson. 1974. "Mechanisms of motion perception." *Psychologia* 17: 38–49.

Teuber, H. L., W. S. Battersby, and M. B. Bender. 1960. *Visual Field Defects After Penetrating Missile Wounds of the Brain*. Cambridge, Mass.: Harvard University Press.

Tyler, C. W., and J. Torres. 1972. "Frequency response characteristics for sinusoidal movement in the fovea and periphery." *Percept. Psychophys.* 12: 232–236.

Van Essen, D. C., and S. M. Zeki. 1978. "The topographic organization of rhesus monkey prestriate cortex." *J. Physiol.* 277: 193–226.

Welch, R. B. 1974. "Research on adaptation to rearranged vision: 1966–1974." *Perception* 3: 367–392.

Woodworth, R. S. 1899. "The accuracy of voluntary movement." *Psychol. Monogr.* 3(suppl.): 54–59.

Note added in proof. Additional results concerning matters discussed in this chapter have been reported in the following: J. Paillard, "The multichanneling of visual cues and the organization of a visually guided response," in *Tutorials in Motor Behavior*, ed. G. E. Stelmach and J. Requin (Amsterdam: North-Holland, 1980); J. Paillard, P. Jordan, and M. Brouchan, "Visual motion cues in prismatic adaptation: Evidence for two separate and additive processes," *Acta Psychol. (1981)*.

13 Visuomotor Mechanisms in Reaching Within Extrapersonal Space

Marc Jeannerod
B. Biguer

Movements directed at visual objects within extrapersonal space represent the net output of complex neural events starting with the detection of visual properties of the objects and ending with the specification of appropriate motor commands. In the case of reaching movements, these commands cannot conceivably be organized on the principle of a pure motor synergy (as is swallowing or walking, for instance). The main argument for this is that the same reaching movement represents the superimposition of several different motor acts, each of which is aimed at a particular aspect of the visual world. Visual objects can be described by a limited number of parameters specifying two main groups of visual properties. Intrinsic properties (such as weight, size, and shape) refer to the object as a physical entity; extrinsic properties (for example, distance in the sagittal plane and location in the frontal plane) refer to its position within the subject's extrapersonal space. Our hypothesis is that these different groups of properties are matched by specific visuomotor mechanisms.

The idea that visuomotor function in reaching is subserved by separate channels is based on the segregation of visual pathways dealing with different types of visual stimuli. The classical view of two visual systems (Trevarthen 1968; Schneider 1969) represents a conceptual basis (probably simplified) for the statement that "shape" and "spatial location," for instance, are not processed in the same brain areas. It is thus conceivable that the motor output of these specialized visual subsystems will also rely upon different and equally specialized motor structures.

The hand can be considered as the effector of the "object" channel (or channels; that is, those matching intrinsic properties of objects). In fact, the primate hand represents a unique achievement for prehensile function. Of particular importance are the improvement of the wrist (which in man is a "ball joint" ensuring a great variety of hand positions with respect to the forearm) and the acquisition of independent finger movements. Together with the opposable thumb, these features are critical for the realization of a precise and fully orientable grip, allowing prehension and manipulation of small objects (Napier 1956, 1960). The

Table 13.1
Schematic description of motor output to hand in reaching for a particular object.

Object Trigger Features	Motor Output Pattern
Orientation: vertical	Forearm: rotated, hand in semipronation
Weight: low	Hand: not dorsiflexed
Size: small	Fingers: Thumb opposed to one finger

great development of hand representation on motor and somatosensory cortices in monkey and man is the physiological correlate of this specialization for prehension.

Simple observation of reaching movements shows that the hand adapts to the shape of the object in anticipation of the grasp. This arrangement has the major advantage of improving the precision of the grasp in minimizing tactile adjustments when the hand comes into contact with the object. In addition, it clearly indicates that relevant visual cues alone are efficient in generating motor commands related to intrinsic object properties. Table 13.1 gives a limited view of possible postures that may combine in shaping the hand according to its goal. For each given object a similar schema could be built, assuming that object features trigger appropriate motor outputs.

"Space" channels have a completely different function: to match the final position of the moving limb with the position of the target within extrapersonal space. For this purpose, central visual processing can be limited to the computation of a set of spatial coordinates establishing the location of the object with respect to the body. This requires that the central "map" used to relate each point of the visual field to the visual system encode body-centered coordinates, rather than retinal coordinates only. Both the position of the eye with respect to the head and the position of the head with respect to the body must be taken into account in determining the direction of the arm movement. In addition, eye, head, and body movements may themselves become part of the act of reaching when it is directed at an object within a peripheral part of the visual field.

Sensory information feeding into "space" channels, and thus generating the arm movement, is mostly gained from the peripheral retina. The role of foveal retina might be limited to optimizing the trajectory at the vicinity of the object, when the hand enters the central visual field. In fact, the analysis of temporal (dynamic) characteristics of reaching arm movements shows the occurrence of systematic variations of velocity during the trajectory. The first part of the trajectory is made ballistically, as indicated by its high velocity and its apparent independence from visual reafference. In contrast, the final part of the movement is marked by a low velocity, and thus might be at least partly under visual guidance.

Intersegmental Coordination in Reaching Arm Movements

Available studies dealing with "voluntary" movements are restricted to simple monoarticular stereotyped movements, such as brisk flexion or extension of the elbow joint. Intrasegmental organization of muscular discharges in these movements is based on simultaneous contraction of synergistic muscles (Wachholder 1928; Livingston et al. 1951) and alternated contraction of the agonist-antagonist pair (Wachholder 1928; Angel 1974). This description applies to the category of rather fast, ballistic-type movements, in which most of reaching movements belong. Cocontraction of agonist and antagonist muscles may be observed if the movement is made more slowly (Hallett et al. 1975) or at the maximum velocity (Freund and Büdinger 1978).

The segmental pattern of muscle contractions in ballistic-type movements is often discussed as reflecting centrally preprogrammed commands (see review in Glencross 1977). In fact, the onset of contraction of the antagonist muscle occurs very shortly after the beginning of the limb displacement, and in some cases even before the displacement (Hallett et al. 1975; Angel 1977). It is thus clear that a reflexive, serial activation of the antagonist by proprioceptive reafferences could not take place efficiently within such short delays.

Coordination of plurisegmental, goal-directed movements, such as reaching for objects, is a more difficult problem, since it involves commands directed at several agonist-antagonist pairs. In agreement with the same "centralist" theory, it might be suggested that segmental routines (such as the biceps-triceps routine) are added over time so as to produce the correct reaching pattern. This would account for a precise timing of the intersegmental sequence. However, according to our previous hypothesis, visual trigger features for the activation of each component of the movement may be very different, as are the command patterns (force, velocity of contraction, etc.) applied to muscles in different segments. The postulated plurality of visuomotor mechanisms thus calls for a more "distributed" programming of central commands.

In this context we have carried out a study of natural movements with multiple degrees of freedom, in normal human subjects. Subjects were seated in front of an apparatus where three-dimensional objects of different shape, texture, and weight could be presented under normal illumination. The apparatus was a variation of that of Held and Gottlieb (1958). Basically, a half-silvered mirror was placed horizontally, midway between the plane of gaze and the hand. According to how the apparatus was illuminated, vision of the hand during the reaching movement could be either allowed ("closed-loop situation") or prevented ("open-loop" situation). In the open-loop case, the mirror reflected a virtual image of the target object, which appeared to be located at the proper reaching distance.

As a rule, subjects kept their eyes closed until the beginning of each trial. A trial involved the following operations: An object was displayed in front of the subject. The subject was required to open his eyes. Shortly afterward (between 500 msec and 1 sec) a small light was flashed, signaling the subject to reach and take the object. He had been asked at the beginning of the experiment to make "fast and accurate" movements. The subject's hand was filmed from the side with a camera operating at 50 frames per second, which was turned on when the subject's eyes opened. Films were processed by hand, by enlarging single frames on a screen. On some occasions, small pieces of tape were attached to the subject's arm in order to facilitate reconstruction of definite parameters of a movement. Finally, electrical activity of the biceps brachii and of the extensor digitorum was recorded through surface electrodes. The electromyogram (EMG) was amplified and displayed on a Grass polygraph.

Film analysis of normal adults' movements directed at three-dimensional objects reveals a definite sequence in the onset of motion of the different arm segments. In cases where the subject keeps a relaxed resting hand position, the first detectable movement generally occurs at the most distal level. As exemplified in figure 13.1, fingers slowly extend and the hand begins to shape according to the object. These distal adjustments start, on the average, some 40 msec before the raising of the hand from the table; in a few cases this delay extended up to 100 or 120 msec. Final finger posture, which will then remain unchanged up to the contact with the target, is acquired by about the halfway point of the trajectory. Rotation of the forearm, which ensures correct orientation of the grasp, also begins early, and is achieved well before the end of the trajectory (figure 13.2).

By synchronizing the recording of muscular events at the arm and forearm level with the film of hand and finger movements, more precise insight into the temporal organization of reaching was obtained. The first contraction from the recorded muscles occurred at the level of the biceps, some 250 msec after the signal to move had been given. Biceps contraction exhibited the typical two-burst pattern (Angel 1974); the end of the first burst was concomitant with the onset of hand raising. This time between the beginning of muscle contraction and the onset of the resulting movement is referred to as isometric tension time. It is worth mentioning, though we have no explanation for this fact, that the resting position of the hand appeared to influence the timing of the movement. When the movement started with the hand in a semipronated position, the delay for the onset of biceps contraction, as well as the biceps isometric tension time, was systematically longer than when the hand was in full pronation (table 13.2).

Contraction of the radial forearm muscles (extensor digitorum),

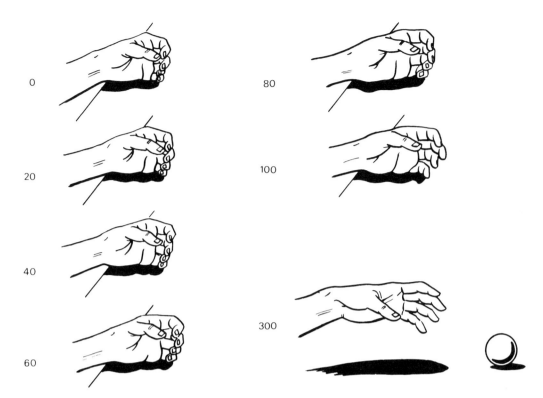

Figure 13.1 Early finger shaping initiating a reaching sequence. In this particular case, the hand is raised from the table only 120 msec after the signal to reach has been given. Time 0 represents the last frame of the film before the first detectable movement, i.e., more than 350 msec after the signal. Note rotation of the forearm and full shaping of the hand later in the sequence (depicted on a frame at 300 msec). (Redrawn from film.)

which occurred slightly later than contraction of the biceps, preceded the first detectable finger movement by an isometric tension time of about 80–100 msec. Muscular activation was represented by a continuously increasing contraction, without evidence of a two-burst pattern. This corresponds to the fact that finger movements performed in this situation were slow, ramplike movements with reduced amplitude. This type of movement has been shown to often involve a slowly rising cocontraction of the agonist-antagonist pair (Hallett et al. 1975).

Figure 13.3 shows a typical example of the intersegmental sequence during initiation of a reaching movement, as reconstructed from film and EMG recordings. This sequence does not merely reflect the delays between the contraction of the two muscular groups. In fact, it appears from EMG recordings that the flexor of the forearm contracts shortly before the extensor of the fingers, although the fingers extend prior to the flexion of the forearm. Hence, it is the load exerted by the different segments that transforms a quasisimultaneous pattern of activation into a sequential movement. This arrangement may have some advantages for the control of the late part of the trajectory.

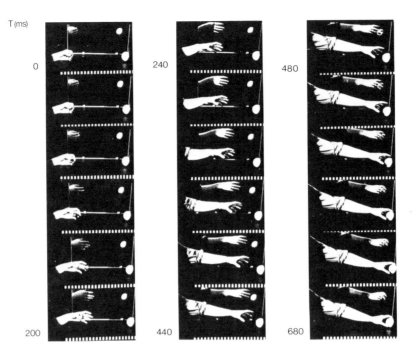

Figure 13.2 Complete film sequence of reaching movement directed at a ball placed in front of the subject. One frame every 40 msec has been represented from a film taken at 50 frames per second. Time 0 represents the last frame before the first detectable movement. Note sequence of finger shaping, rotation of forearm, fast transportation of arm, and slowing down at vicinity of the target.

Table 13.2
Delays, in milliseconds, for six subjects.

	Normal		Open-Loop	
	One-half Pronation	Pronation	One-half Pronation	Pronation
Onset of biceps	256.47	240.87	270.83	264.90
contraction	(82.19)[a]	(58.90)	(90.66)	(94.34)
Biceps isometric	137.84	126.66	140.42	139.20
tension time	(32.12)	(24.95)	(37.01)	(33.97)
Onset of extensor	278.43	261.31	290.64	288.98
contraction	(95.14)	(58.33)	(92.41)	(97.96)

a. Standard deviations are given in parentheses.

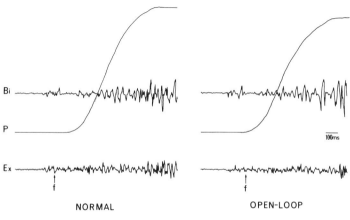

Bi

P

Ex
↑
f

NORMAL

100ms

↑
f

OPEN-LOOP

Figure 13.3 Reconstruction of intersegmental reaching sequences from EMG recordings and film analysis. Bi: EMG recorded from biceps brachii. Ex: EMG from extensor digitorum. P: Hand position redrawn from successive single frames (film taken at 50 frames/sec). f: First finger movement detected on film. Two sequences from the same subject are reproduced, one in the normal and one in the open-loop situation. Beginning of records coincide with signal to move. Note typical two-burst pattern on biceps EMG, as opposed to continuously rising contraction of extensor.

The final part of the movement—transportation of the arm as a whole from resting position to homing position at target—is due to the action of proximal joints. Simultaneous coactivation of muscles at several joints produces a complex pattern where the movement is not just a linear displacement between the initial and the final position. In fact, the hand describes a curved trajectory with a y component that peaks during the first one-third of the movement (figure 13.4, part A). This pattern was not apparent in most previous studies, where the arm movement was either restricted to a single joint or constrained by the displacement of a lever or a joystick.

In order to reconstruct position and velocity profiles of the movements as a function of time, we measured the distance covered by a given point of the forearm on successive single frames (Annett et al. 1958). Part B of figure 13.4 exemplifies the resulting curves for one typical movement, where only the distance covered along the x dimension has been taken into account. All movements exhibited the same stereotyped pattern, with a fast-rise acceleration followed by a slower deceleration (Woodworth 1899; Vince 1948; Beggs and Howarth 1972). In spite of the deceleration, the hand velocity was still relatively high at the vicinity of the object, and the movement came to a sudden stop at its contact.

Role of Visual Reafferences in the Control of Reaching

The same parameters of reaching movements can also be studied in situations involving the absence of visual reafferences from the moving

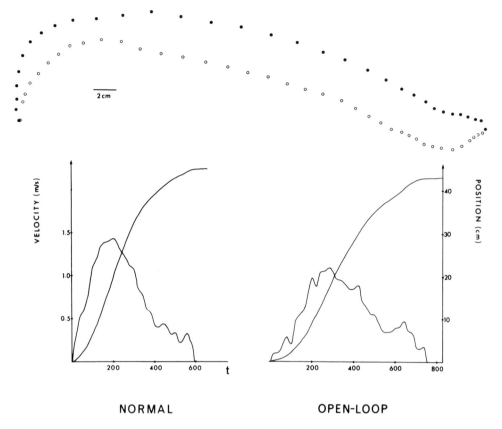

Figure 13.4 Pattern of hand displacement from resting position to target (same subject; normal and open-loop situations). Top: x,y plot of hand trajectory from resting position (left) to target (right). (●) Normal situation; (○) open-loop situation. Each dot represents the position of the hand on a single frame, from a film taken at 50 frames per second. The distance between two dots thus represents the distance covered in 20 msec. Bottom: Reconstruction of velocity (meters per second) and position profiles of the same two movements. Only the x component of the movements has been considered, by projecting each dot on a line joining the resting position and the final position of the hand.

limb before or during the movement (visual "open-loop" situations). The rationale for doing such an experiment is to disclose the possible role of visual cues (issuing from the respective position of the limb and of the object, from the comparison between finger posture and object shape, etc.) in controlling the movement.

EMG-film analysis of reaching movements aimed at virtual objects in the open-loop situation revealed only little difference with respect to movements performed in the normal (closed-loop) situation.

Movement Initiation
Intersegmental coordination during movement initiation appeared to be unchanged. The biceps-extensor delay, the biceps isometric tension time, and the time of occurrence of finger movements were in the same range as with full view of the limb (table 13.2). Posturing of the fingers

and orientation of the hand were also completed within the same time limits. The only significant difference observed during movement initiation in the open-loop situation was an increase in reaction time by about 20 msec on the average (table 13.2). We can only speculate on this difference, which might be due to factors such as an increase in "uncertainty" about completion of the task after suppression of the relevant visual cues or the anticipated processing of nonvisual signals for the control of the trajectory.

Movement Trajectory
Movement pattern was different in the open-loop situation, in that the y component was strikingly reduced (figure 13.4, part A). However, on the basis of the velocity profile of the x component, the difference between movements performed in either situation was much less apparent (compare parts B and C of figure 13.4). To a certain extent, the characteristics of the late part of the trajectory were accentuated in open-loop movements. Deceleration began earlier, and was more pronounced than in the normal situation.

Precision of Movement
Precision of visually goal-directed movements can be better studied in pointing at punctiform targets, a simplified case of reaching. Such a task has been used extensively in visuomotor experiments, in normal as well as in open-loop situations. Preventing vision of the arm during the movement clearly results in pointing errors (Merton 1961; Held and Freedman 1963; Foley and Held 1972; Jeannerod and Prablanc 1978). The subject's pointings are scattered around the target; the amplitude of the scatter, which determines terminal accuracy, is related to the distance between subject and target. Systematic trends in the distribution of errors may also be observed; the most common is undershooting (Beggs et al. 1972). Figure 13.5, drawn from a study done at our laboratory by Prablanc et al. (1979), exemplifies both phenomena (scatter and undershoot) in subjects pointing in the open-loop mode at targets appearing randomly in the peripheral visual field ipsilateral to the hand used for pointing.

Characteristics of reaching movements thus seem to be differentially affected by suppression of direct visual control of the moving limb. Considering movement initiation, it appears that mere inspection of the target is sufficient for specifying motor commands related to intrinsic and extrinsic object properties. This is true for most reaching movements performed in everyday life, where the starting position of the hand is usually out of view. On the other hand, errors in the completion of the movement, though they remain of a small amplitude (a few degrees of arc), may reflect disruption of normal control mechanisms exerted during the movement. The time at which reafferent visual con-

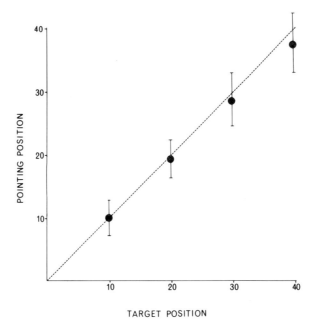

Figure 13.5 Errors in pointing at visual targets in open-loop situation. Targets are small luminous dots appearing in the peripheral visual field (from 10° to 40°). Subjects have their heads fixed, but can move their eyes freely. Dots represent average pointing position of the hand; vertical bars, standard deviation from mean. Note increasing undershoot and scatter when target eccentricity increases.

trol would become effective in adjusting the trajectory is difficult to determine. Classically, visual feedback from the arm takes more than 190–260 msec to process (Keele and Posner 1968). This delay is compatible with an action of vision anywhere in the trajectory during the deceleration phase. Particular emphasis has been put on visual control of the late part of the trajectory for two reasons: because the velocity of the arm becomes notably reduced and because the posture of distal segments anticipates the final position of the limb. The pattern of visual reafference that is available when the hand enters the central visual field at the end of the movement may thus be subjected to fine adjustments in position, velocity, and finger posture according to the corresponding parameters of the object to be grasped. There are strong arguments in favor of this hypothesis.

Other considerations, however, suggest that visual control might not be exerted specifically on a given part of the movement. Reaching in the open-loop condition, instead of being a modified case of normal movements, could represent a different type of motor behavior. If this were the case, the subject would commute from a "visual" to a "non-visual" strategy according to whether visual reafference were available. This alternative hypothesis could account for otherwise difficult-to-explain experimental facts, such as the overall increase in latency of open-loop movements and the modified pattern of their trajectory. In other words, it cannot be excluded that the whole programming of reaching movements might depend on the available cues at the moment of performance.

Eye-Hand Coordination During Reaching

Most, if not all, visually goal-directed actions imply a shift of the gaze in the direction of the goal. The reaching sequence would thus imply the occurrence of a saccadic eye movement, bringing the fovea to the target in the same time as the hand reaches for it. This has been recognized by several authors who have shown that the ocular saccade precedes the arm movements such that the eye is already at target before the hand begins to move (Angel et al. 1970; Megaw and Armstrong 1973; Jeannerod and Prablanc 1978).

In a more detailed analysis of the eye-hand sequence, we simultaneously recorded eye movements and biceps EMG in subjects pointing at targets appearing in the peripheral visual field. The beginning and the end of the arm displacements were also recorded. It was found that the onset of the ocular saccade was statistically synchronous with the onset of biceps contraction. This could have been suspected from the previous experiment, in which the delay of biceps contraction from the signal was around 250 msec. This value corresponds to a typical oculomotor reaction time in a randomized situation. In concordance to this point, oculomotor reaction time has been found to increase slightly in the open-loop situation (Prablanc et al., 1979). Consequently, the saccade-biceps synchrony is preserved.

Figure 13.6 shows the averaging of biceps EMG, eye-movement records, and electrical signals indicating the beginning and the end of the corresponding arm movements over 10 trials. Although the saccade and the onset of the first EMG burst appear to coincide, the hand does not leave the table before the end of the EMG burst, as shown in figure 13.3. Moreover, the hand reaches the target after a rather fixed transportation time of about 600 msec, some 750 msec after the eye begins to move.

A similar relationship would also hold for head movements (which we did not record). Bartz (1966), after others, had shown that in the eye-head sequence during fixation of a target in the peripheral visual field, the eye would start first and would then make a backward compensation movement in order to keep the target fixated during the movement of the head. This means that the resulting eye position in space (gaze) would be at target before the head movement would start. However, a recent study by Warabi (1977), which followed the monkey studies of Bizzi et al. (1971), shows that the neck muscles responsible for head turning are activated some 30 msec before the onset of the ocular saccade.

Again, the phenomenal eye-head-arm sequence is not due to delays in activation of muscles controlling the different mobile segments. From

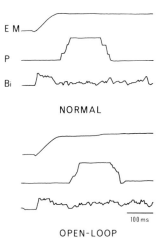

NORMAL

OPEN-LOOP

100 ms

Figure 13.6 Eye-hand sequence during reaching at visual targets. Computer-averaged records of four reaching movements, in open-loop situation, from same subject. Movements are directed at a luminous target appearing at 30° within the right part of the visual field. EM: Averaged eye movements. P: Averaged signals indicating the beginning and the end of the hand displacement. Bi: Averaged EMG from the biceps. Averaging has been triggered by the stimulus (onset of the target), not represented on the records.

the above-mentioned results it appears that eye, neck, arm, and forearm muscles for a given action of reaching contract within less than 50 msec of one another, although the subsequent resulting displacements are interspersed over more than 400 msec (the time at which the hand begins to move) or up to 700 msec (the time at which the hand reaches the target).

Several conclusions can be drawn from these data:

• The fact that the eye fixates the target before the hand starts may account for the use of a reafferent visual control on the movement of the arm. If this control were to be exerted on the basis of information gained during the early part of the movement, this information would have to be channeled via the peripheral retina (specifically that part subserving the lower periphery of the visual field). However, visual control using reafferent input from the terminal part of the movement can be exerted via the central retina.

• Activation of muscles responsible for finger movements also begins prior to foveation of the object, which indicates that at least part of the information for finger posturing originates from the peripheral retina. This is only a logical possibility, however. It remains to be determined whether hand shaping occurs within the same time limits in reaching toward objects in the peripheral visual field and in reaching toward those in the central visual field.

• If the position of the fovea in space is assumed to represent a "goal" to be matched by the arm (see Festinger et al. 1967; Paillard and Beaubaton 1978), then the determination of the nature of the relevant

Jeannerod, Biguer

signal to guide the arm is critical. Obviously, this signal cannot be some reafference (visual or proprioceptive) from the eye movement. Since eye and arm movement seem to be computed at about the same time, even a signal derived from the oculomotor command would not be able to influence the commands forwarded to arm muscles, at least during the first muscle burst. It is conceivable that eye and arm commands are generated independently, and that later on the oculomotor signal (whatever it would be from an "outflow" or "inflow" origin) would be used among others to optimize the trajectory of the arm. This hypothesis tends to minimize the role of eye-position monitoring in visually guided behavior. Tentatively, it can be suggested that the main source of information about "eye position" during the resting periods (that is, between the movements) is the position of the head in relation to the body. Under this assumption, simple mechanisms such as the vestibulo-ocular reflex would automatically ensure a constant "primary position" of the gaze with respect to the head. Accordingly, monitoring of head position would allow the encoding of body-centered coordinates in the internal representation of visual space.

Visuomotor Channels

Input-output relationships in reaching movements could reflect activation of several independent visuomotor channels. Each channel would deal with a specific aspect of the visual stimulus, and when activated would release a motor program adapted to the input pattern. Theoretically, error-correcting mechanisms (using visual feedback, or other reafferent sources) should also be channel-specific. Error signals issuing from a given aspect of the movement (for example, inadequate finger posture relative to object size) would be detected only by the channel specialized for the processing of the relevant visual cue and for generating the proper correction.

Timing of the different channels in producing an ordered output might be a problem if they were to be activated serially. On the other hand, if one speculates that they are activated in parallel and also accepts the centralist theory, then synchronization might rely on a mechanism similar to that already postulated by Bernstein (1967). According to Bernstein, the ordering of motor programs would rely upon a particular structure, lying outside the programs themselves and directing their order by some hierarchical principle. In our opinion, postulation of a hierarchy appears unnecessary. If the commands were released simultaneously, as they appear to be, that temporal ordering would be produced automatically, as a passive consequence of peripheral inbuilt constraints such as conduction time or duration of isometric tension.

Arguments for a plurality of visuomotor channels can be drawn from several studies in man and in animals. For instance, visuomotor development seems to show that "object" and "space" channels mature differentially. Young infants have been shown to be able to reach for visual objects ("pre-reaching"; Trevarthen 1975 and Trevarthan et al. 1975). However, their movement, though it can be directed in the proper direction, is far from complete; the fine coordination of distal segments is usually lacking, the fingers not always shaped, and the hand not oriented properly. According to Halverson (1931), it is not before 4–5 months of age that prehensile responses appear, and not before one year that precise grip can be observed. The late occurrence of hand shaping is controversial; some babies are reputed to show it within the first weeks of age (Bower et al. 1970), though such early shaping is always of an immature style (Di Franco et al. 1978).

One can hardly determine whether this inability (or poor ability) to match intrinsic properties of an object by a motor response is due to a lack of maturation at the input or at the output level of the channel. In favor of the input level, Di Franco et al. (1978) observed that an infant at the prereaching stage will reach as actively for a two-dimensional picture of an object as for the real object. As for the output level, it is worth stressing the deleterious effects on hand shaping of early reflex reactions and synergies. In Twitchell's (1970) terms, the "initial prehension appears more accidental than intentional" and "the emergence of the instinctive avoiding response at the time of these early attempts at voluntary prehension causes ataxia of reach and overpronation of the hand . . . ." Dexterous prehension appears late, and small objects remain difficult for the infant to handle; "the required thumb-fingers apposition does not appear until the grasp reflex can be fractioned during the second half-year of life." Later, the "instinctive avoiding response can contaminate activity and the fingers may abduct or dorsiflex too much as the hand is extended towards the object." These instinctive reactions may reflect an immaturity of the cortico-motoneuronal synapses controlling independent hand and finger movements. In the monkey this pathway does not develop until the eighth month, and not until then can these animals make a precise grip (Kuypers 1962). Monkeys with a lesion in cortical area 4 made in infancy will fail to develop a precise grip when tested up to 33 months after the lesion (Passingham et al. 1978).

Developmental studies of reaching movements also indicate separate maturation of the visual mechanisms initiating the ballistic transportation of the arm and those responsible for its precise guidance at the target. According to Dodwell et al. (1976), presentation of a graspable object to a young infant will first elicit only a visual response (looking at the object). Tentative reach will occur at a later stage (White et al. 1964). Full precision, however, will be the last acquisition. It seems that a second, corrective, component is progressively added to the ballistic movement through the use of visual cues based on the relative position of the hand and of the object within the central visual field (McDonnell 1975). This component is incorporated into the trajectory, and in the adult continuous smooth visual braking replaces terminal correction. Animal experiments provide further evidence for the same separate development of the triggered and the guided component in visual reaching (Hein and Held 1967; see also Jeannerod and Vital-Durand 1975).

A second argument for a plurality of visuomotor channels is based on the fact that pathological conditions may reveal similar dissociations of visuomotor channels. This is the case in humans with unilateral lesions in the visual cortex, who appear to be blind in the half-field opposite the lesion (by perimetric standards) in the sense that they cannot identify or even become aware of the presence of visual items presented on that side. However, by teaching some of them to "guess" at the locations of these stimuli, it has been possible to reveal remarkable residual capacities within their hemianopic field. These capacities, which reflect subcortical visual functions, were expressed in the "visuomotor mode"; that is, as motor responses directed at visual stimuli. Thus, by studying eye movements (Poeppel et al. 1973) or hand pointing (Weiskrantz et al. 1974; Perenin and Jeannerod 1975) in response to targets (bright spots of light) appearing in the blind area, a definite covariation of directional motor output and target position in space can be observed. In other words, the subjects were able to locate what they could not "perceive." In another series of patients with large cortical lesions (hemispherectomy) we have been able to study the same phenomenon more thoroughly (Perenin and Jeannerod 1978). Figure 13.7 shows an average distribution of hand pointings at targets presented for short durations (100–500

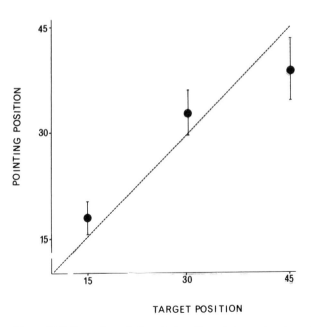

Figure 13.7 Errors in pointing at visual targets in one hemianopic patient. Same situation as in figure 13.5. Note overshooting for targets up to 30° and undershooting for the 45° target.

Visuomotor Mechanisms in Extrapersonal Reaching

msec) in the blind hemifield of one subject. Considering that the patient operates in the open-loop mode, since he cannot see his hand during movements toward that part of his visual field, he compares rather favorably with normals in the open-loop situation (figure 13.5). In both cases, however, precision of movements is affected by residual errors (primarily undershootings), which remain uncorrected.

These results are congruent with the role of central vision in improving terminal accuracy of reaching movements. Visual-cortex lesions, which permanently impair central vision (at the same time as they grossly increase the attentional threshold for the whole hemifield), disrupt fine-error detection and correction mechanisms. This may not be true only of arm movements. Eye movements also are affected when no longer controlled by the central retina. In normal subjects, ocular saccades directed at targets appearing in the peripheral visual field usually fall short of the target position. This "error" is then corrected within a short delay by a secondary correction saccade. When the target is turned off by the time the eyes begin to move (that is, when the saccade is performed and stopped in the open-loop mode), the correction saccade is lacking and the error remains uncorrected (Prablanc and Jeannerod 1975). Absence of correction saccades has also been observed in one monkey with a quasicomplete destruction of the visual cortex (Humphrey 1974).

On the other hand, the fact that reaching responses (though relatively inaccurate ones) can be generated in the absence of visual cortex stresses the role of subcortical pathways as a substrate for "space" channels (Feinberg et al. 1978). Superior colliculus, for instance, seems to be appropriately connected for directing eye and head movements at the absolute spatial locus where the target has been detected. However, unilateral colliculectomy in man (Heywood and Ratcliff 1975) and monkey (Wurtz and Goldberg 1972) does not impair the ability to make correctly oriented saccades in the corresponding hemifield. The main deficit that follows these lesions is a paucity of spontaneous saccades toward that part of the visual field and an increase in latency of saccades induced by a stimulus presented on that side. Collicular function regarding spatial detection would be better described in terms of shifting attention to a given part of the visual field—in terms of gating rather than generating motor commands.

It remains to be determined where these commands originate. Cortical areas outside the visual cortex might be involved. Posterior parietal lesions produce a loss of directional accuracy of reaching movements. In the monkey, the hand contralateral to the lesions misses the target by a large amount, and the error may remain uncorrected. During the first postoperative days, the deficit may even be more severe. Animals consistently avoid using the affected limb in visually directed movements, even though it is devoid of any paralysis and may be used for other purposes such as walking or climbing (Faugier-Grimaud et al. 1978;

Lamotte and Acuna 1978). In man, misreaching with the hand contralateral to the affected hemisphere (usually limited to the contralateral hemifield) is a commonly observed symptom of posterior parietal lesions (optic ataxia) (Rondot and de Recondo 1974; Perenin et al. 1979). Bilateral lesions at this location produce an even more dramatic deficit. Visual attention to the peripheral visual field is decreased. Furthermore, the whole reaching sequence is disrupted. In attempting to reach for an object, patients throw their arms erratically in various directions. At the same time, the gaze wanders and cannot fixate at the target (Balint syndrome; see full description in Hécaen and Albert 1978; also see Michel et al. 1965).

According to Mountcastle (1978), posterior parietal cortex subserves neural mechanisms that appear "to be correlated with the internal state of the organism in terms of needs and interests, . . . and from time to time generate commands for action, for the selective and directed visual attention into the immediate behavioral surround, for the visual grasping of objects, and for skilled, coordinated actions of hand and eye." This contention is based on the pattern of afferent and efferent connections of areas 5 and 7, and on neurophysiological experiments in monkeys in which the activity of parietal neurons has been correlated with reaching behavior (Hyvarinen and Poranen 1974; Mountcastle et al. 1975). It is concordant with the role of subcortical visuomotor structures participating in the control of spatial responses, since these regions (the superior colliculus and the pretectum) are known to be connected (via posterior thalamus) to the posterior parietal areas. In fact, in the hemispherectomized patients described above, subcortical structures probably took over reaching behavior, which is normally controlled by parietal cortex. This is another example of a "layered" system where the upper layer, which normally issues the commands, introduces additional parameters into the generation of these commands. In the case of the "space" channels, subcortical structures might subserve mere "orientation" toward stimuli (spatial detection, gating of attention, and release of reflex commands), while parietal cortex, the higher-level controller, might account for the integration of subcortical mechanisms into complex spatial behavior (involving motivation, decision processes, etc.).

The "object" channel can also tentatively be ascribed to particular structures. In monkeys, lesions in the visual cortex alter hand shaping. During reaching, the arm is fully pronated, the hand remains wide open (Brinkman and Kuypers 1973; Humphrey 1974). This might be a "strategy" to obviate the lack of terminal accuracy by increasing the probability of the hand coming into contact with the object. However, visual cortex is one of the steps on the way to detection of intrinsic properties of objects. It is thus not surprising that lesions at this level alter the whole process leading to motor output to the hand.

This is also the case for the already mentioned effects of area-4 lesions, at the other extremity of the channel, where independent finger movements are no longer observed. In other experiments, area 4 can be functionally excluded without being lesioned. After section of the corpus callosum and the optic chiasm in the monkey, presentation of a graspable object to one eye (and thus to only one hemisphere) elicits shaping of the reaching hand only if the hand contralateral to the eye is used. In this case, the hand is controlled by area 4 and crossed cortico-motoneuronal fibers. If the hand ipsilateral to the stimulated eye is used, no shaping occurs. In this case, the hand cannot be controlled by area 4; only ipsilateral, subcortical pathways are available. Hand movements are awkward and independent finger movements are abolished (Brinkman and Kuypers 1972). Similar observations were made by Gazzaniga et al. (1967) in commissurotomized patients when they had to use the ipsilateral eye-hand combination, especially on the right side.

Intrahemispheric disconnection between visual cortex and area 4 (with visual cortex and area 4 themselves spared) also produces a lack of adaptation of finger movements. Of course, this is a subtle deficit that requires appropriate testing to become apparent, and may not be seen on casual observation (Myers et al. 1962). In the Haaxma-Kuypers (1975) monkey experiment, the occipito-frontal pathways were interrupted at the posterior parietal level. After this lesion, the animals were unable to pick up little pieces of food placed on a board requiring precise orientation of the finger grip. The effect was limited to the hand contralateral to the lesion.

With larger posterior parietal lesions, a complete disruption of con-tralateral reaching movements can be observed. In man (Tzavaras 1978; Perenin et al. 1979) as well as in monkey (Faugier-Grimaud et al. 1978), in addition to the misreaching, the hand is clearly misshaped. Hence, the posterior parietal lobule is located within the visual pathways in such a way that making lesions in it alters both channels involved in reaching toward objects.

Conclusion

Vision is not only devoted to building up an internal representation of the external world. It also has a motor function, which permanently reflects the various levels of sensory processing. Visually directed motion is a continuous motor transform of incoming visual stimuli. Central visual processing is distributed across specialized areas, each one extracting a particular parameter from the milieu and activating the corresponding motor output. Perceptual unity, that is, building up a representation of an object as a sum of many different visual properties, would thus be a pure cognitive concept, not a visuomotor one.

On the other hand, although actions directed to the external world may be organized on the basis of rather simple visuomotor units, the release of these units implies dependence on higher levels. Upper layers of control may specify certain goals to action. Gating mechanisms, for instance, direct attention to a given part of the visual field or can release as well as prevent the output of a given visuomotor channel. A similar view has been cogently conceptualized by Arbib (1980). In our own view, the perceiving subject makes assumptions about external events and verifies them through action. The external world can be seen (at least under certain circumstances) as a flux of sequential events with causal relationships; it progressively becomes for each subject an "assumptive world" (Ittelson 1960) determining his own perceptual experience. Hence, the actual significance of a given external event will be disclosed only through the available consequences of the resulting action, that is, through the possibility of verifying the perceptual assumption made about this event. This is a perception-action cycle, which would rotate in the same direction as the one postulated by Neisser (1976).

Many arguments for this theory can be drawn from experiments involving visuomotor conflicts. Conflict situations produce a mismatch between the assumption and the experienced result of the action. The observed "error" is a cue for elaborating a new set of assumptions (the adaptation process). Adaptation would thus be a progressive change in the law of subjective probability attributed to perceptual assumptions in a given situation. It can be suggested that, even in the normal situation, the external world is always affected by an assumptive probability, and that active verification is always required.

These considerations stress the role of error-correcting feedback as a necessary factor in regulating actions. A purely "centralist" conception could thus not hold for such processes as visuomotor adaptation, acquisition of visuomotor skills, or recovery from the effects of lesions. Central lesions create a permanent disruption between assumptions and verifications, and may also alter the use of feedback loops. Recovery will be based on substitutive strategies, involving introduction of new feedbacks into the perception-action cycle.

References

Angel, R. W. 1974. "Electromyography during voluntary movement: The two-burst pattern." *Electroenceph. Clin. Neurophysiol.* 36: 493–498.

———. 1977. "Antagonist muscle activity during rapid arm movements: Central versus proprioceptive influences." *J. Neurol. Neurosurg. Psychiatr.* 40: 683–686.

Angel, R. W., W. Alston, and H. Garland. 1970. "Functional relations between the manual and oculomotor control signals." *Exp. Neurol.* 27: 248–257.

Annett, J., C. W. Golby, and H. Kay. 1958. "The measurement of elements in an assembly task: The information output of the human motor system." *Q. J. Exp. Psychol.* 10: 1–11.

Arbib, M. A. 1980. "Interacting schemas for motor control." In *Tutorials in Motor Behavior*, G. Stelmach and J. Requin, eds. Amsterdam: North-Holland.

Bartz, A. E. 1966. "Eye and head movements in peripheral vision: Nature of compensatory eye movements." *Science* 152: 1644–1645.

Beggs, W. D. A., and C. I. Howarth. 1972. "The movement of the hand towards a target." *Q. J. Exp. Psychol.* 24: 448–453.

Beggs, W. D. A., J. A. Andrew, M. L. Baker, S. R. Dove, I. Fairclough, and C. I. Howarth. 1972. "The accuracy of non-visual aiming." *Q. J. Exp. Psychol.* 24: 515–523.

Bernstein, N. 1967. *The Coordination and Regulation of Movements.* Oxford: Pergamon.

Bizzi, E., R. E. Kalil, and V. Tagliasco. 1971. "Eye-head coordination in monkeys: Evidence for centrally patterned organization." *Science* 173: 452–454.

Bower, T. G. R., J. M. Brougton, and M. K. Moore. 1970. "The coordination of visual and tactual inputs in infants." *Percept. Psychophys.* 8: 51–53.

Brinkman, J., and H. G. J. M. Kuypers. 1972. "Split-brain monkeys: Cerebral control of ipsilateral and contralateral arm, hand, and finger movements." *Science* 176: 536–539.

———. 1973. "Cerebral control of contralateral and ipsilateral arm, hand, and finger movements in the split-brain rhesus monkey." *Brain* 96: 653–674.

Di Franco, D., D. W. Muir, and P. C. Dodwell. 1978. "Reaching in very young infants." *Perception* 7: 385–392.

Dodwell, P. C., D. Muir, and D. Di Franco. 1976. "Responses of infants to visually presented objects." *Science* 194: 209–211.

Faugier-Grimaud, S., C. Frenois, and D. G. Stein. 1978. "Effects of posterior parietal lesions on visually guided behavior in monkeys." *Neuropsychologia* 16: 151–168.

Feinberg, T. E., T. Pasik, and P. Pasik. 1978. "Extrageniculostriate vision in the monkey. VI. Visually guided accurate reaching behavior." *Brain Res.* 152: 422–428.

Festinger, L., C. A. Burnham, H. Ono, and D. Bamber. 1967. "Efference and the conscious experience of perception." *J. Exp. Psychol.* 74: 1–36.

Foley, J. M., and R. Held. 1972. "Visually directed pointing as a function of target distance, direction, and available cues." *Percept. Psychophys.* 12: 263–268.

Freund, H. J., and H. J. Budinger. 1978. "The relationship between speed and amplitude of the fastest voluntary contractions of human arm muscles." *Exp. Brain Res.* 31: 1–12.

Gazzaniga, M. S., J. E. Bogen, and R. W. Sperry. 1967. "Dyspraxia following division of the cerebral commissures." *Arch. Neurol.* 16: 606–612.

Glencross, D. J. 1977. "Control of skilled movements." *Psychol. Bull.* 84: 14–29.

Haaxma, H., and H. G. J. M. Kuypers. 1975. "Intrahemispheric cortical connections and visual guidance of hand and finger movements in the rhesus monkey." *Brain* 98: 239–260.

Hallett, M., B. J. Shamani, and R. R. Young. 1975. "EMG analysis of stereotyped voluntary movements in man." *J. Neurol. Neurosurg. Psychiatr.* 38: 1154–1162.

Halverson, H. M. 1931. "An experimental study of prehension in infants by means of systematic cinema records." *Genet. Psychol. Monogr.* 10: 110–286.

Hécaen, H., and M. L. Albert. 1978. *Human Neuropsychology*. New York: Wiley.

Hein, A., and R. Held. 1967. "Dissociation of the visual placing response into elicited and guided components." *Science* 158: 390–392.

Held, R., and S. Freedman. 1963. "Plasticity in human sensorimotor control." *Science* 142: 455–462.

Held, R., and N. Gottlieb. 1958. "Technique for studying adaptation to disarranged hand-eye coordination." *Percept. Mot. Skills* 8: 83–86.

Heywood, S., and G. Ratcliff. 1975. "Long-term oculomotor consequences of unilateral colliculectomy in man." In *Basic Mechanisms of Ocular Motility and their Clinical Implications*, G. Lennerstrands and P. Bach y Rita, eds. Oxford: Pergamon.

Humphrey, N. K. 1974. "Vision in a monkey without striate cortex." *Perception* 3: 241–255.

Hyvärinen, J., and A. Poranen. 1974. "Function of the parietal associative area 7 as revealed from cellular discharges in the alert monkey." *Brain* 97: 673–692.

Ittelson, W. H. 1960. *Visual Space Perception*. Berlin: Springer.

Jeannerod, M., and C. Prablanc. 1978. "Résolution et plasticité de la coordination oeil-main." In *Du contrôle moteur à l'organisation du geste*, H. Hécaen and M. Jeannerod, eds. Paris: Masson.

Jeannerod, M., and F. Vital-Durand. 1975. "Les deux étapes du développement visuomoteur." *Lyon Médical* 236: 725–734.

Keele, S. W., and M. I. Posner. 1968. "Processing of visual feedback in rapid movements." *J. Exp. Psychol.* 77: 155–158.

Kuypers, H. G. J. M. 1962. "Corticospinal connections: Postnatal development in rhesus monkey." *Science* 138: 678–680.

Lamotte, R. H., and C. Acuna. 1978. "Defects in accuracy of reaching after removal of posterior parietal cortex in monkeys." *Brain Res.* 139: 309–326.

Livingstone, R. B., J. Paillard, A. Tournay, and A. Fessard. 1951. "Plasticité d'une synergie musculaire dans l'exécution d'un mouvement volontaire chez l'homme." *J. Physiol.* (Paris) 43: 605–619.

McDonnell, P. M. 1975. "The development of visually guided reaching." *Percept. Psychophys.* 18: 181–185.

Megaw, E. D., and W. Armstrong. 1973. "Individual and simultaneous tracking of a step input by the horizontal saccadic eye movement and manual control systems." *J. Exp. Psychol.* 100: 18–28.

Merton, P. A. 1961. "The accuracy of directing the eyes and the hand in the dark." *J. Physiol.* 156: 555–577.

Michel, F., M. Jeannerod, and M. Devic. 1965. "Trouble de l'orientation visuelle dans les trois dimensions de l'espace." *Cortex* 1: 441–466.

Mountcastle, V. B. 1978. "Some neural mechanisms for directed attention." In *Cerebral Correlates of Conscious Experience*, P. Buser and A. Buser-Rougeul, eds. Amsterdam: North-Holland.

Mountcastle, V. B., J. C. Lynch, A. Georgopoulos, H. Sakata, and C. Acuna. 1975. "Posterior parietal association cortex in the monkey: Command functions for operations within extrapersonal space." *J. Neurophysiol.* 38: 871–908.

Myers, R. E., R. W. Sperry, and N. M. McCurdy. 1962. "Neural mechanisms in visual guidance of limb movement." *Arch. Neurol.* 7: 195–202.

Napier, J. R. 1956. "The prehensile movement of the human hand." *J. Bone Joint Surg.* 38B: 902–913.

————. 1960. "Studies of the hands of living primates." *Proc. Zool. Soc. Lond.* 134: 647–657.

Neisser, U. 1976. *Cognition and Reality.* San Francisco: Freeman.

Paillard, J., and D. Beaubaton. 1978. "De la coordination visuo-motrice à l'organisation de la saisie manuelle." In *Du contrôle moteur à l'organisation du geste*, H. Hécaen and M. Jeannerod, eds. Paris: Masson.

Passingham, R., H. Perry, and F. Wilkinson. 1978. "Failure to develop a precision grip in monkeys with unilateral neocortical lesions made in infancy." *Brain Res.* 145: 410–414.

Perenin, M. T., and M. Jeannerod. 1975. "Residual vision in cortically blind hemifields." *Neuropsychologia* 13: 1–7.

————. 1978. "Visual function within the hemianopic field following early cerebral hemidecortication in man. I. spatial localisation." *Neuropsychologia* 16: 1–13.

Perenin, M. T., A. Vighetto, F. Mauguiere, and G. Fischer. 1979. "L'ataxie optique et son intérêt dans l'étude de la coordination oeil-main. A propos de quatre observations." *Lyon Médical* 242: 349–358.

Poeppel, E., R. Held, and D. Frost. 1973. "Residual visual function after brain wounds involving the central visual pathways in man." *Nature* 243: 295–296.

Prablanc, C., and M. Jeannerod. 1975. "Corrective saccades: Dependence on retinal reafferent signals." *Vision Res.* 15: 465–469.

Prablanc, C., J. F. Echallier, E. Komilis, and M. Jeannerod. 1979. "Optimal response of eye and hand motor systems in pointing at a visual target. I. Spatio-temporal characteristics of eye and hand movements and their relationships when varying the amount of visual information." *Biol. Cybernet.* 35: 113–124.

Rondot, P., and J. de Recondo. 1974. "Ataxie optique: Trouble de la coordination visuo-motrice." *Brain Res.* 71: 367–375.

Schneider, G. E. 1969. "Two visual systems." *Science* 163: 895–902.

Trevarthen, C. B. 1968. "Two mechanisms of vision in primates." *Psychol. Forsch.* 31: 299–337.

————. 1975. "Growth of visuo-motor coordination in infants." *J. Human Movement Stud.* 1: 57.

Trevarthen, C. B., P. Hubley, and L. Sheeran. 1975. "Les activités innées du norrisson." *La Recherche* 6: 447–458.

Twitchell, T. E. 1970. "Reflex mechanisms and the development of prehension." In *Mechanisms of Motor Skill Development*, K. Connolly, ed. New York: Academic.

Tzavaras, A. 1978. "Les apraxies unilatérales." In *Du contrôle moteur à l'organisation du geste*, H. Hécaen and M. Jeannerod, eds. Paris: Masson.

Vince, M. A. 1948. "Corrective movements in a pursuit task." *Q. J. Exp. Psychol.* 1: 85–106.

Von Hofsten, C. 1977. "Binocular convergence as a determinant of reaching behavior in infancy." *Perception* 6: 139–144.

Wachholder, K. 1928. *Willkürliche Haltung und Bewegung ins besondere im Lichte electrophysiologische Untersuchungen.* Munich: Bergmann.

Warabi, T. 1977. "The reaction time of eye-head coordination in man." *Neurosci. Lett.* 6: 47–51.

Weiskrantz, L., E. K. Warrington, M. D. Sanders, and J. Marshall. 1974. "Visual capacity in the hemianopic field following a restricted occipital ablation." *Brain* 97: 709–728.

White, B. L., P. Castle, and R. Held. 1964. "Observations on the development of visually directed reaching." *Child Develop.* 35: 349–364.

Woodworth, R. S. 1899. "The accuracy of voluntary movements." *Psychol. Rev. Monogr.* Suppl. 3.

Wurtz, R. H., and M. E. Goldberg. 1972. "Activity of superior colliculus in behaving monkeys. IV. Effects of lesions on eye movements." *J. Neurophysiol.* 35: 587–596.

14 Vision in Action: The Control of Locomotion

David N. Lee
James A. Thomson

In order to get around in the environment and act appropriately with respect to objects and events, an animal needs perceptual information to guide its activities. Conversely, in order to obtain information about the environment, an animal has to move around. The perceptual systems and the motor system are not, therefore, functionally separable; they are components of a unified perceptuomotor system.

Any theory of how the visual component of the perceptuomotor system works needs to address two questions: What information is available through vision for controlling activity? How is the visual information actually used? In this chapter we seek to outline answers to these questions.

We start by considering what general types of information are involved in controlling activity. Classically, the receptor systems of the body have been divided into two distinct groups: exteroceptors and proprioceptors. The exteroceptors—the eyes, ears, nose, mouth, and skin—were taken to be responsible for sensations of external origin, whereas sensations of bodily movement were assumed to arise through stimulation of the proprioceptors—the receptors in the joints, muscles, and inner ear. This classification of the perceptual systems was strongly criticized by Gibson (1966), who pointed out that exteroception, considered as the obtaining of information about extrinsic events, does not depend on exteroceptors, and that proprioception considered as the obtaining of information about one's own actions, does not depend on proprioceptors. Most of the perceptual systems are both exteroceptive and proprioceptive. Therefore, the terms "exteroceptor" and "proprioceptor" are misleading and would best be dropped from the literature.

Gibson's distinction in terms of type of information rather than type of perceptual system is extremely important. However, his categorization of information is arguably too broad. It is important to distinguish between sensing one's bodily actions as such and sensing one's changing relationship to the environment which those actions bring about. Gibson (1958) himself made the distinction clear by considering the case of a bird flying in a headwind; sensing bodily actions per se does not yield information about movement relative to the environment. In other words, there are two types of information about action: body-relative and environment-relative. Lee (1978) suggested that the term "proprioceptive" be used in the more restricted classical sense to refer to the body-relative information, and proposed a new term, *exproprioceptive,* to refer to the environment-relative information.

In short, there are three types of information involved in controlling activity:

• *exteroceptive* information about the layout of the environment and about external objects and events,

• *proprioceptive* information about the positions, orientations, and movements of parts of the body relative to each other, and

• *exproprioceptive* information about the position, orientation, and movement of the body as a whole, or part of the body, relative to the environment.

In this chapter our main concern will be with the exproprioceptive function of vision.

The Optic Flow Field

Information about Movement of the Body Relative to the Environment
What exproprioceptive information is available through vision for controlling activity? In seeking an answer to this question we need to start by examining what the optical input to the eye is.

It is commonly held that the fundamental input to the eye is some form of image or two-dimensional spatial representation of what is being looked at. In other words, the eye may be considered to be rather like a camera. The image has been described in a variety of ways, most recently in terms of spatial frequency spectra (see, for example, Sekuler 1974). The image approach to visual perception has a long history. However, the approach has, from its earliest days, been confronted with a major problem: How can visual perception of the three-dimensional world be at all possible if all that is available to the eye is a two-dimensional image? Where does the missing depth dimension come from? As Gibson (1950), Neisser (1977), and others have pointed out, this problem of the missing depth dimension in the image has lead image theorists into a paradoxical position: In order to account for three-dimensional perception, they have been forced to postulate that there must be embodied in the visual system quite detailed prior knowledge or "assumptions" about what it is that is actually being viewed. These hypothesized built-in assumptions have ranged from the classical pictorial depth cues to the more explicit assumptions recently postulated as necessary to perception by artificial-intelligence theorists (see, for example, Minsky 1975 and Marr 1978). But where does such prior knowledge come from if not through perception? When a

Lee, Thomson

theory logically leads to a paradox, it is time to question its premises.

Is the input to the eye indeed an image? Consider the ecological circumstances in which vision has to operate. An animal is constantly active, moving around its environment, interacting with objects and other organisms. Even when it is sitting or standing still, its body is always swaying slightly, and the sway has to be actively kept in check. The result of this continual activity is that the animal's head is always moving, and so its view of the environment is constantly changing. Thus, the ecological input to the eye is a time-varying, spatially structured array of light, not a time-frozen camera-type image.

Gibson (1950) was the first to recognize the significance of this fundamental ecological fact about vision, and his theory of ecological optics (Gibson 1966, 1979; see also Neisser 1977) provided a radically new approach to understanding visual perception. As we will show, the fact that the input to the eye is a spatiotemporal optical structure—an *optic flow field*—is of threefold significance. The optic flow field contains three types of information: spatiotemporal information about the animal's movement relative to the environment, temporal information necessary for timing its actions, and three-dimensional spatial information about the environment. The problem of the "missing depth dimension" that has so puzzled image theorists does not exist for an animal that is free to move.

Let us consider how the optic flow field is generated. The environment consists of material substances bounded by surfaces, and it is the light reflected from these surfaces that forms the input to the eye. Surfaces contain patches of differing pigmentation, facets, and so on (unless the surface is mirrorlike). That is, surfaces are in general densely covered with texture elements that reflect light differently from their neighbors. This means that the light reflected from the surfaces in the environment to any particular point of observation forms a densely structured optic array, a bundle of cones of light each based on a different surface-texture element and each differentiable from its neighbors in terms of the intensity and/or the spectral composition of the light it contains. Furthermore, at each point of observation in the environment there is a different optic array. Thus, movement of the head relative to the environment generates a continuously changing optic array or optic flow field at the eye.

In order to determine what exproprioceptive information is contained in the optic flow field, we need a way of describing its structure. A convenient way of mathematically describing the optic flow field is in terms of the changing pattern of light formed on a projection surface that intercepts the time-varying bundle of light cones. The choice of projection surface is simply a matter of convenience, since the description of the optic flow field in terms of its projection on one surface is mathematically transformable into a description of its projection on any

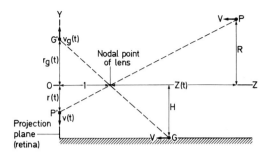

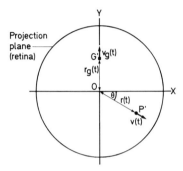

Figure 14.1 Illustration of how rectilinear movement of the point of observation relative to the environment generates an optic flow field. The schematic eye is considered to be stationary, with the environment moving toward it with velocity V in the direction $Z \rightarrow O$. P and G denote texture elements on surfaces in the environment; G is on the ground surface. Light reflected from the moving environmental texture elements passes through the nodal point of the lens, giving rise to the moving optic texture elements P' and G' on the "retina." The densely textured environment gives rise to a densely textured optic flow field wherein all optic texture elements move outward along radial flow lines emanating from O. Adapted from Lee 1974.

other surface. For clarity of exposition, we will consider the projection of the optic flow field onto a plane surface behind the point of observation, rather like the film plane of a camera.

Let us examine the general structure of the rectilinear optic flow field (that which results when the eye is moving along a straight path). It is easier to explain, and equivalent geometrically, if we consider the point of observation to be stationary and the environment moving relative to it. Figure 14.1 shows the environment moving with velocity V toward the point of observation in a direction perpendicular to the projection plane. P and G denote any two environmental texture elements, G lying on the ground surface. Corresponding to P and G are the optic-texture elements P' and G'. It will be seen that each optic element is moving directly away from O.

This brings us to the first invariant property of the rectilinear optic flow field: No matter what the layout of surfaces in the environment, all

Lee, Thomson

optic-texture elements move outward along radial flow lines emanating from the center of the projection plane (see also figure 14.6).

A second important invariant property of the optic flow field pertains to the fact that, during movement of the point of observation, surfaces progressively go out of view and come into view as they are occluded and disoccluded by nearer surfaces. The reflection of this fact in the optic flow field is such that whenever an optic-texture element moving along a radial flow line catches up with a slower-moving element, it "occludes" or replaces it. This is because the faster-moving element corresponds to a nearer texture element in the environment.

It has been proved mathematically (Lee 1974) that if the optic flow field at the eye has these two properties then this specifies that the eye is moving rectilinearly relative to a rigid environment; that is, the optic flow field could not have been generated in any other way (except by artifice). In other words, there is available in the optic flow field information about the physical state of affairs. Of course, the fact that the information is available does not mean that it is actually picked up by the eye and used. To test whether the visual information is in fact picked up, and if so how potent it is in comparison with the information from the other senses, two series of experiments were run. The first series was concerned with what gives rise to the conscious experience of how one is moving relative to the environment. The second series was concerned with the role of vision, at a subconscious level, in controlling balance.

Experiments on the Perception of Ego Movement
The experiments (Lishman and Lee 1973) consisted in giving human subjects visual information about how they were moving relative to their environment that conflicted with the information available through their other senses. The apparatus was very simple. It comprised a "swinging room," a floorless $4 \times 2 \times 2$ m box suspended just above the floor from a high ceiling so that it could be swung noiselessly forward and back along a virtually straight arc. Subjects stood in a trolley inside the swinging room, with their view of the floor outside the trolley occluded by the sides of the trolley. Thus, for example, with the trolley actually stationary, by swinging the room forward optic flow fields could be produced at the subjects' eyes that visually specified that subject and trolley were moving backward, in conflict with the information available through the other senses.

Two sets of experiments, involving passive and active movement, were run. The trolley was either stationary or moving forward and back, and the swinging room was moved forward and back in different ways. Part A of figure 14.2 illustrates one of the passive-movement experiments, in which the subjects simply stood on the trolley floor while the room and trolley were moved mechanically. Part B illustrates one of the active-movement experiments, in which the subjects either walked

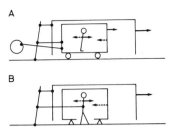

A

B

Figure 14.2 Experiments on visual perception of ego movement. (A) Passive movement: Crank moves subject on trolley and surrounding floorless "room" (at twice the speed) forward and backward. (B) Active movement: With trolley floor removed, subject steps forward and back on laboratory floor, moving "room" with him at twice his speed. Only one direction of movement is illustrated. (← — -) Movement relative to laboratory floor; (——→) visually specified movement, the movement relative to the "room." Lengths of arrows are proportional to speed. In each experiment, 13 out of 16 subjects reported that they and the trolley were moving in the visually specified way shown; they perceived the "room" to be stationary. The remaining subjects' reports were confused; none apprehended what was actually happening. Adapted from Lishman and Lee 1973.

back and forth on the trolley floor or, with that removed, walked on the laboratory floor beneath the trolley and thus moved trolley and room. The results were that vision dominated in 75% of the 112 active-movement trials (7 experiments, 16 subjects) and in 86% of the 64 passive-movement trials (4 experiments, 16 subjects). That is, the ego movement that the subjects experienced was not their movement relative to the earth but the visually specified movement relative to the swinging room, which they perceived to be stationary.

That vision dominated in the active-movement experiments may seem particularly surprising. Surely the subjects knew when they were stepping forward and when they were stepping back! How then could they feel that they were moving backward when they were stepping forward, and vice versa (figure 14.2B)? The experiments, in fact, nicely illustrate the distinction we made at the beginning of this chapter between proprioception and exproprioception—between sensing one's bodily actions as such and sensing how one is moving relative to the environment as a result of those actions—for performing actions designed to move oneself forward docs not guarantee that one will move forward, as walking up sand dunes or on slippery ground and swimming against a current clearly show. In short, locomotion depends not only on how the limbs are moved but also on the nature of the surface or medium of support, and vision is often the only reliable source of exproprioceptive information about how the body is actually moving relative to the stable environment. It is understandable, therefore, that the subjects in the swinging room should have perceived themselves moving in accord with the visual information available to them and felt that they were walking on a sort of conveyor belt.

Let us now turn to the second set of experiments, where the exproprioceptive power of vision is again evidenced, at a much finer level.

Experiments on the Role of Vision in Balance Control

We spend so much time upright on our feet that we do not often realize what a precisely controlled activity balance is. In fact, it is the most fundamental activity—a prerequisite for most other activities, as anyone who has suffered a balance disorder will know.

In maintaining a stance, muscular adjustments have to be continually made on the basis of exproprioceptive information about the orientation and sway of the body relative to the environment. What are the sources of the information? According to the classical view, which is still current in many textbooks, the information is obtained primarily through the vestibular system (often called the organ of balance) and through the mechanoreceptors in the feet and ankles. The experiments described above, however, lead us to suspect that vision might play a major role in balance control. If vision is that powerful in giving rise to the conscious experience of how one is moving, might it not be equally powerful in affording information for balance control, in this case at a subconscious level?

Experiments confirmed this suspicion. The swinging room was again used (this time without the trolley) to visually simulate body sway. The experiment is illustrated in figure 14.3. Both toddlers and adults served as subjects. The control of anterior-posterior and lateral body sway was investigated. (For details of the experiments see Lee and Aronson 1974 and Lee and Lishman 1975.)

The main conclusion drawn from the experiments was that vision generally affords the most sensitive and reliable exproprioceptive information for balance and is an integral component of the control system. For example, oscillating the swinging room through as little as 6 mm caused adult subjects to sway forward and back approximately in phase with this movement (see figure 14.3B). The subjects were like puppets visually hooked to the swinging room. They were quite unaware that the room was oscillating and that they were swaying rhythmically, which indicated that the nonvisual exproprioceptive information available through their vestibular systems and their ankles and feet was not sensitive enough.

Vision was found to be especially important (often crucial) for balance control when the support surface was compliant, unsteady, or narrow, which renders unreliable the information obtainable through the feet, and when the subject was learning a new stance, which requires attunement to unfamiliar afference from the feet and ankles. Toddlers and adults in unpracticed stances could readily be knocked off balance by moving the swinging room. However, while balance was often impossible without vision when subjects were first learning a new stance, with practice vision often became noncritical; this suggested that visually guided practice facilitates attunement to information available

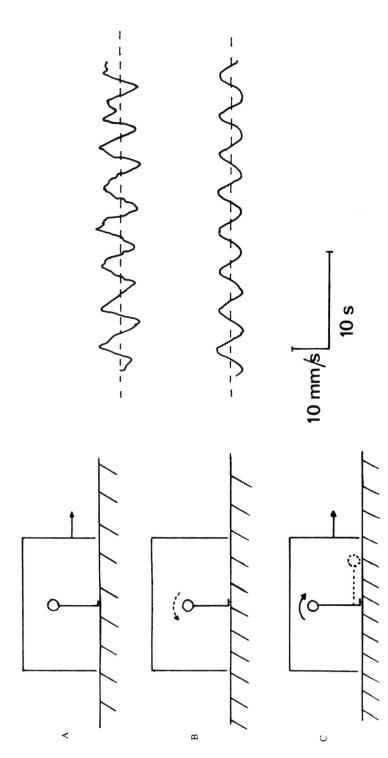

Figure 14.3 Experiments on the role of vision in balance control. Left: The experimental paradigm. Forward motion of the floorless experimental "room" (A) produces optic flow fields at standing subject's eyes corresponding to backward sway of the body (B). Postural adjustment to counteract this apparent backward body sway would result in subject swaying forward and possibly losing balance (C). The opposite holds for backward motion of the room. Right: The result of sinusoidally oscillating the experimental room forward and back through 6 mm with a period of 4 sec. Upper trace: velocity of adult subject's trunk. Lower trace: velocity of room. Positive velocities are forward. The subject was standing on foam pads. Sources: Lee and Aronson 1974 and Lee and Lishman 1975.

through the feet. This is supported by experiments with blind people, who were found to sway twice as much as blindfolded sighted people (Edwards 1946).

Information for Controlling Locomotion
The importance of vision for movement control is particularly evident in locomotion through the normal cluttered environment. Let us now analyze the optic flow field in more detail to determine the information it contains for controlling locomotion.

Information About Distance, Size, and Slant
As figure 14.1 shows, the position of an environmental texture element P relative to the eye is defined by the distance coordinates $Z(t)$ and R, together with the angle between the OZP and OZX planes. This angle is specified in the optic flow field by the angular coordinate θ of the optic-texture element P'. But are the distance coordinates $Z(t)$ and R optically specified? From similar triangles,

$$Z(t)/R = 1/r(t). \tag{1}$$

This equation is an expression of the well-known problem of the missing depth dimension, discussed above, which arises when the visual stimulus is treated as an image (a time-independent spatial structure). The problem is that the position of an optic texture element specifies only the direction in which an environmental texture element lies, not its distance away. However, if we examine the spatiotemporal structure of the optic flow field, we find that the depth dimension is not in fact missing.

Differentiating equation 1 with respect to time, we obtain

$$R/V = r(t)^2/v(t), \tag{2}$$

where $V = -dZ(t)/dt$ is the velocity of the environmental texture element P and $v(t) = dr(t)/dt$ is the velocity of the corresponding optic-texture element P' (see figure 14.1). Eliminating R between equations 1 and 2 gives

$$Z(t)/V = r(t)/v(t). \tag{3}$$

Equations 2 and 3 mean that the distance coordinates $(R,Z(t))$ of all visible texture elements are optically specified to within a scale factor of V. In other words, there is information available in the optic flow field about the relative distances, sizes, and orientations of surfaces and objects in the environment (see also Koenderink and van Doorn 1977 and Nakayama and Loomis 1974). We will show below that there is also available in the optic flow field body-scaled information about the environment, which is the type of information essential for controlling activity. But first let us consider another important type of information given in the optic flow field: temporal information.

Information About Time to Contact

Equation 3 states that the time remaining before the eye will be level with the texture element P if the current approach velocity V is maintained is directly specified by the value of the optic variable $r(t)/v(t)$. This temporal exproprioceptive information is essential for timing actions relative to the environment. For example, in an approach to a surface the optic variable specifies the time to contact, which is the type of information that a bird needs in preparing to land. Another simple example is catching a ball, which requires starting to close the hand at a precise time before the ball hits it (Alderson et al. 1974).

We will show below that the optic variable $r(t)/v(t)$ affords information for controlling a variety of locomotor activities. Since the variable appears to be a particularly informative property of the optic flow field, there is reason to consider it one of the elementary variables. Accordingly, we will designate it by the single symbol $\tau(t)$, where

$$\tau(t) = r(t)/v(t). \tag{4}$$

From this point on we will treat $\tau(t)$ as the basic variable associated with an element of the optic flow field, rather than its velocity $v(t)$. Hence, using equation 4, we rewrite equations 2 and 3 as

$$R/V = r(t)\tau(t), \tag{5}$$

$$Z(t)/V = \tau(t). \tag{6}$$

Body-Scaled Information

As discussed above, there is available in the optic flow field information about the relative sizes and distances of objects and surfaces in the environment. However, this information is, by itself, of little functional value to an animal. What it needs is exproprioceptive information that is relevant to controlling its activity—for example, that a hurdle is a certain fraction of its own height, so many strides away, and so on. That is, an animal needs *body-scaled* information about its environment. The following are two ways that body-scaled information is available in the optic flow field.

Consider an animal running straight over a level stretch of ground. Suppose at a particular time t its speed is V. One bodily yardstick it could use is the height H of its eye above the ground, which will be more or less constant. Since H is the R coordinate of any texture element on the ground over which the animal's eye will pass, applying equation 5 to the ground texture element G depicted in figure 14.1,

$$H/V = r_g(t)\,\tau_g(t), \tag{7}$$

and eliminating V through equations 5–7 yields

$$R/H = r(t)\,\tau(t)/r_g(t)\,\tau_g(t), \tag{8}$$

$$Z(t)/H = \tau(t)/r_g(t)\,\tau_g(t). \tag{9}$$

In other words, a particular relation between the optic flow from the line of ground ahead and the optic flow from other environmental texture elements specifies the distances and sizes of surfaces and objects in the environment in units of the animal's eye height.

Another potential bodily yardstick is stride length. Suppose that at time t the animal's stride length is L and the duration of its strides is t_s. Then for any environmental texture element P (see figure 14.1)

$$Z(t - t_s) - Z(t) = L. \tag{10}$$

Thus, from equation 6

$$[Z(t - t_s) - Z(t)]/V = L/V = \tau(t - t_s) - \tau(t), \tag{11}$$

and so, eliminating V by equations 5, 6, and 11, we obtain

$$R/L = r(t)\,\tau(t)/[\tau(t - t_s) - \tau(t)] = r(t)\,\tau(t)/t_s, \tag{12}$$

$$Z(t)/L = \tau(t)/[\tau(t - t_s) - \tau(t)] = \tau(t)/t_s. \tag{13}$$

That is, the distances and sizes of surfaces and objects in the environment are optically specified in units of the animal's stride length.

Visual Guidance in the Long Jump

How is visual body-scaled information used in regulating locomotor behavior? To investigate this we (Lee et al. 1977) carried out a film analysis of three long jumpers. Sporting activities like the long jump offer the student of visuomotor control excellent classes of behavior for analysis. First, the activities are natural in the sense that no sensory or motor restrictions are imposed on the performer. Second, since athletes are trained to maximize the efficiency of their movements by eliminating random and unnecessary aspects, it is much easier to see the fundamental features of their behavior than with unskilled subjects.

Long jumping requires very precise control of gait directed toward securing accurate footing. After a run of some 40 m, the jumper not only has to strike a 20-cm-wide takeoff board but also has to do so in the right posture for the jump. Most coaches hold that the skill depends on the athlete developing a consistent runup and starting at a measured distance from the board. In their view the skilled jumper's approach to the board is a preprogrammed act performed without visual guidance. This idea might seem plausible, since it is not apparent that skilled long jumpers adjust their strides to zero in on the board, nor do the jumpers normally report doing so. It seemed to us, however, that long jumping must require extremely fine visual regulation of gait, albeit at a subconscious level.

Our film analyses showed that this was indeed the case. For example, figure 14.4 (top) shows the pattern of stride lengths over six jumps of an Olympic jumper as she ran down the track toward the board. It can be seen that up to three strides from the board she produced a reasonably consistent stride pattern, which reflected her training. She progres-

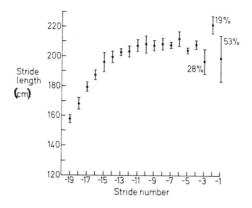

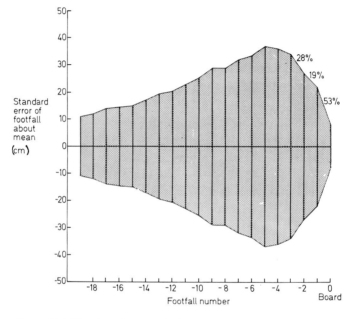

Figure 14.4 Visual guidance in the long jump. Performance of an Olympic athlete over six jumps with a 40-m, 21-stride runup. Top: means and standard errors of stride lengths for last 19 strides to takeoff board. Note how high the standard error is over the last three visually adjusted strides. Bottom: standard errors of successive footfall positions down track. Standard error progressively increases up to a peak of 37 cm and then suddenly drops to 8 cm over last three strides. Percentages shown are estimates of proportion of total adjustment made on each stride, derived from linear regression analyses. Adapted from Lee et al. 1977.

sively lengthened her strides as she accelerated to a steady speed. However, she varied the lengths of her last three strides considerably across her six runs. The reason for this is clear from the bottom part of figure 14.4. As she progressed down the track the small inconsistencies in her successive stride lengths had a cumulative effect, so that by the time she was five strides from the board the standard error of her footfall position was 37 cm. Clearly, if she had continued blindly her standard error at the board would have been even larger. However, this is not what happened. Her standard error rapidly decreased to 8 cm over her last three strides.

Clearly, the jumper was visually adjusting her last three strides to zero in on the board. Furthermore, she appeared to be regulating the strides as a single functional unit, tending to distribute her total adjustment in a consistent way over the three strides (as indicated by the percentages in figure 14.4, top). This makes sense, since during her last three strides she had to "gather" herself for the jump and this involved (among other things) generating the stride pattern "short, long, short, JUMP" seen in figure 14.4.

In the preceding section we described two types of visual information long jumpers might use in adjusting their final strides: distance from the board body-scaled in terms of eye height or stride length. Another possibility is information about time to reach the board, specified by the time-to-contact optic variable τ for the board and body-scaled in terms of the athlete's stride duration. That is, rather than adjusting the *lengths* of strides to just span the remaining *distance* to the board, the jumper's task may be considered to consist in regulating the *durations* of strides so as to just fill the *time* gap between him and the board. This temporal conception of the task is probably more appropriate than the spatial one, since jumpers have direct control over the durations of their strides by how hard they thrust vertically on the ground, whereas stride length is a function also of the jumper's speed.

Driving

The time-to-contact optic variable τ also affords valuable information for driving, in a rather subtle way. Consider the problem confronting a driver approaching an obstacle in the road. How does the driver manage to stop safely? He not only has to start braking early enough, but also has to adjust the vehicle's deceleration to an adequate level during the stop; if he brakes too lightly at the beginning he can run out of braking power. In other words, the driver can get into a "crash state" (speed too high in relation to the distance from the obstacle) well before an actual collision, either before starting to brake or while braking.

Drivers could use time-to-contact information (given by the value of the optic variable $\tau(t)$ corresponding to the obstacle) in judging when to start braking (Lee 1976). But how do drivers avoid getting into a crash state when braking? They clearly need visual exproprioceptive information about how they are closing on the obstacle so that they can appropriately adjust their braking. It might seem that they need to obtain information about distance from obstacle, closing velocity, and deceleration, and then perform complicated mental calculations. However, this is not necessary. It turns out that the value of the time derivative of the optic variable $\tau(t)$ corresponding to the obstacle affords sufficient information.

Suppose that at time t a driver is a distance $Z(t)$ from an obstacle, is traveling with velocity $V(t)$, and is braking with constant deceleration D. The deceleration D is adequate if and only if the distance needed to stop is less than or equal to the current distance from the obstacle, that is, if and only if

$$V(t)^2/2D \leqslant Z(t),$$

or

$$Z(t)D/V(t)^2 \geqslant 0.5. \tag{14}$$

$Z(t)/V(t)$ is specified by the value of the optic variable $\tau(t)$ for the obstacle (see equation 4); that is,

$$Z(t)/V(t) = \tau(t), \tag{15}$$

and differentiating this equation with respect to time we obtain

$$\frac{Z(t)D}{V(t)^2} = 1 + \frac{d\tau(t)}{dt}. \tag{16}$$

Hence, from equations 14 and 16, the value of the time derivative of the optic variable $\tau(t)$ specifies that the driver's current deceleration is adequate if and only if

$$\frac{d\tau(t)}{dt} \geqslant -0.5. \tag{17}$$

In other words, a driver has available visual exproprioceptive information about his potential future state were he to maintain his current braking level. A safe braking strategy would consist in adjusting braking so that $d\tau(t)/dt$ remained at a safe value. The deceleration profiles produced by this hypothetical braking strategy (Lee 1976) matched quite closely those of test drivers recorded by Spurr (1969), the only data on visually controlled braking found in the literature (see figure 14.5).

Steering poses a similar problem, in that a driver can get into a "crash state" well before he actually runs off the road. This can occur not only if he takes a bend too fast, but also if he does not adjust the steering early enough on a bend and so gets into the situation where he needs to steer an impossibly tight curve. What he needs is visual exproprioceptive information not so much about his current position on the road but about his potential future course were he to maintain the current steering angle. This is, in fact, specified in the optic flow field at the driver's eye,

Lee, Thomson

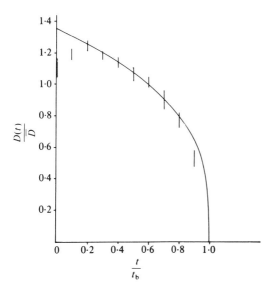

Figure 14.5 Visual control of braking. The data, taken from figure 2 of Spurr 1969, are for a driver stopping at a nominated point from various speeds up to 100 km (62 miles) per hour. They are plotted in a dimensionless form: $D(t)$ is deceleration at time t after initiation of braking; $\bar{D}$ is mean deceleration during stop; t_b is braking time. Curve is derived from theory given in text; it is what would have obtained if the driver had been controlling braking by maintaining the optic variable $d\tau(t)/dt$ at a safe value of -0.425. Deceleration decreases monotonically during stop. In other words, driver is allowing a safety margin: At each point in time, deceleration is higher than it strictly needs to be—if deceleration were maintained, car would stop short. Source: Lee 1976.

as is explained in figure 14.6. McLean and Hoffmann (1973) found that drivers do use such visual information in straight-lane driving. They found that steering adjustments were made primarily on the basis of heading angle (corresponding to potential future course) rather than current lateral position on the road.

The Projective Nature of Visual Guidance
The three skills we have discussed—long jumping, braking, and steering—reveal a basic feature of visual guidance: the projective use of visual information to control future activity. Long jumpers have to regulate their last three strides to the board as a single functional unit on the basis of visual information picked up some time beforehand. In order to stop safely, drivers have to visually adjust their braking to an adequate level well in advance of reaching the obstacle. In negotiating a bend, they have to adjust their steering on the basis of visual information about their potential future course.

The fact that visual information can and must be used projectively suggests that visual guidance does not require the continuous pickup of visual information. Common experience indicates that this is so. For example, drivers often turn around to talk to their passengers, and

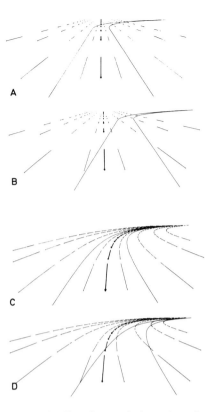

Figure 14.6 Visual control of steering: the optic flow field, projected onto a plane as in figure 14.1, for driving down a road. Solid lines represent edges of road; broken ones represent optic flow lines. Central heavily drawn flow line is "locomotor flow line," which specifies potential future course of vehicle if current steering angle were to be maintained. (A) Steering straight toward a bend on course. (B) Steering straight, but off course. (C) Steering a bend of uniform curvature on course. (D) The same, but off course. To see how the optic flow fields are generated, consider the vehicle to be stationary and the ground moving under it. In A and B, points on the ground are moving along parallel straight paths; the radiating optic flow lines are the projections of these paths. In C and D points on the ground are moving along concentric circular paths, the center corresponding to the center of curvature of the vehicle's path; the hyperbolic optic flow lines are the projections of these paths. Adapted from Lee and Lishman 1977.

people frequently walk along looking in shop windows and at passersby.

To investigate the question, a set of experiments was carried out by Thomson (1977). The basic technique was very simple. In the simplest case, subjects were presented with target marks on the ground at distances varying up to 21 m and, after looking for about 5 seconds, were required to walk to the target with their eyes closed. Other potential sources of distance information, such as echo, were controlled. Thus, if subjects could accurately locate the targets their only source of information would be that acquired through vision before they started to walk.

The results of the experiment were very clear. Figure 14.7 shows that for distances up to 9 m performance was very accurate. The errors were

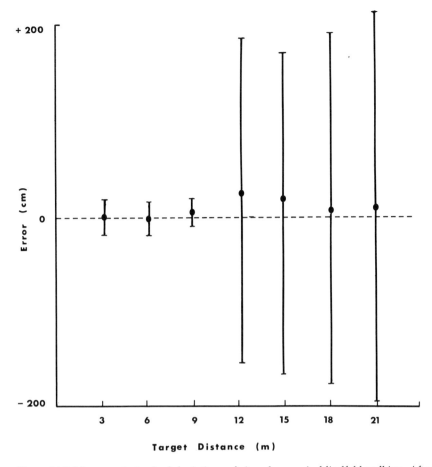

Figure 14.7 Means and standard deviations of signed errors in blindfold walking. After viewing target marks on the ground, subjects had to walk to them with their eyes closed. Accuracy broke down beyond 9 m. Directional errors (walking to left or right of target) were very small. Adapted from Thomson 1977.

no greater than when the subjects did the same task in the normal way with their eyes open and not looking directly down at their feet. Furthermore, within this distance range the addition of up to four obstacles had no effect on ability to circumvent the obstacles or to reach the final target, though performance deteriorated if major reorientations of the body were required.

However, the subjects' accuracy fell dramatically for distances beyond 9 m. Why should this be? Was it that they could not judge these longer distances accurately, or was it that they had picked up accurate information about the location of the target but could only use this information to guide themselves for a certain period of time? The latter possibility fitted the subjects' introspections. Many claimed that at first they could "see" themselves moving toward the target, but that as they got nearer to it the mental image evaporated.

The hypothesis was easily tested. Requiring subjects to take longer to reach the targets (by getting them to stand for a period with their eyes closed before walking) should decrease the distance at which accuracy breaks down. Conversely, requiring them to run should increase the threshold distance. This is indeed what happened, as figure 14.8 shows. Furthermore, the plot of accuracy versus time interval between closing the eyes and reaching the target (figure 14.9) indicates that the visual information picked up before locomoting could be used for accurate guidance for up to about 8 seconds, but not beyond that time.

A further experiment examined how the subjects were using the visual information they had acquired to get themselves to the target. Were they really using the visual information to continuously guide themselves (with their eyes closed), or were they visually determining what action was required to get to the target and then walking or running there in a "ballistic" way without further reference to the visual information, in much the same manner as they would throw a ball? To test this, subjects were required, as before, to walk with their eyes closed towards a target 10 m away, but they were stopped at random places and told to throw a block of wood the rest of the way to the target (with their eyes still closed). If they were carrying out a preformed plan of action, then disrupting the plan should put them at a loss. The finding that they could throw almost as accurately to the target as they could under normal sighted conditions indicated that they really were visually guiding themselves with their eyes closed.

This is not to imply, of course, that they were not also using nonvisual information about how far they had progressed over the ground, obtained for example through the mechanoreceptors in their joints, muscles, and skin. Indeed, it would seem that the most likely explanation of their performance is that they were continuously integrating the visual and nonvisual information into exproprioceptive information about their changing relationship to the target. In this case, the sudden drop in their accuracy after they had had their eyes closed for 8 seconds (figure 14.9) could well have been due to impoverishment of the nonvisual information, since the mechanoreceptors appear to be subject to considerable drift if not kept in tune by vision (Lee 1978).

Quo Vadamus?

How should research on visuomotor control proceed in the future? We may abstract two basic starting points from the present chapter. First, any theory of the visuomotor system (or any other system) must be founded on an adequate description of the input available to the system. Not only do most current theories of motor control fail to recognize this, they even tend to ignore vision completely (see, for example, Shik and Orlovskii 1976 and Stelmach 1976). By analyzing the optic flow field available at the eye of a moving organism we have shown some of the riches of information it contains. But this is only a start. The search for

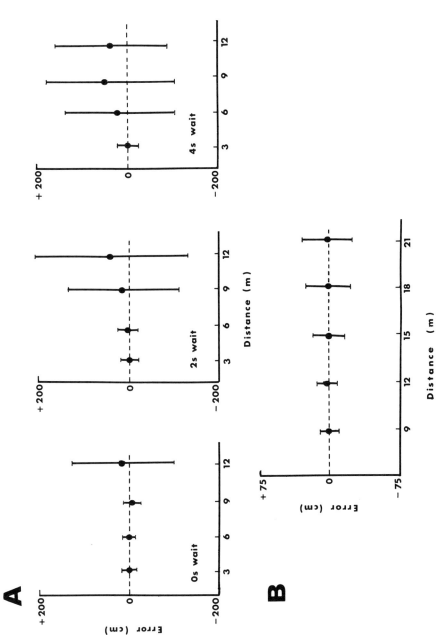

Figure 14.8 (A) Graphs demonstrating that having to wait for a period with eyes closed before walking reduced distance range over which subjects were accurate. (B) Graphs showing that shortening time to reach targets (by running) increased distance range over which subjects were accurate. Adapted from Thomson 1977.

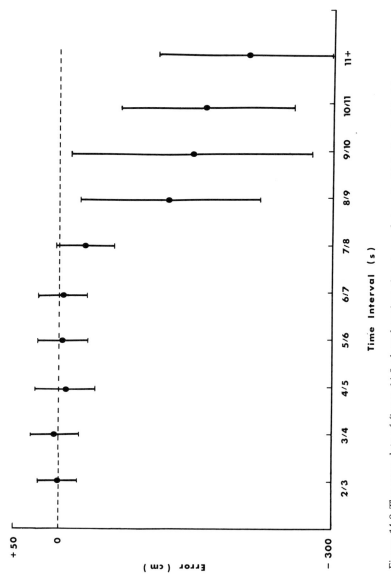

Figure 14.9 The error data of figure 14.8 plotted against time to reach targets. Note threshold at 8 sec. Accuracy appears to be a function of time subjects had eyes closed rather than of target distance. Adapted from Thomson 1977.

meaningful information in the visual stimulus must continue. This brings us to the second point. For the search to be successful we need to understand the functions vision performs in the service of activity. This means taking a much closer look at what animals and people actually do in their everyday lives. Restrictive laboratory experiments are no substitute for an ecological approach. Indeed, to be meaningful, they must be predicated on an understanding of the animal's natural behavior.

Sports and driving skills offer excellent opportunities for studying visuomotor coordination, but the opportunities have been largely neglected, probably because of the laboratory mentality of the times. The skills are demanding and the tasks clearly defined; as a bonus, there is a ready supply of skilled and trainee performers. We have shown how a simple film analysis of long jumpers can yield information about how vision is used to control locomotion. Monitoring the athletes' limb movements and the forces they exert on the ground would yield more detailed information. Such a study needs to be done.

The range of skills waiting to be studied is very wide. High jumping, pole vaulting, and hurdling, for example, involve a rather similar problem of locomotor control as long jumping; however, they also involve particular problems of their own and no doubt would reveal further features of visuomotor control. In such studies the use of vision in the timing of actions deserves particular attention. Ball games, too, offer excellent opportunities to investigate timing, but only a very few studies have been carried out (for example, Hubbard and Seng 1954; Alderson et al. 1974; Sharp and Whiting 1975). Analysis of these skills would probably greatly elaborate the use of the time-to-contact optic variable.

Driving offers a rather simple means of investigating visual guidance, since the motion of the vehicle relative to the road and obstacles and the driver's field of view and actions on the controls can be readily monitored. However, despite the importance of road safety, there have been very few studies concerned with the essential visual component of driving (for a review of some of the literature see Lee 1976). In this chapter we have delineated some of the control problems of driving and shown how vision affords information for control. The ideas now need to be tested on the road.

Finally, there are the locomotive-control problems of the blind. Most of the work on blind mobility has been directed toward developing devices to substitute for vision—a worthy but extremely difficult enterprise. However, since the great majority of visually impaired people are partially sighted, it would make sense to devote much more effort to training them to use what vision they have. To do this, or to develop devices that will substitute for vision, requires an understanding of the functions vision performs in the service of activity. It is to be hoped, therefore, that the future will see much more collaboration between students of visuomotor control and those engaged in helping the blind.

Acknowledgments

The work reported in this chapter was supported by grants from the Medical Research Council and the Science Research Council.

References

Alderson, G. J. K., D. J. Sully, and H. G. Sully. 1974. "An operational analysis of a one-handed catching task using high speed photography." *J. Mot. Behav.* 6: 217–226.

Edwards, A. S. 1946. "Body sway and vision." *J. Exp. Psychol.* 36: 526–535.

Gibson, J. J. 1950. *The Perception of the Visual World.* Boston: Houghton Mifflin.

———. 1958. "Visually controlled locomotion and visual orientation in animals." *Brit. J. Psychol.* 49: 182–194.

———. 1966. *The Senses Considered as Perceptual Systems.* Boston: Houghton Mifflin.

———. 1979. *The Ecological Approach to Visual Perception.* Boston: Houghton Mifflin.

Hubbard, A. W., and Seng, C. N. 1954. "Visual movements of batters." *Res. Q.* 25: 42–57.

Koenderink, J. J., and A. J. van Doorn. 1977. "How an ambulant observer can construct a model of the environment from the geometrical structure of the visual inflow." In *Kybernetik 1977*, G. Hauske and E. Butenandt, eds. Munich: Oldenburg.

Lee, D. N. 1974. "Visual information during locomotion." In *Perception: Essays in Honor of J. J. Gibson*, R. McLeod and H. Pick, eds. Ithaca, N.Y.: Cornell University Press.

———. 1976. "A theory of visual control of braking based on information about time to collision." *Perception* 5: 437–459.

———. 1978. "The functions of vision." In *Modes of Perceiving and Processing Information*, H. Pick and E. Salzmann, eds. Hillsdale, N.J.: Erlbaum.

Lee, D. N., and E. Aronson. 1974. "Visual proprioceptive control of standing in human infants." *Percept. Psychophys.* 15: 529–532.

Lee, D. N., and J. R. Lishman. 1975. "Visual proprioceptive control of stance." *J. Hum. Movement Stud.* 1: 87–95.

———. 1977. "Visual control of locomotion." *Scand. J. Psychol.* 18: 224–230.

Lee, D. N., J. R. Lishman, and J. Thomson. 1977. "Visual guidance in the long jump." *Athletics Coach* 11: 26–30 and 12: 17–23.

Lishman, J. R., and D. N. Lee. 1973. "The autonomy of visual kinaesthesis." *Perception* 2: 287–294.

McLean, J. R., and E. R. Hoffmann. 1973. "The effects of restricted preview on driver steering control and performance." *Hum. Factors* 15: 421–430.

Marr, D. 1978. "Early processing of visual information." In *Computer Vision Systems,* M. A. Hanson and E. Riseman, eds. New York: Academic.

Minsky, M. 1975. "A framework for representing knowledge." In *The Psychology of Computer Vision,* P. H. Winston, ed. New York: McGraw-Hill.

Nakayama, K., and J. M. Loomis. 1974. "Optical velocity patterns, velocity-sensitive neurons, and space perception: A hypothesis." *Perception* 3: 63–80.

Neisser, U. 1977. "Gibson's ecological optics: Consequences of a different stimulus description." *J. Theor. Soc. Behav.* 7: 17–28.

Sekuler, R. 1974. "Spatial vision." *Annu. Rev. Psychol.* 25: 195–232.

Sharp, R. H., and H. T. A. Whiting. 1975. "Information-processing and eye movement behaviour in a ball-catching skill." *J. Hum. Movement Stud.* 1: 124–131.

Shik, M. L., and G. N. Orlovskii. 1976. "Neurophysiology of locomotor automatism." *Physiol. Rev.* 56: 465–501.

Spurr, R. T. 1969. "Subjective aspects of braking." *Automobile Eng.* 59: 58–61.

Stelmach, G. (ed.). 1976. *Motor Control: Issues and Trends.* New York: Academic.

Thomson, J. A. 1977. "Maps, programs, and the visual control of locomotion." Ph.D. diss., Edinburgh University.

Recognition and Transfer Processes

Introduction

Richard J. W. Mansfield

One of the most challenging problems in the study of vision today is determining the mechanisms of central visual function. The chapters in this section, which detail new major advances, provide strong evidence that the solution will not prove elusive. More researchers are focusing on this problem than ever before in all of the history of science, and progress has been marked by steady advance as well as by breakthroughs. In neuroanatomy alone, the number of known neural circuits has more than doubled in the past 10 years. Yet no one approach or technique holds the key to understanding central vision; indeed, some of the most rapid advances in understanding are achieved by combined approaches to a single question. One emerging organizational concept that has been highlighted by combined anatomical, physiological, and behavioral studies is that the brain—even within a given modality—processes information in a distributive manner. The classical notion, dating back to the British associationists and reified by early twentieth-century neuroanatomists into a tripartite division of the cerebral cortex into sensory, association, and motor areas, is that a single neural representation of the visual world generated in the context of a serial processor underlies the multiple proficiencies of visually guided behavior. This concept has proved too simplistic to guide modern research, and does not explain many apparent paradoxes. One such paradox resolved by the new concept of distributive processing concerns the fact that lesions of area 17 produce an isolated and subtle deficit in the cat but one that is obvious and devastating (but, curiously, spares visually guided locomotion) in the monkey. Although the evolutionary lines leading to present-day primates and carnivores diverged over 400 million years ago, it is only recently that the neural substrates underlying the behavioral adaptations that facilitate survival in particular ecological niches are beginning to be known in cellular detail, as these chapters illustrate.

My chapter, 15, gives an overview of the primate visual system detailing the neuroanatomical wiring diagram and neuronal receptive field and response properties, all of which serve as the basis for distributive processing, and contains a specific analysis of a computational model of the spatial filtering subserving spatial acuity carried out by supragranular neurons in area 17.

Mitchell and Timney (chapter 16) address the important issue of how early visual experience can affect subsequent performance, specifically in terms of spatial visual acuity. Using an ingenious forced-choice procedure to measure acuity for grating patterns in kittens, they obtain acuity development curves that, by comparison with those obtained in single-unit electrophysiological studies, imply that the development of acuity is limited by the spatial resolving power of the X cells of the lateral geniculate nucleus. From studies of monocular deprivation and subsequent recovery they infer that binocularity of cortical neurons in area 17 recovers in two stages: one that is noncompetitive and depends only on patterned stimulation and a second that is competitive since it depends on removal of patterned stimulation from the contralateral eye. They report the remarkable discovery that the critical period can be prolonged beyond 4 months by rearing in the dark.

Berkley and Sprague (chapter 17), who used a battery of tests of visual capacity (grating acuity, vernier acuity, orientation acuity, and form discrimination for simple geometric patterns) together with sophisticated psychophysical testing, resolve the paradox of the role of area 17 in the cat. They point out that for the cat the serial-processing model of vision is considerably in conflict with recently accumulated anatomical data revealing multiple visual pathways, and conclude from their lesion studies that area 17–18 is uniquely important for vernier acuity but that the X cells, which are unique to these areas, are not necessary for discriminating simple patterns.

In chapter 18 Ungerleider and Mishkin take up the theme of multiple visual pathways in primates and focus on the functional significance of two separate pathways projecting from the striate cortex (an occipitoparietal pathway to the posterior parietal cortex and an occipitotemporal pathway to the inferotemporal cortex). Reviewing behavioral, single-unit electrophysiological, and anatomical experiments including metabolic mapping of the central visual areas with labeled 2-deoxyglucose, they conclude that a functional dichotomy exists: Inferotemporal cortex handles the recognition of objects, but parietal cortex handles the spatial localization of objects in extrapersonal space. One particularly powerful approach employed is the crossed-lesion technique, which involves behavioral testing of residual visual capacity after each of three stages of lesioning: ablation of a higher visual area, ablation of the contralateral striate cortex, and sectioning of the corpus callosum to eliminate interhemispheric transfer. For inferotemporal cortex, the first two lesions do not eliminate pattern discrimination, but callosal sectioning does. This indicates that the inferotemporal cortex is a station processing pattern information that receives considerable contralateral input—a conclusion consistent with the large receptive fields of inferotemporal units that invariably cross the midline and include the fovea. On the other hand, for the parietal cortex, the striate lesion has an effect on a spatial-localization task more devastating than the terminal callosal lesion. Thus, the posterior parietal cortex is a critical stage in processing action-oriented spatial-localization information, but receives little interhemispheric information—a conclusion consistent with the mainly unilateral, peripheral receptive fields of parietal units and the contralateral-neglect syndrome observed after unilateral parietal destruction. Ungerleider and Mishkin also resolve a puzzling paradox concerning the ineffectiveness of total prestriate lesions on inferotemporal cortical function by finding a new striate projection area deep in the superior temporal sulcus that serves as a relay to inferotemporal cortex.

In chapter 19 Dean pursues in depth the nature of the deficit in visual performance produced by inferotemporal lesions. He takes the increasingly accepted viewpoint that an explanatory description of brain-behavior relations consists of a wiring diagram of the brain with anatomical connections and electrophysiological properties and a map of these modular components and their interactions onto the functional components of behavior. Dean's point is that, despite the absence of a complete computational theory of object recognition, the lesion method can make important contributions because of the modular organization of the visual system and its consequent mode of distributive processing. Cogently reviewing his recent experimental studies, Dean concludes that the basic deficit that follows inferotemporal lesions is one of imprecise categorization (a term he analyzes logically), and that such an impairment underlies both the marked retroactive inference shown by inferotemporal animals having otherwise normal retention and their apparent lack of global stereopsis.

Levine (chapter 20), also analyzing the role of the occipitotemporal pathway and drawing on his clinical expertise, presents the case for cross-species similarity between humans and monkeys with visual agnosia. First, he points out the remarkably similar anatomical substrates: foveal occipital cortex, foveal region of the ventrolateral visual-association cortex, and inferobasal temporal neocortex. He also notes the correspondence in the nature of the visual deficits produced by lesions at each of these sites, such as the important distinction between the foveal prestriate syndrome ("perceptual") and the inferotemporal syndrome ("associative"). One problem in making such comparisons is that human patients are mainly tested using verbal naming tasks; Levine deftly circumvents this difficulty by using nonverbal forced-choice discriminations modeled on those used with monkeys. Levine shows that for both man and monkey the visual agnosia syndromes are dissociable from five other syndromes arising from lesions in adjacent structures. The only dissimilarity for visual agnosia in man and monkey arises when unilateral lesions are considered, because of the left hemisphere's dominance for language and the right hemisphere's advantage for complex familiar and unfamiliar forms (such as faces) in man and the limited channel capacity of the corpus callosum in monkey.

Since vision is an action-oriented sense, special insight into the distributive nature of visual processing can be obtained by a comparative analysis of the several subsystems influencing visual attention, eye movements, and visually guided reaching into extrapersonal space. Latto (chapter 21) makes such a valuable comparison for three regions: frontal eye fields, posterior parietal cortex, and superior colliculus. From his lesion experiments, Latto concludes that the frontal cortex is more concerned with response modulation than with direct execution (a role played by parietal cortex). On the other hand, the superior colliculus has input as well as output functions: Lesions of the deeper layers affect visual search and reaching and produce contralateral neglect, as do lesions to the more central structures; but lesions confined to the superficial layers raise brightness thresholds—a remarkable and controversial result, since confirmed by other laboratories.

One fruitful approach to the localization of processing stages, given the distributive nature of the visual system, is the analysis of interocular equivalence. Hamilton with the monkey and Berlucchi and Marzi with the cat describe how studies of interocular equivalence can be combined with physiological, environmental, and surgical manipulations to determine the functions of different visual areas. The basic rationale is that if interocular equivalence is present, then the locus of function is downstream from the place where information from the two eyes is combined. Two mechanisms, convergence (based on anatomical connections) and active transfer (a necessary inference from interhemispheric transfer of learning sets) suffice to account for the experimental findings of interocular-equivalance studies.

In primates, Hamilton points out in chapter 22, manipulations that reduce or eliminate binocular units at lower levels in the central visual system, such as environmental deprivation early in life or splitting of the optic chiasm, result in the elimination of the interocular transfer of perceptual aftereffects of motion or tilt but leave intact transfer of visual-discrimination learning. Further surgery on the monkey's corpus callosum that retains the integrity of only the anterior commissures preserves the transfer of discrimination learning (but in cat the anterior commissures alone are insufficient to mediate transfer). Similar experiments indicate that the splenium conveys perceptual information identical to that carried by the anterior commissures, but that bilateral memories are laid down only when the anterior commissures are left intact. An exciting prospect for fine-grained analysis of such experiments is emerging from anatomical experiments designed to reveal the cortical terminals and the callosal pathways of the interhemispheric connections in monkey. It should then be possible to separate commissural connections of those areas near primary visual cortex from those deeper in the processing hierarchy.

Berlucchi and Marzi (chapter 23) analyze in detail the interocular transfer of visual discrimination in the cat and conclude that the mechanism for the tasks examined is innately determined convergence, since it survives binocular deprivation, monocular deprivation, strabismus, and segregated input to the two eyes. For example, in Siamese cats with genetic absence of binocular units in area 17 and 18, transfer still occurs because of remaining binocular units in the suprasylvian gyrus (area PMLS of Palmer, Tusa, and Rosenquist) since section of the corpus callosum, which eliminates binocularity in such units, abolishes transfer. In normal cats the units in the suprasylvian gyrus have large binocular receptive fields extending across the midline, but bilateral ablation of this region abolishes transfer. Since in cats neurons with large bilateral receptive fields are not limited to the suprasylvian gyrus, Berlucchi and Marzi raise the interesting and experimentally testable possibility that the subcortical commissures linking the superior colliculi may also participate in interocular transfer, as they do in the goldfish.

The stage is now set for consideration of a difficult problem in the analysis of a distributive processing system: What is the nature of the different internal representations? Cornwell and Warren (chapter 24) scrutinize the methodology of assessing stimulus generalization by transfer and equivalence experiments, and then suggest how the processes underlying impaired visual learning resulting from posterior extrastriate ablations can be specified experimentally. Hughes (chapter 25) marshals evidence that the transcortical serial processing model generating a single internal representation is incorrect or oversimplified, and describes a series of experiments utilizing a series of dot patterns with quantitatively variable local features (high-spatial-frequency components) as well as global or contextual cues (low-spatial-frequency components) to identify the internal representations formed by striate as well as extrastriate visual cortex. In cats, ablation of area 17–18 leaves discrimination of global features intact but abolishes or impairs discrimination based on local features; conversely, it is probable that ablation of area 20 leaves discrimination based on local features intact but impairs discrimination of global features. Hughes raises the intriguing possibility that the striate cortex of the cat serves as a high-pass spatial-frequency filter to a central processor that may depend on the integrity of extrastriate visual pathways for basic figural synthesis subserved by low-pass spatial-frequency filtering. In the final chapter, Dodwell takes the viewpoint that deeper insights into the mechanisms of global analysis by the visual system require the exploitation of geometrical models of visual processing. Four such models are presented, with considerable attention paid to the Lie transformation group theory. A network filtering approach having a more detailed but restricted scope is also stressed. Dodwell argues convincingly that the concepts of vector field, manifold, and differential operator all have natural interpretations and uses in the analysis of visual processing, and presents experimental evidence from visual-deprivation studies in kittens and from apparent-motion studies in humans supporting strong predictions of the latter two models. From such approaches viable computational models of object recognition may well emerge.

15 Role of the Striate Cortex in Pattern Perception in Primates

Richard J. W. Mansfield

A central problem in the study of vision is the determination of the cellular neural mechanisms underlying behaviorally relevant functions. No single approach or technique has proved sufficient to unravel all the complexities and to provide a complete understanding of visual behavior. Precise assessment of behavioral performance with well-defined tasks in the intact animal can yield a quantitative description of visual behavior, but any inferences concerning causal neural mechanisms must remain speculative, since the central nervous system of any interesting species is too complicated for a unique mechanism to be thus identified. As Dean describes in his discussion of the inferotemporal cortex (chapter 19), the same limitation applies to behavioral studies of lesioned animals. At the other end of the spectrum of approaches, neuronal activity can be measured directly at some level of the visual pathway by fracturing the integrity of the central nervous system with anesthesia, drugs, or surgery, and with the application of techniques from the armamentarium of modern neurobiology specific cellular mechanisms can often be identified. As several chapters in this volume emphasize, visual processing in the central nervous system is distributive (that is, there are multiple visual systems), so a particular candidate mechanism may be irrelevant to the behavior of interest or may operate differently in the intact animal. However, converging evidence from different approaches to the study of the visual system can offer an effective strategy.

The primate striate cortex provides a model system in which to illustrate such a strategy. Considerable information has accumulated concerning the anatomy and physiology of the striate cortex and antecedent structures in the retino-geniculo-cortical pathway. In contrast to the afferent visual pathways of other species (such as the afferent pathways in the cat, as illustrated in figure 17.1 in the chapter by Berkley and Sprague), the striate cortex in Old World anthropoids is the major if not exclusive recipient of fine-grain visual information. For Old World anthropoids the assumption of cross-species similarity appears valid; behavioral and neural data obtained for one species can be assumed to hold for all others. To explain visual performance in general requires a reasonably complete if not exhaustive knowledge of the central nervous system, but for optimal performance in (for example) a visual-acuity task we need only focus on the relevant mechanisms that provide constraints on optimality. Strong evidence places the locus of such constraints in the striate cortex, and the quantification of striate-cortex receptive fields and visual performance in acuity tasks makes possible the development of a computational theory relating causal neural mechanisms to behavior. In this chapter a combined approach of neural and behavioral analyses and mathematical synthesis is applied to the role of the striate cortex in visual acuity. First, those critical features of the primate visual system that are pertinent to the discussion are reviewed.

Primate Anatomy and Physiology

Existing Old World anthropoids (a category that includes a range of species from monkeys such as *Macaca mulatta* to man) exhibit a number of special evolutionary adaptations in their visual systems. These species, which probably evolved from nocturnal, solitary visual predators (Polyak 1957; Cartmill 1974) into diurnal, social frugivores or omnivores, possess frontally directed eyes with a large degree of binocular overlap (150°), high central visual acuity (Blakemore 1970), well-developed trichromacy (DeValois and Jacobs 1968) and stereopsis that baffle natural camouflage and promote precise eye-hand coordination, excellent visuospatial memory (Menzell 1973), and acute perception of visual social cues (Savage-Rumbaugh et al. 1978). Subserving these behavioral adaptations are a number of neural adaptations, not all of which are known or understood.

Foveal Specialization
The retina in Old World anthropoids and in a number of other primates possesses a specialization corresponding to the center of visual fixation, the fovea. As in all vertebrate retinas, the photoreceptors are located at the inner margin of the retina, facing away from the light. However, in the primate retina, at the region corresponding to the center of gaze the proximal neurons in the inner nuclear layer (amacrine and ganglion cells) are displaced laterally, forming the foveal pit and allowing light more direct access to the transduction elements. Reduction in light scattering is also achieved by avoidance of the foveal region by the blood vessels on the vitreal surface of the retina, leaving intact only the choroidal circulation. Associated with the ciliary stocks of the photoreceptors in the fovea (cones) is the dense yellow macular pigment, which serves to absorb short-wave light and reduce the chromatic aberration inherent in the simple lens system of the eye. Flicker-photometry measurements (Wooten et al. 1980) indicate that the macular

pigment is densest in the center of the fovea, and that its distribution drops off sharply outside the fovea in parallel with the cone distribution. Microspectrophotometric measurements of cone outer segments have revealed a basis for trichromatic color vision in the differing absorption spectra with λ_{max} values of 430, 535, and 565 nm (Marks et al. 1964; Wald and Brown 1965; Bowmaker et al. 1980) described by a modified vitamin A_1 nomogram (Ebrey and Honig 1977). Although the center of the fovea is rod-free, the dense packing of the cones distorts the outer segment into a slender rodlike shape (Dowling and Boycott 1966) from their normal pyramidal shape in the periphery. Although rods and cones probably do not differ significantly in intrinsic sensitivity (Fain and Dowling 1973), they do exhibit substantially different saturation characteristics (Boynton and Whitten 1970). The convergence necessitated by the existence of 200 million photoreceptors but only 1 million optic-nerve fibers is not uniform, but is graded across the retina, with the central region receiving a disproportionate share as reflected in the magnification factor in the retinotopic map of the dorsal lateral geniculate (Malpeli and Baker 1975). In contrast to other vertebrate species, Old World anthropoids have simple retinal circuitry containing relatively few classes of amacrine cells; correspondingly, the retinal ganglion cells fall into relatively few classes. According to terminology developed for the cat retina, ganglion cells have been classified into three classes termed X, Y, and (when specified) W (DeMonasterio and Gouras 1975; Schiller and Malpeli 1977; DeMonasterio 1978a). X cells have small receptive fields, summate linearly across these fields, often have a sustained response to standing contrast, and are found predominantly in the foveal region. Y cells have large receptive fields, summate nonlinearly across these fields, mainly exhibit a transient response to standing contrast, and are found predominantly in the periphery. The central fovea (1°) yields 90% X cells and 10% Y cells, but in the near periphery (11°–20°) only 45%–70% X cells and 30%–50% Y cells (Schiller and Malpeli 1977; DeMonasterio 1978a). W cells—clearly not a homogeneous class—lack well-defined center-surround organization, are distributed across the retina but mainly outside the foveal region, and project directly to the superior colliculus rather than through the geniculate, with some exceptions (Wiesel and Hubel 1966; Dreher et al. 1976; DeMonasterio 1978b). In the primates (in contrast to other mammals), W cells provide a complete representation of only the contralateral half of the visual field. The actual proportion of the retinal ganglion-cell distribution consisting of W cells is difficult to estimate, since the recording techniques tend to sample from the larger or more responsive X and Y cells. Rodieck (1979) estimated that 50% of cat retinal ganglion cells might be W cells, so in monkeys the proportion may be larger than the 10% encountered experimentally. The retinal segregation into separate functional classes undoubtedly has adaptive significance, since it is retained at higher levels in the visual system.

Functional Segregation in the Dorsal Lateral Geniculate Nucleus

The axons of the retinal ganglion cells form the optic nerve, which decussates at the optic chiasm before reaching its targets in the dorsal region of the lateral geniculate of the thalamus and the superior colliculus in the brain stem. Each hemiretina has a geniculocortical target in a separate hemisphere except for a 1° vertical strip along the midline and a few ganglion cells around the rim of the fovea (Stone et al. 1973; Bunt et al. 1977). The primate dorsal lateral geniculate nucleus (dLGN) is a six-layered structure receiving in precise retinotopic register the inputs from the two retinas and maintaining them in separate laminae with a discrete functional organization (Dreher et al. 1976; Schiller and Malpeli 1978). The four parvocellular layers receive X-cell input and have cells with concentric color-opponent receptive field organization (R+G− or G+R−); the blue-selective "on" cells are found predominantly in the ventral pairs of parvocellular layers. The two magnocellular layers receive Y-cell input and have cells that are concentrically organized, but as might be expected they have spectrally broad-band or achromatic receptive fields.

Cortical Organization

In contrast to other mammalian species (see the chapter by Berkley and Sprague for a discussion of the multiple and parallel input pathways into the cortex of the cat), in catarrhine monkeys, apes, and man the dLGN forms the major if not exclusive input into the cortex, and that entry is confined to the striate cortex (area 17 of Brodman), as shown in figure 15.1. Moreover, the functional segregation present in the geniculate is maintained (though less strictly) in this initial cortical stage. The axons of the X cells of the parvocellular layers terminate mainly in layer 4A and layer 4Cβ of this six-layered koniocortical region (Hubel and Wiesel 1972). The field of termination is the spiny stellate neurons that project to layer 3B and 5A as well as within 4C (Lund 1973; Lund and Boothe 1975). The axons of the Y cells emerging from the magnocellular layers project mainly to layer 4Cα (Hubel and Wiesel 1972); however, they also bifurcate and terminate in the superior colliculus (Schiller et al. 1979). The field of termination is the spiny stellate neurons that contribute to laminae 4B and 5A but also distribute widely within 4Cα (Lund 1973; Lund and Boothe 1975). Both anatomical evidence based on Golgi staining (Lund and Boothe 1975) and electrophysiological evidence based on selective anesthetization of a specific geniculate lamina (Schiller et al. 1979) suggest that the input from the parvocellular and magnocellular geniculate neurons remains largely segregated, with X cells the dominant influence for cortical neurons with cell somas in the upper or supragranular layers and Y cells the major influence for the infragranular layers. Such a distinction has a definite functional significance, since the efferent projections of the two regions are quite different.

Area 17 or V1 projects and receives reciprocal projections from both cortical and subcortical targets, with the cortical targets receiving efferents predominantly from the supragranular layers and the subcortical targets receiving efferents from the infragranular layers (Lund et al. 1975). The pyramidal cells in the upper regions of layer 6 project back to the parvocellular layers of the dLGN, but the pyramidal cells in the lower portion of layer 6 project back to the magnocellular layers (Marrocco et al. 1979). The pyramidal cells in layer 5 project retinotopically to both superior colliculus and pulvinar, but receive a reciprocal projection to layer 2 only from pulvinar which also projects widely to the rest of the visual field (Diamond 1979). However, the projection from the pulvinar to the more rostral regions of cortex does not contribute to the spatial selectivity of the cells in these regions (Gross and Mishkin 1977). The pyramidal and stellate cells in layer 4A (which exhibit strong directional selectivity [Dow 1974]), as well as the largest pyramidal cells in the upper portion of layer 6, project to MT, which in turn sends a reciprocal projection to layers 4A and 6. This circuit may contribute to visually guided movement rather than object vision, since MT sends a heavy projection to the pontine nucleus, which then relays the information to the cerebellum (Glickstein et al. 1980). The pyramidal cells in the supragranular layers project transcortically to V2 and V3 both ipsilaterally and (near the V1/V2 border) contralaterally; these regions reciprocate with a projection to layer 1 (Spatz and Tigges 1972; Wong-Riley 1979). In addition there is a recently discovered third cortical projection site in the most caudal portion of the interparietal sulcus (see the chapter by Ungerleider and Mishkin for a review of the evidence for the site). The corticocortical pathways from the supragranular layers conveying the transformed input from the parvocellular X cells are undoubtedly the critical ones for pattern vision.

Since in primates but not other mammals V1 is the dominant if not exclusive entry site of visual information into the cortex, the supragranular layers of V1, as indicated in figure 15.1, are the gateway to an arena of rich cortical integration. The terminal target regions in the visual projection field have been implicated in different classes of visual behavior: MT could play a role in visually guided movement (Glickstein et al. 1980). Area TE of von Bonin and Bailey, on the inferior convexity of the temporal lobe, appears to mediate the acquisition of complex visual discriminations (Gross and Mishkin 1977; see Dean, this volume). Area 8 in the frontal lobe serves as a second control center for saccadic eye movements in addition to the superior colliculus (Schiller et al. 1979; see Latto, this volume). Area 7 in the parietal lobe appears to direct visual attention to objects in extrapersonal space (Lynch et al. 1977; Mountcastle 1978; see the chapter by Ungerleider and Mishkin for evidence from lesion studies implicating this region in object localization). However, these more rostral regions of the visual system may be irrelevant or at least redundant for certain basic pattern-detection tasks, since

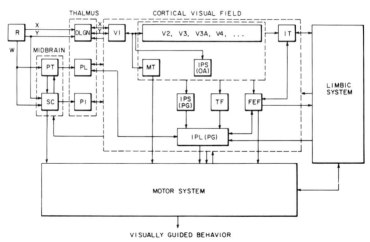

Figure 15.1 Anatomical organization of the primate visual system. R: retina. X, Y, W: axons of three classes of ganglion cells, or (at the geniculate) projections to and from parvocellular and magnocellular layers; see part B. dLGN: dorsal lateral geniculate nucleus. PT: pretectum. SC: superior colliculus. PL: lateral pulvinar. PI: inferior pulvinar. V1: striate cortex. V2, etc.: prestriate visual areas (See Van Essen 1979). IT: inferotemporal cortex (cytoarchitectonic areas TEO and TE of von Bonin and Bailey). FEF: frontal eye fields in the region of the arcuate sulcus consisting of cytoarchitectonic areas 8A, 8B, and posterior 9 in the terminology of Brodmann. MT: a region at the posterior end of the temporal sulcus described as the visual area of the superior temporal sulcus by Zeki (1971) but probably homologous to the middle temporal area in the owl monkey. IPS (OA): a region delineated by Ungerleider and Mishkin (this volume) using anterograde transport from V1 of radioactive amino acids in area OA (von Bonin and Bailey) at the depth of the posterior extent of the intraparietal sulcus. IPS(PG): a region delineated by Mesulam et al. (1977) using retrograde transport of horseradish peroxidase from PG, located in the intraparietal sulcus anterior to IPS(OA). TF: a region described by von Bonin and Bailey on the medial surface of the temporal lobe adjacent to TE and OA. IPL(PG): the region PG described by von Bonin and Bailey on the inferior parietal lobule, which corresponds approximately to the posterior half of area 7 of Brodmann. (A) Schematic representation of the main structures, showing the principal pathways, with arrows indicating direction of flow of information and control. For clarity of presentation only ipsilateral connections have been shown and the detailed structures of the limbic and motor systems and of the prestriate visual cortical field have been omitted. For purposes of this diagram area 6aβ (premotor cortex), which receives a projection from area 9p in FEF, is included in the motor system along with the pons (which receives input from IPL(PG), MT, and intervening regions [Glickstein et al. 1980]) and the interstitial nucleus of Cajal (which receives input from the superior colliculus). Major sources of evidence for the newly described projections in the cortical visual field: Projections to IPL(PG) are based on Mesulam et al. 1977 and Divac et al. 1977; projections to and from limbic system are based on Nauta and Domesick 1980; projection from motor system to visual cortical field is based on Johnston et al. 1979; projections from FEF (areas 8 and 9p) are based on Kunzle and Akert 1977 and Kunzle 1978; projection to IPS(OA) is based on Ungerleider and Mishkin (this volume); connections to and from V1 are based on Lund et al. 1975. Dashed rectangles indicate structures whose substructures are not indicated or not known in detail, such as prestriate visual areas. Arrows to dashed rectangles indicate projections to more than one substructure; arrows from such rectangles indicate projections from more than one substructure. (B) Schematic representation of the main structures in the retinocortical pathways, with arrows indicating direction of information flow. For clarity of presentation only the connections from one hemiretina and to one hemisphere are shown. Brackets on left side of lateral geniculate nucleus indicate feedback distributed over parvocellular (upper bracket) or magnocellular (lower bracket) layers. Major sources of evidence: for the existence and distribution of X, Y, and W type retinal ganglion cells, DeMonasterio 1978a, b; Schiller and Malpeli 1977; for the projections from retina to superior colliculus and pretectum, (Schiller and Malpeli 1977; Bunt et al. 1975; for the connections from retina to dorsal lateral geniculate nucleus, Dreher et al. 1976; Schiller and Malpeli 1978; for the projections from geniculate to striate cortex, Hubel and Wiesel 1972; Rezak and Benevento 1979; for the projections from striate cortex to cortical and subcortical structures, Lund et al. 1975; Van Essen 1979; Maunsell et al. 1979; Ogren and Hendrickson 1976; Rezak and Benevento 1979; for feedback connections to striate cortex, Van Essen 1979; Ogren and Hendrickson 1977; Wong-Riley 1979); for connections among superior colliculus, pretectum, and their projections, Benevento and Fallon 1975; Rezak and Benevento 1977; Berman 1977 (cat); Benevento 1980; for feedback from V2 and MT to geniculate, Lin and Kaas 1977 (owl monkey). (C) Schematic diagram of anatomically segregated visual processing in striate cortex (adapted from Mansfield 1979).

all regions of the cerebral cortex project to the basal ganglia, the major motor-control system. In particular, visual acuity is unimpaired by inferotemporal lesions (Cowey and Weiskrantz 1967; Blake et al. 1977; Weiskrantz, pers. comm.) or by parietal-lobe ablation (Ungerleider and Mishkin, this volume); this suggests that some relatively simple detection mechanism applied to the output of area V1 may be sufficient for high-level performance in a visual-acuity task. It is of interest, then, to consider how visual information is processed in area V1 and represented in the population profile of its efferents.

Cortical Processing and Representation in Area V1

In order to analyze the contribution of area V1 it is necessary first to consider visual information that remains essentially invariant so as to define more clearly the intrinsic processes.

Invariants of Retinal and Geniculate Origin

The receptive fields of individual neurons recorded in the supragranular layers of striate cortex preserve invariant or only slightly transformed several properties encoded in more distal portions of the visual pathway. The minimum-discharge receptive fields are discrete, with excitatory cores only slightly greater in size than those of the dLGN (Schiller et al. 1976a) resulting in approximately the same magnification factor as in the geniculate (Hubel and Wiesel 1974) but with less precise retinotopic organization. Light and dark adaptation have their major locus of control in the retina, so for example the inverse relation between discharge latency and the luminance of the background or target is determined up to an additive constant by retinal mechanisms and in the dark-adapted case more likely by photoreceptor mechanisms (Mansfield 1976b; Mansfield and Daugman 1978). Object brightness, which appears to be encoded by relative impulse density, is established for isolated targets at the retinal-ganglion-cell level in a relation that is initially linear up to one log unit above the background level but nonlinear at higher levels (Mansfield 1975; 1976a,b). In the dark-adapted case, simple reaction time and the latency of cortical evoked potentials decrease as the inverse cube root of target luminance, approaching asymptotically different but constant values (Mansfield 1976b). In the dark-adapted case, retinal ganglion cells, lateral geniculate neurons, and supragranular striate cells (although only 10% of the population) increase their average discharge rate as the cube root of target luminance, a relation also found with numerical magnitude estimates in psychophysical tasks. However, the brightness or lightness of targets in more complex scenes is influenced by large-range spatial interactions (see, for example, Land and McCann 1971) that undoubtedly reflect central convergence. Such global interactions of lightness may underlie object color perception, but local spectral and achromatic differences are present more distally. The majority of neurons in the supragranular layers of the foveal projection region maintain a differential sensitivity to wavelength (Poggio et al. 1975; Crawford et al. 1979), but in the representations of the more peripheral regions the proportion drops sharply relative to that of the geniculate, indicating some convergence. In addition, the constant-criterion thresholds of the monkey's spectrally broad-band cells in both the foveal and parafoveal projection regions reflect the same balanced input from the three classes of cone photoreceptors exhibited by behavioral measures of luminance threshold in the whole animal (Poggio et al. 1975). On the other hand, four attributes of cortical receptive fields represent intrinsic contributions of striate cortex.

Intrinsic Cortical Transformations

There are four major transformations carried out on the incoming visual information by the synaptic circuitry of area V1 and reflected in the discharge patterns of efferent neurons: binocular convergence, spatial-frequency filtering, pattern-specific adaptation, and orientation and direction selectivity. For each visual hemifield area V1 represents the first site of binocular convergence. With increasing distance from layer 4 the proportion of binocularly driven neurons increases, in a manner suggesting independence of connectivity for orientation selectivity and ocular dominance (Poggio 1972). Although the degree of orientation-preference disparity (the amount to which the axis of orientation differs in the two eyes) is small in primates—it is estimated to be 4° in the macaque monkey, compared with 9° in the cat (Mansfield et al. 1980)—there is also a binocular disparity in the retinal locus of receptive fields that probably serves a role in local stereopsis. Poggio and Fischer (1977) found that in the foveal striate projection region, where receptive fields were small, the majority of neurons were sensitive to location of stimuli in depth, with the maximum discharge or shift in discharge occurring within ±0.4° of disparity, but the total range of binocular interaction could extend to more than ±1° of disparity. A second transformation that takes place is that the spatial-frequency filtering present in the geniculate neurons becomes sharper (Poggio et al. 1977; DeValois et al. 1978). The third transformation is the adaptation or habituation of neural discharges to specific patterned stimuli (Mansfield et al. 1978), a feature that is not observed in the lateral geniculate (Wiesel and Hubel 1966; Schiller and Malpeli 1978; De Valois et al. 1978) but whose appearance at the cortical level reflects the initiation of a temporal filtering mechanism for visual salience and novelty. A fourth and major transformation is that instead of isotropic, concentrically organized receptive fields, such as those found in the dLGN, the majority of striate neurons in the supragranular layers exhibit an orientation preference and are directionally selective (Hubel and Wiesel 1968; Poggio 1972; Dow and Gouras 1973; Dow 1974; Poggio et al. 1975; Schiller et al. 1976a,b; Mansfield et al. 1980). From recent experiments quantitative estimates can be made of the population response characteristics in the efferent projection from area V1.

Quantitative Measurements of Receptive-Field Properties

In order to predict behavioral performance it is necessary to know in quantitative detail the excitability profiles of the receptive fields of the efferent neurons in the supragranular layer and to understand, at least to a first approximation, how the neural information is utilized in a particular behavioral task.

For the first of these problems the main obstacle is that of sample bias when recording from only a small proportion of neurons in the layers. One source of sample bias arises from the columnar organization for the attributes of orientation selectivity and ocular dominance (Hubel and Wiesel 1968). The invariance of particular receptive-field properties with cortical depth is a general principle of neocortical organization (Mountcastle 1957, 1978) probably reflecting the conservation of genetic information required for synaptic organization of the cortex using modular subunits. From the experimenter's point of view, however, microelectrode penetrations normal or nearly normal to the pial surface encounter neurons whose receptive-field properties are not independent; oblique penetrations encounter similar problems of statistical dependence when the attributes like orientation selectivity exhibit sequential regularity (Hubel and Wiesel 1974). One strategy for obtaining statistical independence is to count as a sample the receptive field of a single neuron (for example, the first neuron encountered that could be studied) in a given penetration and to separate penetrations by at least some fixed distance (at a separation of 500 microns the mean serial correlation coefficient for orientation preference has dropped to chance levels) (Mansfield 1974; Mansfield et al. 1980). A second potential source of sample bias with microelectrode penetrations is that the neurons with large cell somas will be more heavily weighted since their action potentials will be more easily isolated (Humphrey and Corrie 1978). However, since the neurons of interest are the pyramidal cells in layers 2 and 3 rather than the stellate cells, such a bias will aid in selecting the population of interest. A third potential source of bias is the stochastic nature of the neuronal discharge; however, multiple and randomized stimulus presentation can provide objective, stable measures (Stryker and Sherk 1975; Schiller et al. 1976a; Mansfield et al. 1980). A fourth potential source of bias is that a particular choice of measure for neuronal discharge may not be relevant for further processing by the more rostral portions of the visual system; however, integrated discharge (pulse number) for brief stimulus presentations is a probable candidate code.

Excitability Profiles

The majority of the receptive fields in the supragranular layers of primate area V1 have discharge characteristics resembling those of X cells in the geniculate but have the spatial characteristics of complex cells (a single activating region) (Hubel and Wiesel 1968; Poggio et al. 1975; Schiller et al. 1976a; Poggio et al. 1977; Mansfield et al. 1980). No receptive-field type, such as simple cells, is distinguishable on the basis of receptive-field size or breadth of orientation tuning (Schiller et al. 1976d; Mansfield et al. 1980). At the descriptive level a central excitatory zone with flanking silent inhibition of cortical origin is a first approximation.

Orientation Tuning

As with orientation tuning curves in the cat when they have been examined in quantitative detail (Henry et al. 1974), a symmetric bell-shaped curve provides an excellent fit to primate data, as illustrated in figure 15.2 for a binocular cell. The shape of the orientation tuning curve does not change systematically with orientation preference or with retinal eccentricity, at least between the central foveal projection (0°–2°) (Mansfield et al. 1980) and the parafoveal region (3°–6°) (Schiller et al. 1976b; Mansfield et al. 1980) where it has been examined. However, as illustrated in figure 15.3, the distribution of tuning bandwidths (half-width at half-height) is broad but skewed toward the narrowly tuned neurons, with a median near 24°, a value that is consistent across electrophysiological experiments (Schiller et al. 1976b; Mansfield et al. 1980). One striking difference illustrated in figure 15.4 between foveal and parafoveal projection regions is that in the parafoveal projection region all orientations are represented equally (Hubel and Wiesel 1968; Finlay et al. 1976) but in the foveal projection region the horizontally and vertically selective receptive fields are overrepresented (Mansfield 1974; Mansfield and Ronner 1978; Mansfield et al. 1980). The orientation anisotropy is probably innate and is not confined to primates (see, for example, the following cat studies: Pettigrew et al. 1968; Levanthal and Hirsch 1977; Kennedy and Orban 1979; Orban and Vandenbussche 1979).

Size or Spatial-Frequency Tuning

The size of receptive fields in the supragranular layers changes with two gradients: It increases with retinal eccentricity (Hubel and Wiesel 1974) and with distance from layer 4, the main input layer. Thus, at each location in the retinotopic projection there is a distribution of receptive field sizes. One method of quantifying the degree of size tuning of the cell is to measure the response of the cell to grating targets. When displayed along a log spatial-frequency axis, the response curves assume a shape that is approximately 1–2 octaves wide at half-maximal response (Schiller et al. 1976c; Poggio et al. 1977) or at half-maximal sensitivity for

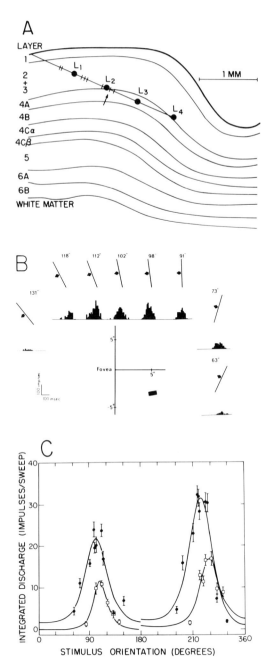

Figure 15.2 Orientation selectivity in binocular bidirectional striate-cortex neuron in monkey. (A) Anatomical reconstruction of microelectrode track (parasaggital oblique penetration T10). On the left side of the diagram the nomenclature for each distinguishable lamina of the striate cortex is indicated. Solid black circles represent the locations of four marking electrolytic lesions. The short line segments perpendicular to solid line marking track represent recording sites of isolated single units. Solid black arrow indicates recording site of unit 8T107, described more fully in parts B and C. (B) Receptive-field map and response profile (unit 8T107). Solid black rectangle represents minimal-discharge receptive field of neuron in right visual hemifield. The receptive field measured

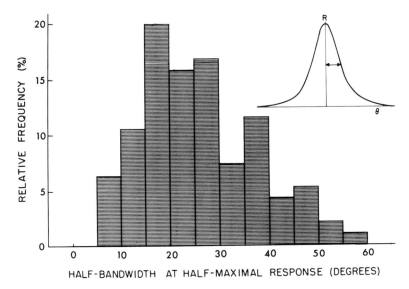

Figure 15.3 Frequency distribution of orientation selectivity in striate-cortex neurons. Orientation selectivity was measured as the half-bandwidth at half-maximal response on an orientation tuning curve, as shown in the insert. The measurements were obtained from 96 independently sampled neurons. Source: Mansfield et al. 1980.

a constant criterion response (DeValois et al. 1978). The available samples of neurons from the supragranular layers for which optimal spatial frequency has been measured are relatively small, but an approximation to the distribution is shown in figure 15.5. An additional factor is the degree to which neurons with different optimal spatial frequencies respond at their optimal frequency to contrast differences; however, there are essentially no quantitative data that can be brought to bear on this question.

Pattern-Specific Adaptation
A common observation is that neurons in the supragranular layers habituate easily, and so are difficult to activate repeatedly (Wurtz 1969). The decrement in response is exponential with time and depends upon the cell's response; that is, nonoptimal stimulation does not produce

---

$\frac{1}{2}°$ by 1° and was located approximately 6° from the fovea. Arranged around the receptive field map are peristimulus-discharge histograms generated by a long slit subtending 0.1° in width moved through the discharge zone at a velocity of 5.6°/sec at the orientation indicated. Each peristimulus histogram is based on 15 stimulus presentations. (C) Orientation tuning curves (unit 8T107) showing integrated discharge as a function of target-slit orientation. Each data point is the average calculated from 15 stimulus presentations. Dots represent responses from the receptive field in right eye; open squares represent responses from receptive field in left eye. Error bars represent ± 1 standard error of mean integrated discharge. Solid lines fitted to data points represent theoretical functions for orientation tuning derived from a simple model of orientation determination. (Adapted from Mansfield et al. 1980.)

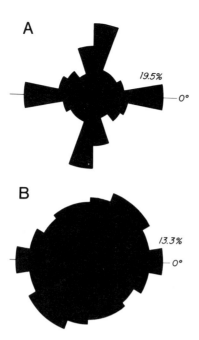

A

19.5%

0°

B

13.3%

0°

Figure 15.4 Distribution of axes of orientation preference for striate-cortex neurons. Each neuron was sampled from a single penetration with multiple penetrations separated by at least 500 μm across the cortical surface. (A) Polar histogram of distribution for receptive fields of independently sampled neurons with foveal receptive fields ($N = 82$). The deviation from uniformity is statistically significant ($\chi^2 = 15.94$, d.f. $= 8$; $P < 0.05$). (B) Polar histogram of a distribution of independently sampled neurons with parafoveal receptive fields ($N = 128$). The χ^2 test for departure from a uniform distribution was not signifi cant at the 5% level ($\chi^2 = 1.66$, d.f. $= 8$; $P > 0.5$). Source: Mansfield et al. 1980.

habituation (Mansfield et al. 1979). The exponential time constants vary across cells, and are smaller for adaptation than for recovery, in which case the average value is on the order of 10 seconds. Direct synaptic activation of the supragranular pyramidal cells produces habituation *in vitro* (Mansfield and Simmons 1979) that follows a similar time scale for recovery.

Quantitative Model of Population Response Profile
In order to develop a computational theory of the function of the supragranular layers of area V1 based on the population response profile, we need to consider the excitation profile of a typical cortical receptive field in the supragranular layers, the local and global distributions of receptive-field parameters, and a candidate algorithm for combining the neural output from area V1 to produce a response.

The experimental studies reviewed above indicate that the cortical receptive fields in the supragranular layers are shape-invariant with changes in preferred orientation and scale multiplicatively in the frequency domain. As a simplified but instructive approach to the question of the construction of a cortical receptive field, we can consider

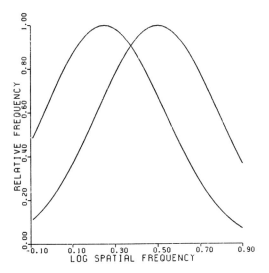

Figure 15.5 Frequency distribution of optimal spatial frequency of striate-cortex neurons in foveal and parafoveal projection regions. Left curve represents parafoveal striate-cortex neurons from projection region 4°–6° from fovea; right curve represents foveal striate-cortex neurons from projection region covering central 2°.

how the receptive fields at successive levels in the primate visual system are generated. For convenience, the on-center retinal-ganglion X cell will be used as an example. The spatial-domain weighting function describing the contributions of the photoreceptor input to the cell can be reasonably well approximated by a circularly symmetric Gaussian distribution, the expected limiting distribution as the number of photoreceptors becomes large assuming isotropic and globally random connectivity. However, lateral inhibition is also contributed by the retinal circuitry and can be represented by the simplest isotropic form of spatial differentiation, the two-dimensional Laplacian operator, ∇^2, defined as

$$\nabla^2 = \frac{\partial^2}{\partial x^2} + \frac{\partial^2}{\partial y^2}, \tag{1}$$

or more simply as

$$\frac{\partial^2}{\partial r^2} + \frac{1}{r}\frac{\partial}{\partial r}$$

for circularly symmetric spatial functions in polar coordinates. The resulting excitation profile $E(r)$ for the retinal ganglion cell is given by the equation

$$\nabla^2 G(r) = 4\left(\frac{r^2}{a^4} - \frac{1}{a^2}\right)e^{-(r/a)^2} \tag{2}$$

representing central excitation with surround inhibition as depicted in figure 15.6. The profile is essentially similar to the empirically derived difference of Gaussians used by Enroth-Cugell and Robson (1966) to

Striate Cortex in Pattern Reception

describe the structure of cat retinal ganglion cells. At the next level in the parvocellular layers of the lateral geniculate, the principal cells receive excitatory input from probably a very small number of retinal ganglion cells of the same type (on-center). No inhibition is directly transmitted; rather it is supplied by inhibitory interneurons. The profiles would remain essentially isotropic and unchanged in shape except for stronger inhibition, since the lateral inhibition is applied on the afferent side rather than on the efferent side. At the cortical level the geniculate input is probably exclusively excitatory, and supragranular pyramidal cells probably receive direct monosynaptic contact from very few geniculate axons. Inhibition is supplied by disynaptic or polysynaptic circuits. Since the intracortical inhibition is polysynaptic, the inhibitory subfields are larger than the excitatory central zone and slightly displaced. The inhibitory-side lobes, although silent, could generate orientation selectivity, and the width of the excitatory central zone could determine the optimal size or spatial-frequency tuning for the cell. A second operation of lateral inhibition is suggested by the recurrent collaterals of pyramidal cells found elsewhere in the neocortex (Marin-Padilla 1969, 1970) and by the difficulty in generating responses by electrical stimulation in the optic radiation below the cortex. Antidromic action potentials would serve to silence cellular activity by recurrent inhibition. Such pericolumnar inhibition, as Mountcastle (1978) has described it, would by the sequential regularity of the orientation columns be directed orthogonal to the preferred orientation. Formally, in terms of the simple theoretical receptive-field profile described by equation 2, the Laplacian operator takes the form ∇_x^2 if for convenience we let the y axis be the axis of preferred orientation. When the operator is applied, the equation for the receptive-field profile becomes

$$\nabla_x^2 \nabla^2 G = 4 \left(\frac{4r^4(\cos\theta)^2}{a^8} - \frac{12r^2(\cos\theta)^2}{a^6} - \frac{2r^2}{a^6} + \frac{4}{a^4} \right) e^{-(r/a)^2}. \tag{3}$$

The receptive-field profile corresponding to equation 3 is shown in figure 15.6. The pericolumnar inhibition would have the effect of sharpening orientation tuning.

At a given point in the cortex, a typical receptive field can be described by an excitation profile such as that given by equation 3. Such a receptive field then acts as a local spatial filter with a particular center frequency and frequency bandwidth and a preferred orientation and orientation bandwidth. Locally, the center frequency is distributed as illustrated in figure 15.5, and the frequency bandwidths at half-height are distributed in a small range between 1 and 2 octaves. Depending on location, the orientation preferences have distributions similar to those shown in figure 15.4, and the orientation bandwidths have a distribution similar to that in figure 15.3. More globally, the distributions of orientation preferences and center frequencies change with retinal eccentricity. An important question concerns the degree to which orien-

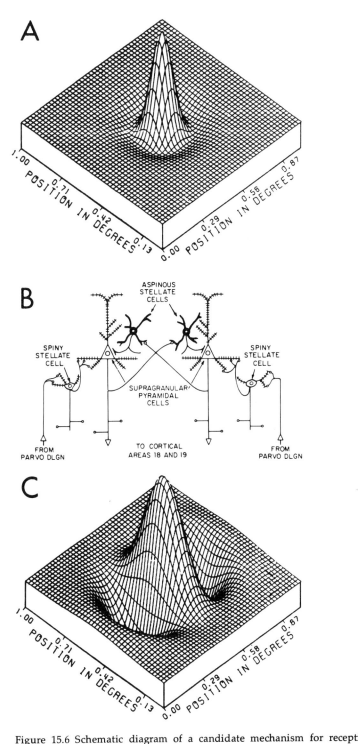

Figure 15.6 Schematic diagram of a candidate mechanism for receptive-field organization of supragranular neurons. (A) Excitatory profile generated by retinal lateral inhibition. (B) Candidate neural network in supragranular region of striate cortex exhibiting recurrent lateral inhibition. (C) Excitatory profile generated by operating on profile in A by neural network in B.

tation tuning and spatial-frequency selectivity are interrelated. These two aspects appear to be functionally separable, since orientation-tuning bandwidth for supragranular neurons (Mansfield et al. 1980) or presumed supragranular neurons (Schiller et al. 1976b) does not appear to change with retinal eccentricity, at least up to 20°. On the other hand, since receptive-field size and center frequency are correlated, band-width is roughly proportional to center frequency (1–2 octaves), and center frequency shifts with retinal eccentricity (Schiller et al. 1976c), there is a change in absolute bandwidth with retinal eccentricity. A deeper question is how the population response profile information is utilized for a particular task.

The ensemble of supragranular pyramidal neurons forms an input array to the rest of the visual system that carries the necessary and suffi-cient information for optimal acuity performance. If $I(x,y)$ denotes the luminance distribution in the retinal image, then the neural response of a particular class of orientation-selective neurons at the corresponding retinotopic point, $N_{\theta,f}(x,y)$, is given by

$$N_{\theta,f}(x,y) = S[I(x,y)],\qquad(4)$$

where S is the visual-system operator. The operator can be considered linear if the input to the neurons is exclusively from X cells, since by definition they respond linearly. The luminance distribution can be written in terms of the Dirac δ function, $\delta(x - \xi, y - \eta)$, as

$$I(x,y) = \int_{-\infty}^{\infty}\int_{-\infty}^{\infty} I(\xi,\eta)\delta(x - \xi, y - \eta)d\xi d\eta.\qquad(5)$$

Then, substituting this definition of $I(x,y)$ in equation 4 and using the linearity of the S operator since the input to the supragranular layers is dominated by X-cell input, we obtain

$$N_{\theta,f}(x,y) = \int_{-\infty}^{\infty}\int_{-\infty}^{\infty} I(\xi,\eta)S[\delta(x - \xi, y - \eta)]d\xi d\eta.\qquad(6)$$

But the operation of S on the δ functions is simply to produce spatial filter functions $h(x,y;\xi,\eta,\theta)$ corresponding to the excitatory profiles pre-viously discussed, so that

$$N_{\theta,f}(x,y) = \int_{-\infty}^{\infty}\int_{-\infty}^{\infty} I(\xi,\eta)h(x - \xi, y - \eta, \xi, \eta, \theta)d\xi d\eta.\qquad(7)$$

That is, the response of the neurons, the integrated discharge, is the convolution of the luminance distribution with the excitatory profile of the receptive field. Such detailed considerations based on the present limited knowledge of synaptic circuitry in the central visual system cannot yet be extended with confidence beyond the striate cortex. How then can visual behavior be dealt with in a predictive quantitative manner? By choosing a visual task that requires optimal performance, such as visual acutiy, and by considering what an optimal information-processing system can do given the output array from V1. As will be seen, optimal or near-optimal detection can be carried out by at least two neurally realizable mechanisms shown schematically in figure 15.7.

In discussing the detection process involved in visual acuity we need to consider the issues of the independence of noise in different neural channels, the coherence or incoherence of the detection, and the nature of the optimal detection mechanism. An important theorem of stochastic processes (Cinlar 1972) states that the sum of point processes that are Poissonian is itself Poissonian. For this reason, the simplest random process—a Poisson process—is a good first approximation to the stochastic nature of impulse trains for cortical neurons when suitably corrected for refractoriness, which is intrinsic to neural activity (Teich and McGill 1976; Teich et al. 1978). On the other hand, the very multiplicity of sources guarantees the statistical independence of the noise in different channels; the noise is uncorrelated. On the issue of coherence of the detection process, the question is whether phase information between channels tuned to the same spatial frequency but at different retinotopic locations is preserved. With a coherent detector spatial-frequency bandwidth and spatial summation area are inversely related, but with an incoherent detector the two are independent. The studies of receptive fields in the more rostral portions of the visual system indicate that the fields are large but in general do not show correspondingly specific spatial-frequency tuning; thus, incoherent detection is the most likely form. In order to detect a pattern of interest presented to the retina, the central mechanisms can be considered to monitor the output of M orthogonal channels and to perform a decision process that is independent of spatial phase except locally. Optimal detection arises from the application of the Neyman-Pearson criterion, which maximizes the probability of detection with the probability of a false alarm constrained to a particular value as described by the theory of signal detectability (Peterson et al. 1954; Green and Swets 1966). The general decision rule can be stated in terms of likelihood ratios, but for a Poisson noise-counting distribution the decision rule reduces to a simple comparison of a pulse count to a criterion number of pulses, a threshold process readily realizable in terms of known neural mechanisms. In the case of small signals, the optimal detector is closely approximated by a multiband detector that simply compares the sum of the outputs of the M channels to a criterion level (Cohn 1978). The optimal detector is also closely approximated by another suboptimal detector in which detection occurs if the activity in at least one channel exceeds a threshold (Nolte and Jaarsma 1967; Green and Weber 1980). Such a probability-summation detector is readily realizable as a neural logical "or" gate, and is particularly convenient for computation. The detection process in

A

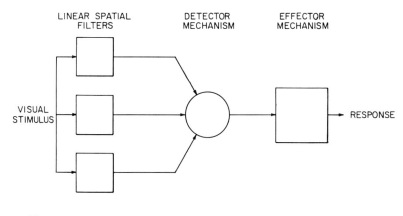

B

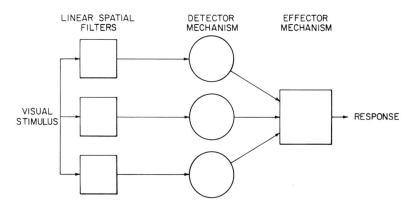

Figure 15.7 Schematic diagram of two quasioptimal neurally realizable pattern-detection mechanisms. (A) Pooling detector. In this case the luminance corresponding to the visual stimulus distribution is processed by linear spatial filters and the responses of the filters are summed or pooled in a detector, which responds if the combined input exceeds some criterion level. For clarity of presentation the independent noise sources for the spatial filters have been omitted. (B) Probability-summation detector. In this case the luminance distribution corresponding to the visual stimulus is processed by linear spatial filters but the response of each is handled by an independent detector, which responds if the input exceeds a criterion level. Thus, the action of the effector depends on the action of one or more of the detectors. For clarity of presentation the independent noise sources for the spatial filters have been omitted.

a visual-acuity task then can be conceptualized in terms of a set of orthogonal spatial filters with uncorrelated Poisson noise followed by an incoherent but quasioptimal detector.

Behavioral Experiments

Behavioral performance reflects organized cellular (mainly neural) activity, and a detailed understanding requires precise information about the underlying neural activity. However, optimal performance can be considered within the context of a computational theory with complete information on not all neural activity, but only the neural activity that sets constraints on optimality. In the case of simple visual acuity, optimal performance is constrained by the spatial filtering carried out by the cortical receptive fields in the supragranular layers of area V1. A comparison can be made between experimentally determined spatial visual acuity and the predictions of a computational theory based on an optimal incoherent detector constrained by M linear orthogonal spatial filters having characteristics like the cortical receptive fields for a particular set of stimulus conditions. However, there is a developing consensus that such a computational theory can be extended to a wide class of stimuli.

Comparisons under Stabilized Viewing

The close correspondence in behavioral capacity among Old World anthropoids makes possible the comparison of visual performance in man with predictions from a computational theory based upon neural parameters derived mainly from the macaque monkey. Visual acuity assessed by contrast sensitivity functions is essentially identical in man and macaque under the same stimulus conditions (DeValois et al. 1974). For oriented grating targets, the assumption of cross-species similarity for neural mechanisms is quite plausible but the experimental conditions under which the neural data were collected must be carefully considered.

Precise assessment of single-unit receptive fields necessitates the control of eye movements—typically by immobilizing the ocular and skeletal musculature with a curarelike drug, gallamine triethiodide (Schiller et al. 1976a; Mansfield et al. 1980). Even with drug-induced paralysis of the neuromuscular synaptic transmission there is some residual eye movement, but it is on the order of only 0.1° over the course of 10 minutes. Similar drug-induced paralysis in human observers results in a reduction in apparent contrast so that large uniformly illuminated objects disappear and edges fade (Stevens et al. 1976), a phenomenon observed in normal vision for peripheral targets viewed with steady eccentric fixation under dim illumination. Such Troxler fading is confined to the periphery, since the eye movements present in steady fixation are small relative to the dimensions of the receptive fields engaged by the stimulus. Steady fixation with foveal viewing does not achieve the same result, since even in the best normal observers the direction of gaze about a fixated target is erratic (mean radius ≈ 0.1°) and microsaccades are not suppressed (Higgins et al. 1981). In order to examine visual performance under comparable conditions to the neurophysiological experiments in any detailed psychophysical task "would have required hours of total paralysis—a very unrealistic requirement" (Stevens et al. 1976). An alternate means of stabilizing the retinal image is to compensate optically for image motion on the surface of the retina.

The conventional approach developed 25 years ago involves the use of a contact lens, to which either a mirror is attached to form one arm of an optical lever (Riggs et al. 1953) or a miniature projector is mounted (Ditchburn and Pritchard 1956). These methods are limited both in the range of stimuli that can be presented and in the ease with which extended and detailed observations can be made in several observers, because of the specially fitted scleral contact lenses involved. Recent innovations in the methods of accurate eyetracking without physical contact have set the stage for a new technique of retinal-image stabilization.

The approach uses infrared light reflected from the eye to determine eye position. As is shown schematically in figure 15.8, the novel feature is that the SRI Dual Purkinje Image Eyetracker monitors infrared reflections from both the cornea (first Purkinje image) and the back of the lens (fourth Purkinje image) (Cornsweet and Crane 1973; Clark 1975). By measuring the optical distance between the two images, the eyetracker achieves an accuracy of greater than 1 minute of arc. Such a high degree of accuracy suggested to B. R. Wooten of Brown University the possibility of developing a means of retinal-image stabilization without contact lenses. In the Wooten modification a pair of fast-response (1–2 msec rise time) galvanometer-driven mirrors use the horizontal and vertical eye-movement signals of the eyetracker to compensate optically for movement of the stimulus across the retina resulting from eye movements.

Figure 15.9 shows an observer in the apparatus and a typical stimulus pattern on the display oscilloscope. The observer views the target monocularly, with his head held steady by means of a hard wax bite bar. Typically, mydriatic and cycloplegic drugs are used to maintain the pupil at maximal dilation and the lens in a relaxed state of accommodation. A televisionlike raster scan on the face of the display oscilloscope can be readily produced by waveform generators to yield a uniformly illuminated surface. The voltage on the z axis of the oscilloscope can be modulated in phase with each frame of the scan to produce patterns on the display face; electronic hardware can be used for simple sinusoidal patterns, or digital-to-analog readout from a computer for more complex patterns.

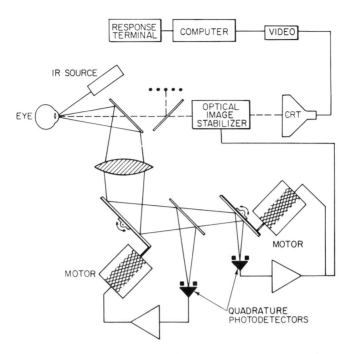

Figure 15.8 Schematic diagram of Dual Purkinje Eyetracker and Retinal Image Stabilization System. In determining eye position, a low-intensity invisible beam of infrared light from a light-emitting diode is reflected from the eye and the images of the first Purkinje image (corneal reflection) and the fourth Purkinje image (posterior surface of the lens) are separately focused on quadrature photodetectors. The error signals from the quadrature photodetectors drive mirrors in the optical path via servomotors. The output of the second quadrature photodetector (fourth Purkinje image) then corresponds to the difference between the first and fourth Purkinje images, a measure of eye position that is relatively invariant under lateral movement of the head. For retinal image stabilization the eye-position signal is used to operate galvanometer-driven mirrors in the optical path of the viewer which compensate for vertical and horizontal eye movements by appropriate movements of the visual target, a computer-controlled video display. Independent of the stabilization, a set of visible LEDs serve as fixation targets. In addition, the observer's responses are accessible to the computer and can be used in interactive psychophysical procedures.

The choice of a target stimulus with which to probe the visual system and test the range of validity of the computational theory outlined above depends on several characteristics of the primate visual system. Using low-contrast patterns keeps the response in the linear range. By choosing patterns with a small spatial extent a piecewise isoplanatic approximation to retinal inhomogeneity can be made. A sinusoidally modulated luminance grating such as that shown in part B of figure 15.9 is easily generated and controlled and can be used to determine the frequency response, which is related to the line-spread function by a Fourier transform. The case of foveal vision is of particular interest; figure 15.10 shows three results obtained for a small grating patch centered at the fovea ($\pm 1°$) surrounded by a broad border of approximately equal

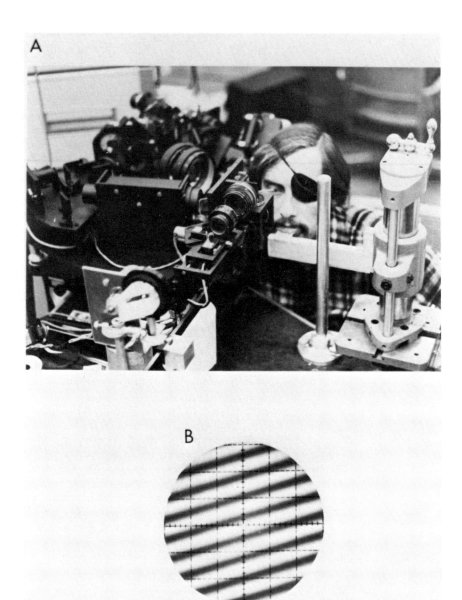

Figure 15.9 (A) Photograph of observer viewing target under retinal image stabilization by using the outputs of the SRI Dual Purkinje Image Eyetracker to drive small mirrors on the stimulus deflector seen in the foreground. (B) Photograph of target on video screen at high contrast.

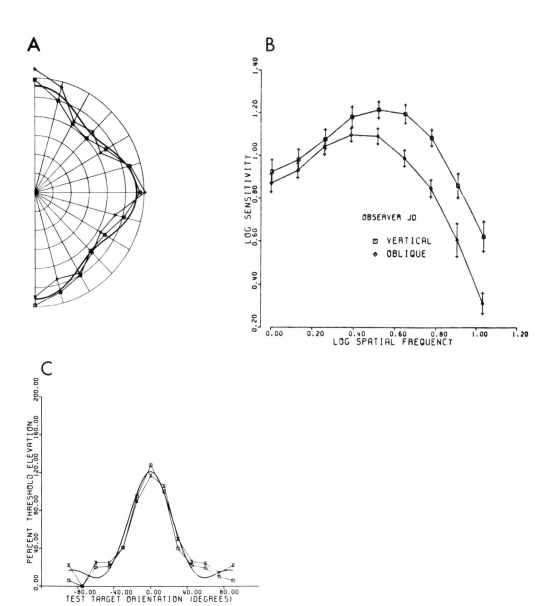

Figure 15.10 Comparison of calculated response from synthesized striate-cortex neuron population with human psychophysical data on orientation acuity. Theoretical curves have been scaled in amplitude to best fit experimental data. (A) Detection threshold as a function of target orientation for an 8 cycle/degree sinusoidally modulated luminance grating subtending 2° foveally. The ● are the data points; the solid line represents the computed values. (B) Detection threshold as a function of grating spatial frequency. (●) Data points for oblique (45°) target; (■) data points for vertical target. Solid lines represent computed values. (C) Detection-threshold elevation after adaptation to 8 cycle/degree grating subtending 2° foveally. (□) Data obtained with adaptation and testing in the same eye; (X) data obtained with testing in the contralateral eye. Solid curve represents computed values with adaptation proportional to pooled neural response.

luminance to eliminate edge effects, which could swamp the response of interest. Temporal transients also are avoided by the use of a relatively long display period.

The basic datum consists of an observer's contrast-threshold setting, but each data point is typically based on several observations to obtain an accurate average. Part A of figure 15.10 displays detection threshold as a function of target orientation for a grating of relatively high spatial frequency. The solid line drawn through the data points is the prediction from the computational theory scaled to the sensitivity level of the observer. The results show how the orientation anisotropy present in the neuronal population translates into a difference in the detectability of targets at different orientations. These results are generalized in part B, which shows detection threshold as a function of spatial frequency of the grating target for both a vertical and an oblique angle (45°). The solid line is again the theoretical prediction from computation. At low spatial frequencies the sensitivity is depressed beyond that inherent in the frequency response of the central fovea because the limited number of grating cycles present in the 2° target decreases the optimization of detection produced by the information pooling that can take place when multiple bars are present. Moreover, the convergence of the vertical and oblique sensitivities reflects the dominance of the response by the Y cells, which do not exhibit orientation anisotropy. The effect on detection threshold of prior exposure to a high-contrast-grating target is shown in part C. The selective elevation of threshold reflects the orientation-selective nature of the pattern adaptation. Since the pooling operation in effect averages over the spectrum of orientation-tuning curves, the resulting curve resembles the tuning curve of the average receptive field.

Implications for Pattern Detection
There is accumulating evidence that the detection of a wide class of monochromatic patterned stimuli can be explained by a computational theory based on quasioptimal incoherent detection applied to the output of an array of independent channels whose medium-bandwidth filter characteristics resemble the receptive-field excitatory profiles of efferent neurons in the supragranular layers of area V1. Incoherent detection is consistent with the localized effects of pattern adaptation. Previous work from this laboratory (Legge 1976) demonstrated that the adaptation is retinotopically localized. Legge found that adaptation to a high-contrast fine line did not affect subsequent detection of spatially extended sinusoidally modulated grating patterns, as would be predicted if the effect of adaptation were local rather than global. Daugman and Mansfield (1979) found that the adaptation is not even transferred interhemispherically, since no adaptation is produced by an eccentric high-contrast-grating patch placed in the homologous location in the opposite hemifield. Graham and her colleagues (Graham and Nachmias

1971; Graham et al. 1978) demonstrated independence of detection of separate components in a compound grating even when it was presented to a relatively homogeneous retinal region—a result consistent with the existence of independent channels. Variation with retinal eccentricity of several measures, such as visual acuity (Aulhorn and Harms 1972), summation area (Westheimer 1967), Mach-band locations (Shipley and Wier 1972), contrast-sensitivity functions (Hilz and Cavonius 1974), and line-spread functions (Hines 1976; Limb and Rubinstein 1977; Wilson 1978) is consistent with an increase in receptive-field size with distance from the fixation point. The work from Wilson's laboratory in particular suggests that the increase is linear, at least up to 4°. The scatter in receptive-field sizes and types at a given retinotopic projection point is reflected in the differences in detectability of different patterns presented at the same retinal locus (Kulikowski and King-Smith 1973; Wilson and Geise 1977). Experiments with frequency-gradient patterns (Wilson and Geise 1977) and with artificial central scotomas (Kelly 1978) indicate that all receptive-field sizes are represented in the foveal projection region but that there is a decline in the relative proportion of small receptive fields with retinal eccentricity. On the important issue of the estimate of the spatial-frequency bandwidth of the receptive fields, a growing consensus indicates that although several studies have been interpreted as implying extremely narrow bandwidths (¼ octave) (Sachs et al. 1971; Kulikowski and King-Smith 1973; Quick and Riechert, 1975), subsequent work has shown that if an incoherent detection process such as probability summation across space is taken into account the data are consistent with medium bandwidths (Stromeyer and Klein 1975; King-Smith and Kulikowski 1975; Graham and Rogowitz 1976; Mostafavi and Sakrison 1976; Wilson 1978; Wilson and Bergen 1979; Bergen et al. 1979). The bandwidth estimates range from 1.5 to 2.1 octaves at half-height. The data of Blakemore and Campbell (1969), obtained using spatially selective frequency adaptation, yield an estimate of 1.5 octaves with full bandwidth at half-height, but the fact that retinal inhomogeneity and probability summation are not taken into account results in an underestimation. Mostafavi and Sakrison (1976) obtained an estimate of 2.1 octaves using band-limited noise displays; they took into account retinal inhomogeneity and spatial probability summation but not summation over channels, which resulted in an overestimation. Wilson and his colleagues (Wilson 1978; Wilson and Bergen 1979; Bergen et al. 1979), using subthreshold summation to determine the line-spread function, found 1.75 octaves, taking into account spatial probability summation over channels and space. The monkey neurophysiology data described earlier yield estimates of 1–2 octaves (Schiller et al. 1976c; Poggio et al. 1977; DeValois et al. 1978). The measurements of DeValois et al., made under conditions comparable to the psychophysical measurements, gave a median value of 1.4 octaves (1.2 for the more orientation-selective neurons). A similar situation holds for orientation selectivity. Early studies, particularly those that employed subthreshold summation (see, for example, Kulikowski et al. 1973), were interpreted as evidence for extremely narrow orientation bandwidths, such as $\pm 3°$. Other measurement procedures using adaptation (Blakemore and Nachmias 1971) and masking (Campbell and Kulikowski 1966) gave estimates of $\pm 7°$ to $\pm 15°$. However, these studies did not take into account spatial probability summation. With this summation taken into account, the data from subthreshold summation experiments are consistent with orientation bandwidths on the order of $\pm 16°$. Mostafavi and Sakrison (1976), using two-dimensional filtered noise and assuming a Gaussian shape, retinal inhomogeneity, and a spatial probability summation index of 6, obtained an orientation bandwidth of $\pm 14.4°$; however, they did not take into account summation across channels or orientation anisotropy. As we have seen, taking into account spatial probability summation, orientation anisotropy, and probability summation across channels yields an estimate of $\pm 20°$ (see part c of figure 15.10). The neurophysiological data uniformly concur on an estimate of average bandwidth of $\pm 20°$ to $\pm 25°$ (Schiller et al. 1976b; Poggio et al. 1977; DeValois et al. 1978; Mansfield et al. 1981). The concordance of results from behavioral and neurophysiological studies, particularly with regard to orientation and frequency bandwidths, carries a number of implications for visual processing. For example, the view that the visual system performs some type of Fourier analysis on a visual scene (Campbell and Robson 1968; Blakemore and Campbell 1969; Pollen et al. 1971) is oversimplified at best, since the channel bandwidths are too broad. The apparent sharp tuning is the result of the statistical nature of the incoherent detection process. An alternative view consistent with experimental facts reviewed here is that the visual system analyzes target objects using an ensemble of medium-bandwidth spatial filters whose characteristics match those of the receptive fields of supragranular striate-cortex neurons.

Future Prospects

The development of a truly computational theory of primate vision is an exciting prospect but one that can be realistically approached. Two guiding principles have emerged from the studies reviewed here: that the primate visual system is organized in terms of functionally significant channels, and that optimal visual performance can be analyzed into a critical transformation site and optimal utilization of the information from that site. The filtering operations performed by the visual system up to the level of the striate cortex and by the striate cortex itself give rise to a filtered neural image of the retinal stimulus represented by the population response profile of the efferent neurons, the transcortical projection of which consists of the pyramidal cells in the supragranular layers. This neural image contains all the fine-grain information available to the more rostral portions of the visual system, which operate upon the information in a manner similar to that of a noncoherent optimal detector. In the detailed example the stimuli were stationary, low-contrast, monochromatic, monocularly presented targets and the task was one of detection. Each of these restrictions can be lifted or modified.

The neurons in the supragranular layers respond differentially to direction of target motion, but no quantitative comparison has been made with the directionally sensitive mechanisms revealed by prior adaptation in psychophysical experiments. One experimental finding that makes reasonable the suggestion that these neurons mediate the detection of target direction is that target detection is superior to detection of direction of motion even when absence of eye movements is achieved through retinal-image stabilization (Mansfield and Nachmias 1981). This result is consistent with the dominance of X input over Y input into these layers. Although the simplest cases to analyze are those at threshold, linearity extends above that level (Mansfield 1976a). Indeed, as Cannon (1979) has shown, the entire contrast-sensitivity curve remains parallel to those generated from equal apparent contrast up to at least a 70% contrast ratio. At suprathreshold levels in the linear range, the spatial filtering imposed by the striate cortex has predictable but not intuitively obvious consequences, such as the distortions from veridicality epitomized by the familiar optical illusions (Ginsburg 1979) or, for instance, the spatial distribution of eye saccades and fixations in free viewing of even a complex target such as a face (Ginsburg, Mansfield, and Higgins, in prep.). Saccadic eye movements also induce an adaptive suppression of retinal blur through spatial-frequency-selective masking that can be modeled in terms of inhibitory interactions between fast- and slow-responding spatial filters (Breitmeyer and Ganz 1976; Rogowitz 1980). Within the same conceptual framework, binocular interactions such as interocular adaptation and masking or even local stereopsis can be usefully analyzed in terms of striate-cortex neuronal activity. On the other hand, successful extension of the approach of the cellular synthesis of behavior described here to more complex visual perceptual and visuomotor behaviors such as global stereopsis, object color determination, and object tracking will require a more detailed understanding of the primate visual system—particularly of regions beyond area V1.

The basis for a new and detailed understanding of the primate extrastriate cortex will rest on three developments in cellular and behavioral neurobiology:

• *An understanding of the general functional subdivision of labor among the different regions.* A recurrent theme of the chapters in this volume is the existence of visual subsystems, and a premier goal of behavior analysis must be that of fractionating an animal's repertoire into units that optimally engage a particular subsystem. Insights into causal sequences in the dynamic unfolding of a behavioral script are now possible with single-unit and multiunit recording from alert animals in well-defined behavioral tasks with eye movements closely monitored, as in the study by Lynch et al. (1977) of the initiation and control of visual grasping of an object of motivational significance. On the other hand, anatomical labeling of a subsystem by its elevated metabolic activity when engaged can be accomplished by using ^{3}H-labeled or ^{14}C-labeled 2-deoxyglucose, which accumulates in cells since it does not substitute for glucose beyond a certain point in the Krebs cycle. Frost (this volume), using this technique, has localized the region of a pigeon's brain that deals with relative motion. In the living animal, positron-emission transverse tomography (PETT scanning) will soon supplement the resolution of metabolic activity on the cortical surface now possible with gamma emission, and will provide resolution in three dimensions (eventually down to the millimeter level).

• *An extensive knowledge of the interconnections between cortical regions and the interactions with subcortical structures.* Area V1 is the simplest visual cortical area in terms of its connections, and yet it is reciprocally interconnected with at least four other structures: dLGN, the pulvinar complex, and areas V2 and MT. Although connections can sometimes be inferred from disconnection syndromes produced by lesions, application of such an approach to a distributed processing system such as the visual system—particularly in advance of anatomical investigations—can often yield paradoxical results (see the chapter by Dean and that of Levine for a discussion of such inferences for inferotemporal lesions). Recent advances in neuroanatomical labeling by axonal transport have brought this question within the range of experimental analysis, so that precise information is accumulating; see the review of Van Essen (1979). Since the function played by a particular region is determined not only by the intrinsic operations it performs but by the pattern of its afferents and efferents in the circuit of which it forms a part, even a first approximation to the necessary circuit diagram such as the one illustrated in part A of figure 15.1 provides a rational basis for identifying functional subsystems (see the chapter by Ungerleider and Mishkin for a discussion of two cortical visual subsystems: one terminated by the inferotemporal region TE and the second by the inferior parietal lobule, area PG).

• *An understanding at the synaptic level of the functioning of cortical columns, the modules of organization of the neocortex.* Although a connectional diagram even as detailed as that in part B of figure 15.1, together with recording and lesion studies, can illuminate the role of a particular region (for example, that of the striate cortex as a preprocessor of object information), the mechanism by which the cortex operates on its afferent input requires a more fine-grain analysis. As Mountcastle (1976) pointed out, "the central problem of the intrinsic physiology of the cerebral cortex is to discover the nature of the neuronal processing within the translaminar chains of interconnected cells (in columns)." Techniques are now available to determine the synaptic circuitry of morphologically identified neurons in functional modules of neocortex using *in vitro* recording from brain slices (Mansfield and Simmons 1979; Mansfield, in prep.). Anatomical and physiological evidence (see Mountcastle 1978 for a review) suggests that cortical modules may have a unitary organization, so detailed knowledge of the synaptic organization in one cortical region could serve as a paradigm for the cortex as a whole. The present and continuing challenge will be to synthesize the results from these three lines of development into a coherent computational theory of visual behavior.

Acknowledgments

The past and present members of the Laboratory of Behavioral Neurobiology, John G. Daugman, Gordon E. Legge, Steven F. Ronner, and Linda K. Simmons, have contributed significantly to the substance of the present report, and I gratefully acknowledge their collaboration. Visiting colleagues on sabbatical from other institutions (Kent E. Higgins from the Pennsylvania College of Optometry and Jacob Nachmias of the University of Pennsylvania) have influenced not only the results but also the perspectives presented here in important ways which I greatly appreciate. I am particularly grateful to Margaret E. Keahey for her skilled editorial assistance. This research was supported in part by The Milton Fund of Harvard University, by the National Science Foundation under research grant BNS 75-08437, and by the Air Force Office of Scientific Research under contract F44-76-C-0109.

References

Aulhorn, E., and H. Harms. 1972. "Visual perimetry." In *Handbook of Sensory Physiology*, vol. VII/4, D. Jameson and L. M. Hurvich, eds. Berlin: Springer.

Benevento, L. A., and J. H. Fallon. 1975. "The ascending projections of the superior colliculus in the rhesus monkey (*Macaca mulatta*)." *J. Comp. Neurol.* 160: 339–362.

Bergen, J. R., H. R. Wilson, and J. D. Cowan. 1979. "Further evidence for four mechanisms mediating vision at threshold: Sensitivities to complex gratings and aperiodic stimuli." *J. Opt. Soc. Am.* 69: 1580–1587.

Berman, N. 1977. "Connections of the pretectum in the cat." *J. Comp. Neurol.* 174: 227–254.

Blake, L., C. D. Jarvis, and M. Mishkin. 1977. "Pattern discrimination thresholds after partial inferior temporal or lateral striate lesions in monkeys." *Brain Res.* 120: 209–220.

Blakemore, C. 1970. "The range and scope of binocular depth discrimination in man." *J. Physiol.* 211: 599–622.

Blakemore, C., and F. W. Campbell. 1969. "On the existence of neurones in the human visual system selectively sensitive to the orientation and size of retinal images." *J. Physiol.* 203: 237–260.

Blakemore, C., and J. Nachmias. 1971. "The orientation specificity of two visual after-effects." *J. Physiol.* 213: 157–174.

Bowmaker, J. K., H. J. A. Dartnall, and J. D. Mollon. 1980. "Microspectrophotometric demonstration of four classes of photoreceptor in an old world primate (*Macaca fascicularis*)." *J. Physiol.* 298: 131–144.

Boynton, R. M., and D. N. Whitten. 1970. "Visual adaptation in monkey cones: Recordings of late receptor potentials." *Science* 170: 1423–1426.

Breitmeyer, B. G., and L. Ganz. 1976. "Implications of sustained and transient channels for theories of visual pattern masking, saccadic suppression, and information processing." *Psychol. Rev.* 83: 1–36.

Bunt, A. H., D. S. Minckler, and G. W. Johanson. 1977. "Demonstration of bilateral projection of the central retina of the monkey with horseradish peroxidase neuronography." *J. Comp. Neurol.* 171: 619–630.

Bunt, A. H., A. E. Hendrickson, J. S. Lund, R. D. Lund, and A. F. Fuchs. 1975. "Monkey retinal ganglion cells: Morphometric analysis and tracing of axonal projections, with a consideration of the peroxidase technique." *J. Comp. Neurol.* 164: 265–286.

Campbell, F. W., and J. J. Kulikowski. 1966. "Orientation selectivity of the human visual system." *J. Physiol.* 187: 437–445.

Campbell, F. W., and J. Robson. 1968. "Application of Fourier analysis to the visibility of gratings." *J. Physiol.* 197: 551–566.

Cannon, M. 1979. "Linearity of perceived contrast as a function of spatial frequency." *Vision Res.* 19: 975–984.

Cartmill, M. 1974. "Rethinking primate origins." *Science* 184: 436–443.

Cinlar, E. 1972. "Superposition of point processes." In *Stochastic Point Processes: Statistical Analysis, Theory, and Applications*, P. A. W. Lewis, ed. New York: Wiley-Interscience.

Clark, M. R. 1975. "A two-dimensional Purkinje eye tracker." *Behav. Res. Methods Instrum.* 7: 215–219.

Cohn, T. E. 1978. "Detection of 1-of-M orthogonal signals: Asymptotic equivalence of likelihood ratio and multiband models." *Optics Lett.* 3: 22–23.

Cornsweet, T. N., and H. D. Crane. 1973. "Accurate two-dimensional eye tracker using first and fourth Purkinje images." *J. Opt. Soc. Am.* 63: 921–928.

Cowey, A., and L. Weiskrantz. 1967. "A comparison of the effects of inferotemporal and striate cortex lesions on the visual behaviour of rhesus monkeys." *Q. J. Exp. Psychol.* 15: 91–115.

Crawford, M. L. J., J. D. Fagan, M. Borchert, A. Heston, and R. E. Marc. 1979. "Metabolic patterns induced in monkey visual cortex by stimulation with colored lights." *Soc. Neurosci. Abstr.* 5: 780.

Daugman, J. G., and R. J. W. Mansfield. 1979. "Adaptation of spatial channels in human vision." *Invest. Ophthalmol. Visual Sci.* Supplement 18: 91.

DeMonasterio, F. M. 1978a. "Properties of concentrically organized X and Y ganglion cells of macaque retina." *J. Neurophysiol.* 41: 1394–1417.

———. 1978b. "Properties of ganglion cells with atypical receptive-field organization in retina of macaques." *J. Neurophysiol.* 41: 1435–1449.

DeMonasterio, F. M., and P. Gouras. 1975. "Functional properties of ganglion cells of the rhesus monkey retina." *J. Physiol.* 251: 167–195.

DeValois, R. L., and G. H. Jacobs. 1968. "Primate color vision." *Science* 162: 533–540.

DeValois, R. L., H. C. Morgan, and D. M. Snodderly. 1974. "Psychophysical studies of monkey vision. III. Spatial luminance contrast sensitivity tests of macaque and human observers." *Vision Res.* 14: 75–81.

DeValois, R. L., D. G. Albrecht, and L. G. Thorell. 1978. "Cortical cells: Bar and edge detectors, or spatial frequency filters?" In *Frontiers in Visual Science,* S. J. Cool and E. L. Smith III, eds. New York: Springer.

Diamond, I. T. 1979. "The subdivisions of neocortex: A proposal to revise the traditional view of sensory, motor, and association areas." In *Progress in Psychobiology and Physiological Psychology,* J. M. Sprague and A. N. Epstein, eds. New York: Academic.

Ditchburn, R. W., and R. M. Pritchard. 1956. "Vision with a stabilized retinal image." *Nature* 177: 434–435.

Divac, I., J. H. LaVail, P. Rakic, and K. R. Winston. 1977. "Heterogeneous afferents to the inferior parietal lobule of the rhesus monkey revealed by the retrograde transport method." *Brain Res.* 123: 197–207.

Dow, B. M. 1974. "Functional classes of cells and their laminar distribution in monkey visual cortex." *J. Neurophysiol.* 37: 927–946.

Dow, B. M., and P. Gouras. 1973. "Color and spatial specificity of single units in rhesus monkey foveal striate cortex." *J. Neurophysiol.* 36: 79–100.

Dowling, J. E., and B. B. Boycott. 1966. "Organization of the primate retina: Electron microscopy." *Proc. Roy. Soc.* B 166: 80–111.

Dreher, B., Y. Fukada, and R. W. Rodieck. 1976. "Identification, classification, and anatomical segregation of cells with X-like and Y-like properties in the lateral geniculate nucleus of old-world primates." *J. Physiol.* 258: 433–452.

Ebrey, T. C., and B. Honig. 1977. "New wavelength dependent visual pigment nomogram." *Vision Res.* 17: 147–157.

Enroth-Cugell, C., and J. G. Robson. 1966. "The contrast sensitivity of retinal ganglion cells of the cat." *J. Physiol.* 187: 517–552.

Fain, G., and J. E. Dowling. 1973. "Intracellular recordings from single rods and cones in the mudpuppy retina." *Science* 180: 1178–1181.

Finlay, B. L., P. H. Schiller, and S. F. Volman. 1976. "Meridional differences in orientation sensitivity in monkey striate cortex." *Brain Res.* 105: 350–352.

Ginsburg, A. P. 1979. "Visual perception based on spatial filters constrained by biological data." In *Proceedings of the International Conference on Cybernetics and Society.* New York: IEEE.

Glickstein, M., J. L. Cohen, B. Dixon, A. Gibson, M. Holland, E. Labossie, and R. Farrel. 1980. "Cortico-pontine visual projections in macaque monkey." *J. Comp. Neurol.* 190.

Graham, N., and J. Nachmias. 1971. "Detection of grating patterns containing two spatial frequencies: A comparison of single-channel and multiple-channel models." *Vision Res.* 11: 251–259.

Graham, N., and B. E. Rogowitz. 1976. "Spatial pooling properties from the detectability of FM and quasi-AM gratings: A reanalysis." *Vision Res.* 16: 1021–1026.

Graham, N., J. G. Robson, and J. Nachmias. 1978. "Grating summation in fovea and periphery." *Vision Res.* 18: 815–825.

Green, D. M., and J. A. Swets. 1966. *Signal Detection Theory and Psychophysics.* New York: Wiley.

Green, D. M., and D. L. Weber. 1980. "Detection of temporally uncertain signals." *J. Acoust. Soc. Am.*

Gross, C. G., and M. Mishkin. 1977. "The neural basis of stimulus equivalence across retinal translation." In *Lateralization in the Nervous System,* S. Harnad *et al.,* eds. New York: Academic.

Henry, G. H., B. Dreher, and P. O. Bishop. 1974. "Orientation specificity of cells in the cat striate cortex." *J. Neurophysiol.* 37: 1394–1409.

Higgins, K. E., J. G. Daugman, and R. J. W. Mansfield. 1980. "Amblyopic contrast sensitivity: Role of unsteady fixation." *Invest. Ophthalmol.*

Hilz, R., and C. R. Cavonius. 1974. "Functional organization of the peripheral retina: Sensitivity to periodic stimuli." *Vision Res.* 14: 1333–1337.

Hines, M. 1976. "Line spread function variation near the fovea." *Vision Res.* 16: 567–572.

Hubel, D. H., and T. N. Wiesel. 1968. "Receptive fields and functional architecture of monkey striate cortex." *J. Physiol.* 195: 215–243.

————. 1972. "Laminar and columnar distribution of geniculo-cortical fibers in the macaque monkey." *J. Comp. Neurol.* 146: 421–450.

————. 1974. "Uniformity of monkey striate cortex: A parallel relationship between field size, scatter, and magnification factor." *J. Comp. Neurol.* 158: 295–305.

Humphrey, D. R., and W. S. Corrie. 1978. "Properties of the pyramidal tract neuron system within a functionally defined subregion of primate motor cortex." *J. Neurophysiol.* 41: 216–243.

Johnston, M. V., M. McKinney, and J. T. Coyle. 1979. "Evidence for a cholinergic projection to neocortex from neurons in basal forebrain." *Proc. Nat. Acad. Sci.* 76: 5392–5396.

Kelly, D. H. 1978. "Photopic contrast sensitivity without foveal vision." *Optics Lett.* 2: 79–81.

Kennedy, H., and G. A. Orban. 1979. "Preferences for horizontal or vertical orientation in cat visual cortical neurones." *J. Physiol.* 296: 61P–62P.

King-Smith, P. E., and J. J. Kulikowski. 1975. "The detection of gratings by independent activation of line detectors." *J. Physiol.* 247: 237–271.

Kulikowski, J. J., and P. E. King-Smith. 1973. "Spatial arrangement of line, edge, and grating detectors revealed by subthreshold summation." *Vision Res.* 13: 1455–1478.

Kulikowski, J. J., R. Abadi, and P. E. King-Smith. 1973. "Orientational selectivity of grating and line detectors in human vision." *Vision Res.* 13: 1479–1486.

Kunzle, H. 1978. "An autoradiographic analysis of the efferent connections from premotor and adjacent prefrontal regions (areas 6 and 9) in *Macaca fascicularis.*" *Brain, Behav., Evol.* 15: 185–234.

Kunzle, H., and K. Akert. 1977. "Efferent connections of cortical area 8 (frontal eye field) in *Macaca fascicularis:* A reinvestigation using autoradiographic technique." *J. Comp. Neurol.* 173: 147–164.

Land, E. H., and J. J. McCann. 1971. "Lightness and retinex theory." *J. Opt. Soc. Am.* 61: 1–11.

Legge, G. E. 1976. "Adaptation to a spatial impulse: Implications for Fourier transform models of visual processing." *Vision Res.* 16: 1407–1418.

Leventhal, A. G., and H. V. B. Hirsch. 1977. "Effects of early experience upon orientation selectivity and binocularity of neurons in visual cortex of cats." *Proc. Nat. Acad. Sci.* 74: 1272–1276.

Limb, J. O., and C. B. Rubenstein. 1977. "A model of threshold vision incorporating inhomogeneity of the visual field." *Vision Res.* 17: 571–584.

Lin, G. S., and J. H. Kaas. 1977. "Projections from cortical visual areas 17, 18, and MT onto the dorsal lateral geniculate nucleus in owl monkey." *J. Comp. Neurol.* 173: 457–474.

Lund, J. S. 1973. "Organization of neurons in the visual cortex area 17 of the monkey (*Macaca mulatta*)." *J. Comp. Neurol.* 147: 455–496.

Lund, J. S., and R. G. Boothe. 1975. "Interlaminar connections and pyramidal neuron organization in the visual cortex, area 17, of the monkey (*Macaca mulatta*)." *J. Comp. Neurol.* 159: 305–334.

Lund, J. S., R. D. Lund, A. E. Hendrickson, A. H. Bunt, and A. F. Fuchs. 1975. "The origin of efferent pathways from the primary visual cortex, area 17, of the macaque monkey as shown by retrograde transport of horseradish peroxidase." *J. Comp. Neurol.* 164: 287–304.

Lynch, J. C., V. B. Mountcastle, W. H. Talbot, and T. C. T. Yin. 1977. "Parietal lobe mechanisms for directed visual attention." *J. Neurophysiol.* 40: 362–389.

Malpeli, J. G., and F. H. Baker. 1975. "The representation of the visual field in the lateral geniculate nucleus of *Macaca mulatta*." *J. Comp. Neurol.* 161: 569–594.

Mansfield, R. J. W. 1974. "Neural basis of orientation perception in primate vision." *Science* 186: 1133–1135.

———. 1975. "Neural mechanisms subserving brightness coding in monkey visual cortex." *Soc. Neurosci. Abstr.* 1: 431.

———. 1976a. "Visual adaptation: Retinal transduction, brightness, and sensitivity." *Vision Res.* 16: 679–690.

———. 1976b. "Psychophysics and the neural basis of information processing." In *Advances in Psychophysics*, H.-G. Geissler and Yu. M. Zabrodin, eds. Berlin: VEB Deutscher Verlag der Wissenschaften.

———. 1979. "Neural information processing in the primate visual system." In *Proceedings of the International Conference on Cybernetics and Society*. New York: IEEE.

Mansfield, R. J. W., and J. G. Daugman. 1978. "Retinal mechanisms of visual latency." *Vision Res.* 18: 1247–1260.

Mansfield, R. J. W., and J. Nachmias. 1981. "Perceived direction of motion under retinal image stabilization." *Vision Res.*

Mansfield, R. J. W., and S. F. Ronner. 1978. "Orientation anisotropy in monkey visual cortex." *Brain Res.* 149: 229–234.

Mansfield, R. J. W., and L. K. Simmons. 1979. "Intrinsic processing in the visual cortex of primates." *Soc. Neurosci. Abstr.* 5: 795.

Mansfield, R. J. W., J. G. Daugman, and S. F. Ronner. 1978. "Cortical mechanism in primate vision for detecting oriented targets." *J. Opt. Soc. Am.* annual meeting abstracts, 68: 1419.

Mansfield, R. J. W., L. K. Simmons, and J. G. Daugman. 1979. "Cortical locus of pattern-specific adaptation in primates." *Invest. Ophthalmol. Vis. Sci.* suppl. 18: 228.

Mansfield, R. J. W., S. F. Ronner, and J. G. Daugman. 1981. "Neural mechanisms of orientation perception: Properties of monkey striate neurons." *J. Neurophysiol.*

Marin-Padilla, M. 1969. "Origin of the pericellular baskets of the pyramidal cells of the human motor cortex: A Golgi study." *Brain Res.* 14: 633–646.

———. 1970. "Prenatal and early postnatal otogenesis of the human motor cortex: A Golgi study. II. The basket-pyramidal system." *Brain Res.* 23: 185–191.

Marks, W. B., W. H. Dobell, and E. F. MacNichol, Jr. 1964. "Visual pigments in single goldfish cones." *Science* 143: 1181–1182.

Marrocco, R. T., J. W. McClurkin, and Z. H. H. Farooqui. 1979. "Modulation of LGN cell responsivity by visual activation of the corticogeniculate pathway." *Soc. Neurosci. Abstr.* 5: 795.

Maunsell, J. H. R., J. L. Bixby, and D. C. van Essen. 1979. "Areal boundaries and topographic organization of visual areas V2 and V3 in the macaque monkey." *Soc. Neurosci. Abstr.* 5: 812.

Menzell, E. W. 1973. "Chimpanzee spatial memory organization." *Science* 182: 943–945.

Mesulam, M.-M., G. W. van Hoesen, D. N. Pandya, and N. Geschwind. 1977. "Limbic and sensory connections of the inferior parietal lobule (area PG) in the rhesus monkey: A study with a new method for horseradish peroxidase histochemistry." *Brain Res.* 136: 393–414.

Mostafavi, H., and D. J. Sakrison. 1976. "Structure and properties of a single channel in the human visual system." *Vision Res.* 16: 957–968.

Mountcastle, V. B. 1957. "Modality and topographic properties of single neurons of cat's somatic sensory cortex." *J. Neurophysiol.* 20: 408–434.

———. 1976. "The world around us: Neural command functions for selective attention." The F. O. Schmitt Lecture for 1975. *Neurosci. Res. Program Bull.* 14, suppl. 1.

———. 1978. "An organizing principle for cerebral function: The unit module and the distributed system." In *The Mindful Brain,* G. E. Edelman and V. B. Mountcastle, eds. Cambridge, Mass.: MIT Press.

Nauta, W. J. H., and V. B. Domesick. 1980. "Neural associations of the limbic system." In *Neural Substrates of Behavior,* A. Beckman, ed. New York: Spectrum.

Nolte, L. N., and D. Jaarsma. 1967. "More on the detection of one of *M* orthogonal signals." *J. Acoust. Soc. Am.* 41: 497–505.

Ogren, M. P., and A. H. Hendrickson. 1976. "Pathways between striate cortex and subcortical regions in *Macaca mulatta* and *Saimiri sciureus:* Evidence for a reciprocal pulvinar connection." *Exp. Neurol.* 53: 780–800.

———. 1977. "The distribution of pulvinar terminals in visual areas 17 and 18 of the monkey." *Brain Res.* 137: 343–350.

Orban, G. A., and E. Vandenbussche. 1979. "Behavioral evidence for the oblique effect in the cat." *J. Physiol.* 295: 15P–16P.

Peterson, W. W., T. G. Birdsall, and W. C. Fox. 1954. "The theory of signal detectability." *IRE Trans. Information Theory* IT-4: 171–212.

Pettigrew, J. D., T. Nikara, and P. O. Bishop. 1968. "Responses to moving slits by single units in cat striate cortex." *Exp. Brain Res.* 6: 373–390.

Poggio, G. F. 1972. "Spatial properties of neurons in striate cortex of unanesthetized macaque monkey." *Invest. Ophthalmol.* 11: 368–376.

Poggio, G. F., and B. Fisher. 1977. "Binocular interaction and depth sensitivity of striate and prestriate cortical neurons of the behaving rhesus monkey." *J. Neurophysiol.* 40: 1392–1405.

Poggio, G. F., F. H. Baker, R. J. W. Mansfield, A. Sillito, and P. Grigg. 1975. "Spatial and chromatic properties of neurons subserving foveal and parafoveal vision in rhesus monkey." *Brain Res.* 100: 25–29.

Poggio, G. F., R. W. Doty, and W. H. Talbot. 1977. "Foveal striate cortex of the behaving monkey. Single neuron responses to square-wave gratings during fixation of gaze." *J. Neurophysiol.* 40: 1369–1391.

Pollen, D. A., J. R. Lee, and J. H. Taylor. 1971. "How does the striate cortex begin the reconstruction of the visual world?" *Science* 173: 74–77.

Polyak, S. D. 1957. *The Vertebrate Visual System.* University of Chicago Press.

Quick, R. F., and T. A. Reichert. 1975. "Spatial-frequency selectivity in contrast detection." *Vision Res.* 15: 637–643.

Rezak, M., and L. A. Benevento. 1977. "A redefinition of pulvinar subdivisions in the macaque monkey: Evidence for three distinct subregions within classically defined lateral pulvinar." *Soc. Neurosci. Abstr.* 3: 574.

———. 1979. "A comparison of the organization of the projections of the dorsal lateral geniculate nucleus, the inferior pulvinar, and adjacent lateral pulvinar to primary striate cortex (Area 17) in the macaque monkey." *Brain Res.* 167: 19–40.

Riggs, L. A., F. Ratliff, J. C. Cornsweet, and T. N. Cornsweet. 1953. "The disappearance of readily fixated visual objects." *J. Opt. Soc. Am.* 43: 495–501.

Rodieck, R. W. 1979. "Visual pathways." *Annu. Rev. Neurosci.* 2: 193–225.

Rogowitz, B. 1980. "Spatial/temporal interactions: Backward and forward metacontrast masking with sine-wave gratings." *Vision Res.*

Sachs, M. B., J. Nachmias, and J. G. Robson. 1971. "Spatial frequency channels in human vision." *J. Opt. Soc. Am.* 61: 1176–1186.

Savage-Rumbaugh, E. S., D. M. Rumbaugh, and S. Boysen. 1978. "Symbolic communication between two chimpanzees (*Pan troglodytes*)." *Science* 201: 641.

Schiller, P. H., and J. G. Malpeli. 1977. "Properties and tectal projections of monkey retinal ganglion cells." *J. Neurophysiol.* 40: 428–445.

―――. 1978. "Functional specificity of lateral geniculate nucleus laminae of the rhesus monkey." *J. Neurophysiol.* 41: 788–797.

Schiller, P. H., B. L. Finlay, and S. F. Volman. 1976a. "Quantitative studies of single-cell properties in monkey striate cortex. I. Spatio-temporal organization of receptive fields." *J. Neurophysiol.* 39: 1288–1319.

―――. 1976b. "Quantitative studies of single-cell properties in monkey striate cortex. II. Orientation specificity and ocular dominance." *J. Neurophysiol.* 39: 1320–1333.

―――. 1976c. "Quantitative studies of single cell properties in monkey striate cortex. III. Spatial frequency." *J. Neurophysiol.* 39: 1334–1351.

―――. 1976d. "Quantitative studies of single-cell properties in monkey striate cortex. V. Multivariate statistical analyses and models." *J. Neurophysiol.* 39: 1362–1374.

Schiller, P. H., J. G. Malpeli, and S. J. Schein. 1979. "Composition of geniculostriate input to superior colliculus of the rhesus monkey." *J. Neurophysiol.* 42: 1124–1133.

Schiller, P. H., S. D. True, and J. L. Conway. 1979. "Effects of frontal eye field and superior colliculus ablations on eye movements." *Science* 206: 590–592.

Shipley, T., and C. Wier. 1972. "Asymmetries in the Mach band phenomena." *Kybernetik* 10: 181–189.

Spatz, W. B., and J. Tigges. 1972. "Species difference between Old World and New World monkeys in the organization of the striate-prestriate association." *Brain Res.* 43: 591–594.

Stevens, J. K., R. C. Emerson, G. L. Gerstein, T. Kallos, G. R. Neufeld, C. W. Nichols, and A. C. Rosenquist. 1976. "Paralysis of the awake human: Visual perceptions." *Vision Res.* 16: 93–98.

Stone, J., J. Leicester, and S. M. Sherman. 1973. "The naso-temporal division of the monkey retina." *J. Comp. Neurol.* 150: 333–348.

Stromeyer, C. E., and S. Klein. 1975. "Evidence against narrow-band spatial frequency channels in human vision: The detectability of frequency modulated gratings." *Vision Res.* 15: 899–910.

Stryker, M. P., and H. Sherk. 1975. "Modification of cortical orientation selectivity in the cat by restricted visual experience: A reexamination." *Science* 190: 904–906.

Teich, M. C., and W. J. McGill. 1976. "Neural counting and photon counting in the presence of dead time." *Phys. Rev. Lett.* 36: 754–758.

Teich, M. C., L. Matin, and B. I. Cantor. 1978. "Refractoriness in the maintained discharge of the cat's retinal ganglion cell." *J. Opt. Soc. Am.* 68: 386–402.

Van Essen, D. C. 1979. "Visual areas of the mammalian cerebral cortex." *Annu. Rev. Neurosci.* 2: 227–263.

Wald, G., and P. K. Brown. 1965. "Human color vision and color blindness." *Cold Spring Harbor Symp. Quant. Biol.* 30: 345–359.

Westheimer, G. 1967. "Spatial interaction in human cone vision." *J. Physiol.* 190: 139–154.

Wiesel, T. N., and D. H. Hubel. 1966. "Spatial and chromatic interactions in the lateral geniculate body of the rhesus monkey." *J. Neurophysiol.* 29: 1115–1156.

Wilson, H. R. 1978. "Quantitative prediction of line spread function measurements: Implications for channel bandwidths." *Vision Res.* 18: 493–496.

Wilson, H. R., and J. R. Bergen. 1979. "A four mechanism model for threshold spatial vision." *Vision Res.* 19: 19–32.

Wilson, H. R., and S. C. Geise. 1977. "Threshold visibility of frequency gradient patterns." *Vision Res.* 17: 1177–1190.

Wong-Riley, M. 1979. "Columnar cortico-cortical interconnections within the visual system of the squirrel and macaque monkeys." *Brain Res.* 162: 201–217.

Wooten, B. R., K. Thawley, and K. Knoblauch. 1980. "Spectral absorbance and retinal distribution of the macular pigment." *J. Opt. Soc. Am.*

Wurtz, R. H. 1969. "Visual receptive fields of striate cortex neurons in awake monkeys." *J. Neurophysiol.* 32: 727–742.

Zeki, S. M. 1971. "Convergent input from the striate cortex (area 17) to the cortex of the superior temporal sulcus in the rhesus monkey." *Brain Res.* 28: 338–340.

16 Behavioral Measurement of Normal and Abnormal Development of Vision in the Cat

Donald E. Mitchell
Brian Timney

The last decade has seen the accumulation of a wealth of information concerning the anatomical and physiological development of the central visual pathways of the cat and the monkey during the first few months of postnatal life. Although there are much earlier clinical antecedents (see, for example, Worth 1903; von Senden 1960), the undoubted stimuli for much of this work were the demonstrations by Hubel and Wiesel in the early 1960s of the remarkable susceptibility of connections in the kitten's visual cortex to modification by various forms of early visual deprivation (Wiesel and Hubel 1963, 1965; Hubel and Wiesel 1963, 1965). The susceptibility of functional connections to modification early in life is particularly exemplified by the plasticity of binocular connections.

In normal animals, the visual signals from the two eyes are usually in concert, so binocular cortical connections present at birth are strengthened and refined during the first few postnatal months (Hubel and Wiesel 1963; Pettigrew 1974; Blakemore and Van Sluyters 1975; Buisseret and Imbert 1976; Frégnac and Imbert 1978). This manifests itself as a gradual refinement in the orientation and disparity tuning of binocular cortical cells (Pettigrew 1974; Blakemore and Van Sluyters 1975; Buisseret and Imbert 1976). In the absence of visual input to either eye, specificity either is not maintained or fails to develop (Blakemore and Van Sluyters 1975; Buisseret and Imbert 1976), so that after several months a high proportion of cortical cells respond only very weakly or not at all to visual stimuli, and those that do respond lack the specificity for orientation or disparity characteristic of normally reared animals (Wiesel and Hubel 1965; Imbert and Buisseret 1975; Kratz and Spear 1976; Cynader et al. 1976). The extreme vulnerability of binocular connections to disruption early in life is exemplified by the consequences of various manipulations of the visual input that either place one eye at a competitive advantage over the other (as in monocular occlusion) or else (as in strabismus) render the visual input to the two eyes discordant. After even brief periods of monocular visual deprivation imposed in the first few months of life, the vast majority of cortical neurons lose their functional connections with that eye so that they may only be activated through the eye that had been open (Wiesel and Hubel 1963; 1965). Disruptions of concordant binocular visual experience (such as results after surgically induced strabismus where one eye is made to deviate) also result in a marked breakdown of binocularity, leaving an approximately equal number of cortical cells that can only be monocularly activated through one or the other eye (Hubel and Wiesel 1965).

It was soon apparent that these marked functional changes in the visual pathway could only be produced by visual deprivation imposed within a certain "sensitive period," since similar effects were not observed in the cortex after comparable periods of deprivation imposed on adult animals (Wiesel and Hubel 1963). The approximate time course of the period of sensitivity to the effects of monocular visual deprivation was later delineated by Hubel and Wiesel (1970) from examination of the effects of varying periods of visual deprivation imposed on kittens at various ages. They concluded that the sensitive of critical period "begins suddenly near the start of the fourth week, remains high until sometime between the sixth and eighth weeks, and then declines, disappearing finally around the end of the third month." This conclusion was later confirmed by Blakemore and Van Sluyters (1974), who in addition showed that during the sensitive period there was a potential for substantial recovery from the effects of previously imposed visual deprivation. They showed that, if at the time the sutured eyelids of the deprived eye were opened those of the other eye were sutured closed (thereby forcing the animal to employ the previously deprived eye), the effect of the initial deprivation could be partly or even completely reversed, depending upon the age at which the reverse suture was performed. However, if the reverse suture was delayed until 14 weeks of age, almost no recovery was evident after the nine weeks that they allowed to elapse before recording.

These very striking early demonstrations of modification of the ocular dominance of cortical cells, as well as the reports that followed of modification of other visual-response characteristics of cortical cells (such as their specificity for orientation and direction of movement) by suitably biased early visual exposure (for reviews see Daniels and Pettigrew 1978; Blakemore 1978), led naturally to investigations of the consequences for visual perception and behavior. One important motive for such studies was the hope that documentation of the behavioral deficits that followed various forms of early visual deprivation might throw some light on the relationship between the visual-response characteristics of cortical neurons and perception. But in addition there was the very real need for behavioral data to complement and even support physiological data on the effects of various forms of early visual deprivation—data that may be prone to error due to sampling bias during recording (Stryker and Sherk 1975; Blasdel et al. 1977; Blakemore 1978).

Until recently, adequate assessment of the perceptual deficits that follow various forms of early visual deprivation was made difficult by the lack of behavioral techniques that would permit rapid measurement of the visual capacities of young kittens. Consequently, it has not been possible to assess the initial effects on perception of early visual deprivation or of the time course of any recovery. However, a behavioral technique developed in our laboratory (Mitchell et al. 1976a, 1977a) that permits the rapid assessment of the visual capacities of kittens as young as 4 weeks now allows measurement not only of the normal development of various visual capacities, but also of the immediate effects on these capacities of various regimens of visual deprivation, as well as the time course of any recovery.

We have employed this technique to assess the effects of various deprivation procedures on the animal's visual acuity for square-wave gratings. The principal reason for favoring this particular measure of visual function over other more traditional measures, such as the ability to perform various complex pattern discriminations (see, for example, Rizzolatti and Tradardi 1971; Ganz et al. 1972) or even simple visuomotor behaviors (see Movshon 1976), was the belief that these measures may be easier to correlate with certain visual-response characteristics of cortical cells (particularly their spatial resolution) than other measures.

After a detailed description of the method and its more recent modifications, we will present a number of examples of its use, concentrating on studies that examine the behavioral effects of early monocular and binocular visual deprivation. These experiments raise a number of important questions concerning the nature of the sensitive period.

The Kitten Jumping Stand

Development

The jumping stand was originally developed in collaboration with F. Giffin and F. Wilkinson after several unsuccessful attempts with conventional techniques to train kittens to make simple pattern discriminations. Although it was possible to train kittens on light-dark discriminations in a T maze after many trials spread over several days, it was obviously impossible for them to learn pattern discriminations sufficiently quickly for the technique to be useful as a means of documenting changes in perceptual abilities during early development. Consequently, we decided to investigate methods that relied on the animal's observational responses rather than methods that required the animal to learn various visual discriminations.

One observational behavior that we attempted to exploit was "preferential looking" (Teller et al. 1974; Atkinson et al. 1974; Leehey et al. 1975), a procedure that has been widely used to study the perceptual capacities of human infants. This procedure exploits a natural tendency of human infants to look preferentially at contoured stimuli over adjacent noncontoured surfaces (especially during the first 6 months of life). None of our kittens exhibited the slightest degree of interest in any display that we presented in front of a slightly elevated platform on which they were placed. Instead, they spent most of the time looking down at the floor. Even when the visual displays were placed below them, the kittens gave no evidence of attending to a contoured stimuli (usually a grating pattern) more than to a noncontoured field. At this juncture we investigated the possibility that animals might preferentially jump to the contoured field beneath them if prompted to leap from the platform. Although our kittens gave no evidence of this, we were so impressed with their willingness to jump small distances and by the grace with which they jumped that we decided to attempt to train them to leap toward the contoured stimulus. To accomplish this we constructed a jumping stand similar to the basic design depicted in figure 16.1. This arrangement immediately proved so successful that we have found it unnecessary to make any other than minor modifications to the original design. Our apparatus turned out to be similar in design to the jumping stand that Lashley (1930) developed for his studies of the visual capabilities of rats. With Lashley's jumping stand, the animal is required to jump horizontally from a platform to one of two trapdoors on which the visual stimuli are placed. If the animal leaps to the correct stimulus, the trapdoor opens to provide the access to a food reward; if it makes the incorrect choice, the trapdoor remains closed so that the rat

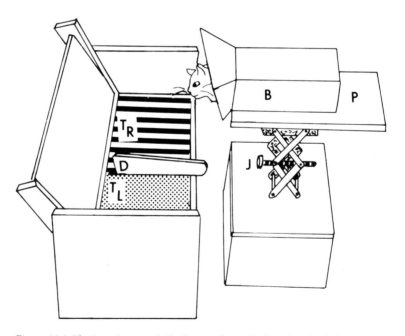

Figure 16.1 The jumping stand. T_L, T_R: trapdoors. D: Divider. P: platform. J: jacks. B: box.

falls to the floor. By contrast, our version of the jumping stand requires kittens to jump down for a food reward toward one of two adjacent stimuli placed beneath them. Although our stimuli are also displayed on trapdoors, these are both normally kept closed. Incorrect choices are only punished by denial of the food reward.

The Basic Apparatus

The basic apparatus consists of a rectangular box about 40 cm high whose top is about 65 cm long and 40 cm wide. Cut into the top of the box are two trapdoors separated by a wedge-shaped divider. The visual stimuli are placed or projected onto the two trapdoors. The kitten is trained to leap from a platform to one of the two stimuli for a food reward and for petting. These rewards are denied if the animal jumps to the incorrect stimulus. A wall about 25 cm high surrounds three sides of the box in order to discourage the animal from attempting to escape. The platform is supported by two laboratory jacks that are bolted together so that the height of the platform can be adjusted to any desired level up to 100 cm. The animals are trained to jump from the platform by inducing them to walk to the edge of the platform through an open-ended box whose dimensions are such that the animal cannot easily turn around once it is inside. Stimuli that are placed on the trapdoors are evenly illuminated by a lamp under the platform.

Measurement of Visual Acuity

The vast majority of experiments to date have involved measurements of the visual acuity for square-wave gratings. In most cases we have employed as stimuli photographic reproductions of commercially prepared high-contrast square-wave gratings, pasted onto thick pieces of cardboard 25 cm square and placed directly on one of the trapdoors. The other stimulus is a grey uniform field of exactly the same mean luminance, so that, to the human eye, the two stimuli appear indistinguishable when the stimuli are sufficiently distant that the grating can no longer be resolved. In order to facilitate measurement of acuity, a finely graded series of gratings of different periods was prepared. The step size between the grating period was made extremely small—about one-eighth of an octave (where an octave represents a difference of a factor of 2 between the periods of two gratings)—so that it was possible to alter the spatial frequency of the gratings in extremely small steps by combining changes in the grating period with small alterations of the elevation of the platform.

Occasional experiments have been run using stimuli projected from above onto matte white paper placed on the two trapdoors. By means of circular neutral-density wedges placed in front of the projector lenses it is possible to precisely match the luminances of the two stimuli as well as to alter their absolute luminance to a maximum of 170 candelas per m². This is somewhat higher than the luminance of the stimuli placed directly on the trapdoors (110 cd/m²) and illuminated by the lamp under the platform. All stimuli have a Michelson contrast ratio (Campbell and Green 1965) close to 1.0.

We initially train animals by setting the platform only a few centimeters above the divider so that the animal can be induced by means of a piece of food held in front of its nose to leave the platform and step onto the stimulus. During training one of the trapdoors is left open so that the animal has virtually no choice but to step onto the closed trapdoor on which is placed the positive stimulus. For measurement of acuity this is a grating with a period of 4 cm. Once a kitten has stepped onto the grating it is rewarded with petting and a small amount of food. The grating is then placed on the other trapdoor, with its companion trapdoor open for the next trial. This procedure is repeated for about ten trials or until the kitten steps promptly onto the closed trapdoor. Once this is achieved, the platform is elevated several centimeters with each trial until the animal freely jumps about 25–30 cm to the correct stimulus. At this juncture the negative stimulus is introduced on the trapdoor adjacent to the grating. The majority of kittens continue to jump toward the grating. The very occasional leaps toward the grey stimulus are punished by denial of the rewards and the animal is immediately placed back on the platform.

After a kitten has made 10 or more consecutive jumps to the grating, which is altered from trial to trial in a quasirandom sequence such as that proposed by Gellerman (1933), the platform is raised once more in small steps to a height that permits measurements of visual acuity to be initiated. For these measurements the platform is placed as high as is consistent with the animal's ability to jump, in order to minimize the effects of minor fluctuations in the distance of the animal from the stimuli prior to each jump.

As indicated in earlier publications (Mitchell et al. 1976a, 1977a), we have employed a number of different procedures to estimate a given animal's acuity. Our earliest procedures (Mitchell et al. 1976a) made extensive use of the trapdoor for punishment of incorrect responses. We have since found that such severe punishment causes most animals to exhibit extreme distress, which in turn detracts from their overall performance. In the last few years we have found it unnecessary to punish the animals by dropping the trapdoor.

In our experience the most consistently reliable procedure has been a modified version of the staircase procedure, a variant of one of the common psychophysical methods utilized for the measurement of human sensory thresholds (Cornsweet 1962). The basic feature of this method is that the stimulus magnitude on any trial or block of trials is determined by the response on the immediately preceding trial(s). After initial training with a grating of period 4 cm, the animal is given a block of trials (usually 5 or 10) with a grating of a slightly smaller period (and thus a somewhat higher spatial frequency). If it achieves a criterion level of performance on this block of trials, which we have arbitrarily set at 4 out of 5 or 7 out of 10 correct, the spatial frequency for the next block of trials is increased once more. This procedure is repeated until a spatial frequency is reached where the animal fails to achieve criterion performance. It is safe to assume that the animal experiences difficulty discriminating a grating of this spatial frequency from the adjacent uniform field. This view is supported by the animal's general behavior. Initially, with gratings having very low spatial frequencies, the animal jumps quickly and without any prompting. However, as the spatial frequency of the gratings is increased, the animal becomes much more hesitant and the latency between jumps increases dramatically. At the point where an animal fails to achieve criterion performance, it often shows signs of distress. Typically, they cry and attempt to back out from the jumping box before reluctantly leaping. After a block of trials in which criterion is not achieved, the spatial frequency of the grating is decreased several steps to a value that enables the animal to perform the task once more without any sign of distress. The procedure is then repeated until a spatial frequency is reached where the animal once again no longer achieves criterion performance. Frequently this is with a grating of a somewhat higher spatial frequency than that on which the animal first failed. Usually the procedure is repeated at least once more, or until either two consecutive threshold estimates agree or the animal's general behavior declines.

Figure 16.2 shows typical results obtained from a normal kitten when four consecutive estimates of acuity were made in a single session. Each horizontal bar indicates the spatial frequency of the grating on a

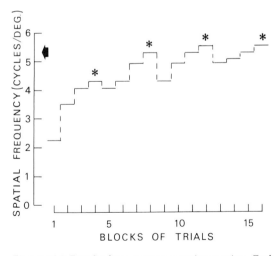

Figure 16.2 Results from a representative session. Each horizontal bar indicates the spatial frequency of the grating employed on successive blocks of 10 trials. Blocks of trials on which the animal failed to attain criterion performance are indicated by asterisks. The highest spatial frequency on which criterion performance was achieved on this series is indicated by an arrow.

Development of Vision in the Cat

particular block of 10 trials, beginning with a grating of 3.4 cycles per degree. While the animal achieved criterion performance (7 or more correct), the spatial frequency of the grating was increased for the next block of trials and so on until eventually the animal failed to make the discrimination. When this happened the spatial frequency of the grating was reduced several steps for the next block of trials to a value at which good performance could be reestablished. The procedure was then repeated until two successive estimates of acuity agreed or else the animal's behavior deteriorated. Thus, the graph adopts a profile with the appearance of an irregular staircase or a series of battlements. The general improvement in the animal's performance during the course of the session is reflected by the fact that the animal failed to achieve criterion performance on only the third block of trials with gratings of 4.3 cycles/degree, but after good visual behavior was reestablished at a lower spatial frequency performance remained above criterion for spatial frequencies up to 5.25 cycles/degree. On the third and fourth replications, performance deteriorated on both occasions at a slightly higher spatial frequency, 5.5 cycles/degree.

As with any psychophysical measurement, the criterion for threshold is somewhat arbitrary. We have usually adopted one of two criteria. The simplest is to define as threshold the grating of the highest spatial frequency for which the animal could achieve criterion performance (7 of 10 correct). In the case of the session depicted in figure 16.2 this would be 5.3 cycles/degree. Alternatively, it is possible to construct a frequency-of-seeing curve by combining all the trials at each spatial frequency. The threshold can be derived from this curve by adopting as criterion the spatial frequency for which the animal was correct on 70% of trials. For the animal of figure 16.2 the threshold defined in this way, 5.3 cycles/degree, was actually the same as that determined by the other criterion.

We have consistently found that measures of acuity obtained on the jumping stand are as high as and sometimes even higher than values obtained on adult animals by means of more time-consuming behavioral measures (Blake et al. 1974; Bisti and Maffei 1974; Jacobson et al. 1976; Bloom and Berkley 1977). Animals behave on the jumping stand as if they are severely taxed by the task. Performance with suprathreshold gratings of low spatial frequency tends to be flawless, and only near threshold do animals tend to make many errors. Consequently, frequency-of-seeing curves obtained from data such as shown in figure 16.2 tend to be quite steep (Mitchell et al. 1977a), so it is possible to titrate thresholds quite precisely. Besides making errors of judgment, animals exhibit a number of other indications that the task is more difficult at higher spatial frequencies. Whereas at low spatial frequencies they will jump almost without hesitation, at higher spatial frequencies the animals often will look from one stimulus to the other many times before they leap. As previously mentioned, near their threshold animals exhibit clear signs of distress.

Thresholds obtained on the jumping stand with older animals, beyond the age where their perceptual capacities are rapidly changing, tend to be highly repeatable. Furthermore, thresholds obtained by changing the period of the grating, with the jumping height held constant, tend to be identical with thresholds obtained by progressively increasing the jumping distance to a grating whose period is held constant. This suggests very strongly that the performance declines when the grating becomes difficult to resolve rather than because of the other changes, such as the different motor demands of the task as the jumping height is altered. In addition, this result is just one of the arguments that rule out the possibility that the animals employ small local or overall luminance differences between the stimuli in making their judgments. The very fact that performance declines to chance levels at high spatial frequencies is a convincing argument in itself. Furthermore, test trials made with small but inconsistent luminance differences introduced between the two stimuli do not disrupt performance, as would be expected if the animals were employing a consistent luminance cue in their judgments.

We have frequently found that it requires considerable perseverance on the part of the experimenter to obtain good acuities in some animals. Certain animals quickly adopt other strategies (frequently position habits) once the task becomes at all difficult. Some of the strategies that we have successfully employed to overcome these problems have been outlined in detail (Mitchell et al. 1977a); here we emphasize only that we have found it helpful during a session to persevere through a number of reversals using the staircase procedure.

Measurement of Other Pattern Discriminations

The jumping stand can be used to train animals in more complex pattern discriminations. Conventional discriminations of this sort (for example, between geometrical shapes such as squares and circles) are frequently learned in only a handful of trials. Although we have not made extensive use of the jumping stand to examine such forms of pattern discrimination, it has been successfully employed by others for this purpose (Wilkinson and Dodwell 1980) as well as for the examination of the ability of animals to discriminate random from nonrandom dot arrays (V. Emerson, pers. comm.).

Measurement of Depth Perception

In collaboration with M. Kaye, we have made some minor modifications to the jumping stand to permit quantitative measures of the abilities of cats to perceive depth (Mitchell et al. 1979). The principle of the method is illustrated by figure 16.3. Kittens are prompted to leap

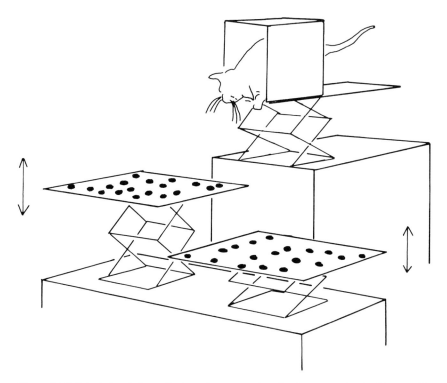

Figure 16.3 Principle of the technique for measuring depth perception in kittens. The animals are trained to jump toward the platform that is closest to them, in this case the platform on the animal's right.

from the platform (P) onto the nearer of two surfaces (S_1 and S_2) located beneath them. The distance separating the two surfaces in depth (d) is made progressively smaller until eventually the animal no longer jumps correctly to the nearer surface; from this it can be concluded that the difference in depth between the surfaces can no longer be discriminated by the animal from the platform.

Normal animals' ability to discriminate much smaller separations in depth using both eyes than under monocular viewing conditions implies the existence of a cue to depth that is available only with binocular viewing. Exactly the same situation obtains in human vision; this fact was exploited by Helmholtz (see Helmholtz 1924) and later by Howard (1919) as a means of efficiently screening human observers for the presence of stereoscopic vision. The assumption was that individuals who performed no better with two eyes than with one lack stereopsis. Thus, by analogy with humans, it is not unreasonable to suppose that the superiority of binocular performance in cats is also due to the additional cue to depth provided by stereopsis.

The Howard-Dolman box (Hirsch and Weymouth 1948) was widely used to exploit the superiority of binocular over monocular depth discrimination as a means of screening individuals for stereopsis. The observer was required to adjust a movable rod in depth until it appeared at the same depth as a flanking rod when both rods were viewed through an aperture that obscured their extremities. Under monocular conditions the task is almost impossible within the range of excursion of the movable rod, since observers have to rely solely on monocular cues to depth (such as differences in size) to make their settings. Consequently, the smallest discriminable displacement in depth with binocular vision is often between one and two orders of magnitude smaller than that which can be observed monocularly.

Our method employs the same principle as the Howard-Dolman test. As with that test, we have also found it necessary to obscure the extremities of the stimuli with masks in order to accentuate the superiority of binocular over monocular performance. Figure 16.4 depicts our apparatus in schematic fashion. The stimuli consist of two transilluminated clear plastic plates on which are randomly placed circular dots of various sizes. Each plate can be raised and lowered beneath a mask (M) placed immediately below a glass plate (P), onto which cats are trained to jump. To the cat, the stimuli appear as two sheets of randomly separated dots floating in space beneath the apertures. The animal's task is to jump to the surface that appears closest. Measurement is made of the smallest separation in depth between the two stimuli that can be discriminated under both monocular and binocular viewing

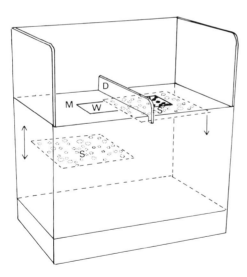

Figure 16.4 Modified jumping stand for measurement of depth perception. Kittens are trained to jump toward the nearer of the two stimuli (S,S') placed on either side of a divider (D). Stimuli consist of plates of clear plastic on which are placed a series of dots of different sizes. The two stimuli are illuminated from below and are viewed through a window (W) cut in a mask (M) that obscures their extremities. The mask lies immediately below the clear sheet of glass onto which the kittens jump.

Figure 16.5 Photographs, taken from above, of a kitten jumping to the nearer of two stimuli separated by the maximum amount (23 cm) that can be achieved with the apparatus of figure 16.4. The photographs provide an indication of the magnitude of the monocular depth cues that are available to the animal.

conditions. Figure 16.5 shows photographs taken from above of a cat jumping to the nearer of two stimuli separated by 23 cm, the maximum amount possible with our apparatus. These photographs illustrate the monocular appearance of the stimuli as well as the magnitude of the monocular cues to depth that are available when the separation of the stimuli in depth is at a maximum.

We have found that normal animals jumping from a distance of 75–85 cm typically discriminate separations between the two stimuli of only 2 cm when using both eyes, but under monocular conditions the animal behaves as if it has to learn a new discrimination. Even after extensive training they are typically only able to discriminate separations of 10–15 cm from the same jumping distance of 75–85 cm. The presence of such clearly superior performance under binocular viewing conditions strongly implies the presence of stereopsis. Nevertheless, to rigorously prove the presence of stereopsis it is necessary to demonstrate that the cat can perceive depth from retinal-disparity cues alone. Because the anaglyphic presentation of stimuli this requires would necessitate that the cat wear goggles or contact lenses so as to allow separate control of the visual stimuli to the two eyes, it is likely that this ultimate proof of the presence of stereopsis may be difficult. The simple test that we have devised can be thought of as a useful short cut for rapid screening.

Application of the Method to Studies of Kitten Visual Development

Normal Development of Visual Acuity

Kittens as young as 4 weeks can be trained on the jumping stand. Consequently, it is possible to trace the development of acuity from about 30 days of age, although it is likely that the measurements at this age may underestimate the animal's true capabilities because of the animal's lack of experience with the task. Figure 16.6 shows the results of measurements of the acuity of a number of different kittens obtained at various ages by Giffin and Mitchell (1978). These findings complement earlier results, obtained with gratings of lower mean luminance (Freeman and Marg 1975; Mitchell et al. 1976a), which show that the acuity of the kitten develops rather slowly, reaching adult values of 6–7 cycles/degree at 3–4 months of age.

There are a number of factors, both peripheral and central, that could be responsible for the gradual improvement in vision. The first of these to consider are changes in the optical media. Ophthalmoscopic observation of the fundus of the eye of a kitten shortly after the eyelids have opened is difficult because of the persistence of embryological vascular material around the crystalline lens. As this vascular network is absorbed over the course of the first 4 weeks of life, there is a marked improvement in the optical quality of the eye (Freeman and Lai 1978; Bonds and Freeman 1978). Subsequently, the optical quality improves only slightly. Quantitative measurements of the optical performance of the eye (Bonds and Freeman 1978) indicate that the optical quality of the

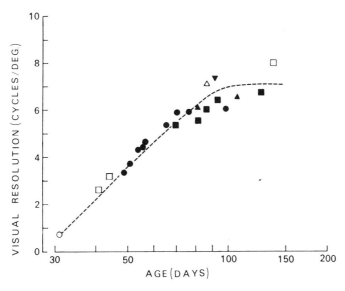

Figure 16.6 Plot of visual acuity of seven normal kittens measured on the jumping stand at various ages. Acuity develops slowly during the second and third months, reaching adult levels at about 4 months of age. Source: Giffin and Mitchell 1978.

eye is very good by the fifth week and thus does not impose a limit to the slow improvement in the kitten's visual capabilities over the next 12 weeks.

Although changes in optical quality do not play a major role in the slow development of acuity, the gradual increase in the dimensions of the eye itself as the animal grows during this period (Donovan 1966) must contribute to the increase in acuity. It is not possible to accurately calculate the influence of this factor on acuity because of the present lack of data on changes in the posterior nodal distance of the eye and on interreceptor separation during the first few postnatal months. However, Donovan's (1966) data on the development of the linear extent of the retina suggest little change in the dimensions of the eye after 6 or 7 weeks of age. Thus, simple changes in the dimensions of the eye, which are most rapid in the first month of life, could only be expected to contribute to the rate of development of acuity during the first 7 weeks.

Since we have eliminated optical and other peripheral factors as major contributing factors to the gradual improvement of acuity during the second and third months of life, it is not unreasonable to attribute much of the slow improvement of acuity to changes occurring more centrally in the visual pathway. A few studies have already attempted to link the development of acuity to changes in the spatial resolving power of neurons at various levels in the visual pathways. This has been done by determining the highest spatial frequency of a grating that, when drifted through the receptive field of a cell, elicits a modulated response from that cell. Figure 16.7 shows the results of two such studies. The

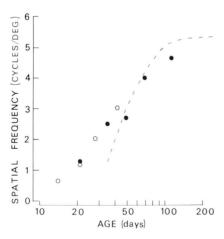

Figure 16.7 Comparison of visual acuity of cells in lateral geniculate nucleus (LGN) and visual cortex of kittens of various ages. ○ depict the cell with the highest visual acuity observed among a sample of cells recorded in the visual cortex of kittens of various ages by Derrington (1978); ● depict the highest acuity observed in cells in the LGN as a function of age by Ikeda and Tremain (1978). Dashed line shows behaviorally measured visual acuities of four kittens at different ages obtained with gratings of luminances comparable to those used in the physiological measures by Mitchell et al. (1976a).

Mitchell, Timney

tacit assumption is made here that behavioral acuity is dictated by the neuron with the highest sensitivity. Although this is not the assumption made by Ikeda and Tremain (1978) (see their figures 5 and 6), it is one of the major assumptions of psychophysics, and it underlies the conventional interpretation of such fundamental functions as dark-adaptation curves as well as color-increment-threshold curves (Stiles 1949).

Although the cortical data are rather limited, the agreement with the geniculate measurements is remarkable. It appears from these data that the gradual development of the resolving power of cortical cells parallels the development of the spatial resolution of their afferent input, the sustained cells in the lateral geniculate nucleus. If the possible inaccuracies in the behaviorally measured acuity of the youngest animals are taken into account, the overall agreement of the behavioral data with the spatial resolution of the geniculate and cortical cells is extremely good.

Effects of Various Forms of Early Visual Deprivation on Development of Visual Acuity

Monocular Visual Deprivation
We have conducted a number of experiments to determine the effects on visual acuity of fixed periods of monocular occlusion imposed on kittens of different ages. These experiments parallel those of Hubel and Wiesel (1970) which delineated the period of susceptibility of visual-cortex neurons to the effects of monocular deprivation. In the main our experiments complement the earlier physiological experiments, although our data strongly suggest that the cortex may be affected by monocular deprivation for a somewhat longer period.

Figure 16.8 shows the effect on visual acuity of a 23-day period of monocular eyelid suture imposed on three kittens at either 42, 64, or 86 days of age (Giffin and Mitchell 1978). At the termination of the deprivation the sutured eyelids were parted to allow visual input to both eyes. In each case filled and open symbols depict the result of measurements of acuity of, respectively, the nondeprived and deprived eyes. The latter measurements were made with an opaque contact lens occluding the nondeprived eye; the former were usually assessed under binocular viewing conditions, as control experiments revealed that the results obtained with both eyes open were no different from those obtained with the deprived eye occluded.

The effects of the monocular occlusion imposed on the kittens at 6 weeks of age were very dramatic. When the vision of the formerly deprived eye was first tested several hours after the eyelids had been opened, the animal showed the severe behavioral defects first described by Wiesel and Hubel (1963), appearing completely blind on all conventional tests of simple visuomotor behavior. To test for any visual capac-

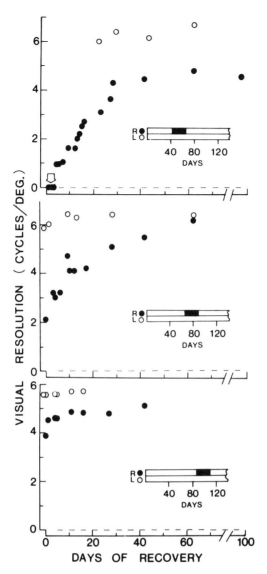

Figure 16.8 Effect on vision of 23 days of monocular occlusion imposed on three kittens, at age 42, 64, and 87 days, respectively. Each graph shows the results of measurements of visual acuity of the formerly deprived eye at various times following termination of the period of deprivation. The visual histories of the three animals are indicated schematically as an inset to each graph. Note the gradation in the severity of the initial deficits and the magnitude of the subsequent recovery. The animal deprived at 42 days of age appeared completely blind when it first was forced to use its deprived eye; however, signs of vision began to emerge within a day as indicated by the arrow. Data from Giffin and Mitchell 1978.

ity on the jumping stand the platform was set very low, only a few cm above the trapdoors. One trapdoor was opened while on the other was placed a grating with a 4-cm period. The kitten, which was quite agitated when placed on the platform, was induced by gentle prodding to leave the starting box. Immediately on emerging from the box it felt for the wooden divider with the front paws. Using the divider as a support, the animal reached out and felt for the closed trapdoor with each paw in turn. Although it could locate the closed trapdoor in this way by tactile cues, it was completely unable to do so by visual cues alone, as could be easily shown by raising the platform a few centimeters so that the animal could no longer reach the divider. However, after 2 days it became quite obvious that the animal could now see with its deprived eye, since it was able to jump to the closed trapdoor on the jumping stand from a height of 20 cm. By the fourth day it was capable of making pattern discriminations between a grating and an opposed uniform field and in fact demonstrated an acuity of 0.9 cycles/degree. The visual acuity of this eye improved rapidly over the next few weeks and eventually attained an acuity of 4.8 cycles/degree after 2 months, a value that was only 30% lower than that of the nondeprived eye.

By contrast, the effects of a similar period of monocular occlusion imposed at later ages were not nearly so profound; instead of being rendered apparently blind by the period of deprivation, the other two animals suffered only a reduction in the visual acuity of their deprived eyes. The initial acuity was lower (only 2 cycles/degree) in the animal deprived at 64 days of age than the animal deprived at 86 days (3.8 cycles/degree). Again the acuity of the deprived eye improved considerably in the next few weeks so that eventually the acuity of the deprived eye was only marginally poorer than that of the nondeprived eye.

Two aspects of these results deserve mention. First, the initial effects of the period of monocular occlusion were graded in severity according to the age at which eyelid suture was imposed. This result complements the neurophysiological finding of a gradual decline in the degree of susceptibility of the visual cortex to the effects of monocular deprivation from a peak around the fourth week of postnatal life. The second point to emphasize is the remarkable degree to which the acuity of the deprived eye recovered in these animals even though no steps were taken (such as reverse suture) to force the animal to use this eye during the period of recovery. In fact, later experiments (Mitchell et al. 1977b; Giffin and Mitchell 1978) indicate that the rates of recovery of vision of the deprived eye after monocular occlusion are remarkably close in the two situations where the animal either has both eyes open after the period of deprivation, or else is forced to use its deprived eye by suturing closed the eyelids of the initially open eye (a reverse suture). Although the acuity eventually attained by the deprived eye was always somewhat higher in the latter case, the initial stages of recovery were almost identical. Physiological recordings made on these animals once behavioral recovery was complete revealed that even in the animals that had both eyes open after the period of monocular occlusion a sizable proportion of cortical cells could be excited through the formerly deprived eye. Nevertheless, the proportion of cells dominated by the formerly deprived eye of the animals that were reverse-sutured after the same period of deprivation was somewhat higher (Mitchell et al. 1977b). The recovery observed in the former situation cannot be attributed to the brief periods of occlusion of the nondeprived eye during behavioral testing of the vision of the deprived eye, since almost identical recovery was observed in animals that received no behavioral training (Mitchell et al. 1977b; Olson and Freeman 1978).

The fact that substantial physiological and behavioral recovery occurs in animals that have both eyes open after the initial period of deprivation suggests that functional connections with the deprived eye are reestablished passively by a noncompetitive mechanism that manifests itself once occlusion is terminated. In addition to the recovery induced by this mechanism, additional recovery must be induced once the initially deprived eye is placed at a competitive advantage over the other by means of reverse suture.

Experiments to Determine Period of Sensitivity to Effects of Monocular Deprivation

Two converging lines of behavioral evidence lead us to suspect that the visual cortex may remain susceptible to the effect of monocular deprivation somewhat longer than was previously thought. The first indication of this arose from studies of the recovery of vision after prolonged periods of monocular deprivation imposed from birth. Figure 16.9 shows the recovery of the acuity of the deprived eye of three kittens that were monocularly occluded until nearly 4 months of age. Two of these (108R and 113R) were reverse-sutured after the period of deprivation; the other (120B) was allowed binocular vision throughout the period of recovery. Although all the animals appeared blind for a long period when using the deprived eye, eventually they all attained an acuity in excess of 2.5 cycles/degree in this eye.

We subsequently examined two cats that had been monocularly sutured from birth for 5 and 6 months and thereafter allowed binocular vision. Both recovered some vision with the deprived eye. Figure 16.10 shows the time course of this recovery for the cat (160B) that was deprived until 5 months of age. Eventually this animal achieved an acuity of 1.4 cycles/degree with its deprived eye. The animal deprived for 6 months eventually achieved somewhat better acuity with its deprived eye: 2.1 cycles/degree. The period of apparent blindness prior to the emergence of any sign of vision in the deprived eyes of these animals was quite long, extending well beyond a month.

With increasing length of deprivation, the period of apparent blindness becomes progressively longer and the acuity eventually attained

Mitchell, Timney

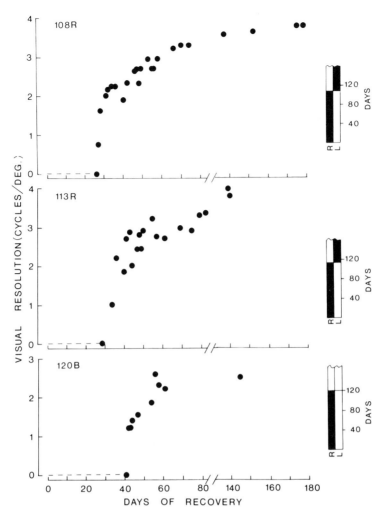

Figure 16.9 Recovery of vision in the deprived eye of three cats, 108 R, 113 R, and 120 B, monocularly deprived from birth for (respectively) 108, 113, or 120 days. After the period of occlusion, either the animal was forced to use the formerly deprived eye by reversal of the eyelid suture (108 R and 113 R), or else, as in the case of 120 B, visual input was allowed to both eyes. Horizontal dashed line indicates period of time animals appeared to be functionally blind when using deprived eye. Visual histories are indicated schematically at right.

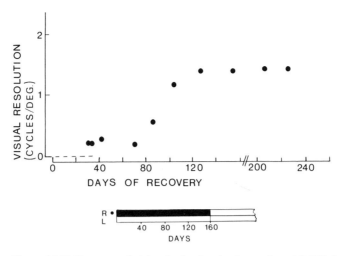

Figure 16.10 Recovery of vision in the deprived eye of a cat (160B) that was monocularly sutured from birth until 5 months of age, after which the sutured eyelids were opened to allow visual input to both eyes. Period during which animal appeared to be functionally blind is indicated by dashed line.

becomes lower. These two facts can be observed in figures 16.10 and 16.11, which plot data from all of our monocularly sutured animals that had been deprived for varying periods of time. In figure 16.11 we have plotted, as a function of the period of deprivation, the number of days that each animal required for its deprived eye to attain a criterion level of vision, which was defined as the ability to resolve gratings of 0.5 cycles/degree. Figure 16.12 shows the final acuities eventually attained by the deprived eye of each animal as a function of the age at which deprivation was terminated. In order to highlight the regular nature of the decline in the final acuities attained with increasing periods of occlusion, the latter graph was plotted on semilogarithmic coordinates.

As with animals deprived for shorter periods of time, it is tempting to conclude that the small degree of behavioral recovery observed in the animals deprived for 4 months or more reflected some limited recovery from the initial effects of monocular deprivation in the visual cortex. The very regular nature of the decline of acuities exhibited by the animals of figure 16.12 suggest that this may be so at least for animals deprived up to 5 months of age. However, if the lines drawn by eye through the data are continued and extrapolated to the abscissa, they suggest that animals deprived beyond a certain age may never recover any vision at all. This certainly is not the case, since there have been reports of animals recovering some pattern vision even after deprivation for a year or more (Chow and Stewart 1972). This means that the curve of figure 16.11 must eventually flatten out, as suggested by the dotted line. Although the regular nature of the decline in the final acuity attained by animals deprived to about 5 months of age is consistent with the notion that the recovered vision reflects some limited recovery

Mitchell, Timney

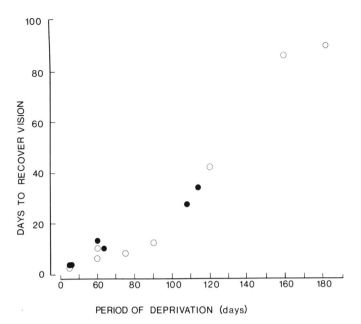

Figure 16.11 Number of days required by a number of monocularly deprived animals to recover a criterion level of vision (defined as the ability to resolve gratings of 0.5 cycles/ degree with the deprived eye. All animals were monocularly sutured from birth to 45–180 days of age. ○ represent animals that had both eyes open after the period of deprivation; ● represent animals that were reverse-sutured.

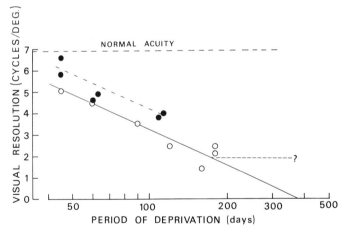

Figure 16.12 Final acuities attained by a number of monocularly deprived animals that had been deprived from birth to the ages shown on the abscissa (logarithmic scale). Symbols are as in figure 16.11. Separate lines of best fit have been fitted by eye through the two sets of data from reverse-sutured animals and those that had both eyes open after the period of monocular eyelid suture. The line drawn through the latter data (○—○) has been extrapolated to the abscissa to define the period of deprivation beyond which recovery may be impossible. An alternative possibility—that the curve flattens out for periods of deprivation extending beyond 6 months—is depicted by the queried horizontal dashed line.

Development of Vision in the Cat

in the visual cortex, it is possible that the vision acquired by animals deprived beyond 6 months of age may be mediated by visual areas other than the striate cortex.

There are at least two experimental ways to ascertain whether the behavioral recovery observed after extended monocular deprivation is mediated by the visual cortex. The most direct approach is to simply record from a sample of units in the visual cortex to determine if there has been any increase in the proportion of cells that can be influenced through the formerly deprived eye. In collaboration with M. Cynader we have recorded from two animals, 120B and 160B (see figures 16.9 and 16.10) that were monocularly sutured until 4 and 5 months of age, respectively (Cynader et al. 1980). One animal was deprived until 4 months of age (120B); then recordings were made 13 months after the sutured eyelid had been opened, long after it had attained an acuity of 2.45 cycles/degree with the formerly deprived eye. A very sizable proportion of cells, 53 out of 84 (63%), could now be influenced through the formerly deprived eye, although no cells were encountered that could be excited exclusively through that eye. Therefore, it is not at all unreasonable to suppose that the acuity eventually attained by the deprived eye of this animal was mediated by those cells in the visual cortex that had recovered their connections with this eye. The results from this animal were unambiguous, but the physiological findings from the other animal (160B), with which recordings were made 6 months after the sutured eyelids were opened, were somewhat equivocal. Of 30 cells encountered in this animal, 4 (13%) could be influenced through the deprived eye, a percentage that is only marginally different from that which would have been encountered in layers other than layer IV (Shatz and Stryker 1978) immediately on termination of monocular deprivation (Wiesel and Hubel 1963; 1965; Kratz et al. 1976). Thus, in this animal, there is the very real possibility that the behavioral recovery was mediated by cortical areas other than the visual cortex. The second approach that can be employed to determine if recovery after monocular deprivation depends upon recovery in the visual cortex is to observe whether the restored vision remains after ablation of area 17. This test was employed by Spear and Ganz (1975) to determine whether the ability of animals using their formerly monocularly deprived eye to make simple visual pattern discriminations was retained after lesions of the visual cortex. Their animals, which were monocularly deprived until 6 months of age, eventually attained the ability to make a number of simple visual pattern discriminations. However, the fact that this ability was completely abolished after lesions involving area 17 implied that this recovery depended upon the visual cortex. Although this experimental approach argues that the vision that was recovered by their animals that were monocularly sutured for 6 months was mediated by the visual cortex, it may be difficult to verify this directly by physiological recording. This is partly due to the need to sample a large region of

Mitchell, Timney

cortex, since it is probable that the proportion of cells that can be influenced through the deprived eye might be only marginally greater than that observed immediately after the period of deprivation. It remains to be seen whether the visual recovery observed after even longer periods of monocular occlusion is mediated by the visual cortex. However, Hoffman and Cynader (1977) were unable to observe any physiological recovery in area 17 in animals that had been monocularly sutured until 8 months of age, despite the fact that the deprived eye subsequently showed substantial recovery in its ability to mediate various pattern discriminations.

Does the Critical Period End at 4 Months of Age?

The considerable physiological recovery observed in cat 120B suggests very strongly that the visual cortex retains a certain degree of plasticity beyond 4 months of age. We have examined this possibility directly (Cynader et al. 1980) by examining both the behavioral and physiological effects of 3 months of monocular occlusion imposed on kittens at 4 months of age. Figure 16.13 shows the results of measurements of the acuity of the deprived eye immediately after the termination of occlusion, when the sutured eyelids were opened to allow vision to both eyes.

There was no doubt that the vision of the deprived eye was reduced on first being tested 2½ hours after the eyelids were opened. Even 24 hours later the acuity was still the same: 2.9 cycles/degree. However, the

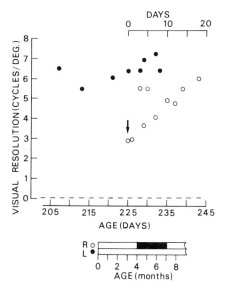

Figure 16.13 Measurements of the visual acuity of the undeprived (●) and the deprived (○) eyes of an animal that was monocularly sutured for 3 months at 4 months of age, after which time the eyelids of the sutured eye were opened to allow visual input to both eyes (see schematic visual history below). Arrow indicates acuity of deprived eye 2½ hours after termination of eyelid closure (zero on upper abscissa).

Development of Vision in the Cat

acuity of the deprived eye improved dramatically; after 3 weeks the acuity of this eye was very close to that of the nondeprived eye. Physiological recordings were made from the visual cortex, in a number of other cats reared in an identical manner, immediately after the period of monocular occlusion (Cynader et al. 1980). Here too there was quite a definite shift in the distribution of ocular dominance of the sample of cells recorded from these animals in favor of the nondeprived eye (see figure 16.14). At first sight, this appears to conflict with the finding Hubel and Wiesel obtained from a single cat that was also monocularly deprived for 3 months at 4 months of age (Hubel and Wiesel 1970, figure 6A). Here, however, physiological recording was made 16 months after the sutured eyelids were opened, thus allowing for the possibility of substantial physiological recovery from any immediate affects of the period of occlusion. The very fast behavioral recovery observed in the animal of figure 16.12 suggests very strongly that this physiological recovery may have occurred very quickly and certainly long before the physiological recordings were made 16 months later.

Taken together, our results indicate that the visual cortex remains susceptible to the effects of monocular deprivation beyond 4 months of age. Although our results do not indicate how much longer the period of susceptibility extends beyond 4 months, the preliminary results from cat 160B as well as results obtained by Hubel and Wiesel (1970) with a cat that was monocularly deprived for 4 months at 6 months of age

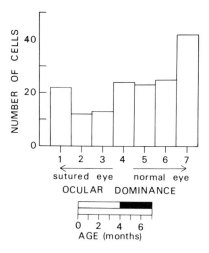

Figure 16.14 Distribution of ocular dominance in a sample of 161 cells recorded from four cats that were monocularly deprived for 3 months at 4 months of age. Recording electrode was located in hemisphere contralateral to formerly sutured eye. Cells have been classified into ocular-dominance groups according to the original scheme of Hubel and Wiesel. Cells in groups 1 and 7 can only be excited by visual stimuli presented to the eye contralateral or ipsilateral to the recording electrode, respectively. Cells classified as group 4 can be influenced approximately equally by both eyes. Binocularly excitable cells were classified in the intermediate group according to the relative influence of the two eyes. Source: Timney et al. 1979.

Mitchell, Timney

suggest that very little measurable plasticity exists beyond 6 months of age.

Recovery from Extended Total Binocular Visual Deprivation

There has been a long history of studies of the effects on vision of total visual deprivation, beginning with the compilation of clinical cases over a 900-year period by von Senden (1960). Behavioral experiments on animals preceded the pioneering physiological investigation of Wiesel and Hubel (1965) on the effects of binocular lid suture or dark rearing on the properties of cortical cells (see Ganz 1978 for a review). Subsequently, a number of studies have examined the immediate effects of long periods of visual deprivation on complex visuomotor behaviors and on the response characteristics of cortical cells, as well as the extent to which there is recovery from the initial deficits on exposure to a normal illuminated environment (Baxter 1966; Pettigrew et al. 1974; Cynader et al. 1976; van Hof–van Duin 1976; Kalil 1978).

Kittens reared with binocular lid suture or in total darkness from birth for 4 months or longer initially appear completely blind on being exposed to illuminated surroundings. Their first explorations of the new surroundings are performed in a characteristic hesitant crouching manner with their bellies pressed close to the ground. They nearly always walk into objects in their path, frequently exhibiting strong startle responses. Formal tests of simple visuomotor behaviors (such as visually triggered placing of the paws toward a horizontal surface, visual following of large moving objects with the head and eyes, or visual startle responses) are all negative, although the pupils constrict to an abnormal degree in response to light. Nevertheless, within a few days the above-mentioned visuomotor behaviors begin to appear in piecemeal fashion over the next 2 weeks in the case of animals deprived for 4 months and over the next 6 weeks in animals deprived for 6 months (van Hof–van Duin 1976; Timney et al. 1978).

There is general agreement that cortical cells in binocularly deprived animals are quite abnormal, although not nearly as abnormal as might be expected if the effect of binocular eyelid closure were the sum of the separate effects of monocular closure (Wiesel and Hubel 1965). Typically, cells respond very poorly to visual stimuli. The vast majority that are visually responsive are extremely broadly tuned for orientation as well as other stimulus parameters, and respond very weakly (Wiesel and Hubel 1965; Blakemore and van Sluyters 1975; Buisseret and Imbert 1976; Cynader et al. 1976; Kratz and Spear 1976; Singer and Tretter 1976). However, even animals that have been binocularly deprived for a year show considerable recovery of stimulus specificity in the visual cortex after having lived in illuminated surroundings for a time (Cynader et al. 1976).

In light of the latter observation, we examined the extent to which visual acuity recovers in animals that had been reared from birth in total

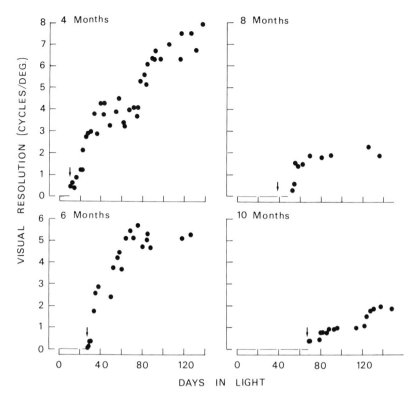

Figure 16.15 Time course of development of visual acuity in four animals introduced to a normally illuminated environment after having been reared in total darkness from birth to the ages shown. Arrows indicate first day on which animal showed evidence of visual behavior on jumping stand, defined as the ability to distinguish a closed from an open trapdoor. Throughout the period indicated by the dashed line, the animals were unable to discriminate between a grating and a uniform field of the same space-average luminance. With the shorter periods of deprivation, this was possible on the first day the animal showed any sign of visual behavior on the jumping stand. On the other hand, the animals that were dark-reared until 8 or 10 months of age showed some signs of vision several days before they were able to discriminate between a grating and a uniform field.

darkness for 4–10 months. The recovery observed in animals deprived for 4 and 6 months have been described in detail elsewhere (Timney et al. 1978). Figure 16.15 shows the results obtained with two representative animals from this study, together with results from two additional animals that were dark-reared until either 8 or 10 months of age. A characteristic pattern of recovery emerged in every case. After a period of apparent blindness that became progressively longer (over 2 months in the case of the animal that was dark-reared for 10 months) with increasing deprivation, signs of vision began to emerge. Concurrent with improved general visuomotor behavior was a long and gradual recovery in visual acuity. Animals that were binocularly deprived until 4 months of age eventually attained, after 4 months, visual acuities that were close to that of normally reared animals. The animal deprived for 6 months attained an acuity of 5.7 cycles/degree, only marginally lower than normal. The two animals that were deprived for longer never achieved acuities in excess of 2 cycles/degree. Nevertheless, this does represent a substantial recovery of vision, which was reflected by the fact that these cats' general behavior in the colony room was to casual inspection not very different from that of the normal animals in the same room.

The gradual increase in the time to recover signs of vision with increasing length of deprivation is plotted in figure 16.16 for all the dark-reared animals that we have studied. The behavioral and physiological recovery that is observed in animals that have been binocularly deprived from birth for long periods poses a challenge to the classical concept of the critical period as being a definite chronologically bounded period in life beyond which the visual system is immutable to the effects of perturbations of its visual input. Although it might just be possible that the recovery observed in the animals that were dark-reared for 4 months simply reflected the residual plasticity that is observed in the cortex of light-reared animals after 4 months of age, it is not possible to explain the high acuity of the animal that was dark-reared for 6 months in such terms.

The same evidence also eliminates an alternative extreme viewpoint: that the period of dark-rearing "freezes" the visual cortex in a condition close to the immature state that existed when the animal was placed in the dark. Although this would account for the normal acuities attained by the animals that were dark-reared for only 4 months, it cannot explain the lowered acuity of the animals deprived for longer periods. Instead, our results, together with the earlier physiological findings of

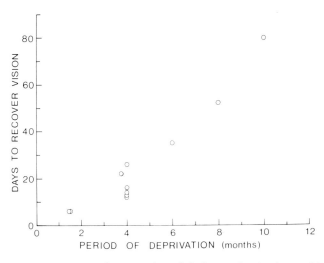

Figure 16.16 Time for a number of dark-reared animals to achieve a criterion level of visual performance (the ability to discriminate a grating of 0.5 cycles of deg from a uniform field) on exposure to illuminated surroundings as a function of the period of initial deprivation.

Development of Vision in the Cat

Cynader et al. (1976), suggest that dark rearing impedes the decline of sensitivity of the visual cortex to modification by its visual input. This possibility was tested directly in the next set of experiments, in which we collaborated with M. Cynader to examine the effects of periods of monocular occlusion on animals that had been dark-reared from birth for 4 months or more. If the visual cortex does indeed retain considerable plasticity after extended periods of binocular visual deprivation, then this would be reflected by a heightened susceptibility to modification by periods of monocular deprivation in comparison to normally reared animals of the same age.

Evidence for Extension of the Critical Period by Dark Rearing

Figure 16.17 shows results obtained from two cats that were reared in total darkness from before the time of natural eye opening until they were either 4 or 6 months of age. The right eyelids of both animals were sutured shut immediately after the animals were removed from the darkroom into illuminated surroundings. Prior to the time the sutured eyelids were opened, either 6 weeks (in the case of the animal that was dark-reared for 4 months) or 3 months (for the animal dark-reared for 6

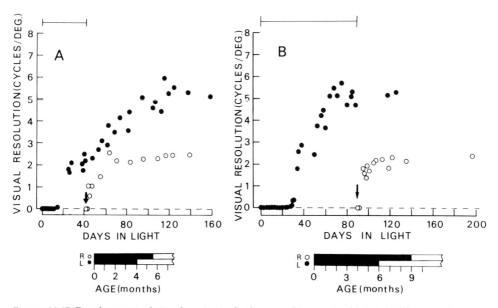

Figure 16.17 Development of visual acuity in both eyes of two animals that had been dark-reared until either 4 (A) or 6 (B) months of age on exposure to illuminated surroundings. As indicated schematically below each graph, both animals were subjected to monocular eyelid closure for a period of either 1½ or 3 months immediately upon introduction to an illuminated environment. Horizontal bar at top of each graph indicates duration of eyelid closure. (●) Acuity of nonoccluded eye; (○) results of measurement of the acuity of the formerly sutured eye after termination of the period of monocular deprivation. Horizontal dashed line through zero on ordinate indicates absence of any sign of form vision on jumping stand, such as ability to discriminate between an open and a closed door or between a very coarse grating and a uniform field. Result obtained on first test of vision mediated by the formerly sutured eye is indicated by arrow.

months) later, the time course of development of vision was regularly documented by measurement of visual acuity on the jumping stand.

Immediately prior to the termination of monocular eyelid closure, both animals had recovered good visual behavior and acuity through the unsutured eye. The cat that was dark-reared for 4 months (DR-4) had achieved an acuity of 2.5 cycles/degree by the end of the 6-week period of occlusion of the other eye. The other cat (DR-6) had attained an acuity of 5.3 cycles/degree with its sutured eye at the conclusion of 3 months of monocular deprivation, a value very close to the acuity that the animal eventually achieved using its unsutured eye. In striking contrast to the good visual behavior mediated through the nonsutured eye, both animals appeared to be functionally blind when first forced to use the monocularly deprived eye. One cat (DR-4) began to show signs of vision with this eye 2 days later, after which there was a gradual improvement to a final value of 2.5 cycles/degree. Physiological recordings made with the other animal 24 hours after termination of monocular occlusion precluded behavioral testing of the acuity of the monocularly deprived eye for another 4 days, at which time there was no doubt that the animal was able to see with this eye. In fact, formal tests of the visual acuity of this eye revealed that it was capable of resolving gratings having spatial frequencies as high as 1.8 cycles/degree. Thereafter, the vision of this eye improved only very slightly, to about 2.3 cycles/degree. Thus, although both animals showed a limited degree of recovery from the initial severe effects of the period of monocular occlusion, the visual acuity of the deprived eye remained very poor.

The results obtained from these animals stand in striking contrast to the effects of similar periods of monocular occlusion imposed on normally reared animals at the same age. For example, both the initial and the final visual deficits exhibited by cat DR-4, which was monocularly sutured for just 6 weeks at 4 months of age, were very much greater than those observed in the animal that was reared in illuminated surroundings until 4 months of age and then monocularly sutured for 3 months (see figure 16.13). In fact, the final acuity that DR-4 eventually attained with its deprived eye was lower than the acuity mediated by the deprived eye of the light-reared animal when it was first measured on opening the sutured eyelids. Moreover, the initial deficit in the vision of the deprived eye of this animal was quite temporary; within 3 weeks the acuity of this eye had improved to near normal levels. The visual deficits observed with the monocularly deprived eye of the dark-reared animals were thus very much greater than those observed after comparable periods of monocular deprivation imposed on light-reared animals of the same chronological age. This conclusion, drawn from behavioral observations, supports the results of physiological recording (Cynader and Mitchell 1980; Cynader et al. 1980) from animals reared in a similar manner which show a striking shift in ocular dominance in favor of the nonsutured eye. For example, the recording made with cat

DR-6 36 hours after the termination of the 3-month period of monocular occlusion imposed at 6 months of age indicated that only 20% of a sample of 40 cells encountered in the recording session were strongly dominated (ocular dominance groups 1 and 2) by the formerly deprived eye. The proportion of cells that could be excited binocularly (27.5%) was also very much reduced. The findings from this animal compare well with results obtained from other animals that had been dark-reared until 4–10 months of age and then monocularly occluded for periods of 3 months immediately upon exposure to normally illuminated surroundings (Cynader and Mitchell 1980). The marked changes in ocular dominance observed in these animals were very much greater than the rather small shifts observed after similar periods of monocular occlusion imposed on light-reared animals at 4 months of age (figure 16.14). Thus, together, the behavioral and physiological observations indicate that the visual cortex of dark-reared animals is far more susceptible to modification by monocular occlusion than that of normally reared animals of the same age.

A preliminary set of experiments have been undertaken to determine the duration of the period during which dark-reared animals remain susceptible to the effects of monocular deprivation. To investigate this question, we reared three cats in darkness until 4 months of age and then monocularly occluded one eye for a period of 3 months after the animals had lived in an illuminated environment for either 1, 2, or 3 months. The results of measurements of the acuity of the monocularly deprived and nondeprived eyes of the first of these animals, which was monocularly deprived after 1 month in the light, are shown in figure 16.18. Prior to the suturing of the eyelids of the right eye, all measurements of acuity were made binocularly, with the single exception of one measurement (indicated by an open symbol) of the monocular acuity of the right eye that was made immediately before the eyelid suture.

Immediately after the termination of monocular occlusion, tests of the vision of the formerly deprived eye revealed that the acuity of this eye was much reduced in comparison with that of the other eye. Indeed, the acuity of this eye was even lower than that it had been 3 months earlier, prior to the period of occlusion. This eye improved rapidly, so that within a few days its acuity was in excess of 3 cycles/degree. Thereafter, there was only a slight further increase in acuity, to about 4.5 cycles/degree (a value much less than that of the nondeprived eye). The magnitudes of the initial and final deficits in the vision of the formerly occluded eye suggest very strongly that the visual cortex of animals that have been reared in darkness until 4 months of age are still highly susceptible to monocular deprivation 1 month after they have been placed in illuminated surroundings.

A striking feature of the results from this animal and those of figure 16.16 was the rapid recovery of the vision of the formerly deprived eye.

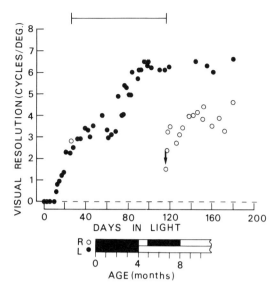

Figure 16.18 Development of vision in both eyes of an animal that had been reared in darkness until 4 months of age and subjected to 3 months of monocular eyelid closure 1 month after introduction to illuminated surroundings (see schematic visual history at bottom). (●) Results of measurements of binocular visual acuity, or that of the nonsutured left eye; (○) acuity of right eye alone. Note that a measurement was made of the acuity of the right eye immediately before it was sutured closed for the period indicated by the horizontal bar at the top of the figure. Arrow indicates first results obtained after 3 months of occlusion.

This recovery was far faster than the time course of development of vision in the nondeprived eye upon introduction of the animal to illuminated surroundings—a finding that at first might seem puzzling. Particularly in the case of animals (like those of figure 16.16) that were monocularly sutured as soon as they were introduced to light, it might be thought that the development of vision in the formerly sutured eye should follow closely the same time course as that of the nondeprived eye. However, this line of argument is erroneous, since it is based on the assumption that the two eyes are essentially independent. If this were the case, the time course of development of vision in the formerly occluded eye would depend only on how long the eye had been deprived of vision, and would be identical to that of an animal that had been reared in darkness for the same total time. However, because binocular connections are present at birth the two eyes are not independent, and so the development of vision in the two eyes reflects quite different processes. The development of vision in the nonoccluded eye undoubtedly reflects the gradual refinement of the tuning characteristics of cortical cells that occurs once the animal is exposed to an illuminated environment (Wiesel and Hubel 1965; Imbert and Buisseret 1975; Cynader et al. 1976). On the other hand, the development of vision in the formerly sutured eye reflects a different chain of events, in which the first step is a passive reinnervation of cortical cells by the deprived eye. But because of the gradual maturation of the cortical network that takes place once the animal is in the light, those cells that are passively reinnervated would already have become more finely tuned to visual stimuli and so could immediately mediate reasonably high acuity through the deprived eye (Timney et al. 1979).

Effects of Various Forms of Early Visual Deprivation on Depth Perception

To date, only a limited number of visually deprived animals have been examined for their ability to perceive depth on the jumping stand of figure 16.4. Nevertheless, the results have been quite unequivocal in revealing a lack of any superiority of binocular over monocular performance on depth discrimination. Two animals reared with surgically induced strabismus, which consequently suffered a substantial reduction in the proportion of binocular cortical cells (Hubel and Wiesel 1965), performed no better on depth discriminations with two eyes than with either eye separately. Furthermore, the thresholds obtained under all three of these viewing conditions were identical to those of normally reared animals using only one eye. This finding supports earlier reports (Packwood and Gordon 1975; Blake and Hirsch 1975) that strabismic cats lack stereopsis.

Finally, we have recently obtained some interesting results with a dark-reared animal that provide further support for the notion that dark rearing maintains plasticity in the visual cortex (Kaye et al. 1980). This particular animal was reared in darkness until 4 months of age. As is common (Cynader et al. 1976; Cynader 1979) with dark-reared animals, on being introduced to illuminated surroundings this animal developed an extreme convergent misalignment of the visual axes during the first month or so it was in the light. Although this animal eventually achieved normal and identical acuities in the two eyes, it showed a severe deficit of binocular depth perception. Under binocular viewing conditions it proved to be unable to perceive smaller separations in depth than with either eye alone, from which it would be concluded that it lacked stereopsis. Subsequently, electrophysiological recordings from the visual cortex of this animal revealed a substantial reduction in the proportion of binocularly excitable cells, a result that is consistent with the behavioral observation. Presumably, the breakdown of binocularity occurred after the animal had been exposed to illuminated surroundings at 4 months of age, possibly as a consequence of the convergent misalignment of the visual axes that developed over the course of the next month. The fact that this breakdown of binocular connections occurred at such a late age provides even more evidence that the visual cortex of this animal retained considerably more plasticity than that of a light-reared animal of the same age.

Overview

The above applications of the jumping stand to measurements of behavioral effects of various forms of early visual deprivation demonstrate the usefulness of the method for quickly documenting the nature of the visual deficits, and the time course of any subsequent recovery. In turn, these findings have in some cases provided an impetus for further physiological studies. A case in point was the finding of an amazing degree of behavioral recovery after monocular deprivation in situations where the eyelids of the deprived eye were simply opened to allow vision to both eyes during recovery (figures 16.8–16.10). This led to experiments to determine the degree of physiological recovery in this situation (Mitchell et al. 1977b). A second example was the finding of considerable behavioral recovery after periods of monocular occlusion from birth that extended throughout the first 4 months of life, a finding that questioned current notions of the duration of the critical period (figures 16.9, 16.10).

The above examples also indicate the close relationship between the behavioral and physiological deficits that follow various forms of early visual deprivation. Much of the compatability of the behavioral and physiological findings in these experiments is in our opinion due to the use of threshold measures such as measurements of visual acuity rather than conventional tests of pattern discrimination. However, prior to a discussion of the benefits that accrue from the use of measurement of visual acuity, it is worthwhile to consider the nature of the relationship between this behavioral measure of the size of the deficit and the physiological measure, which conventionally is defined in terms of changes in the proportion of cells driven by the two eyes.

One of the underlying assumptions of psychophysics is that thresholds are determined by the most sensitive neurons, or channel, that responds along the particular stimulus dimension under examination. Such an assumption, for example, underlies the interpretation of dark-adaptation curves and the increment threshold curves of Stiles (1949). On the basis of this assumption, the visual acuity of an eye might be expected to depend upon the spatial resolution of cells that can be excited by that eye, and specifically those cells with the highest spatial resolution. Though the spatial resolution of cells excited by a given eye must be related in some way to the number or proportion of cortical cells that can be influenced by that eye, the nature of the relationship is not well understood. Consequently, for purposes of comparing the behavioral and physiological effects of a given regimen of early visual deprivation it may be more appropriate to measure the spatial resolution of cells excited by the two eyes, in addition to their numbers—a practice that is becoming more common (Eggers and Blakemore 1978).

Benefits of the Use of Visual-Acuity Measures

One obvious advantage of the use of threshold measures in general and of visual acuity in particular over the conventional measurement of the ability of animals to make various pattern discriminations is the fact that the former are continuous (rather than discrete) measures and so permit the severity of any defect to be accurately graded. Also, of course, such measures allow for accurate documentation of the time course of the development of various perceptual capacities, as well as the recovery from any deficit induced by procedures such as early visual deprivation or surgical ablation of neural tissue. On the other hand, the interpretation of the results depends to a considerable extent on the accuracy with which these behavioral measures reflect the animal's true sensory capacities. We have previously described our trust in the power of the behavioral technique itself as a means of probing the animal's sensory limits (Mitchell et al. 1977a). Our confidence in the method is enhanced by the results obtained from our older cats, which are as good as the highest estimates of acuity obtained by other, more time-consuming behavioral procedures with adult cats (Blake et al. 1976; Jacobson et al. 1978) or the estimates obtained from evoked potentials (Harris 1978). In addition to this confidence in the technique itself, it is highly unlikely that the measures of acuity are contaminated by refractive errors. Retinoscopy is routinely performed on all of our animals after the instillation of cycloplegic drops. We have never observed in any of our animals a difference in the refractive state of the two eyes of more than 1D; indeed, the refractions have proved remarkably consistent in revealing a negligible difference of less than 0.50D between the two eyes. The absolute refraction is usually about +1D, a value close to that observed in normal animals by Hughes (1977). Such small refractive errors have been shown to produce no effects at all on the contrast thresholds of single cortical cells (C. Blakemore, pers. comm.), and so would be very unlikely to exert any influence on our behavioral measures.

Although measurements of acuity are useful for the reasons just described, we wish to emphasize here a potentially far greater benefit: the possibility that in certain cases measures of acuity may assist in defining either the locus or the type of cells that mediate the animal's performance.

During the last decade considerable evidence has accumulated for the existence of at least three classes of cat retinal-ganglion cells showing concentric receptive fields (see Levick 1975 for review). These three classes can be distinguished not only on the basis of physiological criteria, but also on morphological and other criteria, including differences in their projection patterns. The same three classes of cells can also be identified in the lateral geniculate nucleus on the basis of the same set of criteria as well as on the basis of the conduction velocities of their axons (Cleland et al. 1971, 1976; Hoffman et al. 1972; Wilson and Stone 1975).

One class of both retinal-ganglion and geniculate cells, referred to as X cells (Enroth-Cugell and Robson 1966) or brisk-sustained cells (Cleland et al. 1971), tend to respond to gratings of higher spatial frequency than those responded to by the other two classes of cells. The most common other class of cells, the Y or brisk-transient cells, have an average cutoff spatial frequency about an octave lower than that of the X cells (Enroth-Cugell and Robson 1966; Ikeda and Wright 1976; Cleland et al. 1979). In the vicinity of the area centralis, Y or brisk transient cells have a cutoff spatial frequency of only 4 cycles/degree or less, with a mean of only 3 cycles/degree (Cleland et al. 1979). On the other hand, X cells can respond to gratings having spatial frequencies comparable to the grating acuities obtained by normal animals on the jumping stand.

Besides this very clear difference, these two classes of cells have quite different projection patterns. Whereas the X or brisk-sustained cells mainly project to area 17, the axons of the Y or brisk-transient cells apparently bifurcate, with one branch innervating area 17 while the other projects to area 18 (Stone and Dreher 1973).

The difference between the cutoff spatial frequencies of these two classes of cells provides a means whereby it might be possible in certain cases to identify at least the class of cell, and possibly the cortical area, that is utilized by a visually deprived or lesioned animal in making a discrimination between a grating and a uniform field on the jumping stand. Animals achieving acuities in excess of 3 cycles/degree most likely retain some X cells and could be utilizing area 17 to make the discrimination. On the other hand, animals that cannot achieve acuities in excess of 3 cycles/degree may be employing Y cells. In general, knowledge of the spatial resolution of cells at various levels within the visual pathways provides some hint as to the visual area that may be mediating visual behavior in certain deprived or lesioned animals. For example, the degree of acuity that was eventually recovered by animals that were monocularly deprived from birth to 4 months of age, which was in excess of 2.5 cycles/degree, virtually eliminates the superior colliculus as a structure that might be mediating visual discriminations by these animals, since in normal animals the cutoff spatial frequencies for collicular cells is less than 2 cycles/degree (Bisti and Sireteanu 1976). Likewise, knowledge of the acuity of an eye that had been rotated 90° shortly after birth may help locate the structure mediating the amazing degree of behavioral recovery observed in such animals (Peck and Crewther 1975; Mitchell et al. 1976b; Timney and Peck 1981).

In summary, we would like to argue that simple behavioral measurement of the ability of lesioned or visually deprived animals to resolve grating patterns may provide much better clues to the pathway or structures mediating the visual perceptions of these animals than conventional tests of their ability to perform complex pattern discriminations. The simplicity of our methodological procedure together with the new twist provided by measurement of visual acuity may help provide fresh clues to the link between function and structure within the visual pathway.

Acknowledgments

The experiments described here were supported by grants from the Medical (MA-5027) and National (AP-7660) Research Councils of Canada. The preparation of this manuscript was aided by grants from The National Health and Medical Research Council of Australia. We thank Dr. F. Wilkinson, Fred Giffin, Melissa Fuller, Evelyn Sutton, and Karl Grantmyre for their many contributions to this work.

References

Atkinson, J., O. Braddick, and F. Braddick. 1974. "Acuity and contrast sensitivity of infant vision." *Nature* 247: 403–404.

Baxter, B. 1966. "The effect of visual deprivation during postnatal development on the electroretinogram of the cat." *Exp. Neurol.* 14: 224–237.

Bisti, S., and L. Maffei. 1974. "Behavioral contrast sensitivity of the cat in various visual meridians." *J. Physiol.* 241: 201–210.

Bisti, S., and R. C. Sireteanu. 1976. "Sensitivity to spatial frequency and contrast of visual cells in the cat superior colliculus." *Vision Res.* 16: 247–251.

Blake, R., and H. V. B. Hirsch. 1975. "Deficits in binocular depth perception in cats after alternating monocular deprivation." *Science* 190: 1114–1116.

Blake, R., Cool, S. J., and M. L. J. Crawford. 1974. "Visual resolution in the cat." *Vision Res.* 14: 1211–1217.

Blakemore, C. 1978. "Maturation and modification in the developing visual system." In *Handbook of Sensory Physiology*, vol. VIII: *Perception*, R. Held et al., eds. Berlin: Springer.

Blakemore, C., and R. C. Van Sluyters. 1974. "Reversal of the physiological effects of monocular deprivation: Further evidence for a sensitive period." *J. Physiol.* 237: 195–216.

———. 1975. "Innate and environmental factors in the development of the kitten's visual cortex." *J. Physiol.* 248: 663–716.

Blasdel, G. G., D. E. Mitchell, D. W. Muir, and J. D. Pettigrew. 1977. "A combined physiological and behavioural study of the effect of early visual experience with contours of a single orientation." *J. Physiol.* 265: 615–636.

Bloom, M., and M. A. Berkley. 1977. "Visual acuity and the near point of accommodation in cats." *Vision Res.* 17: 723–730.

Bonds, A. B., and R. D. Freeman. 1978. "Development of optical quality in the kitten eye." *Vision Res.* 18: 391–398.

Buisseret, P., and M. Imbert. 1976. "Visual cortical cells: Their developmental properties in normal and dark reared kittens." *J. Physiol.* 255: 511–525.

Campbell, F. W., and D. G. Green. 1965. "Optical and retinal factors affecting visual resolution." *J. Physiol.* 181: 576–593.

Chow, K. L., and D. L. Stewart. 1972. "Reversal of structural and functional effects of long-term visual deprivation in cats." *Exp. Neurol.* 34: 409–433.

Cleland, B. G., M. W. Dubin, and W. R. Levick. 1971. "Sustained and transient neurones in the cat's retina and lateral geniculate nucleus." *J. Physiol.* 217: 473–496.

Cleland, B. G., W. R. Levick, R. Morstyn, and H. G. Wagner. 1976. "Lateral geniculate relay of slowly conducting retinal afferents to cat visual cortex." *J. Physiol.* 255: 299–320.

Cleland, B. G., T. H. Harding, and U. Tulunay-Keesey. 1979. "Visual resolution and receptive field sizes: Examination of two kinds of cat retinal ganglion cells." *Science* 205: 1015–1017.

Cornsweet, T. N. 1962. "The staircase method in psychophysics." *Amer. J. Psychol.* 75: 485–491.

Cynader, M. 1979. "Interocular alignment following visual deprivation in the cat." *Invest. Ophthal. Vis. Sci.* 18: 726–741.

Cynader, M., and D. E. Mitchell. 1980. "Prolonged sensitivity to monocular deprivation in dark-reared cats." *J. Neurophysiol.* 43: 1026–1040.

Cynader, M., N. Berman, and A. Hein. 1976. "Recovery of function in cat visual cortex following prolonged deprivation." *Exp. Brain Res.* 25: 139–156.

Cynader, M., B. Timney, and D. Mitchell. 1980. "Period of susceptibility of kitten visual cortex to the effect of monocular deprivation extends beyond 6 months of age." *Brain Res.* 191: 545–550.

Daniels, J. D., and J. D. Pettigrew. 1978. "Development of neural responses in the visual system of cats." In *Studies on the Development of Behavior and the Nervous System*, vol. 3: *Neural and Behavioral Specificity*, G. Gottlieb, ed. New York: Academic.

Derrington, A. M. 1978. "Development of selectivity in kitten striate cortex." *J. Physiol.* 276.

Donovan, A. 1966. "The postnatal development of the cat retina." *Exp. Eye Res.* 5: 249–254.

Eggers, H. M., and C. Blakemore. 1978. "Physiological basis of anisometropic amblyopia." *Science* 201: 264–267.

Enroth-Cugell, C., and J. G. Robson. 1966. "The contrast sensitivity of retinal ganglion cells of the cat." *J. Physiol.* 187: 517–552.

Freeman, D. N., and E. Marg. 1975. "Visual acuity development coincides with the sensitive period in kittens." *Nature* 254: 614–615.

Freeman, R. D., and C. E. Lai. 1978. "Development of the optical surfaces of the kitten eye." *Vision Res.* 18: 399–408.

Frégnac, Y., and M. Imbert. 1978. "Early development of visual cortical cells in normal and dark-reared kittens: Relationship between orientation selectivity and ocular dominance." *J. Physiol.* 278: 27–44.

Ganz, L. 1978. "Innate and environmental factors in the development of visual form perception." In *Handbook of Sensory Physiology,* vol. VIII: *Perception,* R. Held et al., eds. Berlin: Springer.

Ganz, L., H. V. B. Hirsch, and S. B. Tieman. 1972. "The nature of perceptual deficits in visually deprived cats." *Brain Res.* 44: 547–568.

Gellerman, L. W. 1933. "Chance orders of alternating stimuli in visual discrimination experiments." *J. Genet. Psychol.* 42: 207–208.

Giffin, F., and D. E. Mitchell. 1978. "The rate of recovery of vision after early monocular deprivation in kittens." *J. Physiol.* 274: 511–537.

Harris, L. 1978. "Contrast sensitivity and acuity of a conscious cat measured by the occipital evoked potential." *Vision Res.* 18: 175–178.

Helmholtz, H. von. 1924. *Treatise on Physiological Optics,* vol. III, translated from the third German edition; J. P. C. Southall, ed. Ithaca, N.Y.: Optical Society of America.

Hirsch, M. J., and F. W. Weymouth. 1948. "Distance discrimination. I. Theoretic considerations." *AMA Arch. Ophthalmol.* 39. 210–223.

Hoffman, K. P., and M. Cynader. 1977. "Functional aspects of plasticity in the visual system of adult cats after early monocular deprivation." *Philos. Trans. Roy. Soc.* B 278: 411–424.

Hoffman, K. P., J. Stone, and S. M. Sherman. 1972. "Relay of receptive field properties in dorsal lateral geniculate nucleus of the cat." *J. Neurophysiol.* 35: 518–531.

Howard, H. J. 1919. "A test for the judgment of distance." *Amer. J. Ophthalmol.* 2: 656–675.

Hubel, D. H., and T. N. Wiesel. 1963. "Receptive fields of cells in striate cortex of very young, visually inexperienced kittens." *J. Neurophysiol.* 28: 994–1002.

———. 1965. "Binocular interaction in striate cortex of kittens reared with artificial squint." *J. Neurophysiol.* 28: 1041–1059.

———. 1970. "The period of susceptibility to the physiological effects of unilateral eye closure in kittens." *J. Physiol.* 206: 419–436.

Hughes, A. 1977. "The topography of vision in animals of contrasting life styles: Comparative optics and retinal organization." In *Handbook of Sensory Physiology,* vol. VII/5: *The Visual System in Evolution,* F. Crescitelli, ed. Berlin: Springer.

Ikeda, H., and K. E. Tremain. 1978. "The development of spatial resolving power of lateral geniculate neurones in kittens." *Exp. Brain Res.* 31: 207–220.

Ikeda, H., and M. J. Wright. 1976. "Properties of LGN cells in kittens reared with convergent squint: A neurophysiological demonstration of amblyopia." *Exp. Brain Res.* 25: 63–77.

Imbert, M., and P. Buisseret. 1975. "Receptive field characteristics and plastic properties of visual cortical cells in kittens reared with or without visual experience." *Exp. Brain Res.* 22: 25–36.

Jacobson, S. G., K. B. J. Franklin, and W. I. McDonald. 1976. "Visual acuity of the cat." *Vision Res.* 16: 1141–1143.

Kalil, R. 1978. "Dark-rearing in the cat: Effect on visuomotor behaviour and cell growth in the dorsal lateral geniculate nucleus." *J. Comp. Neurol.* 178: 451–468.

Kaye, M., D. E. Mitchell, and M. Cynader. 1980. "Depth perception, eye alignment and cortical ocular dominance of dark-reared cats." *Dev. Brain Res.*

Kratz, K. E., and P. D. Spear. 1976. "Effects of visual deprivation and alterations in binocular competition on responses of striate cortex neurons in the cat." *J. Comp. Neurol.* 170: 141–152.

Kratz, K. E., P. D. Spear, and D. C. Smith. 1976. "Postcritical-period reversal of effects of monocular deprivation on striate cortex cells in the cat." *J. Neurophysiol.* 3: 501–511.

Lashley, K. S. 1930. "The mechanism of vision. I. A method for the rapid analysis of pattern in vision in the rat." *J. Genet. Psychol.* 37: 453–460.

Leehey, S. C., A. Moskowitz-Cook, S. Brill, and R. Held. 1975. "Orientational anisotropy in infant vision." *Science* 190: 900–902.

Levick, W. R. 1975. "Form and function of cat retinal ganglion cells." *Nature* 254: 659–662.

Mitchell, D. E., F. Giffin, F. Wilkinson, P. Anderson, and M. L. Smith. 1976a. "Visual resolution in young kittens." *Vision Res.* 16: 363–366.

Mitchell, D. E., F. Giffin, D. Muir, C. Blakemore, and R. C. van Sluyters. 1976b. "Behavioral compensation of cats after early rotation of one eye." *Exp. Brain Res.* 25: 109–113.

Mitchell, D. E., F. Griffin, and B. Timney. 1977a. "A behavioral technique for the rapid assessment of the visual capabilities of kittens." *Perception* 6: 181–193.

Mitchell, D. E., M. Cynader, and J. A. Movshon. 1977b. "Recovery from the effects of monocular deprivation in kittens." *J. Comp. Neurol.* 176: 53–64.

Mitchell, D. E., M. Kaye, and B. Timney. 1979. "Assessment of depth perception in cats." *Perception* 8: 389–396.

Movshon, J. A. 1976. "Reversal of the behavioural effects of monocular deprivation in the kitten." *J. Physiol.* 261: 175–187.

Olson, C., and R. D. Freeman. 1978. "Monocular deprivation and recovery during sensitive period in kittens." *J. Neurophysiol.* 41: 65–74.

Packwood, J., and B. Gordon. 1975. "Stereopsis in normal domestic cat, Siamese cat, and cat raised with alternating monocular occlusion." *J. Neurophysiol.* 38: 1485–1499.

Peck, C. K., and S. G. Crewther. 1975. "Perceptual effects of surgical rotation of the eye in kittens." *Brain Res.* 99: 213–219.

Pettigrew, J. D. 1974. "The effect of visual experience on the development of stimulus specificity by kitten cortical neurones." *J. Physiol.* 237: 49–74.

Rizzolatti, G., and V. Tradardi. 1971. "Pattern discrimination in monocularly reared cats." *Exp. Neurol.* 33: 181–194.

Shatz, C. J., and M. P. Stryker. 1978. "Ocular dominance in layer IV of the cat's visual cortex and the effects of monocular deprivation." *J. Physiol.* 281: 267–283.

Singer, W., and F. Tretter. 1976. "Receptive-field properties and neuronal connectivity in striate and parastriate cortex of contour-deprived cats." *J. Neurophysiol.* 39: 613–630.

Spear, P. D., and L. Ganz. 1975. "Effects of visual cortex lesions following recovery from monocular deprivation in the cat." *Exp. Brain Res.* 23: 181–201.

Stiles, W. S. 1949. "Increment thresholds and the mechanisms of color vision." *Docum. Ophthalmol.* 3: 138–165.

Stone, J., and B. Dreher. 1973. "Projection of X- and Y-cells of the cat's lateral geniculate nucleus to areas 17 and 18 of visual cortex." *J. Neurophysiol.* 36: 551–567.

Stryker, M. P., and H. Sherk. 1975. "Modification of cortical orientation selectivity in the cat by restricted visual experience: A re-examination." *Science* 190: 904–906.

Teller, D. Y., R. Morse, R. Borton, and D. Regal. 1974. "Visual acuity for vertical and diagonal gratings in human infants." *Vision Res.* 14: 1433–1439.

Timney, B., and C. K. Peck. 1981. "Visual acuity in cats following surgically induced cyclotropia." *Behav. Brain Res.*

Timney, B., D. E. Mitchell, and F. Giffin. 1978. "The development of vision in cats after extended periods of dark-rearing." *Exp. Brain Res.* 31: 547–560.

Timney, B., D. E. Mitchell, and M. Cynader. 1980. "Behavioral evidence for prolonged sensitivity to the effects of monocular deprivation in dark-reared cats." *J. Neurophysiol.* 43: 1041–1054.

van Hof–van Duin, J. 1976. "Development of visuomotor behavior in normal and dark-reared cats." *Brain Res.* 104: 233–241.

von Senden, M. 1960. *Space and Sight,* P. Heath, tr. London: Methuen.

Wiesel, T. N., and D. H. Hubel. 1963. "Single cell responses in striate cortex of kittens deprived of vision in one eye." *J. Neurophysiol.* 26: 1003–1017.

Wiesel, T. N., and D. H. Hubel. 1965. "Comparison of the effects of unilateral and bilateral eye closure on cortical unit responses in kittens." *J. Neurophysiol.* 28: 1029–1040.

Wilkinson, F., and P. C. Dodwell. 1980. "Young kittens can learn complex visual pattern discriminations." *Nature* 284: 258–259.

Wilson, P. D., and J. Stone. 1975. "Evidence of W-cell input to the cat's visual cortex via the C laminae of the lateral geniculate nucleus." *Brain Res.* 92: 472–478.

Worth, C. A. 1903. *Squint: Its Causes, Pathology and Treatment.* Philadelphia: Blakiston.

17 The Role of the Geniculocortical System in Spatial Vision

Mark A. Berkley
James M. Sprague

In a series of early studies, Lashley (1950) carried out experiments on the role of neocortex in behavior. He used an ablation-behavior paradigm in which he measured behavior before and after removing various portions of cortex. As is well known, the animals that he tested were marvelously resistant to revealing deficits after rather extensive cortical lesions. These puzzling findings, which seemed to run counter to the idea of localization of function, forced Lashley to put forward a rather weak hypothesis to account for his results: that specific behaviors may be mediated by structures that are widely dispersed through the brain, perhaps through all of it. Recent discoveries in the physiology and anatomy of the brain have not supported such a view, but rather have reaffirmed the idea: those parts of the brain that have distinct structural and functional organization make somewhat different or even unique contributions to specific behaviors.

One contemporary view of the visual system in which functional localization is assumed is the simple serial processing model. This model is based on the classic experiments of Hubel and Wiesel (1962) in which the receptive-field properties of visual-cortex neurons were determined. Hubel and Wiesel discovered that the receptive fields in the cortex are elongated and respond best to linear contours of a specific orientation. They found other cells with larger and more complicated receptive fields which also responded best to elongated contours. Since the receptive fields of retinal ganglion cells and the neurons in lateral geniculate were known to be essentially round, with center-surround organization, Hubel and Wiesel suggested that the elongated receptive fields in cortex were built up by a process of convergence (stringing together) of the concentric receptive fields from the lateral geniculate nucleus (LGN). They further suggested that the larger, more complicated receptive fields of neurons in the cortex might be built up by a process of convergence of input from the more simple, elongated fields of other cortical cells. One implication of such a model is that the shape of any object might be specified by a process of successive convergence. In its simplest formulation, this model suggests that visual stimulus elements or features are strung together in ever greater complexity to form a visual precept (Barlow 1972). Those cells signaling the locus and orientation of the edges of the stimulus object ultimately converge onto a single cell and thus uniquely define the stimulus object. Clearly, intervention anywhere in such a serially organized system should produce total and permanent deficits in form vision—something which the studies of Lashley (1950) and others (Doty 1971; Berkley et al. 1976; Diamond 1976; Sprague et al. 1977) did not find in some animals in which visual cortex was ablated. However, simple variants of a serially ordered system could explain these negative results. If the visual system were organized with *several* parallel serial systems (Berkley 1978), it would not be surprising that function can be recovered or retained after an ablation or interruption of one subsystem if the various subsystems shared some functions. The results of our ablation-behavior studies of the visual system of the cat suggest the operation of such mixed parallel-serial systems.

Is there anatomical evidence that the visual system is organized in parallel-serial fashion? In the monkey, the geniculocortical system ends exclusively or chiefly in area 17 (Hubel and Wiesel 1962; Wong-Riley 1976). Widespread divergence of this system does not appear to occur until beyond striate cortex in the monkey. However, there is a second route from retina to cortex: the tecto-thalamic route, in which the tectum, which receives direct retinal input, projects to portions of the pulvinar, which in turn projects chiefly to areas 18 and 19 of the cortex (Benevento and Rezak 1976). Removal of striate cortex (area 17) results in marked, retrograde atrophy of all parts of the dorsal lateral geniculate nucleus (LGN), presumably rendering it functionally inactive, with no effect on the tecto-thalamocortical route (Pasik and Pasik 1971; Humphrey 1974; Denny-Brown and Chambers 1976). Indeed, behavioral tests indicate severe spatial-vision deficits in destriate monkeys. Recent studies have also indicated that the destriate monkey may have more spatial visual capacity than originally thought. The residual vision is probably mediated via the tecto-thalamocortical pathway (see Keating, this volume).

In the cat, the situation is significantly different. There is great divergence of output from thalamus. Thus, LGN_d and the tecto-thalamocortical system project to cortical areas 17–19 and parts of the lateral suprasylvian cortex (LSA) (Garey and Powell 1967; Glickstein et al. 1967; Niimi and Sprague 1970; Burrows and Hayhow 1971; Rosenquist et al. 1974; LeVay and Gilbert 1976; Hollander and Vanegas 1977). On the basis of these differences between cat and monkey, one might expect to find a greater effect of ablation of areas 17 and 18 on vision in the monkey than in the cat—a prediction entirely consistent with the ablation-behavior studies of both species (Denny-Brown and Chambers 1976; Keating, this volume). Figure 17.1 is a summary of current knowledge about the anatomical pathways in the cat from retina to cerebral cortex, and depicts the known multiple visual pathways in the cat.

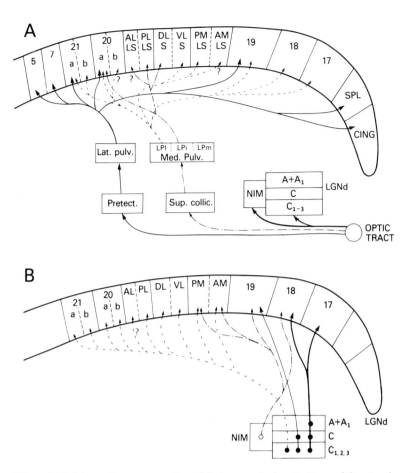

Figure 17.1 Schematic representation of thalamocortical projections of the visual systems in the cat. (A) Projections of the pulvinar complex; (B) projections of the dorsolateral geniculate complex. Lat. pulv.: lateral pulvinar; Med. pulv.: medial pulvinar (terminology of Niimi and Kuwahara 1973). LPl: lateral posterior, lateral division; LPi: lateral posterior, intermediate division; LPm: lateral posterior, medial division (terminology of Updike 1977). Pretect.: pretectum; Sup. collic.: superior colliculus; LGN$_d$: lateral geniculate, dorsal part; laminae A + A$_1$, C and C$_1$, C$_2$ and C$_3$; NIM: medial interlaminar nucleus; cortical areas 17, 18, 19, 20a, 20b, 21a, 21b; ALLS: anterior lateral, lateral suprasylvian; PLLS: posterior lateral, lateral suprasylvian; DLS: dorsal lateral suprasylvian; VLS: ventral lateral suprasylvian; PMLS: posterior medial, lateral suprasyl; AMLS: anterior medial suprasylvian (terminology of Tusa 1978; Palmer et al. 1978). SPL: splenial visual area (terminology of Kallia and Whitteridge, 1973). cing: cingular cortex. Summary includes unpublished work by Symonds and Rosenquist in A and by Raczkowski and Rosenquist in B. Figure adapted from Sprague et al. 1977.

Geniculocortical System in Spatial Vision

One important question raised by the presence of multiple parallel systems in the cat is: Are these various pathways equipotential; for example, can they perform the same function interchangeably? Though some degree of redundancy is to be expected in an organ as complex as the brain, it seems unlikely that each of the various visual subsystems has the same processing capabilities. More likely is the possibility that each individual pathway, while sharing peripheral input with its partners, makes a unique contribution to visually guided behavior.

How can a system that has multiple pathways sharing peripheral information be studied? First, one must decide on an anatomical locus. Second, one must have a working hypothesis about the function of the structure. In this case, the hypothesis may be stated simply: Each pathway, in addition to participating in functions shared with other pathways, has a unique function that cannot be performed by other portions of the system. Finally, tests must be selected that make it possible to evaluate the hypothesis.

We have been studying the cortical portions of one subsystem of the cat visual system, cortical areas 17 and 18, in an attempt to discover their unique functions. This region receives not only the classic input from LGN_d, but input from other thalamic regions as well. As can be seen in figure 17.1, the area 17–18 input from the LGN actually is only a small part of a large and complex visual thalamic projection to cortex. Not knowing area 17–18's unique function, but assuming that it has one, we selected a battery of tests of visual capacities derived from consideration of the extensive anatomical and physiological studies of these cortical areas. The anatomical and physiological literature suggests that such stimulus dimensions as movement, size, orientation, and topographic contour alignment are important in activating neurons in areas 17 and 18. Thus, we have attempted to determine to what extent areas 17 and 18 contribute to the detection of these stimulus attributes by devising test stimuli that contain them. Having selected the structure and stimulus dimensions, we turned our attention to the problem of ascertaining whether the animals were actually utilizing the selected stimulus dimensions. Because communication with animal subjects is limited, determining whether an intended cue is being attended to is a serious and often overlooked problem. Regardless of how fundamental or simple a visual stimulus may appear to be to us, only the animal's behavior can be used to determine what dimensions are actually being used by the animal. The rationale for using the animal's behavior to tell us what cue the animal is using is as follows: If variations of the selected stimulus dimension (with everything else held constant) produce concomitant changes in the animal's behavior—that is, if they control the animal's behavior—we have evidence that the manipulated stimulus dimension is the one the animal is attending to and utilizing in its discriminations.

Thus, the present studies were designed to combine the anatomical knowledge of the cat's visual system with tests of visual capacity determined with the behavioral techniques derived from human psychophysics and modern animal behavior-control methodology (Berkley 1976).

Specifically, the present study examined the effects of ablations restricted to areas 17 and 18, or of the remainder of the visual cortices sparing 17 and 18, in the cat, on the following visual capacities: grating acuity, topographic alignment (vernier offset), orientation acuity, and complex form vision.

Summary of Experimental Details

Randomly bred adult cats were used; the only criterion for selection was that a cat be gentle and cooperative and able to learn a simple black-white discrimination in the test apparatus.

The test apparatus was a two-choice discrimination chamber, described in detail elsewhere (Berkley 1970). The test chamber is a box with a small round opening at one end. Covering this opening from the outside is a small plexiglas chamber with two transparent plexiglas response keys at its outer end. Below the keys is a pneumatically operated reward terminal that dispenses small portions of a pureed-beef baby food (Berkley et al. 1971). The cats are trained to sit or stand in the chamber, extend their heads into the plastic chamber, inspect the stimuli by peering through the response keys, and make choices by pressing the left or right response key with their noses. A sketch of the apparatus is shown in figure 17.2.

Data collection and stimulus manipulation were performed by a special-purpose computer interfaced with several stimulus-display devices. Daily test sessions consisted of 200 trials, each consisting of the presentation of a pair of visual stimuli, one appearing on the left side of the display device and one on the right. A trial was terminated when the animal indicated its choice by pressing the left or the right nose key. If the choice was correct, a small amount of pureed beef was delivered. If the choice was incorrect, no reward was delivered and a "time out" (no stimuli presented or response permitted) ensued. In all cases, the animals were considered to have learned the discrimination problem when they achieved a criterion of either 90% correct choices in all 200 trials on one day or 85%–90% correct in 600 trials on three days. (Further details of the training procedures may be found in Bloom and Berkley 1977, Tunkl and Berkley 1977, and Berkley and Sprague 1979).

The test stimuli used in the visual discriminations fell into two categories: form or shape stimuli, and threshold-detection stimuli (gratings, line orientation, and vernier offset). Examples of stimuli are shown in figure 17.3.

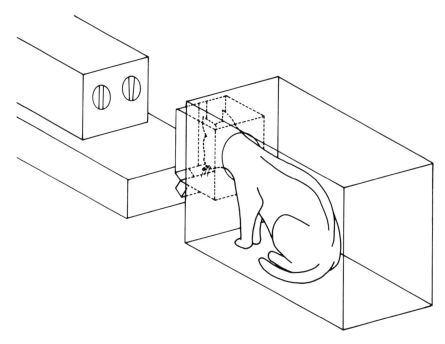

Figure 17.2 Sketch of apparatus used to test cats. The chamber in which the cat places its head has two response keys, which can be operated by the cat by pushing them with its nose. A food-delivery terminal is located just below the keys. The visual target display device (CRT, TV, or projection screen) is placed on a bench and can be located at various distances from the cat. The entire apparatus is enclosed in a larger chamber to exclude extraneous stimuli.

After preoperative testing was completed, the cat was deeply anesthetized with sodium pentabarbital and lesions were induced by gentle subpial aspiration under strict aseptic precautions. Surface landmarks described by Otsuka and Hassler (1962), Sanides and Hoffman (1969), Tusa et al. (1978), and Palmer et al. (1978) were used to guide the placement of the lesions. All animals were permitted to recover for 30 days before retesting was begun.

All postoperative threshold testing was preceded by testing on a light-dark discrimination task. When this test was completed, the animals were evaluated for any changes in threshold. (Threshold tests were given in the same order used before operation.)

In addition to these tests, all animals were tested after operation on a variety of visuomotor tasks, such as visual placing, visual localization of objects, visual tracking of moving objects, jumping across gaps, and measuring the extent of visual fields in a perimetry test (Sprague and Meikle 1965; Sherman 1973).

At the completion of behavioral testing, all animals were killed and their brains were prepared for histologic examination. The extent of each cortical lesion was reconstructed in detail by means of projection drawings of selected coronal sections through the cortex and the

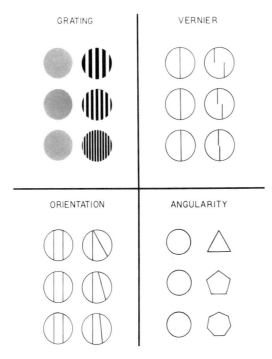

GRATING   VERNIER

ORIENTATION   ANGULARITY

Figure 17.3 Pictorial table showing configurations of stimulus pairs used in present study. Three values of each stimulus dimension are shown, to demonstrate how that stimulus dimension was varied. Left-right positions of stimulus pairs, shown in one configuration for clarity, were varied haphazardly during testing. The size of the target field (aperture) was 8°, and the width of the lines making up the vernier, orientation, and angularity targets was 10'.

thalamus (see Sprague et al. 1977 for details) and evaluation of the state of degeneration in LGN.

Results and Discussion

Lesions

The extent of seven lesions is plotted in surface views in figures 17.4 and 17.5. Striate-cortex lesions are shown in figure 17.4 and extrastriate lesions in figure 17.5. In two of the area 17–18 ablations (cats Zelda and Scarlet) the lesions were large and involved most of areas 17 and 18; in the other two (Francis, Streak) the lesions were chiefly in area 17 and involved 18 only in and near the representation of the vertical meridian. (More detailed descriptions and documentation of the lesions can be found in Berkley and Sprague 1979).

In three cats the lesions were placed outside of area 17–18, as shown in figure 17.5. In one (Arlo) the lesion was large, approximating the volume of tissue removed in the "17–18" cats, and included areas 19, 21, 7, and part of 20. In two others (Earl and BJ) the lesions were smaller, involving chiefly areas 19 and 21.

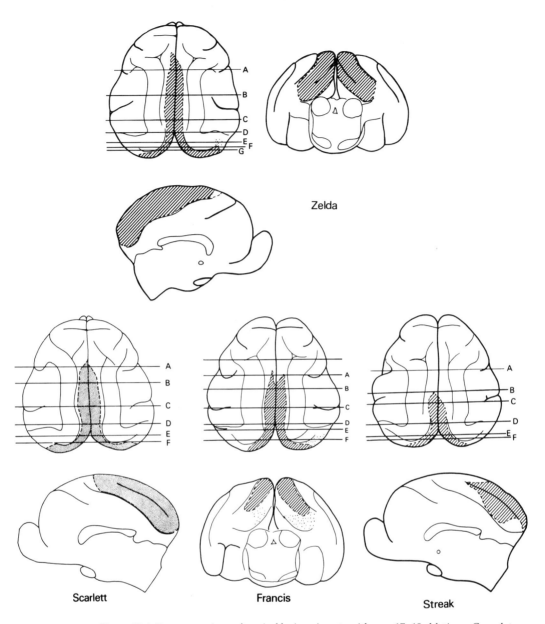

Zelda

Scarlett   Francis   Streak

Figure 17.4 Reconstructions of cortical lesions in cats with area 17–18 ablations. Complete anatomical details are given in Berkley and Sprague 1979.

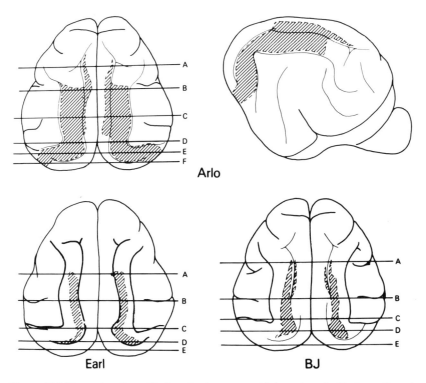

Figure 17.5 Reconstruction of lesions in three cats with lateral cortical lesions. Complete anatomical details are given in Berkley and Sprague 1979.

Visuomotor Behavior

As has been reported previously, none of the cats with area 17–18 lesions showed any lasting deficits in general neurological tests (Sprague 1966; Doty 1971; Sprague et al. 1970). The performance on many of these tests strongly suggested that these animals were using the central visual representation present in intact areas of cortex (19, 20, 21, or LSA) and/or superior colliculi, rather than operating with peripheral vision mediated by small, residual islands of tissue in area 17 or 18 (a suggestion often made to account for residual spatial vision after such lesions). That is, all animals were able to move about without bumping into obstacles; freely jumped up or down from or to platforms; were able to strike small moving targets with their forepaws; made accurate head and eye movements in attending to interesting moving objects, and, in general, showed no overt signs of visual deficits.

Contour-Discrimination Behavior

To ascertain whether the lesions employed in this study permitted the same level of visual-discrimination performance reported in previous ablation-behavior studies of area 17–18 in cats, we used a form-discrimination task (○ versus +) that has been used in other studies. The ○+ test confirmed earlier findings and showed that animals with

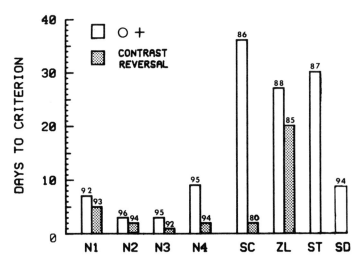

Figure 17.6 Histogram showing number of days needed to reach criterion in discrimination of 0 versus + for four normal cats (N1–N4) and four cats with cortical ablations. Cats SC, ZL, ST, and SD received ablations in area 17–18. Open bars show performance when stimuli consisted of black lines against a white background; shaded bars show performance when contrast of figures was reversed. Number at top of each column indicates terminal performance level in percentage of correct choices.

area 17–18 ablations could learn this discrimination. The rate of acquisition, however, was somewhat slower than normal. These results are summarized in figure 17.6, which shows in histogram form the number of days needed to reach criterion by four inexperienced normal animals (N1–N4) and four animals with area 17–18 ablations (SC, AL, ST, SD). As a control for local flux cues, the contrast was reversed after criterion was reached. As can be seen, this maneuver had little effect on one animal (SC) and a great effect on another (Zelda). This is typical even in normal animals (see N1 and N4). From these results we conclude that animals with area 17–18 ablations can learn discriminations based on small forms (<7°; line thickness 1.8°), but are somewhat slower than normal in doing so. At this stage, we had learned nothing new and had simply confirmed what is now a well-documented fact. All of the arguments previously raised to account for the apparently normal form vision in cats after area 17–18 ablations can be raised in this case as well. Thus, one could argue that the form vision observed was mediated by the residual portions of area 17–18, or that the animals employed some "trick" that did not require spatial vision (such as scanning or the use of a local flux cue) to perform the discriminations. We do not know, for example, what the shapes look like to the animals, only that they can distinguish one from another. The results of the form-vision tests, by themselves, do not permit us to address these issues.

To deal with these arguments as well as to attempt to determine the unique functions of area 17–18 requires the other vision tests we had devised.

As mentioned earlier, a series of stimuli were selected that had two attributes: they had at least one dimension that would stimulate known classes of cortical neurons, and they could be varied along one stimulus dimension. The acuity stimuli consisted of three different types of targets: gratings, vernier offset lines, and line orientations. Figure 17.3 shows examples of the stimulus configuration for each of the stimulus dimensions used; easily discriminable values are shown at the top, more difficult ones below.

Details of how the stimuli were generated, calibration procedures, and other matters can be found in Berkley and Sprague 1979.

Contour Acuities
The effects of the lesions on grating acuity are shown in figure 17.7. At large grating sizes (low spatial frequencies) the operated animals were essentially unimpaired; at the higher spatial frequencies they exhibited only mild deficits in grating acuity. This modest deficit is surprising, because it has been assumed that area 17 is critical for detail vision. Though there was significant impairment, the animals obviously could see relatively fine gratings. Preoperative threshold levels were never recovered, however, even after considerable efforts at retraining. As a control for general visual deficits, animals with lesions lateral to areas 17 and 18 were tested. Their postoperative acuity performance did not show any impairment (figure 17.7, lower right, cat Arlo).

Orientation acuity showed a somewhat greater reduction after area 17–18 ablations. In this task, one stimulus was a pair of parallel lines and the other was a pair of nonparallel lines (figure 17.3). Before operation, the cats rapidly learned the discrimination to about a 5° angular difference without difficulty. In figure 17.8 the effects of an area 17–18 ablation on orientation acuity can be seen. With 56% as a threshold criterion, a threshold of about 5° (varying somewhat between animals, with an occasional cat with a threshold as low as 2°–3°) is observed. After operation, the animals with the area 17–18 lesions (Francis and Zelda) showed a substantial deficit in their ability to discriminate contour orientation. Nevertheless, they performed reasonably well. Recall that these animals were able to perform a classical form discrimination, such as upright versus inverted triangle or circle versus plus (figure 17.4). In an attempt to relate contour-orientation acuity to shape discrimination, we also tested the animals in a polygon (angularity) discrimination task (figure 17.3). As can be seen, the discrimination of the presence of angled contours is an important component in this discrimination task. The performance of the cats with area 17–18 lesions was consistent with the results of the contour-orientation findings (figure 17.9). Thus, these cats (Francis, Streak) were impaired in their ability to discriminate polygons from a circle. Cats with ablations sparing area 17–18 (Earl, BJ) showed no such impairment (figure 17.9); they were able to discriminate 9–11-sided figures from circles.

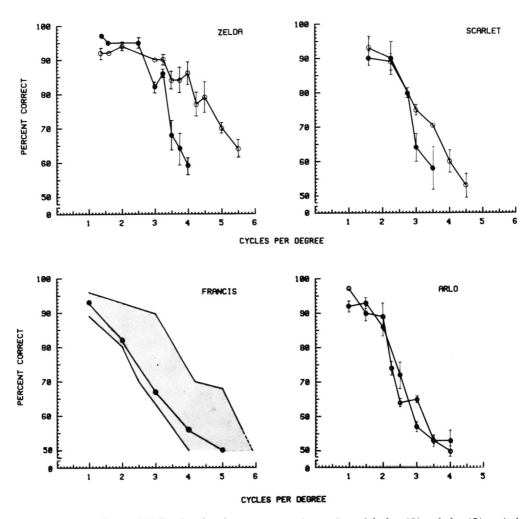

Figure 17.7 Graphs of performance on grating acuity task before (○) and after (●) cortical ablations in four cats. The percentage of correct choices is plotted on the ordinate, and grating size (spatial frequency) on the abscissa. Cats Zelda, Scarlet, and Francis received area 17–18 ablations (see figure 17.4); Arlo had a lateral cortical ablation. Vertical bars at each point indicate ± 1 S.D. Shaded area on graph for Francis depicts range of normal performance. Chance performance as determined empirically is 56% correct.

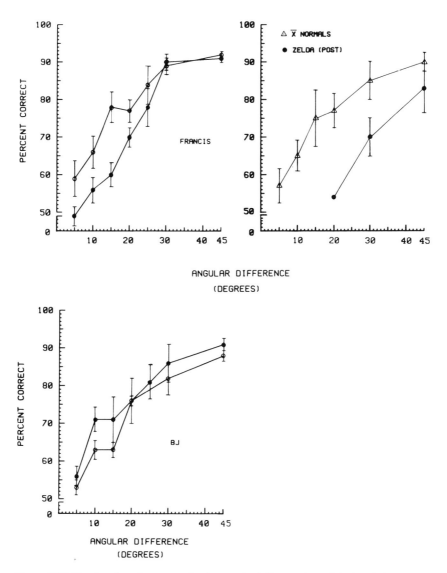

Figure 17.8 Graphs of performance of cats on orientation acuity task before (○) and after (●) cortical ablations. △ represents X̄ normals. Cats Francis and Zelda received area 17–18 ablations; BJ received a lateral cortical ablation (see figures 17.4 and 17.5). Abscissae represent angular difference between nonparallel lines and parallel lines as shown in figure 17.3. Vertical bars indicate ±1 S.D.

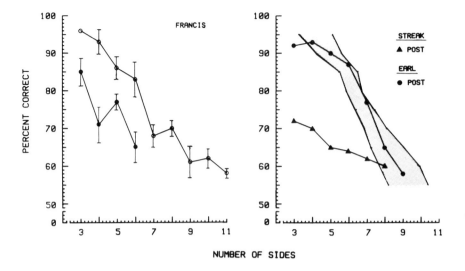

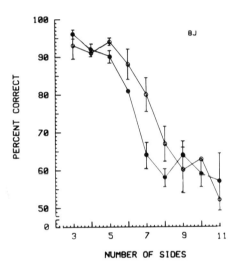

Figure 17.9 Graphs of performance of cats on angularity task before (○) and after (●, ▲) cortical ablations. Cats Francis and Streak received area 17–18 ablations; BJ and Earl received lateral cortical ablations. Shaded portion of upper right graph depicts range of behavioral normal cats tested on this task. Abscissae denote number of sides of polygon figures used in discrimination task (see angularity targets shown in figure 17.3). Vertical bars at each point indicate ±1 S.D.

The last acuity task we selected in our evaluation of the role of area 17–18 in vision was a vernier-offset acuity problem. The stimulus configuration used in this test is shown in figure 17.3. The targets were generated on an oscilloscope, so the lines were luminous—not an ideal target for this kind of discrimination, but an easy one to control. During our first attempts at teaching cats this discrimination it was extremely difficult to teach them to respond to the line offset. However, by placing a small annulus around the offset we were able to speed the acquisition of the discrimination at which time the larger (normal size) annulus was reintroduced. Four animals were tested on this task (figure 17.10). The details of the lesion of one cat (Streak) are worth noting in considering the results of the vernier-offset test: the lesion was the smallest of the lesions in the cats tested, and a considerable portion of area 17 (which represents the peripheral visual field) remained intact. Figure 17.10 shows pre- and postoperative performance on the vernier-offset task. The cats with area 17–18 ablations were severely impaired on this task. For example, after operation we could not find an offset that cat Streak could discriminate. The largest offset that we could produce was about 7°, and yet this animal could not perform on his task. Cat Francis was able only to discriminate large offsets (>2°). Control cats with a lateral lesion were, however, essentially normal (figure 17.10, Earl and BJ).

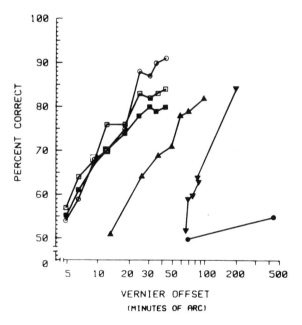

Figure 17.10. Graph of performance of cats on vernier-offset discrimination task before (open symbols) and after (filled symbols) cortical ablations. Cats Streak (○, ●) and Francis (▼; no preoperative test) received area 17–18 ablations; Earl (□, ■) and BJ (▲; no preoperative test) received lateral cortical ablations. NT indicates that a preoperative test was not performed. Abscissa denotes horizontal difference (offset), in visual angle, between broken line segments (see figure 17.3 for depiction of stimulus). Note logarithmic scale on abscissa.

Geniculocortical System in Spatial Vision

General Discussion

Anatomical Issues

A valid criticism often raised in ablation studies when little or no post-operative deficit in behavior is found is that the lesion was incomplete. In visual studies, this argument is supported by numerous experiments showing that only a small amount of cortex is necessary for performance at criterion levels on a variety of visual problems (see, for example, Lashley and Frank 1934). This argument could also be invoked in the case of the present study, because most of the area 17–18 ablations were not complete. We believe, however, that our behavioral tests preclude this argument. First, the residual portions of area 17–18 in our operated cats were in regions of cortex where the peripheral visual field is represented. Even the smallest area 17–18 lesion included at least the representation of the central 5° in these areas. The intact cells we observed in portions of LGN_d concerned with central vision were located in those portions of the nucleus (NIM, C_1, and C_2) that project primarily to visual areas lying outside of area 17–18 (Burrows and Hayhow 1971; LeVay and Ferster 1977). If the ablated cats were using those portions of area 17–18 in which the peripheral visual field is represented, then one might expect the observed decline in acuity to be predicted by the retina-cortex projection ratio (magnification factor) or the acuity–retinal eccentricity function. Although this function is not available for the cat, if one assumes acuity to fall concomitantly with ganglion-cell density, the postoperative acuity observed in the present study is better than that predicted from the ganglion-cell density gradient observed in the cat (Berkley 1976). Second, careful observation of the cats during behavioral and neurological testing gave no hint that the cats were using the peripheral visual fields; for example, their gaze was always centered on the small moving objects we used to test visual tracking. More direct evidence comes from the results of acuity testing in cats with small retinal lesions (made with a laser) of the central 2°–3° of the visual field. These animals showed a larger loss in acuity than we observed in the present study (Berkley et al. 1978), even though the extent of visual field impaired was smaller than the field deficit due to our area 17–18 lesions. Thus, we believe that the residual acuity observed in our animals was not mediated by residual portions of area 17–18.

Behavioral Issues

Another argument often raised in interpreting ablation-behavior studies of the visual system in which little or no deficit is found is that after operation the animals were able to use cues that they had not used before, and that this confounds experimental results. Although in many cases this is a valid criticism, we do not consider it applicable to the present study. In general, using complex visual stimuli (such as shapes) in a discrimination task permits an animal to use any one of a number of dimensions contained in the stimulus (size, orientation, texture, contour, local flux cues, etc.) to differentiate it from other stimuli. Most form stimuli, even those labeled as simple, contain many uncontrolled or unspecified attributes. Thus, what is described as a simple geometric form—for example, a triangle—contains a variety of dimensions that differentiate it from other relatively simple geometric forms. It is not possible from examination of the figure or the behavior of the animal to discover which of the various dimensions available in the stimulus are being used to make the discrimination. Thus, use of such stimuli would permit the shifting of cues between pre- and postoperative tests. The stimuli used in the present study were selected to eliminate or control all except one cue. This was achieved by using a threshold testing procedure in which one dimension of the stimulus complex was varied. As mentioned earlier, the rationale for doing this was that, if the manipulation of the dimension controlled the animal's behavior (for example, by producing systematic changes in choice performance), that was taken as evidence that it was the cue to which the animal was attending. A second and equally important advantage of a threshold testing procedure is that it contains a simple performance decrement test. That is, discrimination performance at suprathreshold values of the stimulus should not be affected by the ablation and thus can be used as a performance control. The postoperative performance data of the present study show that both conditions were fulfilled: At suprathreshold stimulus values, the animals in most cases performed near normal levels and their choice behavior varied systematically with changes in the specified stimulus dimension. Thus, we conclude that the cats did not have an overall performance deficit and that the cue they were attending to after operation was the same cue they had used in preoperative tests.

Finally, it could be argued that our preoperative acuity estimates were too low, permitting the animals room for postoperative improvement and thus reducing the relative magnitude of the "true" postoperative deficits in grating and orientation acuity. This seems unlikely, since our preoperative acuity estimates were similar to those observed in other studies of normal cats (Berkley and Watkins 1971; Blake et al. 1974; Bloom and Berkley 1977) and there is no reason to believe that the cats used in this study deviated from a normal sample.

Theoretical Implications
The results of the present study have implications for current speculative models in vision. A simple hierarchical (series) model of the visual system is not supported by our results; this confirms the expectations derived from anatomical studies that the visual system is not a simple serial processor (Garey and Powell 1967; Glickstein et al. 1967; Niimi and Sprague 1970; Burrows and Hayhow 1971; Diamond 1976). Those studies demonstrated that there are multiple parallel pathways from

LGN to cortex in the cat. Further, the recent demonstrations of a route to cortex via the tectum in both cat and monkey raise the possibility that this pathway may also participate in higher visual processing. Recent studies of the visual capacities of destriate monkeys suggest that this pathway may be more important in spatial vision than was originally thought (see Keating, this volume).

A second and perhaps more important implication of the present results concerns the putative role of one physiological class of neurons (X cells) in the visual system. Numerous investigators have distinguished retinal neurons into several functional classes (for example, X, Y, W) on the basis of their physiological properties (see Rowe and Stone 1977 for review and comments). There is also evidence that classes X, Y, and W remain essentially segregated in their projections to cortex (see, for example, Stone and Dreher 1973; LeVay and Ferster 1977). Current anatomical and physiological evidence indicates that in the cat X cells project only to area 17 and possibly to area 18, whereas the other cell classes project more widely (Stone and Dreher 1973; LeVay and Ferster 1977). Several investigators have suggested or implied functional roles for these physiological cell classes (see, for example, Tolhurst 1973; Kulikowski and Tolhurst 1973; King-Smith and Kulikowski 1975) such that X cells mediate pattern vision and Y cells mediate temporal aspects of vision. Since the present study showed that considerable detail vision (and form vision) is possible without area 17–18, and thus essentially without an X-cell contribution to the neural processing of visual information, it seems reasonable to conclude that X cells are not essential for form vision (see also Hughes, this volume). Since X cells have the finest spatial resolution, then one must also conclude that resolution of high spatial frequencies is not essential for pattern vision. This is not to say that X cells contribute nothing to vision, but rather that other cell classes are capable of signaling the shapes of objects.

If one assumes that X cells mediate the vision of the finest details, then clinical data from humans with impaired optics in which the visibility of high spatial frequencies is impaired but low-spatial-frequency sensitivity is normal also support the view that high spatial frequencies are not essential for seeing shapes (Hess and Garner 1977; Hess and Woo 1978). That is, defocus errors, which produce losses in high spatial frequencies, are less destructive to visual shape perception than optical fogging or clouding, which reduces the visibility of all spatial frequencies (Campbell and Green 1965). (This effect may be easily observed by looking at a contoured visual scene first through a mild defocusing lens and then through frosted glass.)

In a broader context, these studies seem to rule out any simple models of visual perception derived from current knowledge of receptive fields in area 17 and 18. It seems likely that in the cat, and possibly in the monkey, the neural mechanisms of visual perception reside not in area

17 and 18 but elsewhere. These mechanisms may be accessible through a variety of pathways.

Though it can be said with some certainty that area 17–18 is not necessary for cats to see shapes and distinguish a variety of visual details, the contributions of the various thalamocortical pathways to spatial vision is still unclear. From the deficits observed in the present study, it is obvious that area 17–18 contributes heavily to the vision of fine details in the cat. This capacity may be important in such capacities as neural "deblurring," distant-object detection, binocular eye-alignment mechanisms, and stereopsis. The question of which remaining neural areas are mediating the visual capacities exhibited without area 17–18 (for example, the remaining geniculocortical pathways or the tectothalamocortical pathways) is not yet answered and awaits further experimental study.

Summary and Conclusions

The results of the present study clearly demonstrate that the cat is capable of processing information of some detail even after extensive damage to areas 17 and 18. It provides further confirmation of the findings that cats can discriminate complex spatial stimuli after loss of area 17 and most of 18, and it rules out explanations based on use of other cues (luminance differences, local flux cues, etc.).

Anatomical studies provide a plausible explanation for such findings in the overlapping connections of the multiple parallel pathways in the cat's visual system. This and other studies (Spear and Braun 1969; Doty 1971; Wood et al. 1974; Sprague et al. 1977) have shown that removal of one pathway (area 17 or 20), two (17 + 18, 19 + 21, 19 + LSA, 21 + LSA, 20 + 21), or three (17 + 18 + 19) does not eliminate the ability to discriminate spatial features, strongly suggesting that the individual visual pathways share some spatial capabilities. The present study lends support to such a view, but also demonstrates that not all the pathways in the cat are equivalent. From the results of the present study it appears that area 17 and adjoining 18 (especially in the region of the area centralis) participates in the mediation of the detection of fine details, as revealed by the modest increase in threshold in grating acuity, the moderate loss in orientation acuity, and the extensive deficit in a task requiring topographic alignment of contours. Similarly, destriate cats show prolonged learning of form discriminations if the discriminanda are small outline figures. Although the neuronal machinery in area 17 (+18) functions in the analysis of fine spatial details it does not appear to be essential for basic discrimination of patterns and shapes in the cat.

The results of this study permit the exclusion of explanations of residual spatial vision based on noncontour information after removal of 17–18, but do not permit specification of which of the various residual pathways cats use in performing the visual discriminations. For example, is the capacity to discriminate forms mediated chiefly by the tectopulvinar-cortical system, or by the extrastriate part of the geniculo-cortical pathway, or equally by both? The answer to this question will come from continued studies of isolated individual pathways.

Acknowledgments

The authors take pleasure in acknowledging the assistance of D. S. Warmath and Jeanne Levy in the conduct of the study, and the secretarial assistance of Kenna Study. This work was supported by research grants EY00953 to M.A.B. and EY00577 to J.M.S. from the National Eye Institute.

References

Barlow, H. B. 1972. "Single units and sensation: A neuron doctrine for perceptual psychology?" *Perception* 1: 371–394.

Benevento, L. A., and M. Rezak. 1976. "The cortical projections of the inferior pulvinar and adjacent lateral pulvinar in the rhesus monkey (*Macaca mulatta*): An autoradiographic study." *Brain Res.* 108: 1–24.

Berkley, M. A. 1970. "Visual discriminations in the cat." In *Animal Psychophysics: The Design and Conduct of Sensory Experiments*, William Stebbins, ed. New York: Appleton-Century-Crofts.

———. 1976. "Cat visual psychophysics: Neural correlates and comparisons with man." In *Progress in Psychobiology and Physiological Psychology*, vol. 6, J. Sprague and A. M. Epstein, eds. New York: Academic.

———. 1978. "Vision: Geniculocortical System." In *Handbook of Behavioral Neurobiology*, vol. 1: *Sensory Integration*, R. B. Masterton, ed. New York: Academic.

Berkley, M. A., and J. M. Sprague. 1979. "Striate cortex and visual acuity functions in the cat." *J. Comp. Neurol.* 187: 679–702.

Berkley, M. A., and D. W. Watkins. 1971. "Visual acuity of the cat estimated from evoked cerebral potentials." *Nature New Biol.* 234: 91–92.

Berkley, M. A., F. T. Crawford, and G. Oliff. 1971. "A universal foodpaste dispenser for use with cats and other animals." *Behav. Res. Meth. Instr.* 3: 259–260.

Berkley, M. A., J. Sprague, and D. S. Warmath. 1976. "The role of the geniculo-cortical system in form vision: Area 17 and 18 and contour acuities." *Soc. Neurosci. Abstr.*

Berkley, M. A., S. M. Sherman, D. S. Warmath, and J. T. Tunkl. 1978. "Visual capacities of adult cats which were reared with a lesion in the retina of one eye and the other occluded." *Soc. Neurosci. Abstr.* 4: 467.

Blake, R., S. J. Cool, and M. L. J. Crawford. 1974. "Visual resolution in the cat." *Vision Res.* 14: 1211–1217.

Bloom, M., and M. A. Berkley. 1977. "Visual acuity and the near point of accommodation in cats." *Vision Res.* 17: 723–730.

Burrows, G. R., and W. R. Hayhow. 1971. "The organization of the thalamo-cortical visual pathways in the cat." *Brain, Behav., Evol.* 4: 220–272.

Campbell, F. W., and D. G. Green. 1965. "Optical and retinal factors affecting visual resolution." *J. Physiol.* 181: 576.

Denny-Brown, D., and R. A. Chambers. 1976. "Physiological aspects of visual perception. I. Functional aspects of visual cortex." *Arch. Neurol.* 33: 219–227.

Diamond, I. T. 1976. "Organization of the visual cortex: Comparative anatomical and behavioral studies." *Fed. Proc.* 35: 60–67.

Doty, R. W. 1971. "Survival of pattern vision after removal of striate cortex in the adult cat." *J. Comp. Neurol.* 143: 341.

Garey, L. J., and T. P. S. Powell. 1967. "The projection of the lateral geniculate nucleus upon the cortex in the cat." *Proc. Roy. Soc.* B 169: 107–126.

Glickstein, M., R. A. King, J. Miller, and M. Berkley. 1967. "Cortical projections from the dorsal lateral geniculate nucleus of cats." *J. Comp. Neurol.* 130: 55–76.

Hess, R., and L. Garner. 1977. "The effect of corneal edema on visual function." *Invest. Ophthalmol.* 16: 5–13.

Hess, R., and G. Woo. 1978. "Vision through cataracts." *Invest. Ophthalmol.* 17: 428–435.

Hollander, A., and H. Vanegas. 1977. "The projection from the lateral geniculate nucleus onto the visual cortex in the cat: A quantitative study with horseradish peroxidase." *J. Comp. Neurol.* 173: 519–536.

Hubel, D. H., and T. N. Wiesel. 1962. "Receptive fields, binocular interaction and functional architecture in the cat's visual cortex." *J. Physiol.* 160: 106–154.

Humphrey, N. R. 1974. "Vision in a monkey without striate cortex: A case study." *Perception* 3: 241–255.

Kalia, M., and D. Whitteridge. 1973. "The visual areas in the spinal splenial sulcus of the cat." *J. Physiol.* 232: 275–283.

King-Smith, P. E., and J. J. Kulikowski. 1975. "Pattern and flicker detection analysed by subthreshold summation." *J. Physiol.* 249: 519–548.

Kulikowski, J. J., and D. J. Tolhurst. 1973. "Psychophysical evidence for sustained and transient detectors in human vision." *J. Physiol.* 232: 149–162.

Lashley, K. S. 1950. "In search of the engram." In *Symposium of Society for Experimental Biology* No. 4. Cambridge University Press.

Lashley, K. S., and M. Frank. 1934. "The mechanism of vision. X. Postoperative disturbances of habits based on detail vision in the rat after lesions in the cerebral visual areas." *J. Comp. Psychol.* 17: 355–380.

LeVay, S., and D. Ferster. 1977. "Relay cell classes in the lateral geniculate nucleus of the cat and the effects of visual deprivation." *J. Comp. Neurol.* 172: 563–584.

LeVay, S., and C. D. Gilbert. 1976. "Laminar patterns of geniculo-cortical projection in the cat." *Brain Res.* 113: 1–19.

Niimi, K., and E. Kuwabara. 1973. "The dorsal thalamus of the cat and comparison with monkey and man." *J. Hirnforsch.* 14: 303–325.

Niimi, K., and J. M. Sprague. 1970. "Thalamo-cortical organization in the visual system of the cat." *J. Comp. Neurol.* 138: 219–249.

Otsuka, R., and R. Hassler. 1962. "Über aufbau und Gliederung der corticalen Sehsphäre bei der Katze." *Arch. Psychiatr. Z. ges. Neurol.* 203: 212–234.

Palmer, L. A., A. C. Rosenquist, and R. J. Tusa. 1978. "The retinotopic organization of lateral suprasylvian visual areas in the cat." *J. Comp. Neurol.* 177: 237–256.

Pasik, T., and P. Pasik. 1971. "The visual world of monkeys deprived of striate cortex: Effective stimulus parameters and the importance of the accessory optic system." *Vision Res.* suppl. 3: 419–435.

Rosenquist, A. C., S. B. Edwards, and L. A. Palmer. 1974. "An autoradiographic study of the projections of the dorsal lateral geniculate nucleus and the posterior nucleus in the cat." *Brain Res.* 80: 71–93.

Rowe, M. H., and J. Stone. 1977. "Naming neurones. Classification and naming of cat retinal ganglion cells." *Brain, Behav., Evol.* 14: 185–216.

Sanides, F., and J. Hoffman. 1969. "Cyto- and myeloarchitecture of the visual cortex of the cat and of the surrounding integration cortices." *J. Hirnforsch.* 11: 79–104.

Sherman, S. M. 1973. "Visual field defects in monocularly and binocularly deprived cats." *Brain Res.* 49: 25–45.

Spear, P. D., and J. J. Braun. 1969. "Pattern discrimination following removal of visual neocortex in the cat." *Exp. Neurol.* 25: 331–348.

Sprague, J. M. 1966. "Visual, acoustic, and somesthetic deficits in the cat after cortical and midbrain lesions." In *The Thalamus*, D. P. Purpura and M. Yahr, eds. New York: Columbia University Press.

Sprague, J. M., and T. H. Meikle, Jr. 1965. "The role of the superior colliculus in visually guided behavior." *Exp. Neurol.* 11: 115.

Sprague, J. M., J. Levy, A. DiBerardino, and B. Berlucchi. 1977. "Visual cortical areas mediating form discrimination in the cat." *J. Comp. Neurol.* 172: 441–488.

Stone, J., and B. Dreher. 1973. "Projection of X- and Y-cells of the cat's lateral geniculate nucleus to areas 17 and 18 of visual cortex." *J. Neurophysiol.* 36: 551–567.

Tolhurst, D. J. 1973. "Separate channels for the analysis of the shape and the movement of a moving visual stimulus." *J. Physiol.* 231: 385–402.

Tunkl, J. E., and M. A. Berkley. 1977. "The role of superior colliculus in vision: Visual form discrimination in cats with superior colliculus ablations." *J. Comp. Neurol.* 176: 575–588.

Tusa, R. 1978. "Retinotopic organization of visual cortex in cat." Ph.D. diss., University of Pennsylvania.

Tusa, R. J., L. A. Palmer, and A. C. Rosenquist. 1978. "The retinotopic organization of area 17 (striate cortex) in the cat." *J. Comp. Neurol.* 177: 213–236.

Updike, B. V. 1977. "Topographic organization of the projections from cortical areas 17, 18, and 19 onto the thalamus, pretectum, and superior colliculus in the cat." *J. Comp. Neurol.* 173: 81–122.

Wong-Riley, M. T. T. 1976. "Projections from the dorsal lateral geniculate nucleus to prestriate cortex in the squirrel monkey as demonstrated by retrograde transport of horseradish peroxidase." *Brain Res.* 109: 595–600.

Wood, C. C., P. D. Spear, and J. J. Braun. 1974. "Effects of sequential lesions of suprasylvian gyri and visual cortex on pattern discrimination in the cat." *Brain Res.* 66: 443–466.

18 Two Cortical Visual Systems

Leslie G. Ungerleider
Mortimer Mishkin

Damage to the primary visual cortex (striate cortex or area 17), either by disease, by acute assault, or through surgical intervention for therapeutic purposes, results in a scotoma or blind region in the visual field. Moreover, because of the precise topographic mapping of the visual field onto striate cortex, the region of the scotoma corresponds to the part of striate cortex damaged. In contrast to this primary sensory impairment, "higher-order" visual dysfunctions ensue when damage occurs more rostrally in the brain, in either the temporal or parietal lobes; however, the effects that follow lesions of these two cortical areas are markedly different (Newcombe and Russell 1969). Whereas damage to temporal cortex produces an impairment in visual recognition (see, for example, Milner 1958, 1968; Kimura 1963; Lansdell 1968; Benson et al. 1974; Meadows 1974), damage to parietal cortex produces a constellation of visual spatial impairments (see, for example, McFie et al. 1950; Semmes et al. 1963; De Renzi and Faglioni 1967; Butters et al. 1972; Ratcliff and Davies-Jones 1972; Ratcliff and Newcombe 1973).

Although it is clear that visual information must reach the temporal and parietal association areas to enable their participation in visual recognition and visual spatial perception, respectively, the complex circuitry through which this information is transmitted has yet to be unraveled. It is known, however, that two major fiber bundles emerge from occipital cortex and project rostrally in the brain (Flechsig 1896, 1920). One, the superior longitudinal fasciculus, follows a dorsal path, traversing the posterior parietal region in its course to the frontal lobe; the other, the inferior longitudinal fasciculus, follows a ventral route into the temporal lobe. It has been our working hypothesis (Mishkin 1972; Pohl 1973) that the ventral or occipitotemporal pathway is specialized for object perception (identifying *what* an object is) whereas the dorsal or occipitoparietal pathway is specialized for spatial perception (locating *where* an object is). This distinction between the two types of visual perception is not new (see, for example, Ingle 1967; Held 1968). In the past, however, the neural mechanisms underlying object and spatial vision were seen as localized in geniculostriate and tectofugal systems, respectively (Schneider 1967; Trevarthen 1968).

By contrast, in the present formulation, corticocortical connections originating in the striate area are viewed as mediating both types of vision, with two diverging cortical systems replacing the geniculostriate-tectofugal dichotomy. The reasons for stressing corticocortical mechanisms for both types of visual perception in primates are developed below, but this emphasis is not meant to deny that the tectofugal system (including the tectofugal pathway to cortex) contributes to spatial vision, particularly to its visuomotor aspects; it is simply that, with regard to the perceptual aspects of spatial vision, the tectofugal system in the primate appears to play a subsidiary role. In our investigations of the two cortical visual systems, we have used the rhesus monkey (*Macaca mulatta*) as our subject and have employed a combination of behavioral, electrophysiological, and anatomical techniques.

Occipitotemporal Mechanisms in Object Perception

The results of numerous behavioral studies have demonstrated that bilateral removal of inferior temporal cortex in monkeys produces a severe impairment in visual-discrimination performance (for reviews see Gross 1973; Dean 1976; Wilson 1978). In brief, inferior temporal lesions produce a deficit that is exclusively visual, affecting both the retention of discriminations acquired prior to surgery and the postoperative acquisition of new discriminations. Among the deficits that have been reported are those involving hue, brightness, two-dimensional patterns, and three-dimensional shapes. More recent work has shown that damage to the posterior part of the inferior temporal cortex (area TEO) interferes mainly with discriminative ability, whereas damage to the anterior part (area TE) affects primarily visual memory (Iwai and Mishkin 1968; Cowey and Gross 1970).

The pathway through which inferior temporal cortex receives visual information was first suggested on the basis of early neuronographic data (von Bonin et al. 1942). It had been shown in both the monkey and chimpanzee that if strychnine is applied to the striate cortex spike discharges can be recorded from a prestriate cortical belt, whereas if strychnine is applied to any part of this prestriate region spikes can be recorded in the inferior part of the temporal lobe. These neuronographic findings were later confirmed in a neuroanatomical study (Kuypers et al. 1965) that employed the Nauta-Gygax (1954) technique for tracing projections by the silver-staining of degenerating axons after removal of their cell bodies. As anticipated, it was found that striate cortex projects to a prestriate cortical belt, which, in turn, projects to the inferior temporal area. It was also confirmed that each prestriate area projects across the splenium of the corpus callosum to reach the prestriate area of the opposite hemisphere.

These neuronographic and neuroanatomical findings are sche-

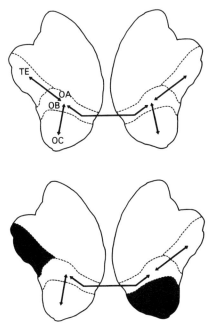

Figure 18.1 Striate-prestriate-temporal pathways. Abbreviations refer to cytoarchitectonic divisions of von Bonin and Bailey (1947). Lower diagram indicates pathways remaining after crossed striate and inferior temporal lesions (in black). Pathways are shown as two-way, in accord with the neuroanatomical evidence that the connections are reciprocal (Kuypers et al. 1965; Rockland and Pandya 1979). Adapted from Mishkin 1966.

matized in figure 18.1. As indicated in the upper diagram, each striate area transmits visual information, relayed through the prestriate cortex, to the ipsilateral inferior temporal area; but, in addition, because of the reciprocal prestriate connections across the corpus callosum, each striate area also transmits visual information to the contralateral inferior temporal area. The behavioral significance of these pathways was first demonstrated in a series of crossed-lesion disconnection experiments (Mishkin 1966). Since such disconnection studies are not commonly employed to investigate brain-behavior relationships, a brief explanation of the underlying logic may be in order.

The usual method of identifying the function of a cortical area is to remove that same area from both hemispheres and then determine what function has been lost. This method does not reveal whether the function was lost because it depended on the area (or station) that was removed or because it depended on a later station along the same pathway that could no longer operate after the pathway leading to it was destroyed. Crossed-lesion disconnection studies make it possible to demonstrate that a pathway with many cortical stations exists. The method is to remove one of the stations in the postulated pathway on one side of the brain, and to remove a different one (later in the pathway) on the other side (see lower diagram of figure 18.1). This leaves one of each pair in the series intact, and it can often be shown that up to this point little or no loss in function occurs. However, if the connections between the two hemispheres are then cut, the functions served by the stations beyond the cut may be lost. This demonstrates that the function in question depends not only on the stations themselves, but also on the connections between them. The same demonstration theoretically could have been achieved by cutting the fiber connections between the stations within each hemisphere, but the complexity of these connections is so great, and they form such a compact network, that the method is technically difficult if not impossible.

In the disconnection study referred to above it was found that if an inferotemporal lesion in one hemisphere was combined serially with total striate removal in the other, animals continued to perform a pattern-discrimination task; however, when these asymmetrical or crossed lesions were followed by transection of the corpus callosum, the performance of the animals fell to chance and they failed to relearn the task. Presumably, the single crossed pathway from the intact striate cortex on one side across the corpus callosum to the intact inferior temporal cortex on the other side is sufficient to mediate pattern-discrimination habits, but if this pathway is cut a severe deficit results. This study thus provided the first behavioral indication that inferior temporal cortex is a late station along a cortical visual pathway running from striate through prestriate cortex.

Electrophysiological experiments (Gross et al. 1972) subsequently revealed that single neurons in inferior temporal cortex, like those in striate (Hubel and Wiesel 1968) and prestriate (Hubel and Wiesel 1970) cortex, have visual receptive fields. The optimal trigger features for inferior temporal neurons, however, are considerably more complex. That visual input to these neurons originates in striate cortex was shown in a combined recording and ablation study by Rocha-Miranda et al. (1975). The method entailed measuring the defect in the visual receptive fields of inferior temporal neurons after selective cerebral lesions or commissural transections. As predicted from earlier ablation studies (Mishkin 1972), the tectofugal pathway from the superior colliculus through the pulvinar to cortex turned out to be unimportant for the receptive field properties of inferior temporal neurons. By contrast, the cortical pathway from striate through prestriate to inferior temporal cortex proved to be essential, since interruption of this pathway by a striate removal or by transection of the forebrain commissures eliminated the corresponding visual input to inferior temporal cells. The results, shown in figure 18.2, indicate that normally over 60% of inferior temporal neurons have bilateral receptive fields. However, after bilateral striate-cortex removal these neurons are totally unresponsive to visual stimulation, after unilateral striate-cortex removal they respond to visual stimulation only in the hemifield opposite the intact striate cortex, and after commissurotomy they respond to stimulation only in the con-

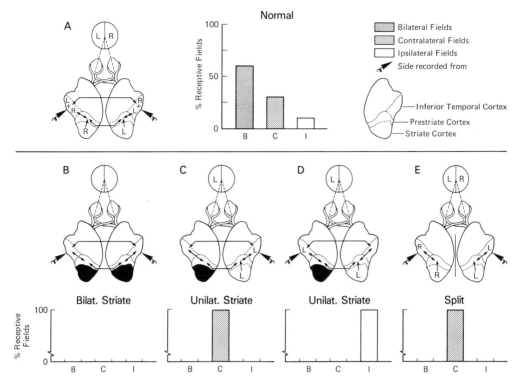

Figure 18.2 Proportion of inferior temporal neurons that had bilateral, contralateral, or ipsilateral receptive fields in (A) normal monkeys, (B) monkeys with bilateral removal of striate cortex, (C, D) monkeys with unilateral removal of striate cortex, and (E) monkeys with transection of the forebrain commissures. The brain diagrams show how information from the right (R) and left (L) visual hemifields could reach inferior temporal cortex along a corticocortical route and how each lesion (in black) interferes with this pathway. Adapted from Rocha-Miranda et al. 1975.

tralateral hemifield. Thus, the dependence of inferior temporal cortex on corticocortical connections arising in striate cortex has now been made evident at the single-cell level.

Most recently, this entire visual pathway was functionally mapped using ^{14}C-labeled 2-deoxyglucose (2-DG) via the method developed by Sokoloff and his colleagues (Kennedy et al. 1975). In this technique 2-DG is used as a marker of local cerebral glucose utilization and hence indicates regions that are metabolically active during the experimental procedure. The visual-mapping studies (Jarvis et al. 1978; Kennedy et al. 1978) were carried out in awake monkeys previously prepared with a unilateral optic-tract section combined with transection of the forebrain commissures (a procedure that visually deafferented one hemisphere while leaving the other intact). On the day of the experiment, the monkeys were presented with visual patterns in a rotating drum or in a discrimination apparatus. In both situations, reduced glucose utilization in the blind as compared with the seeing hemisphere was seen cortically, not only in the geniculostriate projection but throughout the entire expanse of prestriate and inferior temporal cortex as far forward as the temporal pole. With one exception, no other cortical area in the deafferented hemisphere showed reduced activity.

The evidence that has been cited favors a sequential-activation model for object vision in which information that reaches the striate cortex is transmitted for further processing to the prestriate cortex, and from there to the inferior temporal area. This system appears to be important for the analysis and coding of the physical dimensions of visual stimuli needed for their identification and recognition. It is unlikely, however, that any part of this system up to and including the inferior temporal area is involved in the still higher-order process of associating visual stimuli with other events, such as motivational and emotional ones. Recordings from single cells in monkeys performing visual discrimination and reversal tasks have shown that although inferior temporal neurons are sensitive to the physical properties of stimuli, they are relatively insensitive to changes in the reward value of the stimuli (Jarvis and Mishkin 1977). Similar results have been reported by others (Rolls et al. 1977; but see Ridley and Ettlinger 1973). Presumably, the process of attaching reward value to a stimulus depends on stations beyond the occipitotemporal pathway (Jones and Mishkin 1972; Sunshine and Mishkin 1975; Spiegler and Mishkin 1978; Rolls et al. 1979b; Sanghera et al. 1979). However, for object vision, the inferior temporal cortex may well be the final station. It is significant in this regard that, by virtue of the extremely large receptive fields of inferior temporal neurons (Gross et al. 1972), this area provides the neural mechanism of stimulus equivalence across retinal translation (Gross and Mishkin 1977; Seacord et al. 1979)—that is, the ability to recognize a stimulus as the same regardless of its position in the visual field and, by extrapolation, regardless of its spatial location. Indeed, a necessary consequence of this equivalence mechanism is that within the occipitotemporal pathway there is a loss of information about the spatial locations of objects.

Occipitoparietal Mechanisms in Spatial Perception

The neural mechanism for the analysis of the spatial locations of objects also entails the transmission of visual information from the striate through the prestriate area; however, the rest of the pathway for spatial vision appears to be quite separate from the ventral pathway into the temporal cortex. Evidence in support of this dichotomy of cortical visual systems comes from recent studies in our laboratory on the parietal lobe.

In the initial study of this series, Pohl (1973) demonstrated a dissociation of visual deficits after inferior temporal and posterior parietal lesions. Whereas the temporal but not the parietal lesion produced a severe impairment on an object discrimination task, just the reverse was found on tests in which the animal was required to choose a response location on the basis of its proximity to a visual "landmark" (see part A of figure 18.5). The results suggested that ". . . the inferior temporal cortex participates mainly in the acts of noticing and remembering an object's qualities, not its position in space," and that "conversely, the posterior parietal cortex seems to be concerned with the perception of spatial relations among objects, and not their intrinsic qualities" (Mishkin 1972).

Accumulating evidence from other laboratories supports this view of posterior parietal function. Not only has the impairment of landmark tasks after posterior parietal lesions been corroborated (Milner et al. 1977; Ungerleider and Brody 1977; Brody and Pribram 1978; however, see Ridley and Ettlinger 1975), but impairments after such lesions have also been found on other visual spatial tasks, including a stylus maze (Milner et al. 1977), patterned-string tests (Ungerleider and Brody 1977), cage finding (Sugishita et al. 1978), and route following (Petrides and Iversen 1979). Visual spatial disorientation, however, is not the only deficit produced by lesions of this region. Indeed, the classical symptoms of posterior parietal dysfunction are misreaching in the dark as well as in the light (Ettlinger and Wegener 1958; Bates and Ettlinger 1960; Ettlinger and Kalsbeck 1962; Hartje and Ettlinger 1973; LaMotte and Acuña 1978; Faugier-Grimaud et al. 1978; Stein 1978), contralateral neglect of auditory and tactile as well as visual stimuli (Denny-Brown and Chambers 1958; Heilman et al. 1970), and impairments of tactile discrimination (Blum 1951; Pribram and Barry 1956; Wilson 1957, 1975; Ettlinger and Wegener 1958; Pasik et al. 1958; Bates and Ettlinger 1960; Wilson et al. 1960; Ettlinger and Kalsbeck 1962; Ettlinger et al. 1966; Moffett et al. 1967; Moffett and Ettlinger 1970; Ridley and Ettlinger 1975).

The posterior parietal lesions in these studies have nearly always included two or even more cytoarchitectonic areas, and so it is natural to assume that the heterogeneity of effects is a consequence of the heterogeneity of the tissue included in the removals. However, the few studies that have directly examined this question (Moffett et al. 1967; Ridley and Ettlinger 1975) failed to uncover any clear-cut instance of functional dissociation. From the evidence at hand, the most parsimonious interpretation is one that has been offered by Semmes (1967), namely, that the multiple deficits are but different reflections of a single, supramodal spatial disorder (see also Ratcliff et al. 1977). According to this view, and in the light of more recent anatomical information (Pandya and Kuypers 1969; Jones and Powell 1970), a supramodal spatial framework could be constructed out of converging inputs to posterior parietal cortex from all exteroceptive sensory modalities, with a significant contribution from vision. In the studies in our laboratory that followed the "landmark" experiment by Pohl (1973), we addressed the questions raised above regarding localization of spatial function within the posterior parietal region and the dependence of this function on visual inputs.

As in most other posterior parietal ablation studies, the lesions in Pohl's study were composites that included not only inferior parietal but also dorsal prestriate tissue. To test for localization of function within this region, we partitioned it into three approximately equal sectors—inferior parietal, lateral preoccipital, and medial parietal-preoccipital—as shown in figure 18.3. Monkeys were then prepared with lesions of either one, two, or all three of these sectors and tested on the landmark, or distance discrimination, task (Mishkin, Lewis, and Ungerleider, in prep.). Like the earlier attempts, this one failed to reveal any evidence of functional specialization within the larger area. Instead, there was an especially clear relationship between severity of effect and extent of lesion, completely independent of locus. This finding is illustrated in figure 18.4.

The fact that performance on the landmark task depends as much on dorsal prestriate tissue as on inferior parietal tissue pointed to the possibility that the visual information on which posterior parietal cortex operates arises in striate cortex. The alternative possibility, that the critical visual input arises in the superior colliculus, found no clear support in a study of the effects of tectal lesions on landmark discrimination (Snyder and Mishkin, unpublished data); even complete destruction of the superior colliculus failed to produce a reliable loss in retention. Therefore, to test for the functional dependence of posterior parietal cortex on striate output, a crossed-lesion disconnection experiment analogous to the striate-temporal disconnection experiment described earlier (Mishkin 1966) was undertaken (Ungerleider and Mishkin 1978a). Monkeys were preoperatively trained on the landmark task, and then received a three-stage operation intended to serially disconnect posterior parietal from striate cortex. At each stage, the monkeys were given a preoperative retention test followed by surgery, allowed 2 weeks to recover, and then retested.

As shown in part B of figure 18.5, the first removal was a unilateral posterior parietal ablation, which included all three sectors investigated in the partial-lesion study to ensure a complete effect. On the hypothesis that each striate area has corticocortical connections with both posterior parietal areas, this first lesion may be viewed as a test of the effects of destroying two of these connections (the remaining two are indicated by the black arrows in the figure). A comparison of preoperative with postoperative retention scores on the task shows that this amount of damage to the system had only minimal effect.

The second-stage removal was a contralateral striate-cortex ablation, illustrated in part C of figure 18.5. It included all of the striate cortex, both laterally and medially. The results of the retest given after this operation revealed a severe impairment; the monkeys required an average of 880 trials to relearn, despite the remaining interhemispheric cor-

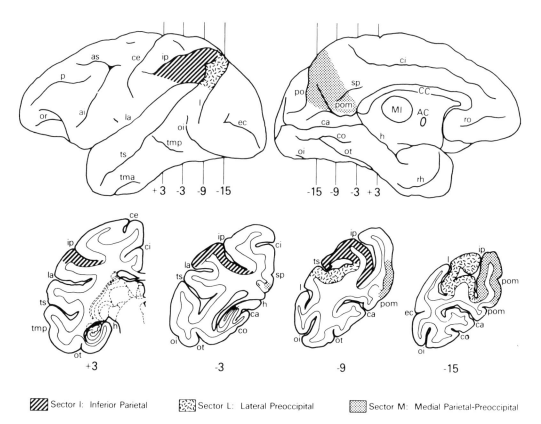

Figure 18.3 Partition of posterior parietal cortex into three sectors: inferior parietal (I), lateral preoccipital (L), and medial parietal-preoccipital (M). Monkeys received bilaterally symmetrical lesions intended to remove one (groups I, L and M), two (groups IL, LM, and IM), or all three (group PP) of these sectors. The sectors are shown on lateral and medial surface views and on representative cross-sections of a standard rhesus monkey brain. Numerals indicate approximate A-P stereotaxic levels. AC: anterior commissure. ai: inferior arcuate sulcus. as: superior arcuate sulcus. ca: calcarine fissure. CC: corpus callosum. ce: central sulcus. ci: cingulate sulcus. co: collateral sulcus. ec: ectocalcarine sulcus. h: hippocampal fissure. ip: intraparietal sulcus. l: lunate sulcus. la: lateral fissure. MI: massa intermedia. oi: inferior occipital sulcus. or: orbital sulcus. ot: occipitotemporal sulcus. p: principal sulcus. po: parieto-occipital incisure. pom: medial parieto-occipital fissure. rh: rhinal fissure. ro: rostral sulcus. sp: subparietal sulcus. tma: anterior middle temporal sulcus. tmp: posterior middle temporal sulcus. ts: superior temporal sulcus.

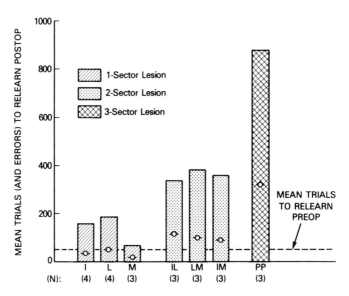

Figure 18.4 Postoperative retention on the landmark task after bilateral lesions of one, two, or three posterior parietal sectors, illustrated in Figure 18.3. Bars indicate mean trials to criterion; circles within bars indicate mean errors. N in this and following figures equals number of monkey in group. Mean number of trials to relearn task before operation (dashed line) is based on scores on a two-week retention test given just prior to surgery and represents the data from all 23 monkeys. The results showed no significant differences among the groups given a one-sector lesion, or among the groups given a two-sector lesion. However, as the number of sectors included in the posterior parietal lesion increased, so did the severity of the impairment. (Source: Mishkin et al., in prep.).

tical connections between the intact striate and parietal areas. This suggested that the interaction between the striate and parietal cortex via uncrossed connections had been of considerable importance for performance. However, whether such interaction was mediated by a corticocortical pathway could only be determined by examining the effects of the third stage of operation, transection of the corpus callosum.

In this operation, the posterior half of the callosum was transected, thereby cortically disconnecting the intact striate and posterior parietal areas. As shown by the postoperative retention score in part D of figure 18.5, the effect of the transection was to produce a partial reinstatement of the impairment. We can infer from these results that performance on the landmark task had indeed been mediated via corticocortical pathways between the striate and parietal areas.

However, the disruption of landmark discrimination that followed this lesion was not complete; all monkeys successfully relearned the task. At this point, the question was whether recovery of the discrimination was mediated by the interaction of the intact posterior parietal cortex with subcortical visual structures or by the interaction of the intact striate cortex with other cortical areas. In subsequent work on this preparation we found that removal of the remaining posterior parietal cortex was completely without effect, which indicated that tectofugal

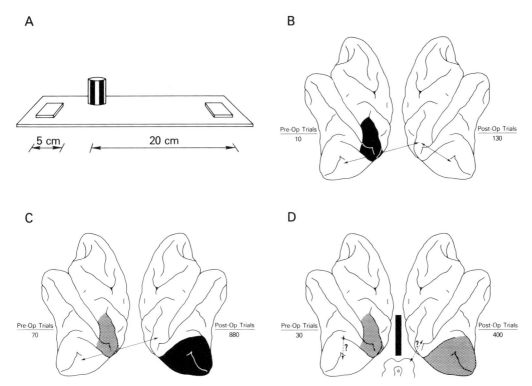

Figure 18.5 Design and results of the striate-parietal crossed-lesion disconnection study. (A) Landmark task. Monkeys were rewarded for choosing the covered foodwell located closer to a striped cylinder (the "landmark"), which was positioned on the left or the right randomly from trial to trial, but always 5 cm from one foodwell and 20 cm from the other. Training was given for 30 trials per day to a criterion of 90 correct responses in 100 consecutive trials. (B) Discrimination retention before and after first-stage lesion (unilateral posterior parietal; $N = 3$); 10 preoperative trials and 130 postoperative trials. (C) Discrimination retention before and after second-stage lesion (contralateral striate; $N = 3$); 70 preoperative and 880 postoperative trials. (D) Discrimination retention before and after third-stage lesion (corpus callosum; $N = 3$); 30 preoperative and 400 postoperative trials. At each stage the lesion is shown in black and the lesions of prior stages are shaded. Arrows denote hypothetical connections left intact by lesions. Adapted from Ungerleider and Miskin 1978a.

inputs to this cortex, which appear to be unnecessary when corticocortical connections are intact, also play no significant role in the recovery after corticocortical disconnection. By contrast, removal of prestriate cortex ipsilateral to the intact striate cortex produced a marked impairment on the task. Taken together, the results of this serial-lesion experiment demonstrate that the visual spatial functions of posterior parietal cortex depend on inputs from striate cortex and that such inputs are transmitted across corticocortical pathways. In this regard, it is significant that the only cortical area apart from those along the occipitotemporal pathway to show differential glucose utilization in the 2-DG experiments described earlier (Jarvis et al. 1978; Kennedy et al. 1978) was located within the posterior parietal region. The affected area extended forward from the dorsal prestriate cortex to include the posterior half of the inferior parietal gyrus, both on its crown and on the ventral bank of the intraparietal sulcus.

The Two Cortical Visual Systems Compared

These results on striate-parietal interaction parallel the earlier finding that corticocortical inputs from striate cortex are crucial for the visual recognition functions of inferior temporal cortex. But, though the results of these two studies are analogous, one important difference emerges (see figure 18.6). On the pattern-discrimination task of the earlier study, only a moderate effect followed the striate lesion, and a severe deficit did not result until the callosal transection. By contrast, on the landmark task, the relative effects of the striate and callosal lesions were reversed; that is, the striate-cortex lesion produced the more severe effect. These data suggest that the posterior parietal cortex, like the inferior temporal, depends heavily on corticocortical inputs from striate cortex for its visual function, but that, unlike the inferior temporal, the posterior parietal cortex in a given hemisphere does not seem to receive a heavy input via the corpus callosum from striate-cortex neurons representing the ipsilateral visual field. Thus, each posterior parietal area may be organized largely as a substrate for contralateral spatial function, and this could account in part for the symptom of contralateral neglect that has so often been reported after unilateral injury to the parietal lobe in man (see, for example, Denny-Brown and Chambers 1958).

A second important difference in the organization of visual inputs to the two systems was uncovered in an experiment comparing the behavioral effects of selective striate-cortex removals. In this experiment, monkeys received bilateral lesions of the striate area representing either central vision (lateral striate) or peripheral vision (medial striate). The lateral lesion included the entire lateral surface of striate cortex, and the medial lesion included the calcarine fissure as well as both banks of its ascending and descending limbs (figure 18.7). The monkeys were tested

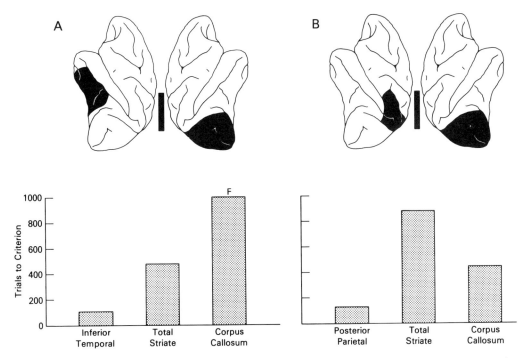

Figure 18.6 Comparison of two crossed-lesion disconnection experiments: (A) pattern discrimination after striate-temporal disconnection (Mishkin 1966) and (B) landmark discrimination after striate-parietal disconnection (Ungerleider and Mishkin 1978a). In each case, brain diagrams illustrate the three lesions (in black) involved in the disconnection and bar graphs show the mean trials to relearn after each lesion. F indicates failure to relearn within the limits of training.

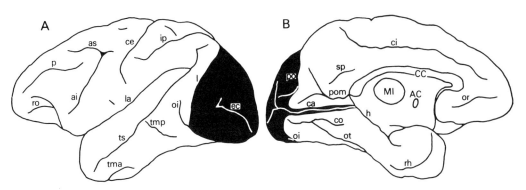

Figure 18.7 Locus of two different bilateral striate-cortex removals (in black): (A) lateral striate, the area representing central vision (N = 5); (B) medial striate, the area representing peripheral vision (N = 5). For details of visuotopic organization see figure 18.13; for abbreviations see figure 18.3.

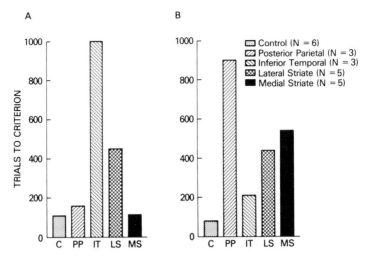

Figure 18.8 Comparison of the effects of lateral striate lesions with those of medial striate lesions on (A) postoperative acquisition of a pattern discrimination and (B) postoperative retention of the landmark task. Adapted from Ungerleider and Mishkin 1978a.

before operation both on a pattern-discrimination task, to assess residual inferior temporal function, and on the landmark task, to assess residual posterior parietal function. For purposes of comparison, their scores are plotted in figure 18.8 together with those of monkeys with either bilateral inferior temporal or bilateral posterior parietal lesions

The data indicate that on the pattern-discrimination task severe impairment was produced by inferior temporal but not by posterior parietal lesions, whereas on the landmark task severe impairment followed posterior parietal but not inferior temporal lesions. This, then, is another instance of the dissociation of visual recognition and visual spatial functions of temporal and parietal cortex, respectively. Of more direct concern for the question of differences in anatomical organization, however, were the effects of the striate lesions. Here the data indicate that on the pattern-discrimination task impairment was produced only by lateral striate-cortex lesions, whereas on the landmark task equally severe impairment followed lateral and medial striate-cortex lesions. Apparently, inputs from central vision are especially important for the visual recognition functions of the inferior temporal cortex, but inputs from central and peripheral vision are equally important for the visual spatial functions of posterior parietal cortex.

Thus, although interactions with striate cortex are critical for the parietal just as for the temporal area, the inputs of the striate cortex to these two regions appear to be organized differently; relative to inferior temporal cortex, posterior parietal cortex receives a smaller contribution from contralateral striate-cortex inputs (representing the ipsilateral visual field) but a greater contribution from medial inputs (representing

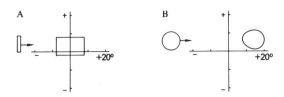

Figure 18.9 Receptive-field properties of neurons in (A) inferior temporal and (B) posterior parietal cortex. Axes represent horizontal and vertical meridians. Plus sign indicates upper or right visual field; minus sign indicates lower or left. Scale is in degrees of visual angle. A slow-moving stimulus typically effective in activating the given population of neurons is shown at the left of each diagram. The enclosed geometric shape within each set of axes indicates the border of the receptive field for a typical neuron in that population. Note that of the inferior temporal neurons over 60% are bilateral, 100% include the fovea, and there is selectivity for stimulus features; of the posterior parietal neurons over 60% are contralateral only, over 60% do not include the fovea, and there is no selectivity for stimulus features. (Part A adapted from Gross et al. 1972; part B adapted from Robinson et al. 1978.)

the peripheral visual field). Interestingly, these differences in the organization of striate-cortex inputs inferred from the ablation studies are clearly reflected in the receptive-field properties of neurons within the inferior temporal (Gross et al. 1972; Jarvis and Mishkin 1975; Rolls et al. 1977) and posterior parietal areas (Hyvärinen and Poranen 1974; Mountcastle et al. 1975; Robinson et al. 1978; Rolls et al. 1979a). The relevant properties are shown diagrammatically in figure 18.9.

First, there is a smaller representation of the ipsilateral visual field in the receptive fields of posterior parietal neurons than in those of inferior temporal neurons: Over 60% of inferior temporal neurons have bilateral receptive fields, while over 60% of posterior parietal neurons have contralateral fields only. Second, there is a relatively greater representation of the peripheral visual field in the receptive fields of posterior parietal neurons than in those of inferior temporal neurons: 100% of inferior temporal neurons have receptive fields that include the fovea, while over 60% of posterior parietal neurons have receptive fields outside the fovea. Finally, and perhaps most important, not only are the striate-cortex inputs to the temporal and parietal areas organized differently, but these inputs must also be carrying different information to the two areas, as is indicated by the behavioral evidence for functional dissociation. This too is clearly seen at the level of single neurons, for, in contrast to the specific and often complex trigger features needed for maximal activation of inferior temporal neurons (see, for example, Gross et al. 1972), most posterior parietal neurons can be maximally driven by simple spots of light (see Robinson et al. 1978). These several differences in the receptive-field properties of inferior temporal and posterior parietal neurons, each of which has a behavioral correlate that was revealed in the ablation studies, presumably reflect differences in the cortical processing required for object versus spatial vision.

Corticocortical Pathways Through Prestriate Cortex

The evidence presented thus far strongly supports the view that neural mechanisms underlying object and spatial perception depend on the relay of different kinds of visual information from striate cortex through prestriate cortex to targets in inferior temporal and posterior parietal areas, respectively. Despite the abundance of positive evidence, however, the support has not been unanimous. The results of prestriate-cortex ablation studies have repeatedly raised serious problems for this conception of corticocortical pathways. If prestriate cortex were an essential relay in either a striate-temporal or a striate-parietal pathway, then damage to this relay should yield effects at least as severe as damage to its target areas. Yet, monkeys sustaining extensive bilateral prestriate lesions commonly exhibit only mild visual effects. In a few instances in which severe visual deficits did follow prestriate removals (Keating and Horel 1972; Keating 1975), the removals included the posterior part of inferior temporal cortex (area TEO), damage to which, by itself, produces severe visual-discrimination impairment (Iwai and Mishkin 1968). Extensive prestriate-cortex removals that spare this posterior temporal region, however, have repeatedly failed to yield appreciable effects in either visual discrimination or in spatial orientation (Lashley 1948; Meyer et al. 1951; Riopelle et al. 1951; Chow 1952; Evarts 1952; Pribram et al. 1969; Ungerleider and Pribram 1977).

Is there a resolution of this paradox? It was suggested by Mishkin (1966) that failures to obtain severe impairments after extensive prestriate-cortex removals simply indicated that spared prestriate remnants continued to serve as effective relays between the striate area and its rostral targets—that is, that the prestriate cortex was invested with a high degree of equipotentiality. According to this proposal, only a complete prestriate lesion would be expected to yield a corticocortical disconnection. The results of a study by Iwai and Mishkin (Mishkin 1972) provided support for this proposal with regard to occipitotemporal transmission. When putatively complete prestriate-cortex lesions were made in a large group of monkeys, none was able to relearn a visual pattern discrimination. By contrast, partial prestriate ablations, irrespective of their location, were nearly without effect. These findings therefore pointed to the existence of multiple pathways through prestriate cortex, any of which can be utilized to convey visual information out of striate cortex. It seemed that only if all these equivalent pathways were destroyed would the transmission be completely disrupted and the anticipated deficit ensue.

Although it has since been found that even the massive removals of prestriate cortex referred to above did not totally disconnect the striate-temporal pathway, the conclusion reached remains valid nonetheless. Indeed, it has now been demonstrated that any sparing, no matter how minimal, of the visual functions of inferior temporal cortex after a prestriate ablation is directly attributable to the continued relay of information across a viable prestriate pathway. This demonstration emerged from a series of interrelated behavioral, electrophysiological, and anatomical experiments (Ungerleider, Iwai, Gross, Bender, Snyder, and Mishkin, in prep.) that began in the following way: In an attempt to verify that the inferior temporal cortex had been functionally disconnected in the Iwai-Mishkin study, two of the monkeys that had failed to show relearning after the prestriate lesions were given prolonged pattern-discrimination training by a method of approximation, in preparation for a second-stage lesion. The method entailed presentation of a series of stimulus pairs that differed first in brightness, then in size, then in contour, and finally in pattern; the successive pattern pairs approximated the originals more and more closely until the original pattern discrimination had been relearned. Over several months both monkeys were successfully retrained by this method, and both were then given bilateral inferior temporal ablations in a second-stage operation. The supposition was that if the initial prestriate removal had produced a total disconnection of inferior temporal from striate cortex, then the slow pattern-discrimination relearning by approximation should have been achieved without the participation of inferior temporal cortex, and, consequently, removal of this cortex should not disrupt the relearned habit. As it turned out, however, the inferior temporal ablation produced a complete reinstatement of the deficit in both monkeys. Thus, unless inferior temporal cortex can participate in visual discrimination learning in the absence of all corticocortical visual input, the only conclusion to be drawn from this unexpected result is that the initial prestriate-cortex removal had not produced a total striate-temporal disconnection after all.

In order to examine directly this question of preserved neural transmission despite massive prestriate removals, an electrophysiological study was undertaken in animals prepared with removals identical to those in the Iwai-Mishkin study except that they were limited to one hemisphere. The distribution of the receptive fields of inferior temporal neurons in the operated animals is summarized in figure 18.10. According to the proposed route of visual information flow, indicated by the arrows in the brain diagrams, these neurons should have responded to visual stimulation only in the hemifield opposite the intact prestriate cortex. That is, inferior temporal neurons in the intact hemisphere should have had strictly contralateral visual fields, while inferior temporal neurons in the hemisphere with the prestriate ablation should have had strictly ipsilateral fields. In short, the results should have duplicated those obtained after unilateral striate removals, illustrated in parts C and D of figure 18.2. In fact, however, the receptive fields of approximately 25% of the neurons sampled did not fit this prediction.

In the normal monkey, stimulation in a given hemifield will activate about 80% of inferior temporal neurons. For example, as indicated in

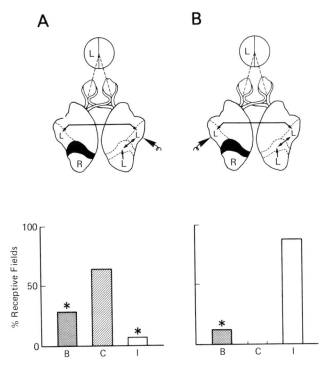

Figure 18.10 Proportion of inferior temporal neurons that had bilateral (B), contralateral (C), or ipsilateral (I) receptive fields after unilateral ablation of prestriate cortex (in black). Data for part A were obtained from inferior temporal neurons contralateral to the lesion, whereas data for part B were obtained from those ipsilateral to the lesion. Asterisks denote the receptive fields deviating from the prediction; that is, deviating from those obtained after unilateral striate-cortex removals, shown in figure 18.2. Adapted from Ungerleider, Snyder, and Mishkin, in prep.

part A of figure 18.2, stimulation in the right hemifield will activate about 90% of the inferior temporal neurons in the left hemisphere (all those with either bilateral or contralateral visual fields) and about 70% of the inferior temporal neurons in the right hemisphere (all those with either bilateral or ipsilateral visual fields). By contrast, in the animals with left prestriate removals, stimulation in the right hemifield activated about 25% of inferior temporal neurons. That is, as indicated in figure 18.10, right-hemifield stimulation activated about 15% of the inferior temporal neurons in the left hemisphere and about 35% in the right. Thus, the massive prestriate removals did greatly reduce the visual input to inferior temporal neurons—in fact, by more than two-thirds (80% to 25%)—thereby accounting for the severe deficit that such lesions produced in pattern discrimination relearning. On the other hand, the removals did not completely eliminate striate input to inferior temporal neurons (25%, as compared with 0% after striate removals), and this fact accounted for the ultimately successful retraining by approximation that was achieved in the animals with such lesions.

Even with this electrophysiological evidence, the conclusion that behavioral sparing resulted from sparing of remnants of the prestriate relay was only inferential. In an attempt to demonstrate the preserved pathway directly, a neuroanatomical experiment was undertaken on two additional monkeys with massive prestriate removals that had been successfully retrained by approximation on the pattern discrimination. More than a year after the prestriate-cortex removals, one of these animals was given a second-stage operation in which the entire occipital lobe posterior to the initial lesion was removed from one hemisphere, and the brain was examined for anterograde degeneration by the Fink-Heimer (1967) technique. The critical question was whether degeneration would appear in the inferior temporal cortex ipsilateral to the occipital lobectomy. Since striate cortex does not project directly to the inferior temporal area, the presence of degeneration here would indicate that the lobectomy had destroyed prestriate-relay tissue spared by the initial lesion—a finding that would account for the functional sparing that had been demonstrated both behaviorally and electrophysiologically. In fact, terminal degeneration did result in the ipsilateral inferior temporal cortex, specifically in the ventral part of area TEO. In addition, degenerating material was found in an unexpected area, the floor of the caudal portion of the superior temporal sulcus, which turned out to be a second spared route through which visual input could reach inferior temporal cortex. This was demonstrated in the other monkey of the anatomical experiment, which was given a second-stage lesion in just this portion of the superior temporal sulcus of one hemisphere, again more than a year after the initial prestriate removal. Terminal degeneration was found in the ipsilateral inferior temporal cortex of this case also, specifically in the dorsal part of area TEO.

The foregoing series of experiments made it clear that, although the massive prestriate-cortex removals had severely disrupted the prestriate relay, they had not totally abolished it. As illustrated in figure 18.11, spared pathways that presumably continued to serve as effective relays were directly demonstrated anatomically. At the same time, however, the anatomical results raised some important new questions. For example, what were the loci of the spared prestriate remnants whose removal gave rise to the degeneration in the ventral part of area TEO? Also, what was the source of the unexpected degeneration in the depth of the caudal portion of the superior temporal sulcus, an area from which a projection had been traced to dorsal TEO? These questions highlighted our lack of knowledge about the precise links through which visual information originating in the striate cortex is transmitted to inferior temporal cortex. Even greater ignorance surrounded the location and arrangement of the links in the striate-parietal pathway. It became apparent that, in order to obtain this information, a comprehensive anatomical investigation was needed, beginning with an analysis of the projections of the striate cortex itself.

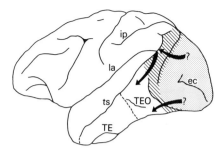

Figure 18.11 Schematic representation of results from two anatomical experiments, demonstrating that striate-prestriate-temporal pathways were preserved despite massive prestriate removals. In the first experiment, an occipital lobectomy (dots) was performed more than a year after the initial prestriate removal (stripes). Degeneration from the lobectomy was found in ventral area TEO and in the floor of the caudal portion of the superior temporal sulcus. The exact source of these two projections, however, was indeterminate (indicated by question mark). In the second experiment, the floor of the superior sulcus was removed, again more than a year after the initial prestriate lesion. Degeneration in this case was found in dorsal area TEO. For abbreviations see figure 18.3. Adapted from Ungerleider et al., in prep.

Clarification of Striate-Prestriate Connections

Early attempts to examine striate-prestriate connections (von Bonin et al. 1942; Bailey et al. 1944) employed the method of strychnine neuronography, described above. Although the map of striate projections obtained with this method has been confirmed by more recent work using degeneration (Kuypers et al. 1965; Cragg and Ainsworth 1969; Zeki 1969, 1971a; Jones and Powell 1970) and autoradiographic (Zeki 1976) and horseradish-peroxidase (Lund et al. 1975) tracing techniques, all these methods, both old and new, have been used to define projections primarily from lateral striate cortex, the part of striate cortex representing central vision. Surprisingly, there has been almost no information regarding the projections from posterior and medial striate cortex, the parts representing peripheral and far-peripheral vision. Indeed, it was undoubtedly the near absence of information about the location of the prestriate tissue serving noncentral vision that accounted for the repeated failure, recounted above, to completely disconnect the higher-order visual areas from their striate input. We therefore undertook a study of the cortical efferents from all parts of striate cortex, with the aim of defining the locus, extent, and topographic organization of the entire striate-prestriate projection system.

To delineate the entire map of striate projections to prestriate cortex, one series of monkeys was prepared with partial striate lesions such that, collectively, they included all of area 17 with little or no invasion of area 18. The brains were then processed by the Fink-Heimer (1967) procedure for silver staining of degenerating axon terminals. Reconstructions of the lesions in three of the five cases from this series are shown in figure 18.12. The lateral lesion involved the lateral surface of area 17 only, with no invasion of area 18; the posterior lesion included the lateral and medial banks of the vertical limbs of the calcarine fissure, again with no invasion of area 18; and the medial striate lesion involved the tissue within the stem of the calcarine fissure, in this case with slight damage to area 18 around the lips of the fissure. According to the electrophysiological map of Daniel and Whitteridge (1961), these three sectors of striate cortex correspond to the central 7° of the contralateral visual field, to the field between 7° and 22° from fixation, and to eccentricities greater than 22° from fixation, respectively (see figure 18.13). Thus, it was anticipated that not only would the entire striate-prestriate projection system emerge from the data on this series of monkeys, but the general visuotopic organization of the system would be apparent as well.

To verify the degeneration results, and also to investigate the details of the visuotopic organization, a second series of monkeys were prepared with injections of radioactively labeled amino acids into selected sites in striate cortex, and the brains processed for autoradiography. A summary of the injection sites is shown in figure 18.14. These loci correspond to positions in the visual field ranging from less than ½° from fixation (site number 1 on the lateral surface) to greater than 45° from fixation (site number 10 in the stem of the calcarine fissure). The loci injected included representations of both the upper and the lower visual fields.

The results from both sets of experiments, degeneration and autoradiographic, indicated that all parts of striate cortex project to three separate visual areas within prestriate cortex (Ungerleider and Mishkin 1979a): The first is a circumstriate cortical belt surrounding area 17 at the 17–18 border, the second is located along the caudal portion of the superior temporal sulcus, and the third is buried deep within the caudal part of the intraparietal sulcus. Each of these three projection areas will be described in turn.

Within the first projection area—the circumstriate cortical belt—the representations of the upper and lower visual fields are entirely separate (figure 18.15). The topographic organization of these separate representations is illustrated in figure 18.16. Progression from central to peripheral to far peripheral vision is represented in the lower field by a progression into the posterior bank and depth of the lunate sulcus, medially along the surface of the buried annectent gyrus into the parieto-occipital incisure, and then rostrally along the upper lip of the calcarine fissure; and in the upper field by a progression into the inferior occipital sulcus, ventromedially into the occipitotemporal and collateral sulci, and then rostrally along the lower lip of the calcarine fissure. The total extent of this projection field corresponds remarkably closely to area OB of von Bonin and Bailey (1947); indeed, the two may be equivalent. Moreover, the autoradiographic evidence indicates that whereas the

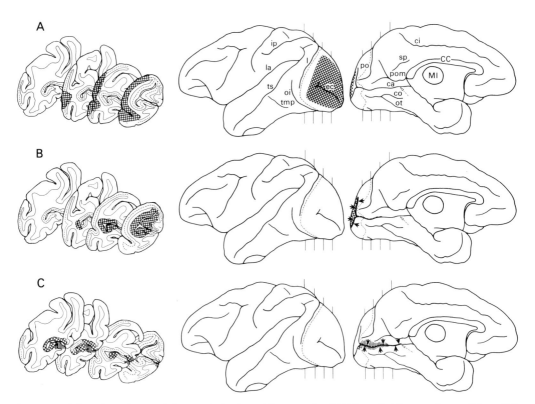

Figure 18.12 Location of cortical damage in brains with removals of (A) lateral, (B) posterior, or (C) medial striate cortex. The lesions are mapped on lateral and medial surface views and on representative cross-sections through a standard rhesus monkey brain. Cross-sections are at levels indicated by vertical lines on surface views. Dashed lines indicate extent of striate cortex on cross-sections but borders of striate cortex on surface views; anterior limit of striate cortex within the calcarine fissure is indicated on medial surface views by dashed arrow. Removal of striate cortex is shown in crosshatch; black arrows on the medial surface views indicate removal of striate tissue from within the banks and depths of the calcarine fissure. Damage to prestriate tissue is shown in stripes. For abbreviations see figure 18.3.

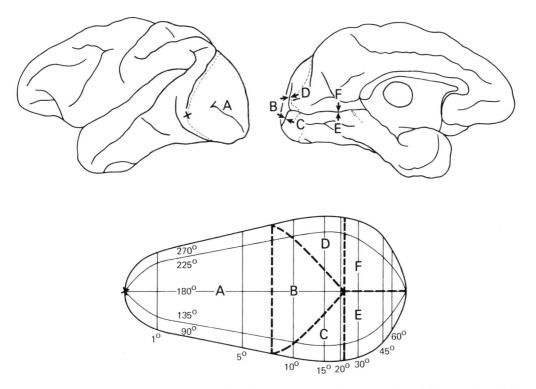

Figure 18.13 Relationship between topography of striate cortex and representation of the contralateral visual field. According to the electrophysiological map of Daniel and Whitteridge (1961), shown at bottom, the center of gaze (fixation) is represented in the far anterolateral part of striate cortex, just below the ventral tip of the lunate sulcus. The representation of the vertical meridian passes through fixation (X) and continues all along the 17–18 border (dashed line). The representation of the horizontal meridian is a horizontal line (not illustrated) passing from fixation across the lateral surface, entering the depth of the calcarine fissure, and continuing to the rostral limit of the medial striate cortex (dashed arrow). The upper visual field is represented below the horizontal meridian and the lower visual field above the horizontal meridian. The lateral surface of striate cortex (A), the lateral bank of the calcarine limbs (B), the medial bank of the calcarine limbs (C, D), and the stem of the calcarine fissure (E, F) represent approximately the center 7°, 7°–13°, 13°–22°, and eccentricities greater than 22° from fixation, respectively. Thus, a progression from lateral to posterior to medial striate cortex corresponds to a progression from central to peripheral to far-peripheral vision.

vertical meridian is represented at the inner boundary of this projection field (that is, at the striate-OB border), the horizontal meridian is represented at its outer boundary (at the OB-OA border).

The second projection field of striate cortex is located within cytoarchitectonic area OA along the caudal portion of the superior temporal sulcus (Ungerleider and Mishkin 1978b, 1979b). As indicated in the lateral view of the brain shown in figure 18.17, the ventral limit of this region can be demarcated by an imaginary line connecting the ventral tips of the lunate and intraparietal sulci; from this limit the region extends dorsocaudally for about 1 cm to the point at which the superior temporal sulcus frequently bifurcates, sending one spur forward into the inferior parietal lobule. Within this portion of the superior temporal sulcus there is again an orderly mapping of the contralateral visual field,

571   Two Cortical Visual Systems

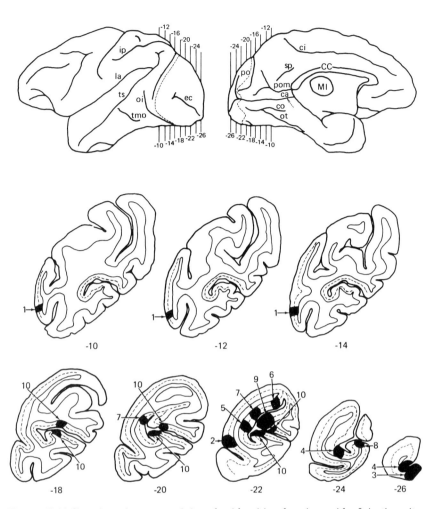

Figure 18.14 Sites in striate cortex injected with tritiated amino acids. Injection sites (shown in black) are numbered according to the eccentricity of their representation in the visual field. These representations ranged from less than one-half degree from fixation (site 1) to greater than 45° from fixation (site 10). Each injection was made in a separate hemisphere. Dashed lines indicate extent of striate cortex on cross-sections, but borders of striate cortex on surface views; anterior limit of striate cortex within the calcarine fissure is indicated on medial surface view by a dashed arrow. Negative numerals indicate approximate A-P stereotaxic levels. For abbreviations see figure 18.3.

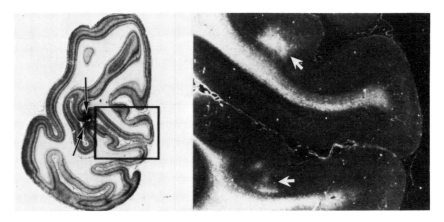

Figure 18.15 Photomicrographs exemplifying autoradiographic results after injections of striate cortex. The bright-field photomicrograph on the left shows an injection (marked by arrows) involving both the dorsal and ventral banks of the calcarine fissure (see figure 18.14, injection site 10). In the dark-field photomicrograph on the right, labeling is seen in two prestriate loci, marked by arrows: dorsally in the medial parieto-occipital fissure and ventrally in the ventral bank of the calcarine fissure. The patchy pattern of label, especially apparent in the ventral locus, is typical of striate projections to the circumstriate cortical belt. Adapted from Ungerleider and Mishkin 1979a.

as shown on the three selected cross sections. Progression from central to peripheral to far-peripheral vision is represented by a progression down the posterior bank of the superior temporal sulcus and up along the lower and then the upper part of the sulcal floor. Figure 18.18 summarizes this topographic arrangement. The finding that the floor of the superior temporal sulcus receives a direct projection from posterior and medial striate cortex provides an answer to one of the anatomical questions posed in the preceding section (see figure 18.11). As yet, however, we do not have an answer to the second question raised there, except the confirmation that the immediate source of the occipital projections to ventral area TEO is not the striate cortex.

The third area that receives direct striate-cortex projections is also located within cytoarchitectonic area OA, in the depth of the most caudal portion of the intraparietal sulcus. A cross-section through the projection shows that it is buried in the foot-shaped part of the sulcus beneath its lateral bank (figure 18.19). This projection field is by far the smallest of the three; its total rostral-caudal length extends only about 2 mm. There is no discernible topographic representation of the visual field within this area, for after a lesion or an injection involving any part of striate cortex a projection is always seen in the same part of the sulcus. Thus, although the dark-field photomicrograph in figure 18.19 shows degeneration in the foot of the intraparietal sulcus after a medial striate lesion, the very same picture could as well represent the projections of either posterior or lateral striate cortex. Indeed, the first evidence of direct striate projections to this area came from studies that

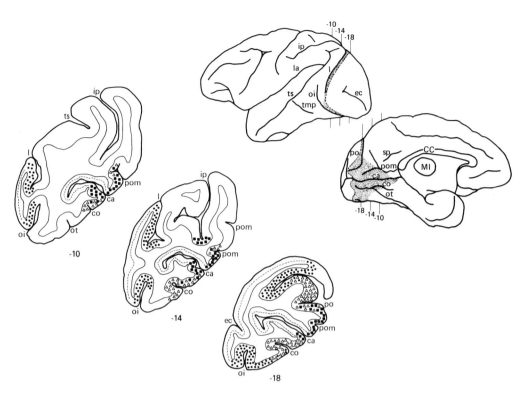

Figure 18.16 Location, extent, and topographic organization of striate projections (●: lateral; △: posterior; ■: medial) to the circumstriate cortical belt. The drawings are composites based on both degeneration and autoradiographic material. Shaded area represent total projection field on surface views. As shown on the surface views of the hemisphere, this projection zone surrounds the striate cortex at the 17–18 border and corresponds remarkably closely to cytoarchitectonic area OB of von Bonin and Bailey (1947). Within the projection zone, there is a precise topographic map of the contralateral visual field (see cross-sections); note, however, that the representations of the upper and lower fields are entirely separate. Dashed lines indicate extent of striate cortex on cross sections, but borders of striate cortex on surface views. Numerals indicate approximate A-P stereotaxic levels. For abbreviations see figure 18.3. Adapted from Ungerleider and Mishkin 1979a.

had examined only lateral striate connections (Kuypers et al. 1965; Jones and Powell 1970), though the area had not been recognized as a separate projection field at that time.

These three projection fields thus comprise all the possible first-stage prestriate relays through which striate output can reach the inferior temporal and posterior parietal cortex. Undoubtedly, second- and third-stage prestriate relays also participate in both corticocortical pathways (Zeki 1971b; Mesulam et al. 1977; Stanton et al. 1977; Desimone et al. 1980), although the exact location and topographic arrangement of these further relays remain to be delineated. But even from the evidence pertaining to the first stage alone, we can begin to understand why attempts to produce striate-temporal disconnection by removal of prestriate cortex have repeatedly failed. In all prestriate-

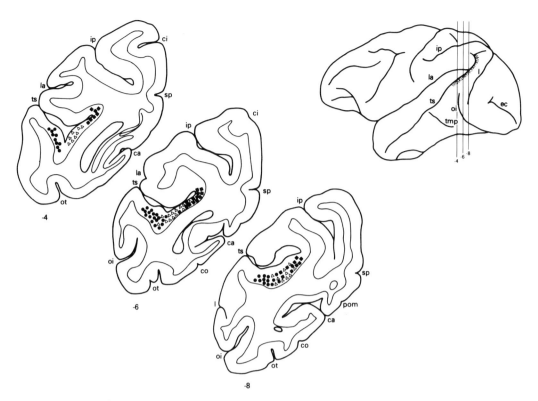

Figure 18.17 Location, extent, and topographic organization of striate projections (●: lateral; △: posterior; ■: medial) to cortex in the superior temporal sulcus. The drawings are composites based on both degeneration and autoradiographic material. Shaded area represents locus of projection on lateral view. All areas of striate cortex project to a restricted region inside the caudal portion of the superior temporal sulcus, its A-P extent indicated on the lateral view of the hemisphere. As indicated on the cross-sections, the topographic arrangement of the contralateral visual field is such that a progression from central to peripheral to far peripheral vision is represented by a progression down the posterior bank of the sulcus and up along the lower and then the lower part of the sulcal floor. Although this topography is clearest in the rostral portion of the projection zone, it is also discernible caudally. Numerals refer to approximate A-P stereotaxic levels. For abbreviations see figure 18.3. Adapted from Ungerleider and Mishkin 1979b.

ablation studies, certain areas have consistently been spared that can now be identified as receiving direct projections from posterior and medial striate cortex. These prestriate areas are the medial portion of area OB (figure 18.16), the floor of the caudal portion of the superior temporal sulcus (figure 18.17), and the depth of the most caudal portion of the intraparietal sulcus (figure 18.19). Each of these three areas has escaped damage for a different reason: the first because of its relative inaccessibility, the second because it was not even recognized to be a part of prestriate cortex, and the third because its precise location had not been described. In light of the nearly complete sparing of these areas in most ablation studies, the finding of only mild visual impairment is no longer surprising. Only when these areas were invaded substantially, as in the massive prestriate removals of the Iwai-Mishkin study, did severe visual impairment result. However, even in those le-

Figure 18.18 Drawings of the lateral and medial views of the left hemisphere and of a cross-section through the superior temporal sulcus to illustrate the topographic arrangement of striate projections to this secondary visual area. The orderly mapping within the sulcus of the entire representation of the contralateral visual field is shown. For abbreviations see figure 18.3. Source: Ungerleider and Mishkin 1979b.

sions, where there was almost total destruction of the prestriate areas that receive direct projections from lateral striate cortex (the part representing central vision), there was no more than 50% destruction (Ungerleider et al., in prep.) of the prestriate areas that receive direct projections from posterior and medial striate cortex (the parts representing peripheral and far peripheral vision). Clearly, it was the sparing of this tissue that accounted for the continued (though reduced) transmission of visual information to inferior temporal cortex that had been demonstrated not only behaviorally but also electrophysiologically. Thus, the present anatomical findings, many of which have now been confirmed by others (Weller and Kaas 1978; Maunsell et al. 1979; Rockland 1979; Van Essen et al. 1979; Weller et al. 1979), provide a coherent explanation for a large body of seemingly paradoxical results regarding the effects of prestriate lesions. That is, just as "peripheral" striate cortex can mediate pattern discrimination in the absence of "central" striate cortex (Blake et

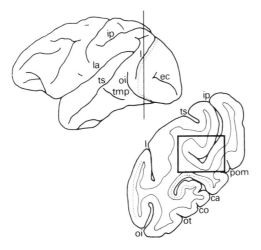

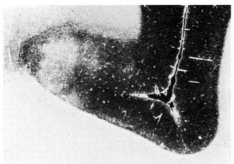

Figure 18.19 The striate projection zone within the intraparietal sulcus. This zone is located in the depth of the most caudal portion of the sulcus. Its approximate A-P level is indicated by the vertical line on the lateral view of the hemisphere, and its depth by the black square on the cross-section. The dark-field photomicrograph shows the locus of degeneration after a medial striate lesion. Projections are seen in this very same region from all parts of striate cortex, which suggests the absence of a visuotopic organization within this projection zone. Source: Ungerleider and Mishkin 1979a.

al. 1977), so can "peripheral" prestriate cortex substitute for "central" prestriate cortex in this function. And if this is the case for the visual recognition functions of inferior temporal cortex, where inputs from peripheral vision are less important than those from central vision, then the same must surely be true for the visual spatial functions of posterior parietal cortex, where inputs from peripheral vision are equal in importance to those from central vision (see figure 18.8).

Now that the total system of striate efferents has been delineated, the effects of completely disconnecting temporal or parietal from striate cortex can finally be investigated. Such disconnection should be achievable by removal of just the three striate projection fields that have been described, without inclusion of any other prestriate tissue. In addition, with special testing methods it may now be possible to study the behavioral effects of prestriate-cortex damage without producing total

Two Cortical Visual Systems

disconnection. In view of the visuotopic organization of the striate projection zones in both area OB and the caudal part of the superior temporal sulcus, the details of which have been corroborated electrophysiologically (Gattass and Gross 1979; Gattass et al. 1979), even limited lesions within these prestriate zones should yield severe visual deficits if the animals are forced to use the part of the visual field corresponding to the area damaged. Testing of visual functions within specified parts of the visual field can be accomplished in monkeys who have been trained to maintain fixation (Wurtz 1969). Thus, the combination of this method and the use of lesions based on our new understanding of prestriate anatomy offers a promising approach for the study of the processing characteristics and functional organization of prestriate cortex.

Summary and Conclusions

The hypothesis that has been guiding our research is that appreciation of an object's qualities and of its spatial location depends on the processing of different kinds of visual information in the inferior temporal and posterior parietal cortex, respectively. From an initial concern with these higher-order visual areas, our research has proceeded backward. Having first obtained strong support from ablation studies for the postulated dichotomy of temporal and parietal function, we next examined the pathways through which visual information reaches these two cortical areas. Converging evidence from a variety of sources—behavioral and electrophysiological disconnection experiments as well as anatomical and metabolic mapping studies—indicated that both the temporal and parietal visual areas depend on corticocortical inputs relayed from striate through prestriate cortex. The results of prestriate ablation studies, however, conflicted with this conclusion, since they commonly indicated only modest deficits in both visual recognition and spatial orientation. The solution to this puzzle emerged from a series of interrelated behavioral, electrophysiological, and anatomical experiments that demonstrated that every case of functional sparing after prestriate damage can be directly attributed to the continued relay of information through viable prestriate remnants. To determine the locus of these prestriate remnants, we turned to a study of the cortical efferents from all parts of striate cortex. The results revealed that all prestriate lesions to date, no matter how massive, have failed to include varying extents of tissue that receive direct projections from posterior and medial striate cortex, the parts representing peripheral vision. Having delineated the projections of striate cortex, which were found to consist of three separate re-representations of the visual field, we are now in a position to proceed in a forward direction and follow these projections, with the inferior temporal and posterior parietal cortex as our targets. In the course of this endeavor, a major goal will be to determine where within the complex of prestriate cortex the two cortical visual systems begin to diverge. On the assumption that both systems can indeed be followed stepwise to our target areas, not only in the temporal but also in the parietal lobe, a major question for the future will be how the object and spatial information carried in these two separated systems are subsequently integrated into a unified visual percept.

Acknowledgments

We gratefully acknowledge the contribution of past members of the Laboratory of Neuropsychology at the National Institute of Mental Health: Eiichi Iwai, Charlene Drew Jarvis, Michael E. Lewis, Walter G. Pohl, and Marvin Snyder. We also wish to thank our collaborators from other laboratories: Charles G. Gross and David B. Bender at Princeton University, Edward G. Jones at Washington University, Charles Kennedy and Louis Sokoloff at NIMH, and Blair H. Turner at Howard University. Skillful technical assistance was provided by Margaret E. Knapp, Bertha McClure, and George Creswell. The research was supported in part by the National Institutes of Health under postdoctoral fellowship EY05009.

References

Bailey, P., G. von Bonin, E. W. Davis, H. W. Garol, and W. S. McCulloch. 1944. "Further observations on association pathways in the brain of *Macaca mulatta*." *J. Neuropathol. Exp. Neurol.* 3: 413–415.

Bates, J. A. V., and G. Ettlinger. 1960. "Posterior biparietal ablations in the monkey." *Arch. Neurol.* 3: 177–192.

Benson, D. F., J. Segarra, and M. L. Albert. 1974. "Visual agnosia-prosopagnosia." *Arch. Neurol.* 30: 307–310.

Blake, L., C. D. Jarvis, and M. Mishkin. 1977. "Pattern discrimination thresholds after partial inferior temporal or lateral striate lesions in monkeys." *Brain Res.* 120: 209–220.

Blum, J. S. 1951. "Cortical organization in somesthesis: Effects of lesions in posterior associative cortex in somatosensory function in *Macaca mulatta*." *Comp. Psychol. Monogr.* 20: 219–249.

Brody, B. A., and K. H. Pribram. 1978. "The role of frontal and parietal cortex in cognitive processing: Tests of spatial and sequence functions." *Brain* 101: 607–633.

Butters, N., C. Soeldner, and P. Fedio. 1972. "Comparison of parietal and frontal lobe spatial deficits in man: Extrapersonal vs. personal (egocentric) space." *Percept. Motor Skills* 34: 27–34.

Chow, K. L. 1952. "Further studies on selective ablation of associative cortex in relation to visually mediated behavior." *J. Comp. Physiol. Psychol.* 45: 109–118.

Cowey, A., and C. G. Gross. 1970. "Effects of foveal prestriate and inferotemporal lesions on visual discrimination by rhesus monkeys." *Exp. Brain Res.* 11: 128–144.

Cragg, B. G., and A. Ainsworth. 1969. "The topography of the afferent projections in the circumstriate visual cortex of the monkey studied by the Nauta method." *Vision Res.* 9: 733–747.

Daniel, P. M., and D. Whitteridge. 1961. "The representation of the visual field on the cerebral cortex in monkeys." *J. Physiol.* 159: 203–221.

Dean, P. 1976. "Effects of inferotemporal lesions on the behavior of monkeys." *Psychol. Bull.* 83: 41–71.

Denny-Brown, D., and R. A. Chambers. 1958. "The parietal lobe and behavior." *Res. Publ. Assoc. Res. Nerv. Ment. Dis.* 36: 35–117.

De Renzi, E., and P. Faglioni. 1967. "The relationship between visuo-spatial impairment and constructional apraxia." *Cortex* 3: 327–342.

Desimone, R., J. Fleming, and C. G. Gross. 1980. "Prestriate afferents to inferior temporal cortex: An HRP study." *Brain Res.* 184: 41–55.

Ettlinger, G., and J. E. Kalsbeck. 1962. "Changes in tactile discrimination and in visual reaching after successive and simultaneous bilateral posterior parietal ablations in the monkey." *J. Neurol. Neurosurg. Psychiatr.* 25: 256–268.

Ettlinger, G., and J. Wegener. 1958. "Somaesthetic alternation, discrimination, and orientation after frontal and parietal lesions in monkeys." *Q. J. Exp. Psychol.* 10: 177–186.

Ettlinger, G., H. B. Morton, and A. Moffett. 1966. "Tactile discrimination performance in the monkey: The effect of bilateral posterior parietal and lateral frontal ablations, and of callosal section." *Cortex* 2: 5–29.

Evarts, E. V. 1952. "Effects of ablation of prestriate cortex in auditory-visual association in monkey." *J. Neurophysiol.* 15: 191–200.

Faugier-Grimaud, S., C. Frenois, and D. G. Stein. 1978. "Effects of posterior parietal lesions on visually guided behavior in monkeys." *Neuropsychologia* 16: 151–168.

Fink, R. P., and L. Heimer. 1967. "Two methods for selective silver impregnation of degenerating axons and their synaptic endings in the central nervous system." *Brain Res.* 4: 369–374.

Flechsig, P. 1896. *Gehirn und Seele.* Leipzig: von Veit.

———. 1920. *Anatomie des menschlichen Gehirns und Rückenmarks auf myelogenetischer Grundlage.* Leipzig: Thieme.

Gattass, R., and C. G. Gross. 1979. "A visuotopically organized area in the posterior superior temporal sulcus of the macaque." *Invest. Ophthalmol.* (ARVO suppl.) 18: 184.

Gattass, R., J. H. Sandell, and C. G. Gross. 1979. "V2 in the macaque: Visuotopic organization and extent." *Soc. Neurosci. Abstr.* 5: 786.

Gross, C. G. 1973. "Visual functions of inferotemporal cortex." In *Handbook of Sensory Physiology*, VII/3, R. Jung, ed. Berlin: Springer.

Gross, C. G., and M. Mishkin. 1977. "The neural basis of stimulus equivalence across retinal translation." In *Lateralization in the Nervous System*, S. Harnad et al., eds. New York: Academic.

Gross, C. G., C. E. Rocha-Miranda, and D. B. Bender. 1972. "Visual properties of neurons in inferotemporal cortex of the macaque." *J. Neurophysiol.* 35: 96–111.

Hartje, W., and G. Ettlinger. 1973. "Reaching in light and dark after unilateral posterior parietal ablations in the monkey." *Cortex* 9: 344–352.

Heilman, K. M., D. N. Pandya, and N. Geschwind. 1970. "Trimodal inattention following parietal lobe ablations." *Trans. Amer. Neurol. Assoc.* 95: 259–261.

Held, R. 1968. "Dissociation of visual functions by deprivation and rearrangement." *Psychol. Forsch.* 31: 338–348.

Hubel, D. H., and T. N. Wiesel. 1968. "Receptive fields and functional architecture of monkey striate cortex." *J. Physiol.* 195: 215–243.

———. 1970. "Stereoscopic vision in macaque monkey." *Nature* 225: 41–42.

Hyvärinen, J., and A. Poranen. 1974. "Function of the parietal associative area 7 as revealed from cellular discharges in alert monkeys." *Brain* 97: 673–692.

Ingle, D. 1967. "Two visual mechanisms underlying the behavior of fish." *Psychol. Forsch.* 31: 44–51.

Iwai, E., and M. Mishkin. 1968. "Two visual foci in the temporal lobe of monkeys." In *Neurophysiological Basis of Learning and Behavior*, N. Yoshii and N. A. Buchwald, eds. Osaka University Press.

Jarvis, C. D., and M. Mishkin. 1975. "Responses of inferior temporal neurons to visual discriminanda." *Soc. Neurosci. Abstr.* 1: 61.

———. 1977. "Responses of cells in the inferior temporal cortex of monkeys during visual discrimination reversal." *Soc. Neurosci. Abstr.* 3: 564.

Jarvis, C. D., M. Mishkin, M. Shinohara, O. Sakurada, M. Miyaoka, and C. Kennedy. 1978. "Mapping the primate visual system with the [^{14}C] 2-deoxyglucose technique." *Soc. Neurosci. Abstr.* 4: 632.

Jones, B., and M. Mishkin. 1972. "Limbic lesions and the problem of stimulus-reinforcement associations." *Exp. Neurol.* 36: 362–377.

Jones, E. G., and T. P. S. Powell. 1970. "An anatomical study of converging sensory pathways within the cerebral cortex of the monkey." *Brain* 93: 793–820.

Keating, E. G. 1975. "Effects of prestriate and striate lesions on the monkey's ability to locate and discriminate visual forms." *Exp. Neurol.* 47: 16–25.

Keating, E. G., and J. A. Horel. 1972. "Effects of prestriate and striate lesions on performance of simple visual tasks." *Exp. Neurol.* 35: 322–336.

Kennedy, C., M. H. Des Rosiers, J. W. Jehle, M. Reivich, F. Sharp, and L. Sokoloff. 1975. "Mapping of functional neural pathways by autoradiographic survey of local metabolic rate with [^{14}C]deoxyglucose." *Science* 187: 850–853.

Kennedy, C., C. D. Jarvis, O. Sakurada, and M. Mishkin. 1978. "A delineation of the visually responsive loci of the temporal lobe by means of [^{14}C]deoxyglucose." *Neurology* 28: 366.

Kimura, D. 1963. "Right temporal-lobe damage." *Arch. Neurol.* 8: 264–271.

Kuypers, H. G. J. M., M. K. Szwarcbart, M. Mishkin, and H. E. Rosvold. 1965. "Occipitotemporal corticocortical connections in the rhesus monkey." *Exp. Neurol.* 11: 245–262.

LaMotte, R. H., and C. Acuña. 1978. "Defects in accuracy in reaching after removal of posterior parietal cortex in monkeys." *Brain Res.* 139: 309–326.

Lansdell, H. 1968. "Effect of extent of temporal lobe ablations on two lateralized deficits." *Physiol. Behav.* 3: 271–273.

Lashley, K. S. 1948. "The mechanism of vision. XVIII. Effects of destroying the visual 'associative areas' of the monkey." *Genet. Psychol. Monogr.* 37: 107–166.

Lund, J. S., R. D. Lund, A. E. Hendrickson, A. H. Bunt, and A. F. Fuchs. 1975. "The origin of efferent pathways from the primary visual cortex, area 17, of the macaque monkey as shown by retrograde transport of horseradish peroxidase." *J. Comp. Neurol.* 164: 287–304.

Maunsell, J. H. R., J. L. Bixby, and D. C. Van Essen. 1979. "The middle temporal area (MT) in the macaque: Architecture, functional properties, and topographic organization." *Soc. Neurosci. Abstr.* 5: 796.

McFie, J., M. F. Piercy, and O. L. Zangwill. 1950. "Visual-spatial agnosia associated with lesions of the right cerebral hemisphere." *Brain* 73: 167–190.

Meadows, J. C. 1974. "The anatomical basis of prosopagnosia." *J. Neurol. Neurosurg. Psychiatr.* 37: 489–501.

Mesulam, M.-M., G. W. Van Hoesen, D. N. Pandya, and N. Geschwind. 1977. "Limbic and sensory connections of the inferior parietal lobule (area PG) in the rhesus monkey: A study with a new method for horseradish peroxidase histochemistry." *Brain Res.* 136: 393–414.

Meyer, D. R., H. F. Harlow, and H. W. Ades. 1951. "Retention of delayed responses and proficiency in oddity problems by monkeys with preoccipital ablations." *Amer. J. Psychol.* 44: 391–396.

Milner, A. D., E. M. Ockleford, and W. Dewar. 1977. "Visuo-spatial performance following posterior parietal and lateral frontal lesions in stumptail macaques." *Cortex* 13: 350–360.

Milner, B. 1958. "Psychological defects produced by temporal-lobe excision." *Res. Publ. Assoc. Res. Nerv. Ment. Dis.* 36: 244–257.

———. 1968. "Visual recognition and recall after right temporal-lobe excision in man." *Neuropsychologia* 6: 191–209.

Mishkin, M. 1966. "Visual mechanisms beyond the striate cortex." In *Frontiers in Physiological Psychology*, R. Russell, ed. New York: Academic.

———. 1972. "Cortical visual areas and their interactions." In *Brain and Human Behavior*, A. G. Karczmar and J. C. Eccles, eds. Berlin: Springer.

Moffett, A. M., and G. Ettlinger. 1970. "Tactile discrimination performance in the monkey: The effect of unilateral posterior parietal ablations." *Cortex* 6: 47–67.

Moffett, A., G. Ettlinger, H. B. Morton, and M. F. Piercy. 1967. "Tactile discrimination performance in the monkey: The effect of ablation of various subdivisions of posterior parietal cortex." *Cortex* 3: 59–96.

Mountcastle, V. B., J. C. Lynch, and A. Georgopoulos. 1975. "Posterior parietal association cortex of the monkey: Command functions for operations within extrapersonal space." *J. Neurophysiol.* 38: 871–908.

Nauta, W. J. H., and P. A. Gygax. 1954. "Silver impregnation of degenerating axons in the central nervous system: A modified technique." *Stain Tech.* 29: 91–94.

Newcombe, F., and W. R. Russell. 1969. "Dissociated visual perceptual and spatial deficits in focal lesions of the right hemisphere." *J. Neurol. Neurosurg. Psychiat.* 32: 73–81.

Pandya, D. N., and H. G. J. M. Kuypers. 1969. "Cortico-cortical connections in the rhesus monkey." *Brain Res.* 13: 13–36.

Pasik, P., T. Pasik, W. S. Battersby, and M. B. Bender. 1958. "Visual and tactual discriminations by macaques with serial temporal and parietal lesions." *J. Comp. Physiol. Psychol.* 51: 427–436.

Petrides, M., and S. D. Iversen. 1979. "Restricted posterior parietal lesions in the rhesus monkey and performance on visuospatial tasks." *Brain Res.* 161: 63–77.

Pohl, W. 1973. "Dissociation of spatial discrimination deficits following frontal and parietal lesions in monkeys." *J. Comp. Physiol. Psychol.* 82: 227–239.

Pribram, H. B., and J. Barry. 1956. "Further behavioral analysis of parieto-temporo-preoccipital cortex." *J. Neurophysiol.* 19: 99–106.

Pribram, K. H., D. N. Spinelli, and S. L. Reitz. 1969. "The effects of radical disconnexion of occipital and temporal cortex on visual behaviour of monkeys." *Brain* 92: 301–312.

Ratcliff, G., and G. A. B. Davies-Jones. 1972. "Defective visual localization in focal brain wounds." *Brain* 95: 49–60.

Ratcliff, G., and F. Newcombe. 1973. "Spatial orientation in man: Effects of left, right, and bilateral posterior cerebral lesions." *J. Neurol. Neurosurg. Psychiat.* 36: 448–454.

Ratcliff, G., R. M. Ridley, and G. Ettlinger. 1977. "Spatial disorientation in the monkey." *Cortex* 13: 62–65.

Ridley, R. M., and G. Ettlinger. 1973. "Visual discrimination performance in the monkey: The activity of single cells in infero-temporal cortex." *Brain Res.* 55: 179–182.

———. 1975. "Tactile and visuo-spatial discrimination performance in the monkey: The effects of total and partial posterior parietal removals." *Neuropsychologia* 13: 191–206.

Riopelle, A. J., H. F. Harlow, P. H. Settlage, and H. W. Ades. 1951. "Performance of normal and operated monkeys on visual learning tests." *J. Comp. Physiol. Psychol.* **44**: 283–289.

Robinson, D. L., M. E. Goldberg, and G. B. Stanton. 1978. "Parietal association cortex in the primate: Sensory mechanisms and behavioral modulations." *J. Neurophysiol.* 41: 910–932.

Rocha-Miranda, C. E., D. B. Bender, C. G. Gross, and M. Mishkin. 1975. "Visual activation of neurons in inferotemporal cortex depends on striate cortex and forebrain commissures." *J. Neurophysiol.* 38: 475–491.

Rockland, K. S. 1979. "Cortical connections of the occipital lobe in the rhesus monkey." Ph.D. diss., Boston University.

Rockland, K. S., and D. N. Pandya. 1979. "Laminar origins and terminations of cortical connections of the occipital lobe in the rhesus monkey." *Brain Res.* 179: 3–20.

Rolls, E. T., S. J. Judge, and M. K. Sanghera. 1977. "Activity of neurones in the inferotemporal cortex of the alert monkey." *Brain Res.* 130: 229–238.

Rolls, E. T., D. Perrett, S. J. Thorpe, D. Puerto, A. Roper-Hall, and S. Maddison. 1979a. "Responses of neurons in area 7 of the parietal cortex to objects of different significance." *Brain Res.* 169: 194–198.

Rolls, E. T., M. K. Sanghera, and A. Roper-Hall, 1979b. "The latency of activation of neurones in the lateral hypothalamus and substantia innominata during feeding in the monkey." *Brain Res.* 164: 121–135.

Sanghera, M. K., E. T. Rolls, and A. Roper-Hall. 1979. "Visual responses of neurons in the dorsolateral amygdala of the alert monkey." *Exp. Neurol.* 63: 610–626.

Schneider, G. E. 1967. "Contrasting visuomotor functions of tectum and cortex in the golden hamster." *Psychol. Forsch.* 31: 52–62.

Seacord, L., C. G. Gross, and M. Mishkin. 1979. "Role of inferior temporal cortex in interhemispheric transfer." *Brain Res.* 167: 259–272.

Semmes, J. 1967. "Manual stereognosis after brain injury." In *Symposium on Oral Sensation and Perception,* J. F. Bosma, ed. Springfield, Ill.: Thomas.

Semmes, J., S. Weinstein, L. Ghent, and H.-L. Teuber. 1963. "Correlates of impaired orientation in personal and extrapersonal space." *Brain* 86: 747–772.

Spiegler, B. J., and M. Mishkin. 1978. "Evidence for the sequential participation of inferior temporal cortex and amygdala in stimulus-reward learning." *Soc. Neurosci. Abstr.* 4: 263.

Stanton, G. B., W. L. R. Cruce, M. E. Goldberg, and D. L. Robinson. 1977. "Some ipsilateral projections to areas PF and PG of the inferior parietal lobule in monkeys." *Neurosci. Lett.* 6: 243–250.

Stein, J. 1978. "Effects of parietal lobe cooling on manipulative behaviour in the conscious monkey." In *Active Touch*, G. Gordon, ed. Oxford: Pergamon.

Sugishita, M., G. Ettlinger, and R. M. Ridley. 1978. "Disturbance of cage finding in the monkey." *Cortex* 14: 431–438.

Sunshine, J., and M. Mishkin. 1975. "A visual-limbic pathway serving visual associative functions in rhesus monkeys." *Fed. Proc.* 34: 440.

Trevarthen, C. B. 1968. "Two mechanisms of vision in primates." *Psychol. Forsch.* 31: 299–337.

Ungerleider, L. G., and B. A. Brody. 1977. "Extrapersonal spatial orientation: The role of posterior parietal, anterior frontal, and inferotemporal cortex." *Exp. Neurol.* 56: 265–280.

Ungerleider, L. G., and M. Mishkin. 1978a. "Interactions of striate and posterior parietal cortex in spatial vision." *Soc. Neurosci. Abstr.* 4: 649.

———. 1978b. "The visual area in the superior temporal sulcus of *Macaca mulatta:* Location and topographic organization." *Anat. Rec.* 190: 568.

———. 1979a. "Three cortical projection fields of area 17 in the rhesus monkey." *Soc. Neurosci. Abstr.* 5: 812.

———. 1979b. "The striate projection zone in the superior temporal sulcus of *Macaca mulatta:* Location and topographic organization." *J. Comp. Neurol.* 188: 347–366.

Ungerleider, L. G., and K. H. Pribram. 1977. "Inferotemporal versus combined pulvinar-prestriate lesions in the rhesus monkey: Effects on color, object, and pattern discrimination." *Neuropsychologia* 15: 481–498.

Van Essen, D. C., J. H. R. Maunsell, and J. L. Bixby. 1979. "Areal boundaries and topographic organization of visual areas V2 and V3 in the macaque monkey." *Soc. Neurosci. Abstr.* 5: 812.

Weller, R. E., and J. H. Kaas. 1978. "Connections of striate cortex with the posterior bank of the superior temporal sulcus in macaque monkeys." *Soc. Neurosci. Abstr.* 4: 650.

Weller, R. E., J. Graham, and J. H. Kaas. 1979. "Cortical connections of striate cortex in macaque monkeys." *Invest. Ophthalmol.* (ARVO suppl.) 18: 157.

von Bonin, G., and P. Bailey. 1947. *The Neocortex of Macaca Mulatta.* Urbana: University of Illinois Press.

von Bonin, G., H. W. Garol, and W. S. McCulloch. 1942. "The functional organization of the occipital lobe." *Biol. Symp.* 7: 165–192.

Wilson, M. 1957. "Effects of circumscribed cortical lesions upon somesthetic and visual discrimination in the monkey." *J. Comp. Physiol. Psychol.* 50: 630–635.

———. 1965. "Tactual discrimination learning in monkeys." *Neuropsychologia* 3: 353–361.

———. 1978. "Visual system: Pulvinar-extrastriate cortex." In *Handbook of Behavioral Neurobiology*, vol. 1: *Sensory Integration*. R. B. Masterton, ed. New York: Plenum.

Wilson, M., J. A. Stamm, and K. H. Pribram. 1960. "Deficits in roughness discrimination after posterior parietal lesions in monkeys." *J. Comp. Physiol. Psychol.* 53: 535–539.

Wurtz, R. H. 1969. "Visual receptive fields of striate cortex neurons in awake monkeys." *J. Neurophysiol.* 32: 727–742.

Zeki, S. M. 1969. "Representation of central visual fields in prestriate cortex of monkey." *Brain Res.* 14: 271–291.

———. 1971a. "Convergent input from the striate cortex (area 17) to the cortex of the superior temporal sulcus in the rhesus monkey." *Brain Res.* 28: 338–340.

———. 1971b. "Cortical projections from two prestriate areas in the monkey." *Brain Res.* 34: 19–35.

———. 1976. "The projections to the superior temporal sulcus from areas 17 and 18 in the rhesus monkey." *Proc. Roy. Soc.* B 193: 199–207.

19 Visual Behavior in Monkeys with Inferotemporal Lesions

P. Dean

Bilateral removal of cortex from the inferior part of the temporal lobe (figure 19.1) affects visual behavior in monkeys in such a way that after surgery they appear to have forgotten visual discriminations learned before the operation, and they learn new discriminations more slowly than normal animals. Since the discovery of these effects (Mishkin 1954; Mishkin and Pribram 1954), many experiments have been performed to try to answer the following question: What are the functions of inferotemporal cortex, as revealed by analysis of changes in behavior occurring after inferotemporal ablation? This problem is examined in the present chapter, both in its own right and as an example of a particular technique in the analysis of visual behavior.

The Nature of the Problem

The general objectives of studying the relationship between brain and behavior can be described very crudely as constructing a "wiring diagram" of the brain in which the connections between components (anatomy) and the properties of those components (electrophysiology) are understood, and mapping this diagram onto a "conceptual" wiring diagram of behavior, thus allowing the functions of different parts of the brain to be explained. One method of helping to generate or confirm such a map is to deduce the function of a component from the changes in behavior that occur when it is damaged.

Meaning of "Function"

What is meant by the word "function" in this context? It has been argued that it is improper to speak of parts of the brain having functions "localized" in them—especially parts of association cortex, such as inferotemporal cortex—since "all the cells in the brain . . . are participating, by a sort of algebraic summation, in every activity" (Lashley 1950, p. 477). However, this argument depends on confusing two uses of the word "function": function as a property of the system as a whole, and the functions of its component parts. The confusion can be illustrated by considering a relatively simple system. The function of a radio set is

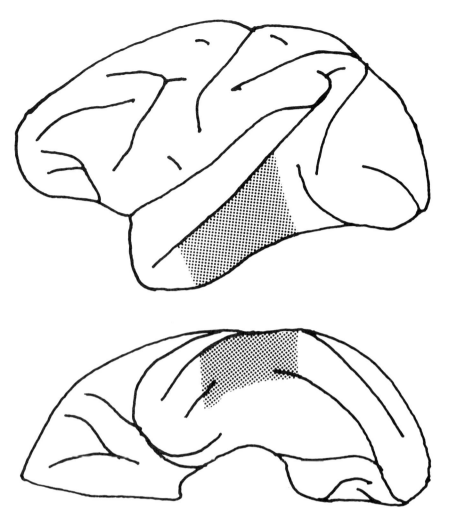

Figure 19.1 Typical inferotemporal lesion in rhesus monkey. Top: lateral view. Bottom: ventral view. (Copyright 1976 by The American Psychological Association. Reprinted by permission).

to convert electromagnetic signals into acoustic signals. The functions of its components are to perform subtasks necessary for the operation as a whole. There is localization of function in the sense that different parts of the set do different things, but there is no single part that carries out the overall function.

It makes no sense to ask in which component of a system the function of the entire system is performed, or to assert that an activity of the entire organization, like feeding, is carried out solely by a particular region of the brain (such as the hypothalamus). This is not true of the question "Which parts of a radio set *contribute* to (or are necessary for) its overall function?"; a particular knob, for example, may be for decoration only and have no functional role. But that question is logically distinct from the issue of whether the parts that do contribute do so in the same or different ways. All the parts may contribute, yet each may be highly specialized. (See also Laurence and Stein 1978; Dean 1980.)

This distinction is particularly important for flexible systems capable of performing many tasks or "functions." The operation that each component is capable of carrying out may be invariant, but the significance of that operation for a particular "function" (of the system as a whole) depends on the nature of the problem that the system is dealing with. Therefore, "functions of inferotemporal cortex" must refer to its functions *as a component*—that is, roughly speaking, to the transformations that inferotemporal cortex performs on its inputs to generate its outputs. These can be investigated and specified even if every cell in the brain does take part in every activity, although they cannot be specified in terms that apply only to the system as a whole, such as "perceiving," "attending," "remembering," and the like. Inferotemporal cortex may perform an operation that is necessary for, e.g., normal perception, but cannot by itself "perceive," and nor can it be said that perception was localized in the temporal lobe.

Ambiguity of Behavioral Tests

When the functions of inferotemporal cortex are viewed in this way, the nature of the problem of deducing those functions from the effects of lesions becomes clearer. As with any other area of the brain, these functions will not be directly expressed as behavior of the whole animal. Other parts of the system—for example, those that provide inputs to inferotemporal cortex and process its outputs—are required before it can influence that behavior. Inferotemporal cortex is embedded in a complex system, in which any behavior displayed by the system as a whole requires the cooperation of several components and, conversely, any components must cooperate with others to produce behavior.

A consequence of this is that failure on any behavioral test after damage to inferotemporal cortex does not by itself unequivocally specify the cause of that failure. Visual discrimination learning, for example, requires a whole series of sensory, cognitive, and motor skills, and the impairment of discrimination learning by inferotemporal ablation might be caused by a failure in any one of these. How, then, is it possible to use behavioral tests to investigate the functions of system components, given the intrinsic ambiguity of such tests?

The Nature of the Solution

In general terms, the solution to this problem is clear: If any single behavioral test is ambiguous, it is necessary to give more than one. But how are the tests to be chosen, and in what order should they be administered? Partial answers to these questions come from consideration of other circumstances in which the nature of a fault in a complex system has to be deduced from the system's aberrant behaviour. The process is one of diagnosis, as used in medicine, analytic chemistry, and engineering, and the general strategy is to systematically eliminate possible (but wrong) explanations of the symptoms displayed. To achieve this systematic elimination, sequences of tests are arranged in a diagnostic tree, such as those used to trace particular classes of faults in an automobile engine. The important feature of this type of diagnostic tree is that the first questions asked about the fault are general ("Does the starter turn the engine?") rather than specific ("Has the coil insulation broken down?"). There are two reasons for this. One is that the first question is about equally informative whether the answer is "yes" or "no" (since if the answer is "no" a large class of potential faults have been eliminated), whereas if the answer to the second question is "no" ignorance has been diminished very slightly since only one potential fault has been eliminated. The situation is analogous to the game of Twenty Questions: Unless one possesses extrasensory perception, it is not a good idea to ask a first question like "Is it an aardvark?" Information theory shows that, with yes-no answers, questions should be chosen at each stage that split the remaining possibilities in two; this ensures that the answer is arrived at with the least number of questions (see, e.g., Garner 1962). The second reason applies particularly to diagnostic trees that employ only behavioral tests. Testing the coil insulation produces an unequivocal solution to the problem if an impairment is found, but involves going inside the system to examine a component directly. Behavioral tests are ambiguous, whether or not they reveal an impairment. However, if a behavioral test is part of a system in which possible explanations of the fault have already been eliminated, the ambiguity of a positive finding can be much reduced. A particularly notorious example: The finding that a rat is slow at learning to traverse a maze gives, on its own, little information, since maze learning can be affected in so many different ways. But if it is already known that the animal's sensory and motor capacities are intact, and that other kinds of learning are normal, a maze-learning impairment is more likely to indicate, say, a specific fault in organizing spatial information.

Consideration of other complex systems, therefore, suggests that a reasonable procedure for investigating the inferotemporal defect is first to ask general questions about the nature of the behavioral disturbances, with more precise questions following later as more and more possible explanations of the defect are eliminated. Before asking how this can be achieved in practice, attention has to be paid to arguments that, because the brain is so complicated, the lesion method is for theoretical reasons incapable of providing useful information about the relations between brain and behavior. There are two reasons for attending to these criticisms. One is that, if they were correct, detailed consideration of the effects of a particular lesion would be otiose. The other is that, even if not correct in their strongest form, these arguments expose the especial difficulties the lesion method has to face—in particular, the difficulty of devising effective diagnostic trees.

Criticisms of the Lesion Method

Objections to the lesion method range from feelings rather like Gandalf's—"He who has to break a thing to find out how it works has departed from the path of wisdom"—to the powerful and detailed arguments of Gregory (1961). They fall into two main classes:

• Experience with man-made complex systems does not suggest that damaging them is a good way to investigate the intricacies of their normal function. The analogy sometimes offered here (not, apparently, in self-parody) is that of trying to find out how a computer works by seeing what happens when you hit it with an axe.

• The brain is so complex that the behavior it produces is often not understood, even in principle. If normal behavior is so poorly understood, disturbances such as those due to brain lesions are impossible to interpret.

The following discussion of these objections relies heavily on the arguments of Weiskrantz (1968). In this discussion, "lesion method" refers to the attempt to understand the behavioral symptoms that occur after any particular lesion. Other uses of lesions—for example, studying the effects of different lesions on one particular behavioral task to trace the parts of the brain responsible for its performance—may not be vulnerable to these criticisms to the same extent.

"Wrecking the Computer"

This attack on the lesion method often conflates two somewhat different points: that some systems are very vulnerable to any damage, no matter how precise and localized, because they are highly interactive; and that any system, no matter how well designed, will give uninterpretable symptoms if the damage to it is sufficiently widespread.

A highly interactive system is one in which each component directly and powerfully interacts with many others. Damaging an individual component therefore has far-reaching effects throughout the entire system, producing symptoms that bear no direct relation to the function. In fact, with such a system any damage is likely to cause complete collapse. Thus, "removal of any of several widely spaced resistors may cause a radio set to emit howls" (Gregory 1961, p. 323), or, as it sometimes appears, an industrial grievance in any part of certain British car-manufacturing companies brings the entire organization to a halt. Very little can be deduced from this sort of symptom about the functions of components in normal operation, and if the brain were a highly interactive system the lesion method would not be an appropriate way of studying it.

In fact it appears that, at least in certain respects, the brain does not behave as a highly interactive system. It is common to find that a brain lesion affects some kinds of behavior and not others, and that different lesions produce different patterns of behavioral change. For example, many lesions that affect visually guided behavior leave hearing intact, and vice versa. The brain appears to contain relatively independent functional subsystems or "modules," each of which directly influences only a small number of other subsystems. In modular systems (provided they are not organized in simple serial fashion), the effects of removing individual modules can give important information about their function and the arrangement.

Apart from this and much other empirical evidence for modular construction in the brain (many anatomical and electrophysiological findings strongly suggest that the brain is parceled into relatively independent functional subareas), there are also theoretical considerations to be taken into account. Simon (1969) vividly described the advantages of modular over interactive design for complex systems. One is superior resistance to disruption, the very reason why lesion studies of the brain are feasible. Localized damage to a modular system need not cause the entire system to malfunction; rather, its behavior undergoes "graceful degradation" (Marr 1976). Another advantage is that modular systems are easier to improve: If one component of an interactive system is changed, the input to all other components in the system is altered as well, giving rise to unwanted side effects which may be very difficult to compensate. It is as if the introduction of color vision substantially altered the properties of neurons throughout the auditory system. The fact that interactive systems are difficult to improve unless they are completely redesigned suggests that systems such as radio sets or some car industries would not survive as products of biological evolution. Conversely, systems that (like the brain) are the products of evolution are likely to be modular (see also Marr and Poggio 1976).

The fact that some areas of the brain directly influence only a small number of others means that the lesion method cannot be dismissed *a priori* as inadequate, but it does not mean that the entire brain is organized in this way. For example, the dorsal noradrenergic system arises from a handful of cells in the hindbrain, yet innervates almost the entire neocortex (Ungerstedt 1971). How any particular system is organized needs to be determined empirically. In the visual system, there is now a substantial amount of evidence for the existence of some relatively independent functional units (see also Mansfield, this volume). Information from the eye travels to (among other destinations) the superior colliculus, and to striate cortex via the lateral geniculate nucleus. Neurones in the superior colliculus and striate cortex have different electrophysiological properties, and the two areas can function, to some extent, independently of one another, at least in some species. For example, collicular lesions affect a hamster's ability to orient to suddenly appearing visual stimuli, but do not alter its ability to tell apart two-dimensional visual patterns; visual-cortex lesions impair pattern discrimination but not orientation (Schneider 1969). Farther on in the visual system, prestriate cortex (to which striate cortex projects) has been found to contain numerous maps of the visual world (see e.g. the review by Kaas [1978]), and current evidence suggest that the neurons in each map have distinctive properties. For example, in rhesus monkeys there is one prestriate area that contains cells responsive primarily to color, while another contains neurons specialized for direction of movement (Zeki 1978). It has been suggested that damage restricted to such specialized areas may underlie the selective perceptual deficits that have been observed clinically, such as impaired color vision with normal form perception (Cowey 1963; Meadows 1974). If this suggestion is correct, then these specialized prestriate areas must be able to function relatively independently of one another.

However, it is important to emphasize that such independence can only be relative. Modules have to get information from somewhere and have to communicate their computational achievements to the rest of the system. Effective behavior of the system as a whole requires smooth cooperation of the component operations. Thus, although striate cortex and superior colliculus behave independently in some respects, there is a projection from striate cortex to colliculus which in monkeys is very important for some aspects of normal tectal functioning: Ablation of striate cortex abolishes the visual responsiveness of cells in the deep layers of the monkey superior colliculus (Schiller et al. 1974). The "interactive system" argument can be overstated, but it does draw attention to the fact that removing a component from a system will produce symptoms that depend not only on the function of that component, but also on the changes produced in the other components with which it interacts. Inference from symptom to function has to take this into account. However, if the connections between components are constrained in the ways modularity suggests, the lesion method of analysis remains feasible for at least some lesions. Although inferotemporal-cortex ablation will affect the operations of some neighboring modules, it is reasonable to suspect that their number will not be too great, and that the nature of the influence might be relatively straightforward.

No matter how modular a system is, sufficiently widespread damage will degrade normal performance too far to allow easy inferences about the functions of components. Whether this is true for any particular lesion again needs to be determined empirically. There are two lines of evidence relevant to the issue of whether inferotemporal lesions cause widespread damage to many components. One is anatomical and electrophysiological, the other behavioral.

Originally it appeared that inferotemporal lesions of the kind illustrated in figure 19.1 destroyed only one cytoarchitectonic area, called area TE in the cytoarchitectonic map of von Bonin and Bailey (1947) (see figure 19.2). However, more recent investigations suggest that this is an

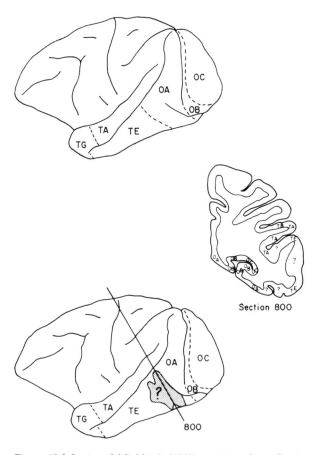

Figure 19.2 Iwai and Mishkin's (1969) revision of von Bonin and Bailey's cytoarchitectonic map of prestriate and inferotemporal cortex in *Macaca mulatta*. Top: drawing based on von Bonin and Bailey's (1947) original map. Bottom: revised drawing with new area between OA and TE, labeled "?". Middle: section 800 of von Bonin and Bailey, showing new area. Iwai and Mishkin suggest TEO as a suitable label for the new area. Reprinted, with permission, from Iwai and Mishkin 1969 (copyright 1969 by Academic Press).

oversimplification. Iwai and Mishkin (1969), using as part of their extensive evidence the original data of von Bonin and Bailey, suggested that intercalated between OA and TE is a further cytoarchitectonically and functionally distinct area, TE(O) (figure 19.2). If so, the lesion in figure 19.1 would destroy both TE and part of TE(O). More recently, Seltzer and Pandya (1978) also reinvestigated the cytoarchitectonics of inferotemporal cortex. They did not discern any new area between OA and TE, but did divide TE itself into five subareas (figure 19.3). In practice, inferotemporal lesions vary somewhat in size and location, so that the number of Seltzer and Pandya's areas that are invaded differs from experiment to experiment; but it is seldom less than three (TE_m, TE_2, and TE_3) and is often all five.

It remains to be seen how far these cytoarchitectonic subdivisions correspond to discrete functional areas with separate connections, each containing neurons with distinctive electrophysiological properties. On the one hand, experience with nearby visual areas in rhesus monkeys suggests that even within cytoarchitectonically uniform regions of prestriate cortex (for example, OA and OB, or in Brodmann's terminology areas 18 and 19) there exist subareas with quite different anatomical connections and electrophysiological properties (Zeki 1978). On the other hand, current evidence suggests that the electrophysiological properties of single units vary only slightly from one part of area TE to another (Gross et al. 1972; Desimone and Gross 1979).

At the moment, then, anatomical and electrophysiological findings are consistent with only a small number of functional areas being destroyed by a typical inferotemporal lesion.

The other reason for thinking that inferotemporal lesions do not cause widespread damage to many components is the specificity of behavioral effects produced. The important findings (see Mishkin 1966, 1972; Gross 1973a,b; Dean 1976) were made quite soon after the initial discovery that inferotemporal lesions impaired visual-discrimination learning—in fact, well before the anatomy and electrophysiology of inferotemporal cortex were at all well understood. One early question about the impairment was whether it was confined to vision. It turned out that discriminations in other modalities were not affected by inferotemporal ablation. This was an important finding since it ruled out many explanations of a discrimination learning defect—for example, that the animal was generally unhealthy or disorientated, had problems with motor coordination, or could not supress some conflicting behavior. Another important issue was whether all aspects of visual performance were affected. This seemed unlikely, since in their home cages monkeys with inferotemporal lesions did not bump into objects or misjudge distances but behaved like normal animals. This lack of a general visual impairment was confirmed by formal testing. For example, the delayed-response task, in which the animal has to remember where it saw a piece of food being hidden, was not impaired by inferotem-

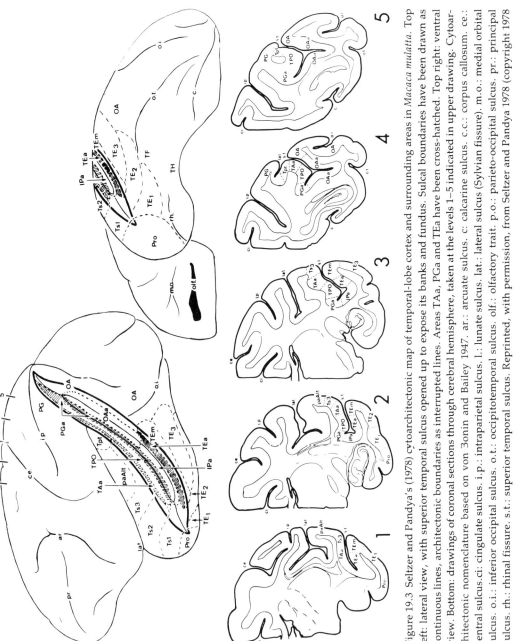

Figure 19.3 Seltzer and Pandya's (1978) cytoarchitectonic map of temporal-lobe cortex and surrounding areas in *Macaca mulatta*. Top left: lateral view, with superior temporal sulcus opened up to expose its banks and fundus. Sulcal boundaries have been drawn as continuous lines, architectonic boundaries as interrupted lines. Areas TAa, PGa and TEa have been cross-hatched. Top right: ventral view. Bottom: drawings of coronal sections through cerebral hemisphere, taken at the levels 1–5 indicated in upper drawing. Cytoarchitectonic nomenclature based on von Bonin and Bailey 1947. ar.: arcuate sulcus. c: calcarine sulcus. c.c.: corpus callosum. ce.: central sulcus.ci: cingulate sulcus. i.p.: intraparietal sulcus. l.: lunate sulcus. lat.: lateral sulcus (Sylvian fissure). m.o.: medial orbital sulcus. o.i.: inferior occipital sulcus. o.t.: occipitotemporal sulcus. olf.: olfactory trait. p.o.: parieto-occipital sulcus. pr.: principal sulcus. rh.: rhinal fissure. s.t.: superior temporal sulcus. Reprinted, with permission, from Seltzer and Pandya 1978 (copyright 1978 by Elsevier/North-Holland Biomedical Press).

poral ablation. In fact, early experiments suggested that the effects of inferotemporal lesions were confined to visual-discrimination learning, though within that paradigm learning was impaired for all kinds of visual discriminanda (except those that normal animals find extremely easy to tell apart, such as certain pairs of junk objects), whether they differed in shape, color, size, orientation, etc.

The relative specificity of the behavioral impairment produced by inferotemporal ablation therefore supports the anatomical and electrophysiological evidence in suggesting that a relatively small number of component modules are destroyed by the operation. Analogies between inferotemporal lesions and wanton damage to vulnerable manmade systems seem unfounded, and certainly not strong enough to warrant *a priori* the dismissal of a careful evaluation of the behavioral effects of inferotemporal lesions.

Poor Understanding of Normal Behavior
The second major criticism of the lesion method is that brain damage affects components of a system so complex that its normal behavior is often very poorly understood. The "conceptual" wiring diagram referred to above exists in, at best, an extremely rudimentary form. This problem with the lesion method has been recognized for at least 40 years:

> . . . the study of the cerebral function has two aspects. The first is the physiological analysis, in which established categories of behavior are taken for granted and their relation to cerebral activity is studied. The second is the analysis of behavior itself, and the attempt to formulate more significant behavioral concepts or categories, from which it may be possible to gain a truer idea of the nature of cerebral action. It is clear that the success of the physiological analysis depends upon the adequacy of the behavioral tests used. Much of the difficulty found in the evaluation of the effects of cerebral destruction, both clinically and with animal experimentation, is due to the fundamental difficulty of the analysis of behavior and to the unsatisfactory available accounts of it. (Hebb 1938, p. 333).

The criticism is sometimes taken to imply that normal behavior must at some time be completely understood before lesion studies can give any information (see Weiskrantz 1968)—a sort of all-or-none argument—but this is to overstate the case. Ability to diagnose faults in a complex system is likely to vary gradually with knowledge of its normal function, or so experience with systems less dauntingly complex than the brain suggests. Some drivers can identify none of the component operations that occur inside their cars' engines (for them the engine is one giant black box), and they can describe faults but not diagnose them. An expert mechanic may know all the component operations (at the appropriate level of description there are no black boxes for him) and be able to conduct tests to characterize any fault. But in between these extremes lies every intermediate stage of knowledge and diagnostic power, in general related by the fact that only if the distinction between component subsystems (e.g., between fuel and electrical systems, or between different parts within the electrical system) is appreciated can one ask in which of the subsystems the fault is located. There is almost always some knowledge available about the behavior disturbed by brain lesions, since we are generally capable of that behavior ourselves. It is not as if the investigator is confronted with some totally unknown device. So the force of this criticism is not that we can say nothing about lesion-induced symptoms, but that what we can say may be very limited. The nature of this limitation can be illustrated by the example of object recognition. We lack even a "computational theory" of object recognition (see, e.g., Marr and Poggio 1976): We cannot say what computation *any* system would need to carry out in order to recognize objects as successfully as people do. Still less can we say how any particular system implements those computations. Trying to understand the operation of a component in an object-recognition device in ignorance of these computational principles may be like trying to understand a part of a thermoregulatory system before the principles of control systems and such concepts as error signal and set point have been elucidated.

The implications of this relationship between knowledge and diagnostic power for analysis of the effects of inferotemporal lesions are serious. Knowledge about many of the processes involved in visual behavior is very incomplete, even at a rather general level. There is, consequently, no widely accepted model of normal visual processing among workers on inferotemporal lesions (though it would be difficult to tell if there were, since the topic is explicitly mentioned so rarely). This lack of a good model of normal visual behavior, with the attendant difficulty of constructing a reasonable diagnostic sequence of behavioral tests, is the major reason for the slow progress that has been made in analysis of the inferotemporal defect since its discovery over 25 years ago. Researchers have been forced to construct their own diagnostic systems (and implied theories of normal function), using such knowledge of normal vision as was available. Since these systems have had great influence on the course of research on inferotemporal lesions, the heuristic methods used for their construction are very important, and will be discussed in a separate section below. One further comment belongs here.

It is a slightly paradoxical point that knowledge of normal visual behavior is so incomplete that the lesion method has a power it would lack in a well-understood system. It was suggested earlier that it was necessary to realize the distinction between two subsystems before particular fault could be located in one or the other. However, as Hebb was aware, the situation may work in reverse. The occurrence of a particular fault in behavior may suggest the existence of hitherto unsuspected subsystems—for example, one for the detection of certain visual attributes without the awareness that such detection has taken place ("blindsight"; see Weiskrantz et al. 1974). In other words, the lesion method can give information not only about brain and behavior, but also about normal behavior as such. It is in this sense that the lesion method can be a method for the analysis of visual behavior. Researchers on the effects of lesions may therefore find themselves asking two questions: "What do these results tell me about the function of part X?" and "What do these results tell me about the organization of behavior Y?" In some cases, such as that of diffuse brain damage in human patients, the second question may predominate; but in research on well-defined lesions in animals attention tends to focus on the first. Indeed, the fact that there is anything still to discover about the behavior in question is often not explicitly stated.

Strategies for Diagnosis

The Problem
It was suggested above that, since any behavioral test is directed at the whole system and therefore requires many component operations for its solution, a diagnostic sequence of tests was required for efficient characterization of a fault in one particular component, by systematic elimination of false possibilities. It now turns out there is a further and much graver difficulty: The nature of the operations tested by any particular task is often not known. The problem becomes one of how to conduct informative tests with the very partial knowledge of visual behavior that is currently available. Since a diagnostic test asks whether the fault is in subsystem (class of operations) A or in subsystem (class of operations) B, part of this problem involves the generation of putative subsystems in the relevant visual behavior. The other part is how to devise suitable behavioral tasks to test these subprocesses. There seem to have been three main heuristic strategies used to generate component processes and behavioral tests for them.

One approach has been to rely on theories of normal visual function and the nature of visual tasks. Such theories may at one extreme be public and explicit (such as the Sutherland-Mackintosh [1971] two-stage theory of visual discrimination learning), or at the other private and intuitive, amounting perhaps to no more than a set of what seem to be reasonable behavioral categories. One possible danger with this method is a bit like the danger of asking precise questions too early in a diagnostic tree: Tests derived from the premature use of a theory that turns out not to be correct may give very little information. This danger tends to promote an attitude of least theoretical commitment (see Marr 1976).

The second strategy, in almost complete opposition to the first, has been to administer many behavioral tests in shotgun fashion, in the hope that interesting dissociations will emerge from the pattern of impairment found and will lead to a *post hoc* identification of component behavioral processes. This is perhaps the strategy that is least worried by ignorance of normal function, but its disadvantage is that of asking questions at random in the game of Twenty Questions: It may take a very long time to find the answer.

The third approach is in a sense a compromise between the first two. A task impaired by a particular lesion is conceptually analyzed to see what processes are necessary for its solution, and further tasks are devised that test some of those processes but not others. This avoids the Scylla of asking questions based on false premises and the Charybdis of asking questions at random, but is inadequate on its own because of the dangers of incompleteness. The processes described may be necessary for performance of a particular task, but taken together need not be sufficient.

Diagnosis and Inferotemporal Lesions
Because each of these "pure" methods of partitioning and testing visual behavior has its disadvantages, real-life strategies in the analysis of lesion effects have tended to be judicious mixtures of all three. This has certainly been true of work on the inferotemporal defect. The fact that no two workers seem to have used quite the same mix makes it hard to arrange the questions that have been asked, and the tests used, into a universally acceptable diagnostic tree. The arrangement in figure 19.4 is therefore to some extent an idiosyncratic one. It is, however, in rough correspondence with the actual sequence of discoveries, and shows how, in accord with effective diagnostic practice, the later questions have tended to be more precise than the earlier ones, as possibilities were more or less systematically eliminated. It can also be used to illustrate the different strategies that have been used in investigating the inferotemporal defect. For example, the early work showing that inferotemporal lesions affected visual discrimination learning rather than (e.g.) delayed response used a battery of visual tasks designed to test a wide variety of visual abilities (Mishkin 1954; Mishkin and Pribram 1954). Once the defect had been traced to discrimination learning, later work was able to concentrate more specifically on the processes that were necessarily involved in such a task; for example, the following:

• The discriminanda must be told apart, which involves the animal's being able to perceive the difference between them and to attend to the discriminanda so that the difference is in fact perceived.

• Some internal representation of these discriminanda must be formed—the stimuli have to be categorized.

• An association has to be formed between these representations and reward or punishment.

• Information has to be stored in memory from trial to trial.

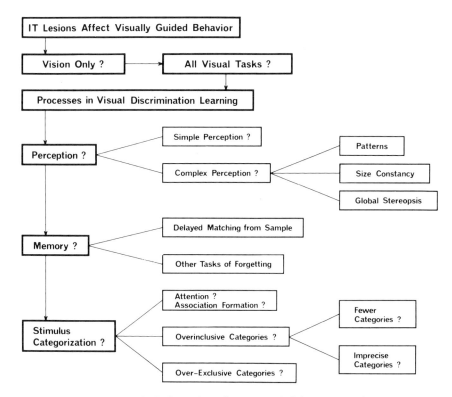

Figure 19.4 Questions asked about the inferotemporal defect, arranged in not too implausible diagnostic tree. ("Attention" and "Association formation" are here included in the "stimulus categorization" section for convenience of exposition only, not because they are component processes of categorization.)

Finally, for the investigation of such a complex but ill-defined process such as perception, rather intuitive notions of the important subprocesses have been used.

To see how far the tree of mixed origin has succeeded in diagnosing the inferotemporal defect, and how well it has grappled with the computational complexities of such processes as object recognition, the detailed evidence relating to the inferotemporal defect will now be examined.

Telling Discriminanda Apart

Perhaps the most obvious way in which a lesion could produce an exclusively visual defect would be to affect sensory or perceptual processes in vision. Animals must be able to tell the discriminanda apart to learn a discrimination, and if for some reason inferotemporal lesions made the discriminanda look more similar one would expect learning to be slower. Although this is perhaps the most obvious explanation of the inferotemporal defect, it is not necessarily the most plausible. In-

ferotemporally lesioned animals are slow at learning visual discriminations in every dimension tested (for example, hue, size, orientation,
"shape," and flicker), and it may not be clear what sort of sensory or
perceptual impairment would affect processing of such a wide range of
stimuli. Nevertheless, it seems a prudent diagnostic strategy to try and
rule out (or confirm) such an obvious explanation before proceeding to
investigate more esoteric hypotheses. It would be very embarrassing to
spend time investigating, say, memory for visual stimuli, only to discover later that the animals could not tell those stimuli apart.

The basic question that is asked in investigations of inferotemporally
lesioned animals' sensory and perceptual capacities is the following:
Are there cases in which a monkey with an inferotemporal lesion cannot learn to tell two simultaneously presented stimuli apart when a
normal animal can? The stimuli must be present simultaneously
(otherwise, memory becomes involved), and the animals must be given
adequate opportunity to learn the task (otherwise an impairment is not
unambiguously sensory). The kind of stimuli about which this question
has been asked can be divided, conveniently though to some extent
arbitrarily, into simple and complex.

Simple Stimuli
How well can monkeys whose vision is affected by inferotemporal lesions detect the difference between simple stimuli (light flash versus no
light flash, presence versus absence of contrast, and so forth)? To put it
another way, do inferotemporal lesions cause any kind of simple "sensory" loss? The obvious way to answer this question is to measure the
thresholds of monkeys with inferotemporal lesions by standard psychophysical methods. There arises immediately an example of the ambiguity of behavioral tests, which is well illustrated by experiments
trying to measure critical flicker-fusion frequency in monkeys with inferotemporal lesions.

Mishkin and Weiskrantz (1959) investigated critical flicker-fusion
frequency by first training animals to discriminate between an apparently steady and an obviously flickering light, and then gradually increasing the rate of flicker. Inferotemporal removal reduced the rate of
flicker at which the monkeys could distinguish the two lights. However, the conclusion that the operation had an effect on sensory capacity
is not warranted, because the threshold of both lesioned animals and
unoperated control animals continued to improve throughout the many
months of testing. In other words, the monkeys continued to learn
better ways of performing the task. As the authors were aware, the
inferotemporal impairment they observed could therefore have been
caused, not only by an alteration in best possible performance, but by
any other way in which visual discrimination *learning* could be affected.
This problem was subsequently resolved by Symmes (1965), who altered the discrimination apparatus to make learning easier, and trained

his monkeys before operation until their thresholds stopped improving; that is, until learning had ceased. In these circumstances, inferotemporal ablation had no effect on flicker-fusion threshold.

These two experiments demonstrate an important general point about measuring sensory and other capacities in monkeys with inferotemporal lesions. Psychophysical methods used with animals frequently involve two kinds of learning. One concerns general instructions about what to do to get reward—for example, "touch the flickering light, not the steady light." The other is an ill-understood kind of perceptual learning, by which the animal becomes able to make finer and finer discriminations. There is no reason why corresponding forms of learning should not occur in tests of other abilities—for example, of how long a stimulus can be remembered. To remove this ambiguity it is necessary to get rid of the learning "contamination" by testing the animals until learning stops, often a long and tedious business. (The inferotemporal *learning* impairment is very long-lasting [Gross 1973b]; so prolonged testing on psychophysical and related tasks does not simply ensure that the animal completely recovers from the effects of the operation.)

When they have been measured properly, the thresholds of animals with inferotemporal lesions turn out to be normal in a variety of tasks. One particularly important demonstration of this kind was carried out by Cowey and Weiskrantz (1967), who showed that monkeys with inferotemporal lesions did not have field defects as tested in Cowey's (1963) perimeter, and had normal acuity. In contrast, monkeys with damage to that part of striate cortex to which the fovea projects (lateral striate lesions) did have field defects and reduced acuity, yet learned pattern discriminations faster than the animals with inferotemporal lesions. This finding very strongly indicates that the effects of inferotemporal ablation on discrimination learning are not mediated by the most obvious forms of decreased sensitivity to simple-stimulus differences. Anatomical and electrophysiological evidence also supports this conclusion. The most important source of visual input for inferotemporal cortex is from striate cortex via a series of synapses in prestriate areas (Gross 1973a; Mishkin 1972), and many of its outputs are to structures that are not exclusively visual in function (for example, basal ganglia, entorhinal cortex, temporal pole cortex). The suggestion is that, insofar as the visual processes with which inferotemporal cortex is concerned are carried out serially, the area represents the final stages in purely visual processing. Similarly, comparison of single unit properties in striate and inferotemporal cortex reveal that inferotemporal units have much larger fields, which always include the fovea and frequently extend into the ipsilateral hemifield; that in some cases they need a much more specific and complex stimulus to produce appreciable firing; and that they respond to such a stimulus in the same way wherever it is situated in the receptive field. It could be argued that these data are consistent with the idea that inferotemporal cortex processes visual information at a much "higher level" than striate cortex; accordingly, inferotemporal lesions would not be expected to affect the detection of simple attributes or features of visual stimuli (provided there were other outlets from striate cortex for information about such stimuli besides the prestriate-inferotemporal connection) (see Mansfield, this volume).

Complex Stimuli

This last argument carries the implication that, if animals with inferotemporal lesions are bad at telling any stimuli apart, those stimuli are likely to be "complex." What is meant by complexity here? In the absence of an accepted theory of object recognition, no very precise meaning can be given to the term. However, for the limited purposes of discussing experiments on inferotemporal lesions, it could be said that stimuli are complex if more than simple brightness or contrast measurement are needed to tell them apart. In practice, three types of this kind of stimulus have been used to investigate the perceptual abilities of monkeys with inferotemporal ablation: relatively simple patterns, disks of varying size and distance, and random-dot stereograms.

Discrimination of Patterns

The patterns used by Blake et al. (1977) were inverted V's with different angles at the vertex. Monkeys had first to learn to pick one pattern with a 90° angle at the vertex in preference to another presented simultaneously with a 10° angle. The angle of the 10° stimulus was then expanded in steps so that an "angle threshold" could be measured. Animals with lesions in TE learned the original task, and reached threshold performance, more slowly than control monkeys, but their final threshold performance was indistinguishable from normal. However, what this result shows about angle discrimination in inferotemporally lesioned animals is uncertain. Other cues besides the angle between the two arms of the inverted V's were available—for example, the exact position on the stimulus plaque of a single arm. There was some evidence that the monkeys, both normal and operated, were using a peculiar cue, because their "angle thresholds" were much higher than a human observer's. (In most visual psychophysical tasks, macaques perform rather like humans.)

The problem of what cues the animals were using also cropped up in an experiment to measure reaction times for pattern discriminations in inferotemporally lesioned animals (Dean 1974a). The rationale for that experiment was that one measure of how discriminable two stimuli are is the time taken to choose one of them; if inferotemporal lesions impair the ability to tell patterns apart, choice reaction times for learned pattern discriminations should be lengthened. In fact, although monkeys with inferotemporal lesions took more trials to learn the pattern problems (which were discriminations between numerals: 4 vs. 2 and 5 vs. 9), they subsequently could perform them as fast as normals (unless the lesion strayed into tissue posterior to TE). An attempt was made to check on the possibility that the animals were using local cues (components of the patterns)—for example, the presence of a horizontal line at the bottom of the figure in the case of the 4 vs. 2 discrimination—to solve individual problems. They were given the two problems mixed together, so that on any trial the monkey would see either 4 (positive) and 2, 4 and 9, 5 (positive) and 2, or 5 and 9. The inferotemporally lesioned animals solved the mixed discrimination immediately, and their choice reaction times were again normal.

As in the experiment of Blake et al. (1977), patterns giving rise to slow learning in inferotemporally lesioned animals produced subsequent normal performance, and in addition there was some reason to believe that local cues were not being used. Thus, the evidence indicates that monkeys with inferotemporal lesions can tell simple patterns of these kinds apart as well as normal animals.

Size Constancy

It is clear that in some circumstances the brain can calculate the physical size of objects given their distance and retinal image size, but it is not known where in the brain the calculation takes place. To investigate this problem, Humphrey and Weiskrantz (1969) trained monkeys to discriminate two disks on the basis of their physical size when their distances varied, and then removed either parietal or inferotemporal cortex. The animals with inferotemporal lesions were unable to relearn the task in the time available; instead they adopted choice strategies based either on the retinal image size of the disks or on their distance. This evidence could be interpreted as showing either that monkeys with inferotemporal lesions were incapable of using image size and distance to calculate physical size or that (as the normal animals' learning scores indicated) the task was very difficult to learn under the conditions present and the inferotemporally lesioned animals were poor because of their learning impairment, which had nothing to do with size constancy as such (Dean 1976). Recent evidence supports the latter interpretation: Ungerleider et al. (1977) improved the apparatus so that normal animals found the task much easier to learn, and found that after subsequent inferotemporal removal all the animals relearned the task (although slowly) and none showed evidence of peculiar strategies.

Global Stereopsis

Recent evidence indicates that there is at least one kind of complex stimulus that animals with inferotemporal lesions are unable to tell apart (Cowey and Porter 1979). Monkeys wearing goggles with a red filter over one eye and a green filter over the other were trained to discriminate two red-green anaglyph forms of random-dot stereograms (Cowey et al. 1975). In both stereograms a central square appeared in depth when the stimuli were viewed with the goggles, but in one stereogram the square appeared in front whereas in the other it appeared behind. After the discrimination had been learned, the percentage of stereogram elements in binocular correspondence was varied in steps from 100% to 40%. Inferotemporal-cortex removal produced an apparently permanent defect in the ability to discriminate depth when binocular correspondence was reduced. Inferotemporally lesioned animals were also impaired when element size and disparity were reduced.

Conclusions

Experiments on the ability of monkeys with inferotemporal lesions to tell complex stimuli apart indicate that pattern and size-constancy problems can be handled normally, but that global stereopsis is impaired. These findings raise two important questions. The first is how far the global stereopsis defect can explain the difficulties that monkeys with inferotemporal lesions find with visual discriminations in general. Although the discrimination of three-dimensional objects might be impaired, it is hard to see why many other tasks that use essentially two-dimensional stimuli, such as colors or line drawings, should be affected. It seems either that inferotemporal lesions disrupt visual behavior in more than one way (perhaps by destroying more than one functional area in the brain) or that the global stereopsis defect is itself the result of some simpler fault that affects other kinds of discrimination training as well. Two candidates for such a fault are suggested below. The other question concerns the previously mentioned problem of what is meant by the complexity of a visual discrimination. This problem is exacerbated when the process underlying visual recognition is not understood. "Complexity" can be obtained in many different ways. Is it a good idea to continue testing monkeys with inferotemporal lesions on all the different sorts of complex discriminations that can be devised? If not, what reliable conclusions can be drawn about the capacity of inferotemporally lesioned animals to solve complex discriminations in general?

The general point made in connection with the whole topic of inferotemporally lesioned animals' sensory and perceptual capacities is worth repeating here: Since their learning impairment is found for all kinds of visual stimuli, a general, widespread perceptual defect would be needed to explain it. But the negative results quoted in this section show that no such general defect exists. There must be something else wrong after inferotemporal removal—for example, a fault in visual memory. Although inferotemporal ablation may cause some yet-undiscovered specific impairments of particular kinds of complex discrimination, it is hard to see how these impairments, like the one found with random-dot stereograms, would be able to explain the general learning difficulties also caused by the operation.

This state of affairs suggests that insofar as the cause of the inferotemporal learning impairment is the problem of interest, other capacities besides perceptual ones should be investigated. It is interesting to note in this context that two of the above experiments on complex discriminations in inferotemporally lesioned monkeys were not attempts to test explanations of their learning impairment: In the case of both size constancy and global stereopsis, the question was rather which parts of the brain are involved in those particular processes. It is likely that any other perceptual impairments that inferotemporal lesions may cause will be discovered in a similar way.

Remembering Visual Information

Besides telling the stimuli apart, another process that must be involved in visual discrimination learning is that of holding information in store from one trial to the next. If inferotemporal lesions make monkeys forget information faster than normal, their learning should be slow (though final asymptotic performance, as in a threshold task, might well be normal). The problem, then, is to measure how fast inferotemporally lesioned (and normal) monkeys forget visual information.

Delayed Matching from Sample

Perhaps the most popular task for investigating visual memory in animals is delayed matching from sample. The task comes in slightly different forms, but common to them all is that the animal is presented first with a "sample" stimulus, then, after a delay, with a stimulus identical to (matching) that sample together with one or more distractor stimuli. The animal is required to respond after the delay to the stimulus that matches the sample on that trial. For example, the sample stimulus might be a red light appearing behind a particular response key (the sample key). When the animal presses the sample key (a requirement often used to make sure the animal notices the sample) the red light goes off; 10 seconds later lights appear behind three new response keys, one red, one blue, and one green. If the animal presses the red key, it is rewarded. If one of the two other keys is pressed no reward is given. Often the incorrect or distractor stimuli that appear with the correct match at the end of the delay are themselves samples on other trials; in the example given, the sample on the next trial might be a green light, and on the trial after that a blue light, and so on. As might be expected, the proportion of trials on which the animal correctly chooses the stimulus that matches the sample declines as the delay is lengthened. If inferotemporal removal increases rate of forgetting, this decline should be steeper in animals with inferotemporal lesions.

An early experiment to investigate this question did indeed suggest that monkeys with inferotemporal lesions could not remember the sample for as long as normal monkeys, at least under some conditions (Buffery 1965). However, since other processes besides visual memory are required to solve delayed matching from sample, the effect of inferotemporal lesions on this task is ambiguous. In particular, the problem that arose in connection with psychophysical tasks arises again here: Perhaps the inferotemporally lesioned monkeys were bad at delayed matching from sample because they had not, in some sense, finished learning it.

Two ways to overcome this problem were tried (Dean 1974b). First, performance was tested at zero delay as well as at longer delays. In zero delay the match and distractor stimuli appear immediately when the sample is turned off, so that zero-delay performance is a measure of how well the basic delayed-matching task has been learned. It was thus possible to see whether inferotemporal removal worsened performance at long but not at short delays (which would indicate faster forgetting). Second, normal and control animals were tested until their performance, averaged over all delays, was relatively stable over 40 sessions, as an indication that learning had finished or was proceeding very slowly.

In other respects the task was fairly standard: The sample was one of four colors (red, yellow, blue, green), all of which could appear at the end of the delay. (On half of the trials only one distractor stimulus was used; since this factor had no effect on the results, it will not be mentioned further.) With the extensive training, some animals performed quite well before operation at delays of up to 50 seconds. However, despite the enormous amount of practice these animals received, inferotemporal removal had a drastic effect: The animals were scarcely able to perform the task after operation at any delay, including zero. (See also Wilson et al. 1972; Ibuka et al. 1974.) This was an interesting finding, since the colors used as samples are very easy for monkeys to discriminate, and because the retention of preoperatively overtrained *simultaneous* visual discriminations is apparently little affected by inferotemporal damage (Chow and Survis 1958; Orbach and Fantz 1958). It appears that some aspect of the delayed-matching-from-sample procedure is very sensitive to the effects of inferotemporal removal. However, since the animals were so bad at all delays immediately after operation, the importance of delay itself could not be investigated. To remedy this, the animals were laboriously retrained to a stringent criterion of performance at zero delay; then longer delays were tested. Preoperative performance at the longer delays was attained very rapidly, in contrast to both the amount of postoperative training necessary at zero delay and the time taken by the animals in learning to perform well at long delays before operation. It seemed that whatever aspect of delayed matching from sample was affected by the lesion, it was not the ability to hold visual information in memory once it had gotten there.

Other Tests of Forgetting
Normal rates of forgetting by monkeys with inferotemporal lesions have been found in other tests, which will only be mentioned briefly here (for fuller treatment, and references, see Dean 1976). The tests include delayed response (which measures among other things how well an animal can remember where it has seen food), memory for previously learned visual discriminations, and visual discrimination learning with manipulation of the intertrial interval. These last two tests are important because of yet another problem caused by ignorance about normal behavior. Perhaps the kind of memory used to remember samples in delayed matching from sample differs in some important way from the kind of memory used in visual discrimination learning (for one possibility see Dean 1976). If it did, normal forgetting in delayed matching from sample would be compatible with abnormal forgetting in discrimination learning itself. But it appears that in fact the lack of effect of inferotemporal removal on forgetting in delayed matching from sample is not restricted to that paradigm.

However, there is at least one circumstance in which animals with inferotemporal lesions do appear to forget rapidly, and that is when discrimination A is learned, followed by discrimination B, then retention of A is tested. Performance on the retention test is worse than normal, in terms of both initial score and number of trials to relearn (Iversen 1970; Passingham 1971; Dean and Cowey 1977). Animals with inferotemporal lesions seem to be abnormally vulnerable to retroactive interference (and possibly to proactive interference also; see references in Dean 1976).

Conclusions
The pattern of performance of monkeys with inferotemporal lesions on tests of forgetting resembles their pattern on tests of sensory or perceptual capacity: A particular set of conditions produces an impairment, although performance on several other tasks is normal. And it appears that, as with their global stereopsis defect, the vulnerability of inferotemporally lesioned animals to retroactive interference cannot easily be invoked to explain their difficulties with a wide variety of discrimination learning problems. In normal discrimination learning situations, retroactive interference is not deliberately provided. As before, this suggests two possibilities: that the vulnerability to interference identifies a further separate function of the cortex removed in inferotemporal lesions and that the basic cause of the learning defect also causes the rapid forgetting when interference is present.

The normal retention of the inferotemporal animals in the other tasks is at least discouraging for attempts to account for their poor learning in terms of rapid forgetting. Since the possibility cannot be ruled out that forgetting in discrimination learning differs in some important way from the forgetting studied in the tests that have been used, it is impossible to be more certain than this. However, the discouragement is sufficient to suggest that at this stage it may be more profitable to ask questions about other processes involved in discrimination learning. At least provisionally, the "leaky memory" interpretation of the inferotemporal learning impairment does not hold water.

Stimulus Categorization

The two examples just discussed illustrate the sort of questions that have been asked in the analytic stage about the effects of inferotemporal lesions. Other questions have been asked about the ability of inferotemporally lesioned animals to attend to visual stimuli, or to form associations between such stimuli and rewards or punishments. These will not be considered further here, beyond saying that at present no unequivocal evidence has emerged that those abilities are directly impaired by the lesion (Dean 1976; Mishkin and Oubré 1976). Though it would be premature to rule out *any* possibility of such an impairment, it seems that a better practical approach is to look for other kinds of ability that animals need to solve visual-discrimination problems, and that therefore might be impaired by inferotemporal removal.

Retrograde Amnesia

One possibility is suggested by the effects of inferotemporal ablation on visual discriminations learned before operation. Saving scores for simultaneous discriminations (unless overtrained before operation) are substantially reduced, often to near zero. In one experiment, where retraining of a series of object discriminations was compared with new learning of similar problems (Dean and Weiskrantz 1974), the inferotemporally lesioned animals learned both kinds of problem at the same speed, showing no signs of having encountered the preoperative problems before. Similarly, their performance on the first retention trial of problems taught prior to operation was no different from chance. Such a result could have occurred if the perceptual world of the inferotemporally lesioned animal was so altered that visual stimuli seen after operation bore no resemblance to the same stimuli seen before, but the evidence reviewed above provides no support for an alteration of this severity. One way that such a postoperative loss could occur is if the lesion destroyed the internal descriptions that the animals had formed of the discriminanda, so that descriptions had to be built up anew. Before this idea can produce testable predictions, some substance must be given to the notion of an internal description.

Internal Descriptions

It is clear introspectively, and can be demonstrated experimentally, that the full richness of our perceptual experience is not generally captured in memory. A simple example: Buying wool to mend a colored sweater will not produce a perfect match unless we take the sweater with us; delaying comparison between colors produces errors that are immediately obvious when the comparison is direct. What appears to be stored in memory is a simplified, impoverished description of the perceptual world. Presumably, simplification of this kind greatly reduces the amount of "unnecessary" information that is stored about an object or scene, with advantages for subsequent manipulation of that information.

"Categorization" has been suggested as a convenient label for this process of simplification (Dean 1976), but there are certainly others—for example, "labeling" or "classification"—that would be appropriate. A flavor of the categorization process can be gotten by considering the sort of description we would give of an unfamiliar fruit: for example, "quite long and thin, yellow, and curved." We do not and could not say exactly how long or precisely what shade of yellow the fruit was, and we would probably ignore small blemishes in its skin. Categorizing is at least logically separate from "identifying"; the above description could be given whether or not we knew the fruit was a banana.

It is presumably a simplified description of this kind—though nonverbal—that monkeys make of stimuli in the visual discriminations they have to learn. It is possible even that making *appropriate* descriptions is what takes most of the time in many discrimination tasks; otherwise, it is certainly puzzling why monkeys should take so many trials to solve tasks that seem perfectly trivial to us. In any event, it is conceivable that inferotemporal ablation disrupts discrimination by impairing categorization. Such an impairment would at least fit with the evidence so far provided, since abnormal internal descriptions need not affect detailed comparison of simultaneously presented stimuli, nor the rate at which information is forgotten (except perhaps in special circumstances). Categorization can be thought of as fitting in between perceptual discrimination and memory.

The results of two recent experiments are consistent with the idea that categorization is impaired by inferotemporal removal (Mishkin and Oubré 1976; Delacour 1977). Both used a variation of delayed matching from sample, which differed in three important ways from the task described above: The samples and distracting stimuli were objects, not colors; each sample object and distracting stimulus was presented for only one trial; and the correct response was to pick the distracting stimulus, not the object that matched the sample. The task was thus "trial-unique delayed non-matching from sample," and it was quite easy for normal monkeys to learn, at least under the conditions used (see Elliott et al. 1977). Nonetheless, as with standard delayed matching from sample, inferotemporal removal produced a severe impairment. The problem requires animals to produce an internal description of each sample sufficient to differentiate it from the subsequently presented distracting object, and the results could be interpreted as showing that the ability of monkeys to produce such a description (that is, to categorize the sample) is impaired by inferotemporal lesions. Other interpretations would be that the operated animals lacked the perceptual ability to tell sample and distractor apart, or that they forgot their description of the sample very rapidly; however, as has been seen, many of their perceptual abilities, and their forgetting of samples in only slightly different situations, are normal. These prior negative results increase the suitability of this task for testing categorizing ability in inferotemporally lesioned monkeys.

Possible Defects of Categorization

The supposition that removing inferotemporal cortex destroys an animal's machinery for categorizing visual stimuli might explain their lack of retention of discriminations encountered before operation, but an additional assumption is needed to explain the slow learning of new problems. If an animal lost its normal categorizing apparatus with inferotemporal removal, it would presumably then have to rely on some "backup" system (perhaps located in prestriate cortex; see Ettlinger et al. 1968). The additional assumption is that this backup system is less powerful in certain respects than the original. This weaker system has difficulty coping with a situation in which the stimulus to be categorized is seen only once; for this reason, it could be argued, inferotemporal lesions produce the defect found in trial-unique nonmatching. The problem is to test whether the categorizing abilities of inferotemporally lesioned monkeys are in fact less powerful than normal, given that, yet again, normal abilities of this kind are very poorly understood. Two approaches to this problem will be outlined.

Overinclusive Categories

One suggestion (Dean 1976) is that inferotemporally lesioned animals confuse visual stimuli under certain conditions because their categories are too broad, so that a description in terms of them conveys less information than the descriptions provided by the more precise categories of normal animals. This idea of broad or imprecise categories might explain why animals with inferotemporal lesions have particular difficulty learning discriminations between stimuli that are rather similar (Mishkin and Hall 1955). If it is assumed that because each category is broad there are fewer of them, it might also be possible to explain why inferotemporally lesioned animals have difficulty when they are required to learn several discriminations at once (Mishkin 1972).

Fewer Categories

The possibility that animals with inferotemporal lesions have fewer categories than normal animals was suggested by the results of experiments on humans using the "absolute judgment" paradigm (Miller 1956; Garner 1962; Corcoran 1971). In these experiments, subjects were shown stimuli one at a time and asked to judge which was which, for example by assigning each a number. Thus, if the stimuli were a series of tones differing in pitch, the highest might be stimulus 1, the next-highest stimulus 2, and so on. A rather curious fact emerged from these studies. If the stimulus set varied along only one dimension (for example, tones varying in pitch), subjects could reliably identify only a small number, around 7 (according to Miller 1956, 7 ± 2). This number is surprisingly unaffected by a number of variables, including practice (for references see Corcoran 1971), and is very much smaller than would be expected from performance with pairs of stimuli. Thus, two tones 5 Hz apart might be discriminably different, but still only 5–7 tones each 100 Hz apart can be reliably identified. Constraints also operate on people's ability to identify stimuli varying along more than one dimension. In general, performance is worse with two or more dimensions than would have been simply predicted from performance in each separately; for example, fewer than $7^2 = 49$ stimuli can be identified from a set varying independently along two dimensions (Corcoran 1971). These results are consistent with rather severe limitations on the information that internal descriptions can carry. On the assumption that there are similar limitations on the ability of monkeys to categorize visual stimuli, it is possible that inferotemporal removal reduces the number of stimuli that the system can handle. In Miller's terminology, the "magical number" of monkeys with inferotemporal lesions will be less than 7 ± 2. A little support for this idea comes from an experiment that used a variant of the matching from sample paradigm (Dean 1972). The matching and distractor cues always appeared in the same positions, so the animals had a choice between remembering the sample as a visual stimulus, or as its position in the response array. The samples were colors, and were increased in number from two to seven during training by the addition of one new color at each stage. The number of possible samples and available responses at the matching stage were always kept the same, and animals had to reach a criterion of successful performance, on both simultaneous and zero-delay conditions. Monkeys with inferotemporal lesions learned two and three alternative tasks as rapidly as controls, were slower but not significantly so for four and five alternatives, and were significantly slower for six and seven alternatives, although the control animals found that the six- and seven-sample tasks were easier to learn than the two- and three-alternative problems. In fact, two of the three animals with inferotemporal lesions were unable to perform reliably at criterion with seven alternatives.

This did not constitute a direct test of the "reduced magical number" idea, for a variety of reasons: The task could be solved by matching rather than absolute judgment; the colors varied along more than one dimension and were not spaced equally apart in discriminability (so that the inferotemporally lesioned animals confused particular pairs of colors); and performance was not measured before and after inferotemporal removal. However, the magnitude of the impairment in two of the animals with inferotemporal lesions in stages of the tasks that caused normal monkeys no particular problem was impressive and suggests that a proper test might well be worth carrying out, despite the daunting amounts of time and effort that might be required.

Imprecise Categories

The broadness or imprecision of individual categories is apparently rather easier to test. Evidence suggesting that the categories of animals with inferotemporal lesions are less precise than normal comes from a study by Butter et al. (1965), who in one experiment trained monkeys to discriminate between successively presented vertical and oblique gratings. The animals were then presented with gratings of intermediate orientations, and their generalization gradients were measured. These gradients were broader than normal in animals with inferotemporal lesions, though the effect failed to reach significance. Howlett (described in Weiskrantz 1974) improved the design by measuring gradients during overtraining of the basic task for long periods of time in the same animals before and after operation. He obtained clearly and significantly broader generalization gradients after inferotemporal ablation. Both these results indicate that monkeys with inferotemporal lesions are more likely to treat stimuli similar to the positive stimulus as if they were positive. This would be expected if the internal representation of the positive stimulus were less precise in such animals.

Although Howlett's results are suggestive for the imprecise-categorization hypothesis, they can be explained in other ways. Perhaps his animals decided whether a stimulus was vertical by comparing it not with some model in its head but with other vertical lines in the apparatus. Perhaps the animals with inferotemporal lesions were still *learning* to discriminate near-vertical from vertical gratings, and would have succeeded given more time. To eliminate these problems and provide a more direct estimation of category precision, a new task was devised in which the successive orientation thresholds of normal and inferotemporal monkeys were measured (Dean 1978).

Normal animals were first trained to discriminate two gratings presented one at a time, using an apparatus with three response keys in a straight line. When a grating at 45° to vertical was projected on the central response key, the animal had to press the left key for reward. If a horizontal grating appeared centrally, the right key gave reward. In the next stage thresholds were measured by altering the angle of the horizontal grating so that, depending on an animal's performance, it approached nearer and nearer to the 45° stimulus on a titration schedule. The 45° stimulus itself remained unchanged as the reference stimulus. It was not parallel to any other feature of the apparatus, to prevent simultaneous comparison. Animals were tested until their threshold performance reached stability over a 30-day period. After operation, the basic task was retaught, if necessary, and thresholds were remeasured. When stable performance was reached, the operated animals were given a period of "overtraining." Their thresholds were measured for a further 30 days or until they had completed as many sessions of threshold testing as they received before operation—whichever took the longer. This was to see whether any effects of surgery would be improved by very extensive practice.

The effects of inferotemporal lesions on successive orientation thresholds were clear. A significant increase was produced, about 75% for three animals and over 400% for a fourth. This rise in threshold was not reduced by overtraining and so, apparently, was not the product of slow learning. Nor did it appear to be a general consequence of task difficulty, since Weiskrantz and Cowey (1963) had measured acuity with an equally "difficult" psychophysical method and found a slight improvement after inferotemporal ablation. But in the Weiskrantz-Cowey experiment the animals could compare the stimuli to be discriminated, since these were presented simultaneously. In the orientation experiment such comparison was impossible. Perhaps, as implied by the imprecise-categorization hypothesis, comparability of stimuli and not general task difficulty was the important differences between the two experiments.

Scope and Generality of the Hypothesis
The results of the "successive orientation threshold measurement" and "generalization gradient" experiments were encouraging for the hypothesis of overinclusive categories, and suggested more detailed examination of what aspects of the inferotemporal learning defect it would in fact explain. An immediate problem is that inferotemporal lesions impair learning of simultaneous discriminations, in which the stimuli can be compared and the need for precise internal representation is reduced. However, it appears that normal monkeys often do not compare simultaneously presented discriminanda. At least in some circumstances they prefer instead to look at one stimulus position only, gradually learning to look at and respond to the other only if the first is incorrect (Oscar-Berman et al. 1971; Schrier and Wing 1973; Geary and Schrier 1975). Looking backward and forward between the stimuli occurs rather rarely. This tendency to treat simultaneous discriminations as if they were successive is, if anything, exaggerated in monkeys with inferotemporal lesions (Bagshaw et al. 1970; Oscar-Berman et al. 1974),

and emphasizes the importance of internal descriptions. The fact that comparison does occur when the stimuli in a learned simultaneous discrimination are made very similar, as in acuity measurement (Cowey 1961), may explain why inferotemporally ablated animals can reach normal performance on some threshold tasks: Comparison would remove the need for precise internal representations. On the other hand, it is possible that the threshold impairment in global stereopsis (Cowey 1978) arises because some feature of the unusual testing conditions prevents effective comparison.

There are three other results, otherwise somewhat anomalous, for which the hypothesis can offer explanations. Manning (1971) found that using electric shock as punishment for incorrect responses abolished the learning impairment for difficult pattern discriminations that typically followed inferotemporal lesions. It is feasible that electric shock, as it were, jolts the inferotemporally lesioned animal out of its lazy habit of not comparing the stimuli. Comparison may also be induced by the circumstances that Brody et al. (1977) found improved postoperative learning of inferotemporally lesioned monkeys: pattern stimuli presented in unpredictable locations. Finally, if descriptions of stimuli are imprecise they are more likely to be confused with descriptions formed subsequently, as in experiments on retroactive interference.

Since the hypothesis appears to have some explanatory power, it is important to establish what kinds of visual stimulus it applies to—especially as the inferotemporal learning defect is so general. Howlett (unpublished results) found broader-than-normal generalization gradients for the dimensions of size and brightness, but as yet no successive-threshold measurements have been carried out for these dimensions. However, successive thresholds have been measured for the dimension of hue (Dean 1979). Monkeys were presented with two identical Munsell cards in a Wisconsin General Test Apparatus. If the cards were a particular shade of red, the right one concealed reward; if they were of any other hue, the left card was rewarded. The initial discrimination was between red and green. Thresholds were then measured by varying the hue of the green stimulus on a titration schedule. Testing was continued until an animal's threshold performance remained stable over 20 sessions (1,200 trials).

Inferotemporal-cortex removal had a clear effect on the basic red-green discrimination: All four animals showed poor retention, and as a group they required slightly longer to relearn the task after operation than before (median: 1,680 trials). For comparison, no animal with a large prestriate-cortex lesion (intended to remove area V4 [Zeki 1973]) took more than 180 trials to relearn the task. However, despite their retention defect, the inferotemporal animals were able to reattain preoperative threshold performance.

This was an unexpected result, and not only because of the previous findings with orientation. Butter et al. (1965) found that inferotemporal lesions broadened generalization gradients for wavelength rather as they had for orientation, and Dean (1972) found that some animals with inferotemporal lesions failed to discriminate two similar reds used as samples in a delayed matching task even though they were given thousands of trials and the stimuli were not particularly difficult for normal animals. Both these experiments used light transmitted through response keys. Perhaps for monkeys with inferotemporal lesions colored lights and colored papers are importantly different. But it cannot be argued that inferotemporally ablated animals have no problem learning discriminations between colored papers, because Gross et al. (1971) have demonstrated that they are clearly impaired. The imprecise-categorization hypothesis must, therefore, predict raised thresholds for successively presented hues.

Perhaps the difference between the effects of inferotemporal lesions on successive hue and orientation thresholds can be explained by the apparently slight difference in procedure between the two experiments that measured them. The basic red-green discrimination was much easier for normal animals to learn than the 45°-versus-horizontal grating problem—indeed, the procedure was changed with this aim in mind. However, inferotemporal removal produced a striking loss for this easy part of the hue-discrimination problem, while hue thresholds (which were measured with a very similar titration schedule to the orientation thresholds and were thus equally "difficult") were not impaired. The poor retention for the basic hue discrimination was in some ways surprising (and recalls the similar loss for delayed matching from sample), since the stimuli used (red and green) are very easy for normal monkeys to discriminate and the task had been extensively overtrained before operation. It is certainly compatible with inferotemporal removal destroying the machinery normally used for generating internal descriptions of such stimuli. Yet when that machinery is reformed, its precision is apparently no less than normal.

It seems, then, that the dimensions of hue and of orientation are affected in different ways by inferotemporal-cortex removal, and that the learning impairments for those dimensions therefore have different explanations. Besides being displeasing from the point of view of parsimony, this argument leaves unsolved the problem of what explanation applies to the dimension of hue. Before pursuing these points, it is convenient to describe the other suggestion concerning a possible weakness in inferotemporally ablated animals' ability to categorize visual stimuli: that, rather than being too broad and therefore overinclusive, their categories are in one important respect too specific and therefore over*exclusive*.

Overexclusive Categories

It is important that internal descriptions of objects in the world be to some extent independent of where those objects are in space. Although it is necessary to know where things are to approach or avoid them, for purposes of recognizing what they are information about location is much less relevant. In fact, to recognize an object independent of which particular set of photoreceptors its image stimulates requires the "loss" of some kinds of position information. Rocha-Miranda et al. (1975) pointed out that inferotemporal cortex might be concerned with this process of "losing" position information so that images of the same object falling on different parts of the retina were treated as equivalent, since electrophysiological findings indicated that the receptive fields of neurones in inferotemporal cortex were large and often extended into the ipsilateral hemifield and that the optimum trigger features of these cells remained constant over their large receptive fields. Gross and Mishkin (1977) provided behavioral evidence to support this idea by showing that interocular transfer in monkeys with split optic chiasms was abolished by removal of inferotemporal cortex. Since in the split-chiasm animal each eye sees only one hemifield, this suggests that inferotemporal cortex is the part of the brain that brings together information from the two hemifields. Another experiment provided further, though indirect, support for this idea: Normal monkeys find patterns that are mirror images of each other particularly confusing, but monkeys with inferotemporal lesions do not (Gross et al. 1975)—in fact, with such patterns the inferotemporal impairment disappears. If patterns that are mirror images are always fixated centrally, the confusing parts will lie in the opposite hemifields. The fact that monkeys with inferotemporal lesions do not find such patterns especially confusing is consistent with their processing each hemifield separately. (However, see Gross 1978.) For what may be related effects in split-chiasm and split-brain monkeys, see (e.g.) Hamilton et al. 1973; Lehman and Spencer 1973.

These experiments provide no information about the ability of inferotemporally ablated animals to transfer information from one location to another *within* a hemifield, but certainly add plausibility to the suggestion that such transfer is deficient. If it were, the inferotemporal learning impairment could be explained, at least in part. On every trial the discriminanda form images on slightly different parts of the retina. This does not matter to normal animals, since they are able to extract from the image a position-independent internal description. Monkeys with inferotemporal lesions, however, extract internal descriptions that are tied to the position of the image and therefore vary from trial to trial. The inferotemporally ablated animal is thus, in effect, trying to learn many discriminations while the normal animals is learning one, and is (not surprisingly) slower.

This explanation has at least two difficulties to overcome. The first is that monkeys appear to fixate at least certain discriminanda with surprising accuracy (Hamilton et al. 1973; Lehman and Spencer 1973)—accuracy that should be sufficient to bring the stimulus on every trial within receptive fields of the size found in parts of prestriate cortex

(Gross et al. 1972). Second, stimuli that are the same over substantial areas of the visual field—such as hues or gratings—should on this hypothesis pose little problem for monkeys with inferotemporal lesions, but the evidence is clear that such stimuli do cause impairment for inferotemporally ablated monkeys. Thus, even if subsequent experiments were to show that inferotemporally ablated animals did not recognize images from one retinal location presented in another, it is far from clear that this would explain the inferotemporal learning impairment in its entirety.

Unsolved Problems

Both suggestions about categorization impairments after inferotemporal-cortex removal have some evidence in their favor, but neither appears to be a full explanation of the discrimination-learning data. This suggests that perhaps inferotemporal removal causes both categorization defects—possibly by destroying two separate functional areas, one of which brings together information from all over the visual field and the other of which then reads out a description of that information. If both areas are destroyed, the animal describes visual stimuli too precisely in respect to their position, but not precisely enough in terms of their orientation, size, and brightness. Consequently, the learning of a wide range of visual discriminations will be impaired.

This agreeable compromise faces difficulties. One is that neither component hypothesis has yet been adequately tested. As already said, transfer between sites within a hemifield in inferotemporally ablated monkeys has not yet been examined—a crucial point for the "overexclusive categorization" hypothesis. And it is very important for the "opposite" hypothesis that although successive orientation thresholds are impaired, simultaneous orientation thresholds when the stimuli are compared should be normal. Current evidence on this point is inconclusive. I have found (unpublished results) that when the animals in the successive-orientation experiment were transferred to a simultaneous discrimination (designed so that the stimuli *had* to be compared), three out of four inferotemporally ablated animals improved, whereas the control animals (with foveal prestriate or lateral striate lesions) did not. However, unoperated animals were not tested. Blake et al. (1977) found normal "angle" thresholds in inferotemporal monkeys (see above), but it is not clear which cue the monkeys used, since their performance was much worse than that of human observers.

One reason why it is particularly important to measure simultaneous orientation thresholds properly is to assess the following possible explanation of the successive orientation impairment. Gross and Mishkin (1977) indicated that monkeys with inferotemporal lesions processed the two hemifields independently. If the inferotemporally ablated monkeys in the successive orientation discrimination had fixated the gratings centrally, they would have had to judge the orientation of the right and left halves of the gratings separately. Each judgment of orientation would then have been based on a stimulus half as long as that used by the normal animals. Evidence from humans suggests that a restriction of this kind would raise orientation thresholds, since accuracy of orientation perception in humans increases with line length (Andrews 1967). It is also conceivable that the global stereopsis problem used by Cowey (1978) was affected by independent processing of the two hemifields, since once again the inferotemporally ablated animal had in effect to deal with a stimulus of smaller area, giving a reduced sample of points for disparity computation. On the other hand, the discrimination of hues might not be much impaired by independent processing of the hemifields unless the stimuli were very small. If Gross and Mishkin's evidence did account for the results of these three threshold experiments, an important puzzle—why inferotemporal lesions have different effects on different dimensions—would be cleared up.

However, even if *both* the categorization hypotheses did survive more testing, a further problem would still have to be solved: Animals with inferotemporal lesions are slow to learn discrimination between different hues (Gross et al. 1971), and between gratings of different orientations even when the orientation difference is as large as 45° (Butter et al. 1965). It is not clear that either hypothesis can explain these impairments. Colored papers and gratings stimulate relatively large areas of the retina in similar fashion, reducing the problems of stimulus equivalence: Successive hue thresholds are unaltered by inferotemporal ablation, and although successive orientation thresholds were increased by the operation, three out of four monkeys with inferotemporal lesions were able to discriminate gratings 12° apart. It appears that there has to be yet another way in which inferotemporal lesions disrupt visual discrimination. One possibility is suggested by the results of the successive-hue-discrimination experiment. It seemed that inferotemporal ablation destroyed the animal's normal machinery for categorizing hues, but that the backup machinery was capable of a similar degree of precision once it was functioning properly. However, it might be that in the backup system new categories take *longer* to form. A suggestion similar to this one was made by Butter et al. (1965) when they speculated about the role of inferotemporal cortex in the "learned identification of visual stimuli." It seems reasonable that in the course of learning, as a stimulus becomes familiar, a particular internal representation of it is set up. This setting up may occur in inferotemporal cortex. But what is meant by "setting up of internal representations" here? Again, there is an urgent need for better understanding of normal vision before ideas about effects of inferotemporal lesions can be formulated properly. Even if the process could be described precisely, how could the rate of formation of such representations be measured experimentally? Such questions seem to be much more hampered by ignorance about normal vision than questions about forgetting stimuli or telling them apart. This may be a measure of the progress that has been made in the study of inferotemporal lesions.

Conclusions

Functions of Inferotemporal Cortex

Deducing the functions of inferotemporal cortex from the effects on behavior of removing it faces the serious problem that the behavior affected (visual discrimination learning) is very poorly understood. We know little for certain about the processes that are involved in such learning, or, in many cases, what are suitable tasks to investigate them. Ways that have been used in attempts to carry out an effective diagnostic procedure in such circumstances were described above. Some of the progress that has been made in the diagnosis can be very briefly summarized as follows.

• As a result of trying to answer general questions about perception and memory in monkeys with inferotemporal lesions, a number of possible explanations of the impairment have been rendered at least less plausible.

• A heterogeneous collection of new symtpoms have been discovered. Monkeys with inferotemporal lesions have been found to have apparently specific impairments connected with global stereopsis, retroactive interference, successive presentation of stimuli, and interhemifield transfer.

• New, more specific questions are now being asked—in particular, about the way in which monkeys with inferotemporal lesions form internal descriptions of visual stimuli—and some attempts to answer those questions have already been made. At present there is some evidence suggesting that, in monkeys with inferotemporal lesions, internal descriptions are less precise than normal for certain aspects of visual stimuli (for example, orientation) and overprecise about others (in particular, the location of their images on the retina). These faults could explain some aspects of the inferotemporal learning impairment, and perhaps some or all of the new symptoms.

What implications do these findings have for our understanding of the functions of inferotemporal cortex? If the above suppositions are correct, it appears that one function of inferotemporal cortex is to extract from the rich "perceptual" representation of the visual world (provided by prestriate cortex?) a much briefer symbolic description suitable for general-purpose manipulation (in, for example, memory and thinking). This description is then handed on to other areas of the brain that are involved in such manipulations. Inferotemporal cortex may thus be indirectly concerned in other kinds of visual processing in memory (where efficient descriptions ensure better storage) and attention (where much of the rich perceptual information has to be thrown away), as well as in learning itself.

It was argued that how well the functions of an area of the brain could be characterized from studying lesion effects depended very closely on how much was known about the organization of the behaviors disrupted by removing the area. The above speculation, if nothing else, illustrates that dependence. What sorts of processes are involved in "extracting a brief symbolic description from a rich perceptual representation"? What behavioral tasks can be used to investigate those processes? It may well be necessary to pursue questions like these to make further progress in deducing the functions of inferotemporal cortex from the effects of removing it.

Inferotemporal Lesions as a Tool for Analyzing Visual Behavior

Because the systems that produce behavior are in certain important respects relatively independent or modular, removing parts of the brain can produce interesting decompositions of normal behavior that can provide information about the nature and arrangement of the underlying component modules. This has been true in the case of inferotemporal lesions and visual discrimination learning. At a gross level, studying the effects of inferotemporal lesions has focused attention on the complexity of discrimination learning. Such learning clearly involves many component processes. This is perhaps not such a trivial point as it may appear, since not so long ago two-process or even one-process models were offered as serious simulations of discrimination learning. More specifically, the inferotemporal impairment indicates that some aspect of learning to discriminate visual stimuli on the basis of their appearance is relatively independent of nonvisual learning, of knowing where the stimuli are, of being able to make fine discriminations, and of holding in memory certain kinds of visual information. Suggestions about how these modules (and others) are arranged have been obtained by comparing the effects of removing inferotemporal cortex with other lesions to the visual system. For example, lesions to foveal striate cortex produce field and acuity defects, but have less effect on learning itself. This evidence indicates that there is not a simple serial arrangement "striate cortex–prestriate cortex–inferotemporal cortex," since information about simple stimulus attributes somehow "leaks out." Other investigations of the effects of removing areas that projects to inferotemporal cortex or receive projections from it have been equally informative (see, for example, Iwai and Mishkin 1969; Cowey and Gross 1970; Jones and Mishkin 1972; Mishkin 1972; Mishkin and Oubré 1976).

However, it is important to mention some limitations of this approach to analyzing normal behavior. One is that destroying relatively large areas of the brain is unlikely to reveal much about *how* a particular module works. Another is that productive questions about lesion effects are often generated not directly by the nature of the effects themselves, but by other techniques and disciplines, such as cognitive psychology, artificial intelligence, or learning theory. As Weiskrantz (1968) pointed out, in one sense this matters not at all; the lesion technique is but one weapon used in a more or less orchestrated assault, from different directions, on the problems of brain and behavior. But from the point of view of understanding normal behavior, as opposed to which bit of the brain does what, the lesion technique often seems uncomfortably parasitic and inefficient in comparison with other methods. Particularly, at the moment, it is at least possible that where visual behavior is concerned advances in understanding the effects of visual-system lesions will depend more on advances in areas such as cognitive psychology and artificial intelligence than the other way round.

References

Andrews, D. P. 1967. "Perception of contour orientation in the central fovea. II. Spatial integration." *Vision Res.* 7: 999–1013.

Bagshaw, M. H., N. H., Mackworth, and K. H. Pribram. 1970. "The effect of inferotemporal cortex ablations on eye movements of monkeys during discrimination training." *Int. J. Neurosci.* 1: 153–158.

Blake, L., C. D. Jarvis, and M. Mishkin. 1977. "Pattern discrimination thresholds after partial inferior temporal or lateral striate lesions in monkeys." *Brain Res.* 120: 209–220.

Brody, B. A., L. G. Ungerleider, and K. H. Pribram. 1977. "The effects of instability of the visual display on pattern discrimination learning by monkeys: Dissociation produced after resections of frontal and inferotemporal cortex." *Neuropsychologia* 15: 439–448.

Buffery, A. W. H. 1965. "Attention and retention following frontal and temporal lesions in the baboon." In Proceedings of the 73rd Annual Convention of the American Psychological Association. Washington, D.C.: APA.

Butter, C. M., M. Mishkin, and H. E. Rosvold. 1965. "Stimulus generalization in monkeys with inferotemporal and lateral occipital lesions." In *Stimulus Generalization*, D. I. Mostofsky, ed. Stanford, Calif.: Stanford University Press.

Chow, K. L., and J. Survis. 1958. "Retention of overlearned visual habit after temporal cortical ablation in monkey." *Arch. Neurol. Psychiatr.* 79: 640–646.

Corcoran, D. W. J. 1971. *Pattern Recognition*. Harmondsworth, Middlesex: Penguin.

Cowey, A. 1961. "Discrimination in monkeys." Doctoral thesis, Oxford University.

———. 1963. "The basis of a method of perimetry with monkeys." *Q. J. Exp. Psychol.* 15: 481–490.

———. 1973. "Brain damage and seeing: A new look at some old problems." *Trans. Ophthal. Socs. U.K.* 93: 409–416.

Cowey, A., and C. G. Gross. 1970. "Effects of foveal prestriate and inferotemporal lesions on visual discrimination by rhesus monkeys." *Exp. Brain Res.* 11: 128–144.

Cowey, A., and J. Porter. 1979. "Brain damage and global stereopsis." *Proc. Roy. Soc.* B 204: 399–407.

Cowey, A., and L. Weiskrantz. 1967. "A comparison of the effects of inferotemporal and striate cortex lesions on the visual behavior of rhesus monkeys." *Q. J. Exp. Psychol.* 15: 91–115.

Cowey, A., A. M. Parkinson, and L. Warnick. 1975. "Global stereopsis in rhesus monkeys." *Q. J. Exp. Psychol.* 27: 93–109.

Dean, P. 1972. "Functions of the temporal lobe in visual discrimination and memory in *Macaca mulatta*." Doctoral thesis, Oxford University.

——. 1974a. "Choice reaction times for pattern discriminations in monkeys with inferotemporal lesions." *Neuropsychologia* 12: 465–476.

——. 1974b. "The effect of inferotemporal lesions on memory for visual stimuli in rhesus monkeys." *Brain Res.* 77: 451–469.

——. 1976. "Effects of inferotemporal lesions on the behavior of monkeys." *Psychol. Bull.* 83: 41–71.

——. 1978. "Visual cortex ablation and thresholds for successively presented stimuli in rhesus monkeys. I. Orientation." *Exp. Brain Res.* 32: 445–458.

——. 1979. "Visual cortex ablation and thresholds for successively presented stimuli in rhesus monkeys. II. Hue." *Exp. Brain Res.* 35: 69–83.

——. 1980. "Recapitulation of a theme by Lashley? Comment on Wood's simulated lesion experiments." *Psych. Rev.* 87: 470–473.

Dean, P., and A. Cowey. 1977. "Inferotemporal lesions and memory for pattern discriminations after visual interference." *Neuropsychologia* 15: 93–98.

Dean, P., and L. Weiskrantz. 1974. "Loss of preoperative habits in rhesus monkeys with inferotemporal lesions: Recognition failure or relearning deficit?" *Neuropsychologia* 12: 299–311.

Delacour, J. 1977. "Role of temporal lobe structures in visual short-term memory, using a new test." *Neuropsychologia* 15: 681–684.

Desimone, R., and C. G. Gross. 1979. "Visual areas in the temporal cortex of the macaque." *Brain Res.* 178: 363–380.

Elliott, R. C., E. Norris, G. Ettlinger, and M. Mishkin. 1977. "Some factors influencing nonmatching to sample in the monkey." *Bull. Psychon. Soc.* 9: 395–396.

Ettlinger, G., E. Iwai, M. Mishkin, and H. E. Rosvold. 1968. "Visual discrimination in the monkey following serial ablation of inferotemporal and preoccipital cortex." *J. Comp. Physiol. Psychol.* 65: 110–117.

Garner, W. R. 1962. *Uncertainty and Structure as Psychological Concepts.* New York: Wiley.

Geary, N. D., and A. M. Schrier. 1975. "Eye movements of monkeys during performance of ambiguous cue problems." *Animal Behav.* 3: 167–171.

Gregory, R. L. 1961. "The brain as an engineering problem." In *Current Problems in Animal Behaviour,* W. H. Thorpe and O. L. Zangwill, eds. Cambridge University Press.

Gross, C. G. 1973a. "Inferotemporal cortex and vision." In *Progress in Physiological Psychology,* vol. 5, E. Stellar and J. M. Sprague, eds. New York: Academic.

————. 1973b. "Visual functions of inferotemporal cortex." In *Handbook of Sensory Physiology,* vol. 7, part 3B: Central Processing of Visual Information, R. Jung, ed. Berlin: Springer.

————. 1978. "Inferior temporal lesions do not impair discrimination of rotated patterns in monkeys." *J. Comp. Physiol. Psychol.* 92: 1095–1109.

Gross, C. G., and M. Mishkin. 1977. "The neural basis of stimulus equivalence across retinal translation." In *Lateralization in the Nervous System.* S. Harnad et al. eds. New York: Academic.

Gross, C. G., A. Cowey, and F. J. Manning. 1971. "Further analysis of visual discrimination deficits following foveal prestriate and inferotemporal lesions in monkeys." *J. Comp. Physiol. Psychol.* 76: 1–7.

Gross, C. G., C. E. Rocha-Miranda, and D. M. Bender. 1972. "Visual properties of neurons in inferotemporal cortex of the macaque." *J. Neurophysiol.* 35: 96–111.

Gross, C. G., M. Lewis, and D. Plaisier. 1975. "Inferior temporal cortex lesions do not impair discrimination of lateral mirror images." *Soc. Neurosci. Abstr.* 1: 74.

Hamilton, C. R., S. B. Tieman, and H. L. Winter. 1973. "Optic chiasm section affects discriminability of asymmetric patterns by monkeys." *Brain Res.* 49: 427–431.

Hebb, D. O. 1938. "Studies of the organization of behavior. I. Behavior of the rat in a field orientation." *J. Comp. Psychol.* 25: 333–351.

Humphrey, N. K., and L. Weiskrantz. 1969. "Size constancy in monkeys with inferotemporal lesions." *Q. J. Exp. Psychol.* 21: 225–238.

Ibuka, N., K. Kubota, and E. Iwai. 1974. "Ablation of a small circumscribed portion of the inferotemporal cortex and a delayed matching-to-sample task." *Abstr. Fifth Cong. Internat. Primatol. Soc.,* pp. 67–68.

Iwai, E., and M. Mishkin. 1969. "Further evidence on the locus of the visual area in the temporal lobe of the monkey." *Exp. Neurol.* 25: 586–594.

Iversen, S. D. 1970. "Interference and inferotemporal memory deficits." *Brain Res.* 19: 277–289.

Jones, B., and M. Mishkin. 1972. "Limbic lesions and the problem of stimulus-reinforcement associations." *Exp. Neurol.* 36: 362–377.

Kaas, J. 1978. "The organization of visual cortex in primates." In *Sensory Systems of Primates,* C. R. Noback, ed. New York: Academic.

Lashley, K. S. 1950. "In search of the engram." *Symp. Soc. Exp. Biol.* no. 4: 454–482.

Laurence, S., and D. G. Stein. 1978. "Recovery after brain damage and the concept of localization of function." In *Recovery from Brain Damage,* S. Finger, ed. New York: Plenum.

Lehman, R. A. W., and D. D. Spencer. 1973. "Mirror-image shape discrimination: Interocular reversal of responses in the optic chiasm sectioned monkey." *Brain Res.* 52: 233–241.

Manning, R. F. J. 1971. "Punishment for errors and visual-discrimination learning by monkeys with inferotemporal cortex lesions." *J. Comp. Physiol. Psychol.* 75: 146–152.

Marr, D. 1976. "Early processing of visual information." *Phil. Trans. Roy. Soc.* B 275: 483–524.

Marr, D., and T. Poggio. 1976. "From understanding computation to understanding neural circuitry." *Neurosci. Res. Prog. Bull.* 15: 470–488.

Meadows, J. C. 1974. "Disturbed perception of colours associated with localized cerebral lesions." *Brain* 97: 615–632.

Miller, G. A. 1956. "The magical number seven, plus or minus two: Some limits on our capacity for processing information." *Psychol. Rev.* 63: 81–97.

Mishkin, M. 1954. "Visual discrimination performance following partial ablations of the temporal lobe. II. Ventral surface versus hippocampus." *J. Comp. Physiol. Psychol.* 47: 187–193.

———. 1966. "Visual mechanisms beyond the striate cortex." In *Frontiers of Physiological Psychology,* R. W. Russell, ed. New York: Academic.

———. 1972. "Cortical visual areas and their interactions." In *The Brain and Human Behaviour,* A. G. Karczmar and J. C. Eccles, eds. Berlin: Springer.

Mishkin, M., and M. Hall. 1955. "Discrimination along a size continuum following ablation of the inferior temporal convexity in monkeys." *J. Comp. Physiol. Psychol.* 48: 97–101.

Mishkin, M., and J. L. Oubré. 1976. "Dissociation of deficits on visual memory tasks after inferior temporal and amygdala lesions in monkeys." *Soc. Neurosci. Abstr.* 2: 1127.

Mishkin, M., and K. H. Pribram. 1954. "Visual discrimination performance following partial ablations of the temporal lobe. I. Ventral versus lateral." *J. Comp. Physiol. Psychol.* 47: 14–20.

Mishkin, M., and L. Weiskrantz. 1959. "Effects of cortical lesions in monkeys on critical fusion frequency." *J. Comp. Physiol. Psychol.* 52: 660–666.

Orbach, J., and R. L. Fantz. 1958. "Differential effects of temporal neocortical resections on overtrained and non-overtrained visual habits in monkeys." *J. Comp. Physiol. Psychol.* 51: 126–129.

Oscar-Berman, M., S. P. Heywood, and C. G. Gross. 1971. "Eye orientation during visual discrimination learning by monkeys." *Neuropsychologia* 9: 351–358.

———. 1974. "The effects of posterior cortical lesions on eye orientation during visual discrimination by monkeys." *Neuropsychologia* 12: 175–182.

Passingham, R. E. 1971. "Behavioural changes after lesions of frontal granular cortex in monkeys *(Macaca mulatta)*." Doctoral thesis, University of London.

Rocha-Miranda, C. E., D. B. Bender, C. G. Gross, and M. Mishkin. 1975. "Visual activation of neurons in inferotemporal cortex depends on striate cortex and forebrain commissures." *J. Neurophysiol.* 38: 375–491.

Schiller, P. H., Stryker, M., Cynader, M., and Berman, N. 1974. "Response characteristics of single cells in the monkey superior colliculus following ablation or cooling of visual cortex." *J. Neurophysiol.* 37: 181–194.

Schneider, G. E. 1969. "Two visual systems. Brain mechanisms for localization and discrimination are dissociated by tectal and cortical lesions." *Science* 163: 895–902.

Schrier, A. M., and T. C. Wing. 1973. "Eye movements of monkeys during brightness discrimination and discrimination reversal." *Animal Learning Behav.* 1: 145–150.

Seltzer, B., and D. N. Pandya. 1978. "Afferent cortical connections and architectonics of the superior temporal sulcus and surrounding cortex in the rhesus monkey." *Brain Res.* 149: 1–24.

Simon, H. A. 1969. *The Sciences of the Artificial.* Cambridge, Mass.: MIT Press.

Sutherland, N. S., and N. J. MacKintosh. 1971. *Mechanisms of Animal Discrimination Learning.* New York: Academic.

Symmes, D. 1965. "Flicker discrimination by brain-damaged monkeys." *J. Comp. Physiol. Psychol.* 60: 470–473.

Ungerleider, L. G., L. Ganz, and K. H. Pribram. 1977. "Size constancy in rhesus monkeys: Effects of pulvinar, prestriate, and inferotemporal lesions." *Exp. Brain Res.* 27: 251–269.

Ungerstedt, U. 1971. "Stereotaxic mapping of the monoamine pathways in the rat brain." *Acta. Physiol. Scand.* suppl. 367: 1–48.

von Bonin, G., and P. Bailey. 1947. *The Neocortex of Macaca Mulatta.* Urbana: University of Illinois Press.

Weiskrantz, L. 1968. "Treatments, inferences, and brain function." In *Analysis of Behavioral Change,* L. Weiskrantz, ed. New York: Harper and Row.

———. 1974. "The interaction between occipital and temporal cortex in vision: An overview." In *The Neurosciences: Third Study Program,* F. Schmitt and F. G. Worden, eds. Cambridge, Mass.: MIT Press.

Weiskrantz, L., and A. Cowey. 1963. "Striate cortex lesions and visual acuity of the rhesus monkey." *J. Comp. Physiol. Psychol.* 56: 225–231.

Weiskrantz, L., E. K. Warrington, M. D. Sanders, and J. Marshall. 1974. "Visual capacity in the hemianopic field following a restricted occipital ablation." *Brain* 97: 709–728.

Wilson, M., H. M. Kaufman, R. E. Zierler, and J. P. Lieb. 1972. "Visual identification and memory in monkeys with circumscribed inferotemporal lesions." *J. Comp. Physiol. Psychol.* 78: 173–183.

Zeki, S. M. 1973. "Colour coding in rhesus monkey prestriate cortex." *Brain Res.* 53: 422–427.

———. 1978. "Uniformity and diversity of structure and function in rhesus monkey prestriate visual cortex." *J. Physiol.* 277: 273–290.

20 Visual Agnosia in Monkey and in Man

David N. Levine

In recent years, there have been few attempts to relate disorders of visual recognition in man to visual disturbances resulting from circumscribed cerebral ablations in animals. Early workers (see Lissauer 1890) had no hesitation in attempting to relate their observations in patients to experimental results. Indeed, the term *Seelenblindheit* (psychic blindness) that was used to describe the syndrome known today as associative visual agnosia was taken directly from Munk's (1881) description of the behavior of dogs after bilateral occipital lesions. In the ensuing decades, however, the laboratory and the clinic drifted apart in the study of cerebral disturbances of vision. Perhaps part of the reason for this drift lay with Lissauer himself. In choosing to give a psychological explanation for his well-defined syndrome he opened the door to a controversy that continues to this day among clinicians as to the psychological nature of visual agnosia. Whether "perception" can be intact but have no "meaning" has been the focus of this controversy and has diverted attention from the study of the comparative anatomy and comparative behavior of the syndrome.

Geschwind (1965) discussed "agnosias" in animals and in man, and pointed out that one cannot understand visual agnosias in man without considering the specialization of the left hemisphere for speech and the effects of disconnecting the nonverbal right hemisphere from the speaking left. He concluded that "visual agnosias" in monkey and in man had entirely different anatomical and psychological substrates. "Visual agnosia" in the monkey was conceived as a disconnection of the calcarine cortex from the "limbic" regions anatomically and as the loss of visual-visceral (olfactory, gustatory) associations psychologically. Visual agnosia in man was conceived as a disconnection of the calcarine cortex from the angular gyrus of the dominant hemisphere anatomically and as loss of visual-language associations psychologically.

The purpose of this chapter is to reexamine the degree to which the syndromes of visual agnosia in man parallel the syndromes of impaired visual discrimination in the monkey. Given Geschwind's cogent arguments, it would be unwise to expect complete correspondence, because the monkey lacks speech and its underlying lateralized cerebral substrate. Nevertheless, I will attempt to show that there are remarkable parallels both in behavioral abnormalities and in pathological anatomy between the visual agnosias in monkey and in man. In both primates, similar impairments in visual discrimination result from interruption of an interconnected system of neurons in striate cortex, visual association cortex, and inferior temporal neocortex. In both man and monkey, these syndromes can be distinguished from those involving other cerebral areas. The appearance of hemispheric specialization in man results in a greater variety of visual-discrimination deficits than are seen in the monkey, but this greater variety can be seen as a differentiation in man of a common behavioral and anatomical pathology that is shared with the monkey.

The Anatomic Basis of Impaired Visual Discrimination in the Monkey

Munk (1881) performed bilateral occipital ablations in dogs. The animals appeared to see in that they walked about without colliding with obstacles. But they did not otherwise behave discriminatively toward visual stimuli, no longer taking food, no longer cringing from fire, and remaining indifferent to the sight of their master though responding to his voice.

Brown and Schafer (1888) were the first to produce a similar picture in the primate. After removal of both temporal lobes, a rhesus monkey oriented its body and limbs properly to visually presented objects, but appeared tame, allowing itself to be handled. It approached and investigated by feeling, tasting, and smelling all sorts of objects, animate or inanimate, edible or inedible. This syndrome was largely forgotten until the experiments of Klüver and Bucy (1937, 1939) confirmed that after bilateral temporal lobectomy monkeys were tame, compulsively examined every visual object in the environment ("hypermetamorphosis"), and tended to chew, lick, or sniff all such objects (oral tendencies). In addition, the monkeys would often eat meat, which a normal monkey will not do, and displayed excessive indiscriminate (auto-, hetero-, and homo-) sexual behavior. Although visual acuity appeared intact, as did orientation of the body and limbs to visual objects, these monkeys had great difficulty (but eventual success) in learning to discriminate visually a circle from a square and also in retaining such discriminations that had been learned before the operation.

Numerous studies followed in an attempt to delimit more precisely the lesions necessary to produce the syndrome. Early studies (Blum et al. 1950; Chow 1951, 1952) demonstrated that impaired acquisition and retention of visual discriminations resulted from lesions involving only neocortex of temporal lobe. Shortly thereafter, studies with less extensive lesions demonstrated that ablation of the neocortex of the inferior temporal lobe, coinciding roughly with von Bonin and Bailey's (1947) area TE, would produce the visual-discrimination deficits, but that lesions of superior temporal gyrus (Mishkin and Pribram 1954), hippocampus, and amygdala (Mishkin 1954) or of temporal pole, anterior insula, and orbital frontal cortex (Pribram and Bagshaw 1953) would not. The inferotemporal lesions, while producing defects in acquiring and retaining visual discriminations, did not as a rule produce the other manifestations of the "Klüver-Bucy syndrome"—the tameness, hypermetamorphosis, oral tendencies, and changes in dietary habits or sexual behavior.

Once it was shown that lesions of inferior temporal neocortex (TE) were crucial in producing impaired acquisition and retention of visual discrimination, further studies attempted to determine which anatomical connections to this area were necessary for adequate function of this region. It seemed possible, on the one hand, that cortico-cortical connections (striate cortex → visual association cortex → inferotemporal cortex) were important; on the other hand, cortico-subcortical relays (striate cortex → pulvinar → inferotemporal cortex) or extrageniculostriate pathways (retina → superior colliculus → pulvinar → inferotemporal cortex) might be crucial. Mishkin (1966) demonstrated the importance of cortico-cortical connections. He ablated inferotemporal cortex in one hemisphere and striate cortex in the other. The monkeys were somewhat impaired in retaining visual discriminations but could be retrained. He then sectioned the corpus callosum, destroying the only remaining cortico-cortical connection between striate and inferotemporal cortex. Severe impairment in retention of visual discriminations ensued. Thus Mishkin demonstrated the importance of striate-inferotemporal connections, both ipsilateral and crossed (via corpus callosum) in ensuring adequate visual discrimination performance in the monkey.

Mishkin's results heightened a paradox that had existed for almost two decades. Modern studies employing the Nauta technique (Kuypers et al. 1965; Jones and Powell 1970) have confirmed earlier conclusions, based on strychnine neuronography, that striate cortex projects to inferotemporal cortex not directly but through visual-association cortex. Thus, if cortico-cortical pathways from striate to inferotemporal regions were necessary for intact visual discrimination, impairment should result from lesions of visual-association cortex as well. Yet earlier investigators (Lashley 1948; Chow 1951, 1952; Evarts 1952) had ablated extensive areas of visual-association cortex in the monkey without producing impairment in the acquisition of visual form discriminations. This failure had prompted either of two alternative conclusions: that cortico-subcortical loops were more important than cortico-cortical connections and capable of maintaining normal visual discrimination when cortico-cortical connections were interrupted (Pribram 1960); or that cortico-cortical paths were indeed important, but that association cortex was equipotential and highly redundant, so that even small areas remaining after large ablations could subserve the acquisition of simple visual discriminations (Lashley 1948). Though experiments such as Mishkin's demonstrated the insufficiency of cortico-subcortical pathways alone, only recently has sufficient anatomic information about the organization of visual-association cortex become available to offer an alternative to Lashley's conclusion.

Investigations (Iwai and Mishkin 1968; Cowey and Gross 1970) taking advantage of new information about the anatomic organization of visual-association cortex have contributed to the resolution of the above paradox. This anatomic information was derived from studies employing the Nauta technique and its variations to trace fiber pathways after focal occipital ablations (Myers 1965a; Kuypers et al. 1965; Zeki 1969, 1970, 1971a, b) and from studies employing microelectrodes to map visual areas by determining receptive fields of single neurons (Allman and Kaas 1971, 1974a, b, 1975, 1976). These studies have demonstrated that visual-association cortex is not a mass of equivalent neurons in which visual information is represented diffusely. Instead, it appears to consist of numerous, geographically distinct representations of the visual fields with complex interconnections.

A detailed review of the results of these still-incomplete studies is beyond the scope of this article (see Allman 1977). Of importance to us is that these studies consistently demonstrate that the representation of the fovea in striate cortex projects to specific, restricted areas of visual-association cortex. The densest projections are to the immediately anterior cortex on the ventrolateral convexity of the occipitotemporal region (figure 20.1). This "foveal prestriate" cortex (Cowey and Gross 1970) appears to correspond to the representations of foveal information in several electrophysiologically mapped visual areas, including Allman and Kaas's V2, DL, and MT (Allman 1977).

When Iwai and Mishkin (1968) and Cowey and Gross (1970) made small bilateral lesions restricted to the foveal prestriate area (much smaller than the extensive lesions of previous investigators), their monkeys were impaired in learning visual form discriminations even more than animals with the usual, more anterior inferotemporal (TE) lesions. These investigators pointed out that all of their predecessors had spared part or all of the foveal prestriate region in their ablations for fear of damage to the underlying visual radiations. Thus the paradox of the apparent dispensability of the visual association areas despite the proven importance of cortico-cortical connections was solved. The solution was that some areas of association cortex were more important than others with respect to the acquisition of visual form discrimination. The area of foveal representation was crucial; when it was destroyed, performance was impaired.

Thus, a coherent anatomic basis for understanding impaired visual form discrimination in the monkey has begun to emerge. At the cortical level, an interconnected system consisting of the portions of striate,

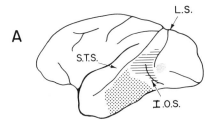

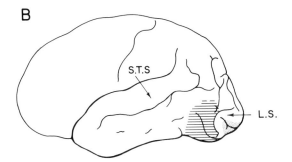

Figure 20.1 (A) The foveal prestriate (horizontal lines) and inferotemporal regions (coarse dots) in the rhesus monkey, as demonstrated by Zeki. (B) The site of lesions resulting in human visual agnosia according to Nielsen (horizontal lines). L.S.: lunate sulcus. I.O.S.: inferior occipital sulcus. S.T.S.: superior temporal sulcus. Fine dots represent anterior extent of foveal representation in striate cortex.

prestriate, and inferotemporal cortex representing central vision is necessary for intact visual form discrimination. This interconnected system, if intact in *either* cerebral hemisphere, will allow visual discrimination learning. Even after certain bihemispheral lesions of this system (e.g., striate cortex in one hemisphere and inferotemporal cortex on the other), discrimination may be relatively preserved because posterior callosal fibers may suffice to interconnect remaining foveal striate, prestriate, and inferotemporal regions. The crossed connections, however, do not appear as strong as the intrahemispheric ones (Myers 1965b; Mishkin 1972).

Although the necessity of the above interconnected cortical system for normal visual discrimination is now clear, a role for the pulvinar in visual discrimination cannot be dismissed on the basis of the experiments to date. As previously indicated, the pulvinar might relay information either from superior colliculus or from striate cortex to the prestriate and inferotemporal regions. Early studies (Chow 1954; Rosvold et al. 1958) demonstrated that bilateral pulvinar ablations did not produce impairment in acquisition of visual discriminations, suggesting that these routes were relatively unimportant. However, modern anatomical studies have confirmed earlier work establishing the anatomic specificity of the different subdivisions of the monkey pulvinar.

In the squirrel monkey (Mathers 1971), as in the macaque (Chow 1950), it is the inferior nucleus of the pulvinar that projects to the foveal prestriate region of occipital cortex. Also, it is only the inferior pulvinar in the squirrel monkey that receives projections from the superficial layers of the superior colliculus (Mathers 1971). Allman et al. (1972) have mapped a representation of the contralateral visual half-field in this region. Thus, a clear-cut route from superior colliculus to foveal prestriate cortex exists via the inferior nucleus of the pulvinar. The existence of striate cortex → pulvinar connections is disputed. Siqueira and Franks (1974), using both Marchi and Nauta techniques, found no such projections in the macaque, whereas Campos-Ortega et al. (1970) found projections from striate cortex to inferior pulvinar. Thus, if a route to inferotemporal cortex from striate cortex via pulvinar exists, it too is probably via the inferior nucleus.

If one now reviews the studies of the effects of pulvinar lesions on visual discrimination learning, it is clear that in no case has there been bilateral destruction of all or even of a substantial portion of the inferior nucleus. In Chow's studies, the published reconstructions of the lesions demonstrate this to be the case. In the work of Rosvold et al., pulvinar lesions were produced by mistake when lesions of superior colliculus were intended, and it was the medial, not the inferior, pulvinar that was affected. Thus, it is possible that appropriately placed lesions, bilaterally destroying nucleus pulvinaris inferior, will produce impairment in acquisition of visual discriminations.

The Anatomic Basis of Visual Agnosia in Man

According to Bodamer (1947), inability to recognize familiar objects and people after lesions of the central nervous system was first recorded by Thuycidides in his description of the typhus epidemic of Athens in 430 B.C. Lissauer (1890) offered the first extensive clinical description of a patient, who, after a stroke in the territory of the left posterior cerebral artery, was unable to name, describe, or demonstrate the use of visually presented objects, though he easily named them when he touched them or heard them. He had a right homonymous hemianopia. Although central visual acuity was about one-third of normal, the patient could draw visually presented objects even when he was unable to name them. To describe the syndrome Lissauer employed the term *Seelenblindheit* (psychic blindness)—the same term Munk had used to describe the vision of his destriate dogs. Freud, in his monograph on aphasia (1891), substituted the term *agnosia* for the then current term *asymbolia* (Spamer 1876) and did not consider the term *psychic blindness* in his discussion. Thus, *psychic blindness* or *visual agnosia* (synonymous in the clinical literature) came to denote the conditions of patients who, when faced with a visual object, were unable to name it, show its use, or sort it into a group of morphologically dissimilar objects with identical functions—that is, showed no "recognition" of the object—but nevertheless showed evidence of preserved vision in the usual tests of central visual acuity and peripheral visual fields. The impairment was specific to the visual modality, and performances were normal when the patient touched the object or heard it make its characteristic sound. The syndrome was well defined and has been employed consistently by clinicians interested in the subject in the past 70 years (for example, Dejerine [1914], von Monakow [1914], Pötzl [1928], Lange [1936], Nielsen [1937], Ajuriaguerra and Hécaen [1960], and Brain [1961]). However, in addition to its referring to the above syndrome, *visual agnosia* has come to have another meaning, defined not operationally but in psychological terms. This notion, originating with Lissauer's (1890) attempt to explain the abnormalities in his patient in psychological terms, is that the visual agnosic has intact "perception" but that his percepts have no "meaning." This theoretical concept of visual agnosia is not a satisfactory explanation of the syndrome, for reasons to be discussed later in the chapter. Here, I emphasize that in this paper the term *visual agnosia*, unless otherwise stated, is used to denote the syndrome defined above and not the theoretical concept.

Although studied since the latter part of the nineteenth century, the anatomic basis of visual agnosia in man is less securely established than is the syndrome of impaired visual discrimination in the monkey. This results, of course, from reliance on postmortem neuropathologic study. The number of cases of well-documented visual agnosia is small. With rare exceptions, cases that have been well studied clinically have not had postmortem examinations, whereas cases examined postmortem have often been studied insufficiently during life.

Despite these difficulties, there has been a remarkable degree of agreement among many authors who have tried intensively to establish the localization of visual agnosia (von Stauffenburg [1914], Pötzl [1928], Lange [1936], Nielsen [1937]). A major review is that of Nielsen (1937). Although Nielsen did not distinguish sufficiently between visual agnosia and (the independent syndrome of) visual disorientation, and although he considered only cases with unilateral cerebral disease, the former difficulty leads to exclusion of at most one or two of his cases and the latter can be supplemented with other reviews considering bilateral lesions as well. The "cortical area seemingly essential for the recognition of objects" that emerged from Nielsen's survey is illustrated in Part B of figure 20.1. This area comprises the ventral convexity portions of occipital and temporo-occipital cortex.

A second review is that of Pötzl (1928), who included cases with both bilateral and unilateral disease and who carefully distinguished visual agnosia from visual disorientation. According to Pötzl, visual agnosia results most frequently from infarction in the territory of the posterior cerebral arteries bilaterally. The crucial cerebral area is the ventral occipital convexity and the fusiform gyrus at the base of the temporo-occipital lobe. Unilateral, left-sided infarction of this area could also produce the syndrome if the corpus callosum was involved. Pötzl emphasized that the crucial area "lies very basally and often reaches far into the inferior portion of the temporal lobes anteriorly." He noted the frequency with which the white matter in these areas is more severely damaged than the overlying cortex.

Subsequent cases confirmed this localization. The majority (Heidenhain 1927; Hoff et al. 1962; Gloning et al. 1970; Lhermitte et al. 1972; Benson et al. 1974; Cohn et al. 1977) resulted from bilateral infarctions in the territory of the posterior cerebral arteries. The lesions involved the basal occipitotemporal regions, extending forward from a centimeter or two anterior to the occipital pole to at least the level of the splenium of the corpus callosum. The infarctions included variable portions of the cortex of lingual, fusiform, and inferior occipital gyri. There was consistent involvement of the white matter deep to these gyri, most often the white matter subjacent to the collateral sulcus and the contiguous lingual and fusiform gyri. This white matter consisted not only of the predominantly radially oriented fibers immediately subjacent to the cortex, but also of the basal portions of the strata saggitalia adjacent to the floor of the lateral ventricle. The affected white matter thus included

• cortico-cortical connections—both short fibers interconnecting gyri locally and longer fibers connecting inferior occipital cortex with temporal and inferior frontal cortex,

• afferent and efferent fibers interconnecting basal occipitotemporal cortex with subcortical structures (these connections include the inferior geniculocalcarine radiations as well as occipitotemporal connections to pulvinar, striatum, midbrain tectum and tegmentum, and basis pontis), and

• commissural fibers interconnecting the basal temporo-occipital regions of each hemisphere (these fibers traverse the ventral portions of the splenium of the corpus callosum).

Although the majority of cases have shown bilateral, roughly symmetric lesions of the basal temporo-occipital regions, one case showed a unilateral, left-sided lesion in this region (Caplan and Hedly-Whyte 1974), one case showed a right-sided temporo-occipital lesion associated with a left-sided lesion of the angular gyrus (Pevzner et al. 1962), and one case (involving a tumor) showed a right temporo-occipital lesion with only minimal involvement of the left hemisphere (Hécaen et al. 1957). The influence of such asymmetric distributions of lesions on the clinical manifestations of visual agnosia will be discussed later in the chapter.

It is thus clear that in man disorders of visual identification can occur in association with bilateral lesions of the cortex and white matter of the basal occipital and basal posterior temporal regions. The precise extent of relevant cortex is still unknown, because few cases are available and they have been relatively uniform in their pathology. For example, the precise lateral extension of relevant cortex beyond fusiform gyrus to inferior occipital and inferior temporal gyrus, or even to more dorsal cortex of the occipitotemporal convexity, remains unknown.

Despite our incomplete knowledge of the relevant human anatomy, the regions of known importance in man are remarkably similar to those in the monkey. In both, the occipital association cortex contiguous with the representation of the fovea in striate cortex appears to be important. In man, where much of the striate cortex has been dislodged from the lateral convexity of the hemisphere, the foveal portion of striate cortex lies at or near the occipital pole. The foveal "prestriate" region, which would correspond roughly with Nielsen's "essential area" (figure 20.1), is therefore more posterior than in the monkey. In both man and monkey inferior temporal neocortex is also important. In man, it is primarily the posterobasal portion of inferotemporal cortex that is known to be relevant, and the role of more lateral and more anterior inferotemporal cortex is still not clear. In the monkey, the important inferotemporal cortex extends laterally to the superior temporal sulcus and anteriorly toward the temporal pole.

Thus (except for the question of unilateral lesions, to be considered shortly), there is a striking correspondence between the lesions associated with human visual agnosia and the ablations that result in impaired visual discrimination in the monkey. It would thus appear that in man an interconnected system consisting of foveal striate, foveal prestriate, and inferotemporal cortex is necessary to prevent visual agnosia.

Behavioral Parallels

We have suggested that visual agnosia in man and impaired acquisition and retention of visual discrimination in monkeys (see Dean, this volume) have, as far as the available data will allow, identical anatomical substrates. It is now time to compare the behavioral deficits produced by comparable lesions in the two primates to determine to what extent they too are comparable.

The Problem
At the outset, the two syndromes appear different because one usually tests man and monkey differently. The patient is asked to "recognize" a visual stimulus. By "recognition" is meant a classification into one of a group of categories that correspond to those of the surrounding normal population. This classification may be verbal (the patient names the visual pattern) or nonverbal (the patient demonstrates the conventional use of the stimulus objects or sorts visual stimuli into different groups). It is obvious that every such classification is also a discrimination. For example, in naming the stimulus, the human patient is discriminating

the object from others that would be labeled by different words in his vocabulary. (If he were severely aphasic so that his vocabulary consisted of a single word, naming would be neither a classification nor a discrimination.) "Recognition" is said to be impaired when no means of response, verbal or nonverbal, yields a set of classifications (discriminations) that corresponds to the conventional. The categories may be too broad, lumping conventionally differently classified patterns together, or they may be too narrow, such that each stimulus pattern is unique to itself. In addition to the above static changes, the classifications may fluctuate excessively, so that a given stimulus pattern, on separate occasions, is classified differently far in excess of normal tendencies. (Bruner [1957] emphasized that some fluctuations occur in the normal individual where classifications are influenced by needs and expectations as well as by the stimulus material.) If such impaired "recognition" of visual stimulus patterns exists in the presence of "sufficient" vision (tested by conventional measures of central acuity, e.g., the Snellen chart, and peripheral fields, e.g., kinetic perimetry) and normal "recognition" by touch and hearing, visual agnosia is said to exist.

The monkey, on the other hand, is usually faced with an apparatus in which there are several visual patterns. He is required to select one of these patterns consistently over successive trials. The number of such trials required to reach a preestablished criterial level of performance is compared with that of the normal monkey, and if it is excessive the discrimination performance is said to be impaired.

Several important differences between the performances demanded of man and monkey are thus evident:

• The monkey is often asked to acquire a new visual discrimination (that is, to form categories that it may never have had occasion to form before), whereas the human is asked to make a classification into conventional, well-established categories.

• The monkey is required to discriminate between only two (or at most among a few) visual patterns, that is, it must form only a twofold categorization system, whereas the human often must assign the pattern to one of a great number of possible categories (for example, in a naming task, all possible namable objects).

• The classification (discrimination) capacities of the monkey are conventionally tested with the very same visual patterns used to form the categories initially, whereas the human is often faced with a visual stimulus pattern not exactly like any he has used to form the category of which this stimulus pattern is an example (for example, he may never have seen a key of this shape before, but is still required to "recognize" it as a key).

• In learning a categorization, the monkey is usually given only the same two patterns, that is, one example per category, whereas the human has usually had a broad range of experience with many members of each category (he has previously classified as "keys" many patterns of different size and shape).

• The monkey is usually faced with all of the choices (in its visual presence is an example of each of the categories among which the discrimination is to be made), whereas the human is often faced with only one visual pattern and must select the proper response (category) from among many of which no examples are present.

• The discriminating (classifying) response of the monkey is a limb or whole-body movement (such as pointing to the correct discriminandum), whereas that requested from the human may be either a limb movement or a specific pattern of oro-linguo-laryngeal movements.

These differences, however, do not present as insurmountable an obstacle to comparing the behavior of man and monkey as might at first be expected. Several attempts to alter the monkey tests to correspond better to the tasks usually given humans demonstrate that deficits persist under the new test conditions. Thus, the work of Klüver and Bucy (1937, 1939) as well as the latter studies of Mishkin and Pribram (1954) established that, in monkeys with inferotemporal lesions, retention of previously acquired visual discriminations is affected in addition to the acquisition of new ones (the first difference listed above). Pribram and Mishkin (1955) showed that deficits also persist when the monkey is presented with only one stimulus pattern at a time and is required to respond discriminatively to it. This "go/no go" task is more similar to the situation of a human confronted with a single visual pattern and asked to respond with its name or its use (the fifth difference above). Butter et al. (1965) have shown that inferotemporally ablated monkeys are impaired in tests of "stimulus generalization," in which the animal, after acquiring a visual discrimination, is asked to respond to a new pair of stimuli resembling, but not identical to, those of the initial training (see the third difference above).

Conversely, if one alters the tests in human agnosics to correspond better to the customary testing situation in the monkey, the behavior remains abnormal. Thus, the agnosic patient reported on in Levine 1978, when tested for learning of a visual discrimination in a two-alternative forced-choice situation, was unable to point to the correct one of two complex visual patterns that differed very subtly (figure 20.2) though she succeeded with easier discriminations.

Once it is realized that many of the apparent differences between the behavior of the patient with visual agnosia and the monkey with impaired visual discrimination can be minimized by variations in testing procedures, many striking parallels emerge between the behavior of the two primates in tasks of visual form discrimination as well as in other types of visual and nonvisual tasks. Before discussing these parallels, however, we must consider the subdivision of both visual agnosia in man and impaired visual discrimination in the monkey.

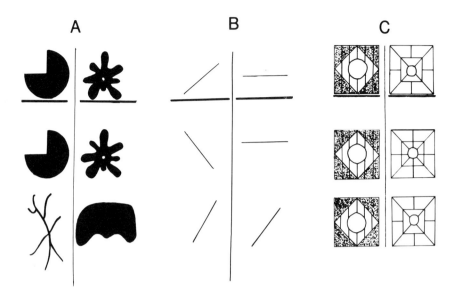

Figure 20.2 Examples of visual discrimination tasks that were quickly mastered (A,B) or never learned (C) by the visual agnosic reported on in Levine 1978. In each task the patient was presented with a card on which two patterns (below the thick horizontal lines) were displayed. The examiner indicated that one of the two shown above the line was correct. On subsequent trials the same two patterns were shown and the patient was asked to indicate the position of the correct choice verbally. (The relative positions of correct and incorrect choices were varied randomly from trial to trial. The number of trials preceeding 5 consecutive correct responses were recorded). Learning was immediate for 14 tasks similar to the 4 shown in A and B. No learning had occurred, even after 30 trials, for 4 tasks similar to the two shown in C.

Subdivisions of the Syndromes

To this point, we have treated the syndrome of impaired visual pattern discrimination in the monkey with foveal-prestriate and/or inferotemporal cortex damage as a single entity. The purpose of this treatment was to demonstrate its anatomical homology to the syndrome of visual agnosia in man, which we also have not subdivided. It is clear, however, that in both man and monkey the syndrome can be fractionated.

In the monkey, Iwai and Mishkin (1968, 1969) and Cowey and Gross (1970) have shown that foveal-prestriate lesions produce a qualitatively different impairment of visual pattern discrimination from inferotemporal lesions. On the one hand, monkeys with foveal-prestriate lesions are more impaired than inferotemporally lesioned animals in acquiring and retaining discriminations between three-dimensional visual objects or two-dimensional visual patterns but are less impaired with color discrimination. Once a discrimination has been acquired, the addition of an extra but irrelevant visual cue distracts (impairs performance of) foveal-prestriate-lesioned but not inferotemporally lesioned animals (Gross et al. 1971). On the other hand, monkeys with inferotemporal lesions are more impaired than those with foveal-prestriate lesions in tasks of "concurrent discrimination." For a procedure of concurrent discrimination only simple pairs of discriminanda are used that would be no trouble, even for the monkey with lesions, if presented alone. However, several such tasks are presented concurrently so that the same task does not appear on successive trials. The greater impairment of inferotemporally lesioned monkeys on tasks of concurrent discrimination has led to the interpretation that inferotemporal lesions produce a deficit in visual "association" or "memory" (Cowey and Gross 1970), whereas the foveal-prestriate deficit is more "perceptual" in nature, amounting to a disturbance in "selective attention" necessary in the orderly construction of a percept.

Wilson et al. (1972) demonstrated that the "memory" defect of the monkey with inferotemporal lesions could not be interpreted in the sense of decay of a trace, because in a matching-to-sample task increasing delay between presentation of the sample and presentation of the choices did not affect the performances of inferotemporal animals any more than it did those of normal or foveal-prestriate monkeys. It is possible that the "memory" deficit may relate to another aspect of the behavior of inferotemporal monkeys demonstrated by Wilson. She showed that these animals have more "intrusion errors" than either foveal-prestriate or normal monkeys in a matching-to-sample problem; they often err by responding in the same manner as on the previous trial. Such proactive interference, or perseveration, in addition to accounting for impaired acquisition of concurrent discrimination, might also explain the beneficial effect of adding punishment for wrong responses to reward for correct responses on the performances of the inferotemporal monkeys (but not the foveal-prestriate animals) in visual discrimination tasks (Manning 1971). The punishment may minimize the tendency to perseverate by approaching the wrong (punished) discriminandum on subsequent trials. A similar explanation may apply to the deterioration in the performances of inferotemporal animals when only intermittent reinforcement is given (Manning et al. 1971).

The perseveration itself, however, may be a result of, or may be associated with, other impairments. Thus, it appears that inferotemporally lesioned animals do not employ a flexible strategy that results in detecting the relevant features in a discrimination problem and in disregarding the irrelevant. That is, the relative values of the component features in determining the response are not those of the normal animal. This distortion of feature values involves not only the various linear and contour features of a shape in a shape-discrimination problem (Butter et al. 1965; Butter 1968) but also a tendency to devalue shape and orientation cues altogether in favor of brightness differences (Iversen and Weiskrantz 1967). Perseveration may at least partially reflect this inflexibility in perceptual strategy, and may represent the stubborn clinging to a distorted set of values of the stimulus features.

In man, cases of visual agnosia have been subdivided into two groups since Lissauer (1890) distinguished "apperceptive" from "associative" visual agnosia on a largely theoretical basis. The former group had an abnormality of "apperception," which Lissauer defined as "the seizing by consciousness of sensory impressions." Abnormal "apperception" could be detected, despite adequate central visual acuity and peripheral visual fields, by requiring the patient to make a "judgment" about his perception. Thus, the patient would be unable to draw a visual stimulus pattern or to match it with an identical pattern in an array of stimulus patterns. The "associative" agnosic, on the other hand, had intact (or nearly intact) "apperception" but was impaired in "association," which to Lissauer meant the ability to relate the trace of the visual stimulus pattern in the occipital cortex to past traces left by the object in other sensory areas of the brain (via corticocortical pathways from occipital lobe to other sensory areas). In actual fact, cases of visual object agnosia do vary between two extremes that in the neurological literature are usually classified as either "apperceptive" or "associative." (The further subdivision of associative agnosia into object agnosia and prosopagnosia will be discussed later.) I shall use these terms as descriptive of syndromes, though I do not feel that the psychological or anatomical formulations of Lissauer best describe them.

The cases of "apperceptive" visual agnosia (Goldstein and Gelb 1918; Adler 1944, Benson and Greenberg 1969) have an extremely severe defect in discriminating shapes, but retain nearly normal absolute and differential brightness thresholds and nearly normal visual fields when tested by kinetic perimetry (Benson and Greenberg 1969). The shape-discrimination deficit is so severe that the patient is unable to distinguish, for example, a circle from a cross. A patient examined by the author was an inveterate card player, yet was unable to distinguish a diamond from a club, though visual acuity by testing detection of black dots of different sizes on a white background was at least 20/70 and the upper quadrants of both right and left homonymous visual fields were full. Such patients are completely unable to copy or match visual stimulus patterns. Despite the severe problem with form discrimination, most of these patients have been able to discriminate and to name colors. They are also highly susceptible to distraction by adjacent (Adler 1944) or overlapping (Goldstein and Gelb 1918) visual objects when attempting a visual discrimination performance. Patients with "associative" visual agnosia (see, for example, Lissauer 1890; Heidenhain 1927; Rubens and Benson 1971), on the other hand, are less impaired in shape discrimination. They are, as a rule, able to discriminate and name simple geometric figures such as circles and squares. However, their performances on more difficult discrimination tasks are not normal, and failure can be elicited when the discriminanda are complex patterns that differ from each other subtly (Levine 1978). This point is important, for since the time of Lissauer it has been claimed that accurate performance in matching a visual sample to one of a set of choices by an associative visual agnosic guaranteed that "perception" was intact. However, in fact, these patients match much more slowly than do normals, accomplishing the task by painstakingly comparing components of the discriminanda serially until they detect differences. Thus, they often become hypercritical; that is, their categories become very narrow. For example, Rubens and Benson's patient said that two identical patterns were different because he included slight smudges of printer's ink or irregularities in the texture of the paper in his comparisons. The same procedure is employed in drawing an object. The end product may be good, but drawing is performed slowly, proceeding serially ("slavishly") feature by feature to build up the final product. The inadequacy in discrimination (categorization) is also revealed in tasks of naming and sorting, where underspecification occurs; that is, the patient names an object with a similar overall shape or an identical inner detail (Levine 1978). These patients, as a rule, are less distracted than the apperceptive agnosics by extraneous visual stimuli, but response perseveration—calling an item by the same name that was given to a previous item—is very common.

From the foregoing discussion, it appears that there are parallels in the behavior of the inferotemporally lesioned monkey and the human associative visual agnosic, and in the behavior of the foveal-prestriate-lesioned monkey and the human apperceptive visual agnosic.

Behavioral Parallels Between Human Associative Visual Agnosia and Simian Inferotemporal Visual Discrimination Impairment

The first parallel between the two syndromes is that, in both, the disturbance is reflected in only one sensory sphere: the visual. Monkeys with inferotemporal lesions and impairment in acquiring visual discriminations have shown normal abilities to acquire olfactory discriminations (Brown 1963; Brown et al. 1963), tactile discriminations (Wilson 1957; Pasik et al. 1958), and auditory discriminations (Weiskrantz and Mishkin 1958). In the human visual agnosic this can also be the case (Pötzl 1928; Rubens and Benson 1971), although often there is mild difficulty in naming through other modalities as well (Ettlinger and Wyke 1961; Levine 1978). However, where nonvisual naming difficulties occur, they are much less severe than the visual disturbance.

A second parallel is that disturbances in "elementary" visual functions are not sufficient to account for the inability to make visual form discriminations. "Elementary" visual functions are difficult to define in a manner that will distinguish them precisely from more "complex" functions such as the discrimination of complex forms. Roughly, they comprise discriminations that utilize one or two points of light and can be plotted as a function of the position of these points in the visual fields—such as visual acuity (discrimination of two points from one), absolute and differential brightness thresholds (which determine the "visual fields" of static or kinetic perimetry and amount to discrimination of one point from none), flicker-fusion frequencies, or the closely related tachistoscopic sensation time and adaptation time.

In the inferotemporally lesioned monkey, "elementary" visual function has been assessed in several ways. Visual fields are normal when tested by a method resembling static perimetry in man (Cowey and Weiskrantz 1967). Central visual acuity is also normal when tested by the ability to discriminate alternating black and white vertical stripes (that can be made progressively narrower) from a homogeneous field of the same total luminous flux (Weiskrantz and Cowey 1963). These studies were performed without normal controls, and one cannot assert that the functions were entirely normal; however, they were in all cases better than in animals with lesions of striate cortex, even though the latter perform better on tests of visual form discrimination. Similarly, critical flicker-fusion thresholds in the central visual fields are either slightly (Mishkin and Weiskrantz 1959) or not at all (Symmes 1965) affected by inferotemporal lesions—again different from the case with lesions of striate cortex, where flicker thresholds are worse though form-discrimination performance is better than in inferotemporal animals.

Humans with associative visual agnosia may have normal central visual acuity (see, e.g., Bay 1952 for a report of tests requiring tracing and matching simple forms of progressively smaller size; see also Rubens and Benson, 1971) and normal visual fields (Bay 1952) when tested by kinetic perimetry. Although Bay (1952, 1953) claimed that impaired tachistoscopic sensation time and shortened local adaptation time (probably very similar to elevated flicker-fusion thresholds) are invariably present in human visual agnosics, Ettlinger (1956) and Ettlinger and Wyke (1961) demonstrated that nonagnosic patients may have more severe impairment on such tests than patients with visual agnosia. Unfortunately, in assessing visuosensory efficiency in his patients, Ettlinger did not distinguish impairment in central visual fields from impairment in the peripheral fields. More recently, however, Levine and Calvanio (1980) demonstrated no correlation between impaired identification of pictures of objects and elevated central flicker-fusion frequency in patients with bilateral lesions of the posterior cerebrum. The situation in man thus appears to parallel that in the monkey.

Although disturbances of "elementary" visual functions are not sufficiently severe to entail either associative visual agnosia in man or the inferotemporal discrimination deficit in monkey, it is apparent that stressing these functions will impair performance far more than in normal man or monkey. The effects of reducing visual angle of the discriminanda (stimulus size) in a form-discrimination task is one example. In the inferotemporally lesioned (Pasik et al. 1960) and in the foveal-prestriate-lesioned (Iversen 1973) monkey, reducing the size of the discriminanda adversely affected the ability to discriminate a triangle from a square or a triangle from a plus, though even the smallest stimulus used was well within the range of normal acuity. In the author's human agnosic (Levine 1978) there was a suggestion that larger objects were better described than small objects of similar shape. Moreover, better naming of large than of small letters may occur in agnosic alexia in man (Rubens and Benson 1971; Woods and Poppel 1974). In the human agnosic, other stresses of elementary visual functions also impair form discrimination at levels not affecting normal performance at all. Reduced background illumination, reduced contrast between stimulus and background, or reduced exposure time of the stimulus patterns markedly impaired form discrimination in an agnosic patient (Levine 1978), whereas control subjects were not at all affected in the range of variables used. These studies should be made in the monkey, where lesions can be better controlled.

In addition to modality specificity and to lack of dependence on disturbances of "elementary" visual function, the syndromes of associative visual agnosia in man and of impaired visual discrimination in inferotemporally lesioned monkeys show marked parallels with respect to the nature of the errors made in tasks of visual discrimination:

• Associative visual agnosics characteristically make errors of "under-specification" (Levine 1978). When asked to name a visually presented object, they often give the name of an object that has a roughly similar outline or topological structure to that of the object presented. Or they may name a prominent feature of the presented object, or give the name of another object that shares this feature (and perhaps no others) with the object presented. Thus, Rubens and Benson's (1971) patient called a key a "violin" (topological similarity) and the author's patient called a harmonica a "comb" because she confined her attention to the spaces separated by wooden partitions, which she took to be the teeth of the comb. In the inferotemporal monkey the experiments of Butter et al. (1965) and of Butter (1968) established that the animal also attends excessively to a single feature of the discriminanda in learning visual discriminations. For example, it attends more than the normal monkey to the horizontal base line of a triangle in learning to choose it instead of a circle. Hence, in an equivalence test it will select a semicircle instead of an inverted triangle, whereas a normal monkey will not. This impairment in pattern equivalence thus resembles the behavior in man that has been called underspecification of the discriminandum.

• A very frequent characteristic of humans with associative visual agnosia is perseveration; that is, these subjects name an object shown them with the name of an object that they had previously been shown or thought that they had been shown (Lissauer 1890, Pötzl 1928; Critchley 1964; Levine 1978). The perseveration is specific to tasks of visual discrimination and does not appear in perceptual tasks in other modalities. Thus, the author's patient called most of the objects shown to her in one series of observations a pen. At times the perseveration is so strong that even salient features of the object held before the eyes are disregarded. More often, however, perseveration will occur when the name perseverated (i.e., that of a previous object) denotes an object that is also an underspecification of the object currently presented, and will not occur when the object presented does not at all resemble previous objects morphologically. In the monkey, as previously mentioned, Wilson et al. (1972) showed the presence of more "intrusion" errors (i.e., perseveration) in animals with inferotemporal lesions than in normal monkeys or monkeys with foveal-prestriate lesions.

The syndromes in man and in monkey are further parallel in the beneficial effects of "overtraining" on the performance of visual discriminations. In the monkey, extensive preoperative experience with a form discrimination minimizes the retention deficit after inferotemporal lesions (Orbach and Fantz 1959). In the human agnosic, although the effects of prelesion "overtraining" cannot be readily studied, the relative lack of impairment of visual identification in familiar surroundings has often been mentioned (Pötzl 1928; Critchley 1964). The author's patient (Levine 1978) was able to identify objects better when she saw them utilized in their natural manner (e.g., a toothbrush brushing teeth) than when they were held motionless or were rotated and moved back and forth in front of her eyes.

Several of the above parallels between inferotemporally lesioned monkey and agnosic human can be summarized by the more general parallel that visual discriminations easily acquired by the normal organism are acquired with little difficulty and eventually to the same level of proficiency by the organism with lesions, while tasks difficult for the normal may never be acquired to the normal level of proficiency by the damaged organism or may require extensive and prolonged training. Gross (1972) reviewed this principle for the inferotemporally lesioned monkey. In man this principle is less well established. However, the author's patient (Levine 1978) easily mastered visual discriminations that were obvious to normals, but could not acquire discrimination between subtly different complex forms (figure 20.2).

Behavioral Parallels Between Human Apperceptive Visual Agnosia and Foveal-Prestriate Visual Discrimination Deficit in the Monkey

Much less can be said about the relationship of these syndromes than about the relationship between associative visual agnosia and the inferotemporal visual discrimination deficit. This ignorance arises from the fact that the significance of the foveal-prestriate region in the monkey has only recently been appreciated and less information about the effects of foveal-prestriate lesions is available than about the effects of inferotemporal (TE) lesions. Even more limiting is the paucity of cases of "apperceptive" visual agnosia in man. Not a single one of the available clinically studied cases has had postmortem examination, in contrast with the situation for associative visual agnosia. Because several of the cases (Adler 1944, 1950; Benson and Greenberg 1969) arose in the context of asphyxiation, where lesions may be widespread, it is unclear to what degree the symptoms reflect foveal-prestriate damage independent of involvement of other areas.

Nevertheless, several parallels between apperceptive visual agnosia in man and foveal-prestriate visual discrimination impairment in monkey are evident when one considers them in relation to associative visual agnosia and inferotemporal visual discrimination deficit, respectively:

• The discrimination deficit is more severe, as mentioned previously. In the monkey, this has been shown for three-dimensional visual objects, for two-dimensional visual patterns, and in pattern-equivalence tests (Gross 1973). Human apperceptive visual agnosics cannot discriminate, name, or draw even simple geometric figures (Benson and Greenberg 1969; Brown 1972), whereas associative visual agnosics can.

• Although form discrimination is more impaired, color discrimination is relatively preserved. Gross et al. (1971) demonstrated this in monkeys with foveal prestrate lesions. Also, the human patients of Adler (1944) and Benson and Greenberg (1969) were able to name colors despite the severe form-discrimination deficit.

• Distraction by irrelevant features of the discriminandum is prominent. Gross et al. (1971) showed that foveal-prestriate-lesioned monkeys performed poorly on a visual form discrimination that they had already mastered when an irrelevant color cue was added to the forms to be discriminated, whereas inferotemporally lesioned animals were less affected. Iversen (1973) showed that irrelevant forms, either surrounding or partially covering the discriminanda, impaired discrimination in foveal-prestriate animals but not in controls. The human apperceptive visual agnosic also is even further impaired in efforts at naming simple visual patterns if a line is drawn over them (Goldstein and Gelb 1918), whereas such need not be the case in associative visual agnosia (Heidenhain 1927). Such distractability may be merely a manifestation of the severity of the form-discrimination deficit in both monkey and man. Thus one might expect similar distractability with inferotemporally lesioned monkeys and human associative visual agnosics if the discriminanda and the overlapping distractions are made more complex.

• The pattern of eye movements during the examination of visual discriminanda is abnormal in foveal-prestriate monkeys (it is also abnormal in inferotemporal animals; see Oscar-Berman et al. 1973), and was abnormal in Benson and Greenberg's (1969) apperceptive visual agnosic. These eye-movement abnormalities may reflect the difficulty of the form discrimination.

The relationship of impaired visual discrimination in foveal-pre-striate-lesioned monkeys and of apperceptive visual agnosia in man to disturbances of "elementary" visual function has not yet been clarified. In the foveal-prestriate monkey, Bender (1973) demonstrated normal brightness thresholds under scotopic conditions, but visual-acuity tests (i.e., two-point discrimination), static perimetry, and tests incorporating temporal factors such as flicker fields have not been done, as they have for the inferotemporal animals. In humans, Efron (1968) demonstrated only modest elevation of brightness thresholds, mild decreases in flicker-fusion thresholds, and full perimetric fields in Benson and Greenberg's patient with apperceptive agnosia. Visual acuity was not measured. Clearly more work must be done in both monkey and man in this area.

Thus, it has been established that numerous parallels exist between the syndromes of impaired visual discrimination in monkeys with foveal-prestriate and inferotemporal lesions and the syndromes of visual agnosia (psychic blindness) in man. The analogous behavior and homologous anatomic localization of the syndromes in the two primates makes it likely that similar physiologic processes are impaired in each. In the following section I shall demonstrate that in both man and monkey these syndromes are at least partially dissociable from other syndromes of the temporal lobe and from other disturbances of visual behavior resulting from parietal and parieto-occipital lesions.

Other Syndromes

In this section a number of syndromes that may be found in association with visual agnosia in man or impaired visual discrimination in the monkey will be discussed. The purpose of this discussion is to show that the syndromes are at least partially independent of visual agnosia and impaired visual discrimination. This partial independence suggests that the anatomic substrates of these syndromes are not congruent with those of visual agnosia, though some overlap may occur. The term "visual agnosia," if it is to be applied in its usual sense, should, therefore, not be used in reference to these other syndromes.

The first syndrome that must be distinguished from impaired visual discrimination in the monkey and visual agnosia in man is the portion of the "Klüver-Bucy syndrome" produced by resection of part or all of the territory occupied by the amygdaloid body and adjacent pyriform cortex, neocortex at the tip of the temporal lobe (von Bonin and Bailey's TG), the anterior insula, and the cortex of the orbital surface of the frontal lobe. Monkeys with bilateral lesions of this entire area (Pribram and Bagshaw 1953) demonstrate marked tameness in that they approach or allow themselves to be approached and touched by human observers without displaying fear or aggression. "Hypermetamorphosis" is common, as is ingestion of inedible, unpalatable, and noxious objects such as burning paper, feces, or sharp metal objects. They often approach other monkeys abnormally frequently to groom or to mount, but do not interact normally with their fellows and lose their social status. With more restricted lesions in this area, such as those of amygdala and pyriform cortex alone (Walker et al. 1953; Weiskrantz 1956), the tameness is prominent but hypermetamorphosis either is absent or disappears after a week or two. Instead the animals appear apathetic, sitting in one place with drooping head and arms for prolonged periods and apparently not attending to their environment. Sexual activity is usually decreased, but recovers at a point when tameness and apathy may still be present. The tameness is prominent for visual stimuli (for example, humans or noxious objects), but snarling may be obtained (in some animals) by tail pinch. Conditioned avoidance may be acquired when the reinforcement is electric shock, but not when it is the sight of a man, although acquisition is slower than normal even to electric shock.

It is beyond the scope of this article (and perhaps not possible) to fractionate the components of this "orbitofrontal–anterior temporal" syndrome according to the site of lesion within this system. The major point to be made here is that the results of tests of visual discrimination (Pribram and Bagshaw 1953; Mishkin 1954; Schwartzbaum 1965) have been normal or nearly normal in all such animals. Hearst and Pribram (1964) demonstrated that stimulus generalization with respect to the dimension of background illumination was also normal in animals with bilateral amygdalectomy. To date, there have been no studies of generalization with respect to pattern discrimination analogous to the experiments of Butter et al. (1965) and of Butter (1968) with inferotemporally lesioned monkeys. However, given the information (see below) from human cases with similar lesions, it is predicted that these will be normal as well. If so, it is clear that these animals can discriminate visual patterns normally and can group visual patterns into the usual categories employed by normal monkeys.

In man, similar cases have been reported (see, for example, Terzian and Dalle Ore 1955; Friedman and Allen 1969; Gascon and Gilles 1973); all these cases included lesions of hippocampal formation as well as part or all of the fronto-temporal regions mentioned above). Gascon and Gilles (1973) reported the most detailed study of behavior. After an encephalitic illness, the patient, a young housewife, was left in a state called by the authors "limbic dementia." She wandered continuously about the ward, was found aimlessly searching through other patients' drawers, and often mouthed inedible objects, at one time even feces. Her attention span was very short. She was highly distractable and was unable to sustain goal-directed activity. Although no overt sexual behavior was observed, a psychiatric examiner suggested sexual connotations in her conversation and that she showed "confused courting" of the examiner. In addition to this "limbic dementia" an amnesic syndrome was present. Despite the above permanent changes, after the first few weeks of her illness she was consistently able to name common objects, pictures of common objects, and body parts. She was able to read aloud and to match a printed noun to the appropriate picture. She was also able to name colors well. In short, there was no evidence of the syndrome of visual agnosia. Postmortem examination showed bilateral destruction of amygdala, pyriform cortex, anterior tips of the temporal lobe, and parts of the posterior orbital cortex and the cingulate gyri. The hippocampus and parahippocampal regions were destroyed bilaterally. In addition, the fusiform gyrus on the left and both fusiform and inferior temporal gyri on the right were partially damaged. The latter incomplete damage to the posterior inferotemporal neocortex probably accounted for the transient misnaming of common objects in the first few weeks of illness. This misnaming at that time showed, from the few examples given by the author, the visual underspecification (e.g., thermometer: "needle") characteristic of visual agnosics.

It appears, then, that the orbitofrontal–anterior temporal syndrome of tameness, "hypermetamorphosis" with oral tendencies, and abnormal sexual behavior may occur without impaired visual discrimination in monkey and without the syndrome of visual agnosia in man. The reverse dissociation is also clear. Monkeys with inferotemporal or foveal-prestriate lesions and impaired visual discrimination do not as a rule show the orbitofrontal–anterior temporal syndrome (see, e.g., Mishkin and Pribram 1954; Cowey and Gross 1970).

The independence of the syndromes of striate-prestriate-inferotemporal cortex from those of orbitofrontal–anterior temporal cortex in both monkey and man makes it unwise to use the term *visual agnosia* to describe both syndromes. Unfortunately, that term has been applied to many of the symptoms of the orbitofrontal–anterior temporal syndrome, such as hypermetamorphosis, oral tendencies, and indiscriminate sexual behavior. When used in this sense, *visual agnosia* is no longer referring to the syndromes of impaired visual discrimination characteristic of striate-prestriate-inferotemporal lesions; instead it is being used in the theoretical sense of "an intact but meaningless percept." The argument is that, since the organism no longer responds discriminatively to visual objects with respect to the behavioral dimension "to be approached/to be avoided," the objects have lost their "meaning." But, as we have just seen, human patients (and probably monkeys) with the orbitofrontal–anterior temporal syndrome may categorize visual patterns into their conventional groups (for example, by naming, by sorting, or by generalization tests), and in this sense the visual patterns do have "meaning." The loss of "meaning" of visual objects in the orbitofrontal–anterior temporal syndrome is only partial; it is restricted to the behavioral dimension of approach/avoidance. If the term *agnosia* is applied here, one might argue that it should be applied to other syndromes with partial loss of the "meaning" of visual stimuli—for example, to apraxia, where the patient has lost the "meaning" of an object with respect to how to use it; or to aphasia where the patient has lost the "meaning" of an object with respect to how to name it. It would thus seem wise to restrict the term *visual agnosia* to the deficits in visual identification (discrimination) resulting from damage to the foveal striate, foveal prestriate, and inferotemporal portions of the cerebral hemispheres.

Although I would discourage the use of the term *visual agnosia* to describe the orbitofrontal–anterior temporal syndrome, I do not wish to deny that this syndrome may be at least partially modality-specific. Several investigations, as previously discussed, have demonstrated more prominent loss of avoidance for visual stimuli than for stimuli such as electric shock to the skin. Perhaps the modality specificity relates to the strong input from inferotemporal cortex to amygdala, as demonstrated by Whitlock and Nauta (1956). However, one must note that the stimuli used to assess modality specificity in animals with orbitofrontal–anterior temporal lesions may not really be comparable. One should present such animals with complex natural sounds (such as the growl of a leopard) instead of a loud noise, or with complex, natural tactile patterns (such as the crawl of a scorpion) instead of electric shock. If the animals were to respond with avoidance to these but not to complex visual patterns such as the human face, modality specificity would be better established than it is at present. Furthermore, although inputs to basolateral amygdala are particularly heavy from inferotemporal (visual) cortex, Druga (1969–1970) demonstrated in the cat that the heaviest connections seemed to arise from an area corresponding most closely to Woolsey's auditory AII area.

A second temporal-lobe syndrome that is at least partially dissociable from visual agnosia in man is the amnesic syndrome. Patients with amnesia show only mild or no impairment in recall of material that is within the normal immediate span, provided no distracting stimuli are introduced in the interval between presentation and recall. But with the introduction of distracting stimuli, or with lists of items exceeding immediate span, marked deficits in recall occur (Scoville and Milner 1957; Drachman and Arbit 1966). The deficit is not specific to any particular sensory modality of stimulus presentation.

It is beyond the scope of this chapter to deal in detail with the behavioral characteristics of the amnesic syndrome. It is sufficient to point out that such patients, although unable to learn to identify new visual stimuli, do not have difficulty identifying even complex visual patterns with which they have had extensive experience prior to the onset of their lesions. Thus, they display no signs of visual agnosia (Milner and Teuber 1968). Conversely, although patients with associative visual agnosia often perform subnormally on standardized tests of memory (e.g., verbal paired-associate learning subtest of the Wechsler Memory Scale), they do better than patients with the amnesic syndrome. Visual agnosics, to the extent that they can discriminate and classify visual patterns, have little difficulty with subsequent recall, provided that they "see it the same way" (i.e., make the same classification) on re-presentation (Pötzl 1927; Levine 1978).

The neuropathology of the amnesic syndrome is still not entirely clear. Those postmortem examinations that have been conducted have shown that cases associated with temporal-lobe lesions have generally resulted from bilateral infarctions in the territory of the posterior cere-bral arteries (Victor et al. 1961; DeJong et al. 1969). Bilateral lesions of Ammon's horn and parahippocampal gyrus have been present, but damage also has extended laterally to the fusiform gyrus and posteriorly to the occipital lobe to a variable degree. Thus, there is considerable overlap with the neuropathology found in associative visual agnosia. Further studies will be required to determine the precise extent to which the amnesic syndrome and the syndrome of associative visual agnosia can be dissociated, both clinically and pathologically.

There remains considerable controversy over the existence of an am-nesic syndrome in the monkey with lesions of corresponding struc-tures. Bilateral hippocampectomy in the monkey has not resulted in the expected deficits in discrimination learning (Orbach et al. 1960; Kimble and Pribram 1963). Two explanations of these results have generally been offered: that hippocampal lesions alone are not the basis for am-nesia either in monkey or in man (Horel 1978), and that hippocampec-tomy in the monkey does produce a deficit in memory if appropriate methods of testing (Gaffan 1974) or appropriate interpretation of ex-perimental results (Douglas 1967) are employed. In any event, the be-havioral syndrome resulting from hippocampectomy, whatever its relationship to the human amnesic syndrome, appears to be dissociable from impairments in acquisition and retention of visual pattern dis-criminations. Monkeys with hippocampal lesions are normal in ac-quiring visual pattern, size, and brightness discriminations (Kimble and Pribram 1963; Douglas and Pribram 1966).

Yet another syndrome that must be distinguished from visual-object agnosia in man and from the syndrome of impaired visual discrimina-tion in the monkey is the syndrome of visual disorientation described initially by Balint (1909), by Holmes (1918), and by Holmes and Horrax (1919). Patients with this syndrome in severe form are unable to localize extrapersonal visual objects by reaching for them with their limbs. Thus, they misreach for objects they wish to grasp, and in drawing or writing they lose the place when the pencil is lifted from the paper. They cannot state which of two objects is nearer to their own bodies or to an external landmark, and cannot traverse a path without colliding with obstacles. Disturbances of eye movement are present. The patients have marked difficulty capturing by fixation objects in the peripheral visual fields, show instability of fixation once it is achieved, and show difficulty in tracking moving objects in a textured visual field. They do not converge or accommodate to approaching visual objects, nor do they blink to an approaching visual threat. Yet visual agnosia is not present. Visual patterns that normally can be recognized without the need for eye movement, including geometric shapes, common objects, and faces, are usually promptly and accurately named (Balint 1909; Holmes 1918; Holmes and Horrax 1919; Hécaen and Ajuriaguerra 1954; Godwin-Austen 1965; Michel et al. 1965). The literature contains six cases of visual disorientation with subsequent postmortem examinations (Balint 1909; Holmes 1981 [2]; Hécaen and Ajuriaguerra 1954 [2]; Michel et al. 1965). The lesions in these cases are more dorsal than those resulting in visual agnosia, and involve the parietal lobes. The cortex of the precuneus, superior parietal lobule, intraparietal fissure, and portions of supramarginal and angular gyri is often affected, and much of the subjacent white matter is usually involved. The area is thus quite distinct from the more ventral temporo-occipital region involved in visual agnosia.

The converse dissociation is also clear—at least with respect to the "associative" type of visual agnosia, in which patients may show no difficulties in orienting their limbs to visual objects, in walking about, in scanning a visual array, or in discriminating lengths of lines (see, e.g., Heidenhain 1927; Rubens and Benson 1971). However, the same may not be true of patients with apperceptive visual agnosia. Although they may navigate adequately in their environment (see Goldstein and Gelb 1918; Adler 1944, 1950; Benson and Greenberg 1969), there are abnormalities of eye movement and disturbances of drawing (Adler 1944; Benson and Greenberg 1969) that resemble features of the syndrome of visual disorientation. Other features of visual disorientation, such as prehension of visual objects and distance discriminations, have not been adequately assessed in these patients. Even if frequently accompanied by features of visual disorientation, the latter may not necessarily be present in apperceptive visual agnosia, for the cases of Adler and of Benson and Greenberg were the result of asphyxiation, in which more widespread lesions were undoubtedly present that those accounting for the visual agnosia. Or the abnormalities in drawing and in eye movement in apperceptive visual agnosia may differ qualitatively from those of visual disorientation. Only further detailed clinical and pathological studies can resolve these questions.

A syndrome resembling visual disorientation can be produced in the monkey by bilateral ablation of Brodmann's areas 5 and 7 on the medial and lateral surfaces of the hemispheres as well as the dorsal portion of the prelunate gyrus (part of areas 18 and 19 of Brodmann). Such animals misreach for visual stimulus objects (Pribram and Barry 1956; Ettlinger and Waegner 1958; Bates and Ettlinger 1960). They have difficulty grading their leaps, so they often collide with obstacles and have trouble finding their own cages among many in the laboratory (Bates and Ettlinger 1960). Pohl (1973), using monkeys with more restricted lesions (sparing area 5 but including area 7 and dorsal 18 and 19), showed impairments in discriminating which of two identical food wells was closer to a given extrapersonal landmark. Inferotemporally lesioned animals show none of these disturbances (Bates and Ettlinger 1960; Pohl 1973). Data are not yet available concerning the effects of foveal-prestriate lesions on these tasks. Conversely, monkeys with superior parietal lesions do not show any impairment in visual shape discrimination when compared with unoperated controls (Bates and Ettlinger 1960; Pohl 1973). Thus, this syndrome is doubly dissociable from the syndrome of impaired acquisition and retention of visual shape discrimination produced by inferotemporal lesions, and is at least singly dissociable from the impaired visual form discrimination produced by foveal-prestriate lesions. Further experiments comparing foveal-prestriate and superior posterior parietal lesions with respect to tasks of visual orientation would be informative (see Ungerleider and Mishkin, this volume).

Finally, visual-object agnosia is also doubly dissociable from the disturbances associated with damage to the angular gyri of either the dominant or the nondominant hemisphere in man. Constructural apraxia, the inability to construct copies of visual patterns, was first described by Kleist (1934). He emphasized the independence of these constructional disturbances from those characteristics of visual disorientation and localized the syndrome to cortex of the angular gyrus. (Deeper lesions involving the strata saggitalia produce visual disorientation.) Later, Paterson and Zangwill (1944) established that, at least with lesions of the nondominant hemisphere, the analysis of spatial relations of elements of a compound visual structure is impaired even when no complex praxic task is required. Thus, the terms *visuo-spatial agnosia* and *apractagnosia for spatial relations* have been used in the English literature to denote this disturbance in the analysis and synthesis of spatial relations. Both early investigations (Lange 1936) and extensive later studies (e.g., Ettlinger et al. 1957) established the lack of visual-object agnosia in patients with severe constructional apraxia. Conversely, a defining feature of associative visual agnosia is the ability of the patient to copy even complex visual patterns accurately, albeit slowly. Thus, associative visual agnosia and constructional apraxia are independent syndromes.

The same is probably true with regard to apperceptive visual agnosia. In apperceptive visual agnosia, although copying of visual patterns is extremely poor (Adler 1944; Benson and Greenberg 1969), the nature of the impairment resembles that of severe visual disorientation (constant losing of place resulting in "piecemeal drawings") far more than it does constructional apraxia (no losing of place, rotational errors prominent). The relationship of the drawing disturbances in the syndromes of apperceptive visual agnosia, visual disorientation, and visual spatial agnosia needs further study.

In the dominant hemisphere, lesions of angular gyrus produce, in addition to constructional disturbances, the elements of Gerstmann's syndrome: right-left confusion, finger agnosia, acalculia, and agraphia. This syndrome may occur in severe form without visual agnosia (Gerstmann 1931; Stengel 1944), and, conversely, associative visual agnosia occurs independent of the Gerstmann syndrome (Rubens and Benson 1971). Data on "apperceptive" visual agnosia are not available.

One feature distinguishing the above syndromes of the angular gyrus from visual-object agnosia is the multimodal nature of the former. Constructional apraxia (visuospatial agnosia) probably involves difficulties in spatial analysis not only of visual patterns but of tactile patterns as well (Ettlinger et al. 1957; Levine, unpublished observations). Finger agnosia is also characteristically a supramodal deficit (Lange 1936; Kinsbourne and Warrington 1962). This lack of modality specificity is understandable in view of anatomical studies (Pandya and Kuypers 1969; Jones and Powell 1970) demonstrating that in the rhesus monkey the cortex in the walls of the superior temporal sulcus—presumably a rudimentary homologue of the human angular gyrus—receives converging projections from visual, auditory, and somesthetic association areas.

Thus, in man, the syndromes of visual agnosia produced by foveal-prestriate and inferotemporal lesions are at least partially dissociable from the orbitofronto-anterior temporal syndrome, the amnesic syndrome, visual disorientation, and spatial apractagnosia. Correspondingly, in the monkey, impaired visual discrimination from foveal-prestriate and inferotemporal lesions is dissociable from the fronto-temporal syndrome, the hippocampal syndrome, and parietal-lobe syndromes.

The Question of Dominance

To this point I have demonstrated the striking anatomic and behavioral parallels between the syndromes of impaired visual discrimination in the monkey and the syndromes of apperceptive and associative visual agnosia in man. A possible exception to this parallelism is that unilateral lesions may produce visual agnosia in man, whereas serious impairment in visual discrimination in the monkey requires bilateral lesions.

Ettlinger and Gautrin (1972) studied the effects of unilateral inferotemporal lesions on retention and acquisition of form discrimination in the monkey. Mild impairment in retention of a circle-versus-triangle discrimination and in acquisition of a plus-versus-square discrimination were found. The performances were worse than those of monkeys with unilateral superior temporal lesions and better than those of monkeys with bilateral inferotemporal lesions. No differences were found between animals with unilateral inferotemporal lesions on the right and left sides.

The overwhelming majority of humans with visual agnosia have bilateral lesions. However, in several cases with postmortem examination (Nielsen 1937) and in many more clinically studied cases visual agnosia resulted from unilateral lesions. In those cases where sufficient data were available (Lissauer 1890; Nodet 1899; Poussepp 1923; Caplan and Hedley-Whyte 1974) the agnosia was found to be of the associative type. In one case (Poussepp 1923) the lesion involved the ventral occipital convexity (in addition to a lesion of the angular gyrus). In the other cases, the medial and basal portions of the occipital and occipitotemporal lobes were involved by infarction in the territories of one posterior cerebral artery (see Hahn's [1895] analysis of Lissauer's case).

In those cases of associative visual agnosia resulting from unilateral temporo-occipital lesions, it was at first not clear which hemisphere was more important. Early authors (Heilbronner 1910; Mingazzini 1922; Pötzl 1928) asserted that the left hemisphere was the one usually affected. In Nielsen's (1937) series of twelve cases with unilateral lesions, eight had left-hemisphere damage.

Modern studies in which the visual-stimulus material and the required responses have been varied systematically have elucidated the nature of hemisphere specialization and cooperation in the perception and memory of forms. Agreement between various investigators, however, is not complete. Milner (1958, 1971) and her colleagues (Kimura 1963), comparing patients with unilateral anterior temporal lobectomy, found poorer recall of printed (or spoken) language in a group with left-side lesions and poorer recognition of nonsense shapes (recurrent figures paradigm) in a group with right-side lesions. In tests of matching after tachistoscopic presentation of a sample stimulus, the two groups did not differ when the stimuli were multiple familiar objects, multiple letters, or overlapping pictures of familiar objects. However, the patients with right-side lesions were more impaired when the stimuli were scattered dots or overlapping nonsense forms. On the basis of these experiments, supplemented by studies demonstrating superiority of the normal right visual field in identifying tachistoscopically presented letters and pictures of familiar objects (Mishkin and Forgays 1952; Heron 1957; Wyke and Ettlinger 1961), these authors concluded that the crucial difference appeared to be whether the visual-stimulus pattern had a short verbal description—a name. If so, its "perception" and "memory" would be affected by left-side lesions; if not, by right-side lesions. Sperry and his colleagues (Levy et al. 1972), studying patients with section of the forebrain commissures, came to a somewhat different conclusion. They demonstrated superiority of the left hemisphere (and thus the right visual field) for identification of tachistoscopically presented chimeric stimuli only in tasks where a verbal response such as naming was required. When the required response was only to point to the correct stimulus with either hand, the right hemisphere was superior even when the stimuli were namable objects.

Levine and Calvanio (1980b) studied identification of tachistoscopically presented arrays of visual stimuli in the right and left visual fields of an adult woman with an acquired lesion of the posterior third of the corpus callosum and evidence of little or no extracallosal damage. They found marked right-visual-field superiority for identification of groups of letters, numbers, and colors, whether the required response was naming, drawing what was seen, or matching. There was, however, no difference between the visual fields when the stimuli were simple geometric shapes or letterlike nonsense forms. These results are more consistent with the conclusions of Milner and Kimura than with those of Levy. In further experiments, utilizing partial report, Levine and Calvanio demonstrated that increasing the number of simultaneously presented letters enhanced right-visual-field superiority, even when response requirements remained constant. In this sense, the right-visual-field superiority was one of letter "perception," and not merely one of generating a response to what was perceived. However, this stimulus-specific perceptual superiority of the right visual field was also task-specific. It was present when a group of letters had to be identified (by naming, drawing, or pointing), but was much less marked when the patient had only to indicate how many different letter shapes were present in the array. The presence of such task specificity begins to resemble some of the results of Levy et al. (1972). However, important differences remain, which may be the result of testing procedures (chimeric stimuli splitting the midline vs. whole stimuli lateralized to one visual field), extent of interhemispheric disconnection, and presence of cerebral lesions with onset in childhood.

The marked superiority of the left hemisphere in the perceptual identification of letters, numbers, and colors is also evident from the differing effects of left and right temporo-occipital lesions on reading and identifying colors. Numerous cases of agnosic alexia and color agnosia with unilateral left temporo-occipital damage have been reported (see, for example, Lissauer 1890; Dejerine 1892; Pötzl 1928; Geschwind and Fusillo 1966; Mohr et al. 1971). Though the deficit in alexia and color agnosia has often been described in terms of a naming disability (Geschwind 1965), recent experiments (Kinsbourne and Warrington 1962; Levine and Calvanio 1978) suggest that letter perception itself is affected in these patients. Thus, they match or draw tachistoscopically presented letter arrays no better than they name them. No such deficits are found in patients with right temporo-occipital lesions.

In contrast to the marked advantage of the left hemisphere in the identification of letters, numbers, and colors, there appears to be a definite though considerably smaller advantage of the right hemisphere in the identification of faces. Numerous studies comparing patients with unilateral lesions of either the right or the left hemisphere (De Renzi et al. 1968), the right or the left temporal lobe (Milner 1968), and the right or the left retrorolandic area (Newcombe and Russell 1969; Yin

1970) demonstrated more impairment in tasks of face recognition with right-hemisphere lesions than with left hemisphere lesions. However, the deficits in these cases were not severe. All postmortem examinations of cases with severe prosopagnosia revealed bilateral lesions (see, for example, Heidenhain 1927; Gloning et al. 1970; Rubens and Benson 1971; Lhermitte et al. 1972). Thus, it would appear that the recognition of faces, though performed better by the right hemisphere than by the left, is shared by the two sides.

These studies allow a better understanding of how the various manifestations of associative visual agnosia are related to the corresponding lateralization of the cerebral lesions. As we have seen, the critical zones within the hemispheres are the inferior occipital and posterior inferotemporal lobes. This localization is consistent with the greater ease of demonstrating hemispheric visual perceptual asymmetries with posterior unilateral lesions than with more anterior lesions, such as anterior temporal lobectomy. The same areas are of importance in the monkey; however, in man one must consider the laterality of the lesion as well as its intrahemispheric localization.

Left temporo-occipital lesions, or lesions that simultaneously disconnect the two hemispheres and deprive the left hemisphere of visual input, impair identification of arrays of letters, numbers, and colors (agnosic alexia and color agnosia). Identification of common objects and faces may be less impaired, although with larger lesions (Lissauer 1890; Caplan and Hedly-Whyte 1974) the identification of common objects may be abnormal (visual object agnosia). Right temporo-occipital lesions produce no overt identification deficits as a rule, although careful testing reveals deficits in face identification with little or no impairment in letter or color identification. Bilateral lesions will greatly intensify the difficulties in face identification (prosopagnosia) and object identification, while deficits in identification of letters, numbers, and colors appear to depend on the extent of the left-hemisphere lesions or disconnection. Thus, bilateral lesions, in which the left-side lesion is not massive enough to produce severe letter, number, and color identification deficits, may result in prosopagnosia and moderate object agnosia without severe alexia or color agnosia (Levine 1978).

The psychological principle (if there is one) underlying the asymmetric perceptual performances of the two human hemispheres has been the subject of much speculation, and is far from clear. Sweeping dichotomies, such as "analytic versus synthetic," "categorical versus appositive," and "serial-temporal versus parallel-spatial," abound. Each may be partially true. I prefer as a point of departure the incontrovertible fact that speech production and comprehension are strongly lateralized to the left hemisphere in the overwhelming majority of right-handed people. The anatomic and physiologic bases for this speech dominance are not clear, though they have been the subject of intensive investigation in recent years (for a review see Harnad et al.

1977). In any event, by the time a child begins to read, he has mastered an elaborate system of audio-orolingual communication that is largely under the control of the left hemisphere. Furthermore, he learns to read by using speech to identify (discriminate) certain visual forms (letters and letter groups). The ability to make such discriminations increases steadily throughout childhood (Hoffman 1927); that is, the visual span of apprehension for letters increases. If this form of perceptual learning were under the control of the left hemisphere, the entire process, including input and output, would be intrahemispheric. If the right hemisphere were involved, input to this hemisphere would have to be integrated with speech output from the left hemisphere across the corpus callosum. Myers (1965b), however, has shown that in animals the corpus callosum is a channel of limited capacity. Learning that is entirely intrahemispheric is stronger than learning that requires transfer across the corpus callosum. Thus, it may be the case that the perceptual learning resulting in an increasing span of visual letter identification occurs more strongly in the left hemisphere than in the right as a direct result of the left hemisphere's superiority for speech. A similar explanation may underlie the left hemisphere's superiority in number and color identification.

The basis for the right hemisphere's more modest advantage in identification of faces is less clear. It is true that the preverbal infant is capable of discriminating familiar from unfamiliar faces. So it is no surprise, in view of the remarks above, that there is no left-hemisphere advantage in face identification, since the latter is not learned in intimate relationship with speech. But the reasons for a right-hemisphere advantage, however modest, remain unknown. The advantage is probably not limited to faces. Prosopagnosic patients have difficulty identifying not only faces, but also species of animals, birds, trees, or landscapes (see Brown 1972 for a brief review). Kimura (1963), as previously mentioned, showed that patients with right anterior temporal lobectomy performed more poorly than patients with left temporal lobectomy in matching tachistoscopically presented nonsense figures and dot patterns. It seems that the right hemisphere's advantage is most evident for the discrimination of complex forms that differ from each other not because of distinctive, salient features, but because there are numerous, small differences, along multiple, not obviously separable dimensions (Garner 1974), that cannot be concisely described with words. Is this right-hemisphere advantage merely the result of commitment of the left hemisphere's perceptual apparatus to the processing of language-facilitated visual material, or is its specialization independent of the left? The answer is still unknown.

It appears that the asymmetry of hemispheric function in human visual perception is intimately associated with the asymmetry in the production and comprehension of speech. It is, therefore, not surprising that such asymmetry has not appeared in the monkey. However, as we have seen, the intrahemispheric locations of the lesions producing visual agnosias in man and disorders of visual discrimination in the monkey are highly similar. The variations in human (associative) visual agnosia are better explained on the basis of how the lesions are distributed *between* left and right temporo-occipital regions than by the assumption (Pötzl 1928) of special regions *within* the left human temporo-occipital lobe for perception of objects, letters, colors, etc.

Conclusions

In man, as in the monkey, a system of corticocortical pathways linking striate cortex, the foveal region of ventrolateral visual association cortex, and the inferobasal temporal neocortex is necessary for intact visual-form classification. Bilateral lesions of this system that allow no connections, ipsilateral or crossed (via corpus callosum), between striate and inferotemporal cortices produce visual agnosia in man and impaired visual-discrimination performance in the monkey. These syndromes are highly comparable with respect to their behavioral features, and it seems entirely appropriate to regard monkeys with bilateral inferotemporal lesions as models of "associative" visual agnosia in man. Bilateral lesions of foveal-prestriate visual-association cortex may also be a model of "apperceptive" visual agnosia in man, but this conclusion must remain tentative until the pathologic anatomy of apperceptive visual agnosia is better established. The major difference between monkey and man concerns the effects of unilateral lesions of the foveal-prestriate inferotemporal system. Because of specialization of the left hemisphere for language in man, disturbances in the visual identification of letters, numbers, colors, and at times common objects can be severely impaired with only a unilateral (left-side) lesion. I postulate that this perceptual specialization is the result of use of previously lateralized speech processes in acquiring more efficient perception of these materials. The right hemisphere is better than the left in visual identification of complex familiar and unfamiliar forms (such as faces), for which facility in discrimination is not dependent on left-lateralized language processes. Unilateral right-sided lesions may modestly impair identification of such forms, while identification of letter and number arrays may remain normal. Because of this asymmetry in man, the comparability of the behavior of man and monkey with lesions in the foveal striate–foveal prestriate–inferotemporal system emerges most clearly when one considers the effects of bilateral lesions in each.

References

Adler, A. 1944. "Disintegration and restoration of optic recognition in visual agnosia." *Arch. Neurol. Psychiatr.* 51: 243–259.

————. 1950. "Course and outcome of visual agnosia." *J. Nerv. Ment. Dis.* 111: 41–51.

Ajuriaguerra, J. de, and H. Hécaen. 1960. *Le cortex cerebrale.* Paris: Masson.

Allman, J. M. 1977. "Evolution of the visual system in the early primates." *Progr. Psychobiol. Physiol. Psychol.* 7: 1–53.

Allman, J. M., and J. H. Kass. 1971. "A representation of the visual field in the caudal third of the middle temporal gyrus of the owl monkey (*Aotus trivirgatus*). *Brain Res.* 31: 85–105.

————. 1974a. "The organization of the second visual area (V2) in the owl monkey: A second order transformation of the visual hemifield." *Brain Res.* 76: 247–265.

————. 1974b. "A crescent-shaped cortical visual area surrounding the middle temporal area (MT) in the owl monkey (*Aotus trivirgatus*)." *Brain Res.* 81: 199–213.

————. 1975. "The dorsomedial cortical visual area: A third tier area in the occipital lobe of the owl monkey (*Aotus trivirgatus*)." *Brain Res.* 100: 473–487.

————. 1976. "Representation of the visual field on the medial wall of occipital-parietal cortex in the owl monkey." *Science* 191: 572–575.

Allman, J. M., J. H. Kass, R. H. Lane, and F. M. Miezin. 1972. "A representation of the visual field in the inferior nucleus of the pulvinar in the owl monkey (*Aotus trivirgatus*)." *Brain Res.* 40: 291–302.

Balint, R. 1909. "Seelenlahmung des 'Schauens,' optische Ataxie, raumliche Storung der Aufmerksamkeit." *Monats. Psych. Neurol.* 25: 57–71.

Bates, J. A. V., and G. Ettlinger. 1960. "Posterior biparietal ablations in the monkey." *Arch. Neurol.* 3: 177–192.

Bay, E. 1952. "Analyse eines Fall von Seelenblindheit." *Deutsche Z. Nervenheilkunde* 168: 1–23.

————. 1953. "Disturbances of visual perception and their examination." *Brain* 76: 515–550.

Bender, D. B. 1973. "Visual sensitivity following infero-temporal and foveal prestriate lesions in the monkey." *J. Comp. Physiol. Psychol.* 84: 613–621.

Benson, D., and J. Greenberg. 1969. "Visual form agnosia." *Arch. Neurol.* 20: 82–89.

Benson, D. F., J. Segarra, and M. L. Albert. 1974. "Visual agnosia-prosopagnosia." *Arch. Neurol.* 30: 307–310.

Blum, J. S., K. L. Chow, and K. H. Pribram. 1950. "A behavioral analysis of the organization of the parieto-temporo-preoccipital cortex." *J. Comp. Neurol.* 93: 53–100.

Bodamer, J. 1947. "Die Prosop-Agnosie (Die Agnosie des Physiognomieerkennens). *Arch. Psych. Z. Neurol.* 179: 6–53.

Brain, W. R. 1961. *Speech disorders: Aphasia, Apraxia, and Agnosia.* London: Butterworths.

Brown, J. W. 1972. *Aphasia, Apraxia, and Agnosia: Clinical and Theoretical Aspects.* Springfield, Ill.: Thomas.

Brown, S., and E. A. Schafer. 1888. "An investigation into the function of the occipital and temporal lobes of the monkey's brain." *Phil. Trans.* 179: 303–327.

Brown, T. S. 1963. "Olfactory and visual discrimination in the monkey after selective lesions of the temporal lobe." *J. Comp. Physiol. Psychol.* 56: 764–768.

Brown, T. S., H. E. Rosvold, and M. Mishkin. 1963. "Olfactory discrimination after temporal lobe lesions in monkeys." *J. Comp. Physiol. Psychol.* 56: 190–195.

Bruner, J. S. 1957. "On perceptual readiness." *Psychol. Rev.* 64: 123–152.

Butter, C. M. 1968. "The effect of discrimination training on pattern equivalence in monkeys with infero-temporal and lateral striate lesions." *Neuropsychologia* 6: 27–40.

Butter, C. M., M. Mishkin, and H. E. Rosvold. 1965. "Stimulus generalization in monkeys with infero-temporal lesions and lateral occipital lesions." In *Stimulus Generalization*, D. J. Mostofsky, ed. Stanford, Calif.: Stanford University Press.

Campos-Ortega, J. A., W. R. Hayhow, and P. F. de V. Cluver. 1970. "The descending projections from the cortical visual fields of macaca mulatta with particular reference to the question of a cortico-lateral geniculate pathway." *Brain, Behav., Evol.* 3: 368–414.

Caplan, L. R., and T. Hedley-Whyte. 1974. "Cuing and memory dysfunction in alexia without agraphia." *Brain* 97: 251–262.

Chow, K. L. 1950. "Retrograde cell degeneration study of the cortical projection field of the pulvinar in the monkey." *J. Comp. Neurol.* 93: 313–340.

———. 1951. "Effects of partial extirpations of the posterior association cortex on visually mediated behavior." *Comp. Psychol. Mongr.* 20: 187–217.

———. 1952. "Further studies on selective ablation of associative cortex in relation to visually mediated behavior." *J. Comp. Physiol. Psychol.* 45: 109–118.

———. 1954. "Lack of behavioral effects following destruction of some thalamic nuclei in monkey." *Arch. Neurol. Psychiatr.* 71: 762–771.

Cohn, R., M. A. Neumann, and D. H. Wood. 1977. "Prosopagnosia: A clinicopathological study." *Ann. Neurol.* 1: 177–182.

Cowey, A., and C. G. Gross. 1970. "Effects of foveal prestriate and infero-temporal lesions on visual discrimination by rhesus monkeys." *Exp. Brain Res.* 11: 128–144.

Cowey, A., and L. Weiskrantz. 1967. "A comparison of the effects of infero-temporal and striate cortex lesions on the visual behavior of rhesus monkeys." *Q. J. Exp. Psychol.* 19: 246–253.

Critchley, M. 1964. "The problem of visual agnosia." *J. Neurol. Sci.* 1: 274–290.

Dejerine, J. 1892. "Contribution a l'etude anatomopathologique et clinique des differente varietes de cecite verbale." *Mem. Soc. Biol.* 4: 61–90.

————. 1914. *Semiologie des affections du systeme nerveux*. Paris: Masson.

DeJong, R. H., H. H. Itabashi, and J. R. Olsen. 1969. "Memory loss due to hippocampal lesions." *Arch. Neurol.* 20: 339–348.

DeRenzi, E., P. Faglioni, and H. Spinnler. 1968. "The performance of patients with unilateral brain damage on face recognition tasks." *Cortex* 4: 17–34.

Douglas, R. J. 1967. "The hippocampus and behavior." *Psychol. Bull.* 67: 416–442.

Douglas, R. J., and K. H. Pribram. 1966. "Learning and limbic lesions." *Neuropsychologia* 4: 197–220.

Drachman, D. A., and J. Arbit. 1966. "Memory and the hippocampal complex." *Arch. Neurol.* 15: 52–61.

Druga, R. 1969–1970. "Neocortical projections to the amygdala. (An experimental study with the Nauta method)." *J. Hirnforsch.* 11: 467–476.

Efron, R. 1968. "What is perception?" In *Boston Studies in the Philosophy of Science*, R. Cohen and M. Wartofsky, eds. Boston: Reidel.

Ettlinger, G. 1956. "Sensory defects in visual agnosia." *J. Neurol. Neurosurg. Psychiatr.* 19: 297–307.

Ettlinger, G., and D. Gautrin. 1972. "Visual discrimination performance in the monkey: The effect of unilateral removal of temporal cortex." *Cortex* 7: 317–331.

Ettlinger, G., and J. Waegner. 1958. "Somaesthetic alternation, discrimination and orientation after frontal and parietal lesions in monkeys." *Q. J. Exp. Psychol.* 10: 177–186.

Ettlinger, G., and M. Wyke. 1961. "Defects in identifying objects visually in a patient with cerebrovascular disease." *J. Neurol. Neurosurg. Psychiatr.* 24: 254–259.

Ettlinger, G., E. Warrington, and O. L. Zangwill. 1957. "A further study of visual spatial agnosia." *Brain* 80: 335–361.

Evarts, L. V. 1952. "Effect of ablation of prestriate cortex on auditory-visual association in monkey." *J. Neurophysiol.* 15: 191–200.

Freud, S. 1891. *On Aphasia*, E. Stengel, tr. New York: International University Press, 1953.

Friedman, H. M., and N. Allen. 1969. "Chronic effects of complete limbic destruction." *Neurology* 19: 679–690.

Gaffan, D. 1974. "Recognition impaired and association intact in the memory of monkeys after transection of the fornix." *J. Comp. Physiol. Psychol.* 86: 1100–1109.

Garner, W. R. 1974. *The Processing of Information and Structure*. Potomac, Md.: Erlbaum.

Gascon, G., and F. Gilles. 1973. "Limbic dementia." *J. Neurol. Neurosurg. Psychiatr.* 36: 421–430.

Gerstmann, J. 1931. "Zur Symptomatologie der Herderkrankungen in der Übergangsregion der unteren Parietal- und mittleren Okzipitalhirnwindung." *Deutsche Arch. Nervenheilk.* 116: 46–49.

Geschwind, N. 1965. "Disconnexion syndromes in animals and man." *Brain* 88: 237–294.

Geschwind, N., and M. Fusillo. 1966. "Color naming defects in association with alexia." *Arch. Neurol.* 15: 137–146.

Gloning, I., K. Gloning, K. Jellinger, and R. Quatember. 1970. "A case of 'prosopagnosia' with necropsy findings." *Neuropsychologia* 8: 199–204.

Godwin-Austen, R. B. 1965. "A case of visual disorientation." *J. Neurol. Neurosurg. Psychiatr.* 28: 453–458.

Goldstein, K., and A. Gelb. 1918. "Psychologische Analysen hirnpathologischer Falle auf Grund von Untersuchungen Hirverletzten." *Z. Ges. Neurol. Psychiatr.* 41: 1–142.

Gross, C. G. 1972. "Infero-temporal cortex and vision." In *Progress in Physiological Psychology*, E. Stellar and J. M. Sprague, eds. New York: Academic.

———. 1973. "Visual function of infero-temporal cortex." In *Handbook of Sensory Physiology*, vol. 1, part 3B, R. Jung, ed. Berlin: Springer.

Gross, C. G., A. Cowey, and F. J. Manning. 1971. "Further analysis of the visual discrimination deficits following foveal prestriate and infero-temporal lesions in rhesus monkeys." *J. Comp. Physiol. Psychol.* 76: 1–7.

Hahn, E. 1895. "Pathologisch-anatomische Untersuchung des Lissauerschen Falles von Seelenblindheit." *Arb. Psychiatr. Klin. Breslau* 2: 107–119.

Harnad, S., R. W. Doty, L. Goldstein, J. Jaynes, and G. Krauthammer, eds. 1977. *Lateralization in the Nervous System*. New York: Academic.

Hearst, E., and K. H. Pribram. 1964. "Appetitive and aversive generalization gradients in amygdalectomized monkeys." *J. Comp. Physiol. Psychol.* 58: 296–298.

Hécaen, H., and J. de Ajuriaguerra. 1954. "Balint's syndrome (psychic paralysis of visual fixation) and its minor forms." *Brain* 77: 373–400.

Hécaen, H., R. Angelergues, C. Bernhardt, and J. Chiarelli. 1957. "Essai de distinction des modalites clinique de l'agnosie des physiognomies." *Rev. Neurologique* 96: 125–144.

Heidenhain, A. 1927. "Beitrag zur Kenntnis der Seelenblindheit." *Monats. Psychiatr. Neurol.* 66: 61–116.

Heilbronner, K. 1910. "Die aphasischen, apraktischen, und agnostischen Storungen." In *Handbuch der Neurologie*, vol. 1, F. Lewandowsky, ed. Berlin: Springer.

Heron, H. 1957. "Perception as a function of retinal locus and attention." *Amer. J. Psychol.* 70: 38–48.

Hoff, H., I. Gloning, and K. Gloning. 1962. "Die zentralen Storungen der optischen Wahrnehmung." *Wiener Med. Wochens.* 112: 450–459.

Hoffman, J. 1927. "Experimentell-psychologische Untersuchungen uber Leseleistungen von Schulkindern." *Arch. ges. Psychol.* 58: 325–388.

Holmes, G. 1918. "Disturbances of visual orientation." *Br. J. Ophthalmol.* 2: 449–468, 506–516.

———. 1938. "The cerebral integration of the ocular movements." *Br. Med. J.* 2: 107–112.

Holmes, G., and G. Horrax. 1919. "Disturbances of spatial orientation and visual attention with loss of stereoscopic vision." *Arch. Neurol. Psychiatr.* 1: 385–407.

Horel, J. A. 1978. "The neuroanatomy of amnesia: A critique of the hippocampal memory hypothesis." *Brain* 101: 403–445.

Iversen, S. D. 1973. "Visual discrimination deficits associated with posterior inferotemporal lesions in monkey." *Brain Res.* 62: 89–101.

Iversen, S. D., and L. Weiskrantz. 1967. "Temporal lobe lesions and memory in the monkey." *Nature* 201: 740–742.

Iwai, E., and M. Mishkin. 1968. "Two visual foci in the temporal lobe of monkey." Japan–U.S. Joint Seminar on Neurophysiological Basis of Learning and Behavior, Kyoto, Japan.

———. 1969. "Further evidence on the locus of visual area in the temporal lobe of the monkey." *Exp. Neurol.* 25: 585–594.

Jones, E. G., and T. P. S. Powell. 1970. "An anatomical study of converging sensory pathways within the cerebral cortex of the monkey." *Brain* 93: 793–820.

Kimble, D. P., and K. H. Pribram. 1963. "Hippocampectomy and behavior sequences." *Science* 139: 824–825.

Kimura, D. 1963. "Right temporal lobe damage: Perception of unfamiliar stimuli after damage." *Arch. Neurol.* 8: 264–271.

Kinsbourne, M., and E. K. Warrington. 1962. "A disorder of simultaneous form perception." *Brain* 85: 461–486.

Kleist, K. 1934. *Gehirnpathologie.* Liepzig: Barth.

Klüver, H., and P. C. Bucy. 1937. " 'Psychic blindness' and other symptoms following bilateral temporal lobectomy in rhesus monkeys." *Amer. J. Physiol.* 119: 352–353.

———. 1939. "Preliminary analysis of functions of the temporal lobes of monkeys." *Arch. Neurol. Psychiatr.* 42: 979–1000.

Kuypers, H. G. J. M., M. K. Szwarczbart, M. Mishkin, and H. I. Rosvold. 1965. "Occipitotemporal cortico-cortical connections in the rhesus monkey." *Exp. Neurol.* 11: 245–262.

Lange, J. 1936. "Agnosien und Apraxien." In *Handbuch der Neurologie,* O. Bumke and O. Foerster, eds. Berlin: Springer.

Lashley, K. S. 1948. "The mechanism of vision. XVIII. Effects of destroying the visual 'associative areas' of the monkey." *Genet. Psychol. Monogr.* 37: 107–166.

Levine, D. 1978. "Prosopagnosia and visual object agnosia: A behavioral study." *Brain Lang.* 5: 341–365.

Levine, D., and R. Calvanio. 1978. "A study of the visual defect in verbal alexia-simultanagnosia." *Brain* 101: 65–81.

———. 1980a. "Disorders of visual behavior following bilateral posterior cerebral lesions." *Psychol. Res.* 41: 217–234.

———. 1980b. "Visual discrimination after lesions of the posterior corpus callosum." *Neurology* 30: 21–30.

Levy, J., C. Trevarthen, and R. W. Sperry. 1972. "Perception of bilateral chimeric figures following hemispheric deconnection." *Brain* 95: 61–78.

Lhermitte, F., F. Chain, R. Escourolle, B. Ducarne, and B. Pillon. 1972. "Etude anatomo-clinique d'un cas de prosopagnosie." *Rev. Neurologique* 126: 329–346.

Lissauer, H. 1890. "Ein Fall von Seelenblindheit nebst Beitrage zur Theorie derselben." *Arch. Psychiatr.* 21: 222–270.

Manning, F. J. 1971. "Punishment for errors and visual discrimination learning by monkeys with infero-temporal cortex lesions." *J. Comp. Physiol. Psychol.* 75: 146–152.

Manning, F. J., C. J. Gross, and A. Cowey. 1971. "Partial re-inforcement; effects on visual learning after foveal prestriate and infero-temporal lesions." *Physiol. Behav.* 6: 61–64.

Mathers, L. H. 1971. "Tectal projection to the posterior thalamus of the squirrel monkey." *Brain Res.* 35: 295–298.

Michel, F., M. Jeannerod, and M. Devic. 1965. "Trouble de l'orientation visuelle dans les trois dimensions de l'espace." *Cortex* 1: 441–466.

Milner, B. 1958. "Psychological defects produced by temporal lobe excision." *Res. Pub. Ass. Res. Nerv. Ment. Dis.* 36: 244–257.

———. 1968. "Visual recognition and recall after right temporal lobe excision in man." *Neuropsychologia* 6: 191–209.

———. 1971. "Interhemispheric differences in the localization of psychological processes in man." *Br. Med. Bull.* 27: 272–277.

Milner, B., and H. Teuber. 1968. "Alteration of perception and memory in man: Reflections on methods." In *Analysis of Behavioral Change,* L. Weiskrantz, ed. New York: Harper and Row.

Mingazzini, G. 1922. *Der Balken.* Berlin: Springer.

Mishkin, M. 1954. "Visual discrimination performance following partial ablations of the temporal lobe. II. Ventral surface versus hippocampus." *J. Comp. Physiol. Psychol.* 47: 187–193.

———. 1966. "Visual mechanisms beyond the striate cortex." In *Frontiers in Physiological Psychology*, R. Russel, ed. New York: Academic.

Mishkin, M., and D. C. Forgays. 1952. "Word recognition as a function of retinal locus." *J. Exp. Psychol.* 43: 43–48.

Mishkin, M., and K. H. Pribram. 1954. "Visual discrimination performance following partial ablations of the temporal lobe. I. Ventral versus lateral." *J. Comp. Physiol. Psychol.* 47: 14–20.

Mishkin, M., and L. Weiskrantz. 1959. "Effects of cortical lesions in monkeys on critical flicker frequency." *J. Comp. Physiol. Psychol.* 52: 660–666.

Mohr, J. P., J. Leicester, L. T. Stoddard, and M. Sidman. 1971. "Right hemianopia with memory and color deficits in circumscribed left posterior cerebral artery infarction." *Neurology* 21: 1104–1113.

Munk, H. 1881. *Über die Funktionen der Grosshirnrinde—Gesammelte Mitteilungen aus den Jahren 1877–1880.* Berlin: Hirschwald.

Myers, R. E. 1965a. "Organization of visual pathways." In *Functions of the Corpus Callosum*, E. G. Ettlinger, ed. Boston: Little, Brown.

———. 1965b. "The neocortical commissures and interhemispheric transmission of information." In *Function of the Corpus Callosum*, E. G. Ettlinger, ed. Boston: Little, Brown.

Newcombe, F., and W. Russell. 1969. "Dissociated visual perceptual and spatial deficits in focal lesions of the right hemisphere." *J. Neurol. Neurosurg. Psychiatr.* 32: 73–81.

Nielsen, J. M. 1937. "Unilateral cerebral dominance as related to mind blindness." *Arch. Neurol. Psychiatr.* 38: 108–135.

Nodet, V. 1899. "Les agnosies; la cecite psychique en particulier." Thesis, Université de Lyon.

Orbach, J., and R. L. Frantz. 1959. "Differential effects of temporal neocortical resection on overtrained and non-overtrained visual habits in monkeys." *J. Comp. Physiol. Psychol.* 51: 126–129.

Orbach, J., B. Milner, and T. Rasmussen. 1960. "Learning and retention in monkeys after amygdala-hippocampus resection." *Arch. Neurol.* 3: 230–251.

Oscar-Berman, M., S. Heywood, and C. G. Gross. 1973. "The effects of posterior cortical lesions on eye orientation during visual discrimination by monkeys." *Neuropsychologia* 12: 175–182.

Pandya, D. N., and H. G. J. M. Kuypers. 1969. "Cortico-cortical connections in the rhesus monkey." *Brain Res.* 13: 13–36.

Pasik, P., T. Pasik, W. S. Battersby, and M. B. Bender. 1958. "Visual and tactual discrimination by macaques with serial temporal and parietal lesions." *J. Comp. Physiol. Psychol.* 51: 427–436.

———. 1960. "Factors influencing visual behavior of monkeys with bilateral temporal lobe lesions." *J. Comp. Neurol.* 115: 89–102.

Paterson, A., and O. L. Zangwill. 1944. "Disorders of visual space perception associated with lesions of the right cerebral hemisphere." *Brain* 67: 331–358.

Pevzner, S., B. Bornstein, and M. Lowenthal. 1962. "Prosopagnosia." *J. Neurol. Neurosurg. Psychiatr.* 25: 336–338.

Pohl, W. 1973. "Dissociation of spatial discrimination deficits following frontal and parietal lesions in monkeys." *J. Comp. Physiol. Psychol.* 82: 227–239.

Pötzl, O. 1928. *Die Aphasielehre vom Standpunkte der klinischen Psychiatrie*, vol. 1. Leipzig: Franz Deuticke.

Poussepp, L. 1923. "Contribution aux recherches sur la localisation de l'aphasie visuelle." *Presse Med.* 31: 564–565.

Pribram, H. B., and J. Barry. 1956. "Further behavioral analysis of parieto-temporo-preoccipital cortex." *J. Neurophysiol.* 19: 99–106.

Pribram, K. H. 1960. "The intrinsic systems of the forebrain." In *Handbook of Physiology: Neurophysiology II,* J. Field et al., eds. Washington, D.C.: American Physiological Society.

Pribram, K. H., and M. H. Bagshaw. 1953. "Further analysis of the temporal lobe syndrome utilizing frontotemporal ablations." *J. Comp. Neurol.* 99: 347–375.

Pribram, K. H., and M. Mishkin. 1955. "Simultaneous and successive visual discrimination by monkeys with infero-temporal lesions." *J. Comp. Physiol. Psychol.* 48: 198–202.

Rosvold, H. E., M. Mishkin, and M. K. Szwarzbart. 1958. "Effects of subcortical lesions in monkeys on visual discrimination and single alternation performance." *J. Comp. Physiol. Psychol.* 51: 437–444.

Rubens, A. B., and D. F. Benson. 1971. "Associative visual agnosia." *Arch. Neurol.* 24: 305–316.

Schwartzbaum, J. S. 1965. "Discrimination behavior after amygdalectomy in monkeys." *J. Comp. Physiol. Psychol.* 60: 314–319.

Scoville, W. B., and B. Milner. 1957. "Loss of recent memory after bilateral hippocampal lesions." *J. Neurol. Neurosurg. Psychiatr.* 20: 11–21.

Siqueira, E. B., and L. Franks. 1974. "Anatomic connections of the pulvinar." In *The Pulvinar-LP Complex*, I. S. Cooper et al., eds. Springfield, Ill.: Thomas.

Spamer, C. 1876. "Über Aphasie und Asymbolie nebst Versuch einer Theorie der Sprachbildung." *Arch. Psychiatr.* 6: 496–542.

Stengel, E. 1944. "Loss of spatial orientation, constructional apraxia, and Gerstmann's syndrome." *J. Ment. Sci.* 90: 753–760.

Symmes, D. 1965. "Flicker discrimination by brain damaged monkeys." *J. Comp. Physiol. Psychol.* 60: 470–473.

Terzian, H., and G. Dalle Ore. 1955. "Syndrome of Klüver and Bucy reproduced in man by bilateral removal of the temporal lobes." *Neurology* 5: 373–380.

Victor, M., J. B. Angevine, E. C. Mancall, and C. M. Fisher. 1961. "Memory loss with lesions of hippocampal formation." *Arch. Neurol.* 5: 244–263.

von Bonin, G., and P. Bailey. 1947. *The Neocortex of Macaca Mulatta.* Urbana: University of Illinois Press.

von Monakow, C. 1914. *Die Lokalisation im Grosshirn und der Abbau der Funktion durch kortikale Herde.* Wiesbaden: J. F. Bergmann.

von Stauffenberg, W. 1914. "Über Seelenblindheit." *Arb. hirnanat. Inst. Zurich* 8: 1–212.

Walker, A. E., A. F. Thompson, and J. D. McQueen. 1953. "Behavior and the temporal rhinencephalon in the monkey." *Johns Hopkins Hosp. Bull.* 93: 65–93.

Weiskrantz, L. 1956. "Behavior changes associated with ablation of the amygdaloid complex in monkeys." *J. Comp. Physiol. Psychol.* 49: 381–391.

Weiskrantz, L., and A. Cowey. 1963. "Striate cortex lesions and visual acuity in the rhesus monkey." *J. Comp. Physiol. Psychol.* 56: 225–231.

Weiskrantz, L., and M. Mishkin. 1958. "Effects of temporal and frontal cortical lesions on auditory discrimination in monkeys." *Brain* 81: 406–414.

Whitlock, D. G., and W. J. H. Nauta. 1956. "Subcortical projections from the temporal neocortex in *Macaca mulatta*." *J. Comp. Neurol.* 106: 183–212.

Wilson, M. 1957. "Effects of circumscribed cortical lesions upon somesthetic and visual discrimination in the monkey." *J. Comp. Physiol. Psychol.* 50: 630–635.

Wilson, M., H. M. Kaufman, R. E. Zieler, and J. P. Lieb. 1972. "Visual identification and memory in monkeys with circumscribed inferotemporal lesions." *J. Comp. Physiol. Psychol.* 78: 173–183.

Woods, B., and E. Poppel. 1974. "Effect of print size on reading time in a patient with verbal alexia." *Neuropsychologia* 12: 31–41.

Wyke, M., and G. Ettlinger. 1961. "Efficiency of recognition in left and right visual fields." *Arch. Neurol.* 5: 659–665.

Yin, R. K. 1970. "Face recognition by brain-injured patients: A dissociable ability?" *Neuropsychologia* 8: 395–402.

Zeki, S. M. 1969. "Representation of central visual fields in prestriate cortex in monkey." *Brain Res.* 14: 271–291.

———. 1970. "Interhemispheric connections of prestriate cortex in monkey." *Brain Res.* 19: 63–75.

———. 1971a. "Convergent input from the striate cortex (area 17) to the cortex of the superior temporal sulcus in the rhesus monkey." *Brain Res.* 28: 338–340.

———. 1971b. "Cortical projections from two prestriate areas in the monkey." *Brain Res.* 34: 19–35.

21 Visual Perception and Oculomotor Areas in the Primate Brain

Richard Latto

Eye movements are intimately involved in many of the processes of visual perception. This is most apparent in the perception of position. The estimation of egocentric position—that is, position with respect to the subject's body—requires knowledge of both the position of the target in the visual field and the direction of gaze. This is true for all animals that possess the ability to move their eyes independently of their heads. As eye movements evolved, perhaps initially to keep the eyes stationary with respect to the environment by means of a vestibulo-oculomotor reflex (Wall 1962), a system of monitoring the direction of gaze and using that information to determine egocentric target position must also have evolved. Since primates move their eyes in different ways and for many different reasons, and since they use several different kinds of information to determine direction of gaze (Stevens et al. 1976), it seems probable that this process involves widespread parts of the oculomotor system.

In animals with well-developed fovea, eye movements also perform a gating function. They are a basic mechanism for spatial attention, enabling one region of the visual world to be attended to by placing it on the fovea while much of the information from the rest of the visual world is lost. The organization of these eye movements into fixed repetitive scanpaths during visual exploration (Noton and Stark 1971) enables such animals to sample the world in a systematic way. Thus, it has been argued (perhaps most influentially by Hebb [1949]) that pattern perception, too, normally involves the oculomotor system, occurring as an additive process integrating the input from a structured series of fixations into a perceptual whole. The scanpath provides the framework for this integration, and without it the information from each fixation would be like the individual pieces of a jumbled jigsaw puzzle. As with position, pattern perception in anything but a tachistoscopic situation is dependent on combining the visual input with information about eye movements.

Where should we look in the brain for structures that, although outside the primary visual system, might nevertheless be concerned with perception through its oculomotor component? Electrical stimulation of

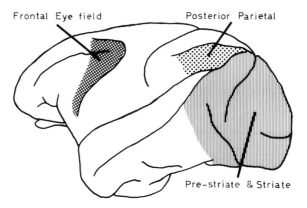

Figure 21.1 Cortical areas where electrical stimulation results in eye movements.

the cerebral cortex of the monkey, first performed systematically by Ferrier (1875, 1886), produces eye movements from surprisingly widespread cortical areas (figure 21.1). All these areas have close homologs in man. Hebb suggested a perceptual role for the frontal eye fields, but Ferrier reported another structure giving eye movements on stimulation: the superior colliculus. Ferrier (1886) concluded his section on the colliculus by saying that "these ganglia form an essential portion of the mechanism of the co-ordination of retinal and general sensory impressions with the mesencephalic motor apparatus" (p. 173).

This chapter is concerned with the possible role in visual perception of three of these oculomotor areas: frontal eye fields, posterior parietal cortex, and superior colliculus. I will not be discussing striate and prestriate cortex, because we know already that these areas are directly involved in visual perception—particularly in visual pattern perception.

Stimulation studies subsequent to Ferrier's (for example, Fleming and Crosby 1955; Wagman 1964; Robinson and Fuchs 1969; Robinson 1972; Schiller and Stryker 1972) demonstrated that all three of these regions produce conjugate, contralateral eye movements in monkeys, with movements in different directions coming consistently from different parts of each region. Single-unit recording also shows considerable similarities between the three areas. All contain units that fire when the monkey moves its eyes (Bizzi 1968; Schiller and Stryker 1972; Wurtz and Goldberg 1972a; Mountcastle et al. 1975; Bushnell and Goldberg 1979), and in each case some of these start firing before the eye movement begins—the traditional definition of a motor unit. All three areas also contain units with visual receptive fields (Mohler et al. 1973; Schiller and Stryker 1972; Robinson et al. 1978).

Stimulation and recording studies tell a good deal about the nature of the information handled in these three regions, and they are beginning to reveal how that information is coded. They say rather little, though, about the kinds of visual behavior these areas are concerned with. Chronic recording of single-unit activity in behaving animals should, in theory, be helpful here. But so far it has only proved possible to correlate unit activity and behavior for very simple visuomotor tasks, and even with these it is not usually possible to deduce much about cause and effect (that is, function) from a simple correlation. For example, units in posterior parietal cortex that fire prior to fixation saccades have been interpreted in two very different ways. Mountcastle et al. (1975) consider them to be motor units, indicating that this region has a high-level command function, while Robinson et al. (1978) consider them sensory units driven by the visual target that also triggers the eye movement, indicating that the region is primarily a sensory association area.

The best source of information for deducing function remains the behavioral analysis of the effects of lesions. This too has many logical problems, not least that of distinguishing the direct behavioral effects of lesions from indirect effects due to disruption of other parts of the brain. But it does allow us to study complex behavior, and it does allow us to draw conclusions about a whole area rather than about a small and not necessarily representative sample of cells within that area. (See also Dean, this volume). This chapter will discuss the effects of lesions in these three higher oculomotor areas—superior colliculus, frontal eye fields, and posterior parietal cortex—on three aspects of visual perception.

Disorders of Visual Perception After Oculomotor-Area Lesions

Visual Neglect
Why did Hebb (1949, p. 35) argue that the frontal eye fields "must have something to do with the elaboration of sensations into visual perceptions"? It was mainly because unilateral lesions of the frontal eye fields had been shown by Kennard (1939) to produce a dramatic, though transient, neglect of contralateral visual stimuli—a spatial neglect. In Latto and Cowey 1971a it was shown that this neglect is not simply a motor deficit—that is, it is not due to the monkey failing to move his eyes or hands into the contralateral half of space, because it was still clearly present when we mapped monkeys' visual fields in a perimeter in which the monkey's eye position was controlled and the responses made by the monkey to signal the presence or absence of a stimulus were not spatially related to the stimulus (figure 21.2). Thus, the neglect is retinotopic.

This deficit is not unique to the frontal eye fields. Small, superficial unilateral superior collicular lesions, made after recovery from frontal-eye-field lesions, also produced contralateral visual-field defects that could be shown by perimetry to be retinotopically organized and independent of head and eye movements (Latto and Cowey 1971a). This effect was strong enough after bilateral lesions to produce a small increase in a brightness discrimination threshold in which the mon-

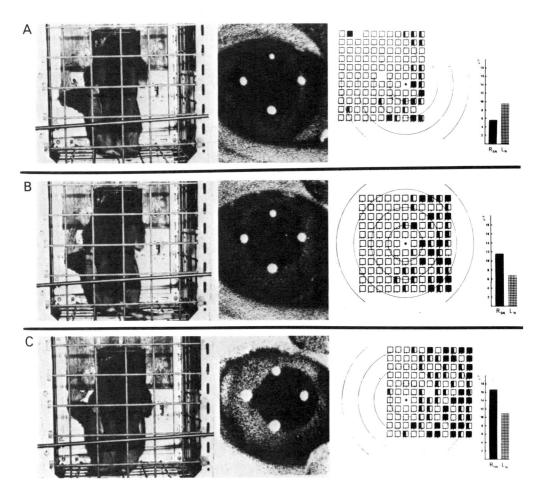

Figure 21.2 The performance of a monkey in the perimeter 15–17 days after a left frontal-eye-field lesion, demonstrating that the contralateral neglect is retinotopic. Visual fields are shown on the right, with circles showing displacement from the monkey's fixation point in 10° intervals. Small squares show the retinal positions at which test flashes occurred, with the actual number of flashes in each position shown by the number of tags along the bottom of each square. The proportion of these flashes of which an incorrect response (R_{SN}) was made is represented by the proportion of the square shaded black. The histogram to the right of each field shows the percentage over all trials of false negative errors (R_{SN}) and false positive errors (L_N). The central column of the figure shows samples of the eye photographs taken on each trial and compared with standard photographs in order to determine the monkey's direction of gaze when the test flash occurred. The column on the left, showing the monkey looking through the peephole at the stimulus display, was also taken synchronously with the test flash. Row b shows the monkey's performance with the perimeter's fixation point straight ahead of him. Row a shows performance with the fixation point moved 15° to the right, and in row c the fixation point is 15° to the left. (For further details of the perimetric technique used see Latto and Cowey 1971a).

keys had time to fixate the discriminanda (Latto 1977). The threshold experiment included a group of monkeys with bilateral posterior parietal lesions, but, unlike the frontal-eye-field-lesion and superior-collicular-lesion groups, they were not impaired. This negative finding after bilateral parietal lesions is interesting, since Heilman et al. (1971) showed that similar, but unilateral, parietal lesions produce a neglect of stimuli presented to the monkey on the contralateral side. Thus, it is possible that, unlike the frontal-eye-field neglect, the parietal neglect is not caused by a genuine field defect or amblyopia, but is the result of a failure to respond to contralateral spatial positions or even of a failure to look spontaneously with the head and eyes to the contralateral side. This leads to the prediction that the parietal neglect would not be present if the monkey's response to a stimulus were not spatially related to that stimulus or if its head and eye movements were controlled. Most of the tests of visual neglect used clinically have a strong motor component. This may explain why, in man, parietal neglect is much the most commonly observed and, clinically, neglect is usually taken as a sign of right parietal damage, although it has occasionally been reported after right frontal lesions as well (Silberpfennig 1941; Jenkner and Kutschera 1965; Heilman and Valenstein 1972) and Zihl and von Cramon (1979) found a small elevation in brightness threshold in the visual field contralateral to a congenital malformation of the superior colliculus.

Visual neglect is more striking after unilateral than after bilateral lesions, but this is probably of little significance. Rather, it is simply because it is easier to demonstrate a lack of response to stimuli in a limited part of space, when the response to the rest of space can be used as a control, than to demonstrate an overall lack of response to all stimuli. In fact, bilateral neglects have been found, in the monkey perimeter after a bilateral frontal-eye-field lesion (Latto and Cowey 1971a), and, clinically, in those patients with bilateral parieto-occipital lesions (or, less commonly, in those with bilateral frontal-lobe lesions) who demonstrate Balint's syndrome (Balint 1909). It has been argued (Hécaen and Albert 1978) that Balint's syndrome is a bilateral version of the contralateral neglect of similar patients with unilateral lesions.

Visual Search
One of the principal functions of the oculomotor system is to explore the visual environment. Therefore, I have looked at the effects of oculomotor-area lesions on the speed and accuracy with which monkeys can conduct a spatially organized search of a visual display (Latto 1978a,b). Monkeys were trained to search a vertical panel for a small circular spot, which appeared in random positions among varying numbers of irrelevant stimuli, and to press the side of the panel on which the target appeared.

The effect of bilateral lesions on the accuracy with which they did this is shown in part A of figure 21.3. Frontal and superior collicular lesions

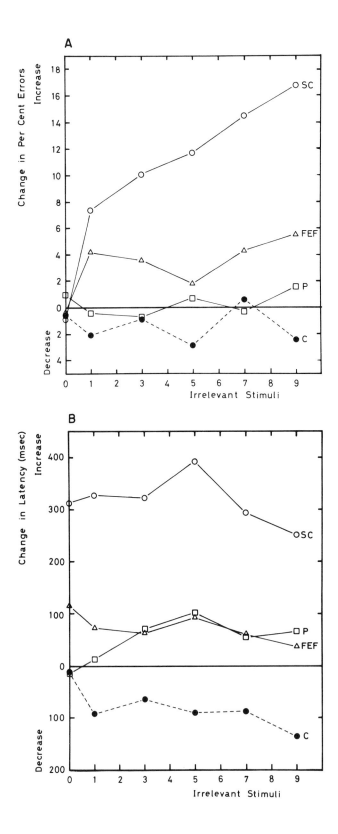

produced a significant increase in errors, whereas the parietal-lesion group did not differ from controls. This again suggests that frontal and collicular deficits have a sensory component whereas the parietal deficit does not. More interesting, part B of figure 21.3 shows the effects of the lesions on the speed with which the monkeys performed the visual search. Now all three groups were impaired, with the collicular-lesion group significantly worse than the parietal-lesion and frontal-eye-field-lesion groups. This picture is largely supported by the clinical evidence. Anterior and posterior cortical lesions and collicular damage have all been found to give impairments of visual search in man (Teuber 1964; Luria et al. 1966; Tyler 1969; Chedru et al. 1973; Kimura et al. 1979), with collicular damage producing the severest deficit (Kimura et al. 1979).

We can conclude that lesions in any of the three oculomotor areas reduce the speed with which primates perform a spatially organized search of their visual environments, but we can say rather little about why the impairment occurs or whether there are qualitative differences between lesions in the three areas. There are three possible causes of the impairment. There might be a disability of basic oculomotor function, such that the subject could no longer fixate or make saccades accurately. There could be a higher-level oculomotor disorder affecting the subject's ability to organize the systematic scanpath that most efficiently extracted the desired information from the visual display. Finally, there could be a disorder in the subject's spatial attention, such that he simply neglected or failed to see stimuli in certain parts of his visual field. It might be expected that different lesions produce these impairments to different extents. A rough analysis of the sequence in which monkeys scanned the visual search display, comparing the latencies for finding targets in different positions over the display (Latto 1978a), found no abnormalities after any of these lesions. So the lesions do not seem to disrupt the ability to organize a systematic scanpath, at least to any great extent. Nor did any of the three lesions impair the visual search for a near-threshold target appearing against a homogeneous background (although one frontal-eye-field-lesioned monkey missed a high proportion of the targets when they fell on the peripheral parts of the display [Latto 1978a]), so it is unlikely that any of the lesions produce an impairment by an overall disruption of saccadic eye movements. Collicular lesions have been found to cause a decrease in saccade accuracy and an increase in saccade latency (Wurtz and Goldberg 1972b; Mohler and Wurtz 1977), so it seems probable that in our experiments

Figure 21.3 Performance of four groups of monkeys on the visual search for a target hidden among varying numbers of irrelevant stimuli. (A) Pre- to postoperative change in response errors, expressed as a percentage of total responses. (B) Pre- to postoperative change in mean latency of correct responses. SC: bilateral superior colliculus lesions. FEF: bilateral frontal-eye-field lesions. P: posterior parietal lesions. C: unoperated controls.

collicular visual-search impairments are due either to a decrease in saccade accuracy that necessitates more correction saccades to achieve fixation or to an increase in the latency of saccades that shift the gaze between stimuli. But the neglect that collicular lesions have been shown to produce in other conditions, such as the perimeter (see above) and the reaching apparatus (see below), suggests that there may be an attentional deficit contributing to the visual-search impairment in addition to the defective fixation. Although there is no direct evidence about the effects of parietal lesions on saccade accuracy and speed, area 7, like the superior colliculus, contains units that fire before the monkey moves its eyes (Mountcastle et al. 1975), so it seems possible that the parietal visual-search deficit, like the parietal neglect, is primarily motor in origin. Since frontal-eye-field neurons do not fire before eye movements unless there is a visual stimulus present in the neuron's receptive field (Bizzi 1968; Bushnell and Goldberg 1979), and since the frontal-eye-field neglect is not due to motor abnormalities (see above), it is probable that the frontal-eye-field visual-search deficit is primarily due to a disorder in spatial attention.

Visually Guided Reaching
The final piece of behavior I want to discuss in this hunt for perceptual deficits after oculomotor-area lesions is the accuracy with which monkeys can reach for a visual target. Reaching is dependent on both visual and eye-position information, and is therefore a likely contender for mediation by one of these areas. Reaching for a brief flash of light in the dark is logically impossible without knowing both the retinal position of the stimulus and the direction of gaze of the eyes at the time the flash occurred. But localizing a target in the light, particularly if it is a persistent target, can be done in another way. It can be localized relative to other visual stimuli in known positions, the subject's hand for example. That is, it can be localized using context information even in the absence of information about the direction of gaze.

Monkeys were trained to press whichever of a horizontal row of 20 levers was, or had been, illuminated. The levers were 2 cm apart, and so, at the minimum distance from which the monkey could view them, they were separated by about 5° of visual angle. The task was self-paced, with the monkey initiating each trial by pressing an observing response lever below the central target lever.

The initial experiment (Latto and Marzi, in prep.) made a pre- to postoperative comparison of reaching under four conditions designed to give a decreasing availability of context information and therefore an increasing dependence on eye-position information: persistent target in the light, persistent target in the dark, brief (62 msec) target in the light, and brief target in the dark.

There were four lesion groups in this experiment, but for simplicity I shall present data only from the bilateral-frontal-eye-field-lesion group (figure 21.1) and the bilateral-superior-collicular-lesion group. There were also two parietal-lesion groups (an area 7 group and a superior temporal sulcus group, both bilateral). Surprisingly, neither of the parietal groups showed any evidence for misreaching in our four basic conditions—perhaps because we did not start retesting until 3 weeks after operation, and parietal misreaching (although commonly reported, at least after unilateral lesions) is often a transient phenomenon (Hartje and Ettlinger 1973; Faugier-Grimaud et al. 1978; LaMotte and Acuna 1978). The collicular lesions were made using a suction pipette under an operating microscope. We were very careful not to penetrate the tentorium cerebelli, but we did have to cut a small amount of the posterior splenium. Thus, we had a control group who had their colliculi exposed, but not damaged, by retracting the hemispheres and cutting the splenium. The four collicular lesions made varied in their completeness and symmetry. This led to some individual differences in the results, to which I shall return later.

The two measures of reaching accuracy, percent errors and standard deviation of responses around the target position, both showed a very similar pattern of pre- to postoperative change, with the collicular-lesion group showing impairments but not the frontal-eye-field-lesion group. Figure 21.4 shows the changes in the standard deviation of response. As with overall errors, the only significant changes were in the collicular-lesion group in both the conditions with a brief target. The

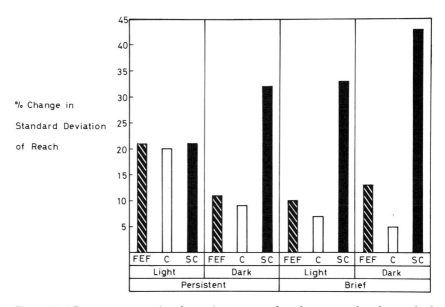

Figure 21.4 Pre- to postoperative change in accuracy of reach, expressed as the standard deviation of responses around the target position, of three groups of monkeys under the four conditions described in the text. FEF: bilateral frontal eye-field lesions. C: sham-operated controls. SC: bilateral superior colliculus lesions.

Perception and Oculomotor Areas in Primates

apparently large change shown by collicular-lesioned monkeys in the condition with a persistent target in the dark was due mainly to one monkey, and so overall the collicular-lesion group was not significantly different from the control group on this condition.

Collicular lesions cause misreaching for a brief target with a larger impairment in the dark. But collicular-lesioned monkeys *are* impaired in the light, as others have shown (Keating 1974; MacKinnon et al. 1976; Butter et al. 1978 and this volume). So maybe context information is impaired as well as information about direction of gaze. I did two experiments to test for this using the accuracy-of-reach paradigm. The first of these looked at the effect of delayed target onset on reaching in the dark. There are two reasons why darkness makes reaching worse in normal monkeys: It removes visual feedback (reafference) from eye movements and limb movements (that is, from fixating the target and moving the hand to the target), and it removes the visual context of the target (the monkey can no longer use the information that the target is 10° to the right of the food cup, or 15° to the left of his hand, as a cue giving more accurate localization). The length of time in the dark before the target appears will not affect the loss of visual feedback information. However, it will affect context information, which we should expect to degrade over time, either through decay of visual memory or through the imposition of errors by unmonitored small drifting eye movements (an effect that Matin et al. [1966] have shown produces errors in context cueing in man). So this experiment compared the effect of varying the interval between the observing response and target onset on reaching for a brief target in the light and in the dark. Figure 21.5 shows the percentage of on-target responses. Again, collicular-lesioned monkeys were worse than control monkeys in both the light and the dark, and the deficit was larger in the dark. The standard deviation of reach also showed a similar pattern. But—and this was the crucial finding—there was no interaction between the deficit in the dark and the delay or between the deficit in the light and the delay. So delay does not make the deficit in the dark less severe, compared with controls. It makes both collicular-lesioned and control monkeys worse to the same extent, presumably because of the decay of context information over time. We can therefore conclude from this experiment that collicular lesions do not affect the use of context information, but produce misreaching by disrupting the coding of the retinal position of the stimulus and/or the direction of gaze, making the system much more dependent on visual reafference and context information. The second experiment testing for an impairment in the use of context information by collicular-lesioned monkeys looked at reaching for a brief target under stroboscopic illumination. It is a well-known phenomenon, described by MacKay (1958) and Gregory (1958), that stroboscopic illumination causes a breakdown in the stability of the visual world. This process, which MacKay says occurs most strongly at 5–6 flashes per second, is vividly

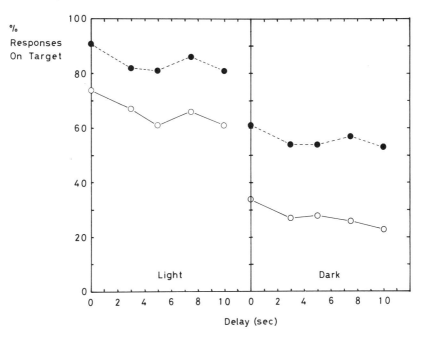

Figure 21.5 Postoperative accuracy of reach, expressed as percentage of responses on target, of two groups of monkeys reaching for a brief target in the light or in the dark with variable delay of target onset. (●) Sham-operated control; (○) bilateral superior colliculus lesions.

demonstrated by MacKay's lighted-cigarette effect. When a glowing cigarette is moved around under stroboscopic illumination, the glowing tip is seen as separating from the body of the cigarette at certain points in the movement. This, I would argue, is because the context (the body of the cigarette) is seen only during the flashes, and therefore gives misleading information about the position of the tip, which is seen continuously. In a situation where the target is self-illuminated and the context is not, stroboscopic light degrades the context information but does not affect information about the retinal position of the target or the direction of gaze. So, if collicular-lesioned monkeys are already impaired in their use of context information, they should be less affected than normals by stroboscopic illumination. After a quick experiment on myself to confirm that stroboscopic illumination does cause misreaching for a self-illuminated target, I tested the two groups of monkeys reaching for brief targets under stroboscopic light. Figure 21.6 shows their performance. The percentage of responses on target for the control group shows a nice MacKay effect, with misreaching increasing down to 6 flashes per second and then improving again. This is shown rather more obviously in their standard deviation of reach, which clearly peaked at around 6–8 flashes per second. But, although the collicular-lesioned group were again considerably worse than the controls, there was no interaction between group difference and flash rate. So the de-

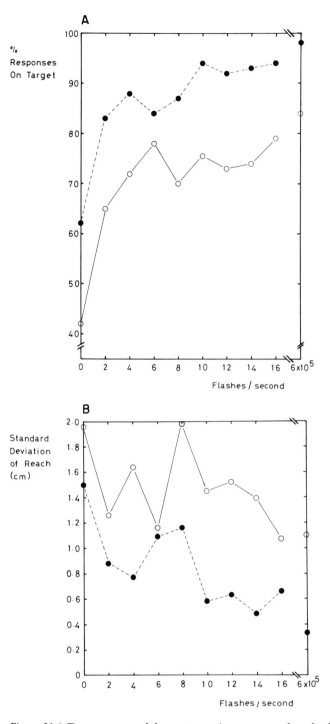

Figure 21.6 Two measures of the postoperative accuracy of reach of two groups of monkeys reaching for a brief target under stroboscopic illumination of varying flash rates. (A) Percentage of responses on target. (B) Standard deviation of responses around target position. (●) Sham-operated control; (○) bilateral superior colliculus lesions.

gradation of context information by stroboscopic illumination affected both groups equally, showing again that the collicular-lesioned monkey is still able to use context information.

Finally, there are two important properties of the collicular deficit that do not show up in the group data but are demonstrated very clearly on a trial-by-trial analysis of individual monkeys' performances. Two out of the four collicular-lesioned monkeys, as well as showing the increase in standard deviation of reach, showed a consistent shift in mean direction of reach for a particular target position. Part A of figure 21.7 shows the performance of one of these monkeys in the basic reaching condition with brief targets. The top part of the figure, giving the mean response position for each target position, shows that this monkey consistently reached 1–2 cm nearer the center of the display than the target, almost as if there was a shrinking of his visual space. The effect was much more severe in the dark than in the light for both monkeys. There was no obvious reason why two monkeys showed this shift in mean response position while two did not. One possibility is that each monkey was adopting a different strategy when reaching for the target and that this resulted in different patterns of head and eye movements or arm use, but nothing could be detected from gross observation. Two monkeys also showed a clear neglect on their trial-by-trial analysis when responding in the light. Most dramatic were the results of the monkey shown in part B of figure 21.7. When the target was to the left of center, this monkey failed to respond on about half the trials and on the other half it made a stereotyped response, pressing a lever about 12 cm to the right of center. The neglect in this experiment was much less transient than the impairment in the earlier experiment in the perimeter. The data in figure 21.7B were collected two months after operation. The two monkeys that showed neglect in the reaching apparatus were the two with slightly asymmetrical lesions, and they both neglected the side contralateral to the larger lesion. However, the neglect is probably not limited to asymmetrical lesions, since one of the monkeys with symmetrical lesions showed quite clear runs of neglecting first one side of the display and then the other. Unlike the misreaching, the neglect was always most striking in the light, perhaps simply because in the light the contrast of the stimulus was much less. This difference between the misreaching and the neglect is important because it suggests that the two deficits are to some extent independent.

Discussion

The problem of distinguishing between sensory and oculomotor correlations in electrophysiology, which was discussed in the introduction, is paralleled by a similar difficulty in discriminating sensory and oculomotor deficits in lesion studies. The three areas investigated in this study are all clearly in some sense oculomotor and are all recipro-

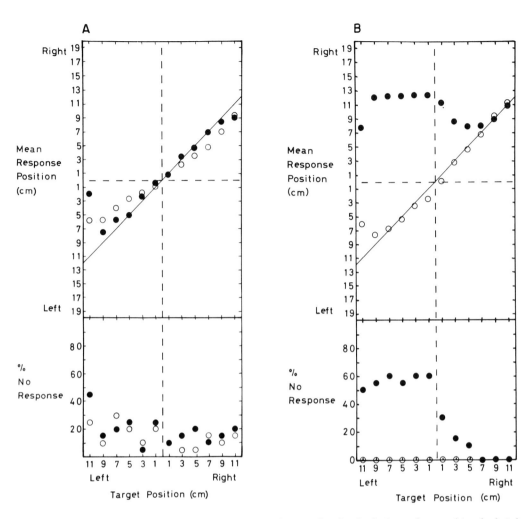

Figure 21.7 Postoperative performance of two of the monkeys with collicular lesions when reaching for brief targets in the light (●) or in the dark (○). Upper section of each figure shows monkey's mean response position for each target position, both positions being measured from the center of the display. Lower section shows percentage of trials at each target position on which no response was made within 1,500 msec. (A) Monkey SC3; (B) monkey SC2.

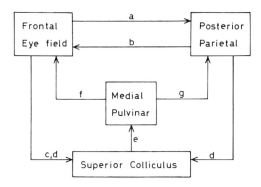

Figure 21.8 The main anatomical pathways between the frontal eyefields, the posterior parietal region, and the superior colliculus. Sources: (a) Pandya and Vignolo 1971; Stanton et al. 1977; (b) Pandya and Vignolo 1969; (c) Kunzle et al. 1976; (d) Kuypers and Lawrence 1967; (e) Benevento and Fallon 1975; (f) Trojanowski and Jacobson 1974; (g) Baleydier and Mauguiere 1977.

cally interconnected (figure 21.8). Lesioning them individually produces rather small effects on simple oculomotor functions (Pasik and Pasik 1964; Latto and Cowey 1971b; Wurtz and Goldberg 1972b; Heywood and Ratcliff 1975; Mohler and Wurtz 1977)—probably mainly because there is considerable redundancy between the areas, so that the functions of a damaged area can be taken over by other areas. This redundancy is powerfully demonstrated by an experiment of Schiller et al. (1979) showing that while frontal-eye-field and superior collicular lesions on their own produce small effects, lesioning both structures at once in the same monkey results in an almost complete loss of all eye movements. But it is most unlikely that these areas are serving the same functions in the normal brain. It may be that the redundancy is only for relatively simple functions, like optokinetic nystagmus (Pasik and Pasik 1964) or fixation of discrete and isolated stimuli (Wurtz and Goldberg 1972b; Mohler and Wurtz 1977), and that these areas *are* specialized for more complex tasks, such as organizing scanpaths to sample the visual world in a meaningful way. Thus, large deficits are found only with more complex oculomotor tasks such as the visual search experiment described above.

This chapter has attempted to show that although both frontal-eye-field and posterior parietal lesions impair visual search, the fact that only frontal-eye-field lesions raise brightness thresholds and (in the visual search situation) increase discrimination errors suggests that frontal deficits are primarily sensory whereas parietal deficits are response-centered.

Collicular lesions result in a deficit in both sensory and motor components. They produce a rise in brightness threshold and a neglect of, or failure to respond to, stimuli occurring in a limited area of the monkey's visual space—a spatial neglect. They result in a severely impaired visual search, with an increase in both latencies and errors. And they cause an impairment in reaching for a brief visual target. Localization of the visual target by context information seems to be normal after collicular lesions; rather, the misreaching is due to errors in the coding of retinal position and/or direction of gaze.

It is logically very difficult to distinguish between impairments in coding retinal position and impairments in coding direction of gaze. Lennie and Sidwell (1978) found that the error in localization following a fixation saccade is matched closely by the error in the size of the saccade, which suggests that the extent of the saccade is not being monitored and that the system assumes the intended fixation saccade was made correctly. So any increase in saccade errors of the kind that have been shown to occur after collicular lesions (Mohler and Wurtz 1977) would necessarily lead to errors in localization. Thus, the reaching deficit, as well as the visual-search deficit, is most simply explained in terms of a decrease in the accuracy with which fixation saccades are made. But we do not know whether this increase in saccade errors is due to an oculomotor impairment, or to inaccuracy in localizing the stimulus to be fixated, or both.

It is similarly difficult to distinguish, in many behavioral situations, between an attentional disorder of the kind argued for by Goodale and Graves (this volume) and a motor or orienting deficit of the kind suggested by earlier experiments (Schneider 1969; Barnes et al. 1970). The two aspects of behavior are commonly causally connected, and this causal link can go in either direction. So "attentional" disorders after collicular lesions in the rat (Goodale, this volume) and the monkey (this chapter) might be due to a disruption in orientational responses, perhaps because of a loss of the process (described at the single-unit level by Goldberg and Wurtz [1972] and at the perceptual level by Crovitz and Daves [1962]) whereby the intention to move the eyes toward a target results in the enhancement of the neural and perceptual response to that target, or perhaps, more simply, because of a disruption in foveation, which, as I argued in the introduction, is an important form of selective attention. Conversely, as Goodale has argued, apparent disorders in orientation after collicular lesions in rodents (Schneider 1969; Barnes et al. 1970) and monkeys (this chapter) might be the result of an attentional failure, with the animal failing to orient to a stimulus because it does not see it. In the reaching experiment described above, the motor deficit (misreaching) and the visual neglect occur most strongly in different conditions, so it seems likely that there are two separate deficits: misreaching due to inadequate monitoring of retinal position or, more probably, eye position; and a spatial neglect, a failure to perceive or respond to stimuli occurring in a limited area of the monkey's visual space. This duality parallels the distinction coming out of recent single-unit studies of the superior colliculus (see, for example, Sparks 1978) between superficial "perceptual" layers and deeper "oculomotor" layers which modulate each other but which can probably function independently, suggesting that the visual neglect may be primarily due to superficial damage while the misreaching and visual-search deficit is primarily due to damage to deeper layers. Certainly the collicular-lesioned monkey tested in the perimeter, whose lesions were both very superficial, showed a neglect but no obvious oculomotor abnormalities.

Attention, in the sense of the active gating of information, is not a unitary function. There is even some doubt about whether it exists at all. Neisser (1976) argued that active filtering out of information is not a necessary function in the explanation of the selectivity of perception, for this can be achieved much more simply by saying that perception is selective because of limitations of sensory and cognitive faculties. Thus, Neisser would argue, we do not perceive ultraviolet light because our sensory systems do not respond to it, and most of us do not perceive the stimuli that differentiate between male and female chicks because we do not possess the necessary perceptual schemata. Similarly, in the classic demonstration of selective attention, the cocktail party problem (Cherry 1957), we ignore irrelevant conversations not because they are filtered out, but because they do not fit the perceptual schemata used to follow the conversation being attended to. But if there is active filtering in the nervous system, then it will be a multiple process occurring at many stages in the passage of information through the brain, from selection by foveation right through to selection by response bias. So it would be naive to suppose that attention could be mediated by a single structure in the brain, such as the superior colliculus. A more likely suggestion is that the colliculus is primarily concerned with other aspects of perception and oculomotor control, such as orientation to and fixation of peripheral stimuli, and that these, almost as a by-product, necessitate the selection of some information and the filtering out of the rest. In any case, until we really understand what attention is, it is probably better for those of us concerned with linking brain function to behavior to avoid using the word at all and stick with behaviorally defineable words such as *orientation* and *neglect*.

References

Baleydier, C., and F. Mauguiere. 1977. "Pulvinar-latero posterior afferents to cortical area 7 in monkeys demonstrated by horseradish peroxidase tracing technique." *Exp. Brain. Res.* 27: 501–507.

Balint, R. 1909. "Die Seelenlahmung des Schauens." *Monats. Psychiatr. Neurol.* 1: 51–81.

Barnes, P. J., L. M. Smith, and R. M. Latto. 1970. "Orientation to visual stimuli and the superior colliculus in the rat." *Q. J. Exp. Psychol.* 22: 239–247.

Benevento, L. A., and J. H. Fallon. 1975. "Ascending projections of superior colliculus in rhesus monkey." *J. Comp. Neurol.* 160: 339–361.

Bizzi, E. 1968. "Discharge of frontal eye field neurons during saccadic and following eye movements in unanaesthetized monkeys." *Exp. Brain Res.* 6: 69–80.

Bushnell, M. C., and M. E. Goldberg. 1979. "The monkey frontal eye fields have a neural signal that precedes visually guided saccades." *Soc. Neurosci. Abstr.* 5: 779.

Butter, C. M., C. Weinstein, D. B. Bender, and C. G. Gross. 1978. "Localization and detection of visual stimuli following superior colliculus lesions in rhesus monkeys." *Brain Res.* 156: 33–49.

Chedru, F., M. Leblanc, and F. Lhermitte. 1973. "Visual searching in normal and brain-damaged subjects (contribution to the study of unilateral attention)." *Cortex* 9: 94–111.

Cherry, C. 1957. *On Human Communication.* Cambridge, Mass.: MIT Press.

Crovitz, H. F., and W. Daves. 1962. "Tendencies to eye movement and perceptual accuracy." *J. Exp. Psychol.* 63: 495–498.

Faugier-Grimaud, S., C. Frenois, and D. G. Stein. 1978. "Effects of posterior parietal lesions on visually guided behavior in monkeys." *Neuropsychologia* 16: 151–168.

Ferrier, D. 1875. "Experiments on the brain of monkeys." *Proc. R. Soc.* B 23: 409–430.

———. 1886. *The Functions of the Brain,* second edition. London: Smith, Elder.

Fleming, J. F. R., and E. C. Crosby. 1955. "The parietal lobe as an additional motor area: The motor effects of electrical stimulation and ablation of cortical areas 5 and 7 in monkeys." *J. Comp. Neurol.* 103: 485–512.

Goldberg, M. E., and R. H. Wurtz. 1972. "Activity of superior colliculus in behaving monkey. II. Effect of attention on neuronal responses." *J. Neurophysiol.* 35: 560–574.

Gregory, R. L. 1958. "Eye movements and the stability of the visual world." *Nature* 182: 1214–1216.

Hartje, W., and G. Ettlinger. 1973. "Reaching in light and dark after unilateral posterior parietal ablations in the monkey." *Cortex* 9: 346–354.

Hebb, D. O. 1949. *The Organization of Behavior.* New York: Wiley.

Hécaen, H., and M. L. Albert. 1978. *Human Neuropsychology.* New York: Wiley.

Heilman, K. M., D. N. Pandya, E. A. Karol, and N. Geschwind. 1971. "Auditory inattention." *Arch. Neurol.* 24: 323–325.

Heilman, K. M., and E. Valenstein. 1972. "Frontal lobe neglect in man." *Neurology* (Minneapolis) 22: 660–664.

Heywood, S., and G. Ratcliff. 1975. "Long-term oculomotor consequences of unilateral colliculectomy in man." In *Basic Mechanisms of Ocular Motility and their Clinical Implications,* G. Lennerstrand, and P. Bach-y-Rita, eds. Oxford: Pergamon.

Jenkner, F. L., and E. Kutschera. 1965. "Frontal lobes and vision." *Confin. Neurol.* 25: 63–78.

Keating, E. G. 1974. "Impaired orientation after primate tectal lesions." *Brain Res.* 67: 538–541.

Kennard, M. A. 1939. "Alterations in response to visual stimuli following lesions of frontal lobes in monkeys." *Arch. Neurol. Psychiatr.* (Chicago) 41: 1153–1165.

Kimura, D., H. J. M. Barnett, and G. Burkhardt. 1979. "The psychological test pattern in progressive supranuclear palsy." *Res. Bull., Dept. Psychol., Univ. Western Ontario* 477: 1–19.

Kunzle, H., K. Akert, and R. H. Wurtz. 1976. "Projection of area 8 (frontal eye field) to superior colliculus in the monkey. An auto-radiographic study." *Brain Res.* 117: 487–492.

Kuypers, H. G. J. M., and D. G. Lawrence. 1967. "Cortical projections to the red nucleus and the brain stem in the rhesus monkey." *Brain Res.* 4: 151–188.

LaMotte, R. H., and C. Acuna. 1978. "Defects in accuracy of reaching after removal of posterior parietal cortex in monkeys." *Brain Res.* 149: 309–326.

Latto, R. 1977. "The effects of bilateral frontal eye-field, posterior parietal or superior collicular lesions on brightness thresholds in the rhesus monkey." *Neuropsychologia* 15: 507–516.

———. 1978a. "The effects of bilateral frontal eye-field, posterior parietal or superior collicular lesions on visual search in the rhesus monkey." *Brain Res.* 146: 35–50.

———. 1978b. "The effects of bilateral frontal eye-field lesions on the learning of a visual search task by rhesus monkeys." *Brain Res.* 147: 370–376.

Latto, R., and A. Cowey. 1971a. "Visual field defects after frontal eye-field lesions in monkeys." *Brain Res.* 30: 1–24.

———. 1971b. "Fixation changes after frontal eye-field lesions in monkeys." *Brain Res.* 30: 25–36.

Lennie, P., and A. Sidwell, 1978. "Saccadic eye movements and visual stability." *Nature* 275: 766–767.

Luria, A. R., B. A. Karpov, and A. L. Yarbus. 1966. "Disturbances of active visual perception with lesions of the frontal lobes." *Cortex* 1: 202–212.

MacKay, D. M. 1958. "Perceptual stability of a stroboscopically lit visual field containing self-luminous objects. *Nature* 181: 507–508.

MacKinnon, D. A., C. G. Gross, and D. B. Bender. 1976. "A visual deficit after superior colliculus lesions in monkeys." *Acta Neurobiol. Exp.* 36: 169–180.

Matin, L., D. Pearce, E., Matin, and G. Kibler. 1966. "Visual perception of direction in the dark: Local signs, eye movements and ocular proprioception." *Vision Res.* 6: 453–469.

Mohler, C. W., M. E. Goldberg, and R. H. Wurtz. 1973. "Visual receptive field of frontal eye field neurons." *Brain Res.* 61: 385–389.

Mohler, C. W., and R. H. Wurtz. 1977. "Role of striate cortex and superior colliculus in visual guidance of saccadic eye movements in monkeys." *J. Neurophysiol.* 40: 74–94.

Mountcastle, V. B., J. C. Lynch, A. Georgopoulos, H. Sakata, and C. Acuna. 1975. "The posterior parietal association cortex of the monkey: Command functions for operations within extrapersonal space." *J. Neurophysiol.* 38: 871–908.

Neisser, U. 1976. *Cognition and Reality.* San Francisco: Freeman.

Noton, D., and L. Stark. 1971. "Scanpaths in saccadic eye movements while viewing and recognizing patterns." *Vision Res.* 11: 929–942.

Pandya, D. N., and L. A. Vignolo. 1969. "Interhemispheric projections of the parietal lobe in the rhesus monkey." *Brain Res.* 15: 49–65.

———. 1971. "Intra- and interhemispheric projections of the precentral, premotor and arcuate areas in the rhesus monkey." *Brain Res.* 26: 217–233.

Pasik, P., and T. Pasik. 1964. "Oculomotor functions in monkeys with lesions of the cerebrum and superior colliculus." In *The Oculomotor System,* M. B. Bender, ed. New York: Harper and Row.

Robinson, D. A. 1972. "Eye movements evoked by collicular stimulation in the alert monkey." *Vision Res.* 12: 1795–1808.

Robinson, D. A., and A. F. Fuchs. 1969. "Eye movements evoked by stimulation of frontal eye fields." *J. Neurophysiol.* 32: 637–648.

Robinson, D. L., M. E. Goldberg, and G. B. Stanton. 1978. "Parietal association cortex in the primate: Sensory mechanisms and behavioral modulations." *J. Neurophysiol.* 41: 910–932.

Schiller, P. H., and M. Stryker. 1972. "Single unit recording and stimulation in superior colliculus of the alert rhesus monkey." *J. Neurophysiol.* 35: 915–924.

Schiller, P. H., S. D. True, and J. L. Conway. 1979. "Effects of frontal eye-field and superior colliculus ablations on eye movements." *Science* 206: 590–592.

Schneider, G. D. 1969. "Two visual systems." *Science* 163: 895–902.

Silberpfennig, J. 1941. "Contributions to the problems of eye movements: III. Frontal lobe tumors." *Confin. Neurol.* 4: 1–13.

Sparks, D. L. 1978. "Functional properties of neurons in the monkey superior colliculus: Coupling of neuronal activity and saccadic onset." *Brain Res.* 156: 1–16.

Stanton, G. B., W. L. R. Cruce, M. E. Goldberg, and D. L. Robinson. 1977. "Some ipsilateral projections to area PF and PG of the inferior parietal lobule in monkeys." *Neurosci. Lett.* 6: 243–250.

Stevens, J. K., R. C. Emerson, G. L. Gerstein, T. Kallos, G. R. Neufeld, C. W. Nichols, and A. C. Rosenquist. 1976. "Paralysis of the awake human: Visual perceptions." *Vision Res.* 16: 93–98.

Teuber, H.-L. 1964. "The riddle of frontal lobe function in man." In *The Frontal Granular Cortex and Behavior*, J. M. Warren and K. Akert, eds. New York: McGraw-Hill.

Trojanowski, J. Q., and S. Jacobson. 1974. "Medial pulvinar afferents to frontal eye fields in rhesus monkey demonstrated by horseradish peroxidase." *Brain Res.* 80: 395–411.

Tyler, H. R. 1969. "Disorders of visual scanning with frontal lobe lesions." In *Modern Neurology*, S. Locke, ed. London: Churchill.

Wagman, I. H. 1964. "Eye movements induced by electrical stimulation of cerebrum in monkeys and their relationship to bodily movements." In *The Oculomotor System*, M. B. Bender, ed. New York: Harper and Row.

Wall, G. L. 1962. "The evolutionary history of eye movements." *Vision Res.* 2: 69–80.

Wurtz, R. H., and M. E. Goldberg. 1972a. "Activity of superior colliculus in behaving monkey. III. Cells discharging before eye movements." *J. Neurophysiol.* 35: 575–586.

———. 1972b. "Activity of superior colliculus in behaving monkey. IV. Effects of lesions on eye movements." *J. Neurophysiol.* 35: 587–596.

Zihl, J., and D. von Cramon. 1979. "Collicular function in human vision." *Exp. Brain Res.* 35: 419–424.

22 Mechanisms of Interocular Equivalence

Charles R. Hamilton

The recent discovery of a dozen or more representations of the visual field in the cerebral cortex of several mammals (reviews: Allman 1977; Van Essen 1979) has dramatically augmented interest in the functional meaning of multiple representations of sensory projection areas. The question of how visual perception results from the differential analysis of the retinal input by these specialized representations poses an exciting challenge that can only be met through coordinated attacks with anatomical, physiological, and behavioral methods. Although intriguing discoveries by anatomists and physiologists have suggested possible functions for some of these areas (Hubel and Wiesel 1965; Gross 1973; Zeki 1978; Van Essen 1979), only behavioral tests can determine if such functional inferences are valid. Often the overly simplified interpretations based on physiological findings have not been supported by behaviorally based measures. For example, discrimination of the orientation of lines by cats is not greatly affected by the imposed absence of appropriately oriented "line detectors" in the primary visual cortex (Hirsch 1972; Muir and Mitchell 1973), nor is pattern discrimination by cats routinely lost after complete removal of visual areas 17 and 18 (Doty, 1973; Sprague et al. 1977). Results of this type emphasize the necessity of testing theories about behavioral functions with behavioral techniques.

In this chapter I will discuss how studies of interocular equivalence may be combined with physiological, environmental, and especially surgical manipulations to determine the functions of different visual areas. I will use the term *interocular equivalence* rather than *interocular transfer* or *convergence* to describe similar performance through the originally exposed and unexposed eyes, because it is relatively devoid of mechanistic interpretations. Transfer and convergence as possible mechanisms for equivalence will be examined in a later section.

It is frequently and reasonably assumed that if interocular equivalence for some function is present, then the locus of that function is downstream from the place where information from the two eyes is combined. This argument has often been used to differentiate retinal processes, which should not show equivalence, from central processes,

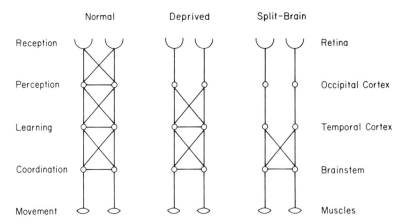

Figure 22.1 Levels of interocular equivalence in three preparations. Circles represent neuronal populations at different levels onto which connections from the two eyes may converge to produce interocular equivalence; lines represent pathways. The site of initial convergence is different in the three preparations. Crossed lines represent intra- and intercortical pathways for the first two examples, but only intercortical ones for the split-brain preparation.

which should (Walls 1953; Day 1958; Anstis and Moulden 1970). It will be extended in this chapter to include differentiating levels of central processing that may be associated with different visual areas. While there are possible pitfalls with this line of logic, they have been identified and can usually be controlled (Day 1958; Barlow and Brindley 1963; Scott and Wood 1966; Anstis and Moulden 1970).

Functional Hierarchy

A few simple examples of functions organized in terms of their sites of interocular equivalence should provide enough confidence in the logic and procedures to warrant the more systematic investigation of equivalence at different cortical levels that will be proposed later. The highly simplified schematic diagrams of the nervous systems of three well-known preparations in figure 22.1 help to illustrate these examples. The small circles represent neuronal populations and the lines represent neural connections. Though the crossed connections of the normal and deprived preparations signify pathways within as well as between the hemispheres, the crossed connections of the split-brain, by necessity, refer only to interhemispheric ones. A plausible hierarchical organization for some gross categories of function is indicated on the left; possible anatomical sites for these functions are suggested on the right. This ordering is derived from results of experiments with these three preparations.

Normal Animals
Mammals usually show a high degree of interocular equivalence for most monocularly experienced events. For example, visual discrimina-

tions learned through one eye are performed nearly as well when tested through the other eye. Interocular equivalence in such cases is thought to occur because the two eyes converge largely on the same population of neurons in the primary visual cortex, as indicated on the left of figure 22.1 Once neural binocularity is established, information from either eye is "short-circuited" to the mechanisms that follow, and it is then irrelevant for most purposes which eye is tested. Incomplete equivalence, if it occurs, may indicate a contribution from monocular mechanisms peripheral to binocular convergence, or from more central mechanisms partly utilizing the small remaining population of monocular neurons; the monocular component of these mechanisms would be unavailable to testing through the unexposed eye. However, one should not fall into the well-populated trap of inferring that convergence at early cortical levels is necessary for interocular equivalence; it is merely sufficient. Many functions continue to show interocular equivalence when binocular convergence is prevented until deeper in the system.

Some experiences do not produce interocular equivalence even in normal subjects. For example, the effects of dark and light adaptation (LeGrand 1957; Woodworth and Schlosberg 1965) and (less clearly) the effects of afterimages (Craik 1940) and brightness contrast (Asher 1950) often remain specific to the exposed eye. Because monocular exposure may also adapt central mechanisms and because afterdischarges from the exposed eye may continue to influence binocular neurons in the cortex, careful controls such as pressure blinding of the exposed eye during stimulation and during testing for interocular equivalence are required (Barlow and Brindley 1963; Scott and Wood 1966). Lack of equivalence in such controlled studies is strong evidence that the mechanisms were located before the normal site of binocular interaction; equivalence indicates that central mechanisms predominate. More important, lack of equivalence shows that functionally useful feedback to preceding monocular mechanisms of the unexposed eye does not occur. It is this fact, which also holds for preparations that artificially delay binocular interactions until later in the visual system, that allows the ordering of functional levels in the visual hierarchy by determining the earliest level for which loss of binocularity can prevent interocular equivalence.

Deprived Animals
The normal complement of binocular neurons in the visual cortex of cats, monkeys, and presumably man can be severely reduced by raising them with discrepant visual input to corresponding points on the two retinae. Siamese cats have a similar, naturally occurring reduction in binocularity. This situation is indicated in the center of figure 22.1. Monocular deprivation alternating from eye to eye on a daily basis and experimentally produced strabismus are particularly useful because they largely destroy binocularity of cells in the visual cortex but leave

relatively intact other properties of these neurons. For what functions is interocular equivalence lost in these preparations?

Aftereffects

After one has viewed a grid of lines tilted about 10° for a few minutes, a vertical line will appear tilted by a few degrees in the opposite direction; this is the well-known "tilt aftereffect." Similarly, inspection of moving stimuli for a time will lead to a perception of movement in the opposite direction when stationary stimuli are viewed. These perceptual aftereffects demonstrate a high degree of interocular equivalence in normal human subjects. However, human subjects with strabismus of early onset show almost no interocular equivalence for these aftereffects, nor do they exhibit good stereopsis (Movshon et al. 1972; Mitchell and Ware 1974; Mitchell et al. 1975). The presumed loss of binocular neurons in visual cortex most likely is responsible for the loss of interocular equivalence and stereopsis. Because it is not yet known how deep into the visual system the loss of binocularity extends, no precise localization of these mechanisms can be inferred, although other evidence strongly suggests that they are located in the early stages of visual processing (Barlow et al. 1967; Nikara et al. 1968; Blakemore and Campbell 1969). The points for now are that interocular equivalence apparently can be destroyed by preventing binocular interactions until after the site of the mechanism in question, and that feedback from later binocular levels does not alter the state of the mechanism for the unexposed eye.

Discrimination Learning

Several experimenters have tested cats with reduced numbers of binocular neurons to see if interocular equivalence for learned visual discriminations would be disrupted (Sherman 1971; Ganz et al. 1972; Hirsch 1972; Marzi et al. 1976). Cats raised with alternating monocular deprivation (Ganz et al. 1972), squint (Sherman 1971), or unrelated input to the two eyes (Hirsch 1972) and Siamese cats (Marzi et al. 1976) show essentially normal interocular equivalence. For monocularly deprived cats equivalence is also quite good if comparisons are made to appropriate control groups to allow compensation for deficits in performance with the deprived eye (Ganz et al. 1972). Even though the specific site of interocular equivalence is not established by these experiments, it seems clear that equivalence for discrimination learning comes after equivalence for perceptual aftereffects of tilt and movement.

Workers in a related field have known for over 15 years that binocular neurons in visual cortex are not needed for interocular equivalence in discriminating visual patterns: Split-brain monkeys with the anterior commissure left intact show good interocular equivalence despite the loss of all binocular neurons in striate and prestriate areas. Because (in contrast with the cat experiments) there are no residual binocular neurons in visual cortex, binocularity must occur at subsequent levels. Thus, the question of the relationship of binocularity and equivalence was answered conclusively with monkeys before it was formally posed for cats! Furthermore, we are reminded again that interpretations based on physiological results often are too simplistic and therefore need behavioral verification before being accepted.

Split-Brain Animals

The level where binocular interactions can first occur may be controlled more precisely by surgical manipulations than by environmental ones. For example, if the optic chiasm and the cerebral commissures are sectioned in the midline, then direct binocular interactions cannot occur at cortical levels. This surgery produces the standard split-brain preparation represented diagrammatically on the right of figure 22.1.

It is well known that split-brain animals show essentially no interocular equivalence for discriminations of visually presented patterns or objects (Sperry 1961; Doty and Negrao 1973). This restriction of learning to the trained hemisphere is most simply interpreted as indicating that the critical learning mechanisms are located before the next available sites of binocular interactions. Once again, backflow of useful information from subsequent binocular levels is not sufficient to establish interocular equivalence.

By contrast, basic coordinational skills, such as guiding an arm to a target by vision through either eye, remain largely intact (Myers et al. 1962; Hamilton 1967; Brinkman and Kuypers 1973). Therefore, subcortical connections that produce interocular equivalence for visuomotor control seem to exist. As a limiting case, at least the motor horn cells in the spinal cord of split-brain animals must be "binocular," because either eye can control the same sets of muscles. Most probably, however, interocular equivalence is established prior to this final level.

Summary of Hierarchy

This brief, simplified overview of prototypical experiments with three preparations differing in the level at which binocular interactions are first achieved indicates that the simplistic, sequential organization given in figure 22.1 has some factual basis. Several phenomena, such as dark adaptation, perceptual aftereffects, discrimination learning, and visuomotor coordination, can be ordered according to the anatomical level at which loss of binocular interactions prevents interocular equivalence. This order fits with that determined from other types of studies and with a logical order of information processing within the visual system. Furthermore, it is apparent that information does not flow backward from later binocular levels to alter monocular mechanisms earlier in the hierarchy. Therefore, it seems that functional mechanisms can be localized within the brain by determining at what level a loss of binocularity prevents interocular equivalence. How detailed a scheme can be made will depend on how well localized are the commissural fibers from different visual areas and on the skill of the surgeon.

Partially Split Brains

The split-brain preparation has been useful in placing anatomical limits on the localization of many behavioral functions (Sperry 1961; Doty and Negrao 1973). A more detailed examination of the hierarchical order of perceptual processing using tests of interocular equivalence, however, depends on the practicality of further fractionating visual perception by only partially disconnecting the cerebral hemispheres. While this approach has already been used to some extent (Black and Myers 1964; Sullivan and Hamilton 1973a; Tieman and Hamilton 1974), its full potential is just beginning to be exploited. A brief review of previous findings on pattern discrimination in partially split-brain animals provides a background for experiments to be discussed.

Most of the dozen or so visual areas of cats and monkeys lie in the occipital cortex. These areas are interhemispherically connected by fibers that pass through the splenium of the corpus callosum. In monkeys, visual areas in the inferotemporal cortex are interconnected by the anterior commissure as well (Zeki 1973; Gross et al. 1976). How efficient are the pathways in producing interocular equivalence? Because I am frequently asked this question, I have collected all the reports I could locate that give data from individual animals for discrimination of symmetrical patterns. These results are summarized in table 22.1 and figure 22.2. (Butler [1979] has recently summarized his work on interocular equivalence in split-brain and partially split-brain monkeys. His results are quite similar to those reported in table 22.1 and figure 22.2, with the qualification that the monkeys with anterior commissure intact did not show quite as good interocular equivalence as those from other laboratories.)

Qualitative Results
After learning a discrimination to criterion with one eye, normal cats and monkeys immediately perform well above chance with the untrained eye. A high level of interocular equivalence is also indicated by the large percent savings. Monkeys with all the forebrain commissures cut but with the optic chiasm intact show similar, high levels of interocular equivalence.

In the remaining groups of animals the optic chiasm is cut, which removes the principal site of binocular convergence within each cortex and leaves only commissural connections as a source of binocular interaction. It is clear that animals with just the optic chiasm split continue to show a high degree of interocular equivalence. Furthermore, if only the splenium in cats, monkeys, or chimpanzees is spared, performance still remains high. The posterior one-third of the corpus

Table 22.1
Interocular equivalence for pattern discriminations.

	Laboratory	Initial Percentage Above Chance						Percent Savings					
		N	SB_{IOC}	SC	IS	IAC	SB	N	SB_{IOC}	SC	IS	IAC	SB
Primates	Hamilton	28_8	38_4	31_5	35_4	29_5	3_{54}	82_8	82_4	76_5	66_4	59_5	3_{54}
	Downer	$\sim 50_9$	$\sim 50_5$		$\sim 50_3$	$\sim 50_1$		100_9	100_5	71_1	100_3	100_1	-11_3
	Myers									98_1	91_1	57_3	3_7
	Gazzaniga									96_3		71_1	0_1
	Butler												1_6
Cats	Hamilton	56_2			38_3	0_1	4_1	91_2			71_3	3_1	4_4
	Berlucchi			28_7		-1_4				66_7		5_4	
	Sechzer	33_2				-12_6		74_2				1_6	
	Myers, Sperry			35_9		0_5						-2_4	
	Webster									52_4			

Notes: Initial percentage above chance is difference in percent correct on the first 20 or 40 trials between the untrained eye and either the trained eye or 50%. Percent savings is calculated according to the expression $100\,(a - b)/(a + b)$, where a and b are, respectively, the trials or errors to criterion for the trained eye and the untrained eye. N: normal. SB_{IOC}: split-brain with intact optic chiasm. SC: split-chiasm. IS: intact splenium. IAC: intact anterior commissure. SB: split-brain. Subscripts indicate numbers of animals. Data are from the following sources: Sperry et al. 1956; Myers 1961; Downer 1962; Black and Myers 1964; Sechzer 1964; Gazzaniga 1966; Butler 1969; Berlucchi and Marzi 1970; Yamaguchi and Myers 1972; Hamilton and Brody 1973; Hamilton and Tieman 1973; Sullivan and Hamilton 1973a; Tieman and Hamilton 1973, 1974; Hranchuk and Webster 1975; Peck et al. 1979; Berlucchi, unpublished; Hamilton, unpublished.

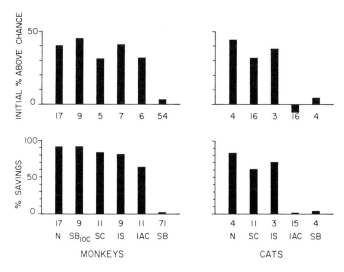

Figure 22.2 Interocular equivalence for pattern discriminations by animals from several laboratories. Bars represent the averages of the numbers in table 22.1 weighted according to the number of animals from each laboratory. Definitions and abbreviations are the same as in table 22.1.

callosum in cats (Myers 1959) and as little as one-fifth in monkeys (Sullivan and Hamilton 1973b) is sufficient to produce equivalence. Similarly, primates with just the anterior commissure left intact can reliably perform pattern discriminations with the untrained eye. Cats, however, show no evidence of transfer of information through the anterior commissure.

Animals with complete division of the forebrain commissures and the optic chiasm show no interocular equivalence. The combined results of 2% savings for 71 split-brain primates and of 2% savings for 19 split-brain cats (with or without the anterior commissure present) attest to the completeness of the isolation of the learning of pattern discriminations to one hemisphere. These results also show that good interocular equivalence in the animals with partially split brains is due to the cerebral commissures and not to subcortical structures.

Results from animals tested with just the anterior commissure left intact deserve special comment. Cats presumably do not show interocular equivalence because the anterior commissure does not interconnect visual areas critical for discrimination learning (Doty and Negrao 1973; Peck et al. 1979). In primates, of course, the anterior commissure extensively interconnects the inferotemporal cortices (Doty and Negrao 1973; Zeki 1973; Gross et al. 1976), which have important visual functions (Gross 1973).

No qualitative differences have been found in the abilities of primates to transfer information concerning discrimination of visual stimuli through the anterior commissure and the splenium. The following types of stimuli have been tested: color (Downer 1962; Hamilton et al. 1968), pattern (Downer 1962; Black and Myers 1964; Gazzaniga 1966; Noble 1968; Sullivan and Hamilton 1973a,b), movement (Hamilton, unpublished), orientation (Hamilton, unpublished; Noble 1968), and position in space (Tieman and Hamilton 1974). The paradigms used in these experiments include two-choice (most of the preceding references), go/no go (Hamilton, unpublished), and matching to sample with one hemisphere receiving the sample and the other the matches (Hamilton et al. 1968; Tieman and Hamilton 1974). The use of electrical stimulation of striate cortex as a conditioned stimulus also produces good interocular equivalence through the anterior commissure or the splenium, but, curiously, bilateral memories apparently are formed only when the anterior commissure is intact (Doty and Overman 1977), in contrast to the formation of bilateral memories via either route when visual stimuli are used (Butler 1969; Hamilton 1977).

It is initially surprising that the anterior commissure, interconnecting areas far into the perceptual systems analyzing visual information, conveys information that is as yet indistinguishable from that carried by the splenium (Sullivan and Hamilton 1973b). The simplest interpretation is that the critical learning and memory mechanisms are located downstream to the first point of entry of the anterior commissure into visual mechanisms in inferotemporal cortex. This must be true unless significant functionally useful feedback to antecedent mechanisms can occur , for which there is no evidence and against which there is the finding that interocular equivalence for many functions may be differentially prevented by deprivation or partial split-brain surgery. Surely the anterior commissure still contains a treasure of information regarding the organization of perceptual mechanisms.

Quantitative Results

The magnitude of interocular equivalence in various preparations may give some indication of the normality and completeness of the transferred information, in addition to providing a background against which the effects of further manipulations may be judged. The data summarized in table 22.1 and figure 22.2 suggest that there is some decrement in equivalence as more extensive brain bisection occurs. However, quantitative comparisons are difficult, because the number of animals and conditions within one experiment are usually small, because various investigators use different procedures, and because most of the experiments are not balanced for critical procedural variables.

The data from my laboratory given in table 22.2 avoid most of these difficulties, although they still suffer from small sample size. To test for interocular equivalence, all monkeys learned two sets of pattern discriminations under identical conditions. Before the first set of six problems was taught (one month after surgery) they had learned a standard series of seven discriminations; by the time the next set of problems was taught (about a year later) they had received training on a variety of similar but not identical tasks. It appears that the additional year's training had no systematic effect on the magnitude of interocular equivalence. Furthermore, the profile of equivalence magnitude versus surgical extent is similar to the data in table 22.1, although perhaps the falloff is more accentuated. (The data from table 22.2 are included in the first line of table 22.1 along with data from other similarly trained monkeys with somewhat more variable backgrounds.)

Table 22.2
Interocular equivalence tested under identical conditions.

	Time after Surgery	4 N	4 SB$_{IOC}$	2 SC	2 IS	2 IAC	20 SB
Initial	~1 month	34	43	32	37	21	4
percentage	~1 year	28	32	33	29	14	3
above chance	Mean	31	38	32	33	18	4
Percent	~1 month	91	88	63	56	39	7
savings	~1 year	91	77	96	72	39	13
	Mean	91	82	80	64	39	10

Note: Abbreviations are the same as in table 22.1.

Overall, the trends in the average degree of interocular equivalence pictured in figure 22.2 seem representative of the less complete data from individual laboratories as given in tables 22.1 and 22.2. A tendency for interocular equivalence to be smaller and better differentiated between surgical groups when the training procedure and apparatus place more demands on the animals seems likely.

The differences between groups, though probably real, still do not clearly indicate that there is less efficient transfer of information through some pathways. The slightly impaired performance often found in tests for interocular equivalence in normal animals most likely reflects the disruptive effects of reversing the occluded eye and, usually for primates, of changing the hand used to respond. However, it is also possible that interocular equivalence is not perfect because monocular cells in the cortex play a significant role, giving a statistical advantage with respect to the number of "learned" neurons available to the trained eye. High levels of performance by split-brain monkeys with optic chiasm intact are not particularly surprising, because the sites of intracortical binocular convergence are still present, but the results do show that cutting the cerebral commissures in itself has no deleterious effects on interocular equivalence.

Performance with the untrained eye of split-chiasm animals is frequently worse than that of normal animals. Though this has been attributed to less efficient transfer of information by the commissures compared with the direct input (Myers 1961), alternative explanations have not been ruled out. For example, deficits in generalization to the untrained eye might result from vision, which is hemianopic, being switched to the opposite half of the visual field or from procedural asymmetries that were present in the transfer designs. Deficits in equivalence for monkeys with just the anterior commissure intact have been reported (Black and Myers 1964; Gazzaniga 1966), although even this result is not universal (Downer 1962; Sullivan and Hamilton 1973a). When found, it may reflect the smaller number of fibers remaining rather than basic impairments in communication. Although not reviewed here, the evidence for weaker memories in the untrained compared to the trained hemispheres, as measured after additional section of the commissures that were intact during training, is considerably stronger (Myers 1961; Cuénod 1972; Doty and Negrao 1973; Hamilton 1977).

Mechanisms of Interocular Equivalence

At least two conceptually different mechanisms of interocular equivalence have been assumed by various researchers, although these assumptions usually have been implicit and not clearly specified. I prefer to call these mechanisms *convergence* and *transfer*; they could also be called *passive transfer* and *active transfer* if one feels that the term *transfer* has already been used too frequently to imply both processes. Convergence is most often assumed by physiologists; in the preceding sections this term has been used only when binocular convergence clearly seemed to occur. Transfer is more likely to be assumed by psychologists, particularly when they are discussing interhemispheric transfer of discrimination learning, memory, or strategies through the cerebral commissures. It is appropriate to specify and distinguish these two possibilities as clearly as possible because such concepts frequently control the framework in which experimental questions are asked, and misunderstanding them often leads to confusion in interpretation of results.

Convergence

The extensive convergence of information onto neurons at successive levels of the visual system is illustrated schematically on the left of figure 22.3. For simplicity, no attempt is made to represent the small proportion of cells that are monocular or the inputs from midbrain mechanisms. Because of continuing convergence, cells deeper in the hierarchy receive visual input from a progressively larger area of visual space, which results in cells of successive levels having larger receptive fields. More important for our purposes, convergence allows the possibility of producing binocular cells at later stages even if binocularity is experimentally prevented at earlier ones. Preserved binocularity of this type, therefore, can explain why deprived cats with few binocular cells in striate cortex still show interocular equivalence in discrimination of patterns. The reason why cells deeper in the hierarchy do not lose binocularity may be that, unlike earlier cells, they continue to receive congruent input from the two eyes because they have relaxed requirements for the position within their receptive fields required for ade-

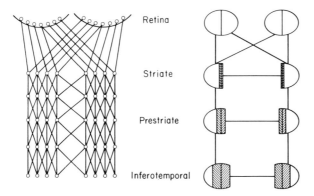

Figure 22.3 Interocular convergence within cortical visual areas. Schematic diagram on the left indicates progressive convergence of neural connections throughout the visual system; picture on the right shows same scheme with the cortical visual areas as visual hemifields. Shaded regions indicate the area around the vertical meridian for which commissural connections contribute to binocularity.

Mechanisms of Interocular Equivalence

quate stimulation. Therefore, I would expect cells deeper in the visual hierarchy to be binocular in such deprived animals and in Siamese cats despite loss of binocularity at earlier visual levels. (Since the writing of this chapter, binocular cells in the lateral suprasylvian area of Siamese cats have been reported [Marzi et al. 1980].)

The preceding description also applies to convergence across the cerebral commissures, which, of course, represents the only source of binocularity in animals with the optic chiasm divided. On the right of figure 22.3, the visual system is redrawn with the visual areas at successive hierarchical levels pictured in terms of the projection of the visual field that they should receive according to the scheme of convergence diagrammed on the left. The two differently shaded regions of each area represent portions of the visual field near the vertical meridian that send or receive information across the cerebral commissures. The progressive increase in the width of the strip of binocular convergence attributable to commissural fibers also follows simply from convergence as pictured on the left. It should be noted that in this model commissural fibers are no different in function than intrahemispheric ones; they are simply more visible because they must bridge the gap between the hemispheres.

Available anatomical and physiological results are at least roughly in accord with the properties of this scheme. Receptive fields are usually larger in cortical areas farther removed from striate cortex (Hubel and Wiesel 1968; Gross 1973; Zeki 1974; Van Essen and Zeki 1978). Interhemispheric connections, by and large, pertain to the vertical meridian of topographically organized visual areas (Berlucchi 1972; Van Essen 1979). The overlap into the ipsilateral visual hemifield is wider in deeper areas (Dubner and Zeki 1971). In the inferotemporal cortex they extend 30° or more into the hemifield that is principally represented in the opposite hemisphere, and this extension of the receptive fields into the ipsilateral visual field is dependent on commissural connections (Gross et al. 1976).

How well can mechanisms based just on convergence explain the results of experiments that test for interocular equivalence? As indicated earlier, the normal convergence from each eye onto binocular neurons in striate cortex is sufficient for the interpretation of interocular equivalence in normal animals and in split-brain animals with the optic chiasm intact, because most functions occur after binocularity is first established and therefore are indifferent to which eye is used. The loss of stereopsis and of interocular equivalence for tilt and motion aftereffects has been interpreted as resulting from the loss of binocular neurons in striate or prestriate cortex, the level at which these functions are probably located (Movshon et al. 1972; Mitchell and Ware 1974). The preservation of interocular equivalence for discrimination in deprived animals eventually may be explained by the predicted preservation of binocular neurons in inferotemporal or suprasylvian visual areas.

Once the optic chiasm is sectioned, however, normal intracortical convergence is lost and only intercortical convergence from cells near the vertical meridian of the visual projection areas can produce binocularity. This may not be too troublesome, because most stimuli are fixated with central vision. Berlucchi et al. (1967) originally noted that even the narrow rectangular window of binocular vision provided by commissural connections at the border between areas 17 and 18 could be sufficient for interocular equivalence. Interhemispheric connections between areas deeper in the hierarchy provide larger sensory-perceptual windows through which binocular convergence could produce interocular equivalence. It should be remembered that midline connections between all of the known visual areas remain intact in split-chiasm and intact-splenium animals; only the inferotemporal areas remain connected in the intact-anterior-commissure monkeys. Finally, the loss of interocular equivalence in split-brain animals would result from the complete loss of binocular convergence at the cortical levels critical for perception, learning, and memory. From these considerations it would appear that interhemispheric equivalence could be subserved by commissural connections at any level in the hierarchy prior to the mechanisms being used. Equivalence at some levels, however, may be less efficient because of the narrower window for binocular convergence or because of specific limitations on the capabilities for transfer of information by intercortical connections of specific visual areas. These are clearly questions for experimental investigation rather than speculation.

Interocular equivalence in partially split-brain animals also has been interpreted by two other groups specifically as resulting from interhemispheric binocular convergence. Berlucchi and his associates have suggested that commissural connections from visual areas in the suprasylvian gyrus rather than areas 17 and 18 are critical for interocular equivalence. They found good discrimination learning and interocular equivalence in split-chiasm cats with large unilateral lesions in areas 17, 18, and parts of 19, and poor learning and equivalence in animals with unilateral lesions in the suprasylvian gyrus (Berlucchi 1972; Berlucchi et al. 1978, 1979). This implies that commissural connections between suprasylvian areas were of primary importance for equivalence. Until split-chiasm cats with bilateral suprasylvian lesions are tested for equivalence, the magnitude of the deficit cannot adequately be assessed. Therefore, although it is clear that commissural connections deep in the hierarchy are sufficient for interocular equivalence in cats, it is not certain that they are necessary.

Gross and his colleagues have more boldly concluded that interocular equivalence for discrimination learning and perceptual constancies in split-chiasm monkeys is critically dependent on commissural connections between the inferotemporal cortices (Gross and Mishkin 1977; Seacord et al. 1979). This is based on a dramatic decrease in interocular equivalence in split-chiasm monkeys with bilateral lesions in inferotemporal cortex. However, the average savings of 23% that remained may reflect the continued utilization of commissural pathways earlier in the hierarchy.

Transfer

Many investigators apparently consider interocular equivalence mediated by commissural connections as resulting from mechanisms specialized for interhemispheric transfer of information. Because the word *transfer* has been used both literally to mean that the function in question (such as memory) is actually transported across the commisures and figuratively to mean an end state for which the function is available to tests through either eye regardless of the mechanism, it is difficult to know precisely what various authors have intended. The following quotations give some indication of how the term has been used:

...the callosum serves to keep each hemisphere up to date on what is new in the other.... (Sperry 1961)

...the corpus callosum has a major mnemonic function. This mnemonic processing is required to achieve full utilization of the capabilities of the cerebral hemispheres by giving access from each hemisphere to information stored in the other (i.e., engram "read out"). (Doty 1973)

...the effect of the performance of bar-pressing, the reinforcement or both, when the two hemispheres are functional is to transfer the trace from the trained to the untrained hemisphere. This indicates that not only is transfer an active process, but that it occurs under fairly restricted conditions. (Russell and Ochs 1963)

The concept of transfer is necessarily vaguer than convergence, because it does not rest on known foundations in anatomy and physiology. The following properties, however, characterize some aspects of transfer that seem in keeping with the thoughts of most investigators:

• Because the commissural connections form part of a mechanism specialized for transfer, they are functionally different from connections within visual areas. This is reflected, for example, in the designation of directly and indirectly trained hemispheres, a distinction that is not relevant for the convergence model.

• Transfer occurs for the results of analyses of visual information rather than for the computational details.

• These results pertain to the analysis of the entire projection of visual space within a hemisphere rather than the portion just near the vertical meridian. The known connections of the vertical midline usually are acknowledged, but considered as part of basic mechanisms of feature extraction, and not for transfer of processed information.

• Transfer of memories need not occur at the time of sensory input.

Whether transfer occurs at many levels in the visual hierarchy or just pertains to deeper areas concerned with highly processed information is not specified, although transfer of sensory, perceptual, conceptual, and mnemonic information are often mentioned.

How well does the mechanism of transfer compare with convergence in explaining the existing data?

Most investigators would agree that convergence at initial levels in the visual system adequately interprets interocular equivalence in animals with an intact optic chiasm. Therefore, this discussion only concerns data from animals with vision through the two eyes lateralized to different hemispheres so that commissural connections must be used. It seems clear that both transfer and convergence can be used to interpret the data on interhemispheric equivalence so far presented. Convergence would be favored, however, because it is more parsimonious to utilize mechanisms known to exist than to postulate new ones. But what of the interocular equivalence for higher-order functions that was originally thought to require sophisticated mechanisms of transfer?

Interhemispheric transfer of memories is often considered to require a process of transfer. But interhemispheric convergence of sensory information during the learning trials could allow memories initially to be formed bilaterally without the need to transfer mnemonic information (Berlucchi et al. 1967; Seacord et al. 1979). Therefore, this example is not critical. Better performance by the untrained hemisphere of monocularly trained split-chiasm cats than by the untrained hemisphere of split-chiasm cats tested for interocular equivalence after additional division of the corpus callosum has been interpreted as showing that the untrained hemisphere in the first instance can tap the better memory in the trained hemisphere (Myers 1961). However, convergence through the callosum at the time of testing the cats with intact commissures could allow the visual input to the untrained hemisphere direct access to the better memory in the trained hemisphere, and therefore no prior transfer of memories need be postulated. Other, more complicated arguments against the existence of tapping can be made, but the preceding logic should suffice for our purposes.

Experiments with spreading depression that in effect create a reversible split brain have provided evidence that memories can be transferred during a single reinforced trial long after the original learning has occurred (Russell and Ochs 1963). If valid, this delayed transfer occurring almost instantaneously could not be explained easily by convergence. Binocular convergence occurs passively as a direct consequence of sensory input and therefore could not operate at later times. Because critics claim that the technique of spreading depression is fraught with interpretative difficulties (Petrinovitch 1976), the existing evidence for delayed transfer of memories does not seem conclusive.

In man, the two hemispheres are differentially specialized for many abilities. To avoid analyzing information from the left and right peripheries differently, some form of active transfer seems necessary. Convergence, even at inferotemporal levels, does not reach to the periphery, and presumably even before that level much important processing occurs in each hemisphere. Indeed, whenever the two hemispheres operate in complementary fashion some form of communication more complex than passive convergence appears necessary. Unfortunately, the problem of how two specialized hemispheres operating in fundamentally different ways manage to communicate has not received sufficient formal analysis to enable a more informed treatment of this intriguing question.

Learning sets for several types of visual discriminations show interocular equivalence in split-brain monkeys if the anterior parts of the corpus callosum remain intact (Sullivan 1971; Noble 1973). This may be thought of as the interhemispheric transfer of a strategy for problem solving. Formulation of a strategy would seem to require analyzing information collected over a number of trials and even over a number of discriminations. Since individual discriminations show no interocular equivalence in this preparation, it is hard to imagine what form of information time-locked to the sensory input could passively converge on the untrained hemisphere to allow parallel development of a learning set. It seems easier to envision the untrained hemisphere tapping the strategy stored in the trained side, or an active transfer of the strategy to the untrained side. The exciting experimental possibilities utilizing interhemispheric transfer of learning sets have hardly been explored.

In summary, both mechanisms of convergence and of transfer are capable of interpreting all forms of interocular equivalence for sensory and perceptual functions in animals with partially split brains. Some higher-order cognitive functions, however, still seem to be more easily explained by the concept of active transfer. It would seem wisest, therefore, to accept the possibility that both types of mechanisms operate, but to favor the structurally based model of convergence whenever it is adequate. Perhaps interhemispheric transfer of higher functions will become interpretable in terms of an expanded concept of convergence as the higher functions are better understood. Alternatively, critical experiments may show one or the other mechanism to be insufficient in particular situations.

Experimental Possibilities

The partially split-brain preparation provides a unique approach to studying the functional roles of visual areas in perception without the necessity of making lesions in the cortex. It is hoped that the overall approach of testing for interocular equivalence in split-chiasm animals with specific commissural connections left intact at different functional levels has become clear. I will now outline some experiments that show

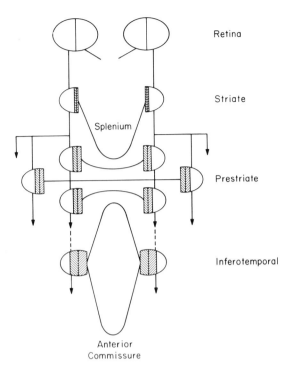

Figure 22.4 Pathways for interocular equivalence in split-chiasm primates. Visual system is drawn as in figure 22.3 to illustrate the multiplicity of visual areas interhemispherically connected by discrete populations of fibers. Commissural connections represent pathways for interocular equivalence occuring because of either convergence or transfer.

how this approach may be used. Figure 22.4 schematically suggests the multiplicity of prestriate areas, organized both serially and in parallel. These areas may have independent outputs and are assumed for this discussion to have spatially separate commissural connections. The interhemispheric connections are intended to represent pathways for mechanisms of convergence or of transfer.

Differentiation of Mechanisms

Humans can verbally describe information presented peripherally in the visual hemifield that projects to the nonspeaking hemisphere. This shows that interhemispheric connections pertaining to the far periphery exist at some level, even though they have not been described anatomically or physiologically. The convergence model predicts that they must arise at the level of the inferotemporal cortex or beyond, because connections for earlier levels are too near the vertical meridian; the active-transfer model predicts that wherever interocular equivalence occurs it will pertain to the entire hemifield. Even if these inferences are not convincing, an experimental demonstration of the location of peripherally related connections is valuable.

Preliminary results from our laboratory have shown that four normal

Mechanisms of Interocular Equivalence

monkeys taught to fixate a central position can accurately determine if two stimuli flashed at 150 msec and up to 90° apart are the same or different. Subsequent section of the splenium in one animal and the anterior commissure in another failed to disrupt performance. Because both the splenium and anterior commissure allow convergence of information onto the inferotemporal cortex from 30° or more off the midline, no critical differentiation of the models has emerged, although one could argue that 45° off the midline may be stretching the capabilities of spatial generalization for the inferotemporal cortex. Increased separation of the stimuli, or, better, additional subdivision of the splenium to allow separation of areas with narrow binocular overlap from those with broader ones, should clarify the interpretation of this result.

Localization of Perceptual Mechanisms

The mechanisms for the perception of stereopsis and of tilt and motion aftereffects are thought to be localized early in the visual hierarchy. As previously noted, however, much of this evidence is indirect, and precise localization on the basis of functional losses in strabismic human patients does not seem feasible. Direct investigation of the locus should be possible in partially split-brain animals if a significant degree of stereopsis or interocular equivalence for aftereffects survives section of the optic chiasm. Subsequent paring down of the commissural pathways could then determine at which levels binocular convergence is sufficient for stereopsis and interocular equivalence for aftereffects. Existing data suggest that connections along the border between areas 17 and 18 would be necessary, but I suspect that interocular equivalence for some aftereffects, such as movement, might survive splenial section to deeper levels, such as the movement area described by Zeki (1974, 1978). Perhaps a hierarchy of perceptual mechanisms could be established in this manner.

Localization of Discrimination Learning

It was suggested previously that learning critically involves levels deep in the visual hierarchy. Interocular equivalence in partially split-brain animals with the anterior commissure intact indicated that interhemispheric communication between the inferotemporal cortices was sufficient (Sullivan and Hamilton 1973a,b). Data from ablation experiments were interpreted by Gross as showing that inferotemporal connections were necessary for interhemispheric equivalence (Gross and Mishkin 1977; Seacord et al. 1979). A direct test of the sufficiency of commissural connections earlier in the hierarchy for interocular equivalence can be made without the confounding effects of cortical ablations by testing for interocular equivalence in animals with different portions of the splenium left intact.

Questions regarding the quality of interocular equivalence in animals with only specific visual areas left interconnected may also be answered with the same preparations. Would, for example, connections between Zeki's "movement" area show preferential ability to support interocular equivalence for discriminations based on movement? In principle, the functions of any area or group of areas could be investigated in this way. At this point it is important to see how precisely fibers from the various areas are localized within the splenium.

Anatomical Considerations

The only separation of cortical visual areas by partial disconnection so far demonstrated is between the inferotemporal areas connected by the anterior commissure and the entire collection of areas connected by the splenium. Even this gross separation has proved valuable in studies of the localization of cortical mechanisms, but more refined disconnection is necessary if this approach is to be pursued.

We have begun to localize the interhemispheric connections in monkeys either by injecting tritiated proline or by making discrete lesions in various visual areas and determining the position of the radioactively labeled or degenerating fibers in the commissures (Tieman et al. 1977). A preliminary summary of these results is given in figure 22.5. In three macaques of different species and ages, injections or lesions on the lateral surface of the cortex along the upper, middle, and lower portions of the representation of the vertical meridian in areas 17 and 18 produced an almost identical, superimposed area of identifiable fibers in

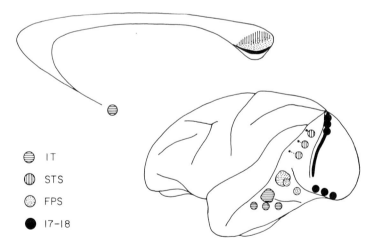

Figure 22.5 Localization of commissural fibers connecting visual cortical areas. Location of fibers in corpus callosum and anterior commissure shown is based either on injections of tritiated proline or on lesions placed in the positions indicated on the diagram of the cortex. Circles represent injections; irregular shapes represent lesions. All are near the surface of the brain except the injections into the STS, which entered at the dots and ended deep in the sulcus. IT: inferotemporal cortex. STS: posterior bank and depths of superior temporal sulcus. FPS: foveal prestriate cortex. 17–18: vertical meridian of areas 17 and 18.

Mechanisms of Interocular Equivalence

the splenium. This region did not overlap with the adjacent region identified from injections or lesions in foveal prestriate cortex of three monkeys. One monkey with injections located in the superior temporal sulcus had fibers in a commissural region partly overlapping with the foveal prestriate region of the splenium. None of these animals had any label or degeneration in the anterior commissure. Three monkeys with injections or lesions in various portions of inferotemporal cortex had labeled or degenerating fibers confined to the anterior commissure. This contrasts with anatomical findings of Zeki (1973) and physiological results of Gross et al. (1976) that demonstrate that the inferotemporal cortex has extensive connections through the splenium. Either large individual variation exists or the interhemispheric connections to inferotemporal cortex may be heterotopic, arising from prestriate areas and terminating in inferotemporal cortex. The latter possibility would account for the findings of Zeki and Gross et al. that show that some input to inferotemporal cortex is received through the splenium and our result that the inferotemporal cortex does not send fibers through the splenium. This possibility will be evaluated by determining the location of label in the uninjected hemispheres of the monkeys with injection of foveal prestriate cortex. (Prestriate cells projecting directly to contralateral inferotemporal cortex have now been demonstrated after HRP injection in inferotemporal cortex [Desimone et al. 1980] and by the tritiated proline injections into foveal prestriate regions mentioned above. We still find no evidence for direct interhemispheric connections through the splenium between the inferotemporal cortices. If this result is confirmed, then Gross's claim that interhemispheric equivalence for visual discrimination learning depends on transfer of information between inferotemporal cortices cannot be correct because such direct connections would not exist in monkeys with only the splenium intact.)

In summary, the localization of interhemispheric connections in the splenium appears precise enough to warrant further selective disconnection experiments in partially split-brain monkeys. At least it should be possible to separate the commissural connections of those areas near the primary visual cortex from those deeper in the hierarchy. If additional brains now being processed validate and extend these anatomical findings, and if the intact commissural segments of partially split-brain animals can be subsequently labeled and their projections adequately localized, then even more sophisticated behavioral tests of the functions of the many visual areas appear plausible.

Acknowledgments

This research was supported by the Public Health Service under grant MH-03372 to R. W. Sperry and by the National Science Foundation under grant BNS 77-12604 to C. R. Hamilton. I thank R. W. Sperry for his support, and M. V. Sullivan, S. B. Tieman, and B. A. Vermeire for participating in some of the experiments and for making many helpful suggestions.

References

Allman, J. 1977. "Evolution of the visual system in the early primates." In *Progress in Psychobiology and Physiological Psychology*, vol. 7, J. M. Sprague and A. N. Epstein, eds. San Francisco: Academic.

Anstis, S. M., and B. P. Moulden. 1970. "Aftereffect of seen movement: Evidence for peripheral and central components." *Q. J. Exp. Psychol.* 22: 222–229.

Asher, H. 1950. "Contrast in eye and brain." *Br. J. Psychol.* 40: 187–194.

Barlow, H. B., and G. S. Brindley. 1963. "Inter-ocular transfer of movement aftereffects during pressure blinding of the stimulated eye." *Nature* 200: 1347.

Barlow, H. B., C. Blakemore, and J. D. Pettigrew. 1967. "The neural mechanism of binocular depth discrimination." *J. Physiol.* 193: 327–342.

Berlucchi, G. 1972. "Anatomical and physiological aspects of visual functions of corpus callosum." *Brain Res.* 37: 371–392.

Berlucchi, G., and C. A. Marzi. 1970. "Veridical interocular transfer of lateral mirror-image discriminations in split-chiasm cats." *J. Comp. Physiol. Psychol.* 72: 1–7.

Berlucchi, G., M. S. Gazzaniga, and G. Rizzolatti. 1967. "Microelectrode analysis of transfer of visual information by the corpus callosum." *Arch. Ital. Biol.* 105: 583–596.

Berlucchi, G., J. M. Sprague, F. Lepore, and G. G. Mascetti. 1978. "Effects of lesions of areas 17, 18 and 19 on interocular transfer of pattern discriminations in split-chiasm cats." *Exp. Brain Res.* 31: 275–297.

Berlucchi, G., J. M. Sprague, A. Antonini, and A. Simoni. 1979. "Learning and interhemispheric transfer of visual pattern discriminations following unilateral suprasylvian lesions in split-chiasm cats." *Exp. Brain Res.* 34: 551–574.

Black, P., and R. E. Myers. 1964. "Visual function of the forebrain commissures in the chimpanzee." *Science* 146: 799–800.

Blakemore, C., and F. W. Campbell. 1969. "Adaptation to spatial stimuli." *J. Physiol.* 200: 11P–13P.

Brinkman, J., and H. G. J. M. Kuypers. 1973. "Cerebral control of contralateral and ipsilateral arm, hand and finger movements in the split-brain rhesus monkey." *Brain* 96: 653–674.

Butler, C. 1969. "A memory-record for visual discrimination habits produced in both cerebral hemispheres of monkey when only one hemisphere has received direct visual information." *Brain Res.* 10: 152–167.

Butler, S. R. 1979. "Interhemispheric transfer of visual information via the corpus callosum and anterior commissure in the monkey." In *Structure and Function of the Cerebral Commissure*, I. S. Russell et al., eds. Baltimore: University Park Press.

Craik, K. J. W. 1940. "Origin of visual afterimages." *Nature* 145: 512.

Cuénod, M. 1972. "Split-brain studies: Functional interaction between bilateral central nervous structures." *Struct. Funct. Nerv. Tiss.* 5: 455–506.

Day, R. H. 1958. "On interocular transfer and the central origin of visual after-effects." *Amer. J. Psychol.* 71: 784–790.

Desimone, R., J. Fleming, and C. G. Gross. 1980. "Prestriate afferents to inferior temporal cortex: An HRP study." *Brain Res.* 184: 41–55.

Doty, R. W. 1973. "Ablation of visual areas in the central nervous system." In *Handbook of Sensory Physiology*, vol. VII/3, part B, R. Jung, ed. Berlin: Springer.

Doty, R. W., and N. Negrao. 1973. "Forebrain commissures and vision." In *Handbook of Sensory Physiology*, vol. VII/3, part B, R. Jung, ed. Berlin: Springer.

Doty, R. W., Sr., and W. J. Overman, Jr. 1977. "Mnemonic role of forebrain commissures in macaques." In *Lateralization in the Nervous System*, S. Harnad et al., eds. New York: Academic.

Downer, J. L. deC. 1962. "Interhemispheric integration in the visual system." In *Interhemispheric Relations and Cerebral Dominance*, V. B. Mountcastle, ed. Baltimore: John Hopkins University Press.

Dubner, R., and S. M. Zeki. 1971. "Response properties and receptive fields of cells in an anatomically defined region of the superior temporal sulcus in the monkey." *Brain Res.* 35: 528–532.

Ganz, L., H. V. B. Hirsch, and S. B. Tieman. 1972. "The nature of perceptual deficits in visually deprived cats." *Brain Res.* 44: 547–568.

Gazzaniga, M. S. 1966. "Interhemispheric communication of visual learning." *Neuropsychologia* 4: 183–189.

Gross, C. G. 1973. "Visual functions of inferotemporal cortex." In *Handbook of Sensory Physiology*, vol. VII/3, part B, R. Jung, ed. Berlin: Springer.

Gross, C. G., and M. Mishkin. 1977. "The neural basis of stimulus equivalence across retinal translation." In *Lateralization in the Nervous System*, S. Harnad et al., eds. New York: Academic.

Gross, C. G., D. B. Bender, and M. Mishkin. 1976. "Contributions of the corpus callosum and the anterior commissure to visual activation of inferior temporal neurons." *Brain Res.* 131: 227–239.

Hamilton, C. R. 1967. "Effects of brain bisection on eye-hand coordination in monkeys wearing prisms." *J. Comp. Physiol. Psychol.* 64: 434–443.

———. 1977. "Investigations of perceptual and mnemonic lateralization in monkeys." In *Lateralization in the Nervous System*, S. Harnad et al., eds. New York: Academic.

Hamilton, C. R., and B. A. Brody. 1973. "Separation of visual functions within the corpus callosum of monkeys." *Brain Res.* 49: 185–189.

Hamilton, C. R., and S. B. Tieman. 1973. "Interocular transfer of mirror image discriminations by chiasm-sectioned monkeys." *Brain Res.* 64: 241–255.

Hamilton, C. R., S. A. Hillyard, and R. W. Sperry. 1968. "Interhemispheric comparison of color in split-brain monkeys." *Exp. Neurol.* 21: 486–494.

Hirsch, H. V. B. 1972. "Visual perception in cats after environmental surgery." *Exp. Brain Res.* 15: 405–423.

Hranchuk, K. B., and W. G. Webster. 1975. "Interocular transfer of lateral mirror-image discriminations by cats." *J. Comp. Physiol. Psychol.* 88: 368–372.

Hubel, D. H., and T. N. Wiesel. 1965. "Receptive fields and functional architecture in two nonstriate visual areas (18 and 19) of the cat." *J. Neurophysiol.* 28: 229–289.

———. 1968. "Receptive fields and functional architecture of monkey striate cortex." *J. Physiol.* 195: 215–243.

LeGrand, Y. 1957. *Light, Colour and Vision.* New York: Wiley.

Marzi, C. A., M. DiStefano, A. Antonini, and C. Legg. 1979. "Binocular interactions in the Clare-Bishop area of Siamese cats which lack binocular neurones in the primary visual cortex." *Neurosci. Lett. Suppl.* 3: 357.

Marzi, C. A., A. Antonini, M. DiStefano, and C. R. Legg. 1980. "Callosum-dependent binocular interactions in the lateral suprasylvian area of Siamese cats which lack binocular neurons in areas 17 and 18." *Brain Res.* 197: 230–235.

Mitchell, D. E., and C. Ware. 1974. "Interocular transfer of a visual after-effect in normal and stereoblind humans." *J. Physiol.* 236: 707–721.

Mitchell, D. E., J. Reardon, and D. W. Muir. 1975. "Interocular transfer of the motion after-effect in normal and stereoblind observers." *Exp. Brain Res.* 22: 163–173.

Movshon, J. A., B. E. I. Chambers, and C. Blakemore. 1972. "Interocular transfer in normal humans and those who lack stereopsis." *Perception* 1: 483–490.

Muir, D. W., and D. E. Mitchell. 1973. "Visual resolution and experience: Acuity deficits in cats following early selective visual deprivation." *Science* 180: 420–422.

Myers, R. E. 1959. "Localization of function in the corpus callosum." *Arch. Neurol.* 1: 44–47.

———. 1961. "Corpus callosum and visual gnosis." In *Brain Mechanisms and Learning*, A. Fessard, ed. Oxford: Blackwell.

Myers, R. E., R. W. Sperry, and N. M. McCurdy. 1962. "Neural mechanisms in visual guidance of limb movements." *Arch. Neurol.* 7: 195–202.

Nikara, T., P. O. Bishop, and J. D. Pettigrew. 1968. "Analysis of retinal correspondence by studying receptive fields of binocular single units in cat striate cortex." *Exp. Brain Res.* 6: 391–410.

Noble, J. 1968. "Paradoxical transfer of mirror-image discriminations in the optic chiasm sectioned monkey." *Brain Res.* 10: 127–151.

———. 1973. "Interocular transfer in the monkey: Rostral corpus callosum mediates transfer of object learning set but not of single-problem learning." *Brain Res.* 50: 147–162.

Peck, C. K., S. G. Crewther, and C. R. Hamilton. 1979. "Partial interocular transfer of brightness and movement discrimination by split-brain cats." *Brain Res.* 163: 61–75.

Petrinovich, L. 1976. "Cortical spreading depression and memory transfer: A methodological critique." *Behav. Biol.* 16: 79–84.

Russell, I. S., and S. Ochs. 1963. "Localization of a memory trace in one cortical hemisphere and transfer to the other hemisphere." *Brain* 86: 37–54.

Scott, T. R., and D. Z. Wood. 1966. "Retinal anoxia and the locus of the after-effect of motion." *Amer. J. Psychol.* 79: 435–442.

Seacord, L., C. G. Gross, and M. Mishkin. 1979. "Role of inferior temporal cortex in interhemispheric transfer." *Brain Res.* 167: 259–272.

Sechzer, J. A. 1964. "Successful interocular transfer of pattern discrimination in "split-brain" cats with shock-avoidance motivation." *J. Comp. Physiol. Psychol.* 58: 76–83.

Sherman, S. M. 1971. "Role of visual cortex in interocular transfer in the cat." *Exp. Neurol.* 30: 34–45.

Sperry, R. W. 1961. "Cerebral organization and behavior." *Science* 133: 1749–1757.

Sperry, R. W., J. S. Stamm, and N. Miner. 1956. "Relearning tests for interocular transfer following division of optic chiasma and corpus callosum in cats." *J. Comp. Physiol. Psychol.* 49: 529–533.

Sprague, J. M., J. Levy, A. DiBerardino, and G. Berlucchi. 1977. "Visual cortical areas mediating form discrimination in the cat." *J. Comp. Neurol.* 172: 441–488.

Sullivan, M. S. 1971. "Interhemispheric transfer of visual discriminations and learning sets via anterior commissure and anterior corpus callosum in monkeys." Ph.D. diss., Stanford University.

Sullivan, M. S., and C. R. Hamilton. 1973a. "Interocular transfer of reversed and non-reversed discriminations via the anterior commissure in monkeys." *Physiol. Behav.* 10: 355–359.

———. 1973b. "Memory establishment via the anterior commissure of monkeys. *Physiol. Behav.* 11: 873–879.

Tieman, S. B., and C. R. Hamilton. 1973. "Interocular transfer in split-brain monkeys following serial disconnection." *Brain Res.* 63: 368–373.

————. 1974. "Interhemispheric communication between extraoccipital visual areas in the monkey." *Brain Res.* 67: 279–287.

Tieman, S. B., C. R. Hamilton, and R. L. Meyer. 1977. "Commissural connections of visual areas in the macaque." *Anat. Rec.* 187: 731.

Van Essen, D. C. 1979. "Visual areas of the mammalian cerebral cortex." *Annu Rev. Neurosci.* 2: 227–263.

Van Essen, D. C., and S. M. Zeki. 1978. "The topographic organization of rhesus monkey prestriate cortex." *J. Physiol.* 227: 193–226.

Walls, G. L. 1953. "Interocular transfer of after-images." *Amer. J. Optom.* 30: 57–64.

Woodworth, R. S., and H. Schlosberg. 1965. *Experimental Psychology.* New York: Holt, Rinehart and Winston.

Yamaguchi, S., and R. E. Myers. 1972. "Age effects on forebrain commissure section and interocular transfer." *Exp. Brain Res.* 15: 225–233.

Zeki, S. M. 1973. "Comparison of the cortical degeneration in the visual regions of the temporal lobe of the monkey following section of the anterior commissure and the splenium." *J. Comp. Neurol.* 148: 167–176.

————. 1974. "The mosaic organization of the visual cortex in the monkey." In *Essays on the Nervous System*, R. Bellairs and E. G. Gray, eds. Oxford University Press.

————. 1978. "Uniformity and diversity of structure and function in rhesus monkey prestriate visual cortex." *J. Physiol.* 277: 273–290.

23

Interocular and Interhemispheric Transfer of Visual Discriminations in the Cat

Giovanni Berlucchi
Carlo Alberto Marzi

Behavioral phenomena subsumed under such terms as stimulus generalization, sensory equipotentiality, and perceptual equivalence are common to all mammalian species, and the clarification of their mechanisms is likely to be of significance to our understanding of the organization of complex neural systems. A specific and limited instance of this large and heterogeneous class of phenomena is the bilateral transfer of training, that is, the carryover of practice effects from one side of the body to the other. For example, an animal that has learned to discriminate between stimuli applied to a set of receptors on one side of the body can usually recognize the same stimuli when they are presented to the corresponding receptors on the other side, even though such stimuli have not been previously delivered to these receptors. Obviously the nervous system must provide a mechanism by which the same significance can be attributed to given sensory information, regardless of whether this reaches the brain via right or left input channels.

In this chapter we will discuss a specific case of bilateral transfer of training: the interocular transfer of visual discriminations. We will limit our consideration to one species, the cat, because of the ample possibility of correlating behavioral phenomena with anatomical and physiological knowledge. However, many of the notions on interocular transfer in cats are also valid for other species, as discussed by Hamilton in this volume.

Methods for Studying Interocular Transfer

Tests of interocular transfer in animals usually involve the following steps:

1. The animal is trained to make discriminatory responses to stimuli presented to one eye only, and training proceeds until a specified criterion of learning is attained.

2. There usually follows some overtraining with the same eye used for learning.

3. The discrimination stimuli are presented to the eye previously occluded and training is continued until the performance reaches the same criterion of learning as attained with the other eye.

A learning score can be assigned to each eye by counting the trials run or errors committed before reaching the criterion of learning, and the presence or absence of interocular transfer can be assessed by comparing the scores for the two eyes. Positive (or veridical) transfer is said to have occurred when the learning score for the second eye is better than the learning score for the first eye, negative (or paradoxical) transfer is said to have occurred when the learning score for the second eye is worse than the learning score for the first eye, and no transfer is said to have occurred when the scores for the two eyes are about equal. By applying the Murdock (1957) formula to the scores for the two eyes, interocular transfer can be evaluated in a quantitative way. The formula is the following:

$$\frac{\text{Score for first eye} - \text{Score for second eye}}{\text{Score for first eye} + \text{Score for second eye}} \times 100$$
$$= \text{Percentage of transfer.}$$

Interocular transfer is expressed by this formula as a percentage measure that will be positive in case of positive transfer, negative in case of negative transfer, and null when the scores for the two eyes are equal. Further, the percentage of transfer increases symmetrically for positive and negative transfers as a function of the absolute value of the difference between the two scores. The extreme values that the percentage of transfer can take are -100% when the score for the first eye is zero (errorless learning) and $+100\%$ when the score for the second eye is zero. The percentage of transfer is indeterminate only when both scores are equal to zero. This is a generally convenient method for measuring interocular transfer, save for the fact that a measure of transfer inferior to 100% does not indicate whether transfer has occurred immediately or only as a saving. In other words, the same measure of interocular transfer may be obtained regardless of whether the initial performance with the second eye is equal to the initial performance of the first eye, but the second eye reaches the criterion of learning faster than the first eye, or the performance with the second eye is better than the performance with the first eye from the very start. Thus, it becomes important to have a measure of initial performance with the second eye. This is generally expressed as percent correct on the initial session for testing transfer, and is compared with the corresponding performance with the first eye (which is usually at chance) or with the average performance with the first eye on the overtraining sessions (which is usually at criterion). Another possibility is that of using a sensitive criterion of learning, such as the first significant deviation of performance from chance (for example, the first significant sequence of correct responses; see Bogartz 1965; Runnels et al. 1968) and to compare the two eyes on this criterion by the formula of Murdock. Figure 23.1 provides examples of the meanings of the various measures of transfer.

Innate and Experiential Factors of Interocular Transfer

Much research has been devoted to analyzing the relative contributions of heredity and environment to the development of visually guided behavior, and the results have clearly indicated that acquisition and maintenance of normal visual abilities depend on complex interactions between genetic and experiential factors. These interactions are especially important during the immediate postnatal period, when the functioning of the visual system can be severely and irreversibly disrupted by lack, reduction, or distortion of the normal visual input. Similar limitations or manipulations of visual experience performed after the critical postnatal period have little or no damaging effect on visual behavior (reviews: Lund 1978; Blakemore 1978; Ganz 1978). In the cat the critical or sensitive period extends from a few days after birth, when opening of the eyelids occurs, to 3–4 months of age (Hubel and Wiesel 1970). Interocular transfer of visual discriminations has often been studied in cats with early abnormal visual experience, and attempts have been made to differentiate the deficits in the capacity for transfer from those in the capacities for perceiving and learning. These experiments on interocular transfer fall into several categories, according to the kind of restraint imposed upon the visual input during the critical period. The capacity for interocular transfer of visual discriminations has been examined in cats raised in complete darkness, or with goggles admitting only diffuse, unpatterned light into the eyes; in cats that during the critical period had received normal visual stimulation through one eye only, with the other eye totally occluded or simply deprived of patterned light stimulation; in cats submitted to an alternate monocular occlusion, such that each eye is exposed to periods of normal visual stimulation but simultaneous normal vision with both eyes is consistently prevented; and in cats allowed to receive patterned stimulation simultaneously through both eyes but experiencing a consistent mismatch between the two monocular inputs, because of either strabismus or the continuous artificial channeling of different information into the two eyes.

The basic questions that can be answered by experiments on interocular transfer of visual discriminations in these groups of cats are immediately apparent. Does the ability for interocular transfer depend on prior visual experience, or can it be present even in the binocularly deprived, visually naive animal? If early visual experience is important for the establishment of the mechanisms of interocular transfer, must such experience be synchronous and congruent for the two eyes in order to be effective? Is the interocular transfer of different classes of visual discriminations, such as brightness and form discriminations,

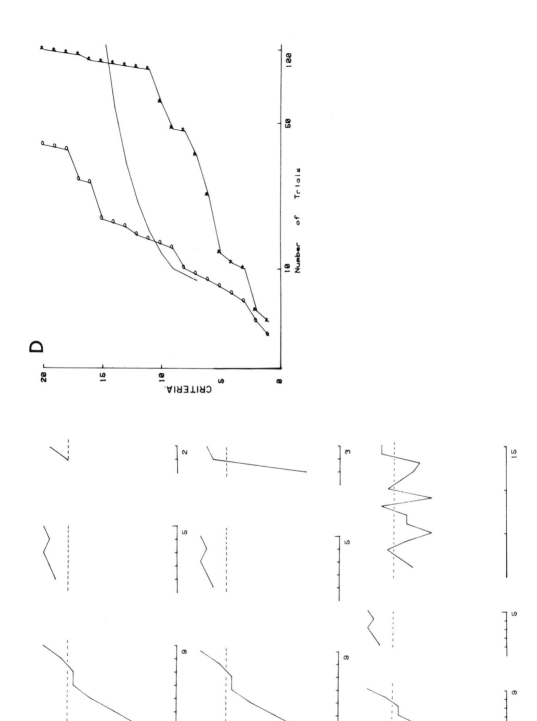

Figure 23.1 Hypothetical learning curves for two eyes, showing some of the difficulties that can be encountered when measuring interocular transfer. In A, B, and C the abscissae represent daily 40-trial training sessions and the ordinates represent percentages of correct responses. Each of the three graphs consists of three parts: learning with the first eye (left), five overtraining sessions with the same eye (center), and learning with the second eye (right). The criterion of learning is fixed at two consecutive sessions with 90% or more correct responses. In A interocular transfer is clear: Initial performance with the second eye is immediately at criterion, and saving is 100% for both errors and trials. In B initial performance with the second eye is not significantly different from chance, but the saving in reattaining criterion is 66% for errors and 75% for trials. In C, although the generally high level of performance with the second eye suggests a successful interocular transfer, the difficulties in meeting the fixed criterion of learning are reflected in negative savings scores for interocular transfer: −30% for trials, −1% for errors. This apparent contradiction can be obviated if the performances of the two eyes are analyzed in terms of successive criteria, as shown in D, where the abscissa is a logarithmic plot of the numbers of trials necessary to meet successive sequences of correct responses allowing for at most one error (Runnels et al. 1968) and the ordinate represents these successive criteria up to the maximum of 20 correct responses out of 21. The criterion for learning is assumed to be at the point where the curves for the two eyes meet the line giving the $P = 0.01$ probability of chance occurrence of that particular sequence within that particular number of trials. This analysis of the data shown in graphs C now indicated a considerable interocular transfer (77%) due to the adoption of a flexible criterion of learning. (○) First eye; (∗) second eye.

differentially affected by different conditions of visual stimulation during rearing? The experimental results that may provide answers to these questions will be reviewed separately for the various groups of cats with different rearing conditions.

Binocular Deprivation

Riesen et al. (1953) reared cats in darkness from birth to 14 weeks. Subsequently these cats were given diffuse light in one eye and patterned light in the other for 30 minutes daily during another 3 or 6 weeks. They were then trained on visual discriminations of geometrical forms, stripes with different directions, and brightness, using the eye that had previously received patterned light. Learning was much prolonged in these cats as compared with control cats reared under conditions of normal visual stimulation, and immediate interocular transfer was totally absent in that performance fell to chance when the eye used for learning was occluded and the other eye was exposed. This absence of immediate interocular transfer contrasted markedly with the virtually perfect carryover of training from the first eye to the second in two normally reared control animals. However, the experimental animals showed marked savings in interocular transfer, usually reattaining the criterion of learning in 20% as many trials as were necessary for learning with the other eye.

These results can be interpreted in two ways. The experiments utilized a two-door discrimination apparatus in which the animal had to walk to the door carrying the positive stimulus and push through it with head and shoulders in order to get the food reward. Riesen et al. (1953) suggested that the savings of training with the second eye could be accounted for by generic practice effects resulting in more effective modes of behavior in the discrimination apparatus, independent of any possible interocular carryover of specific information about the discriminanda. Yet these authors emphasized the clumsy behavior shown by the experimental animals when first tested with the second eye. In opposition to the above suggestion of Riesen et al., one could argue that the initial inability of the animal to perform the discrimination using the second eye was due largely or exclusively to an incapacity to use the apparatus efficiently, and that this incapacity had to be corrected before interocular transfer of the discrimination became apparent. In other words, immediate interocular transfer might have been absent not because the animal failed to recognize the stimuli with the second eye, but because it could not make the appropriate visuomotor response.

Meyers and McCleary (1964) provided a partial test of this hypothesis by using a much simpler discriminatory response. Cats were raised in darkness until they were 7 weeks old, after which they received binocular experience with diffuse, unstructured light for 1 hour daily until 30–35 weeks of age. Training was then begun on a cross/circle discrimination, with the animal restricted in a hammock with one eye occluded. The stimuli were presented one at a time in front of the animal, which received a shock on the right hind leg in association with a given stimulus. The shock could be avoided by lifting the same leg, and training continued until the animal flexed the leg upon the appearance of the positive stimulus, but not the neutral stimulus, on at least 85% of the trials in two consecutive 20-trial sessions. Substantial interocular transfer was observed in 3 cats that, on the first 20-trial session with the second eye, showed a performance ranging from 70% to 95%, in agreement with the scores of 4 normal control cats trained in the same way (range 65%–95%). The 2 other experimental cats, however, performed at chance on the first session with the second eye, and began to exhibit differential responding only on later sessions.

Meyers and McCleary (1964) attributed the successful interocular transfer in 3 of their animals to the simplicity of the discriminatory response, or to the use of a shock avoidance paradigm rather than a food-reward paradigm, or to both factors.

However, immediate interocular transfer of a simple light-intensity discrimination (luminance ratio between stimuli: 32:1) was observed by Riesen and Aarons (1959) in kittens that had lived in total darkness from birth to the sixth week of age, and that were subsequently trained in a two-door discrimination apparatus with food as the incentive for response. A similar result was obtained by Aarons et al. (1963) in kittens reared in darkness until the sixteenth week of age. While indicating that a shock-avoidance motivation is not indispensable for interocular transfer to a visually naive eye, these results suggest that the ease of the discrimination may be important for the presence or absence of interocular transfer in binocularly deprived cats. In one cat in the experiment of Riesen et al. (1953), immediate interocular transfer of a rather difficult intensity discrimination (luminance ratio: about 2.9:1) was totally absent, in agreement with the results on the other discriminations. Therefore, there is no compelling evidence that the mechanisms for interocular transfer of intensity discriminations and those for interocular transfer of form discriminations are differentially affected by binocular deprivation.

Ganz et al. (1972) tested interocular transfer of an intensity (flux) discrimination (luminance ratio: about 105:1), an orientation discrimination (vertical versus horizontal stripes) and a form discrimination (upright versus inverted triangle) in cats raised in complete darkness from birth to 13–25 weeks of age. A locomotor and door-pushing discriminatory response was required, and correct responses were reinforced with food. The three discriminations were learned monocularly in the sequence flux-orientation-form, and high-level immediate interocular transfer was obtained in all cases. These results confirm that interocular transfer of an easy intensity discrimination can be successful even if the eye receiving the transfer has never experienced light stimulation before the transfer. Successful interocular transfer on the subsequent orientation and form discriminations may have been helped by the fact that the animal had previous practice with the discrimination apparatus using either eye, although never with both eyes at the same time. Such experience was obviously lacking in the animals of the previous experiments of Riesen et al. (1953), which had been tested on a single problem. In conclusion, normal binocular experience does not appear to be a prerequisite for a successful interocular transfer of visual discriminations, although the expression of transfer through a complex visuomotor response may be difficult when the eye receiving the transfer has not been previously used for guiding behavior.

Monocular Deprivation

In all of the above experiments on binocularly deprived cats, learning of visual discriminations was greatly retarded as compared with that of control cats with normal visual experience; the only exception was the learning of easy intensity discriminations (Riesen et al. 1953; Riesen and Aarons 1959; Aarons et al. 1963; Meyers and McCleary 1964; Ganz et al. 1972). An even greater deficit in visual learning can be observed in an eye submitted to early visual deprivation if the other eye is concurrently given normal visual experience, possibly because of interocular competition effects. Kittens raised in a normal visual environment with the eyelids of one eye sutured show very little capacity for visually guided behavior when the deprived eye is opened and the other eye is occluded.

The initial results of Ganz and Fitch (1968) and Dews and Wiesel (1970) suggested that a massive incapacity of the deprived eye in pattern perception makes the testing of interocular transfer of form and orientation discriminations difficult or impossible. However, Rizzolatti and Tradardi (1971) showed that 4 cats monocularly deprived for 18–21 weeks after birth could be trained on form and orientation discriminations using the deprived eye. The success may have been due to prolonged training and to the occasional use of shock punishment combined with food reward. Learning with the deprived eye took about 6–20 times as many trials as learning with the normal eye. Interocular transfer from the normal eye to the deprived eye was totally absent, since learning with the deprived eye was not helped at all by previous learning of the same problem with the normal eye. When a problem was learned with the deprived eye first, transfer to the normal eye was not immediately apparent (Rizzolatti, pers. comm.). Although the savings of relearning with the second, undeprived eye were very conspicuous, they may have resulted from the difference in learning rates between the two eyes rather than from interocular transfer effects.

Ganz et al. (1972) produced monocular deprivation in two groups of cats for 13–25 weeks starting 10 days after birth. One group of cats were subsequently trained on a sequence of flux, orientation, and form discriminations (as described above) using the deprived eye; the other group were trained on the same sequence of discriminations using the experienced eye. In the first group (learning with the deprived eye), immediate interocular transfer was generally observed in the flux and orientation discrimination, although the transfer scores were somewhat lower than those of control animals. Of the two cats that were tested on the form discrimination, one showed good interocular transfer after prolonged learning; the other could not learn with the deprived eye. An entirely different picture emerged from the results from the second groups of cats, trained with the experienced eye first. On both flux and orientation discriminations there was no direct evidence for interocular transfer, since initial performance with the second eye was generally at chance and relearning with that eye generally took longer than learning with the other eye. However, a comparison between the two groups of cats showed that learning of the flux and orientation discriminations with the deprived eye was faster for the group using this eye for transfer. This may be taken as an indirect evidence for a small specific interocular-transfer effect from the experienced eye to the deprived eye; yet an unspecific practice factor cannot be ruled out completely, since on each problem using the deprived eye the cats in the second group had more general practice than those of the first group. No cat in the second group learned the form discrimination with the deprived eye, although the same discrimination had been easily learned with the experienced eye.

As already indicated, the interpretation of these asymmetrical interocular-transfer effects in monocularly deprived cats is complicated by the enormous difference in perception and learning between the two eyes. The deprived eye may be deficient not only in processing visual information related to the stimuli to be discriminated, but also in guiding motor behavior in the discrimination apparatus. Effective visuomotor control during visual discriminations may include coordinated head and eye movements for inspecting the discriminatory stimuli, and Rizzolatti and Tradardi (1971) reported that these inspecting movements may be very different depending on the eye used for testing. When using the deprived eye for discrimination, their cats showed large scanning head movements which were absent when the normal eye was used.

Interocular-transfer effects in monocularly deprived cats might be more pronounced if simpler stimuli and simpler discriminatory responses were adopted, as partially suggested by the Meyers-McCleary experiment (1964) on binocularly deprived cats (see above and Held 1968). Even very simple visuomotor responses, however, are not transferred between the eyes of monocularly deprived cats. Thus, for example, visually guided precise placing of the forelimbs does not transfer from the experienced eye to the deprived eye, and this effect has been shown to depend critically not on deprivation *per se* but on the fact that during the critical postnatal period the animal cannot use its deprived eye for watching its forelimbs during spontaneous locomotion (Hein et al. 1970; Hein and Diamond 1971).

On tests of interocular transfer of pattern discriminations from a visually experienced eye to a visually deprived eye some information about the discriminanda may be available to the latter eye, but good discriminatory performance may be hindered by deficits in both pattern analysis and visuomotor coordination, even when elementary motor responses are involved. The successful though imperfect transfer from the deprived eye to the experienced eye suggests that the two monocular perceptual systems are not entirely dissociated, and that the apparent absence of transfer from the experienced eye to the deprived eye may at least in part be attributable to testing procedures.

Ganz and Haffner (1974) studied interocular transfer of a form discrimination (upright versus inverted triangle) in cats raised with monocular deprivation from 5–8 days to 90–115 days after birth. Subsequently, the initially deprived eye was opened and the eyelids of the other eye were sutured in an attempt to improve the visual capacities of the initially deprived eye. All cats learned a flux discrimination, an orientation discrimination, and a form discrimination sequentially. The ability to learn the form discrimination differentiated these "cross-sutured" animals from cats with simple monocular deprivation, which could not learn the same discrimination with the deprived eye (Ganz et al. 1972), and supports previous findings on the usefulness of the reverse deprivation procedure for improving the visual capacities of the initially deprived eye (Dews and Wiesel 1970; Chow and Stewart 1972). Interocular transfer of this form discrimination was tested in the last stage of the experiment by reopening the initially experienced eye and occluding the other eye. There was evidence for some degree of immediate transfer in only 2 animals out of 5, but all animals showed considerable savings of learning with the second eye. Although Ganz and Haffner (1974) concluded that visual information acquired through the initially deprived eye is scarcely accessible to the other eye, such a conclusion is clearly vitiated by the neglect of the fact that the initially experienced eye had been deprived for 95–148 days, and had never been used for performing in the discrimination apparatus, before the interocular transfer test. Again, defective perceptual and/or visuomotor abilities of the eye receiving the transfer might have obscured positive transfer effects. It is therefore interesting to look at the effects of monocular visual deprivation on interocular transfer when the perceptual systems of the two eyes have equal capacities for controlling behavior. This can be achieved with alternate monocular deprivation.

Alternate Monocular Deprivation

Riesen and Mellinger (1956) described good interocular transfer in kittens that were raised in darkness for 1–2 months and then received normal visual stimulation through either eye for an hour daily, with the two eyes exposed at different times. Interocular transfer was imperfect (although clearly successful) in most animals on the first discrimination (a triangle versus a circle), but it was perfect in all animals on the second discrimination (horizontal versus vertical stripes), which was trained through the eye that had received the transfer on the first discrimination. By contrast, another cat that had been raised with monocular pattern experience as in the study of Riesen et al. (1953), except for a shorter stay in complete darkness, failed to show any indication of interocular transfer on the circle/triangle discrimination trained through the pattern-experienced eye. However, this same animal subsequently showed perfect interocular transfer of the horizontal/vertical stripe discrimination trained through the eye that had not received patterned light during the deprivation period.

Ganz et al. (1972) tested interocular transfer of a flux discrimination, an orientation discrimination, and a pattern discrimination in cats raised from birth to 13 weeks of age under conditions of alternating monocular occlusion, with each eye deprived of vision on alternate days. Transfer was virtually perfect and undistinguishable from that of normally raised controls. A similar result was obtained by von Grünau and Singer (1979), who reared kittens in total darkness until 26 days of age. Subsequently these cats were given visual experience for 8 hours a day, using one eye only and alternating eyes from day to day. When 19 weeks old they were trained to discriminate an array of intermixed horizontal and vertical bars from similar arrays containing only horizontal or vertical bars. Learning was monocular, and interocular transfer was perfect. Further, these animals were capable of performing these discriminations binocularly when, by means of color filters, the components of the positive stimulus were presented to different eyes and the negative stimulus was presented to one eye only.

Interocular transfer and simultaneous cross-integration can therefore occur normally in spite of the absence of any prior binocular experience.

Strabismus

Sherman (1971) produced experimental strabismus in kittens by cutting the medial rectus of one eye at 9–10 days of age, or those of both eyes, one at 8 and the other at 23 days of age. Some of these cats were also submitted in adulthood to lesions of visual cortex or to section of telencephalic and mesencephalic commissures. The optic chiasm was intact in all animals. Tests of interocular transfer carried out in these cats involved a flux discrimination and three pattern discriminations. On the average, interocular transfer was not as good in these cats as in intact control cats or in a cat rendered strabismic when adult. However, interocular transfer was consistently high (sometimes perfect) in all experimental subjects, irrespective of whether the strabismic or the normal eye was trained first, and independent of the presence or absence of additional brain damage.

Marzi et al. (1976) found that Siamese cats could transfer pattern discriminations between their eyes in a manner comparable to that of intact ordinary cats. Like many Siamese cats, two out of five of their Siamese subjects were strabismic, but these did not differ from the controls or the nonstrabismic Siamese cats in their capacity for transfer.

Peck et al. (1979a) altered the normal relations between the eyes of kittens by surgical intorsion of one eye, which varied between 10° and 180° and was performed 8–10 days after birth. In some cats the normal eye was left open; in some others it was closed at 6 weeks of age by suturing the eyelids and reopened at the beginning of behavioral testing at 3–4 months of age. In the kittens exposed to binocular experience, forced use of the rotated eye was ensured by occluding the normal eye 3–5 hours each day. Interocular transfer was tested on a brightness discrimination (luminance ratio 27:1), two orientation discriminations (horizontal versus vertical gratings, right oblique versus left oblique gratings), and a pattern discrimination (upright versus inverted triangle). Transfer was usually positive regardless of whether the rotated eye or the normal eye was used for learning and irrespective of the angle of eye rotation. However, the animals with 180° eye rotation were unable to learn the orientation or the pattern discriminations with the rotated eye, so interocular transfer could not be tested on these discriminations. Like the other cats, however, these cats showed good interocular transfer on the brightness discrimination. Bilateral rotations of the eyes, differing in degree, or unilateral eye rotations performed in adult, visually experienced cats, also failed to disturb interocular transfer.

In conclusion, the mismatch between the inputs to the two eyes consistently experienced by a cat with early-induced strabismus does not compete with the mechanisms of perceptual equivalence involved in interocular transfer.

Segregation of Visual Inputs to the Two Eyes
Hirsch (1972) raised cats under conditions of restricted visual experience from birth to 12 weeks of age. These animals wore a mask that continuously presented a field with three black horizontal lines to one eye and a field with three black vertical lines to the other eye. Interocular transfer was tested on three problems: a flux discrimination (luminance ratio between the stimuli about 1:100), a pattern discrimination (outlined square versus ×), and an orientation discrimination (vertical or horizontal grating versus a grating oriented at 45°). Interocular transfer was generally successful: two animals out of six showed poor transfer on the flux discrimination, one animal out of four showed poor transfer on the pattern discrimination, and one animal out of four showed poor transfer on the orientation discrimination. However, with the exception of a cat that was tested only on the flux discrimination, the other animals with transfer difficulties on a given problem trans-

ferred successfully the other two problems. In conclusion, the capacity for interocular transfer was preserved in spite of the fact that the two eyes had consistently received conflicting visual experience prior to testing.

Environmental Manipulations and the Neural Bases of Interocular Transfer

From the above data it is clear that the neural bases of interocular equivalence, at least insofar as it concerns the ability to transfer visual discriminations from one eye to the other, are largely innate. Further, they cannot be disrupted by forcing the two eyes to receive asynchronous and incongruous information during the critical maturation period of the visual system. Elsewhere in this volume Hamilton discusses how convergence of information from the two eyes onto central neuronal systems may contribute to behavioral phenomena of interocular equivalence. Our discussion of the possible neural mechanisms of interocular transfer will be based on the economical assumption that binocularly activated neurons are indeed the basic substrate for interocular transfer.

There are many levels in the central nervous system of the cat whose neurons receive a combined input from both eyes; a schematic description is given in figure 23.2. One of the most studied levels of binocular interaction is the complex of areas 17 and 18 (Hubel and Wiesel 1959, 1962, 1965a; Bishop 1973). These areas receive projections from the main laminae of the thalamic target of the optic tract, the dorsal lateral geniculate nucleus (LGN_d). However, while virtually all neurons in the LGN_d are connected to one eye or the other, but not to both, neurons in areas 17 and 18 receive converging projections from both types of LGN_d neurons, and can therefore be activated from each eye.

Monocular deprivation greatly reduces the number of binocularly activated neurons in areas 17 and 18, simply because the deprived eye becomes largely unable to influence these neurons (Wiesel and Hubel 1965; Hubel and Wiesel 1970). A drastic drop of binocular interactions in areas 17 and 18, accompanied by changes in receptive-field properties, is similarly observed in cats reared with continuous exposure of the two eyes to different visual stimuli (Hirsch and Spinelli 1971).

Binocular interactions in neurons of areas 17 and 18 are also largely eliminated by alternating monocular deprivation or experimental squint during the critical postnatal period. Under these conditions neurons in areas 17 and 18 respond normally to visual stimulation except for the fact that the ratio of binocular to monocular neurons changes from the normal value of 4:1 to about 1:4 or even 1:9 (Hubel and Wiesel 1965b). Behaviorally, the reduction in number of binocular neurons in areas 17 and 18 corresponds to the loss of at least 2 binocular

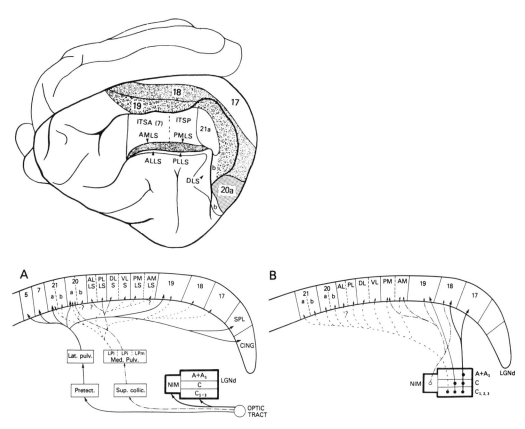

Figure 23.2 Top: Visual cortical areas in the cat. From Sprague et al. 1977, based on work by Palmer et al. (1978) and Tusa et al. (1978, 1979). A and B give a schematic representation of the thalamocortical projections of the visual systems of the cat, with an indication of the centers with predominantly or exclusively monocular neurons (various nuclei of the dorsal lateral geniculate complex, enclosed within a thick outline) and those with predominantly binocular neurons (pretectum, superior colliculus, pulvinar nuclei, and various cortical areas). (A) Projections of the pulvinar complex; (B) projections of the dorsolateral geniculate complex. Lat. pulv.: lateral pulvinar. Med. pulv.: medial pulvinar. LPl: lateral posterior, lateral division. LPi = lateral posterior, intermediate division. LPm: lateral posterior, medial division. Pretect.: pretectum. Sup. collic.: superior colliculus. LGN_d: lateral geniculate, dorsal part, laminae $A+A_1$, C and C_1, C_2 and C_3; NIM: medial interlaminar nucleus. Cortical areas 17, 18, 19, 20a, 20b, 21a, 21b. ALLS: anterior lateral, lateral suprasylvian. PLLS: posterior lateral, lateral suprasylvian. DLS: dorsal lateral suprasylvian. VLS: ventral lateral suprasylvian. PMLS: posterior medial, lateral suprasylvian. AMLS: anterior medial suprasylvian. SPL: splenial visual area. cing: cingular cortex. Summary includes unpublished work by Symonds and Rosenquist in A and by Raczkowski and Rosenquist in B.

processes: fine stereopsis (Packwood and Gordon 1975; Blake and Hirsch 1975) and binocular summation in contrast sensitivity (von Grünau 1979). These deficits are in sharp contrast with the near normal capacity of these animals for interocular transfer of monocularly learned brightness and pattern discriminations. It could be argued that the small number of binocular neurons remaining in areas 17 and 18 is sufficient for interocular transfer but not for binocular stereopsis and summation. However, this hypothesis is highly unlikely, because interocular transfer is successful even after removal of these areas (Sherman 1971; Berlucchi et al. 1978c). It is thus reasonable to look at other brain regions as possible sites of binocular convergence underlying interocular transfer. A search in this direction has been made in the Siamese cat.

Interocular Transfer in Siamese Cats

Because of a genetic abnormality, most of the retinal projections in Siamese cats are crossed, so that each eye is connected primarily with the contralateral half of the brain (see Guillery et al. 1974 and Marzi 1980 for review). As a consequence, neurons in areas 17 and 18 of these cats are almost exclusively responsive to stimulation of the contralateral eye (Hubel and Wiesel 1971; Kaas and Guillery 1973; Shatz 1977a). Shatz (1977b,c) has shown that in Siamese cats areas 17 and 18 receive callosal projections, which in principle could provide neurons in this area with an input from the ipsilateral eye. Yet the almost total absence of responses to ipsilateral eye stimulation in areas 17 and 18 suggests that the callosal input has a negligible influence on its target neurons (Shatz 1977b). In accord with results in ordinary cats reared with abnormal visual experiences, Siamese cats show, in parallel with a strongly reduced number of binocularly driven neurons in areas 17 and 18, an incapacity for combining information from the two eyes in tests of fine stereopsis (Packwood and Gordon 1975) and contrast sensitivity (von Grünau 1979). On the other hand, Siamese cats can transfer visual discriminations from one eye to the other as well as ordinary cats (Marzi et al. 1976; see also Meyers and McCleary 1964). Marzi et al. (1979, 1981) have now found electrophysiologically that binocular interactions do occur in the Siamese cat in at least a cortical area beyond areas 17 and 18. The majority of neurons in a visual area of the suprasylvian gyrus, corresponding to area PMLS of Palmer et al. (1978), receive a binocular input, in contrast with the strong monocular responsiveness of neurons in areas 17 and 18. Section of the corpus callosum virtually eliminates binocularity in the lateral suprasylvian area by suppressing the input from the ipsilateral eye (figure 23.3). In ordinary cats with the normal partial decussation of the optic pathways, callosotomy does not affect binocular interactions in the visual lateral suprasylvian area; it simply

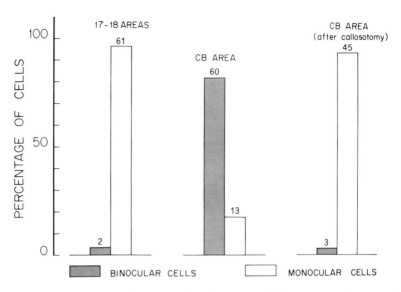

Figure 23.3 Percentages of binocularly and monocularly driven neurons in areas 17 and 18 and in the lateral suprasylvian area (CB area) of Siamese cats. Numbers on tops of bars indicate total number of neurons investigated. In intact Siamese cats virtually 100% of neurons in areas 17 and 18 could be driven only from the opposite eye, while more than 80% of neurons in the lateral suprasylvian area were activated from both eyes. After callosotomy, however, an almost total domination by the contralateral eye was observed with neurons in the latter area. Data from Marzi et al. 1979.

abolishes the representation of the ipsilateral half of the visual field (Legg et al. 1979; Marzi et al. 1981).

The relation between these electrophysiological findings and the capacity of Siamese cats for interocular transfer remains to be assessed, since the latter capacity is not abolished by callosotomy (Marzi and Di Stefano 1978; Marzi et al. 1979). However, these studies provide a clear indication that, in the absence of binocular interactions in areas 17 and 18, such interactions can occur in other cortical areas. It is reasonable to assume that rearing procedures disrupting binocularity in areas 17 and 18 of ordinary cats may leave binocular interactions in other brain regions unaffected. This appears to be the case for an important sub-cortical visual center—the superior colliculus—in cats reared with experimental squint (Gordon and Gummow 1975) or with alternating monocular deprivation (Gordon and Presson 1977). In these cats the superior colliculus retains a high percentage of binocularly activated neurons, in spite of the usual drop of binocular interactions in areas 17 and 18. No recordings from the visual lateral suprasylvian area have been reported in these cats, but it seems probable that, as in Siamese cats, neurons in this area should be mostly binocular. Conversely, there is evidence that the superior colliculus of Siamese cats contains many neurons receiving binocular input (Antonini et al. 1978b, 1981). Whether these binocular interactions are possible because of a direct uncrossed retinal input to the SC, or whether they are due to commissural connections as in the lateral suprasylvanian area, remains to be determined.

The Siamese cat can be considered a simplified preparation for the study of the neural substrates of interocular transfer because its optic input is largely, though incompletely, lateralized. An even more simplified preparation is obtained by sectioning the optic chiasm in the ordinary cat, thus restricting the projections from each retina exclusively to the ipsilateral cerebral hemisphere.

Interocular Transfer in Split-Chiasm Cats

Since the optic input is lateralized in split-chiasm cats, convergence of information from the two eyes onto single cerebral neurons can only occur by way of interhemispheric connections. Interocular transfer of visual discriminations should therefore become dependent on an interhemispheric transmission of visual information. The classic studies of Myers and Sperry (1953), Myers (1955, 1956) and Sperry et al. (1956) showed that split-chiasm cats can transfer pattern discriminations from one eye to the other, but no interocular transfer is possible if a section of the corpus callosum is added to that of the chiasm ("split-brain" preparation, figure 23.4). Split-brain cats are typically unable to recognize

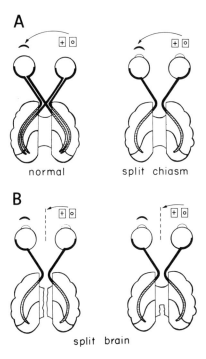

Figure 23.4 Summary of the experiments of Myers and Sperry (1953) and Myers (1955, 1956, 1959), showing that split-chiasm cats can transfer pattern discriminations from one eye to the other (A) but that this interocular transfer is absent if the corpus callosum, or even a posterior portion of it, is sectioned (B).

Transfer of Visual Discrimination in Cats

with one eye visual stimuli that they have previously learned to discriminate with the other eye, and relearning the discrimination with the naive eye on the average takes as long as the original learning with the other eye. This remarkable dissociation of the two eye-hemisphere systems of the split-brain cat in visual learning has subsequently been reproduced in split-chiasm cats with a callosal section limited to the posterior two-thirds of the structure—the part of the corpus callosum that contains the interhemispheric projections of all or most visual cortical areas (Myers 1959).

Certain imperfections in the callosally mediated interocular and interhemispheric transfer of visual discriminations have been described and attributed to a limited capacity of the corpus callosum for transmitting visual information. Myers (1962) reported that in split-chiasm cats easy discriminations show a better interocular transfer than difficult discriminations (the degree of difficulty of the discriminations was assessed on the basis of the number of trials needed to learn with the first eye). Berlucchi et al. (1978b) have shown that interocular transfer of form discriminations improves in split-chiasm cats as a function of practice with interocular transfer tasks, independent of the difficulty of the discrimination (see figure 23.5). This "learning set" for interocular transfer may be accounted for either by a gradual adaptation to the change in visual field that takes place on interocular transfer (the visual fields of the two eyes of the split-chiasm cat are on opposite sides of the vertical midline) or by an improvement with practice in the information-transmitting capacity of the corpus callosum.

Radical distortions of visual information during the transfer in the corpus callosum, such as the right-left reversal of visual stimuli assumed to occur in monkeys by Noble (1968), are probably simulated by testing conditions that encourage the animal to adopt discrimination strategies based on selective and local attention (Hamilton and Tieman 1973; Starr 1979). At any rate, paradoxical interocular transfer of lateral mirror-image discriminations—the preference by the second eye of the negative stimulus for the first eye—has not been observed in split-chiasm cats, which show instead a veridical interocular transfer of these discriminations suggesting a faithful spatial as well as formal representation of visual stimuli by the corpus callosum (Berlucchi and Marzi 1970; Hranchuck and Webster 1975).

Another question of general interest concerns the possible contributions of commissures other than the corpus callosum to interhemispheric transfer of visual discriminations. Meikle and Sechzer (1960) described a virtually perfect interocular transfer of an intensity discrimination (luminance ratio between stimuli 13.6:1) in split-brain cats (which, however, in agreement with the previous findings of Myers and Sperry, could not transfer pattern discriminations between the eyes). Meikle (1964) reiterated this finding and reported that interocular transfer of the above brightness discrimination was absent in split-

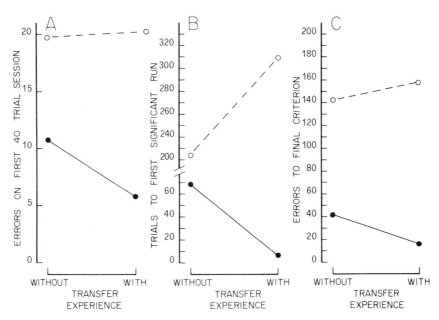

Figure 23.5 Improvement of interocular transfer with practice in split-chiasm cats. (O) First eye; (●) second eye. Four split-chiasm cats were tested for interocular transfer on four pattern discriminations of increasing difficulty. Difficulty of the discrimination was assessed on the basis of number of errors necessary to reach learning criterion in a larger group of control cats. Of the four split-chiasm cats, two used the right eye to learn the first and third discriminations and the left eye to learn the second and fourth; the other two cats followed the reverse order. Therefore, for each cat there were two discriminations that were transferred to an eye without experience of transfer (first and second discrimination) and two discriminations that were transferred to an eye with prior transfer experience (third and fourth discrimination). The three diagrams show the change in performance with the two eyes from the discriminations without prior experience of transfer to the discriminations preceded by such an experience. (A) Number of errors in the first 40-trial training session. (B) Number of trials necessary for performing the first significant run with 0.01 probability of chance occurrence. (C) Number of errors committed before reaching the final criterion of learning (two 40-trial sessions with 90% or more correct responses). Data are means across cats, but in each individual cat the same pattern was observed. As shown by B and C, the first eye showed a decrement in performance on the later discriminations, due to their greater difficulty; in spite of this, the performance with the second eye improved on the later discriminations with respect to the earlier discriminations, both in the initial session and in meeting the two learning criteria. This change in the performance with the second eye is best accounted for by an increase with practice of the ability for interocular transfer in split-chiasm cats. Source: Berlucchi et al. 1978b.

brain cats with an additional section of anterior, posterior, habenular, hippocampal, and intertectal commissures and of the thalamic massa intermedia. These findings have often been cited to support the simplistic "encephalization" theory, which allocates complex visual discriminations (such as pattern discriminations) to the cortex and simple visual discriminations (such as brightness discriminations) to subcortical centers like the superior colliculus. The argument is that the function of a cortical area or a subcortical center in visual learning can be inferred from the effect of the section of their respective commissures on interhemispheric transfer of different classes of visual discriminations. Accordingly, the abolition of interhemispheric transfer of pattern discriminations by section of the neocortical commissure, the corpus callosum, is attributed to a neocortical mediation of these discriminations, whereas the survival of interhemispheric transfer of brightness discriminations after a callosal section and its disappearance after adding the section of other commissures are attributed to a subcortical mediation of such discriminations.

These views must be revised considerably on the basis of more recent experiments by Peck et al. (1979), who have failed to find significant interocular transfer of an easy brightness discrimination (intensity ratio between stimuli 25:1) in split-chiasm cats with a complete callosal section. By contrast, the same discrimination could be transferred interocularly not only by normal cats, but also by split-chiasm cats with a callosal section sparing the splenium. The splenium of the corpus callosum was also essential for interhemispheric transfer of discriminations of directions of moving visual targets, whereas the anterior commissure did not appear to play any manifest role in interhemispheric transfer of either brightness or direction of movement.

Peck et al. (1979b) argue that the previous results of Meikle and Sechzer (1960) and Meikle (1964) may have been contaminated by procedures of eye occlusion that allowed some light to reach the covered eye during training or testing for interocular transfer. The absence of interocular transfer in the cats with radical midbrain surgery in the experiments of Meikle (1964) might have depended on the inability of these animals to make use of the small amount of light leaking through the occluding mask, in keeping with their quite poor learning performances. Meikle himself (1960) reported absence of interocular transfer of near-threshold brightness discriminations in split-chiasm, split-callosum cats.

The conclusion of Peck et al. (1979b) that the corpus callosum is essential for a high degree of interhemispheric transfer not only of pattern discriminations but also of brightness and movement discriminations is in accord with electrophysiological evidence that the corpus callosum is by far the most important, if not the only, cross-midline pathway for transmission of visual information to both cortex and subcortical centers. There is no clear-cut evidence that subcortical commissures are crucially and exclusively involved in the interhemispheric transfer of any type of visual discrimination, since cats reported to have deficits in transfer after section of these commissures had also had callosal sections (Voneida 1963). Section of the posterior and intertectal commissures leaving the corpus callosum intact has no effect on the interhemispheric transfer of pattern discriminations (Berlucchi et al. 1978a). No replication has as yet been made of the experiment by Sechzer (1964) reporting a successful interocular transfer of an orientation discrimination in split-brain cats trained with shock avoidance rather than with food reward. The general significance of this finding remains unclear, but, if confirmed, this capacity for interhemispheric transfer after callosotomy would definitely implicate noncallosal commissures. Berlucchi et al. (1978b) have attempted to potentiate a possible participation of subcortical commissure in interhemispheric transfer of pattern discriminations by giving split-chiasm cats a considerable amount of practice with interocular transfer before sectioning the corpus callosum and the anterior commissure. Some savings of learning with the second eye were found on tests of interocular transfer after commissurotomy, but by and large the magnitude of such savings was not sufficient for concluding that the precommissurotomy practice had been effective in improving postcommissurotomy transfer to any substantial extent.

Possible Neural Bases of Interocular Transfer in Split-Chiasm Cats
After section of the optic chiasm, neurons at the border between areas 17 and 18 can still be activated from both eyes (Berlucchi and Rizzolatti 1968; Berlucchi 1972). The input from the ipsilateral eye is conveyed to these neurons by the corpus callosum (Berlucchi et al. 1967; Hubel and Wiesel 1967; Shatz, 1977b). Binocular interactions in areas 17 and 18 after splitting of the chiasm are limited to a small portion of the visual field adjoining the vertical meridian (Berlucchi 1972), whereas other visual cortical areas in the suprasylvian gyri may contain neurons receiving binocular information from larger expanses of the visual field. This is suggested by the fact that in cats with intact optic pathways neurons in the suprasylvian gyri have a binocular receptive field extending 20° or more into the ipsilateral visual field, this ipsilateral component of the receptive field being abolished by a section of the corpus callosum (Dow and Dubner 1971; Legg et al. 1979; Marzi et al. 1981). As shown in figure 23.6, the callosal input would be necessary for conveying information from the contralateral eye to these neurons after section of the optic chiasm.

Attempts at differentiating the relative roles of the various cortical areas in interocular transfer of pattern discriminations in split-chiasm cats have involved the selective removal of some of these areas, as summarized in figure 23.7.

Unilateral or bilateral lesions of areas 17 and 18, alone or in combination with area 19, do not interfere with interocular transfer of pattern

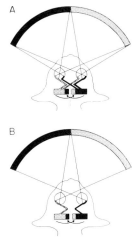

Figure 23.6 Relationship between the two halves of the visual fields, the retinae, and the two halves of the brain in normal and split-chiasm cats. In normal cats (A), each half of the visual field projects to the opposite halves of the retinae and to the opposite half of the brain. The representation of each half of the visual field in the ipsilateral half of the brain can occur by way of interhemispheric connections. In split-chiasm cats (B), each eye is connected solely with the ipsilateral side of the brain, so that interhemispheric connections are necessary for the representation of both the ipsilateral visual field and the contralateral eye in each half of the brain.

discriminations in split-chiasm cats. Since interocular transfer can subsequently be abolished in these animals by section of the corpus callosum, it follows that prior to callosotomy the transfer depended on the callosal connections of other cortical areas (Berlucchi et al. 1978c).

Unilateral lesions sparing areas 17 and 18 and removing large portions of the middle and suprasylvian gyri in split-chiasm cats produce large learning deficits in the eye ipsilateral to the section; in addition, interocular transfer from the eye on the intact side to the eye on the injured side is poor or absent, whereas transfer in the opposite direction is normal (Berlucchi et al. 1979; see figure 23.7). The asymmetry in transfer may be accounted for by the differences in learning rates between the two sides, but preliminary results with bilateral suprasylvian lesions and symmetrical learning deficits suggest a real impairment of the transfer mechanisms (A. Antonini, G. Berlucchi, and J. M. Sprague, unpublished). Thus, as in the split-chiasm monkey (see Hamilton, this volume), the mechanisms for interhemispheric transfer of visual discriminations in the split-chiasm cats are likely to reside beyond area 17.

A drastic impairment of interhemispheric transfer of visual discriminations has been observed in split-chiasm monkeys with bilateral removal of the inferior temporal cortex (Gross and Mishkin 1977; Seacord et al. 1979). Neurons in these cortical areas have large bilateral receptive fields and respond to the same triggering stimuli, which are often highly specific and complex, throughout their receptive field (Gross

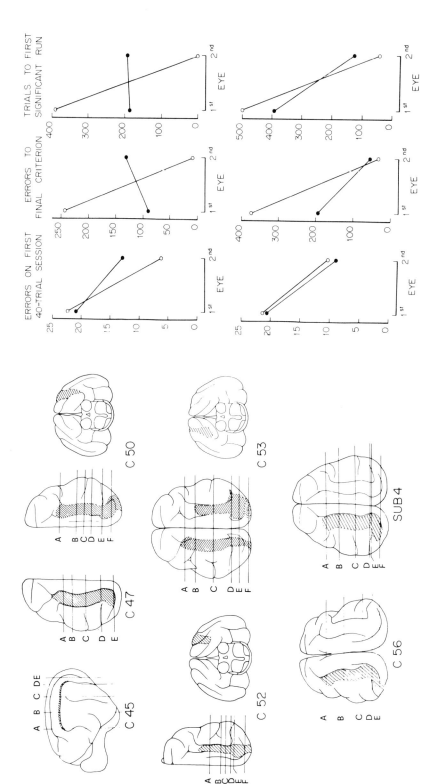

Figure 23.7 Top: Surface view of the brains of split-chiasm cats with unilateral suprasylvian lesions. Bottom: Mean performances using the eye on the intact and on the injured side in original learning and interocular transfer for split-chiasm cats with unilateral suprasylvian lesions (upper part) and split-chiasm cats with unilateral 17-18-19 lesions. (○) Lesion on first side; (●) lesion on second side. Interocular transfer is assessed on the basis of three measures: errors on first 40 trials, number of trials to final criterion (two consecutive 40-trial sessions with at least 90% correct responses), and number of trials to first significant run (first sequence of correct responses, allowing for at most one error, having a chance probability of occurrence equal to or lower than 0.01). Note absence of transfer to injured side in cats with suprasylvian lesions and presence of transfer in both directions in cats with 17-18-19 lesions. Source: Berlucchi et al. 1979.

1973). The ipsilateral portion of these receptive fields, and the input from the contralateral eye after chiasm splitting, depend on the corpus callosum and the anterior commissure (Gross et al. 1977). Gross and Mishkin (1977) and Seacord et al. (1979) argue that the neural substrates of perceptual equivalence between the right and left visual fields, or between the eyes of split-chiasm monkeys, are uniquely localized in the inferior temporal cortex because of the bilaterality and the high specificity for stimulus quality of neurons in this cortical region. Yet other cortical areas with large, bilateral, but relatively unspecific receptive fields, such as the inferior parietal lobule (Robinson et al. 1978) or the frontal eye fields (Mohler et al. 1973), may contribute significantly to interocular and interhemispheric transfer in presence of an intact inferior temporal cortex. The effects of lesions of these areas on interocular transfer have not yet been studied.

Receptive fields of neurons in visual suprasylvian areas of the cat are often large and bilateral, but do not show the stimulus specificity typical of neurons in the inferior temporal cortex of the monkey (Hubel and Wiesel 1969; Wright 1969; Spear and Baumann 1975; Camarda and Rizzolatti 1976). However, highly structured visual information might be encoded by neuronal assemblies irrespective of the absence of selectivity for stimulus quality in the response of single neurons. Many neurons with large bilateral receptive fields in several cortical areas may thus participate in interhemispheric transfer of visual discriminations. (Ganz et al. [1972; see also Ganz 1978] have suggested that the lack of interocular transfer in monocularly deprived cats may depend on the fact that the few binocular neurons remaining in areas 17 and 18 have different receptive fields in the two eyes. The receptive field in the non-deprived eye is usually stimulus-specific for size, orientation, etc.; that in the deprived eye is not. Even if one neglects the evidence against areas 17 and 18 being the locus for interocular transfer, the above hypothesis is weakened by the finding that these animals show good interocular transfer to the undeprived eye [Ganz et al. 1972]. It seems unlikely that the difference in receptive fields of binocular neurons in areas 17 and 18 prevents transfer in one direction but not in the other.)

But neurons with bilateral receptive fields are not limited to cortical areas. In the anterior portion of the superior colliculus many receptive fields extend considerably into the ipsilateral visual field; further, neurons in these regions retain a receptive field in the contralateral eye after splitting of the optic chiasm (Antonini et al. 1978a) (see figure 23.8). Section of the corpus callosum abolishes the response of superior-colliculus neurons to stimulation of the contralateral eye in split-chiasm cats, thus implying that indirect, cross-midline transmission of visual information to the superior colliculus relies on corticotectal projections from cortical areas activated by the corpus callosum (Antonini et al. 1979) (see figure 23.9). Lack of interocular and in-

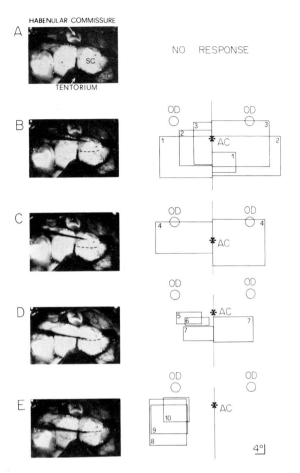

Figure 23.8 Binocularly activated portion of superior colliculus of split-chiasm cat as assessed by combination of photographic and electrophysiological methods. Photographs of superior colliculus were taken through microscope after killing animal and removing hemispheres, with microelectrode repositioned at beginning of tracks along which recordings were previously obtained. In illustration of receptive fields on the right, the vertical meridian of the two eyes is brought into register so that the receptive fields in the ipsilateral (right) eye are to the left of the midline (contralateral visual field) and the receptive fields in the contralateral (left) eye are to the right of the midline (ipsilateral visual field). In track A (Horsley-Clarke coordinates of the electrode: A4, L2) no units responsive to visual stimuli were found. In track B (A, L2) and in track C (A3, L2) binocular units were consistently recorded from. The receptive fields in the two eyes for three of these units are shown in B, and the receptive fields in the two eyes for other units are shown in C. Unit 1 in B was somewhat exceptional because of the large difference is size between the receptive fields in the two eyes. In track D (A2, L2) two superficial units had receptive fields only in the ipsilateral eye, and these fields were somewhat detached from the vertical meridian (units 5 and 6). Unit 7, which was encountered deeper in the track, was binocular and had receptive fields bordering on the vertical meridian in both eyes. In track E (A, L5, L2) all units from which recordings could be obtained had receptive fields only in the ipsilateral eye. These receptive fields were considerably distant from the vertical meridian (see units 8–10). The interrupted line traced on the superior colliculus surface is based on similar explorations with more medial or lateral sequences of tracks. It indicates the posterior border of the binocular portion of the superior colliculus. Source: Antonini et al. 1978a.

Transfer of Visual Discrimination in Cats

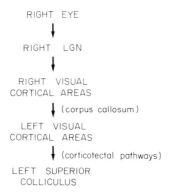

RIGHT EYE

RIGHT LGN

RIGHT VISUAL
CORTICAL AREAS
⬇ (corpus callosum)
LEFT VISUAL
CORTICAL AREAS
⬇ (corticotectal pathways)
LEFT SUPERIOR
COLLICULUS

Figure 23.9 Possible pathway for transmission of information from one retina to the contralateral superior colliculus in split-chiasm cats. The crucial importance of the corpus callosum for this crossed transmission of information is indicated by the lack of superior-colliculus responses to stimulation of the contralateral eye in split-chiasm cats with a section of the posterior two-thirds of the corpus callosum (see Antonini et al. 1978a, 1979).

terhemispheric transfer of visual discriminations in split-chiasm cats with an additional section of the corpus callosum may therefore depend not only on the disconnection of cortical areas, but also on the abolition of cross-midline transmission of visual information to the superior colliculus. An interplay between cortical and subcortical centers in visual learning and interhemispheric transfer had been suggested by Thompson (1965), who postulated that subcortical commissures are not apt to transmit highly structured visual information and that the interhemispheric communication of this information to subcortical centers for storage and analysis must rely on the corpus callosum and on corticofugal projections. A test of the hypothesis that the midbrain is involved in interhemispheric transfer of visual discrimination requires the study of the effects of superior-colliculus lesions on interocular transfer in split-chiasm cats.

References

Aarons, L., H. K. Halasz, and A. H. Riesen. 1963. "Interocular transfer of visual intensity discrimination after ablation of striate cortex in dark-reared kittens." *J. Comp. Physiol. Psychol.* 56: 196–199.

Antonini, A., G. Berlucchi, and J. M. Sprague. 1978a. "Indirect, across-the-midline retinotectal projections and representation of the ipsilateral visual field in superior colliculus of the cat." *J. Neurophysiol.* 41: 295–304.

Antonini, A., G. Berlucchi, M. Di Stefano, and C. A. Marzi. 1978b. "Binocular convergence in visual cortex and superior colliculus of Siamese cat." *Neurosci. Lett.* Suppl. 1: S367.

———. 1981. "Differences in binocular interactions between cortical areas 17–18 and superior colliculus in Siamese cats." *J. Comp. Neurol.*

Antonini, A., G. Berlucchi, C. A. Marzi, and J. M. Sprague. 1979. "Importance of corpus callosum for visual receptive fields of single neurons in cat superior colliculus." *J. Neurophysiol.* 42: 137–152.

Berlucchi, G. 1972. "Anatomical and physiological aspects of visual functions of corpus callosum." *Brain Res.* 37: 371–392.

Berlucchi, G., and C. A. Marzi. 1970. "Veridical interocular transfer of lateral mirror-image discriminations in split-chiasm cats." *J. Comp. Physiol. Psychol.* 72: 1–7.

Berlucchi, G., and G. Rizzolatti. 1968. "Binocularly driven neurons in visual cortex of split-chiasm cats." *Science* 159: 308–310.

Berlucchi, G., M. S. Gazzaniga, and G. Rizzolatti. 1967. "Microelectrode analysis of transfer of visual information by the corpus callosum." *Arch. Ital. Biol.* 105: 583–596.

Berlucchi, G., H. A. Buchtel, and F. Lepore. 1978a. "Successful interocular transfer of visual pattern discriminations in split-chiasm cats with section of the intertectal and posterior commissures." *Physiol. Behav.* 20: 331–338.

Berlucchi, G., E. Buchtel, C. A. Marzi, G. G. Mascetti, and A. Simoni. 1978b. "Effects of experience on interocular transfer of pattern discriminations in split-chiasm and split-brain cats." *J. Comp. Physiol. Psychol.* 92: 532–543.

Berlucchi, G., J. M. Sprague, F. Lepore, and G. G. Mascetti. 1978c. "Effects of lesion of areas 17, 18 and 19 on interocular transfer of pattern discriminations in split-chiasm cats." *Exp. Brain Res.* 31: 275–297.

Berlucchi, G., J. M. Sprague, A. Antonini, and A. Simoni. 1979. "Learning and interhemispheric transfer of visual pattern discriminations following unilateral suprasylvian lesions in split-chiasm cats." *Exp. Brain Res.* 34: 551–574.

Bishop, P. O. 1973. "Neurophysiology of binocular single vision and stereopsis." In *Handbook of Sensory Physiology*, vol. VII/3A. Berlin: Springer.

Blake, R., and H. V. B. Hirsch. 1975. "Deficits in binocular depth perception in cats after alternating monocular deprivation." *Science* 190: 1114–1116.

Blakemore, C. 1978. "Maturation and modification in the developing visual system." In *Handbook of Sensory Physiology*, vol. VIII: *Perception*, R. Held et al. eds. Berlin: Springer.

Bogartz, R. S. 1965. "The criterion method: Some analyses and remarks." *Psychol. Bull.* 64: 1–14.

Camarda, R., and G. Rizzolatti. 1976. "Visual receptive fields in the lateral suprasylvian area (Clare-Bishop area) of the cat." *Brain Res.* 101: 427–443.

Chow, K. L., and D. L. Stewart. 1972. "Reversal of structural and functional effects of long-term visual deprivation in cats." *Exp. Neurol.* 34: 409–433.

Dews, P. B., and T. N. Wiesel. 1970. "Consequences of monocular deprivation on visual behaviour in kittens." *J. Physiol.* 206: 437–455.

Dow, B. M., and R. Dubner. 1971. "Single-unit responses to moving visual stimuli in middle suprasylvian gyrus of the cat." *J. Neurophysiol.* 34: 47–55.

Ganz, L. 1978. "Innate and environmental factors in the development of visual form perception." In *Handbook of Sensory Physiology*, vol. VIII: *Perception*, C. R. Held et al., eds. Berlin: Springer.

Ganz, L., and M. Fitch. 1968. "The effect of visual deprivation on perceptual behavior." *Exp. Neurol.* 22: 638–660.

Ganz, L., and M. E. Haffner. 1974. "Permanent perceptual and neurophysiological effects of visual deprivation in the cat." *Exp. Brain Res.* 20: 67–87.

Ganz, L., H. V. B. Hirsch, and S. B. Tieman. 1972. "The nature of perceptual deficits in visually deprived cats." *Brain Res.* 44: 547–568.

Gordon, B., and L. Gummow. 1975. "Effects of extraocular muscle section on receptive field in cat superior colliculus." *Vision Res.* 15: 1011–1019.

Gordon, B., and J. Presson. 1977. "Effects of alternating occlusion of receptive fields in cat superior colliculus." *J. Neurophysiol.* 40: 1046–1414.

Gross, C. G. 1973. "Visual functions of inferotemporal cortex." In *Handbook of Sensory Physiology*, vol. VII/3B: *Central Visual Information*, R. Jung, ed. Berlin: Springer.

Gross, C. G., and M. Mishkin. 1977. "The neural basis of stimulus equivalence across retinal translation." In *Lateralization in the Nervous System*, S. Harnad et al., eds. New York: Academic.

Gross, C. G., D. B. Bender, and M. Mishkin. 1977. "Contributions of the corpus callosum and the anterior commissure to visual activation of inferior temporal neurons." *Brain Res.* 131: 227–239.

Guillery, R. W., V. A. Casagrande, and M. D. Oberdorfer. 1974. "Congenitally abnormal vision in Siamese cats." *Nature* 252: 195–199.

Hamilton, C. R., and S. B. Tieman. 1973. "Interocular transfer of mirror image discriminations by chiasm-sectioned monkeys." *Brain Res.* 64: 241–255.

Hein, A., and R. M. Diamond. 1971. "Contrasting development of visually triggered and guided movements in kitttens with respect to interocular and interlimb equivalence." *J. Comp. Physiol. Psychol.* 76: 219–224.

Hein, A., R. Held, and E. C. Gower. 1970. "Development and segmentation of visually controlled movement by selective exposure during rearing." *J. Comp. Physiol. Psychol.* 73: 181–187.

Held, R. 1968. "Dissociation of visual functions by deprivation and rearrangement." *Psychol. Forsch.* 31: 338–348.

Hirsch, H. V. B. 1972. "Visual perception in cats after environmental surgery." *Exp. Brain Res.* 15: 405–423.

Hirsch, H. V. B., and D. N. Spinelli. 1971. "Modification of the distribution of receptive field orientation in cats by selective visual exposure during development." *Exp. Brain Res.* 13: 509–527.

Hranchuk, K. B., and W. G. Webster. 1975. "Interocular transfer of lateral mirror-image discriminations by cats: Evidence of species differences." *J. Comp. Physiol. Psychol.* 88: 368–372.

Hubel, D. H., and T. N. Wiesel. 1959. "Receptive fields of single neurons in the cat's striate cortex." *J. Physiol.* 148: 574–591.

Hubel, D. H., and T. N. Wiesel. 1962. "Receptive fields, binocular interaction and functional architecture in the cat's visual cortex." *J. Physiol.* 160: 106–154.

————. 1965a. "Receptive fields and functional architecture in two nonstriate visual areas (18 and 19) of the cat." *J. Neurophysiol.* 28: 229–289.

————. 1965b. "Binocular interaction in striate cortex of kittens reared with artificial squint." *J. Neurophysiol.* 28: 1041–1059.

————. 1967. "Cortical and callosal connections concerned with the vertical meridian of visual fields in the cat." *J. Neurophysiol.* 30: 1561–1573.

————. 1969. "Visual area of the lateral suprasylvian gyrus (Clare-Bishop area) of the cat." *J. Physiol.* 202: 251–260.

————. 1970. "The period of susceptibility to the physiological effects of unilateral eye closure in kittens." *J. Physiol.* 206: 419–436.

————. 1971. "Aberrant visual projections in the Siamese cat." *J. Physiol.* (Lond.) 218: 33–62.

Kaas, J. H., and R. W. Guillery. 1973. "The transfer of abnormal visual field representations from the dorsal lateral geniculate nucleus to the visual cortex in Siamese cats." *Brain Res.* 59: 61–95.

Legg, C. R., A. Antonini, M. Di Stefano, and C. A. Marzi. 1979. "Role of corpus callosum for ipsilateral visual field representation in the Clare-Bishop area of common and Siamese cats." *Exp. Brain Res.* 36: R16–17.

Lund, R. D. 1978. *Development and Plasticity of the Brain.* New York: Oxford University Press.

Marzi, C. A. 1980. "Vision in Siamese cats." *Trends in Neuroscience* 3: 165–169.

Marzi, C. A., and M. Di Stefano. 1978. "Role of Siamese cat's crossed and uncrossed retinal fibres in pattern discrimination and interocular transfer." *Arch. Ital. Biol.* 116: 330–337.

Marzi, C. A., A. Simoni, and M. Di Stefano. 1976. "Lack of binocularly driven neurones in the Siamese cat's visual cortex does not prevent interocular transfer of visual form discriminations." *Brain Res.* 105: 353–357.

Marzi, C. A., M. Di Stefano, A. Antonini, and C. Legg. 1979. "Binocular interactions in the Clare-Bishop area of Siamese cats which lack binocular neurones in the primary visual cortex." *Neurosci. Lett.* Suppl. 3: 357.

Marzi, C. A., A. Antonini, M. Di Stefano, and C. R. Legg. 1981. "Role of the corpus callosum for ipsilateral visual field representation and binocular convergence in lateral suprasylvian area of common and Siamese cats." *Behav. Brain Res.*

Meikle, T. H. 1960. "Role of corpus callosum in transfer of visual discriminations in the cat." *Science* 132: 1496.

———. 1964. "Failure of interocular transfer of brightness discrimination." *Nature* 202: 1243–1244.

Meikle, T. H., and J. A. Sechzer. 1960. "Interocular transfer of brightness discrimination in split-brain cats." *Science* 132: 734–735.

Meyers, B., and R. A. McCleary. 1964. "Interocular transfer of a pattern discrimination in pattern deprived cats." *J. Comp. Physiol. Psychol.* 57: 16–21.

Mohler, C. W., M. E. Goldberg, and R. H. Wurtz. 1973. "Visual receptive fields of frontal eye field neurons." *Brain Res.* 61: 385–389.

Murdock, B. B., Jr. 1957. "Transfer designs and formulas." *Psychol. Bull.* 54: 313–326.

Myers, R. E. 1955. "Interocular transfer of pattern discriminations in cats following section of crossed optic fibers." *J. Comp. Physiol. Psychol.* 48: 470–473.

———. 1956. "Function of corpus callosum in interocular transfer." *Brain* 79: 358–363.

———. 1959. "Localization of function in the corpus callosum: Visual gnostic transfer." *Arch. Neurol.* 1: 74–77.

———. 1962. "Transmission of visual information within and between the hemispheres: A behavioral study." In *Interhemispheric Relations and Cerebral Dominance*, V. B. Mountcastle, ed. Baltimore: Johns Hopkins University Press.

Myers, R. E., and R. W. Sperry. 1953. "Interocular transfer of a visual form discrimination habit in cats after section of the optic chiasma and corpus callosum." *Anat. Rec.* 115: 351–352.

Noble, J. 1968. "Paradoxical interocular transfer of mirror-image discrimination in the optic chiasm sectioned monkey." *Brain Res.* 10: 127–151.

Packwood, J., and B. Gordon. 1975. "Stereopsis in normal domestic cat, Siamese cat, and cat raised with alternating monocular occlusion." *J. Neurophysiol.* 38: 1485–1499.

Palmer, L. A., A. C. Rosenquist, and R. J. Tusa. 1978. "The retinotopic organization of lateral suprasylvian visual areas in the cat." *J. Comp. Neurol.* 177: 237–256.

Peck, C. K., S. G. Crewther, G. Barber, and C. J. Johansen. 1979a. "Pattern discrimination and visuomotor behavior following rotation of one or both eyes in kittens and in adult cats." *Exp. Brain Res.* 34: 401–418.

Peck, C. K., S. G. Crewther, and C. R. Hamilton. 1979b. "Partial interocular transfer of brightness and movement discrimination by split-brain cats." *Brain Res.* 163: 61–75.

Riesen, A. H., and L. Aarons. 1959. "Visual movement and intensity discrimination in cats after early deprivation of pattern vision." *J. Comp. Physiol. Psychol.* 52: 142–149.

Riesen, A. H., and J. C. Mellinger. 1956. "Interocular transfer of habits in cats after alternating monocular experience." *J. Comp. Physiol. Psychol.* 49: 516–520.

Riesen, A. H., M. I. Kurke, and J. C. Mellinger. 1953. "Interocular transfer of habits learned monocularly in visually naive and visually experienced cats." *J. Comp. Physiol. Psychol.* 46: 166–172.

Rizzolatti, G., and V. Tradardi. 1971. "Pattern discriminations in monocularly reared cats." *Exp. Neurol.* 33: 181–194.

Robinson, D. L., M. E. Goldberg, and G. Stanton. 1978. "Parietal association cortex in the primate: Sensory mechanisms and behavioral modifications." *J. Neurophysiol.* 41: 910–932.

Runnels, L. K., R. Thompson, and P. Runnels. 1968. "Near-perfect runs as a learning criterion." *J. Math. Psychol.* 5: 362–368.

Seacord, L., C. G. Gross, and M. Mishkin. 1979. "Role of inferior temporal cortex in interhemispheric transfer." *Brain Res.* 167: 259–272.

Sechzer, J. A. 1964. "Successful interocular transfer of pattern discrimination in split-brain cats with shock-avoidance motivation." *J. Comp. Physiol. Psychol.* 58: 76–83.

Shatz, C. 1977a. "A comparison of visual pathways in Boston and Midwestern Siamese cats." *J. Comp. Neurol.* 17: 205–228.

———. 1977b. "Abnormal interhemispheric connections in the visual system of Boston Siamese cats: A physiological study." *J. Comp. Neurol.* 171: 229–246.

———. 1977c. "Anatomy of interhemispheric connections in the visual system of Boston Siamese and ordinary cats." *J. Comp. Neurol.* 173: 497–511.

Sherman, S. M. 1971. "Role of the visual cortex in interocular transfer in the cat." *Exp. Neurol.* 30: 34–45.

Spear, P. D., and T. P. Baumann. 1975. "Receptive-field characteristics of single neurons in the lateral suprasylvian visual area of the cat." *J. Neurophysiol.* 38: 1403–1420.

Sperry, R. W., J. Stamm, and N. Miner. 1956. "Relearning tests for interocular transfer following division of optic chiasma and corpus callosum in cats." *J. Comp. Physiol. Psychol.* 49: 529–533.

Sprague, J. M., J. Levy, A. Di Berardino, and G. Berlucchi. 1977. "Visual cortical areas mediating form discrimination in the cat." *J. Comp. Neurol.* 172: 441–488.

Starr, B. 1979. "Mirror image transfer in optic chiasm sectioned monkeys." In *Structure and Function of Cerebral Commissures*, I. Steele-Russel et al., eds. London: MacMillan.

Thompson, R. 1965. "Centrencephalic theory and interhemispheric transfer of visual habits." *Psychol. Rev.* 72: 385–398.

Tusa, R. J., L. A. Palmer, and A. C. Rosenquist. 1978. "The retinotopic organization of area 17 (striate cortex) in the cat." *J. Comp. Neurol.* 177: 213–236.

Tusa, R. J., A. C. Rosenquist, and L. A. Palmer. 1979. "Retinotopic organization of areas 18 and 19 in the cat." *J. Comp. Neurol.* 185: 657–678.

Voneida, T. 1963. "Performance of a visual conditioned response in split-brain cats." *Exp. Neurol.* 8: 473–504.

Von Grünau, M. 1979. "Binocular summation and the binocularity of cat visual cortex." *Vision Res.* 19: 813–816.

Von Grünau, M. V., and W. Singer. 1979. "The role of binocular neurons in the cat striate cortex in combining information from the two eyes." *Exp. Brain Res.* 34: 133–142.

Wiesel, T. N., and D. H. Hubel. 1965. "Comparison of the effects of unilateral and bilateral eye closure on cortical unit responses in kittens." *J. Neurophysiol.* 28: 1029–1040.

Wright, M. J. 1969. "Visual receptive fields of cells in a cortical area remote from the striate cortex in the cat." *Nature* 223: 973–975.

24 Evaluation of Visual Performance in Cats with Posterior Extrastriate Lesions

Paul Cornwell
J. M. Warren

Lesions in the posterior extrastriate cortex produce selective subtotal impairments in visual discrimination learning by cats, but the perceptual defects responsible for these losses have not been identified (Hara et al. 1974; Cornwell et al. 1976; Sprague et al. 1977; Campbell 1978; Berlucchi et al. 1977). In this chapter we describe some of the problems that must be considered in research efforts to specify the perceptual troubles that result from posterior extrastriate lesions in cats. Current methods for determining the cues intact animals use in learning visual discrimination problems are critically reviewed, and the implication of this information for studies of the visual functions of the posterior extrastriate cortex are discussed.

Investigators of learning and perception in animals must deal with two basic questions: What stimuli can the animal detect and discriminate under optimal conditions? What cues does the animal actually use in solving a particular discrimination problem? The first question concerns the definition of the animal's sensory capacities, its potential ability to resolve differences between visual stimuli; the second concerns the role of central factors such as selective attention in determining which of the potentially available cues the subject uses in learning a discrimination task.

The disjunction between sensory capacity and attention is important. Animals often fail to use all the sensory data available to them. It often appears that before an animal can learn a discrimination problem it must learn which dimensions to attend to and which to ignore (Sutherland and Mackintosh 1971; Mackintosh 1974). Of the several experimental effects that support this view (Sutherland and Mackintosh 1971), overshadowing has been the most thoroughly investigated and seems most applicable to brain research (see, for example, Nonneman and Warren 1977). Overshadowing was first described by Pavlov (1927), who trained dogs to salivate to a combination of a tactile and a heat stimulus and determined their responses to each stimulus presented separately. The dogs salivated to touch, but not to the thermal stimulus, even though control animals learned if heat was the only conditioned stimulus presented. Thus, the tactile stimulus overshadowed (suppressed learning about) the thermal stimulus in the compound.

Overshadowing has also been shown in instrumental learning experiments with several species. When (for example) groups of goldfish trained to discriminate lines that differ only in orientation, or in both orientation and color, were tested with a new set of lines differing only in orientation, the fish trained with two cues were grossly inferior to those initially trained on orientation only. The presence of color in original learning thus prevented fish trained on both cues from learning about the potentially discriminable orientation cues (Bitterman 1975).

Powerful overshadowing effects may be observed in mammals. Granit (1947) described populations of neural elements in cat retinae that are selectively responsive to red, green, and blue, indicating that cats have the neural basis for color vision. For many years, however, competent psychologists were unable to obtain behavioral evidence of color vision in cats (for a review see Meyer and Anderson 1965). The reason is now clear. Color is not an initially salient dimension for cats. Without rigorous controls, cats trained on color discriminations respond preferentially to irrelevant brightness and position cues (Daw 1973). However, once cats learn to attend to color cues, responses to color may interfere with other tasks, such as heterochromatic brightness matching, in which color is irrelevant (Brown et al. 1973). The attention value of specific cues is thus susceptible to modification by reinforced training.

Ablations of cortex beyond the primary sensory projection areas do not ordinarily result in massive losses in sensory capacity. Since the posterior extrastriate region is not primary visual cortex, our review concentrates on questions regarding cue utilization and attention rather than those concerning sensory capacity, the primary focus of research on the posterior extrastriate cortex until now (Cornwell et al. 1976; Sprague et al. 1977; Berkley and Sprague 1978; Campbell 1978).

Since little has been done to determine to which elements of complex stimuli cats with extrastriate decortications attend, a major part of this chapter is devoted to evaluating the two principal behavioral methods of measuring attention in visual perception and learning. Both of these, the equivalence method and the transfer method, are based on the same premise: One can gain insight into the organization of an animal's perceptual world by determining which features of stimuli it has learned to discriminate may and may not be altered without disrupting discrimination performance. The two methods differ in their use of reinforcement. In the equivalence method tests for the equivalence or nonequivalence of transformed stimuli are carried out with nondifferential reinforcement; in the transfer method such tests are done using differential reinforcement.

Stimulus Equivalence

In the classic studies with the method of equivalence (Klüver 1933; Spence 1937, 1942; Lashley 1938, 1942) animals were trained with differential reinforcement to discriminate a pair of stimuli differing in one or more visual dimensions. They then were tested in a series of sessions in which the experimenter presented intercurrently trials with the original training stimuli, with the initial differential-reinforcement contingencies in force, and trials with an unfamiliar pair of test stimuli, on which choices of either stimulus were reinforced. The test pairs were distortions of the training figures in shape, size, completeness, relative brightness, and the like, or pairs that presented cues in only one of the dimensions present in discrimination training. If the subject chose the test stimulus more like the training figure on a substantial majority of the test trials, it was regarded as equivalent to the rewarded training figure in the sense that it evoked the same response from the subject. The following experiments illustrate the use of the method of equivalence and kind of results it may yield under favorable conditions.

Klüver (1933) trained a Java macaque with differential reinforcement to choose the smaller of two rectangles with areas of 125 and 500 cm². Then, under conditions of nondifferential reward, he observed the monkey's response to changes in the relative and absolute size, shape, brightness, and hue of the discriminanda. He reported data for 57 different equivalence tests, and on 51 of these the monkey responded to the smaller stimulus 75% or more of the time. Some of the altered pairs used in these equivalence tests are shown in figure 24.1, which indicates that when the stimuli were equal (I, J, K, L) or nearly equal (E, F) in area the monkey often chose consistently but responded on the basis of the relative length of the test figures' bases or heights. These findings

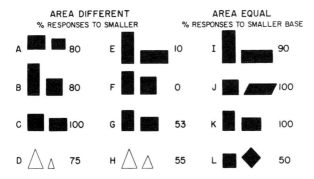

Figure 24.1 Percentages of response to various stimuli by Klüver's monkey D.-L. (Klüver 1933, pp. 162–167). In A–H, areas of stimuli are different and figure gives percentage of responses to smaller stimulus; in I–L, areas are equal and figure gives percentage of responses to stimulus with smaller base.

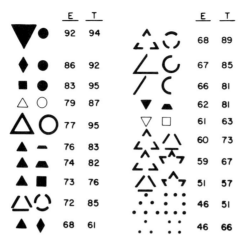

Figure 24.2 Percentages of appropriate responses in equivalence (E) and transfer (T) tests after training with upright triangles versus circles of equal area (from Zimmerman and Torrey 1965). Upper left pair was presented as illustrated for equivalence tests, but stimuli were equal in area for transfer tests. All other pairs were presented as illustrated for both equivalence and transfer tests.

suggest that the monkey had learned something about the relative lengths, as well as about the sizes, of the original training stimuli.

Zimmerman and Torrey (1965) studied the range of equivalence responses by infant rhesus monkeys who had learned to discriminate a circle from a triangle under two conditions. One group was tested for equivalence with nondifferential reinforcement (the equivalence group) and the other was rewarded only for responses to the more triangular figure (the transfer group). The performance of the two sets of infant macaques is compared in figure 24.2. The equivalence and transfer methods yielded comparable results; the correlation between the groups' relative levels of performance on the 20 pairs is significant at the 1% level of confidence (Kendall's $\tau = 0.69$).

The method of stimulus equivalence has been used to advantage in research on learning as well as perception. Humans and other species appear to learn discriminations relativistically, that is, to choose "brighter" or "larger" rather than one absolute value on the size or brightness continuum. Thus, a subject who learns to approach a stimulus with an area of 100 cm² rather than 50 cm² would prefer the stimulus still larger than 100 in equivalence tests with, for example, 200 versus 100. This transposition phenomenon was a strong argument for gestalt psychology until Spence (1937) predicted and confirmed by means of equivalence tests that transposition of differential responses to size would fail with values rather far removed from the training pair. His findings were influential in supplanting gestalt views of transposition with a rigorous treatment of discrimination in terms of stimulus generalization.

Limitations

At first glance, it might seem that after animals have learned a visual discrimination one might present an indefinitely large number of transforms of the original discriminanda and quickly determine precisely what stimulus attributes mediate successful discrimination performance; however, this is not the case. Two serious limitations of the method of stimulus equivalence were recognized early in its history of use (Lashley 1938, 1942).

Response to Novelty
Substitution of novel test stimuli for familiar training stimuli is often disturbing to animal subjects. Lashley's (1938) rats frequently vacillated between the two test figures for a long time and failed to manifest an immediate preference for either stimulus when confronted with an unfamiliar pair of stimuli in the jumping stand.

Extinction of Choice Behavior
Rats and other mammals eventually learn that the differential reinforcement contingencies of discrimination training are suspended in favor of nondifferential reward on equivalence test trials, and cease to respond differentially to the test stimuli. Animals often adopt position habits which are rewarded on every trial. Pairs of test figures that elicit strong preferential responses early in a series of multiple equivalence tests often fail to evoke any preference if presented after the subjects have discovered that all responses on test trials are reinforced (Lashley 1938). Further evidence of the deleterious effects of nondifferential reinforcement on equivalence-test performance is provided by the observations of Lashley and others that transfer tests with differential reward of the test figure physically more like the rewarded training stimulus often reveal immediate and strong generalization when equivalence tests suggest none.

Transfer

The transfer method differs from the method of equivalence in that it does not entail test trials with nondifferential reinforcement. The method of transfer is generally more sensitive than the method of equivalence, but it is far more cumbersome. The central idea is simple enough, but controlling for various threats to internal validity (Campbell and Stanley 1966) may require several groups of animals.

Designs
The basic paradigm involves training an animal with one pair of stimuli and testing for transfer to a second set of stimuli, using differential reinforcement in both training and transfer phases (figure 24.3, design A). Substantial positive transfer in learning the second task may mean

Cats with Posterior Extrastriate Lesions

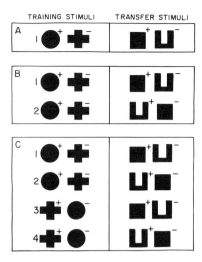

Figure 24.3 Basic experimental designs for tests of stimulus generalization by method of transfer.

that the basis for making the initial discrimination has generalized to the second set of stimuli. Other reasonable hypotheses for the positive transfer are that the animal is now more careful about looking at both stimuli, that it has developed some nonspecific strategy like "win-stay/lose-shift," and that it might be more tame. Thus, at its simplest, the transfer paradigm does not control for the effects of proactive history and is an equivocal method for determining which stimuli are perceived as similar or equivalent.

Design B illustrated in figure 24.3 is a considerable improvement over design A. Here, the subjects are trained on a single discrimination task and then divided into equated groups, matched for performance in initial learning and for critical experimental treatments, and then tested for transfer of their discrimination training. The stimuli presented in transfer differ in potentially crucial aspects abstracted from the discriminanda used in original learning. The significance of these features is tested by comparison of groups for whom the reinforcement contingencies in transfer learning are compatible or incompatible with the contingencies that prevailed in original learning. Transfer is inferred if subjects that must reverse their responses to the putative critical dimension make significantly more errors in transfer learning than animals that are not reversed.

Nonspecific learning effects cannot account for any differences between the subgroups trained with compatible and incompatible reward contingencies in the transfer phase. It is possible, though, that a difference between the two subgroups is due to a specific stimulus preference. If the animal prefers closed shapes to open shapes, this would be confounded with generalizations in the example used to illustrate design B in figure 24.3.

Design C of figure 24.3 controls for both proactive history and specific stimulus preferences, but it requires four groups. Variations of design C are frequently used in experiments with normal animals (see, for example, Sutherland and Mackintosh 1971), but this design is unattractive for work with brain damage because it necessitates four independent subgroups to study each location of lesion and because an internal analysis of the behavioral results within each lesion group is highly desirable, but in design C the effects of individual lesions and training subgroups are confounded. Fortunately, it is often possible to select stimuli that occasion no strong preferences, and when this is so design B offers a valid test of the strength of perceived similarity of the transfer and training stimuli.

Limitations
Although the transfer method does not suffer from the same limitations as the method of equivalence, it has two limitations of its own.

Necessity for More Than One Subgroup
Transfer design B of figure 24.3 is appropriate for many tests of stimulus generalization, but even the two subgroups it requires are too many for most studies of brain damage (for the reasons given above). Hence, tests of transfer by brain-damaged subjects typically employ design A (see, for example, Butter 1972, 1979; Atencio et al. 1975; Cornwell et al., 1980), with the result that the amount of saving in learning the transfer tasks can be due to factors other than stimulus generalization along the dimension investigated. It must be noted, however, that the first step in such research is to find any defect in lesioned animals. Elegant designs are appropriate only after one has an effect to analyze.

Sometimes there is little interest in rigorously separating the effects of stimulus generalization from those of less specific transfer and of problem difficulty. In such instances transfer design A is acceptable, and it has been used by several investigators to study the effects of visual-system damage on the discrimination of spatially masked stimuli. The animal is first trained to discriminate between two stimuli, and in the following transfer phase equivalent and irrelevant lines are added to each stimulus. In these studies interest is focused on the capacity to discriminate patterns in the presence of masking, not on determining which cues controlled choices in the initial pattern discrimination.

Figure 24.4 illustrates some of the results obtained in six experiments of this kind, three with monkeys and three with cats. Each of the experiments with monkeys contained a group of animals with lesions in the lateral striate (LS) cortex, an area that represents the central part of the visual field (Daniel and Whitteridge 1961; Cowey and Ellis 1969). The experiments with cats reported in panels 5 and 6 of figure 24.4 also included subjects with lesions in the central part of the geniculocortical representation of the visual field in the marginal gyrus (M). The lesions

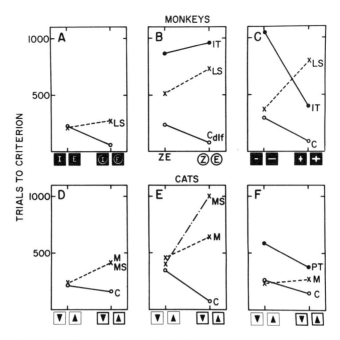

Figure 24.4 Mean trials to criterion in six experiments involving spatial masking. Top three panels from work of Butter with monkeys: (A) 1979; (B) 1972; (C) 1969. Bottom three panels from work by Cornwell and colleagues with cats: (D) Cornwell and Herbein, unpublished; (E) Cornwell et al., in press; (F) Cornwell and Warren 1981. C: control. Cdlf: dorsolateral frontal. LS: lateral striate. IT: inferotemporal. M: marginal gyrus. MS: marginal and splenial gyri. PT: posterior temporal (areas 20 and 21). Group M, MS in panel 4 has not yet been analyzed histologically.

included parts of areas 18 and 19, as well as 17, and so were larger than those incurred by the monkeys. Some cats were studied whose lesions included the medial wall of the marginal gyrus and the depths of the splenial sulcus (groups MS, panel 5 of figure 24.4); for these subjects most of the cortical target of the trilaminar part of the lateral geniculate was removed. Finally, panel 4 of figure 24.4 shows behavioral findings from a study in which the histological results have not yet been analyzed, and consequently it is not possible to distinguish cats in that study with subtotal lesions (M) from those with the more complete lesions (MS).

The data given in figure 24.4 reveal a consistent pattern for controls and animals with striate-cortex (SC) lesions. These two kinds of preparations differ little in initial acquisition, but the control animals made fewer and the SC-lesioned animals more errors in transfer than in original training. Animals with IT or PT ablations are inferior to controls in both acquisition and transfer, but may make more or fewer errors in transfer than in initial learning.

It is very likely that the excellent performance of the controls on the transfer problem was due to a substantial amount of generalization from the original discrimination (that is, positive transfer). After all, it would seem that if other things were equal à discrimination problem with additional irrelevant lines should be more difficult to learn than one without such lines. However, it is possible that this supposition is wrong, and that for normal monkeys and cats problems with masking are somehow easier than those without it. Design A does not allow this possibility to be ruled out.

The difficulties in making inferences about specific transfer effects are clearer in the case of animals with geniculocortical damage. Figure 24.4 suggests a complete absence of stimulus generalization from the unmasked to the masked stimuli by the animals with lesions in the visual cortex (groups LS, M, and MS). However, these subjects may well have performed even more poorly at discriminating masked patterns if they had not had previous experience at discriminating the same patterns without masking.

Figure 24.4 (panels 2, 3, and 6) also shows the performance with unmasked and masked stimuli by animals with lesions in the temporal part of the visual system (inferotemporal in monkeys; posterior temporal in cats). Both monkeys and cats with temporal damage were impaired at learning to discriminate the unmasked stimuli, as well as the masked ones. This difficulty with even fairly simple patterns contrasts sharply to the excellent performance with the same stimuli by animals with geniculocortical damage. Among the subjects with temporal lesions there was no consistent trend for acquisition of the transfer task to be either better or worse than that during the initial problem. But even if there were consistencies, the use of design A would limit conclusions about the effects of temporal lesions on stimulus generalization.

Limits on Number of Stimuli Tested

The differential reinforcement used in generalization testing with the transfer method limits the number of valid generalization tests, because during each such test the subject can learn something about the stimuli in addition to what it learned during the initial training phase. This additional learning is acceptable, and even desirable, if the question asked concerns whether or not the subject can learn to respond to a certain aspect of the stimuli. For instance, such transfer tests have been used to train pigeons to respond to the general class of visual displays containing human figures (Herrnstein and Loveland 1964). The learning produced by each transfer test is unacceptable, however, if the question asked involves what dimensions of the original training display control responding, for by using differential reinforcement it is the investigator who controls the behavior of the subject along these very dimensions. Thus, in many circumstances the number of stimulus pairs that can be used for transfer tests with the same animal is severely limited.

It is possible, of course, to conduct transfer tests with several new stimulus pairs by presenting them concurrently rather than sequentially. When this is done the effects of learning accomplished during the transfer phase are distributed equally across all of the new stimulus pairs. We recently used this design to study the importance of stimulus-response continuity in cats with posterior temporal and insular temporal lesions (Cornwell and Straussfogel, work in progress). The proximity of the response site to the discriminative stimuli is an important determinant of the ease with which normal animals learn visual discriminations (Meyer et al. 1965), and Butter and Hirtzel (1970) suggested that inferior temporal lesions in monkeys restrict the attention in discrimination learning to the region very close to the location of the response. We wanted to see if a similar restriction in attention is produced by posterior temporal lesions in cats.

Sixteen experimentally sophisticated cats were first trained in a Wisconsin General Test Apparatus (WGTA) to discriminate a U-shaped from a crown-shaped figure (figure 24.5) to a criterion of 23 correct responses in 25 trials for two consecutive sessions. The training stimuli of equal area were cut from black construction paper and mounted on white plaques of equal area such that the bottom of each stimulus was 1 cm from the base of the plaque. The cat could obtain an incentive covered by the correct stimulus, which varied from cat to cat, by pushing the stimulus plaque with its paw. Almost all responses by all cats were directed to the bottom of the plaque; the middle of the base became worn and dirty from contact with the cats' paws. After learning the initial discrimination the cats were given a series of transfer tests, each of which began and ended with eight trials with the original training stimuli to ensure that performance of the basic discrimination remained high. The middle 32 trials of each transfer session consisted of eight

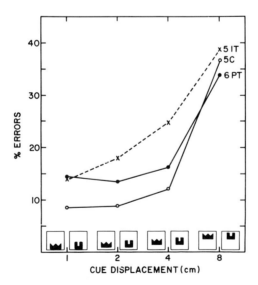

Figure 24.5 Median percent errors by cats with posterior temporal (PT) and insular temporal (IT) lesions and by controls (C) as a function of amount of stimulus-response separation (Cornwell and Straussfogel, work in progress).

Cornwell, Warren

4-trial blocks, in each of which a different pair of stimuli was presented. One pair was the original training stimuli; the other three pairs differed from each other and the training stimuli only in the distance from the bottom of the stimuli to the base of the plaque (figure 24.5). The behavioral results from the transfer phase of this experiment are plotted in figure 24.5, which shows that for each group of cats there was a pronounced effect of displacing the stimulus from the site of the response. Cats with posterior temporal damage, however, were no more affected by the stimulus-response separation than the controls or the cats with lesions in the auditory association cortex (the insular temporal group).

Comparisons of Equivalence and Transfer

The methods of equivalence and of transfer have distinct and different limitations, so the strength of inferences about shape classification by animals is increased when both methods produce the same results. For example, on the basis of equivalence tests with octopi and rats, Sutherland (1960) and Sutherland and Carr (1962) hypothesized that animals classify shapes as open (high ratio of perimeter to area, many sides) or closed (low ratio of perimeter to area, few sides). The generality of their thesis was tested in a transfer experiment with cats (Warren 1972). Separate groups were trained on two discriminations ○ versus ⊔ and △ versus I, with half the animals rewarded for choosing the open and half for choosing the closed figure in each pair. In the transfer task, the cats were switched to the discrimination on which they had not been trained initially; half were rewarded for continuing to respond to open and closed figures as in original training (positive transfer) and half for reversing their responses to the open-closed dimension (negative transfer). The cats that were obliged to choose the open figure on one task and the closed on the other made more than twice as many errors as those that were consistently rewarded for selecting open or closed figures in both tasks. The view that animals classify shapes by perimeter and sidedness was thus supported (Warren 1972). Another example of equivalence and transfer tests leading to the same inferences comes from the experiments done by Zimmerman and Torrey (1965) on pattern discrimination by infant monkeys, described earlier. Still another comes from Hughes's work (this volume) on the salience of local and global cues in discriminating the orientation of stripes by normal cats and by cats with lesions of areas 17 and 18.

The equivalence and transfer methods do not always yield results as consonant as those obtained in the experiments described above. Equivalence tests often fail to disclose evidence of learning about features of the training stimuli when transfer tests do. For example, Warren and McGonigle (1969a) taught cats to discriminate upright and inverted triangles and then tested two groups on the figures illustrated in figure 24.6. The equivalence (E) group was nondifferentially rein-

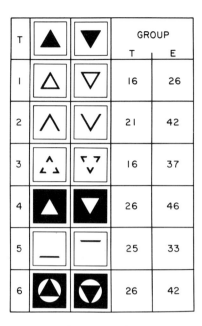

T	▲	▼	GROUP	
			T	E
1	△	▽	16	26
2	∧	∨	21	42
3	^ʌʌ	˅˅˅	16	37
4	▲	▼	26	46
5	—	—	25	33
6	◓	◒	26	42

Figure 24.6 Percentage of inappropriate responses to test figures by cats of the transfer (T) and equivalence (E) groups. Source: Warren and McGonigle 1969a.

forced and the transfer (T) group was differentially rewarded for responses to figures oriented in the same direction as the reinforced training shape. The data of figure 24.6 speak for themselves. On all six test pairs the performance of the equivalence group was inferior to that of the transfer group, by 8%–21%.

A rather homogeneous series of comparisons between equivalence and transfer tests was provoked by Sutherland and Holgate's (1966) analysis of two-cue discrimination learning by animals. One of their experimental designs is represented in figure 24.7 (phases 1 and 2). Different groups of rats first learned to discriminate rectangles differing in brightness and orientation (pair 1 or 2 of phase 1 in figure 24.7) and were then tested, with nondifferential reinforcement, for preferences with four pairs of figures differing in only one dimension (pairs A–D, phase 2). These pairs consisted of the old positive (+) or old negative (−) versus a novel stimulus. Another group was tested with all novel stimuli: black (B) and white (W) squares, and horizontal (H) and vertical (V) gray rectangles. In both cases, preferences for orientation and brightness were negatively correlated, which suggested not only that rats learn multiple-cue problems largely on the basis of single cues, but also that the more a rat learns about one cue the less it learns about the other.

Replications of these experiments with cats and monkeys yielded discordant results. On equivalence trials like those shown in figure 24.7, cats respond to novelty more strongly than to brightness and orientation, manifesting only 62% preferences on these dimensions. Yet in

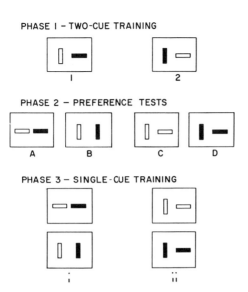

PHASE 1 – TWO-CUE TRAINING

1   2

PHASE 2 – PREFERENCE TESTS

A   B   C   D

PHASE 3 – SINGLE-CUE TRAINING

i   ii

Figure 24.7 Design of experiments on two-cue discrimination learning with tests for preferences (phase 2) between novel and familiar stimuli. Source: Mumma and Warren 1968.

transfer learning on single cues, cats that had to reverse their initial responses made more than twice as many errors as nonreversed cats on both orientation and brightness, which unequivocally indicates that they had learned the reward value of cues in both brightness and orientation in initial learning (Mumma and Warren 1968).

The preference scores for rats tested with two novel stimuli are higher than those of rats tested with one familiar and one novel stimulus (Sutherland and Holgate 1966). The possibility that this relation holds for cats and monkeys was tested in experiments that used the design shown in figure 24.8. The results of the preference tests with cats are given in table 24.1, which shows that the number of appropriate responses to brightness declined progressively while the frequency of responses to the cats' preferred position increased progressively during preference testing, indicating a rather rapid attenuation of responses to brightness cues. There was no evidence of differential response to orientation at any time during preference testing, which suggests that the cats had learned little or nothing about orientation during training on two cues. And yet in subsequent learning on single cues the cats that had the original cue-reinforcement relation reversed relative to initial learning made significantly more errors than nonreversed on both brightness (87 versus 36 errors) and orientation (85 versus 38). Note that, with transfer condition constant, discrimination performance on orientation and brightness cues when presented on their own was almost exactly equivalent. This experiment (Warren, unpublished) shows that, at least under some conditions, performance on preference and transfer tests may be totally uncorrelated. Results very much like those

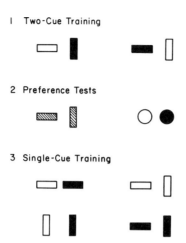

1 Two-Cue Training

2 Preference Tests

3 Single-Cue Training

Figure 24.8 Design of experiments on two-cue discrimination learning with tests for preferences (phase 2) between two novel stimuli. Source: Warren and Warren 1969.

Table 24.1
Percent responses to appropriate visual cues (V) and to preferred position (P) on preference tests.

	Brightness		Orientation	
Trials	V	P	V	P
5	81	61	46	67
10	76	57	54	72
15	69	68	43	68
20	64	72	52	69

from cats were obtained in a similar study with monkeys (Warren and Warren 1969).

The fact that equivalence tests may show no learning about some feature of the discriminanda animals have learned to differentiate while transfer trials reveal clear evidence of learning about the same cue means that equivalence tests often seriously underestimate what experimental animals have learned in discrimination training. One can accept as meaningful positive results from a preference test, but cannot regard negative results as definitive.

The force with which this stricture applies varies, for unknown reasons, among different species of nonhuman animals. The results of the preference tests in the experiments on two-cue learning that are described in the preceding paragraphs are summarized in table 24.2. Cats and monkeys almost always have substantially lower preference scores than rats. Warren and Warren (1969) speculated that rats may be peculiarly less vulnerable than monkeys to generalization decrement on preference tests, but conclusive evidence is lacking.

An experiment by Warren and McGonigle (1969b) epitomizes both

Table 24.2
Performance in preference testing by animals after learning to discriminate stimuli differing in both brightness and orientation.

Species	Experiment[a]	Stimuli	Percent Responses		
			Brightness	Orientation	Training
Rats	1	Old +, − vs. novel	72	86	96
	1	All novel	90	92	97
Cats	2	Old, +, − vs. novel	62	62	80
	3	All novel	78	54	93
Monkeys	4	All novel	72	63	95

a. (1) Sutherland and Holgate 1966; (2) Mumma and Warren 1968; (3) Warren (unpublished); (4) Warren and Warren 1969.

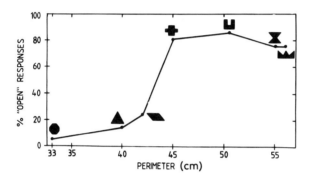

Figure 24.9 Percentage of responses to the position associated with reinforcement of the open training figure on generalization tests. Source: Warren and McGonigle 1969b.

the advantages and the disadvantages of the equivalence method in comparison with the transfer method. Cats learned a successive discrimination between open and closed shapes; when two I-shaped figures were presented they were reinforced for responses to the right, and when two circles were displayed they were reinforced for responses to the left. In preference tests given after the subjects had mastered the initial discriminations, two identical shapes that were intermediate between the ○ and the I in perimeter relative to area and in number of sides were presented and nondifferentially rewarded. Responses to the left implied equivalence to the circle and responses to the right equivalence to the I-shape; 7 test shapes were presented for 21 trials each. The cats' preferences were strongly determined by perimeter and sidedness, as may be seen in figure 24.9. The correlation between the figures' mean rank on perimeter and complexity and responses to the open side was + 0.95 ($P < 0.001$).

The great advantage of the equivalence method is that in relatively few trials one can obtain information on a large number of stimuli which are potentially equivalent to the figures used in original learning,

Cats with Posterior Extrastriate Lesions

and in favorable circumstances very orderly and meaningful relations may be derived from the preference tests. The transfer method is more expensive in time and effort, and yields data on a far more restricted set of stimuli. But the equivalence method as currently used suffers from a unique and serious shortcoming: Animals very quickly stop responding differentially when they are not differentially reinforced. The function given in figure 24.7 is based on the behavior of only 7 of 16 cats that were tested; the data from 9 cats were discarded because they adopted persistent habits of responding to the same side on all preference trials, no matter what stimulus was presented. It was reasonable in a study of stimulus correlates of visual perception to drop those cats that failed to respond to visual cues, but position biases often make it difficult to interpret performance on preference trials in learning experiments where it is less defensible to pick and choose subjects as Warren and McGonigle did.

In summary, equivalence tests are potentially a far more economical way to ascertain what animals have learned in prior training, but they are often insensitive because animals stop responding to visual cues. Equivalence tests might be made more valuable sources of information if one could reduce the subject's disposition to respond to irrelevant cues, such as position.

Suggestions for Improving the Equivalence Method

In order to apply the equivalence method successfully to the study of posterior extrastriate damage in cats, methods are needed to prevent or forestall the extinction of choice behavior. The solution to this problem would seem to involve presenting the novel stimuli during the equivalence phase in such a way that the presence or absence of reinforcement does not allow the cat to discriminate the training conditions from the tests for generalization or equivalence. The use of intermittent reinforcement during the training phase satisfies this requirement nicely, and has been the usual approach to the problem in research on stimulus generalization with free operants (Honig 1969; Honig and Staddon 1977). Intermittent reinforcement on a variable ratio or a variable interval schedule for responses to one stimulus (S+), combined with no reinforcement for responses to a different stimulus (S−), produces excellent control of behavior by the discriminative stimuli. It also allows for multiple tests of stimulus generalization under conditions of extinction before the animal learns that different reinforcement contingencies are in effect in the learning and generalization (equivalence) phases.

Far more work using this technique has been done with rats and pigeons than with cats, but several investigators have shown that with the proper training methods cats will work long and regularly for intermittent positive reinforcement (for reviews see Berkley 1966, 1970), and that they generate orderly generalization gradients in subsequent equivalence tests under conditions of extinction (Mello 1968). We know of no studies that have used such operant-conditioning methods to evaluate attentional processes by cats with posterior extrastriate lesions, but the use of these methods seems both feasible and potentially valuable.

The problems appear to be somewhat less tractable for applying intermittent schedules of reinforcement to discrete-trial situations, such as those using a Wisconsin General Test Apparatus or various discrimination alleys. Here even moderately low ratios of reinforcement seem likely to produce inconsistent responding.

However, studies of probability learning with normal animals suggest a way that might provide a high level of differential responding on equivalence trials. A procedure used in probability learning differs from that used in the typical discrimination task with free operants in two ways: The ratio of reinforcement for responses to one stimulus is typically quite high (usually 60%–90%), and some smaller fraction of responses to the other stimulus is also reinforced.

Poland and Warren (1967) showed that if cats were rewarded 70% of the time for choosing one stimulus and 30% of the time for choosing the other they came to choose the more consistently rewarded stimulus about 85% of the time. The fact that this technique results in a high level of differential responding, despite occasional reinforcements for choosing the other stimulus, could make it very useful as a training method for assessing stimulus generalization in discrete-trial situations. On the average, animals trained in a 70:30 probability task will be accustomed to receiving reinforcement on 60%–65% of their choices, and even the relatively rare choices of the less frequently rewarded stimulus will have been reinforced occasionally. Thus, there should be little in the way of change in reinforcement contingencies between the training and the equivalence phases if reinforcement is made available for both stimuli during about 70% of the equivalence trials. Under these conditions it should take a very long time for even a sophisticated cat to use feedback from differential or nondifferential reinforcement to discriminate between the training stimuli and the stimuli used for the equivalence tests.

References

Atencio, F., I. T. Diamond, and J. Ward. 1975. "Behavioral study of the visual cortex of *Galago senegalensis*." *J. Comp. Physiol. Psychol.* 89: 1109–1135.

Berkley, M. A. 1966. "Cat visual psychophysics: Neural correlates and comparison with man." In *Progress in Psychobiology and Physiological Psychology*, J. M. Sprague and A. N. Epstein, eds. New York: Academic.

———. 1970. "Visual discriminations in the cat." In *Animal Psychophysics*, W. C. Stebbins, ed. New York: Appleton.

Berkley, M. A., and J. M. Sprague. 1978. "Behavioral analysis of the role of the geniculocortical system in form vision." In *Frontiers in Visual Science*, S. J. Cool and E. L. Smith, eds. New York: Springer.

Berlucchi, G., J. M. Sprague, A. Antonini, and A. Simoni. 1979. "Learning and interhemispheric transfer of visual pattern discriminations following unilateral suprasylvian lesions in split-chiasm cats." *Exp. Brain Res.* 34: 551–574.

Bitterman, M. E. 1975. "The comparative analysis of learning." *Science* 188: 699–709.

Brown, J. L., F. D. Shively, R. H. LaMotte, and J. A. Sechzer. 1973. "Color discrimination in the cat." *J. Comp. Physiol. Psychol.* 84: 534–544.

Butter, C. M. 1969. "Impairments in selective attention to visual stimuli in monkeys with inferotemporal and lateral striate lesions." *Brain Res.* 12: 374–383.

———. 1972. "Detection of masked patterns in monkeys with inferotemporal striate or dorsolateral frontal lesions." *Neuropsychologia* 10: 241–243.

———. 1979. "Contrasting effects of lateral striate and superior colliculus lesions on visual discrimination performance in rhesus monkeys." *J. Comp. Physiol. Psychol.* 93: 522–537.

Butter, C. M., and M. Hirtzel. 1970. "Impairment in sampling visual stimuli in monkeys with inferotemporal lesions." *Physiol. Behav.* 5: 369–370.

Campbell, A. L. 1978. "Deficits in visual learning produced by posterior temporal lesions in cats." *J. Comp. Physiol. Psychol.* 92: 45–57.

Campbell, D. T., and J. C. Stanley. 1966. *Experimental and Quasi-experimental Designs for Research.* Chicago: Rand McNally.

Cornwell, P., and J. Warren. 1981. "Visual discrimination defects in cats with temporal and occipital lesions." *J. Comp. Physiol. Psychol.*

Cornwell, P., J. M. Warren, and A. J. Nonneman. 1976. "Marginal and extramarginal cortical lesions and visual discrimination by cats." *J. Comp. Physiol. Psychol.* 90: 986–995.

Cornwell, P., W. Overman, and A. Campbell. 1980. "Subtotal lesions of the visual cortex impair the discrimination of hidden figures by cats." *J. Comp. Physiol. Psychol.* 94: 289–304.

Cowey, A., and C. M. Ellis. 1969. "The cortical representation of the retina in squirrel and rhesus monkey and its relation to visual acuity." *Exp. Neurol.* 24: 374–385.

Daniel, P. M., and D. Whitteridge. 1961. "The representation of the visual field on the cerebral cortex in monkeys." *J. Physiol.* 159: 203–221.

Daw, N. W. 1973. "Neurophysiology of colour vision." *Physiol. Rev.* 53: 571–611.

Granit, R. 1947. *Sensory Mechanisms of the Retina.* Oxford University Press.

Hara, H., P. Cornwell, J. M. Warren, and I. H. Webster. 1974. "Posterior extramarginal cortex and visual learning by cats." *J. Comp. Physiol. Psychol.* 87: 884–904.

Herrnstein, R. J., and B. H. Loveland. 1964. "Complex visual concept in the pigeon." *Science* 146: 549–551.

Honig, W. 1969. *Operant Behavior: Areas of Research and Application.* New York: Appleton-Century-Crofts.

Honig, W., and J. E. R. Staddon, eds. 1977. *Handbook of Operant Behavior.* Englewood Cliffs, N.J.: Prentice-Hall.

Klüver, H. 1933. *Behavior Mechanisms in Monkeys.* University of Chicago Press.

Lashley, K. S. 1938. "The mechanism of vision. XV. Preliminary studies of the rat's capacity for detail vision." *J. Genet. Psychol.* 18: 123–193.

———. 1942. "The mechanism of vision. XVII. Autonomy of the visual cortex." *J. Genet. Psychol.* 60: 197–221.

Mackintosh, N. J. 1974. *The Psychology of Animal Learning.* New York: Academic.

Mello, N. 1968. "Color generalization in cat following discrimination training on achromatic intensity and on wavelength." *Neuropsychologia* 6: 341–354.

Meyer, D. R., and R. A. Anderson. 1965. "Color discrimination in cats." In *Color Vision: Physiology and Experimental Psychology*, A. V. S. deReuck and J. Knight, eds. Boston: Little, Brown.

Meyer, D. R., F. R. Treichler, and P. M. Meyer. 1965. "Discrete-trial training techniques and stimulus variable." In *Behavior of Nonhuman Primates*, vol. 1, A. M. Schrier et al., eds. New York: Academic.

Mumma, R., and J. M. Warren. 1968. "Two-cue discrimination learning by cats." *J. Comp. Physiol. Psychol.* 66: 116–121.

Nonneman, A. J., and J. M. Warren. 1977. "Two-cue learning by brain-damaged cats." *Physiol. Psychol.* 5: 397–402.

Pavlov, I. P. 1927. *Conditional Reflexes.* Oxford University Press.

Poland, W., and J. M. Warren. 1967. "Spatial probability learning by cats." *Psychonom. Sci.* 8: 487–488.

Spence, K. W. 1937. "Analysis of the formation of visual discrimination habits in the chimpanzee." *J. Comp. Psychol.* 23: 77–100.

———. 1942. "The basis of solution by chimpanzees of the intermediate size problem." *J. Exp. Psychol.* 31: 257–271.

Sprague, J. M., J. Levy, A. DiBerardino, and G. Berlucchi. 1977. "Visual cortical areas mediating form discrimination in the cat." *J. Comp. Neurol.* 172: 441–488.

Sutherland, N. S. 1960. "Visual discrimination of shape by *Octopus*: Open and closed forms." *J. Comp. Physiol. Psychol.* 53: 104–112.

Sutherland, N. S., and A. E. Carr. 1962. "Visual discrimination of open and closed shapes by rats. II. Transfer tests." *Q. J. Exp. Psychol.* 14: 140–156.

Sutherland, N. S., and V. Holgate. 1966. "Two-cue discrimination in rats." *J. Comp. Physiol. Psychol.* 61: 198–207.

Sutherland, N. S., and N. J. Mackintosh. 1971. *Mechanisms of Animal Discrimination Learning.* New York: Academic.

Warren, J. M. 1972."Transfer of responses to open and closed shapes in discrimination learning by cats." *Percept. Psychophys.* 12: 449–452.

Warren, J. M., and B. McGonigle. 1969a. "Effects of differential and nondifferential reinforcement on generalization test performance by cats." *J. Comp. Physiol. Psychol.* 69: 709–712.

———. 1969b. "Perimeter, complexity and generalization of form discrimination by cats." *Psychonom. Sci.* 17: 16–17.

Warren, J. M., and H. B. Warren. 1969. "Two-cue discrimination learning by rhesus monkeys." *J. Comp. Physiol. Psychol.* 69: 688–691.

Zimmerman, R. R., and C. C. Torrey. 1965. "Ontogeny of learning." In *Behavior of Nonhuman Primates: Modern Research Trends*, vol. 2, A. M. Schrier et al., eds. New York: Academic.

25 Search for the Neural Mechanisms Essential to Basic Figural Synthesis in the Cat

Howard C. Hughes

Contemporary theories of visual perception often view pattern recognition as an essentially serial process involving two complimentary operations. According to most theorists, the initial processing stages are analytic in nature; that is, feature-extracting mechanisms decompose the image into elementary components whose exact nature varies according to the theory in question. In the later stages of processing, a complimentary process recombines the outputs of feature analyzers and thus creates an internal representation of the image.

Though a variety of theoretical approaches are fairly explicit when dealing with the mechanisms of image analysis, they become much less so when considering the constructive, synthetic aspects of pattern perception. In computerized pattern recognition, for example, elementary features are first detected and are then often compiled into a list. One form of "recognition" process simply compares all the entries in this "feature list" against a set of previously stored "templates," which are simply the feature lists for the set of patterns that the program is able to identify. When the computer finds a match, recognition has occurred. In a sense, then, this algorithm bypasses the problem of pattern reconstruction altogether, since throughout all processing stages the pattern remains little more than a collection of features.

This is most certainly not the case with biological visual systems, which are often able to recognize patterns on a level that transcends the identity of their constituent features. Some examples are shown in figure 25.1. Despite this admittedly anecdotal evidence against the primary importance of elementary features, most theories of biological pattern analysis also tend to emphasize serial processing strategies that begin with feature analysis. The generalized neural implementation of these serial models might be called the "transcortical model" of visual information processing, and is schematically illustrated in figure 25.2. Utilizing easily demonstrated anatomical pathways, the model emphasizes a serial flow of information from the geniculostrate system through successive cortical stages which ultimately lead to the "association cortex," where the classical wisdom of comparative neurology and

Basic Figural Synthesis in the Cat

Figure 25.1 Illustration that letter recognition depends not on analysis of features, but on analysis of their configuration.

Figure 25.2 Illustration that the "transcortical" model of visual information processing suggests a pivotal role for area 17 in vision.

neuropsychology tells us that complex percepts are constructed from the more "elementary" visual sensations processed in area 17.

As in artificial pattern recognition, the neural mechanisms of feature analysis usually receive a much more thorough theoretical treatment than the complimentary mechanisms of figural construction. One of the most explicit of these neural formulations is the serial model of cortical pattern analysis originally invoked by Hubel and Wiesel (1962, 1965) to account for their now-classic observations on the receptive-field organization of cells in the visual cortex. Hubel and Wiesel discovered that cells with "simple" receptive fields are most frequently encountered in layer IV of area 17, whereas "complex" cells are more frequent in the remaining layers of area 17 and area 18. Cells with "hypercomplex" receptive fields were said to be common in area 19. Hubel and Wiesel pointed out that their results could be accounted for by the suggestion that specific patterns of convergence in the outputs of cells with one type of receptive field result in the construction of a new, higher-order type of receptive field. For example, specific patterns of convergence of geniculate cell outputs onto layer IV stellate cells in area 17 could account for the transformation from the concentrically organized receptive fields of the lateral geniculate nucleus into the elongated "on" and "off" regions characteristic of cells with simple receptive fields. In a similar manner, the appropriate convergence of specific simple-cell outputs was suggested to create complex receptive fields.

Although this model has been modified to incorporate more recently discovered features of parallel processing in the geniculocortical system (Hoffman and Stone 1971), the classical notion that cortical processing begins in area 17 has gained widespread acceptance.

Over the years, however, evidence against the transcortical model has gradually accumulated. The most direct negative evidence has come from behavioral demonstrations of substantial capacities for complex spatial vision in destriate animals representing a variety of mammalian species (Spear and Braun 1969; Doty 1971; Killackey et al. 1971; Pasik and Pasik 1971; Winans 1971; Humphrey 1974; Keating 1975; Hughes 1977; Sprague et al. 1977; Dineen and Keating 1979). Perhaps most dramatic are the reports of substantial visual abilities within the visual-field defects that accompany damage to area 17 in man (Pöppel et al. 1973; Weiskrantz et al. 1974). Since these neurological patients seem unaware of their own visual capacities, Weiskrantz et al. (1974) termed the condition one of "blind-sight."

In contrast to the undeniably severe visual consequences of damage to area 17 in primates, including man, a number of experiments have shown quite clearly that removal of both areas 17 and 18 has virtually no effect on learned discriminations of simple patterns or forms in the cat (Doty 1971; Sprague et al. 1977; Berlucchi and Sprague 1980).

A plausible anatomical explanation for the visual abilities that remain after destruction of striate cortex was provided by the discovery of multiple ascending thalamic projections to the visual cortex (Altman and Carpenter 1961; Diamond and Hall 1969). Thus, it is now well established in a number of species that the projection of the superficial layers of the superior colliculus onto specific subdivisions of the pulvinar constitutes a source of input to the visual cortex independent of the well-known geniculocortical system (for review see Jones 1974).

These developments have contributed to an increased interest in the anatomical, physiological, and functional organization of the extrastriate visual cortex. In an important development, electrophysiological mapping experiments have revealed a truly astonishing multiplicity of retinotopically defined visual areas within those parts of the parietal and temporal lobes generally considered "association" cortex. In the cat, for example, ten areas beyond area 19 (or VIII) have been identified

(Tusa et al. 1975; Palmer et al. 1978; Tusa 1978; Tusa et al. 1978, 1979; Tusa and Palmer 1980). Each cortical area receives inputs from the pulvinar complex and, in many cases, a converging input from parts of the geniculate complex. At least some of these extrastriate projections are organized topographically (Hughes 1980) and provide an afferent pathway sufficient to account for many of the receptive-field properties of extrastriate cortical units (Olavarria and Torrealba 1978; Spear and Baumann 1979). For example, removal of areas 17, 18, and 19 does not abolish the visual responsiveness, the retinotopic organization, or the spatial organization of receptive fields of cells in the lateral suprasylvian sulcus. Movement sensitivity and directional selectivity are lost, however (Spear and Baumann 1979). All of these findings serve to emphasize the inadequacies of the transcortical model and raise important questions concerning the possibilities for the localization of visual functions in the extrastriate cortex. Simply put, does the multiplicity of retinotopic organization in the visual cortex imply that each area makes a unique contribution to the central processing of visual information? If so, along what dimensions might the functions of each area differ from one another?

Some evidence for functional specialization has recently been provided by Sprague and co-workers. For example, Sprague et al. (1977) and Berlucchi and Sprague (1980) have shown that, contrary to the classical view, removal of areas 17 and 18 results in virtually no change in a cat's ability to discriminate simple forms and patterns. Such preparations do show modest but reliable losses of visual acuity, however (Berkley and Sprague, 1979 and this volume). This pattern of results contrasts with those obtained from cats with large extrastriate lesions; such cats are deficient in pattern and form discrimination (Cornwell et al. 1976; Sprague et al. 1977; Berlucchi and Sprague 1980) but have normal acuity (Berkley and Sprague, 1979 and this volume).

These results represent a "double dissociation" of the visual disturbances resulting from damage to the striate and extrastriate visual systems, and therefore provide strong evidence for functional specialization within the visual cortex. In addition, the results on extrastriate-lesioned cats raise an important conceptual problem: What types of visual impairments can produce deficiencies in pattern and form recognition without any loss of visual acuity? This problem is not confined to animal studies; lesions of the parietal cortex in humans also result in failures in the recognition of certain types of complex patterns while leaving "more elementary" processes such as increment thresholds and spatial resolution unaffected (Warrington and Taylor 1973).

The work on cats serves to emphasize one important fact, however: Because failures of pattern recognition do not result from removal of the geniculocortical pathways, the inputs relevant to the parietal lobe deficit must be transmitted via pathways other than the geniculostriate pathway, probably including projections from the geniculate and the pulvinar complex.

In order to further explore the cortical mechanisms mediating pattern perception in the cat, J. M. Sprague and I have begun an examination of the visual abilities of cats with ablations of certain subdivisions of the visual cortex using several rigorous tests of pattern vision. The logic of the stimuli we have used grew in part from the demonstration by Sprague et al. (1977) of a form-discrimination deficit without an acuity deficit; specifically, we wanted to explore the possibility that the recognition failures of extrastriate-lesioned cats result from an inability to organize the various features within a simple pattern into a unified percept—that is, from a deficit in "figural synthesis" (Neisser 1966). Such a deficit would leave the detection of features unaffected, but might well lead to a form of pattern "agnosia." Of course, the type of holistic spatial analysis being suggested here is precisely the type of process that was so heavily emphasized by the gestalt psychologists but has largely been overlooked by contemporary theorists. In addition, we wanted to know whether it was possbile to detect any perceptual impairment after removal of areas 17 and 18 beyond a reduction in visual acuity, since recent work by Berlucchi and Sprague (1980) indicates that destriate cats are actually comparable to normal cats in terms of their acquisition and/or retention of simple planemetric pattern and form discriminations. Thus, we required a test in which discrimination of test patterns did not depend on the limits of visual resolution (because we did not want the performance of destriate animals to be limited by their poor acuity), in which the test patterns could be easily metricized and manipulated so as to allow discrimination difficulty to vary widely and thereby provide a rigorous test of pattern vision, and in which the patterns were devoid of any local flux cues that could provide pattern-independent solutions to a discrimination problem, especially in animals with lesions in the visual cortex (Winans 1971; Ritchie et al. 1976). Finally, we wanted a test of pattern vision that required a parallel analysis of the spatial structure of the image; that is, the solution had to be based on spatial cues alone.

A useful class of patterns that fulfills all of these requirements are patterns constructed from arrays of dots. The prototypical pattern we have worked with is shown in figure 25.3. The only cue enabling discrimination of these patterns lies in the relative spacing between neighboring dots, which we might call the "global structure" of the pattern. Thus, these patterns emphasize figural synthesis rather than feature analysis. On the basis of their global structure, dots are clustered into rows that can be seen in either of two orientations, 45° or 135°. The possible mechanisms by which these patterns appear oriented poses a significant theoretical problem, which I will address in the discussion. For the moment, however, it is useful to consider these stimuli

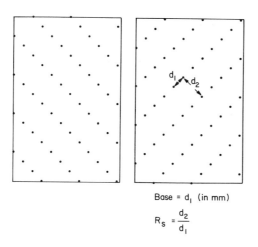

Base = d_1 (in mm)

$$R_s = \frac{d_2}{d_1}$$

Figure 25.3 Two rectilinear arrays of dots, illustrating how two parameters, R_s and a base, specify any particular pattern of this type. Notice that as R_s approaches 1.0, perceived orientation approaches complete ambiguity (i.e., oriented rows at 45° or 135° are equally probable). Variation in d_2 can be used to determine the maximal spatial extent of the cohesive interactions underlying the perception of rows.

simply as a potentially useful tool in the study of feline pattern perception and its dysfunctions after ablation of different subdivisions of the visual cortex.

Procedures

Test Stimuli and Psychometric Scaling

In order to maximize the sensitivity of these tests, we have manipulated discrimination difficulty in two ways. First, the two parameters that specify any pattern of this class are varied in such a way as to generate a set of patterns that are similar but also vary widely with respect to the ease with which they generate a perceived orientation. Thus, as the magnitude of the proximity cue (R_s in figure 25.3) approaches 1.0, perceived orientation approaches complete ambiguity. Furthermore, at any given value of R_s, the value d_1 can also vary, so the distances over which the perceptual grouping of elements is possible can also be measured. The set of patterns generated by manipulating these interdot spacing parameters is shown in figure 25.4. These patterns were then used to generate "frequency-of-seeing" curves, which relate discrimination performance to the magnitude of the proximity cue with interdot spacing as a parameter. In a second series of tests, rectilinear dot patterns are embedded in visual noise, which simply consists of additional dots in random positions. In this case, discrimination performance is measured as a function of the density of noise in the pattern, and threshold noise densities are obtained using a tracking procedure.

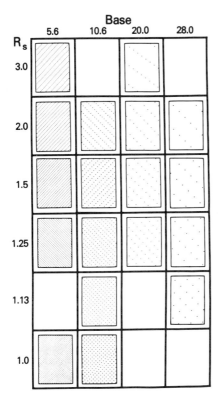

Figure 25.4 The complete set of dot patterns.

Tests of the Relative Importance of Pattern Features Versus Pattern Configuration

The patterns described above differ only in "global structure," that is, the spatial arrangement of their elements. By using line segments instead of dots as pattern elements, we have also examined the influence of the local features themselves in determining perceived orientation. In one set of patterns, the orientation of the features (oriented line segments) conflicts with the global or configurational cue. We refer to these patterns as "structured arrays of line segments," because they contain both local (feature orientation) and global (configurational) cues (see figure 25.8).

The salience of these local orientation cues is measured independent of any configurational cues, using random patterns of identically oriented line segments. We refer to these patterns as "unstructured arrays of line segments," since they are devoid of configurational cues (see figure 25.9).

Apparatus

The test apparatus is a two-choice runway, and the high-contrast stimuli are projected onto panels from the rear. To respond, the animal

Basic Figural Synthesis in the Cat

presses either stimulus panel with its nose. A correct response allows the animal to push open the panel and gain access to a food reward. Touching the wrong panel results in a mild foot shock and no reward. Though a specific viewing distance was not enforced, observation of many cats shows that the animals consistently scan the stimuli at a distance that ranges between 5 and 10 cm.

Results

Normal Cats

The performance of normal cats on these pattern discriminations serves to make several points:

• Perceptual clustering of these rectilinear dot arrays is immediate, as judged by the near-perfect transfer of discriminations of oblique square-wave gratings to an exemplary dot pattern. This is illustrated in figure 25.5.

• The cohesive strength between adjacent dots is an inverse function of the distance between them and a direct function of the magnitude of the proximity cue R_s. These relationships are shown in figure 25.6. The weakness of the interactions between widely separated signal elements is also clearly reflected in the data on random-dot masking, in that threshold noise densities drop precipitously as the spacing between signal dots increases (figure 25.7). Notice that when the signal dots are separated by 24 mm, the presence of as few as ten noise elements leads to a dramatic drop in performance (from 80% to 60%). Since control experiments have demonstrated that the cats only attend to the bottom half of the stimulus panel (masking the top half has no effect), this result

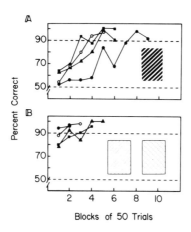

Figure 25.5 (A) Acquisition curves for oblique grating discrimination for four normal cats: (▲) cat 1, (●) cat 2, (■) cat 3, and (○) cat 4. B. Discrimination performance on first exposure to a dot pattern for the same cats. For cats 1 and 2, the pattern B 5.6 has the parameters for R_s:d_2 of 3:1; for cats 3 and 4, the pattern B 10.6 has the parameters for R_s:d_2 of 2.3:1.

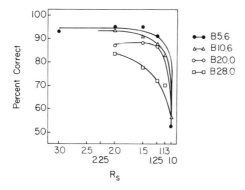

Figure 25.6 Discrimination of performance as a function of R_s-averaged data for four normal cats. (●) B5.6, (△) B10.6, (○) B20.0, (□) B28.0.

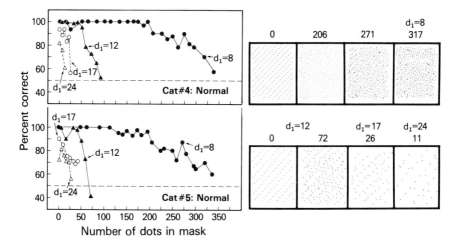

Figure 25.7 Random-dot masking of global pattern structure in two normal cats. As spatial separation between pattern elements increases, threshold noise density is reduced dramatically. Insets show examples of stimuli.

indicates that as few as four or five dots can completely mask an otherwise detectable global structure. This indicates that the individual pattern elements are easily visible to these animals.

• The "local structure" (the orientation of the pattern elements) appears to have little influence on perceived orientation when embedded in a strong global structure. This is inferred from measurements of the animal's stimulus preferences when they are presented with structured arrays of line segments, both before and after they learn to discriminate the orientation of the line segments themselves (see figure 25.8 for details).

As figure 25.9 shows, normal cats vary somewhat in the ease with which they are able to transfer their global orientation discriminations to discriminations based entirely on the orientation of the local features. Control tests indicate that the cats are ultimately able to perform at

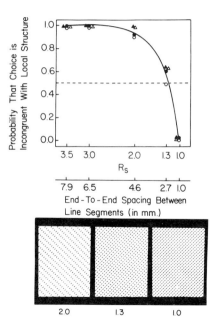

Figure 25.8 Stimulus preferences of two normal cats on five pairs of structures arrays of line segments. The R_s value differed for each stimulus pair in the set; three examples are shown. Upper scale on abscissa is for R_s; lower scale shows end-to-end spacing between adjacent line segments. All patterns oriented at 45° are rewarded, and thus are preferred by cats. Since these patterns present conflicting orientation cues, subjects are rewarded regardless of choice. The probability that the animals selected the pattern that contained line segments in the nonpreferred orientation (135°) is indicated on the ordinate. Filled symbols indicate results; open symbols represent results obtained after training on unstructured arrays of line segments, which requires sensitivity to feature orientation *per se*. (●, ○) Cat 3; (▲, △) cat 4.

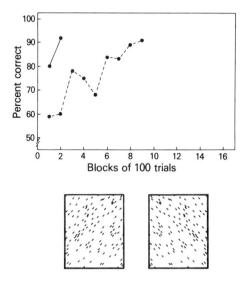

Figure 25.9 Acquisition curves for discrimination of an array of 120 randomly distributed line segments (see left example, figure 25.10) by two normal cats. (●—●) Cat 4; (●---●) cat 3.

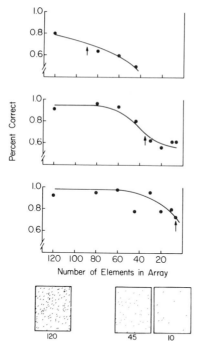

Figure 25.10 Effect of redundancy on discrimination of differences in local structure. Abscissa: numbers of randomly distributed line segments in the arrays. Ordinate: percentage of correct responses. Upper curve presents data from a 17–18 destriate cat (cat 2); lower two curves were obtained with normal cats (3 and 4, respectively). Bottom: Examples of the patterns used.

above-chance levels when as few as four or five line segments appear as the stimulus, which again suggests that the features are individually visible (figure 25.10).

Effects of Removing Areas 17 and 18

Lesion of areas 17 and 18 has virtually no effect on discrimination of global spatial structure, but feature discrimination is adversely affected. Thus, the frequency-of-seeing curves of destriate cats are comparable to those of normal animals for dot patterns (figures 25.11, 25.12) and structured arrays of line segments (figure 25.13), but not for unstructured arrays of line segments (figure 25.14). However, destriate cats do seem incapable of discriminating large-base dot patterns (figure 25.11).

The sensitivity of destriate cats to slight differences in global structure is striking—especially in view of the fact that, even near threshold, striate-lesioned cats are comparable to normal cats. We must conclude that the striate-peristriate visual cortex, ordinarily viewed as the site of the visual system's most detailed spatial analysis, is not an important contributor to the kind of spatial analysis required to discriminate these patterns. We do know, however, that the visually decorticate cat is form-blind (Meyer 1963; Dalby et al. 1970; Ritchie et al. 1976; Berlucchi,

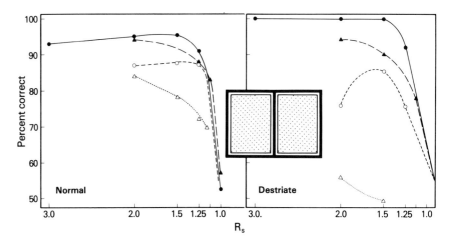

Figure 25.11 Discrimination performance of cats 2 and 3 on structured arrays of dots, before and after removal of areas 17 and 18, as a function of the proximity cue R_s. ($\bullet$) B5.6, ($\blacktriangle$) B10.6, ($\bigcirc$) B20.0, (—) B28.0. Inset shows a difficult discrimination ($R_s = 1.13$, $d_2 = 10$) unaffected by the lesion.

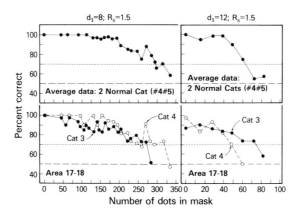

Figure 25.12 Random-dot masking of global pattern structure in two destriate cats (areas 17 and 18). Comparison with two normal subjects shows that the lesion does not alter masking thresholds. (A) $d_1 = 8$, $R_s = 1.5$; (B) $d_1 = 12$, $R_s = 1.5$.

unpublished observations), so the possibility remains that the integrity of some part of the extrastriate cortex is a necessary condition for success in discriminations of global configuration.

Consider for a moment the possible outcomes. If the distinction between global and local spatial processing has validity insofar as the actual neural processing of visual information is concerned, and if these two types of analysis are performed primarily by different cortical areas, it should be possible to make lesions that selectively affect either process. We have already seen how lesions of areas 17 and 18 do not impair discriminations of global structure but do impair discriminations based on local pattern features. The preceding considerations suggest that it

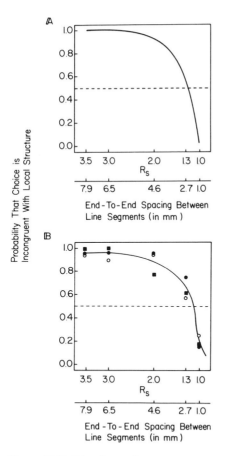

Figure 25.13 Stimulus preferences on structured arrays of line segments. Upper scale on abscissa: R_s. Lower scale: end-to-end spacing of admacent segments. Curve A is redrawn from figure 25.8 and shows the preferences of two normal cats. (B) Stimulus preferences of two destriate cats. ● (cat 2) and ■ (cat 3) represent results obtained prior to discrimination training on unstructured arrays of line segments (see figure 25.8); ○ represent the results of a retest performed with cat 3 after training on the unstructured arrays.

may also be possible to find extrastriate areas whose removal selectively impairs discrimination of global pattern structure.

During the initial testing of the first extrastriate-lesioned animal of this series (one with bilateral removal of area 20), we obtained a pattern of results that seemed to suggest a deficit in figural synthesis. Thus, in comparison with the destriate cats, this animal performed rather poorly when discriminating noise-free dot patterns (figure 25.15) and also showed an increased susceptibility to random-dot masking (figure 25.16). This difficulty was further reflected by apparently increased salience of the local features when the cat was tested for its stimulus preferences on the structured arrays of line segments (figures 25.17). Admittedly, these effects are modest; they seemed somewhat more impressive, however, when this cat subsequently demonstrated perfect

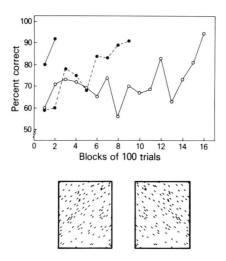

Figure 25.14 Acquisition curves of two normal cats (●—●: cat 4; ●---●: cat 3) compared with postoperative acquisition of a destriate cat (○—○: cat 2) on a discrimination of an array of 120 randomly distributed line segments, as shown at bottom.

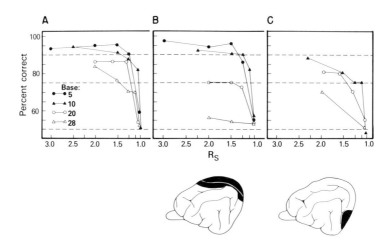

Figure 25.15 Discrimination of structured dot patterns as a function of magnitude of proximity cue R_s. (A) Five normal cats; (B) four cats with area 17–18 destriation; (C) one cat with bilateral ablation of area 20. Bases: (●) 5.6, (▲) 10.6, (○) 20.0, (△) 28.0.

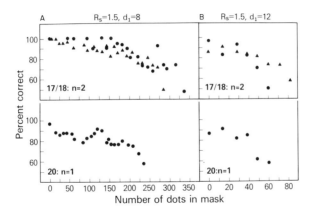

Figure 25.16 Data on random-dot masking of global pattern structure in two destriate cats and one with bilateral lesion of area 20. (A) $d_1 = 8$, $R_s = 1.5$; (B) $d_1 = 12$, $R_s = 1.5$.

transfer to the unstructured arrays of line segments, a problem that most normal and destriate cats find exceedingly difficult (figure 25.18).

These effects were at best temporary, however, since this animal performed extremely well in later retention tests of configurational discriminations. The cause of this improvement is unclear, and we are currently testing additional subjects with somewhat larger extrastriate lesions in an attempt to produce stable and more substantial perceptual impairments. I present these data mainly to illustrate the pattern of results that could serve to dissociate the processes of local and global visuospatial analysis through brain damage. Such a dissociation would indicate that, in addition to providing mechanisms sufficient for normal pattern recognition, the extrastriate cortex contains the mechanisms necessary for global figural analysis.

Discussion

In keeping with the tenor of the conference, the following discussion is wide-ranging and, in some places, highly speculative. Certain arguments may, in light of additional experimentation, require modification. These ideas are presented, however, in the firm belief that speculation and eclectic synthesis can make important contributions to any scientific endeavor.

Consequences of Striate-Cortex Ablations

The success of destriate cats in these tests of pattern perception extend the results of previous reports of pattern and form vision in destriate subjects of a variety of mammalian species (rats, tree shrews, cats, and monkeys) (see, for example, Spear and Braun 1969; Doty 1971; Killackey et al. 1971; Pasik and Pasik 1971; Humphrey 1974; Keating 1975; Hughes 1977; Sprague et al. 1977; Dineen and Keating 1979). Successful discrimination of the discontinuous dot patterns employed in the present

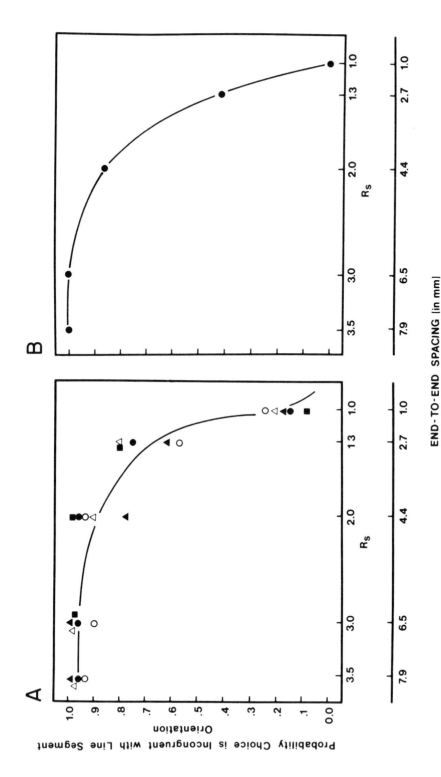

Figure 25.17 Stimulus preferences on structured arrays of line segments. (A) Three destriate cats; (B) one cat with area 20 removed. The lower probability of choosing patterns with features in the nonpreferred orientations (when $R_s = 1.3$) suggests a greater salience of the pattern features in the cat with area 20 removed.

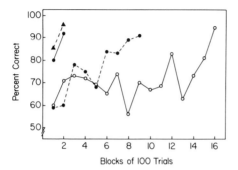

Figure 25.18 Comparison of postoperative acquisition of discrimination of local structural cues in unstructured array of line segments. (▲---▲) One cat with lesion in area 20; (●—●) cat 4 (normal); (●---●) cat 3 (normal); (○—○) cat 2 after destriation of area 17–18.

study requires the ability to generate a precise representation of Euclidean space (or its central equivalent) at some point in the visual pathway. We initially hypothesized that the striate-peristriate cortex (areas 17 and 18) should process information concerning the spatial relationships that distinguish these patterns with more precision and fidelity than any other visuotopic area, since of all the visual areas striate cortex contains the most complete representation of the visual field and has the largest cortical magnification factor (Tusa 1978; Tusa et al. 1978, 1979). Accordingly, removal of this cortex was expected to impair configurational discriminations, at least when the judgments are based on weak proximity cues (low R_s patterns).

This prediction was not confirmed, however. Comparisons between psychometric curves obtained before and after operation show that removal of areas 17 and 18 has little if any effect on configurational discriminations, even when guided by near-threshold proximity cues. It seems clear, moreover, that the destriate cat's capacity to make these discriminations does not represent some "residual" visual ability or one that is somehow "compensating" for impairments caused by the lesion, because these remarkable abilities are demonstrable immediately upon the resumption of testing after operation.

In contrast, more substantial deficits are apparent in destriate cats when the discrimination requires a sensitivity to differences in the local features. These deficits are revealed by slowed rates of acquisition and the poor postoperative retention of discriminations of local structure. It is not known whether the cat's visual system contains any orientation-selective cells after areas 17 and 18 are removed, although Kimura et al. (1980) have demonstrated that cortical cooling of areas 17 and 18 has little effect on receptive-field properties, such as orientation selectivity, in ipsilateral area 19. These results make it clear, however, that removing the entire population of simple cells and a substantial proportion of the complex-cell population does not preclude ultimate success in discriminating local features, which demonstrates that these particular "feature detectors" are not necessary for "feature detection."

Global Configuration versus Local Features in Pattern Discrimination
It appears that normal cats readily transfer a learned discrimination of oblique grating to large-R_s dot patterns and to structured line-segment arrays, but that subsequent transfer from structured to unstructured arrays of line segments is less reliable.

Strong positive transfer can be thought of as indicating perceived similarity, while a lack of transfer indicates perceived dissimilarity between the original training pattern and the transfer pattern. In this context, the results on transfer of training can be interpreted as an indication that similarity between global configurations is a more important determinant of perceived similarity to square-wave gratings than is similarity of the local features.

Additional evidence for a predominance of global structural cues in pattern perception can be found in the results of tests using structured arrays of line segments. We have observed that the R_s value associated with orientational ambiguity in structured line-segment arrays does not change after explicit training on discriminating the orientation of line segments, despite the indications that cats are able to resolve the orientation of individual line segments. This result is observed consistently, and emphasizes the importance of contextual factors in feature recognition. (LeVere 1966). This indication that the processing of local features is suppressed in the presence of a more global pattern structure may explain why destriate and normal cats interpret the orientationally ambiguous structured arrays of line segments in the same way, even though normal cats are able to resolve the local features much better than destriate animals. We might, therefore, suggest that the "grain" of the spatial analysis performed by the visual system can be dependent on certain aspects of the input pattern. This suggestion carries with it the implication that the finely tuned feature analyzers in area 17 and 18 are required only when visual pattern analysis requires high resolution (Berkley and Sprague 1979). This apparent change in the visual system's operating mode is a characteristic of a state-determined system in which one of the controlling variables is a step function (Ashby 1960).

Additional indications of a change in processing strategy from local feature analysis to global configuration based on the nature of the visual input can be found in human pattern-discrimination experiments. The psychophysical experiments of Olson and Attneave (1970), for example, were interpreted as supporting a "feature detection" theory of pattern discrimination. Arrays of randomly distributed pattern elements (oriented line segments, oriented angles, etc.; see part A of figure 25.19) were presented to human observers. The orientation of the pattern elements in one quadrant of the array differed from those in the other three quadrants. The subject's task was to report the location of the disparate

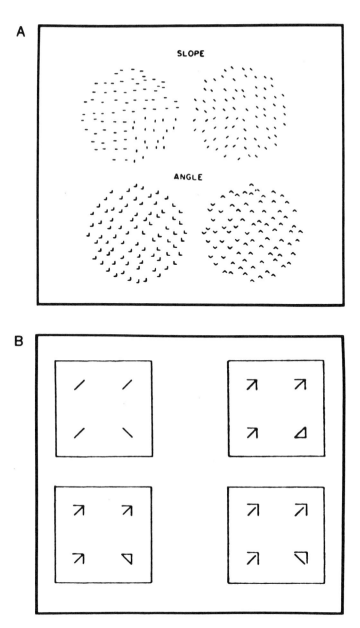

Figure 25.19 (A) Examples of stimuli used by Olson and Attneave (1970, reproduced with permission). Reaction times to detect disparate quadrant were shorter when elements differed in slope rather than in the orientation of angles formed by similarly oriented line segments. (B) Examples of the stimuli used by Pomerantz (1978, reproduced with permission). In this situation, reaction times were longest when stimuli appeared as in 1.

Basic Figural Synthesis in the Cat

quadrant as quickly as possible. The results showed that discriminations between random arrays of line segments with different orientations produced the fastest reaction times, and that patterns whose elements differed in higher-order structural features (such as orientation of angles formed by combinations of similarly oriented pairs of line segments; see part A of figure 25.19) had much longer reaction times. Olson and Attneave interpreted these results as being consistent with a serial, hierarchical discrimination ~ ˙˙˙˙ess in which the most primitive feature (the one extracted first) was the oriented line segment. Pattern elements differing only in higher-order structural information produced longer reaction times, presumably because their discrimination required additional processing ˙˙˙˙ch as scrutinization of the individual features.

Pomerantz (1978) has shown, however, that this result is dependent on the number of elements in the visual display. If the number of elements is reduced to four (one in each quadrant), then the time required to detect the disparate quadrant is longer when the elements are line segments; the shortest reaction times are obtained when the angles formed by similarly oriented line segments differ in orientation (figure 25.19B). Pomerantz pointed out that the apparent discrepancy in the results of the two studies can be reconciled by suggesting that angles are the first feature extracted by the perceptual process; the proximity of dissimilarly oriented line segments in the Olson-Attneave study is assumed to activate angle detectors only at the interface of the disparate quadrant, thus providing a discriminative cue. However, this explanation overlooks a conspicuous difference in the displays used in each experiment: the difference in the number of figural elements. Olson and Attneave's patterns consisted of a large number of randomly distributed elements, and since these patterns contain no global cues they are not conducive to the formation of a figure-ground relationship. These patterns are perhaps better regarded as textures, a suggestion that is supported by the observation that the ease with which they can be discriminated is predicted quite well by the statistical measures shown by Julesz et al. (1973) to govern the discriminability of a variety of textures.

In contrast, the stimuli used by Pomerantz, because they are presented one to a quadrant, clearly result in the formation of figure-ground relationships. Pomerantz's data may, then, relate more to the rules of form discrimination, whereas Olson and Attneave's data relate more to the rules of texture discrimination. If this is the case, it implies that the principles governing these processes may be drastically different—that higher-order configurational cues are dominant in form discrimination, but texture discrimination depends on local processes that act to extract information about the statistical properties of a pattern's microstructure. This again suggests that the visual system may operate in several "modes" whose implementation could be determined by certain aspects of the input pattern. According to this view, the presence of a global structure leads to the separation of figure from ground and that, once embedded in a "figure," features are processed differently than when they are presented in random arrays. Thus, in contrast to the suggestion of most serial-processing approaches that configurational analysis depends on feature analysis, the preceding discussion implies that the perception of stimulus configuration precedes detailed feature analysis. Before exploring this idea further, it may be useful to consider the manner in which these dot patterns may be represented within the visual system.

Relation to Spatial-Frequency Models of Spatial Vision

Spatial-frequency theories suggest that the processes underlying spatial vision are in many ways analogous to Fourier analysis in that, in each case, complex waveforms are decomposed into elementary sine-wave components. This approach has had a great deal of success in explaining a wide variety of visual phenomena, many of which deal with the visibility thresholds of sine-wave-modulated gratings.

For example, Graham and Nachmias (1971) pointed out that when two sine-wave gratings are added the peak-to-peak modulation of the sum (a "complex" grating) varies according to the phase differences between the components. Despite this fact, Graham and Nachmias showed that the visibility of the complex grating is determined only by the amplitude of the components and is independent of their phase difference (the amplitude of the complex). These results, along with many others (see, for example, Robson 1975; Campbell 1977; Cowan 1977), are best explained by postulating that at an early stage of processing the visual system decomposes the image into its orientation-specific spatial-frequency components via orientation-specific "channels" sensitive to a particular band of spatial frequencies. Evidence has been provided that simple and complex cells may provide the neural substrate for these channels by operating as orientation-selective spatial-frequency filters (Pollen et al. 1971; Maffei and Fiorentini 1973). The theory requires that the various frequency-specific channels be independent of one another; this is precisely what the results of the Graham-Nachmias experiment suggest, in that detection of a two-component complex grating is possible if the amplitude of either component exceeds its threshold. Since detection does not depend on the waveform of the complex, the channels corresponding to each component sine wave must be independent.

Using the techniques of Fourier synthesis, any two-dimensional pattern can be synthesized by summing properly chosen sine-wave gratings. In the case of the rectilinear patterns used in the present study, high-frequency components carry information about the size and shape of the features ("element frequencies") and lower-frequency components determine the spacing between features (see figure 25.20).

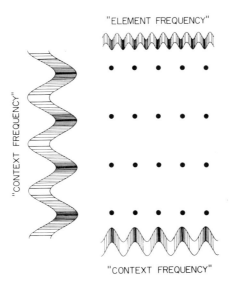

"ELEMENT FREQUENCY"

"CONTEXT FREQUENCY"

"CONTEXT FREQUENCY"

Figure 25.20 Schematic illustration of fundamental frequency components that make up a rectilinear dot pattern.

These lower-frequency components might be called the "context" frequencies. An array that contains a strong proximity cue will contain higher "context" frequencies in the orientation orthogonal to the orientation of the "rows" of dots, and this suggests that if the orientation-specific information is preserved decisions could be mediated by a mechanism that simply selects the pattern on the basis of the orientation of the highest "context" frequencies.

Figure 25.21 shows the relative positions of the "element frequencies" and the "context frequencies" with respect to the cat's contrast-sensitivity curve. The curves are based on two estimates of viewing distance that bracket the observed range of viewing distances chosen by all animals tested thus far. Notice that, at either viewing distance, all fundamental components fall within the range of spatial frequencies visible to normal cats. Notice, however, that the components for an array in which $d_1 = 28$ mm presents something of a problem in that the high and low-frequency components are so widely separated that they nearly exceed the range of visible frequencies. Any constriction of the visible range will result in at least one component falling outside of the visible spectrum. This fact may provide an explanation for the destriate cat's inability to discriminate base 28 dot arrays (see figure 25.11). The results presented by Berkley and Sprague in this volume show that de-striate cats are incapable of resolving spatial frequencies above 3.5–4.0 cycles per degree. Thus, if the cat gets close enough to the stimuli to bring the "dot frequencies" into the visible range, the "context fre-quencies" become so low that they fall beyond the low-frequency cutoff of the curve. Similarly, if the animal increases the viewing distances, to

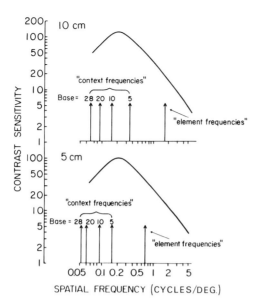

Figure 25.21 Fundamental frequency components of structured dot patterns as a function of viewing distance (10 or 5 cm) plotted on the cat's contrast-sensitivity curve (adapted from Bisti and Maffei 1974). For the sake of clarity, only the "context" frequency which determines the baseline spacing is included. As expected, all components fall within the normal cat's visible range of spatial frequencies.

improve visibility of the context frequencies, then the "element frequencies" move beyond the high-frequency cutoff.

Of course, if two stimuli differ only in frequency components lying beyond the system's resolution, the patterns cannot be discriminated (Campbell 1977). But this does not necessarily imply that stimuli constructed from sine-wave components falling in the visible range will be easily discriminable. For example, an experiment by Mayhew and Frisby (1978) showed quite clearly that textures with different Fourier descriptions can, in effect, be indistinguishable. That is, texture composed of differently oriented and individually visible sine-wave gratings of the same spatial frequency are extremely difficult to discriminate. It seems that information about the sine-wave components of a complex, high-contrast image may be unavailable to subsequent recognition processes.

It is important to realize that, because Mayhew and Frisby's patterns contain components with identical spatial frequencies, the dot patterns formed by the summation of these components present ambiguous proximity cues (in effect, $R_s = 1$). Indeed, it would seem that this is why the patterns are hard to discriminate. Conversely, the presence of a proximity cue is unavoidably confounded with differences in the "context frequency" in various orientations. However, since patterns similar to those used in these experiments that *are* represented by different Fourier descriptions are indistinguishable unless they contain a proximity cue, the suggestion that patterns containing proximity cues are discriminated on the basis of their different Fourier descriptions seems unwarranted at this point. Thus, the supposition that destriate cats are sensitive to the low-frequency components that determine the global configuration of a pattern does not by itself constitute an explanation of how the patterns are actually discriminated. Obviously, an understanding of the relationship between spatial-frequency-specific channels and the perception of global pattern configuration will require experiments specifically designed to explore this important issue.

Some Alternatives to the Spatial-Frequency Model

As an alternative to the spatial-frequency approach, we might consider processes that are somewhat more isomorphic with the pattern structure they analyze. The idea that the perception of a spatial structure is mediated by a topologically similar brain mechanism was a fundamental assumption of the classical gestalt theory of perceptual organization. Certainly the presence of thirteen retinotopically organized cortical areas could provide an ample neural substrate for an analysis of stimulus structure that is based on an internal representation of Euclidean space. Although the evidence from 30 years of single-cell electrophysiology has cast considerable doubt on the plausibility of the "field theories" of gestalt psychology, recent studies of electronic transmission in dendrites and anatomical demonstrations of dendro-dendritic synapses may one day serve to revitalize some form of neural "field theory" to account for the perceived cohesion between pattern elements as a function of their position in Euclidean space. Uttal (1975) has proposed that the autocorrelation function may provide a good model for the detection of ordered dot patterns embedded in visual noise, and has presented evidence in support of this proposal. A noteworthy feature of the autocorrelation model is that the process could be mediated by neuronal networks that are topographically organized but have otherwise undifferentiated receptive fields (Uttal 1975, p. 126), such as might be found in the pulvinar and extrastriate visual cortex.

Global Structure and Preattentive Vision

Successful discrimination of global stimulus configuration can also be viewed as demonstrating the integrity of what Niesser (1967) has called *preattentive vision*. The impetus for hypothesizing the process of preattentive vision comes from the observation that the nature of a visual scene determines to a large extent the scan path used to study it (Niesser 1967). This apparent determinism requires a mechanism that does not rely on focal attention but can analyze the extrafoveal visual field in sufficient detail to extract enough information to decide what in the visual field should be fixated next. Since the process occurs prior to the shifting of focal attention from one point in the visual field to another, Niesser calls it preattentive. In preattentive vision, it is thought that the visual field is analyzed in parallel and that this analysis results in the perceptual structure that underlies the figure-ground dichotomy. The process is based on classic gestalt factors of perceptual organization, such as proximity and similarity.

The present results indicate that this type of analysis is performed outside of areas 17 and 18, and may depend on cortices in the temporal and parietal lobes which we know are retinotopically organized and do not depend on inputs from striate cortex for these properties (Spear and Baumann 1979; Olavarria and Torrealba 1978).

It would seem necessary to suggest that preattentive vision constitutes an early stage in visual processing, one that actually precedes the formation of figure-ground relationships. Indeed, work in the field of artificial intelligence suggests that gestalt-type grouping actually results in the formation of a figure-ground dichotomy (Marr 1976).

In this sense, these behavioral studies are interpreted as support for the thesis first suggested by Diamond and Hall (1969) that, rather than being phylogenetically "new" association cortex, the cortical targets of the pulvinar (or at least its colliculo-recipient divisions) may be phylogenetically older than the geniculostriate system. Diamond and Hall's concepts of the evolution of the central visual pathways fit nicely with the idea that the extrastriate areas perform an early and much more fundamental visual function than the geniculocortical system: that of a low-resolution system that segregates the retinal image into figure and ground. According to this scheme, the geniculocortical system developed to provide a high-resolution system capable of a detailed analysis of objects *after* they were fixated. The frequent observation that striate cortex is most highly developed in foveate animals is also consistent with this speculation.

Elements of this type of processing scheme have received some impressive support from work in artificial intelligence. For example, the relationship between gestalt-type grouping and form vision was highlighted by Marr's (1976) demonstration that image-analyzing algorithms often fail to correctly identify the boundaries of forms unless they make use of gestalt-like grouping processes at an early stage of processing. These grouping operations act on data containing information about the various figural primitives ("blobs," line segments, and various types of edges) and their "place tokens" in the image. This information is in the form of a symbolic description called the *primal sketch*, which is obtained by convolving the image with a large number of orientation-specific spatial-frequency channels. ("Convolving the image" refers to a process that generates a representation of a scene through recursive scanning of the image with elongated slits which are not unlike simple-cell receptive fields. This process effectively detects the first and second derivatives of luminance changes over space; that is, it extracts

Basic Figural Synthesis in the Cat

edges and lines.) The primal sketch is laden with an enormous amount of information, which has very little structure. The grouping processes serve to reduce the amount of information by organizing the data base in ways that aid in subsequent object recognition by identifying the continuity of objects whose borders are interrupted for various reasons (for example, superposition by near objects). Without these grouping operations, feature extraction remains intact, but recognition fails because the contours are not "parsed" appropriately; as a result, the boundaries of different forms are not properly identified. This is precisely the type of deficit we hope to demonstrate in animals with extrastriate cortical lesions: impaired discrimination of stimulus configuration but excellent discrimination of local stimulus features. Thus, based on the demontration that the extrastriate visual cortex is *sufficient* for the analysis of global stimulus configuration, we now wish to pursue the possibility that this cortex is *necessary* for the perception of global structure.

Comparison of Cat and Primate Visual Cortices

The evidence indicates that, in the cat, the formation of a linear gesalt is immediate and unlearned and local pattern features have little influence on perceived pattern structure even though this local structural information is available to the recognition process. Furthermore, and in contrast to our expectations, removal of areas 17 and 18 has no effect on configurational discrimination, although the capacity for detailed feature analysis is impaired (but not abolished). This result indicates that processes that may mark the initial stages of form and pattern recognition do not require the participation of striate-peristriate cortex, but that these cortices are important for more detailed spatial analysis. The widely accepted serial models of pattern recognition, which involve a detailed feature analysis performed in area 17 as the initial processing stage, simply cannot account for these observations. I have argued that, rather than local feature analysis, the analysis of global pattern configuration represents one of the central issues in pattern and form recognition (also see Uttal 1975), and that the perception of global pattern structure can be directly encoded by neurons in the extrastriate visual cortex, since this cortex is sufficient for normal recognition in the cat. The visual impairments of destriate cats seem to be accounted for entirely by the simple loss of visual acuity demonstrated by Berkley and Sprague (this volume). Although primates deprived of area 17 show more severe impairments than subprimate species, they too are capable of complex spatial vision. The relatively more severe effects of striate ablations in primates as compared with cats may relate to the extrastriate projections of the LGN in the cat or to a process of progressive encephalization which favors the dominance of extrastriate visual centers by the geniculostriate system. The endpoint of such a process might conceivably lead to "blindsight" in man, a situation in which the information available via extrastriate pathways is unavailable to consciousness (Weiskrantz et al. 1974). It would appear, then, that in the cat much of the spatial processing that underlies visual pattern perception is carried out by neurons with large and somewhat amorphous receptive fields, and does not require the participation of the finely tuned receptive fields characteristic of the striate cortex. According to this view, striate cortex in the cat serves to provide a high-pass spatial-frequency filter to a central processor that may depend on the integrity of the extrastriate visual pathways for basic figural analysis.

Acknowledgments

This work was supported by grant EY00577 awarded to J. M. Sprague. It is a pleasure to acknowledge the colleagueship of J. M. Sprague, who, in addition to his support and invaluable collaborative efforts, improved the manuscript a great deal through critical readings of earlier drafts. I also thank Mrs. Mary Jack for her skilled and patient secretarial work. The author was supported by a National Research Service Award (EY05130) and a University of Pennsylvania Plan Scholarship.

References

Altman, J., and M. B. Carpenter. 1961 "Fiber projections of the superior colliculus in the cat." *J. Comp. Neurol.* 116: 157–178.

Ashby, W. R. 1960. *Design For a Brain*. London: Chapman and Hall.

Berkley, M. A., and J. M. Sprague. 1979. "Striate cortex and visual acuity functions in the cat." *J. Comp. Neurol.* 187: 679–702.

Berlucchi, G., and J. M. Sprague. 1980. "The cerebral cortex in visual learning and memory, and in interhemispheric transfer in the cat." In *The Organization of the Cerebral Cortex*, F. O. Schmitt et al., eds. Cambridge, Mass.: MIT Press.

Bisti, S., and L. Maffei. 1974. "Behavioural contrast sensitivity of the cat in various visual meridians." *J. Physiol.* 241: 201–210.

Campbell, F. W. 1977. "Sometimes a biologist has to make a noise like a mathematician." *Neurosci. Res. Prog. Bull.* 15: 417–423.

Cornwell, P., J. M. Warren, and A. J. Nonneman. 1976. "Marginal and extramarginal cortical lesions and visual discriminations by cats." *J. Comp. Physiol. Psychol.* 90: 986–995.

Cowan, J. D. 1977. "Some remarks on channel bandwidths for visual contrast detection." *Neurosci. Res. Prog. Bull.* 15: 492–515.

Dalby, D. A., D. R. Meyer, and P. M. Meyer. 1970. "Effect of occipital neocortical lesions upon visual discriminations in the cat." *Physiol. Behav.* 5: 727–734.

Diamond, I. T., and W. C. Hall. 1969. "Evolution of neocortex." *Science* 164: 251–262.

Dineen, J., and E. G. Keating. 1979. "Demonstration of complex pattern vision in the monkey completely lacking striate cortex." *Soc. Neurosci. Abstr.* 5: 782.

Doty, R. W. 1971. "Survival of pattern vision after removal of striate cortex in the adult cat." *J. Comp. Neurol.* 143: 341–370.

Graham, N., and J. Nachmias. 1971. "Detection of grating patterns containing two spatial frequencies: A comparison of single-channel and multiple-channels models." *Vision Res.* 11: 251–259.

Hoffman, K.-P., and J. Stone. 1971. "Conduction velocity of afferents to cat visual cortex: A correlation with cortical receptive field properties." *Brain Res.* 32: 460–466.

Hubel, D. H., and T. N. Wiesel. 1962. "Receptive fields, binocular interaction and functional architecture in the cat's visual cortex." *J. Physiol.* 160: 106–154.

———. 1965. "Receptive fields and functional architecture in two nonstriate visual areas (18 and 19) of the cat." *J. Neurophysiol.* 28: 229–289.

Hughes, H. C. 1977. "Anatomical and neurobehavioral investigations concerning the thalamo-cortical organization of the rat's visual system." *J. Comp. Neurol.* 175: 311–336.

———. 1980. "Efferent organization of the cat pulvinar complex, with a note on bilateral claustro-cortical and reticulo-cortical connections." *J. Comp. Neurol.* 193: 937–963.

Humphrey, N. R. 1974. "Vision in a monkey without striate cortex: A case study." *Perception* 3: 241–255.

Jones, E. G. 1974. "The anatomy of extra-geniculo striate visual mechanisms." In *The Neurosciences: Third Study Volume*, F. O. Schmitt and F. G. Worden, eds. Cambridge, Mass.: MIT Press.

Julesz, B., E. N. Gilbert, L. A. Shepp, and H. L. Frisch. 1973. "Inability of humans to discriminate between visual textures that agree in second-order statistics—Revisited." *Perception* 2: 391–405.

Keating, E. G. 1975. "Effects of prestriate striate lesions on the monkey's ability to locate and discriminate visual forms." *Exp. Neurol.* 47: 16–25.

Killackey, H. M., M. Synder, and I. T. Diamond. 1971. "Function of striate and temporal cortex in the tree shrew." *J. Comp. Physiol. Psychol. Monogr.* 74: 1–29.

Kimura, N., T. Shiaa, K. Tanaka, and K. Toyama. 1980. "Three classes of area 19 cortical cells of the cat classified by their neuronal connectivity and photic responsiveness." *Vision Res.* 20: 169–177.

LeVere, T. E. 1966. "Linear pattern completion by chimpanzees." *Psychonom. Sci.* 5: 15–16.

Maffei, L., and A. Fiorentini. 1973. "The visual cortex as a spatial frequency analyser." *Vision Res.* 13: 1255–1268.

Marr, D. 1976. "Early processing of visual information." *Phil. Trans. Roy. Soc. Lond.* B 275: 483–524.

Mayhew, J. E. W., and J. P. Frisby. 1978. "Texture discrimination and Fourier analysis in human vision." *Nature* 275: 438–439.

Meyer, P. M. 1963. "Analysis of visual behavior in the cats with extensive neocortical ablations." *J. Comp. Physiol. Psychol.* 56: 397–401.

Niesser, U. 1967. *Cognitive Psychology*. New York: Appleton-Century-Crofts.

Olavarria, J., and F. Torrealba. 1978. "The effect of acute lesions of the striate cortex on the retinotopic organization of the lateral peristriate cortex in the rat." *Brain Res.* 151: 386–391.

Olson, R. K., and F. Attneave. 1970. "What variables produce similarity grouping?" *Amer. J. Psychol.* 83: 1–21.

Palmer, L. A., A. C. Rosenquist, and R. J. Tusa. 1978. "The retinotopic organization of lateral suprasylvian visual areas in the cat." *J. Comp. Neurol.* 177: 237–256.

Pasik, T., and P. Pasik. 1971. "The visual world of monkeys deprived of striate cortex: Effective stimulus parameters and importance of the accessory optic system." *Vision Res.* Suppl. 3: 419–435.

Pollen, D. A., J. R. Lee, and J. H. Taylor. 1971. "How does the striate cortex begin the reconstruction of the visual world?" *Science* 173: 74–77.

Pomerantz, J. R. 1978. "Are complex visual features derived from simple ones?" In *Formal Theories of Visual Perception*, E. L. J. Leeuwenberg and H. F. J. M. Buffart, eds. New York: Wiley.

Pöppel, E., R. Held, and D. Frost. 1973. "Residual visual function after brain wounds involving the central visual pathways in man." *Nature* 243: 295–296.

Ritchie, G. D., P. M. Meyer, and D. R. Meyer. 1976. "Residual spatial vision of cats with lesions of the visual cortex." *Exp. Neurol.* 53: 227–253.

Robson, J. G. 1975. "Receptive fields, neural representation of the spatial and intensive attributes of the visual images." In *Handbook of Perception*, vol. 5, E. C. Carterette and M. P. Friedman, eds. New York: Academic.

Spear, P. D., and T. P. Baumann. 1979. "Effects of visual cortex removal on receptive field properties of neurons in lateral suprasylvian visual area of the cat." *J. Neurophysiol.* 42: 31–56.

Spear, P. D., and J. J. Braun. 1969. "Pattern discrimination following removal of visual neocortex in the cat." *Exp. Neurol.* 25: 331–348.

Sprague, J. M., J. Levy, A. DiBerardino, and G. Berlucchi. 1977. "Visual cortical areas mediating form discrimination in the cat." *J. Comp. Neurol.* 172: 441–488.

Tusa, R. 1978. "Retinotopic organization of the visual cortex in the cat." Ph. D. diss., University of Pennsylvania.

Tusa, R. J., and L. A. Palmer. 1980. "Retinotopic organization of areas 20 and 21 in the cat." *J. Comp. Neurol.* 193: 147–164.

Tusa, R. J., L. A. Palmer, and A. C. Rosenquist. 1975. "The retinotopic organization of the visual cortex in the cat." *Neurosci. Abstr.* 1: 52.

————. 1978. "The retinotopic organization of area 17 (striate cortex) in the cat." *J. Comp. Neurol.* 177: 213–236.

Tusa, R. J., A. C. Rosenquist, and L. A. Palmer. 1979. "Retinotopic organization of areas 18 and 19 in the cat." *J. Comp. Neurol.* 185: 657–678.

Uttal, W. R. 1975. *An Autocorrelation Theory of Form Detection.* Hillsdale, N. J.: Erlbaum.

Warrington, E. K., and A. M. Taylor. 1973. "The contribution of the right parietal lobe to object recognition." *Cortex* 9: 152–164.

Weiskrantz, L., E. K. Warrington, M. D. Sanders, and J. Marshall. 1974. "Visual capacity in the hemianopic field following a restricted occipital ablation." *Brain* 97: 709–728.

Winans, S. S. 1971. "Visual cues used by normal and visual-destriate cats to discriminate figures of equal luminance flux." *J. Comp. Physiol. Psychol.* 74: 167–178.

26 Geometrical Approaches to Visual Processing

P. C. Dodwell

The major advances in our understanding of visual processes in recent years have tended to occur at the level of local feature detection and the nature of the elements of a pattern code, as many chapters in this volume demonstrate. This is laudable and in the best tradition of neuropsychology, yet it is clear that there is more to visual perception, to pattern and object recognition and the apprehension of spatial relationships, than the detection of local features. A second approach, also well represented in this volume, is concerned with spatial orientation and visual guidance. Without disputing the importance of this approach either, one may point out that it too fails to capture some of the important attributes of visual perception. What is missing in both cases is a credible treatment of perceptual organization.

The major problem, if one agrees that feature detection is an important first step in visual processing, is how feature information is integrated into perceptual units—the old problem of the perceptual gestalt. Neither pattern recognition nor visual orientation and guidance would be possible in the absence of such organization. I shall argue that, important as the gestalt phenomena are, we have until recently lacked both an appropriate conceptual framework within which to comprehend them and a language (or languages) adequate to their further analysis. That framework and those languages, I shall maintain, are provided by the application of geometrical concepts and operations to vision.

Some theorists (J. J. Gibson foremost among them) have maintained that the very notion of feature detection and the active integration of features into patterns is misguided, so that a theory of perceptual integration is neither necessary nor helpful. Uttal (1975), for example, presented a modern "global" theory in which the necessity for local feature processing is expressly denied. This theory is based on the notion that the autocorrelation functions of patterns are the primary determinants of pattern recognition. The idea that autocorrelations play some role in visual signal processing is appealing, but I believe it is more appropriately applied at the level of separating signal from noise (Dodwell 1971) than at that of global processing of pattern information. At the very least, one has the problem of determining how different autocorrelation functions are themselves classified and recognized. There is no uniquely specifiable mapping from stimulus arrays to perceptual categories. Other global theories tend also to suffer from the inadequate characterization of the operation of recognition (Dodwell 1970).

Another, rather different argument has been brought against the view that feature detection and analysis are fundamental to the process of pattern recognition. Figure 26.1 shows the well-known demonstration by Kolers (1972) that a perceptual category may be defined independently of any particular feature. His point needs no verbal elaboration; obviously "chairs" come in an enormous variety of shapes and sizes. However, it should be pointed out that what Kolers is talking about is a category of use, or an *affordance*, as Gibson (1979) would say, and is at a very high level of cognitive abstraction. I argued elsewhere (Dodwell 1977) that the full understanding of the operations of a perceptual system requires at least that we study it at different levels of processing and different levels of understanding. There I identified three such levels as

1. the level of detection of signals in a noisy environment,
2. the level of organization of visual features into patterns, and
3. the cognitive understanding or use of perceptual categories.

Though these levels are not logically independent of each other, each has properties of its own that place limits on the sorts of explanatory schemes appropriate to it. It should be clear that Kolers's chairs are a category of use, which is to be understood at level 3, and many instances will quickly come to mind of such categorizations that are independent of specific features. However, it is equally certain that there is a level of perceptual organization below this, the one I have identified as level 2, where categories are defined in terms of (perhaps nonexhaustive) specific features or feature concatenations and the organization of such features into perceptual entities. Examples are a texture and a face. Conceptual categorizations such as "chair" are of course necessary, but need not be close to the actual operations of visual processing mechanisms.

I have also argued (Dodwell 1977) that it is at the second level, that of visual organization, that the most theoretical and practical work is needed if we are to understand fully the nature of pattern recognition. Many different approaches to this problem have been proposed, and various analogies have been exploited in attempting to understand the nature of visual integration and organization. The three most prominent approaches to the problem have been the motor integration theories, more abstract sorts of feature analysis and synthesis, and field theories. The first of these, motor integration theories, are well represented by the theorization of Hebb (1949) and, more recently, that of Hochberg (1968). These theorists, although they started from rather different

Figure 26.1 A collection of chairs, each composed of an assortment of different features often not shared with other members of the class. Source: Kolers 1972.

premises, both expounded the view that there are certain visually primitive elements, probably coded by innate mechanisms, which are built into "integrated wholes" by the sequencing of eye movements and (particularly in Hebb's version) the neuronal structures supposedly organized thereby. Whether we call these structures "cell assemblies" or "schematic maps," the basic concept is the same. Integration is brought about by repetitive sequential fixations of spatially extended elements. Though this view has generated a great deal of research, particularly following Hebb's formulation, and though it sits comfortably with our general views about the organization of the nervous system and of the visual field, there is as yet no compelling evidence, I believe, that such processes of motor integration are either necessary or sufficient to explain any great range of configurational properties in vision. (See, however, Noton 1970; Noton and Stark 1971). The second approach, that of abstract feature analysis and synthesis, has enjoyed a great vogue in recent years, particularly in the form of a postulated frequency-analysis mechanism (Campbell 1974). It has even been proposed that some form of Fourier analysis may occur (Campbell and Robson 1968). Experimental work inspired by this conception of visual processing is still appearing with great regularity; however, it is generally believed that any such frequency analysis (and specifically any form of Fourier analysis) could operate only in a very crude way (Sekuler 1974; Robson 1975). Moreover, if such an analysis did occur it would almost certainly operate only at a local level, specifically by the activation of cortical units with specific spatial frequency tuning (Glezer et al. 1973). That being the case, then, the spatial-frequency description of a pattern clearly is being made in terms of local, albeit very abstract, properties; the problem of synthesis remains to be solved. (For other limitations of the Fourier viewpoint see Julesz and Caelli 1979.) Third are the field theories, of which the gestalt theory is classical and the most prominent. The weaknesses of the gestalt formulation, which caused it to fall into disrepute for several decades, are well known, but the gestalt description of visual phenomena remains the only reasonably comprehensive description of organizational principles in the visual field. In my view the gestalt theory is an attempt to use a type of geometrical reasoning in the understanding of vision, but does not rise above the level of metaphor. More specifically, one can argue that although the gestalt theory was on the correct lines at a descriptive level, it lacked any tools for the deeper analysis of visual phenomena. My argument is that the language and tools for this deeper analysis lie in the exploitation of geometrical models for visual processing. The geometrical concepts of vector, vector field, manifold, and differential operator all have their natural interpretation and use in the analysis and understanding of visual organization. The fundamental thesis of gestalt psychology was that "field forces" interact with visual input patterns to determine "output." The trouble with gestalt psychology is that it had no plausible mechanism to handle such interactions. In the remainder of this chapter I shall expound on what the nature of those mechanisms might be.

Geometrical Models of Visual Space

A number of geometrical models of vision have been postulated. I shall describe only four of them, and in no great detail. My intention is simply to show how such models yield rather naturally an organizational and integrative account of visual processing, and by describing four different sorts of models to show that a wide range of phenomena and problems can be tackled with the tools of geometrical analysis.

Luneburg's Theory of Binocular Visual Space

The basic notion of this model (Luneberg 1947) is that there exists a mapping from three-dimensional physical (Euclidean) space to a different space in which the binocular perception of directions and distances occurs. There is a known mapping from the visual directions of single points in a visual field to points on the retina (of an unmoving eye), specified by retinal "local sign" (Hering 1864), that allows us to comprehend the transduction of spatial position into the visual nervous system. So too the mapping from physical to binocular space will be dependent on some form of retinal signature. However, in the binocular case there are two eyes with a fixed interocular distance and the ability to converge and diverge at different fixation distances. The problem is to describe a much more complicated mapping in a way that can account for such things as single vision with two eyes, binocular visual directions, and consistent judgments of deviation from the horizontal plane and the vertical meridian. In Luneburg's model (here much simplified) the properties of binocular visual space are determined by two basic parameters: measures of the state of convergence of the eyes and their elevation from the horizontal plane of regard. Specifically, the simplest mapping is from the bipolar coordinate system having the centers of rotation of the eyes as reference points to the polar coordinate system based on the "egocenter," as shown in figure 26.2.

Points on the Vieth-Müller circle in the horizontal plane defined by the equal-convergence condition of figure 26.2a are mapped into the horopter (the locus of points apparently equidistant from the observer) with the egocenter as origin, as shown in figure 26.2b. Points of "equal-appearing elevation" are defined by rotating the horizontal plane upward about a line through the centers of the two eyes, but maintaining the locus of consistent convergence on it (figure 26.3). A consistent mapping is thus obtained, so that for every point in physical space there is just one point in binocular space, and vice versa, defined by the two measures as described above.

It follows from this model that binocular visual space is non-

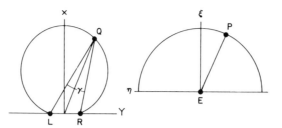

Figure 26.2 The fundamental transform from physical binocular space to "Cyclopean" space of Luneburg's model. The bipolar coordinate system shows the familiar Vieth-Müller circle through the two eyes and the fixation point Q. The circle defines points of equal convergence angle, γ, which are mapped onto the "equidistant" circle centered on the egocenter, as shown at right.

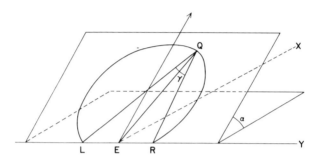

Figure 26.3 The Luneburg mapping outside the horizontal plane of regard, elevated through the angle α. Points on the (elevated) Vieth-Müller circle are still mapped into points on the similarly elevated "equidistant" circle, which is not shown here.

Euclidean and, in fact, hyperbolic (with negative curvature; Euclidean space has zero curvature). That is, it necessarily entails the idea that distances between points in binocular space cannot be measured by standard rigid measuring instruments in the same way as in the ordinary physical space of our local surroundings. Empirical support for the Luneburg model comes from extensive measurements of apparent distances and directions in binocular space, such as the so-called Blumenfeld alley experiments, in which judgments as to the equidistance and parallelism of sets of points are made (Zajaczkowska 1956; Blank 1958). Without going into the details of such experiments one can simply state that, whereas in Euclidean space two distances that are equal to a third distance are necessarily equal to each other, this is not true in hyperbolic space; and, whereas two sets of collinear points that are equidistant from each other are also on lines that are parallel to each other in Euclidean space, they are not so in hyperbolic space. The experiments give general confirmation to the model. The evidence is not totally clear-cut and consistent, but there is no doubt that the space of binocular judgments is not a simple Euclidean space.

Luneburg's model is essentially geometric: It shows us how, from an abstract consideration of the way in which stimulus points must be transduced from the physical world to the retina (and hence into the nervous system) to yield a consistent mapping, a perceptual structure or framework can be deduced. Though the model as such cannot be faulted, the experimental tests have all been performed under highly artificial conditions. The subjects in such experiments are always well trained to make a particular type of judgment; they sit with the head in a headrest, so that little or no movement is possible, and judge the positions of point sources of light in an otherwise dark room. There is no external visual framework, and no points of reference, against which the judgments can be made. Such conditions are necessary if one wants to expose the "pure" nature of binocular processing, but they are far from the situation in which we ordinarily judge position and depth using our two eyes simultaneously. There have been some experiments demonstrating a similar non-Euclidean basis for judgments of visual depth and distance in less artificial conditions (for example, Battro et al. 1976), but the normal space of visual judgments still appears in a certain sense to be four-square Euclidean. The judgment that our ordinary perceptual space is Euclidean is at least partially a cognitive judgement, based on understanding of the nature of physical objects and events, and thus demonstrates that the analysis of processes at different levels (in this case, level 2 for pure binocular processing and level 3 for the judgment of physical characteristics) can seldom work fully independently of one another. The same problem arises in the judgments made in the classical constancy experiments. A compromise is reached between retinal information and physical information—a sort of irrational compromise, but one that acknowledges our sensitivity to different sources of knowledge about the world and the need for them to be simultaneously integrated. The fact that judgments about the physical nature of things can override a mapping and the geometry of visual space it implies does not in itself mean that that mapping and the processes that give rise to it are unimportant. In fact, they can yield a considerable degree of insight into the nature of visual processing.

Although it was the hope of Luneburg and those who followed up his ideas theoretically and experimentally to extend his model to situations where both stimuli and observer are in motion, the model has only been worked out fully for the "static" case. This is obviously a limitation. Any realistic general model of visual processing must take movement as a primary factor, and the failure to do so may explain why so little account has been taken of Luneburg's system in visual science generally.

Variable Geometry and Visual Illusions

The classical visual illusions, such as those of Poggendorf, Ponzo, and Mueller-Lyer, are well known and are usually displayed as line drawings on a flat surface. They have attracted the interest of philosophers and scientists from the earliest times because, apparently, they demonstrate a lack of veridicality in our perceptions. A visual illusion is so named because there is a discrepancy between the "true" properties of a figure and its appearance. The discrepancy is usually in terms of the apparent sizes, distances, angles, and positions of elements in the display. What is meant here by veridicality is correspondence with the measurement paradigms of Euclidean geometry; one would commonly expect a congruence between what is seen and what is known about physical lengths, angles, and the like. A large number of different sorts of theories have been put forward to attempt explanation of such illusions (Robinson 1972). No one of them, however, has enjoyed complete success. In every case an explanation that covers satisfactorily a particular illusion and most of its variants is found not to be adequate for other illusions, even other illusions of apparently similar form. The inappropriate constancy scaling model of Gregory (1966) and Day's (1972) theory of the role of spatial distance cues in illusions are good cases in point. These, and most other attempts to explain visual illusions, have in common the notion that visual space has a fixed geometry—that at least in the local sense it is Euclidean—and that the illusions are merely abberrations from that "true" state of affairs. Watson (1978) considered the possibility that visual space does not have a fixed geometry, but that it is inherently variable in the sense of the variable metric that characterizes certain geometries. Specifically, he postulated that the variable geometry of visual space depends in a precise way on the configuration of elements in different parts of the visual field. These are held to interact with each other as functions primarily of local contrast and separation between elements. The equations that express this interaction are complicated, but the idea underlying them is not; in fact it is akin to the gestalt notion of field forces.

The "Euclidean metric" is familiar to us in the form of the theorem of Pythagoras, which states that the distances between two points with rectangular coordinates (x_1, y_1), (x_2, y_2) can be expressed by the formula

$$s^2 = (x_1 - x_2)^2 + (y_1 - y_2)^2. \tag{1}$$

On a two-dimensional plane surface it would thus be true that, for a small element of distance ds

$$ds^2 = dx^2 + dy^2. \tag{2}$$

This equation expresses a local condition. Where the geometry is non-Euclidean, the departure from the Pythagorean formula can be expressed as

$$ds^2 = (1 + h_1)dx^2 + (1 + h_2)dy^2. \tag{3}$$

This may be termed a "local departure" from the Euclidean metric, provided that h_1 and h_2 are functions of the coordinates x and y and thus vary from point to point. In this case it may be possible to express the new geometry quite explicitly, as that of a so-called Riemannian manifold. Watson's hypothesis is that ds in formula 3 corresponds to perceived distance, so that visual space is indeed a Riemann space.

One could say that h_1 and h_2 represent the action of "field forces" that distort the normal relationships of Euclidean space, just as envisaged in the gestalt theory of field forces. However, the virtue of Watson's approach is that the forces have now been pinned down to a definite form and expressed in a way that leads to explicit, quantifiable predictions about the illusions as functions of the displays in which they occur. Watson's formulae are

$$h_1 = \frac{\partial^2}{\partial x^2} f_1, \quad h_2 = \frac{\partial^2}{\partial x^2} f_2,$$

where f_1 is of the form

$$\iint me^{-x^2}e^{-y^2}\, du\ dv$$

and f_2 is of the form

$$\iint ne^{-x^2}e^{-y^2}\, du\ dv$$

under suitable conditions of normalization, with (u,v) representing the position of elements of a given intensity.

Watson assumes that h_1 and h_2 are "inhibitory" in nature and decay exponentially as functions of the physical separation between elements. This is a reasonable enough assumption. From much of what we know of interactions in the visual field measured psychophysically, and interactions in the visual nervous system measured physiologically, such functions would be suitable candidates for the underlying visual mechanism.

The idea that perceptual space has a variable geometry has two important consequences. First, illusions will appear whenever that geometry departs strongly from the Euclidean form. This will be determined entirely by configurational factors, because of the nature of h_1 and h_2 as explained above. Second, illusions of distance and of path (distortions of contours) will be explainable in terms of the variable geometry. The latter will in fact be geodesics of the visual space. A geodesic on any surface is a "shortest distance," or the path of propagation of a unit vector parallel to itself. (A familiar example would be the lines on a sphere known as great circles.) Geodesics generally define the "shape" or principal characteristics of a surface. Examples of the theoretical predictions and experimental results obtained by Watson for several different illusions are shown in figure 26.4.

The notion that visual space has its own intrinsic geometry, which is variable and dependent on configurations in the field, is very appeal-

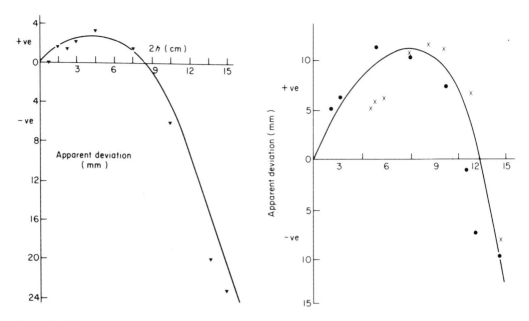

Figure 26.4 Some representative results from Watson's experiments. Data points lie close to the theoretically derived functions for geometrical illusions. The left panel is for the Müller-Lyer illusion, the right for the Ponzo illusion.

ing. It suggests that to use the term "illusion" in this context is wrong, or at least misleading, because there is nothing wrong or illusionary about the perceptions. Rather, they are to be understood as natural consequences of the operation of a rather general processing mechanism—again, a notion congruent with the gestalt tradition. The approach does have some inherent problems, however. First, as with the Luneburg model, the geometry is probably overridden by other aspects of visual processing in most real situations. That is, the nature of the geometry will only be revealed in special situations where the visual input is constrained, as happens in the typical presentation of the classical visual illusions. Thus, while the notion may be correct and yield insight into the nature of visual processing, its application to the understanding of a great array of visual phenomena may be rather limited. It is also not known what range of illusions it can explain. Moreover, the approach has only been worked out for static configurations; thus, it shares another limiting feature with the Luneburg approach. On a more positive note, although Watson presents his ideas as a formal model, it seems that the underlying mechanisms could well be realized in already known neurophysiological processes of lateral inhibition, as was pointed out earlier.

The Lie Transformation Group Theory of Neuropsychology
Hoffman (1966, 1971, 1977) has approached the question of the geometrical nature of the visual field from a somewhat different point of view. He starts with the following question: How do the local, microscopic

processes in the visual field generate macroscopic events that yield the perception of contours, objects, and visual space generally? In parallel with this, one can ask: How do the microscopic events in the visual nervous system (processing by individual neurons) generate macroscopic physiological processes, which must underlie our normal visual perceptions? How is the manifold we call the visual field generated from the discrete processing activities of individual neurons? It happens that a mathematics for proceeding from local infinitesimal operations of a certain sort to the generation of macroscopic events on a manifold was worked out by Sophus Lie late in the nineteenth century (Lie and Scheffers 1893): the theory of continuous transformation groups. It also happens that both the local and the global operations of that theory have a natural interpretation in visual terms, at the levels of both neurophysiology and perception. Hoffman exploited these facts by applying Lie's theory directly to the questions and problems of visual processing.

A transformation is an operation that maps a geometrical object (such as a point, a line, or a figure) into some other object. The simplest possible transformation would be a translation from one place to another. A continuous transformation does this operation smoothly, that is, by proceeding through all intermediate points between the initial and final positions. A set of transformations form a group in the technical sense, provided they fulfil certain simple conditions: that any two transformations applied successively are themselves a transformation, and that there are an inverse (canceling) operation and a null (zero) operation in the set. Movements of a point back and forth along a line of indefinite extent would constitute a group of transformations. If the movements occurred smoothly, by sliding the point along the line, rather than jumping from place to place, this would also be a continuous transformation group. Lie's theory explores the ramifications of such ideas; here we shall be concerned only with their simplest applications.

Operations of this sort on a manifold (and it seems reasonable to identify both the visual field and its cortical representation as manifolds—roughly, smooth surfaces on which continuous transformations can occur) are normally characterized in terms of the actions of vectors. A vector has both position and direction, and thus expresses a purely local property. However, a field of vectors that covers the manifold gives expression not only to all the local properties, but—by the manner in which they can be combined—to macroscopic or global properties as well. The idea that the tangent to a simple curve at any point can be used to express the instantaneous rate of change dy/dx at that point is a familiar one. The tangent used in this way is itself a vector. Conversely, we may conceive of the generation of a curve by the successive set of vectors (local operators) which are tangent to the curve at all possible positions. Generalizing this, one can think of a vector field as potentially generating all the paths, orbits, or *contours* as we may say in the visual case, across the whole manifold. The structure of the manifold may well determine which orbits are most easily generated upon it.

The cortical cells with simple and complex retinal receptive fields discovered by Hubel and Wiesel (1962), with which we are all so familiar, have obvious vectorlike properties. That is, each cell has a positional and directional component to it. Hoffman identified these cells as the components of a cortical vector field which is the basic mechanism for generating orbits (visual contours) on the relevant manifold. Though it has not been possible to identify physiologically how the component vectors are integrated to generate orbits (this is still the great weakness of the single-unit recording approach), the mathematical criterion is understood. It involves a so-called *contact structure,* which ensures that the vectors line up "head to tail" in an appropriate way. We shall see below how, from an ecological point of view, we might expect the simplest form of such a structure to develop in the visual brain. Hoffman (1968) discussed the possible physiological embodiment of various orbit generators in different cell types of the visual cortex, but this remains speculative.

Hoffman noticed that many visual phenomena, in particular the MacKay (1957) complementary afterimages, occur in such a way that pairs of processes are orthogonal to one another. Not only this, but it is apparent that the processes "flow" in natural orbits. He postulated that these orbits are indeed basic characteristics of the visual manifold and express vectorlike operations of the type described above. This notion turns out to be highly compatible with the mathematical structure for continuous transformation groups developed by Lie.

To fix the ideas about "natural flows" on a manifold, consider the situation of an observer in a normal environment. As the head and eyes move around, the visual scene flows across the retina; yet the observer discounts all this activity and maintains pattern and object recognition and a stable frame of reference. A major component in this visual flow is horizontal, because of the way we are constructed and move about. Although we seldom think of it in this way, it is clear that a great deal of neural computation must go on during such transformations—computations of a sort that leave some characteristics of the visual scene invariant under transformation (von Holst and Mittelstaedt 1950). A postulate of the Lie transformation group theory asserts that the "natural flows" across the visual manifold are, initially, those that describe the orbits of transformations under which perception remains invariant. Consider first horizontal translations. The Lie operator that is the infinitesimal generator of a horizontal translation has the symbol $\mathcal{L}_x$ and in differential notation is $\partial/\partial x$, partial differentiation with respect to x. It can also be thought of as a vector moving a geometric object along the horizontal direction on the manifold by some very small amount. Repeated application of this generator will produce continuous (smooth) horizontal translations, or, in the simplest case, orbits that are horizontal straight lines. Vertical translations will be accomplished initially by the operator $\mathfrak{L}_y$, $\partial/\partial y$, repeated application of which represents excursion on the vertical dimension of the manifold, and the orbits will be vertical straight lines. So we get the idea that the infinitesimal operators will, by repeated application, alone or in combination, generate all possible simple translations across the manifold. The two sets of trajectories generated by these transforming operators constitute a net of horizontal and vertical straight lines covering the manifold. As these orbits are basic to visual processing in the theory, we can expect, for example, that patterns involving just these orbits will be particularly simple to discriminate, and that the orbits may even be manifested under special conditions of stimulation, such as those presented by the horizontal/vertical MacKay patterns.

The two sets of orbits cut each other orthogonally, and this is a common property of all the orbits to be discussed. It is well known in visual science that horizontal and vertical visual orientations play a special role in acuity and more generally in the psychophysics of thresholds, as well as in the suprathreshold phenomena of pattern recognition (see Mansfield, this volume). The orbits have been introduced from the point of view of their generation of invariance under an ecologically imposed condition, namely movement of the organism in its visual environment. If there is anything to the theory, we should expect that other basic infinitesimal generators, and their corresponding orbits, should be manifested in visual processing. Some of the early work of Gibson (1950) made it clear that in moving toward an object, or through a visual world, visual flow takes the form of expansion from a point, as shown in figure 26.5. The infinitesimal generators of such flow would be those for dilation, and the orbits would be a star of radial lines. It is not difficult to see that the orthogonal orbits to such a star will be a set of concentric circles, and these would be the generators for transformations that leave patterns invariant under frontal rotation.

So we have two basic orthogonal orbit pairs: one pair for horizontal and vertical translations, and one pair for dilation and rotation, which consists of a star of radial lines and a set of concentric circles. These are illustrated in figure 26.6, together with the Lie operators that generate them and the vectorial representation of the orbit generators. (The second pair also correspond to a pair of MacKay's complementary afterimages.)

It is natural to ask what other orbit pairs, if any, exist. One could consider further the perceptual ecology of the natural environment for clues, but it is equally possible to use a mathematical criterion. It turns out that, under a certain set of simple restrictions, Lie operators such as those shown in figure 26.6 will form a closed system (in fact a group) called a Lie algebra. Under these restrictions there is a limit on the

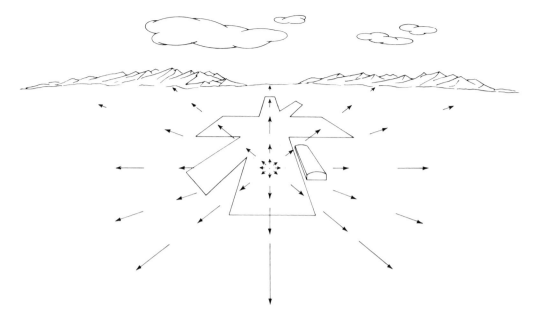

Figure 26.5 Gibson's (1950) well-known illustration of the visual flow experienced during approach to a fixed position in space—in this case an airport runway.

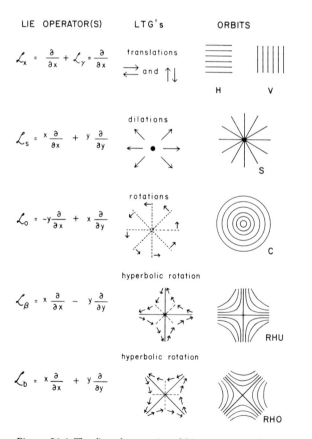

Figure 26.6 The first three pairs of Lie operators, the corresponding Lie transformation groups, and the orbits they generate.

number of possible Lie operators, and all of them have a particularly simple form. According to this criterion, the next orbit pair consists of sets of rectangular hyperbolas asymptotic to a horizontal and a vertical line in the one case and to a pair of perpendicular lines at 45° to these in the other, as is shown in figure 26.6. These hyperbolas are closely related, according to Hoffman, to certain well-known hyperbolic properties of binocular space described by Luneburg (1946) and discussed above. The consequences for perception are not so obvious here, but the mathematical criterion is compelling. (Hoffman predicted, and found, a complementary pair of afterimages for these orbits, similar to those of MacKay.) There are of course other Lie operators, particularly those involving time, but for the moment we will neglect them. The theory has been presented here in terms of the "orbits of invariance," a fundamental part of the theory, but it is held to apply to all aspects of pattern organization and perception, not just those involving "constancy" processing. The general postulate—and this is really the heart of the Lie transformation group theory of neuropsychology—is that the visual system seeks those differential operators that will reduce the output for any path (visual contour) to zero. The simpler the operators (as, for instance, $\mathcal{L}_x$, and $\mathcal{L}_y$), the easier discriminations involving them will be. A recent exposition and discussion of the fuller theory is to be found in Hoffman 1977; Paillard 1977 is the best single source for an introduction and discussion of this approach to visual processing.

What is the practical value of the Lie transformation group theory? If the orbits really are fundamental to visual processing, there should be consequences in visual development and in discrimination and pattern recognition. So far there have been comparatively few explicit tests of the theory, although a good deal of evidence has been marshaled by Hoffman (1966, 1971, 1977) in its support. The first experimental work was reported by Caelli (1974, 1977), who confirmed a number of its predictions in human psychophysical experiments. Recently I have been working with Frances Wilkinson on visual discriminations in young kittens which involve some of the Lie patterns (Wilkinson and Dodwell 1980). As it happens, very little information is available on the pattern-discrimination ability of young kittens, particularly at the time of the physiological "sensitive period." Our results are summarized in table 26.1. The Lie transformation group theory of neuropsychology makes the rather strong prediction that the three orbit pairs shown in figure 26.6 should be simple to discriminate, particularly for the young organism. Horizontal and vertical grids have been used very frequently in research on pattern recognition in animals, and for a variety of reasons have always been considered to be the simplest patternlike discrimination it is possible to devise. A new reason for this doctrine can now be proposed: Horizontal and vertical grids are easy to discriminate because these grids constitute a fundamental Lie orbit pair. It turns out

Table 26.1
Summary of pattern-recognition results of Wilkinson and Dodwell (1980) for young kittens. The actual patterns used are shown in figure 26.6.

Problem	No. of Kittens	Trials to Criterion	Age at Criterion (days)
Horizontal vs. vertical grid	4	271 (97–435)	48 (38–58)
Radial lines vs. concentric circles	9	115 (40–218)	43 (38–58)
Upright vs. oblique hyperbolas	4	327 (165–562)	48 (46–51)

that this discrimination is indeed easy for young kittens, but more surprisingly the radial line/concentric circle discrimination is even easier. Such a result is counterintuitive, and—more importantly—is quite opposite to the prediction one would make on the basis of a simple interpretation of cortical function in terms of line and edge detectors, or indeed in terms of a spatial-frequency analyzer. The sets of rectangular hyperbolas oriented at 45° to each other also turn out to constitute a relatively simple pattern pair for the kitten; perhaps this is an even more surprising result from the orthodox point of view. (We do not at present have a good criterion of what is a "simple discrimination" for a kitten. We have so far established that, compared to some others, the patterns used in this experiment are among the easiest for a kitten to learn to discriminate; but we have not yet demonstrated that it is an exclusive set.) At all events this constitutes, I think, fairly strong evidence that the vectorial visual processing mechanisms for generating elementary orbits proposed in the Lie transformation group theory of neuropsychology are real. We are working on further ways of confirming this notion. There is certainly plenty of scope for additional experimental tests of the theory.

It is clear that the Lie transformation group theory, like the other geometrical systems described above, has its sphere of operation at level 2 (the level at which visual organization naturally occurs). I have spent some time describing it partly because the system is virtually unknown amongst psychologists and partly because it has an obvious application to processing in the normal visual environment, in contrast to the other two theories, which at present are only of value in rather constrained laboratory situations. I am not suggesting that this theory can explain all visual phenomena of interest, and perhaps not even all those that can be analyzed at level 2. However, it strikes me as the deepest and broadest attempt yet to apply geometrical insight and reasoning to the understanding of visual processing.

A Network/Filtering Approach to Visual Organization

The phenomena of visual organization and pattern recognition are so complex that we can scarcely hope to comprehend them all, even at level 2, with a single model or approach. I should like finally to discuss an approach originated by Caelli et al. (1978) which he and I have recently started to apply to the phenomena of spatiotemporal visual events—in particular, apparent movement. It is, I believe, an approach which is not incompatible with the Lie group ideas, although a formal rapprochement has not been attempted. The fundamental idea has been around for a long time, but can be traced in a more or less direct way to Uttley's (1954) proposals about classification of signals in the nervous system. His idea was that interaction between points in the nervous system must be determined by the probability of neural connections between them, and he presented arguments, based on the anatomical studies of Scholl (1953), to show that the probability of connection between any pair of neurons falls off exponentailly with distance. This idea can be formalized by proposing that the output from a single cell consists of at least two components: one due to the present state of the cell, which may be determined by external stimulation and its own internal properties, and one due to inputs from other cells in its neighborhood. These inputs are filtered or weighted, in terms of distance from the cell in question, and the inputs are summed in a simple fashion. Formally, the equation is

$$\phi_i = \eta_i + \sum_j w_{ij} g(\phi_j),$$

where ϕ_i is the output of cell i, η_i is the external input to cell i (e.g., from afferent fibers), $g(\phi_j)$ is some function of cell j's output that is an input to other cells, and w_{ij} is the weighting function between cells i and j. Applying this to the generation of patterns in static displays, Caelli et al. (1978) were able to show a fairly good fit between predictions from the equation (given a particular weighting function) and the global perception by normal human observers of patterns in dot displays. Here a simple filtering and network idea is used to generate, once again, "field properties" for an extended visual display. Earlier gestalt notions of a related sort, such as those of Brown and Voth (1937), simply stated that there must be "forces of interaction" between visual stimuli occurring in different positions, without giving a specific quantitative form to them.

In Caelli and Dodwell 1980 we propose that the system can be extended to the temporal domain by an appropriate choice of filter. Interactions thus occur in the network that are generated by both spatial and temporal separations. The weighting function used is

$$w_{ij} = \alpha\, e^{-\alpha d_{ij}}$$

where

$$d_{ij} = (r_{ij}^2 + \beta t_{ij}^2)^{1/2}.$$

Here r_{ij} is spatial separation, t_{ij} is the temporal separation, and β is a constant of proportionality. The weighting is thus a Poisson process in which α corresponds to the average propagation rate.

Part A of figure 26.7 shows the experimental results of Kolers (1972) for the report of perceived motion between two point sources. The two graphs show the probability of apparent motion as a function of temporal separation at constant visual separation (A) and as a function of visual separation at a constant temporal interval (B). These results clearly suggest a Poisson type of process, in which spatial and temporal separation play complementary or even interchangeable roles. They are the primary reason for postulating the form of filter just described.

We have considered the implications of such a filtering mechanism for apparent motion in multielement configurational displays such as those shown in figure 26.8. The network model has been computer simulated, and the resulting apparent-motion paths or contours have been computed for specific displays. Figure 26.8 shows a comparison between some of the theoretical paths and those actually obtained psychophysically. The fit is quite good.

We also considered the implications of the model for temporal modulation of some of the classical visual illusions. Caelli et al. (1978) had demonstrated the relevance of this approach to the understanding of static visual illusions. We now find that the predictions for spatiotemporal modulation of illusions also fit well with the psychophysical data.

Perhaps the most striking phenomenon to report is that under conditions of apparent movement certain of the classical illusions show very strong enhancement. Specifically, we have studied the Poggendorf illu-

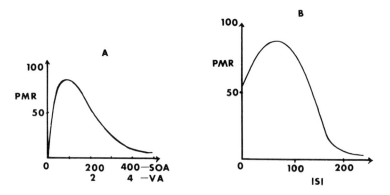

Figure 26.7 (A) Predicted perceived-motion reports as a function of visual angle and stimulus onset asynchrony (SOA) according to the Poisson model. (B) Representative data from Kolers and Pomerantz 1971 showing perceived-motion reports as a function of interstimulus interval.

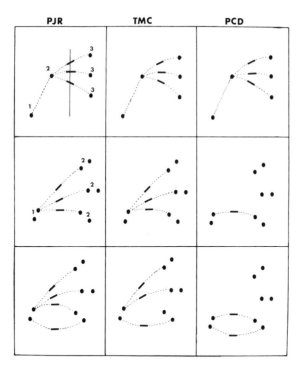

Figure 26.8 Paths of apparent motion as a function of stimulus configuration for three observers. Numbers refer (for each row) to the sequence in which disks or spots were displayed (bottom row same as middle).

sion, which is illustrated in figure 26.9. Part A shows the static configuration, in which it is well known that the two diagonal parts appear to be noncollinear, although in fact they are collinear. The model simulation shows that when the left vertical and diagonal segment are shown a short time before the right segment the degree of noncollinearity grows larger. In fact, under reasonable conditions of exposure duration and interstimulus interval the illusory change in the apparent position of the diagonal is about doubled (Caelli and Dodwell 1980). Here again we have an example where the generation of an illusion, just as in Watson's treatment, is seen not as a distortion of some veridical display or an aberration in the processing system, but rather as a natural consequence of the mechanisms of perceptual organization.

Discussion

I have presented here four approaches to the understanding of visual organization which employ geometrical concepts and operations. One of them—the Lie transformation group theory of neuropsychology—has been discussed at rather greater length than the others, because I feel that this theory has potentially rather wide application and that it is scarcely known among vision researchers. One may well ask what the relationships may be between these various attempts to apply geomet-

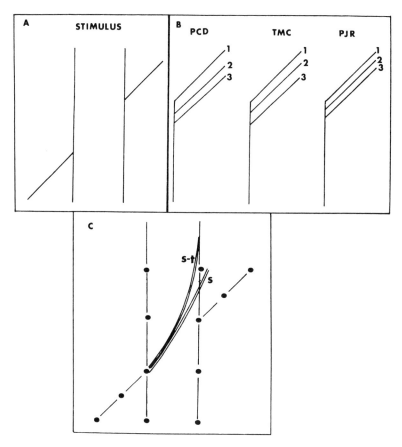

Figure 26.9 Enhancement of the Poggendorf illusion by temporal modulation. (A) The static configuration. (B) Settings for three observers: 1 = veridical, 2 = setting to nullify static illusion, 3 = setting to nullify dynamic illusion. (C) The network simulation of the apparent-motion path for the static (s) and spatiotemporal (s-t) cases. Note that the s-t condition enhances the illusion.

rical intuition and reasoning to the visual system. Are the models mutually consistent? Are they in certain instances interchangeable? What are the incompatibilities, if any? My own belief is that at this comparatively early stage in the application of geometrical tools to vision we should not be too restrictive. That is, the systems under consideration and the phenomena of vision are so multifarious and complex that it would be surprising indeed if a single system or model of the world were adequate to explain them, even at level 2, the level of organization (Dodwell 1978). Rather, we should be willing to consider any variety of modeling or any formal mathematical system that shows promise of giving us some further insight into visual processes. In the end some grand design may be forthcoming which integrates the various approaches, but for now it seems that a wary but conscious eclecticism is the best line to pursue. Certainly, theoretical work to investigate the possible inconsistencies and redundancies between the various models would be in order.

The more modest present aim is to discuss some of the four models' strengths and weaknesses. The Luneburg model has been in existence for quite a long time and has had relatively little general impact on our understanding of visual processes. This I am sure is partly because the mathematics involved is somewhat beyond the knowledge and interest of most psychologists, but more importantly no doubt the neglect stems from the fact that, however correct the theory may be, its practical consequences for visual research or for vision in the normal environment are not tremendously obvious or compelling. Following our earlier argument, "pure" binocular processing may have very frail effects, difficult to detect and usually overridden by other factors. However, it is also true that Luneburg's model describes, in a manner more consistent and clear than has been achieved before, the conditions under which mapping from the physical world into the visual nervous system, via the retinas of the two eyes and their local signatures, can be achieved. As with the theory of retinal "local sign," it may be that the major contribution is one of conceptual clarification rather than practical significance. In a sense, this model or one very like it has to be correct. The question is whether the model is of much consequence for the understanding of a wide variety of other visual phenomena.

Watson's application of Riemannian geometry to the understanding of visual illusions likewise shows great insight into the possiblities of applying a geometrical analysis. However, here too we have to ask how general such a model can be. Watson has shown that it can be applied to a variety of different visual illusions, and in this sense it is perhaps as successful as any other attempt to explain them. But we face once again the overriding problem of the interaction between our cognitive understanding of the nature of the physical world (level 3 analysis) and the organizational constraints imposed by the action of the visual system itself (Dodwell 1978). If the field forces proposed in Watson's model act so weakly or are so easily overridden by other factors that their influence is difficult to detect, again we have the question of how useful such modeling may be for understanding vision in general.

The Caelli-Dodwell approach to modeling apparent motion might also be considered somewhat specialized; on the other hand, it is certainly true that movement and movement-related phenomena are ubiquitous and fundamental in visual processing. The filter or network approach can actually be a very general one, because different model properties (usually readily testable) can be deduced depending on the exact nature of the filter proposed. It is also a sort of statistical approach to the action of the visual manifold, and this can be justified on the grounds that the properties of such a complicated system as the visual cortex (the site, presumably, of the action of filtering and summation) can best be understood, at least to a first approximation, in statistical terms. That is, given the extraordinary level and complexity of local interactions between neurons, statistical characterization may be a more profitable approach than one that attempts modeling at a molecular level. As with other highly complex systems (for example, weather prediction), we do not expect to be able to understand much by looking at local processes like currents of air around a building or the temperature gradient above a particular house or city. This does not mean, however, that we cannot predict and understand weather phenomena on a more global and macroscopic scale. It is simply a matter of deciding on the level of analysis that yields the most reasonable results.

The Lie group transformation approach to visual processing is intriguing because it makes use of some intrinsically interesting mathematics (in common with the other models), but also because it claims a very high degree of generality. It has received surprisingly little comment or attention among visual scientists, no doubt again because the mathematics is a bit difficult for most of us. Also, I believe, it has not been easy for those trained in the life-sciences and neuroscience traditions to accept the possibility that a formal mathematical theory could illuminate so many different aspects of the visual processes. Indeed, a number of individuals have argued that certain aspects of the neuropsychological proposals are quite implausible (see, for example, Frégnac 1977). Given the rate at which understanding of the visual brain is increasing, it seems likely that before too long definite evidence on Hoffman's identification of vector field properties and Lie operators within certain neural structures must be forthcoming. Until that happens, the only rational strategy is to keep an open mind.

In other respects, Lie transformation group theory stands up rather well against a reasonable set of criteria for a neuropsychological theory. I have argued the case in detail elsewhere (Dodwell 1977), so here I will simply state that, in terms of depth, generality, testability and "world-view compatibility," the theory is rather solid and deserves serious attention.

I have tried to show how organizational problems in vision can be tackled with the aid of geometrical tools of analysis and understanding. Problems of the structure of visual space, of field forces that constrain perceptual processing (both spatial and temporal), and of relationships between perceptual organization, stability, and action were all of interest to psychologists of the classical gestalt school. I have tried to show how a deeper and more precise understanding of the phenomena can be attained with the help of geometrical reasoning. No doubt other proposals of this sort will occur with increasing frequency in the future. The geometrical analogies to visual processing are so natural and compelling that it is difficult to imagine that they will not soon dominate the field.

References

Battro, A., S. Netto, and R. Rozestraten. 1976. "Riemannian geometries of variable curvature in visual space: Visual alleys, horopters, and triangles in big open fields." *Perception* 5: 9–23.

Blank, A. A. 1958. "Analysis of experiments in binocular space perception." *J. Opt. Soc. Am.* 48: 911–925.

Brown, J. F., and A. C. Voth. 1937. "The path of seen movement as a function of the vector field." *Amer. J. Psychol.* 49: 543–563.

Caelli, T. M. 1974. "Models, Lie algebras and visual pattern perception." Ph.D. diss., University of Newcastle. Australia.

———. 1977. "Psychophysical interpretations and experimental evidence for the LTG/NP theory of perception." *Cahiers Psychol.* 20: 107–134.

Caelli, T. M., and P. C. Dodwell. 1980. "On the contours of apparent motion: A new perspective in visual space-time." *Biol. Cybernetics* 39: 27–35.

Caelli, T. M., G. Preston, and E. Howell. 1978. "Implications of spatial summation models for processes of contour perception: A geometric perspective. *Vision Res.* 18: 723–734.

Campbell, F. W. 1974. "The transmission of spatial information through the visual system." In *The Neurosciences: Third Study Volume*, F. O. Schmitt and F. G. Worden, eds. Cambridge, Mass.: MIT Press.

Campbell, F. W., and J. G. Robson. 1968. "Application of Fourier analysis to the visibility of gratings." *J. Physiol.* 197: 557–566.

Day, R. H. 1972. "Visual spatial illusions: A general explanation." *Science* 175:1335–1340.

Dodwell, P. C. 1970. *Visual Pattern Recognition*. New York: Holt, Rinehart and Winston.

———. 1971. "On perceptual clarity." *Psychol. Rev.* 78: 275–289.

———. 1977. "Criteria for a neuropsychological theory of perception." *Cahiers Psychol.* 20: 175–182.

———. 1978. "Human perception of patterns and objects." In *Handbook of Sensory Physiology*, vol. VIII, R. Held et al., eds. New York: Springer.

Frégnac, Y. 1977. "Contradictions between L.T.G. model and neurophysiology." *Cahiers Psychol.* 20: 209–211.

Gibson, J. J. 1950. *The Perception of the Visual World*. Boston: Houghton Mifflin.

———. 1979. *The Ecological Approach to Visual Perception*. Boston: Houghton Mifflin.

Glezer, V. D., V. A. Ivanoff, and T. A. Tscherbach. 1973. "Investigation of complex and hypercomplex receptive fields of visual cortex of the cat as spatial frequency filters." *Vision Res.* 13: 1875–1904.

Gregory, R. L. 1966. *Eye and Brain*. London: Weidenfeld & Nicholson.

Hebb, D. O. 1949. *The Organization of Behavior*. New York: Wiley.

Hering, E. 1864. *Beiträge zur Physiologie*. Leipzig: Engelmann.

Hoffman, W. C. 1966. "The Lie algebra of visual perception." *J. Math. Psychol.* 3: 65–98; erratum ibid. 4 (1967): 348–349.

———. 1968. "The neuron as a Lie group germ and a Lie product." *Q. J. Appl. Math.* 25: 423–440.

———. 1971. "Visual illusions of angle as an application of Lie transformation groups." *S.I.A.M. Rev.* 13:169–184.

———. 1977. "An informal, historical description (with bibliography) of the 'L.T.G./N.P.' " *Cahiers Psychol.* 20: 135–174.

Hochberg, J. 1968. "In the mind's eye." In *Contemporary Theory and Research in Visual Perception*, R. N. Haber, ed. New York: Holt, Rinehart & Winston.

Hubel, D. H., and T. N. Wiesel. 1962. "Receptive fields, binocular interaction and functional architecture in the cat's visual cortex." *J. Physiol.* 160: 106–154.

Julesz, B., and T. M. Caelli. 1979. "On the limits of Fourier decompositions in visual texture perception." *Perception* 8: 69–73.

Kolers, P. 1970. "Perception of objects from a category." *Perception* 1:23–45.

———. 1972. "The role of shape and geometry in picture recognition." In *Picture Processing and Psychopictorics*, B. S. Lipkin and A. Rosenfeld, eds. New York: Academic.

Kolers, P., and J. R. Pomerantz. 1971. "Figural changes in apparent motion." *J. Exp. Psychol.* 87: 99–108.

Lie, S., and G. Scheffers. 1893. *Vorlesungen über continuerliche Gruppen*. Leipzig: Teubner.

Luneberg, G. 1947. *Mathematical Analysis of Binocular Vision*. Princeton, N.J.: Princeton University Press.

MacKay, D. M. 1957. "Moving images produced by regular stationary patterns." *Nature* 180: 849–850.

Noton, D. 1970. "A theory of visual pattern recognition." *IEEE Trans. in Systems Sci. Cybernet.* 6: 349–357.

Noton, D., and L. Stark. 1971. "Scan patterns in eye movements during pattern perception." *Science* 171: 308–311.

Paillard, J., ed. 1977. "The Lie transformation group model for perceptive and cognitive psychology." *Cahiers Psychol.* 20 (no. 2/3).

Robinson, J. O. 1972. *The Psychology of Visual Illusion*. London: Hutchinson.

Robson, J. G. 1975. "Receptive fields: Neural representation of the spatial and intensive attributes of the visual image." In *Handbook of Perception,* vol. 5, E. C. Carterette and M. P. Friedman, eds. New York: Academic.

Scholl, D. A. 1953. "Dendritic organization in the neurons of the visual and motor cortices of the cat." *J. Anat.* 87: 387–406.

Sekuler, R. 1974. "Spatial vision." *Annu. Rev. Psychol.* 25: 195–232.

Uttall, W. R. 1975. *An Autocorrelation Theory of Form Detection.* Hillsdale, N.J.: Erlbaum.

Uttley, A. M. 1954. "The classification of signals in the nervous system." *EEG Clin. Neurophysiol.* 5: 479–494.

Von Holst, E., and H. Mittelstaedt. 1950. "Das Reafferenzprinzip." *Naturwissenschaften* 37: 464–476. Translated as "The principle of reafference," in *Perceptual Processing,* P. C. Dodwell, ed. New York: Appleton-Century-Crofts (1971).

Watson, A. 1978. "A Riemann geometric explanation of the visual illusions and figural after effects." In *Formal Theories of Visual Perception,* E. Leeuwenberg and H. Buffart, eds. New York: Wiley.

Wilkinson, F. E., and P. C. Dodwell. 1980. "Young kittens can learn complex visual pattern discriminations." *Nature* 284: 258–259.

Zajaczkowska, A. 1956. "Experimental determination of Luneburg's constants σ and K." *Q. J. Exp. Psychol.* 8: 66–78.

Contributors

Richard J. Andrew
Ethology and Neurophysiology Group
School of Biological Sciences
University of Sussex

Mark A. Berkley
Department of Psychology
Florida State University

Giovanni Berlucchi
Istituto di Fisiologia
Università di Pisa

B. Biguer
Laboratoire de Neuropsychologie Expérimentale
Institute National de la Sante et de la Recherche Medicale
Bron, France

Charles M. Butter
Department of Psychology
University of Michigan

Alphonso Campbell, Jr.
Department of Psychology
University of Michigan

John Cerella
Foundation for Research on the Nervous System
Boston

Thomas Collett
School of Biological Sciences
University of Sussex

Paul Cornwell
Department of Psychology
The Pennsylvania State University

Paul Dean
Department of Psychology
University of Sheffield

John Dineen
Department of Anatomy
Upstate Medical Center
Syracuse, N.Y.

Peter C. Dodwell
Department of Psychology
Queens University
Kingston

Jörg-Peter Ewert
Neuroethology and Biocybernetics Laboratories
University of Kassel

Barrie J. Frost
Department of Psychology
Queens University
Kingston

Melvyn A. Goodale
Department of Psychology
University of Western Ontario

J. A. Graves
Department of Psychology
University of Western Ontario

Charles R. Hamilton
Department of Biology
California Institute of Technology

Lindesay I. K. Harkness
The Biological Laboratories
Harvard University

Werner Himstedt
Institut für Zoologie
Technische Hochschule Darmstadt

Howard C. Hughes
Department of Anatomy
The School of Medicine
University of Pennsylvania

David J. Ingle
Department of Psychology
Brandeis University

Marc Jeannerod
Laboratoire de Neuropsychologie Expérimentale
Institute National de la Sante et de la Recherche Medicale
Bron, France

E. Gregory Keating
Departments of Anatomy and Neurology
Upstate Medical Center
Syracuse, N.Y.

Daniel Kurtz
Department of Psychology
University of Michigan

Richard Latto
Department of Psychology
University of Liverpool

David N. Lee
Department of Psychology
University of Edinburgh

C. C. Leiby III
Department of Psychology
University of Michigan

David N. Levine
Department of Neurology
Massachusetts General Hospital
Boston

Richard J. W. Mansfield
Department of Psychology
Harvard University

Carlo Alberto Marzi
Laboratorio di Neurofisiologia
Consiglio Nazionale delle Richerche
Pisa

J. Mench
Ethology and Neurophysiology Group
School of Biological Sciences
University of Sussex

A. D. Milner
Department of Psychology
University of St. Andrews

Mortimer Mishkin
Laboratory of Neuropsychology
National Institute of Mental Health
Bethesda, Maryland

Donald E. Mitchell
Department of Psychology
Dalhousie University
Halifax

Jacques Paillard
Institut de Neurophysiologie et Psychophysiologie
Centre National de la Recherche Scientifique
Marseille

C. Rainey
Ethology and Neurophysiology Group
School of Biological Sciencies
University of Sussex

James M. Sprague
Department of Anatomy
School of Medicine
University of Pennsylvania

James A. Thomson
Department of Psychology
University of Strathclyde

Brian Timney
Department of Psychology
University of Western Ontario

Leslie G. Ungerleider
Laboratory of Neuropsychology
National Institute of Mental Health
Bethesda, Maryland

J. M. Warren
Department of Psychology
The Pennsylvania State University

Index

Accommodation, 129–138
in chameleon, 136–137, 159, 166, 167
in frog, 136–137, 157
in man, 135–136
in toad, 136–137, 159, 166, 168
Acuity, contour. *See* Contour acuity
Angular gyrus lesions, 654–655
Autocorrelation theory, 794, 801–807
Autoradiography, 2-deoxyglucose method of
in monkey cerebral cortex, 553–554
in pigeon tectum, 190

Barrier avoidance
by frog, 75–91, 289
by gerbil, 264, 286–289
by toad, 120–123, 289
Binocular vision. *See also* Stereopsis
binocular integration pathways in pigeon, 229–230
binocular visual space (Luneburg), 805–807
depth vision in mantid (*Tenodera australasiae*), 151
disparity-sensitive neurons in visual cortex, 146
Burrowing owl, stereopsis in, 147–148

Cat
acuity in, 485–491, 495–514, 535–539
depth discrimination in, 491–494, 514–515
interocular equivalence in, 695–702
interocular transfer in, 721–731, 733–744
pattern-discrimination learning in, 532–534, 696, 721–731, 733–744, 757–767, 778–781, 815–816
Categorization in visual processing, 611–619
Chameleon
accommodation in, 136–137, 159, 166–167
route planning by, 118–120
Chicken
avoidance behavior of, 204
food discrimination in, 201
habituation of feeding in, 197–201
hemispheric specialization in, 197–207
interocular transfer in, 213, 230, 233
sexual differences in, 201–203

Commissures, cerebral
connections of, in monkey, 709, 711–712
lesions of, in cat and monkey, 696–702
Contour acuity
development of, in cat, 495–514
measurement of
in cat, 485–491
in man, 463–468
theoretical basis of, 461–463, 468–470
and visual-cortex lesions, in cat, 535–539

Depth vision
and accommodation, 129–138, 157, 159, 166–168
binocular, in mantid (*Tenodera australasiae*), 151
development of, in visually deprived cat, 514–515
and image size as cue to depth, 138–140, 164
in inferotemporal-lesioned monkey, 605–606
measurement of, in cat, 491–494
and motion parallax, 152–156, 163, 164
and size constancy, 55–57, 89–90, 123–128
and stereopsis, 140–152, 157, 159, 160, 166, 168
Distributive processing in vision, 471–473, 549–550, 592–594
Driving, 423–425, 431. *See also* Parallel-serial systems model
Dysconnection syndrome, 629

Exproprioception, 411–431
Eye movements
in man, 397–399
in monkey, 307–320, 338–339, 353–360, 671–687
in salamander, 63

Feature detection, 788–791, 801–802
Frog
avoidance by, 72–74
accommodation in, 136–137, 157
barrier avoidance by, 75–91
eye separation in, and size of frog, 144, 150
head orientation of, 149

Frog (cont.)
  jumping by, 112, 158
  monocular depth judgments by, 158,
    163
  optokinetic nystagmus of, 82
  phototaxis in, 74–75
  prey catching by, 51–52, 68–71, 83–88
  and size constancy, 89–90, 126
Frontal-eye-field lesions, 673–679

Geniculate, dorsolateral
  connections of
    in cat, 526–528, 732
    in monkey, 446–449
  receptive field characteristics of, 446,
    450–451, 496–497
Gerbil
  barrier avoidance by, 286–289
  locomotion of toward targets, 282–289
  orienting responses of, 92–105, 282–
    289
Gestalt theory, 794, 804–805

Hamster
  locomotion toward targets by, 270, 274,
    275
  orienting responses of, 69, 270, 295
  scanning by, 295
Hippocampal lesions, 651–652
Humans
  accommodation in, 135–136
  driving by, 423–425, 431
  eye movements in, 397–399
  jumping by, 421–423, 431
  locomotion by, 265, 411–431
  posture of, 417–419
  reaching by, 265, 367–373, 377–383,
    387–405
  and size constancy, 123–124
  stereopsis in, 146–147, 160

Image size, as cue to depth, 138–140, 164
Inferotemporal cortex. See also Visual cor-
    tex; Temporal lobe
  electrophysiology of, 403, 552–553, 563,
    565–566
  lesions of, 347, 600–623, 634–637, 641,
    643–646
Insects
  backswimmer (Notonecta glauca), 125–
    126
  bulldog ant (Myrmecia gulosa), 170–171
  horsefly (Syritta pipiens), 113–118, 163
  housefly (Musca domestica), 164
  locust (Schistocerca gregaria), 154–155,
    163–164
  mantid (Tenodera australasiae), 151
Interocular equivalence
  in cat, 695–697
  in man, 696
  in split-brain cat and monkey, 697–702
Interocular transfer
  in cat, 721–731
  in pigeon, 212–235
  in Siamese cat, 733–735
  in split-chiasm cat, 735–744

Jumping
  in frog, 112, 158
  in man, 421–423, 431

Lesion method, rationale for, 587–600
Lie transformation group, 810–816
Locomotion, visual control of,
  in gerbil, 282–289
  in hamster, 270, 275
  in man, 411–431
  in pigeon, 184–188
  in rat, 271–276

Man. See Humans
Monkeys
  acuity in, 461–463
  depth discrimination in, 605–606
  discrimination-training methods for,
    302–312
  eye movements in, 338–339, 353–360
  orienting responses by, 312–324, 328–330
  pattern-discrimination learning by,
    341–353, 554–562, 600–623, 696–702
  reaching by, 373–377, 402–403, 678–684
  spatial localization in, 312–324, 554–563
  visual electrophysiology of
    inferotemporal cortex, 552–553, 563,
      565–566
    lateral geniculate, 450–451
    parietal cortex, 403, 563
    retina, 445, 450
    superior colliculus, 336–337
    visual cortex, 451–456
  visual neglect in. 673–675
  visual search in, 675–678
Motion parallax, 152–156, 163–164
Motion perception, 338, 347–353, 817–820
Motor integration theory, 671, 802–804

Network filtering, 817–820

Optic flow field, 413–415
Optic tectum
  connections in
    in cat, 732
    in monkey, 335–336, 684–685
  function of
    in frog, 68–73, 82–88
    in rodents, 289–291
    in toad, 22–41
  lesions of
    in fish, 68, 73
    in frog, 68–73, 76, 82
    in gerbil, 95–105, 282–289
    in hamster, 270, 274, 275
    in rat, 271–282, 291–295
    in monkey, 292, 307–321, 328–330,
      338–339, 352–354, 360, 673–687
  visual electrophysiology of
    in monkey, 336–337
    in pigeon, 179–184, 188–190
    in salamander, 57–60
    in split-chiasm cat, 742–744
    in toad, 27–31
Orbitofrontal–anterior temporal lesions,
    648–651. See also Visual cortex

Orienting responses
in frog, 68–91
in gerbil, 92–105, 282–289
in goldfish, 73
in hamster, 69, 270, 295
in iguana, 72
in monkey, 312–324, 328–330
and prey recognition, in toads, 8–10

Parallel-serial systems model, 526–528, 541–543. *See also* Distributive processing
Parietal lobe. *See also* Visual cortex
electrophysiology of, 403, 563
lesions of, 402–403, 652–654, 675–679
Pattern-discrimination learning
in cat, 532–534, 696, 721–731, 733–744, 757–767, 778–781, 815–816
and hemispheric differences, in man, 655–660
and hippocampal lesions, in man, 651–652
and inferotemporal lesions, in monkey, 561–562, 600–623
and interhemispheric transfer, 706–708, 710–711
and orbitofrontal-anterior temporal lesions, in man, 649–651
and parietal-lobe lesions, in man, 652–653
and parietal lesions, in monkey, 561–562, 554–555
in pigeon, 241–257
in rat, 755, 762–765
in Siamese cat, 733–735
in split-chiasm cat, 735–744
and striate lesions, in monkey, 322, 324–330, 341–353, 560–562
and superior-colliculus lesions, 269, 307, 338–339
and visual agnosia, in man, 637–648
and visual-cortex lesions, in cat, 532–534
Pigeon
accommodation in, 223
binocular integration pathways in, 229–230
interocular transfer in, 212–235
optokinetic response in, 179–182
pattern recognition by, 241–257
visual guidance of locomotion in, 184–188
Preattentive vision, 794–796
Posture, visual control of, in man, 417–419
Pretectum
and barrier avoidance by gerbil, 264, 286–289
electrophysiology of, in toad, 27, 77
function of, in frog, 77–86
and gerbil locomotion, 275
and toad prey selection, 34–41
Prey catching
in frog, 51–52, 68–71, 83–88
in insects, 170–171
in salamander, 48–57, 61–62
in toad, 10–41, 51–52
Pulvinar complex, connections in
in cat, 527
in monkey, 447–449

Rat
locomotion toward targets by, 271–276
orienting responses by, 277–282
pattern-discrimination learning in, 755, 762–765
scanning by, 291–295
Reaching
in man, 367–373, 377–383, 387–405
in monkeys, 373–377, 402–403, 678–684
Retina, monkey, 444–446, 448–450
Retinal ganglion cells, toad, 24–26
Retinal locus, as factor in interocular transfer in pigeon, 212–235

Salamander
eye movements of, 63
prey catching by, 48–57, 61–62
and size constancy, 55–57
Scanning
by hamster, 295
by monkey, 353–360, 675–678
by rat, 291–295
Selective attention, 312–313, 686–687
Sequential-activation model, 554
Serial-processing model, 525–526
Shape recognition, in toad, 10–22
Size constancy, 55–57, 89–90, 123–128
Spatial-filter model, 456–463, 468–470
Spatial-frequency analysis and synthesis, 791–794, 804
Stereopsis, 140–152
in burrowing owl, 147–148
in man, 146–147, 160
in toad, 157, 159, 166, 168
Superior colliculus. *See* Optic tectum
Suprasylvian lesions, in split-chiasm cat, 740–742. *See also* visual cortex

Temporal lobe, monkey. *See also* Inferotemporal cortex; Visual cortex
anatomical organization of, 594–596
effect of lesion on, 630–631
Thalamic function, in frog, 73–75
Toad
accommodation in, 136–137, 159, 166, 168
and depth constancy, 122
head orientation of, 148, 150
negotiation of obstacles by, 120–123
prey catching by, 51–52
and size constancy, 126–128
stereopsis in, 157, 159, 166, 168
Two-visual-systems model, 263–266, 269–270, 301–302, 328–330, 387

Variable geometry, 808–810
Visual cortex
connections of
in monkey, 446–450, 550–554, 568–578, 684–685
in Siamese cat, 733–735
thalamocortical, in cat, 526–528, 731–732
electrophysiology of
in cat, 483–484, 496–497, 504–506, 511–514, 517
in monkey, 451–456

Visual cortex (cont.)

  comparison of role of in cat and monkey, 797–797

  lesions of

    area 17 and 18, in cat, 531–540, 781–788

    area 17, in monkey, 560–562, 757–759, 264, 301–302, 322–324, 328–330, 339–353, 560–562, 757–759

    crossed striate-parietal, in monkey, 558–560

    extrastriate, in cat, 531–540, 740–742, 757–761

    foveal striate, foveal prestriate, and inferotemporal, in man, 634–637

    posterior cortex, in gerbil, 97–98, 103–104, 284–289

    posterior cortex, in rat, 275–276

    prestriate, in monkey, 564–568, 632–633, 640–641, 646–648

    and reaching, in man, 401–403

    unilateral, in man, 656–660

    unilateral, in monkey, 655

Visual-discrimination test. *See also* Pattern-discrimination learning

  criticism of methodology, 267–268

  role of orientation mechanisms in, 291–295, 307–321

  testing procedure for monkeys, 302–306

Visual neglect, in monkey, 673–675

Visual search, 353–360, 675–678. *See also* Scanning